Drafting for Industry

Walter C. Brown
Professor Emeritus, Division of Technology
Arizona State University, Tempe, AZ

Clois E. Kicklighter
Dean, School of Technology
Indiana State University, Terre Haute, IN

South Holland, Illinois
The Goodheart-Willcox Company, Inc.
Publishers

Lake Superior College Library

Copyright 1995

by

The Goodheart-Willcox Company, Inc.

Previous editions copyright 1990, 1984, 1981, 1978, 1974

Library of Congress Catalog Card Number 93-50718

International Standard Book Number 1-55637-048-5

2 3 4 5 6 7 8 9 10 - 95 - 98 97 96 95

Libary of Congress Catalog-in-Publication Date

Brown, Walter Charles.
    Drafting for Industry / by Walter C. Brown and Clois E. Kicklighter.

        p.   cm.
    Includes index.
    ISBN 1-56637-048-5

    1. Mechanical drawing. I. Kicklighter, Clois E. II. Title.

T353.B873   1995
604.2--dc20   93-50718
                                                        CIP

AutoCAD is registered in the U.S. Patent and Trademark Office by Autodesk, Inc.

INTRODUCTION

As technology has changed, drafting has changed with it. New and improved ways of describing emerging technologies had to be developed. The way that drawings are created has changed greatly with the advent of computer-assisted design and drafting. With all of the changes in technology and drafting, the fundamentals of drafting have remained the same.

Drafting for Industry has been revised and updated to keep pace with the changes in industry and in drafting. A new Chapter 1 has been developed around Drafting as a Communication, Problem-Solving, and Design Tool. Chapter 2, *Using Drafting Tools and Systems,* has been updated to include CADD equipment. *Drafting Careers* has been moved to the last section of the text. The chapters from earlier editions of the text that covered various applications of technical drafting have been condensed into a single chapter. The number of major sections in the text have been expanded from five to six, and the arrangement of chapters has been revised to group related information together.

In addition, new photos and illustrations have replaced obsolete or dated ones. The font size has been increased slightly to improve the readability. Also, important terms appear in **bold-face italics** throughout the text. Each chapter begins with a list of key concepts that will be presented in that chapter. Every aspect of the text has been examined to provide a modern drafting text that is current in every respect.

Even though this book has been updated, the strengths of the previous editions have been retained. Some of these strengths include: true-to-life problems from industry, complete step-by-step procedures for complex drafting procedures, an exemplary model of drafting communication, problem-solving approach used throughout the text, clear and concise examples, relevant career information, and adherence to ANSI and industry standards throughout.

The student *Worksheets* has been updated with many new problems to provide a broader selection. The majority of the *Worksheets* problems are included in the *Drafting for Industry* software. Many of these problems also appear in the text as end of chapter problems. The problems from the text that also appear in the software can be easily identified by their light yellow background. In addition, the problem numbers that correlate to software problems have the same light yellow background. The software problems, provided in AutoCAD® format, can be purchased directly from the publisher.

Drafting for Industry is truly a comprehensive text presented in an easy-to-understand and well-illustrated style.

Walter C. Brown
Clois E. Kicklighter

About the Authors

Dr. Walter C. Brown

Dr. Walter C. Brown's professional background includes teaching and administrative experience of sixteen years at the secondary level and thirty years at the university level. In addition, he has taught evening adult classes and served as a visiting professor at four universities. He has held a variety of professional offices of state and national associations. He has served as a director of an instructional materials laboratory and conducted workshops at the university level. He is the author of texts in drafting, print reading, and mathematics. He received his undergraduate degree from Northwest Missouri State University and his masters and doctorate from the University of Missouri-Columbia.

Dr. Brown is the author or co-author of *Print Reading for Industry, Basic Mathematics, Blueprint Reading for Construction,* and *Drafting for Industry.*

Dr. Clois E. Kicklighter

Dr. Clois E. Kicklighter is Dean of the School of Technology and Professor of Construction Technology at Indiana State University. He is a nationally known educator and has held the highest leadership positions in the National Association of Industrial Technology including Chair of the National Board of Accreditation, Chair of the Executive Board, President, and Regional Director. Dr. Kicklighter was recently awarded the respected Charles Keith Medal for exceptional leadership in the technology profession.

Dr. Kicklighter is the author or co-author of *Architecture: Residential Drawing and Design; Modern Masonry: Brick, Block, and Stone; Residential Housing; Modern Woodworking;* and *Drafting for Industry.*

Dr. Kicklighter's educational background includes a baccalaureate degree from the University of Florida, a master's degree from Indiana State University, and a doctorate from the University of Maryland. His thirty-two years of experience includes industrial, teaching, and administrative positions.

Contents

Acknowledgments

The authors and the publisher wish to thank the following companies, individuals, and organizations for providing illustrations for this text.

Allen-Bradley Company
Altium
Aluminum Association
American Hoist & Derrick Co.
American National Standards Institute
American Welding Society
Amigo Mobility International, Inc.
Ron Anderson
Arco Automation Systems, Inc.
Artisto
Arthur Baker
Baldor
Bell Laboratories
Boston Gear Div.
Briggs & Stratton Corporation
CADAM
Calcomp, Inc.
Larry Campbell
Carboloy Division, GE Co.
Central Foundry Division, GMC
CH Products
Chambersburg Engineering Co.
Chemcut Corp.
Cincinnati Gear Co.
Cincinnati Milacron
Cities Service Oil Co.
Colorado Department of Highways
Convair Aerospace Div., General Dynamics
Cummins Engine Co., Inc.
John Deere & Company
Diazit
DLoG-Remex
Dow Chemical U.S.A.
DP Technology, Inc.
DuPont
Durez Div.
Eastman Kodak Co.
Eaton Corp.
Eberhard Faber, Inc.
ELCO Industries, Inc.
Epson American, Inc.
Esna Corp.
Eugene Dietzgen Co.
Federal Products Co.
Ferguson Machine Co.
FHA
Fisher Body Division, GMC
Ford Motor Company
Formatt
Freightliner/Heil

Garlinghouse, Inc.
Garrett-AiResearch Casting Div.
GE Plastics, Inc.
General Dynamics, Engineering Dept.
General Motors Corp.
General Motors Drafting Standards
General Motors Engineering Standards
Gerber Scientific, Inc.
The Gillette Company
The Gleason Works
Globe Engineering Documentation Services, IBM
B. F. Goodrich Chemical Co.
Goodyear Aerospace Corp.
Gramercy
Groov-Pin Corp.
Grumman Aerospace Corp.
Ken Hawk
Hearlihy & Co.
Hewlett-Packard Co.
Honeywell, Inc.
Houston Instrument, A Summagraphics Company
Hyster Company
IBM
Intergraph Corporation
International Aero Service Corp.
International Harvester Co.
Iomega
ITT Harper, Inc.
Chet Johnson
Kearney & Trecker Corp.
William Kemeny
Keuffel & Esser Co.
Jack Klasey
Klein Tools, Inc.
Koh-I-Noor Rapidograph, Inc.
Kohler Co.
Kubota, Ltd.
Kurta Corporation
Landcadd International, Inc.
LeBlond Makino
Lincoln Electric
Lockheed Aircraft Corp.
Macbeth
Mack Trucks, Inc.
Martin-Marietta Corporation
Mazak Corporation
Mikrosa
Milwaukee Gear
3M Co.
Mitsubishi

Continued

Modulux Div.
Motoman
Motorola, Inc.
NASA
Norton Co.
Paasche Airbrush Co.
Panama Canal Commission
Parametric Technology Corporation
Polaroid
Proctor & Gamble Co.
RapiDesign
Raytheon Co.
Rockwell International Corp.
Rohm and Haas
Scale Models Unlimited
Seagate Technology
SEATCASE,Inc.
James Shaw
SI Handling Systems, Inc.
SoftSource, Inc.
Sperry Flight Systems Div.
Sporlan Valve Co.
Standard Oil of California
Standard Pressed Steel Co.
Stanley Motor Carriage Company
Summagraphics
Superior Electric Co.
Swanson Analysis Systems, Inc.

Tallgrass Technologies Corporation
Tecnomatix
Teledyne Post
Teledyne Rotolite
Trumpf, Inc.
Union Carbide Corp.
United Aircraft Corp.
United States Patent Office, Stanley Motor Carriage
    Company
United States Pipe and Foundry Co.
Urban Investment and Development Co.
U.S. Department of the Interior
U.S. Department of the Interior, Bureau of
    Reclamation
U.S. Department of Transportation
VEMCO
VersaCad
Virginia Dept. of Highways
Western Gear Corp.
Clay Willits
Wilson Instrument Div. ACCO
Winnebago
Wisconsin Department of Transportation
Wood-Regan Instrument Co.
Xerox
Lyle Zeigler
Zenith Data Systems

The publisher would also like to extend a special thanks to the companies Arthur Baker, VEMCO, Artisto, and
Hearlihy & Co. for providing equipment for additional illustrations.

Drafting encompasses many different areas. CADD is an important part of drafting and allows the drafter much versatility. (IBM)

Part I
Drafting Fundamentals

Drafting fundamentals covers all of the basic skills required to produce technical drawings. This part of the text covers *Drafting as a Communication, Problem Solving, and Design Tool; Using Drafting Tools and Systems; Technical Sketching for Communication; Technical Lettering; Basic Geometry Used in Drafting;* and *Advanced Geometry Used in Drafting.*

Drafting is the graphic language used by industry to communicate ideas and plans from the creative-design stage, through production, to service and use. Anyone associated with the processes of creation, production, or use of products will benefit from the study of technical drafting.

The purpose of this text is to aid you in a systematic study of drafting. Whether your goal is to develop an understanding of drafting in a technical world, or enter one of the career fields related to drafting, you will find this an interesting and challenging area of study.

The industrial drawing represents the most effective way to communicate ideas about complex mechanisms and assemblies.

An organized work area is the first step to efficient drafting.

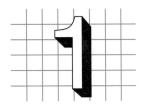

Drafting as a Communication, Problem Solving, and Design Tool

KEY CONCEPTS

- [ ] Sketches serve as a communication tool.
- [ ] Sketches are different from technical drawings.
- [ ] Presentation drawings are used to communicate concepts and ideas.
- [ ] The design problem solving method can be used to solve a variety of problems.
- [ ] There are four steps in design and problem solving: problem definition, preliminary solutions, preliminary solution refinement, and decision and implementation.
- [ ] Models are used in industry to help visualize the final product.
- [ ] Design groups can be used very effectively to solve problems.

Never in history has the world been so technically oriented. With so many new developments in industry, science, medicine, technology, and space, the universal language of drafting has become essential in solving problems, creating designs, and communicating ideas to others, Fig. 1-1.

Drafting is the process of creating technical drawings. A technical drawing is a *graphic repre-sentation* of a real thing–an idea, an object, a process, or a system, Fig. 1-2. Graphic representation has evolved along two related, but separate, directions according to purpose: artistic and technical.

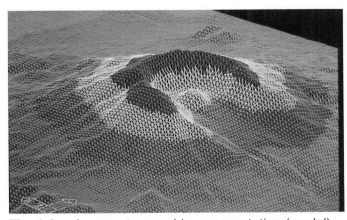

Fig. 1-2. A computer graphic representation (model) can create a clear picture of actual terrain. (Landcadd International, Inc.)

Artists have long expressed their ideas through drawings. Drawings allow abstract concepts to be communicated in ways that people can understand. Frequently, pictures are better understood than words. Historically, images were preserved through pictures or drawings. There were no photographs or videos until relatively recently.

From the earliest times, technical drawings were used as the method of representing the design of objects to be made or constructed. Evidence that drawings existed in ancient times can be seen in the ruins of complex structures such as aqueducts, bridges, fortresses, pyramids, and palaces. These structures could not have been built without technical drawings to serve as a guide for the work. Over time, the "universal graphic language" evolved. Today, the basic principles of this universal language are known throughout the world. Even though people around the world *speak* different languages, the graphic language has remained common. Graphic language is a basic and natural form of communication that is universal and timeless.

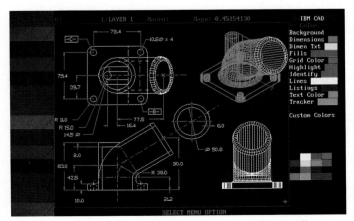

Fig. 1-1. A technical drawing that correctly applies the rules and conventions of the graphic language is the most efficient method of communication between designers, engineers, and technologists. (Altium)

The technology for making technical drawings has advanced far beyond hand sketching and typical mechanical techniques. However, the knowledge of the basic concepts of representing objects graphically still provides the foundation for clear communication. These concepts or principles are well-established and must be applied whether drawing by hand or using a state-of-the-art CADD (Computer-assisted Design and Drafting) system. Without a command of the graphic language, designers, engineers, and technologists would not be able to adequately communicate their ideas to other professionals, Fig. 1-3. All people involved in the design, manufacture, construction, installation, or repair of products need some technical knowledge of this universal language.

DRAFTING AS A COMMUNICATION TOOL

The primary purpose of drafting is to communicate an idea, plan, or object to some other person, Fig. 1-4. However, this is not the only reason for making technical drawings. The process also serves as a problem solving experience, a design tool, and a method of recording acceptable solutions. However, drafting is a basic form of communication and the drafter should always remember that the drawings produced are for someone else. Drawings must be clear and specific. They must follow accepted rules and utilize standard symbols and conventions. Standards developed by the American National Standards

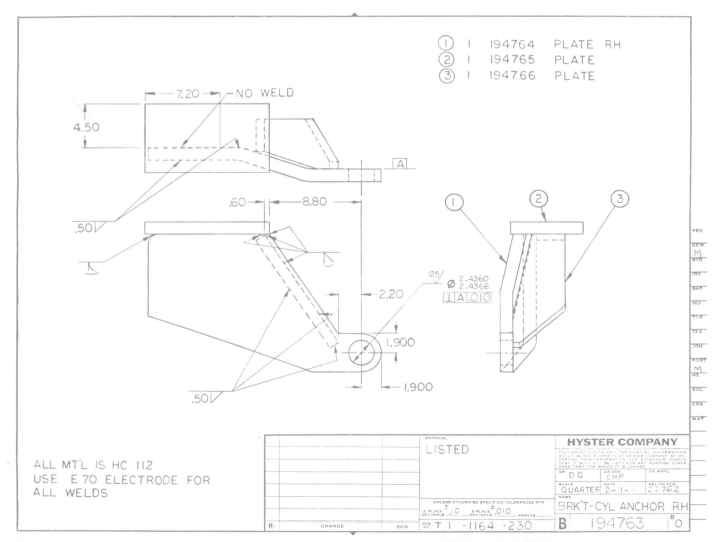

Fig. 1-3.   Drawings are used by many professionals in the manufacturing process. (Hyster Company)

Fig. 1-4. Technical drafting techniques enable ideas to be communicated and expressed in an understandable, efficient, and accurate manner. (Standard Oil of California)

Institute (ANSI) serve as the basic guide for technical drafting. Some large companies have adopted standards that suit their own needs. These company standards may not conform in every respect to the ANSI standards. However, most companies do follow ANSI standards.

Sketches

One form of technical drawing is called freehand sketches. *Freehand sketches* are very useful in communicating undeveloped ideas, Fig. 1-5. A product usually begins as an idea in the mind. The form, dimensions, and details of the product are not yet clear, but the basic concept is there. By sketching the mental image on paper, the idea can be communicated, manipulated, and refined. Others can participate in the process by adding their ideas, or the basic

idea can be saved for future reference. This is the creative phase of technical communication.

Sketches may be detailed or schematic in form. The type of sketch will depend on the nature of the idea and degree of visualization associated with the idea. However, the sketch will provide a quick image of the concept "contained in the head" of the designer when words alone cannot describe something new or unfamiliar, Fig. 1-6.

Mechanical or Computer-generated Drawings

Mechanical or *computer-generated* drawings are other forms of technical drawing. These drawings show an idea or product in a more refined or improved state than sketches. These drawings represent ideas that have moved from the idea or

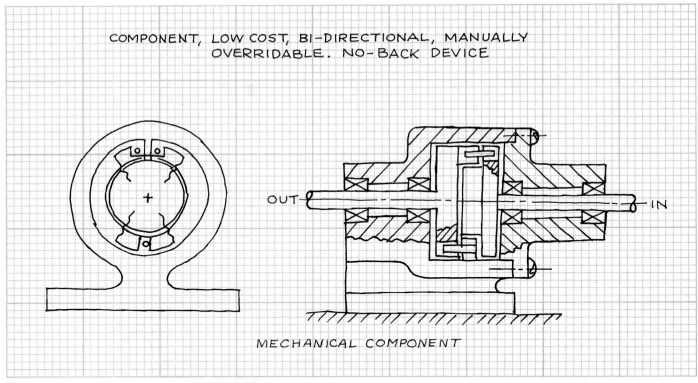

COMPONENT, LOW COST, BI-DIRECTIONAL, MANUALLY OVERRIDABLE. NO-BACK DEVICE

OUT — IN

MECHANICAL COMPONENT

Fig. 1-5. Design engineers use preliminary sketches to get their ideas on paper. (Sperry Flight Systems Div.)

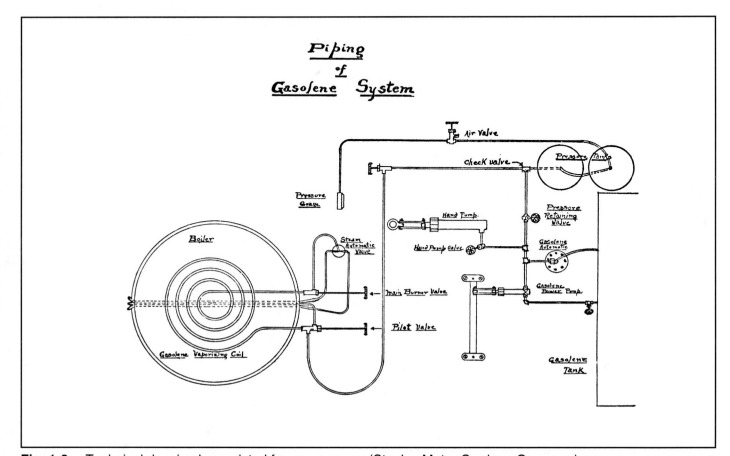

Piping of Gasolene System

Fig. 1-6. Technical drawing has existed for many years. (Stanley Motor Carriage Company)

conceptual stage to a more practical solution. Size, shape, and form have been developed to provide a scaled representation of the object.

Set of Drawings

A third use of technical drawings is called **manufacturing** or **construction communication.** This is not a specific drawing, but rather a group of drawings that communicate the complete product or idea. At this level, the idea has been refined still further from a sketch. Evaluation of the design has been completed, details have been added, and costs figured. Instructions for the production of the item are included. This form of communication is usually referred to as a **set of drawings.** A set of drawings generally contains all of the information required for production, Fig. 1-7.

Presentation Communication

Another use of technical drawing is called **presentation communication,** Fig. 1-8. Like manufacturing communication, presentation communication is not a specific drawing but rather several drawings and other information combined together. Drawings and technical descriptions of products are usually combined with other forms of communication for presentation to the prospective buyer, client, financier, or management. Verbal descriptions of the basic features or specifications are accompanied by pictorial drawings of the product. The purpose of presentation communication is typically to secure acceptance of the proposed product. The goal is to provide a true-to-life image of the finished object that highlights the primary features and specifications.

Each technical drawing, from a sketch to a pictorial, plays a significant role in the development,

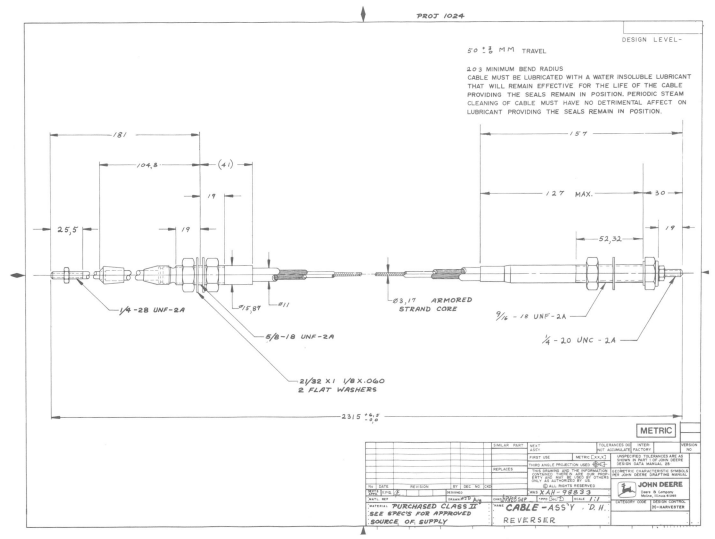

Fig. 1-7. A detail working drawing provides complete information necessary to fabricate, finish, and inspect the part. (John Deere & Company)

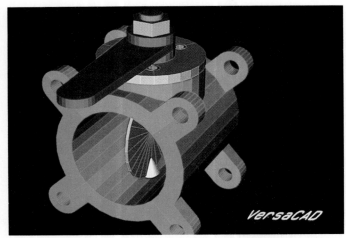

Fig. 1-8. A computer-generated presentation drawing can be used to communicate the physical features of an object. (VersaCad)

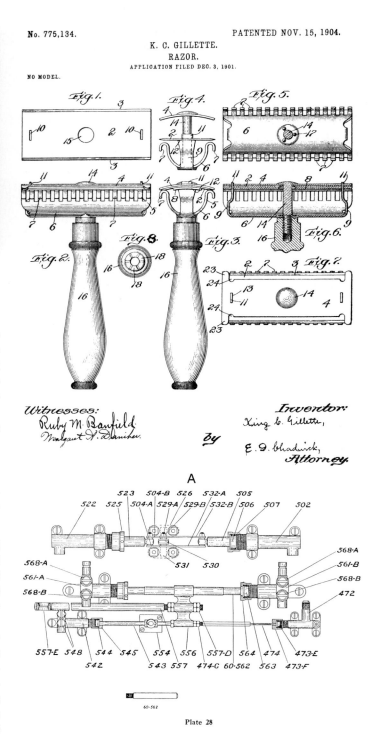

A

B

Fig. 1-9. A–Patent drawings are a form of technical communication. B–Technical literature drawing is useful in identifying the various parts of an object. (United States Patent Office, Stanley Motor Carriage Company)

manufacture, and sale of products. Other specialized or unique forms of technical drawings are sometimes needed as well. For example, patent drawings, reference drawings, advertising drawings, technical literature drawings, and maintenance and repair drawings may also be required, Fig. 1-9. However, the main purpose of all of these drawings is to communicate ideas.

DRAFTING AS A PROBLEM-SOLVING TOOL

Problem solving is the process of seeking practical solutions to a problem. The process varies depending on the type and complexity of the problem that is to be solved. A useful way of thinking about the types of problems that may be confronted is to classify them from well-structured to non-structured.

The most **well-structured problems** are generally rather narrow in scope and have only one correct answer. The process of arriving at a solution for a well-structured problem is usually accomplished through **convergent thinking.** This is where the situation is examined to arrive at the one best solution. **Algorithms** (mathematical equations) are useful in solving well-structured problems. For example, if a person were confronted with a problem of selecting the proper size bearing for a particular shaft, the solution could be found by simply using standard specifications or tables. These specifications or tables were created using algorithms.

The process of making technical drawings involves the use of standards, charts, tables, and references. Experience in solving well-structured problems helps to develop the skills of a problem solver. Know-

ing where to look for the answers to technical questions is a must in learning the graphic language.

Semi-structured problems are more complex than well-structured ones. They may have more than one correct answer and are solved by heuristics. **Heuristics** are guidelines that usually lead to acceptable solutions. You use heuristics in your everyday life without really thinking much about it. An example might be to perform a set of tasks in order of importance. Another might be to perform a group of manufacturing processes in a logical successive order–locate, drill, ream. Solving semi-structured problems requires the identification of the factors that bear on the possible solutions.

In technical drafting, many types of semi-structured problems need to be solved when converting an idea from a sketch to a set of drawings. Decisions on the strength and type of materials, function and use of a part, operations required to manufacture the part, and cost of different manufacturing processes are all part of the drafting and design of products.

Beyond semi-structured problems are **non-structured problems.** These are problems that have a variety of potentially correct solutions. Non-structured problems differ from semi-structured problems in that they require **divergent thinking.** Divergent thinking is examining a very specific problem to come up with several correct solutions. For example, creating a better method of controlling the speed of a motor is a problem that may have many acceptable solutions. However, this type of problem is seldom solved by referring to an algorithm or applying heuristics to produce an acceptable solution. **Creative problem solving** is required to provide the solutions to these types of problems.

Problems that logically fall within the category of non-structured problems clearly are dealt with by designers, engineers, and technologists. Experience in solving more structured problems helps to prepare professionals to solve the more creative problems, but will not necessarily mean success. However, ability to use the graphic language will aid in communicating the creative process of the mind. It will also aid in encouraging others to become involved in the thinking process and thereby help move toward a creative solution.

Problem solving is a process of seeking realistic solutions to a problem. Well-structured problems may be solved with algorithms and convergent thinking. Semi-structured problems may be solved through the use of heuristics, or guidelines. Divergent thinking and creative problem solving techniques are required to solve non-structured problems. Ability to solve all types of problems is necessary for success as a drafter or designer.

DRAFTING AS A DESIGN TOOL

Engineers, industrial designers, and drafters make extensive use of "the design method" in arriving at a final solution to a design problem. An example of a final solution used is shown in Fig. 1-10. From basic sketches come the machines and products of our technological age.

Space vehicles such as the lunar and Mars "Rovers" were first created by designers using the design method. The various components of these items were developed through the use of sketches, Fig. 1-11. Then a prototype of the lunar vehicle was constructed and tested on earth. In Fig. 1-12, a lunar Rover is shown on the moon's surface during the third lunar mission.

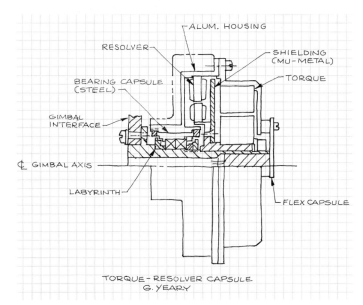

Fig. 1-10. Sketching is one of the most useful techniques to a designer in "thinking through" a solution to a problem.

Fig. 1-11. An artist's rendition of a Mars "Rover" helps communicate the idea to the public. (NASA)

Fig. 1-12. The design method was used extensively in the development of the lunar "Rover." (NASA)

What is the Design Method?

Many of the problems you will solve as a student of design drafting will be presented to you in a clearcut fashion and the solutions apparent, for example, supplying the third view when two views are given. All that is left for you to do is solve for the missing view.

While these experiences are necessary in learning the basics of drafting and other subjects, there are other problems you will need to solve that are not clearly identified. The design for a way of carrying the family camping equipment on a camping trip is a problem that does not have a clear solution. Solving this problem is an example of *creative problem solving.*

Definition of Terms

Some of the terms used in design and problem solving activities are as follows:

- **Problem:** A situation, question, or matter requiring choices and action for solution.
- **Creativity:** The ability to develop original or unique ideas that could solve, or contribute to the solution, of a problem.
- **Design:** As used in this text, design is the result of creative imagination that forms ideas as preliminary solutions to problems. Evaluation of these ideas provides functional solutions to problems. Technical design is not necessarily void of art and beauty, but the emphasis is on inventive and ingenious ways of solving technical problems.

An example of design problem solving is the bicycle cable lock designed by a group of students, Fig. 1-13. The hardened flexible cable, stored on a spring-loaded reel inside the housing, is pulled out and around the bicycle rack, pole, or tree. The end snaps into a lock contained in the housing.

- **Problem Solving Method:** The procedure of systematically approaching a problem and arriving at a solution.

Fig. 1-13. A bicycle locking mechanism solved the problem presented to the designers. The lock secures a bicycle while parked and eliminates the necessity of carrying a chain and padlock. (Ron Anderson, James Shaw, Clay Willits, and Lyle Zeigler)

Steps in Design Problem Solving

In order to systematically approach the solution to a problem, it is necessary to break the solution into four basic steps. These steps are listed and explained here to provide a clear understanding of their nature and use. For this explanation, the example of creating a way to carry a family's camping equipment will be used.

Step 1: Problem definition

Before much progress can be made toward the solution of a problem, the problem must be clearly defined. This step is divided into four parts:

Statement. The statement of the problem should be clear. The statement should describe the need that is the basis of the problem. "Design a device for carrying the family camping equipment." Note that there are no stated requirements or limitations such as quantity or size.

Requirements. The requirements of the problem should be listed. Requirements include needed and desired features. "The space provided must be outside the passenger seating space and trunk space of the car. Space is needed for the equipment of five persons–bed rolls, cooking utensils, dishes, food supply, hiking equipment, and recreational equipment."

Limitations or **Restrictions.** The limitations or restrictions on the design should be specified. In this case, two things that might need to be considered are limiting physical factors (size, weight, and materials) and limiting monetary factors. "The space provided should have easy access, protect the contents from weather and road dust, and cause only minimal drag on the car while driving. Total cost should be kept to a minimum."

Research. The research of the problem should further aid in defining the problem. Information must be gathered relative to the number and size of items to be included and what weather protection is necessary. Questions need to be asked that will reveal the various requirements of the problem statement. "How many bed rolls and what space is required? What cooking utensils, equipment, and dishes are needed and what space is required? How much space will be required for food supplies? What recreational equipment is to be taken and what space requirements are needed?"

Answers for these questions are obtained from previous experience, actual measurements, research in technical publications, and by discussions with knowledgeable persons. "Five bed rolls, 15 inches in diameter by 20 inches in length; cooking utensils, equipment, and dishes approximately 8 cubic feet; food supply approximately 4 cubic feet; and, recreational equipment approximately 3 cubic feet with a length of at least 36 inches."

Step 2: Preliminary solutions

The second step in the design problem solving process is the most creative and will be influenced by the backgrounds (experiences and personal knowledge) of the designers. If the designers have had considerable experience with similar problems, and if the problem has been carefully researched, they will be able to develop more creative and useful solutions.

There are two methods by which this step may be developed–individual methods and group methods. An individual may work alone and list all of the solutions that come to mind, no matter how "far-out" they may seem. The designer should try to think of unique uses of existing items for possible solutions as well as standard or customary ways of solving the problem.

Small groups (up to 10 or 12 individuals) can also work effectively in the above manner. **Brainstorming** is a means of people working together for creative solutions to problems. Each person makes suggestions as they come to mind. New and unusual solutions often come forward as various individuals are stimulated by the suggestions of others.

Some preliminary solutions which may be offered for the family camping equipment carrier are:

- Trailer.
- Waterproof box.
- Plastic pouch.
- Device mounted on the trunk lid.
- Car-top carrier.
- Carrier fastened to front bumper.
- Pouches mounted on the sides of the car.
- Carrier on each front fender.
- Device fastened to rear bumper and free of trunk lid.

It is important to get as many preliminary solutions as possible. No attempt should be made during brainstorming to evaluate the various solutions suggested. Evaluation takes place in the next step.

Step 3: Preliminary solution refinement

In this step, all preliminary ideas are combined or resolved into as few solutions as possible. The better preliminary solutions are reviewed and evaluated in terms of the problem definition. Rough sketches are made to further analyze each solution. Those ideas that do not show promise are eliminated. The remaining preliminary solutions are refined and analyzed until only three or four are left. These remaining ideas are evaluated in terms of the general problem statement.

"Design a device for carrying the family camping equipment. The device must be located outside the car, have easy access, protect contents against the weather, provide minimal drag on car, and cost is a factor."

When all factors are considered, the following three preliminary solutions may be selected for final consideration: a trailer, a car-top carrier, and a device fastened to the rear bumper free of the trunk lid, Fig. 1-14.

Step 4: Decision and implementation

When the best preliminary solutions have been selected, a decision chart is prepared showing comparison of the solutions on the main requirements and limitations of the problem, Fig. 1-15. A rating system of weighing each factor for comparison on a scale of 1 to 3 may be used, with 1 being the best rating. Where two or more preliminary solutions seem equally desirable on a particular factor of comparison, an equal rating should be given. The car-top carrier solution may be given the best rating of the three preliminary solutions considered. This solution is implemented by preparing a working drawing, and building a model and/or prototype of the design solution, Fig. 1-16.

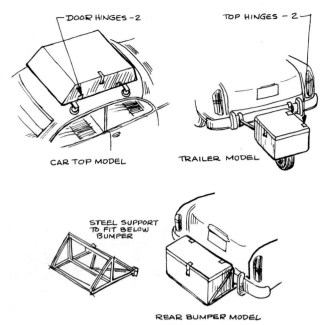

Fig. 1-14. Rough sketches of preliminary solutions are the first step in solving the problem.

MODELS

The use of models in the design process is increasing in industry, Fig. 1-17. Two important ad-

DECISION CHART

FACTORS OF COMPARISON	RATINGS OF BEST PRELIMINARY SOLUTIONS		
	NO. 1 TRAILER	NO. 2 CAR-TOP	NO. 3 REAR BUMPER
Located Outside Car	1	1	1
Easy Access	1	2	1
Weather Protect	2	1	2
Drag on Car	3	1	2
Cost	3	1	2
Difficulty of Construction	2	1	3
RATING TOTALS	12	7	11

Fig. 1-15. A decision chart used in rating preliminary solutions to problems helps narrow the possible solutions.

vantages in using models are improved communication between technical personnel and greater visualization of problems and their solution by nontechnical and management personnel, Fig. 1-18.

Models are also used extensively in the presentation and promotion of a product or design solution. In addition, they are useful in training

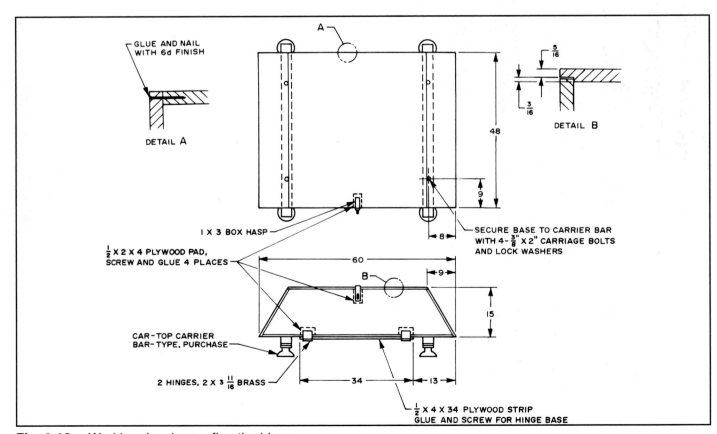

Fig. 1-16. Working drawings refine the ideas.

programs for personnel who will use the equipment. The term **model** is used to refer to three-dimensional scale replicas, mock-ups, and prototypes.

Scaled Models

A **scaled model** is a replica of the actual, or proposed, object. It is made smaller or larger to show proper proportion, relative size of parts, and general overall appearance. Some scaled models are working models used to aid engineers and designers in their analysis of the function and value of certain design features, Fig. 1-19. The size, or **scale,** of models may vary depending on the size of the actual object and the purpose of the model. Construction of models in industry is the work of professional model makers, Fig. 1-20. Materials used for model making include balsa wood, clay, plaster, and aluminum. Standard parts such as wheels, scaled furniture, machine equipment, decals, and simulated construction materials are available from model supply dealers.

Mock-ups

A **mock-up** is a full-size model that simulates an actual machine or part, Fig. 1-21. The full-size mock-up presents a more realistic appearance than a scale model and aids in checking design appearance and function. The mock-up may be used as a simulator for training purposes, but it is not an operational model.

Fig. 1-17. Model makers use clay models to describe new automobile designs. (General Motors Corp.)

Fig. 1-18. A design model helps management and others to "see" the final product. (Dow Chemical U.S.A.)

Fig. 1-19. A working model of the Glen Canyon Dam water spillway outlets helps show the flow of water. (U.S. Department of the Interior, Bureau of Reclamation)

Fig. 1-20. Professional model makers construct full-scale models to help future products be visualized. (General Motors Corp.)

Fig. 1-21. Full-size mock-ups appear the same size as the final product. Everything that will be in the final product will also be in the mock-up. However, a mock-up is not functional. (Scale Models Unlimited)

Prototype

The *prototype* is a full-size operating model of the actual object, Fig. 1-22. It is usually the first full-size working model that has been constructed by craftsmen making each part individually. Its purpose is to correct design and operational flaws before starting mass production of the object.

Virtual Reality

A very modern, experimental type of model is called *virtual reality.* Fig. 1-23 shows a Lockheed

Fig. 1-22. Computers can be used to design products. Some computer systems can be used to "test run" a product. (The Gillette Company)

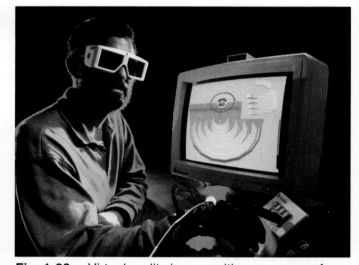

Fig. 1-23. Virtual reality is an exciting new area of modeling which holds promise for a broad range of applications. (Lockheed Aircraft Corp.)

Advanced Computing Laboratory programmer engaged in research in virtual reality to help analysts understand three-dimensional acoustic data for antisubmarine warfare applications. The goggles transform computer display images into three dimensions while the data glove translates the position and shape of his right hand into the graphics on the screen. This allows the operator to "reach into" the screen and directly modify parameters while watching in three dimensions.

Stereolithography

Design prototypes of complex parts can be produced in a few hours by a new technology called *stereolithography.* The process produces a hard plastic model that can be studied to determine whether or not changes need to be made to the part.

Design data that are created on a CADD system supply the necessary information to guide the stereolithography machine. This machine operates similar to a CNC machine tool. A low-power laser beam is used to harden a liquid photo-curable polymer plastic in the shape defined by the computer. This is accomplished by curing thin layers of the polymer to form the desired shape. Construction is begun at the bottom of the object and progresses layer by layer until the object is completed. Ultraviolet curing is required to complete the process.

PROBLEMS AND ACTIVITIES

1. Collect examples of drawings that represent as many of the following types or purposes as possible. Be prepared to discuss with the rest of the class why you believe each drawing represents the category you designated.
    A. Idea sketch.
    B. Technical drawing.
    C. Manufacturing or construction drawing.
    D. Presentation drawing.
    E. Reference drawing.
    F. Patent drawing.
    G. Advertising drawing.
    H. Service or maintenance drawing.

2. Make a list of several problems for each of the following categories:
    A. Well-structured problems.
    B. Semi-structured problems.
    C. Nonstructured problems.

3. List as many useful items as you can that can be made from each of the following items. Let your imagination be your guide. Include all items regardless of practicality for production.
    A. Bale of straw.
    B. One square yard of heavy canvas.
    C. Piece of white pine lumber 1″ x 10″ x 12′.
    D. Sheet of vinyl plastic .005 inch in thickness and 10 feet square.
    E. Unfinished interior flush panel door 2′6″ x 6′8″.
    F. Empty five-gallon bucket.
    G. An automobile radiator core.
    H. Sheet of window glass 30″ x 40″.
    I. Rubber garden hose 50 feet long.
    J. Several glass bottles.

4. Creative thinking is a matter of developing the mind to be resourceful and imaginative. Test your creativity by listing as many different ways you can think of to solve the following:
    A. Paint a board 2″ x 10″ x 10″ on all surfaces.
    B. Transfer a gallon of water to another location 10 feet away and 45° above its present location.
    C. Make a 1 inch square hole in the side of a tin can.
    D. Design a clamping device for gluing wood edge-to-edge without the use of a threaded piece.

5. Make a list of six or more items not currently available on the market that you think have a market or use. Be sure the items can be produced by your class, using materials available to you commercially and using equipment at school or at home.

6. Select one item from Problem 5 and try to interest 3 to 5 members of your class in the design and production of the item. With the approval of your instructor, follow the project through the design stage and development of the prototype. Follow the steps outlined in this chapter on the design method and evaluate the design and production problems. Establish a selling price for which you could manufacture the item and make a reasonable profit.

7. Select any one of the following problems and develop a solution for it. Make use of the design method of problem solving outlined in this chapter to solve the problem. Report your work and activities in writing through each of the steps. Develop sketches for at least two of the best possible solutions and a dimensioned instrument drawing for the proposed final solution.
    A. Portable shoe shine equipment with folding seat and storage space for tools and supplies.
    B. A system for separating gravel (rock) into various sizes such as 1/2″, 1″, 1 1/2″, and 2″.
    C. A study area that provides space and lighting for writing, reading, and storage for materials commonly used in the area.
    D. Means of utilizing discarded plastic containers.

8. One of the most difficult types of problems to solve is one that calls for new uses or unique applications of commonly used items in ways they have not been used before. For example, a small gasoline engine is commonly used as a power unit for such items as lawn mowers, motor bikes, snow blowers, and tree saws.

How many ideas can you suggest for the small gasoline engine that would be useful and yet unique? Test your imagination and list some possible uses. You may add other pieces of equipment or material as needed.

9. Try the same approach used in Problem 8 with the following:
   A. An electric motor from a cordless toothbrush, shaver, or hedge trimmer.
   B. A spare tire and wheel from an automobile and a tire pump.
   C. Bicycle wheels without tires or inner tubes.
   D. The parts from several radios.

10. Ask your instructor, a member of your family, or an acquaintance to help you get in touch with an engineer or industrial designer. Make an appointment for an interview to find out all you can about the nature of the work and the design methods used to solve problems. Ask to be shown sketches used in the solution of problems. Report your findings to your class.

11. Select one or more famous creative inventors or designers and read their biography. See if you can identify things that caused them to develop their creativity. Try to sum up these characteristics in two or three statements that could serve as a guide to others who want to develop their creative abilities. Report your findings to your class.

12. Review the local newspaper for news articles relating to a school or community need that could be solved using the design method. Take one such problem and develop a solution by applying the design method. Prepare a written report of each step in development of a solution.

2 Using Drafting Tools and Systems

KEY CONCEPTS

☐ Basic tools used by drafters to complete mechanical drawings include pencils, pens, erasers, templates, and compasses.

☐ Drawings are created to scale, or in proportion to the actual object.

☐ Industrial drawings use the "alphabet of lines."

☐ There are several basic elements to a CADD system, including software and hardware components.

The purpose of an engineering or technical drawing is to convey information about a part or assembly as clearly and simply as possible. The process of making technical drawings using instruments, templates, scales, and other mechanical equipment is called *instrument drafting.* Electronic or automatic drafting is accomplished using a *computer-assisted drafting and design (CADD) system.*

Because of the need for accuracy in design and clarity in communicating necessary information, most drawings used in technical work are instrument drawings, Fig. 2-1. Today, the production and servicing of industrial components requires careful preparation of drawings.

The purpose of this chapter is to enable you to gain a knowledge of basic drawing instruments and to instruct you in their proper use and care. After you have become familiar with these basic "tools," you can easily acquire skill with more specialized instruments used in drafting. Some of the instruments needed for technical drawing are shown in Fig. 2-2.

DRAFTING EQUIPMENT

Some schools furnish drafting equipment for their students. Others require that students purchase their own. Good instruments are expensive, but it is wise to invest in quality items. Unless you are familiar with drafting equipment, get the advice of your instructor before making a purchase. The equipment and supplies listed below are adequate for most drafting work.

Instruments

- Small compass with pen attachment
- Large compass with pen attachment
- Lengthening bar
- Friction divider
- Lead holder
- Box of leads
- Ruling pen
- Screwdriver

Other Equipment

- Drawing board (approximately 18″ x 24″)
- T-square 24″ plastic edge
- 30-60 degree triangle–10″
- 45 degree triangle–8″
- Architect's and engineer's scales
- Lettering instrument, Ames or Braddock
- Protractor
- Irregular or French curve
- Circle and ellipse templates
- Erasing shield
- Soft rubber eraser
- Vinyl eraser
- Dusting brush
- Sketch pad

Fig. 2-1. An engineering design drafter makes use of numerous drafting instruments and machines.

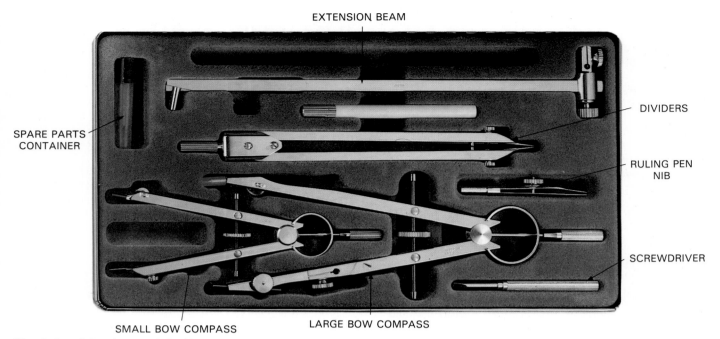

EXTENSION BEAM

DIVIDERS

SPARE PARTS
CONTAINER

RULING PEN
NIB

SCREWDRIVER

SMALL BOW COMPASS   LARGE BOW COMPASS

Fig. 2-2.   A basic set of drafting instruments. (Teledyne Post)

- Drawing paper, tracing paper, or vellum
- Drafting tape
- Drafting pencils or mechanical lead holder
- Sandpaper pad, file, or pencil pointer

Drawing Boards

Drawing boards are usually made of selected softwoods with a smooth, firm surface. The ends are usually cleated to prevent warping, Fig. 2-3. Care should be exercised to avoid marring the surface.

Also available is a vinyl board-top material with excellent resiliency. It mounts over a regular drawing board or drafting table top to improve the drawing surface.

In addition to a smooth surface, the drawing board must have a straight working edge to serve as a base for the T-square and a reference line for drawing. Check the working edge for trueness with a framing square or other true, straight-line tool, Fig. 2-4. Correct a wooden edge that is not true by planing with a hand plane or jointer.

Most drafting tables come with a drafting board surface, eliminating the need for a separate board. Generally, these tables are vinyl covered.

T-Square

The *T-square* consists of two parts, a head and a blade fastened together rigidly, Fig. 2-5.

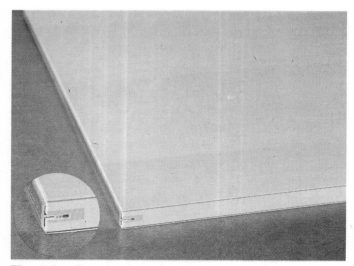

Fig. 2-3.   Drawing boards may have wood and metal cleated ends to restrain warping.

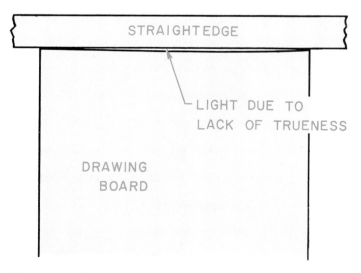

STRAIGHTEDGE

LIGHT DUE TO
LACK OF TRUENESS

DRAWING
BOARD

Fig. 2-4.   Check the working edge of a drawing board for trueness.

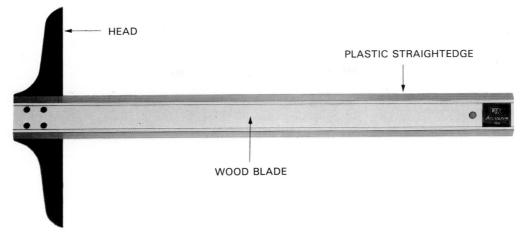

Fig. 2-5. A T-square with clear plastic edges provides a better view of the work. (Teledyne Post)

Some T-squares have transparent plastic edges that are relieved to allow lines to be seen under the edge of the blade. This also avoids the tendency for ink to run under the blade when inking a drawing.

T-squares range in length from 18 to 60 inches in 6 inch intervals. The length should be approximately the same as the width of the drawing board being used.

The three principal uses of the T-square are: (1) to align the paper on the board, (2) serve as a guide in drawing horizontal lines, and (3) serve as a base for triangles. In addition, the T- square is used in conjunction with lettering instruments and various drafting templates.

Occasionally check the T-square for rigidity of the head and for straightness of the blade. If the head is loose, remove the screws and clean off the old glue. Add new glue and firmly reset the screws.

To check the straightness of the blade, mark two points on a sheet of paper at a distance equal to the approximate length of the T-square blade, Fig. 2-6. Join these two points with a sharp line along the edge of the T-square.

Next, revolve the paper 180° so that the points are reversed. Then join the points with a sharp line. If the lines coincide, the blade is true. If they do not, the error is one-half the difference between the lines.

This error can be corrected by lightly sanding the blade at the high points. Use a piece of fine abrasive paper wrapped around a block of wood, and check your progress frequently to insure an accurate edge.

Be careful in handling the T-square so that the edge does not strike other instruments or sharp edges that may dent or nick the straightedge. Never use the T-square as a guide for a knife when cutting material. This could nick the straightedge.

Parallel Straightedge

Drawing boards and tables may be equipped with a ***parallel-ruling straightedge,*** Fig. 2-7. The straightedge is preferred for large drawings and for vertical board work. It is easily manipulated and retains its parallel position.

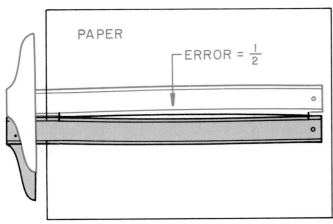

Fig. 2-6. Check the T-square for trueness.

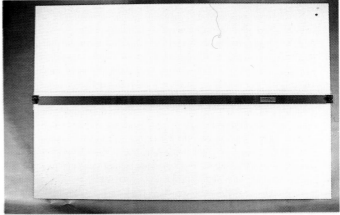

Fig. 2-7. Drafting boards may be equipped with parallel-ruling straightedge.

The straightedge may be moved up or down the board. It is supported at both ends by a cable that operates over a series of pulleys, keeping the straightedge parallel.

Triangles

The most commonly used drafting triangles are **45 degree** and **30-60 degree** types, Fig. 2-8. They are made of transparent plastic, and their size is designated by the height of the triangle. Most popular sizes are the 8 inch for the 45 degree type and the 10 inch for the 30-60 degree. Both may be purchased in sizes from 4 inches to 18 inches.

Special purpose triangles are made to help draw guidelines for lettering, laying out roof pitches, and intricate angular work, Fig. 2-9. An adjustable triangle permits the laying off of any angle from 0° to 90°.

Occasionally check all triangles for nicks by running your fingernail along the edges. Such defects are caused by hitting the triangles against sharp edges, dropping them, or using them as a guide for a knife in cutting or trimming. Minor defects, nicks, or misalignment can be corrected by sanding the edge with fine abrasive paper wrapped around a block of wood.

Test triangles for accuracy by drawing a vertical line. Reverse the triangle and draw a second line, starting at the same point, Fig. 2-10. If the lines coincide, the triangles are true. If not, the error is equal to one-half the distance between the lines.

Drafting Machines

Manual drafting machines have become very popular in industry as well as in many schools, Fig. 2-11. The manual drafting machine combines the functions of the T-square, straightedge, triangles, protractor, and scales into one machine.

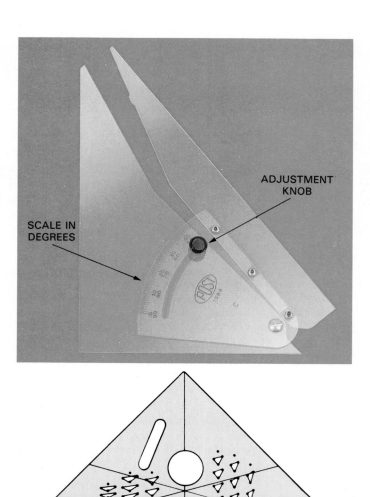

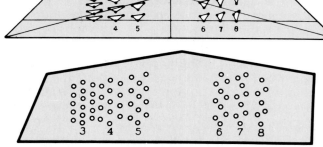

Fig. 2-9. Some triangles are adjustable (above). Some lettering guides have special alignment angles (below).

There are two types of manual drafting machines in use–the **X-Y coordinate** or **track-type** and the **arm-type,** Fig. 2-12. (Refer to Fig. 2-11 for an example of the track-type.) The arm-type is the least expensive. The track-type has the advantage of being more versatile and less troublesome in maintaining accuracy. Both types of drafting machines can be found in industry and in schools.

Essentially, basic operating principles are the same for both types of drafting machines. However, control mechanisms may vary from machine to machine. If an instruction manual is available, review it to become familiar with the controls and adjustments of the machine. If no manual is available, ask your instructor for a demonstration.

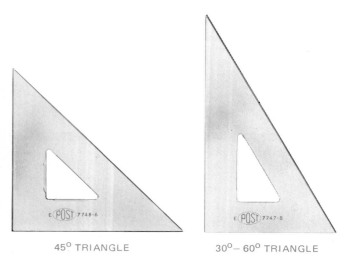

45° TRIANGLE   30°– 60° TRIANGLE

Fig. 2-8.   Transparent plastic triangles are available in many sizes.

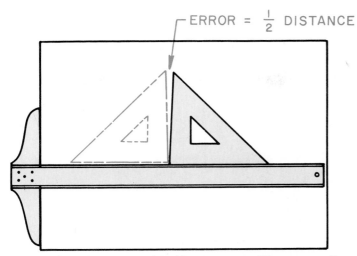

Fig. 2-10. Check triangles for trueness. The error will be 1/2 of the distance between the lines.

Fig. 2-11. A typical industrial track-type drafting machine.

Fig. 2-12. An arm-type drafting machine can be used in many different applications. (Keuffel & Esser Co.)

Alignment of the drafting machine should be checked periodically by comparing results with previously drawn horizontal and vertical lines. When differences occur, check board mounting clamps, scale chuck clamps, and base-line protractor clamp. If these are in order, check the maintenance manual for the machine or check with your instructor.

When using the drafting machine–regardless of make–the following procedures should be observed.

Base-line Alignment

To establish a base line in the desired position:

1. Set protractor indexing mechanism at "zero," then release wing nut or lever securing protractor, Fig. 2-13.
2. Adjust horizontal scale to desired base line. (This may be horizontal or any other angle for necessity or comfort.)
3. Tighten protractor wing nut or screw to "fix" zero setting.

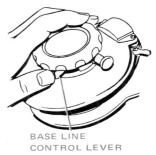

BASE LINE
CONTROL LEVER

Fig. 2-13. Release the base line control lever to align base line.

Vertical Scale Alignment

To align the vertical scale at 90° from the horizontal:

1. Make certain screws holding scale chucks are tight, then firmly insert scales in base plate.
2. With protractor head at zero, draw a reference line along edge of horizontal scale, Fig. 2-14. Hold scale with one hand so pencil pressure does not deflect scale at free end.
3. Index protractor head 90° clockwise to check alignment along reference line. (Refer to Fig. 2-14.)
4. If necessary, align scale by releasing scale chuck clamping screws, adjusting scale, and tightening screws.

Reading the Vernier Scale

To read or set an angle on the vernier scale:

1. Read number of full degrees vernier zero line shows on protractor. In Fig. 2-15, the reading is 12° plus a fraction more counterclockwise from zero.
2. Locate line on counterclockwise side of vernier that aligns exactly with a line on protractor. In Fig. 2-15, the 15' (minute) line aligns.

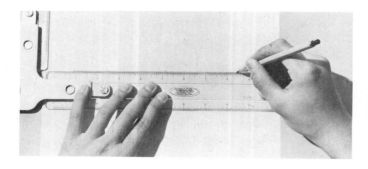

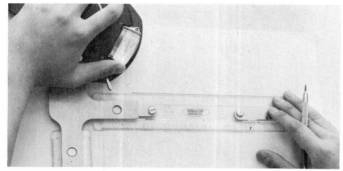

Fig. 2-14. Drawing reference line with head set at zero (above). Checking alignment with head set at 90° (below). (VEMCO)

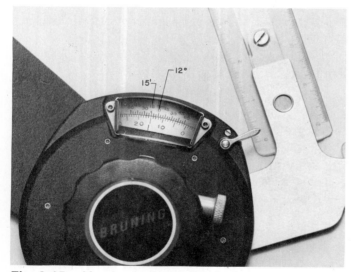

Fig. 2-15. Vernier scale reading in degrees and minutes above horizontal (counterclockwise).

3. The reading of the angle in Fig. 2-15 is 12° 15'.

4. To read a negative angle (below protractor zero), read number of full degrees between zero on the protractor and zero on the vernier. Then locate line clockwise from zero on vernier that exactly aligns with a line on the protractor. Note that this is 5° 35' in Fig. 2-16.

5. To set scales for a certain angle, release protractor clamp and revolve scales until protractor head and vernier align with required number of degrees and minutes.

Fig. 2-16. Vernier scale reading on negative (below horizontal or clockwise) side of protractor zero.

DRAFTING MEDIA

There are several types of *drafting media* (materials) available for use in the preparation of drawings or tracings. These may be classified in three groups—*paper, cloth,* and *film.* They differ in qualities of strength, translucency, erasability, permanence, and stability. The characteristics of each are covered here to assist you in selecting the proper material for a particular drawing.

Paper

Industrial drafting practices have changed in recent years. Now the use of opaque drawing papers is limited to high grade permanent papers for the preparation of maps and master drawings intended to be photographed. Less expensive papers are used for beginning drafting classes in schools. Even here, however, the tendency is to use inexpensive types of translucent materials.

Opaque drawing papers are available in cream, light green, blue tint, and white. White papers are preferable for drawings to be photographed later. Light colored papers reduce eye strain and are less likely to soil.

Media size is important too. Two series of drawing sheet sizes are recognized as standard. The sizes are 8 1/2 x 11 and 9 x 12 inches and multiples of each, designated by letter sizes, Fig. 2-17A and B. Drawing sheet sizes in the metric system are shown Fig. 2-17C.

Tracing Paper

Tracing paper is a translucent drawing paper. Its name was derived from the practice of first making

ENGLISH SYSTEM

8 1/2 × 11 MULTIPLES		9 × 12 MULTIPLES	
LETTER DESIGNATION	SHEET SIZE	LETTER DESIGNATION	SHEET SIZE
A	8 1/2 × 11	A	9 × 12
B	11 × 17	B	12 × 18
C	17 × 22	C	18 × 24
D	22 × 34	D	24 × 36
E	34 × 44	E	36 × 48

A   B

METRIC SYSTEM

DESIGNATION	MILLIMETERS	INCHES
A4	210 × 297	8.27 × 11.69
A3	297 × 420	11.69 × 16.54
A2	420 × 594	16.54 × 23.39
A1	594 × 841	23.39 × 33.11
A0	841 × 1189	33.11 × 46.81

C

Fig. 2-17. These are standard flat sheet sizes.

a drawing in pencil on opaque paper, then "tracing" it in ink on an overlay sheet of translucent paper.

Today, however, the practice in industry is to develop the master drawing in pencil directly on tracing paper that reproductions can be made from (eliminating time and expense of preparing a "tracing"). *Inking* is reserved primarily for permanent drawings or for photographic reproduction work.

Tracing papers (as differentiated from vellums described below) are natural papers that have no additives to make the paper transparent. Natural papers made fairly strong and durable are not very transparent. Papers with high transparency are only moderately durable.

Vellum

Vellums are referred to as *transparentized* or *prepared* tracing papers. They provide strength, transparency, durability (handling and folding), and erasability without *ghosting* (erased lines showing through).

Vellum papers are made of 100 percent rag content and impregnated with a synthetic resin to provide high transparency. They are available in white or blue tint and are highly resistive to discoloration and brittleness due to age. The working qualities of vellum make it the most commonly used in industry even though it is more expensive than tracing paper.

Tracing Cloth

Tracing cloth is considerably more expensive than paper media. However, tracings are made on cloth when greater transparency and permanence are desired.

Tracing cloth is a finely woven cotton fiber material that has been sized with starch to provide a surface that takes pencil and ink. It comes in white for pencil tracings and blue tint for ink. The working

side is dull or frosted. You can erase pencil and ink lines without damaging the surface by using a soft rubber or vinyl eraser.

Tracing cloth is subject to expansion and shrinkage due to changes in moisture content of the air. Therefore, one section or part of a drawing should be completed at a time.

Polyester Film

The latest development in the drafting media field is *polyester film.* With the exception of cost, it has the best qualifications for a drawing medium. It provides dimensional stability, great resistance to tearing, easy erasing (with soft eraser or erasing fluid), and high transparency. It is waterproof and will not discolor or become brittle with age.

Polyester film has an excellent working surface for pencil, ink, or computer printers and typewriters. Many industries feel that the added cost of this medium is offset by its advantages, plus the ease that changes can be made on the tracing from time to time.

Fastening Drawing Sheet to Board

The drawing sheet should be positioned on the drawing board about 2 inches from the working edge. The working edge is the edge of the board that is opposite your drawing hand. The sheet should be placed up from the bottom edge approximately 4 to 6 inches to allow for the manipulation of the T-square and triangles or drafting machine.

Next, place the T-square near the bottom of the board and position the drawing sheet on its blade near the working edge, Fig. 2-18. Fasten the upper corners of the sheet with drafting tape and remove the T-square. Smooth the sheet out to the corners and fasten the lower corners.

Transparent cellophane tape should not be used to attach the drawing sheet. It frequently tears

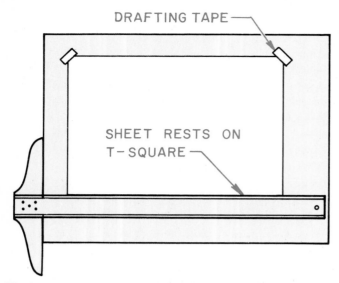

Fig. 2-18. Fasten the drawing sheet to the board using the T-square or drafting machine for alignment.

the sheet upon removal. ***Drafting tape*** is tape that is not as "sticky" as cellophane tape. This prevents the sheet from being torn when the tape is removed.

When a drafting machine is being used, the drawing sheet may be placed at a slight angle. This provides a more "natural" drawing and lettering position than the horizontal position necessary when using a T-square or a parallel rule.

Drawing Pencils

Drawing pencils are manufactured in a variety of types, Fig. 2-19. The following list shows some examples of different types of drawing pencils.

- Standard wood-case.
- Rectangular leads for drawing long lines with a wedge point.
- Mechanical lead holders in standard lead sizes.
- Fractional millimeter lead sizes called ***thin leads.***

Although more expensive, refill pencils are convenient to use, remain the same length, and save the time required in sharpening wood-case pencils.

Fig. 2-19. Wood-case drafting pencils and mechanical lead holders are popular in drafting. Several different types and styles are available.

Leads used in drawing pencils are manufactured by a special process designed to make them strong and capable of producing sharp, even density lines. Drawing leads are graded in eighteen ***degrees of hardness*** from 7B (very soft) to 9H (very hard).

The softer grades of pencil leads (2H, 3H, and 4H) deposit more graphite on the paper and produce more opaque lines. However, many drafters continue to use the harder grades because they produce sharper lines and do not smudge as readily during the drafting process.

Special pencil leads with a plastic base are manufactured for use on polyester drafting films. They come in five grades of hardness from K1 (very soft) to K5 (very hard).

Sharpening the Pencil

When sharpening wood-case pencils, first remove enough wood to expose 3/8 inch of lead on the end opposite the grade marking, Fig. 2-20. Use

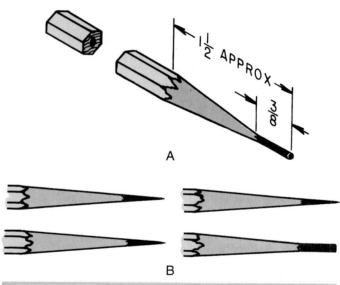

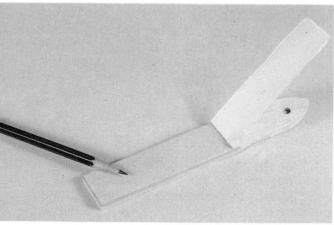

Fig. 2-20. Sharpening the wood-case pencil. C–A sanding pad can be used for final shaping.

a knife or a drafter's pencil sharpener with special cutters that will remove wood only, Fig. 2-21. If a knife is used, exercise care to prevent nicking the lead, causing it to break under pressure of use. For final shaping, finish the point on a piece of scrap paper. This is known as *pointing.*

Lead in standard size lead holders should be extended slightly beyond normal use position for pointing. Thin leads do not require pointing. Finally, remove all excess graphite dust from the pencil point by wiping it on a felt pad or soft cloth.

Two types of points are used on drafting pencils—conical and wedge. The *conical point* is used for general line work and lettering. It is shaped in a lead pointer, on a sandpaper pad, or with a file, Fig. 2-22. (Refer back to Fig. 2-20 for an example of a sandpaper pad.)

The *wedge point* is used for drawing long straight lines because it holds a point (edge) longer than the conical point. Shape the point on a sandpaper pad or file by dressing the two sides to produce a sharp edge. Finish on scrap paper and remove the excess graphite dust.

To maintain a neat work area and produce clean drawings, do not sharpen the pencil over your drawing or instruments. Before storing, remove the graphite dust from the sandpaper pad or file by tapping it lightly against the inside of a wastebasket.

COMPUTER-ASSISTED DRAFTING AND DESIGN EQUIPMENT

Technical drafting and design, as well as generating schedules and analyses, can be performed using a *computer-assisted drafting and design system (CADD).* The entire process—from conceptual design to the actual production of a set of drawings—can be completed on a computer system configured for CADD applications, Fig. 2-23.

New software designed specifically for technical drawing applications makes the drafting and design process easy and efficient. These systems completely replace the equipment used by the traditional drafter.

Using a computer system to design parts or assemblies and generate drawings still requires a knowledge of the graphic language. You must know orthographic projection, sections, auxiliary views, the use of standard symbols and conventions, and dimensioning procedures. Knowledge of contemporary manufacturing practices is also useful. Drafters are still responsible for communicating their ideas in a manner that is recognized by people in the field, Fig. 2-24. The computer system merely replaces the traditional tools for recording, manipulating, and reproducing ideas.

Fig. 2-21. Drafter's pencil sharpeners look like regular pencil sharpeners, however they remove only the wood and leave the lead.

Fig. 2-22. Pencil leads can be shaped to a conical point in lead pointers. (Eberhard Faber, Inc.)

Fig. 2-23. A PC CADD system can be used to test ideas, refine design possibilities, generate complete drawings, and make hard copies of the final design. (Altium)

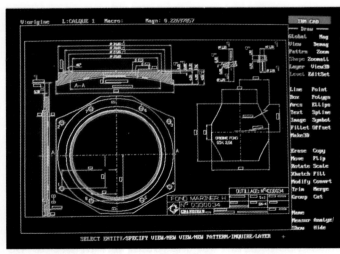

Fig. 2-24. Using accepted symbols, conventions, and arrangement of views is necessary for clear communication in technical drafting. (Altium)

A CADD system generally consists of the following components: computer or processor, monitor, graphics adapter, input and pointing device, software, and hard copy device, Fig. 2-25.

Computer

The **computer** or **processor** is probably the most critical component in a CADD system. The computer is composed of three components: A **data processing** or **logic** section, a **memory** or **control** section, and a **data transfer mechanism** or **data bus.** Each component serves a specific function. **Peripheral devices** (plotters, digitizing tablets, and monitors) are connected to the computer.

The data processing unit executes the commands given the computer by the operator or software program. The type of operations performed generally include:

- Selecting memory locations for programs.
- Transferring instructions and data to and from input and output devices such as disk drives, displays, plotters, digitizers, and printers.
- Performing arithmetic and logic functions.

The memory or control section of the computer contains programs that are currently in use or on-line. Memory locations are assigned for data storage. The data in a memory location frequently changes, but the location itself remains constant.

The transfer mechanism or data bus is the circuits that data travel over. Data travel to and from the processor and peripherals. Data also travel to and from the processing unit and memory section. The transfer mechanism is similar to a pipeline.

Monitors

The monitor, or display, allows the computer to communicate with the operator. Monitors for micro-CADD applications can be divided into two main categories, color or monochrome. **Monochrome monitors** are generally used for word processing and similar applications where color is not important. **Color monitors** are popular for CADD since color can perform a function and is supported by many CADD software programs, Fig. 2-26. Monitors range in size from 5 inches to over 25 inches, but size is generally not as important as resolution. **Resolution** refers to the sharpness of details displayed on the screen. High resolution monitors produce smooth lines at any angle. Generally, a resolution of at least 640 X 480 pixels is desirable. (A **pixel** is one dot of light emitted from the screen.)

Graphics Adapters

The graphics monitor requires a special circuit board in the computer called a graphics adapter or

Fig. 2-25. A typical PC CADD system including processor, monitor, keyboard, mouse, and software. Not shown is a device for making hard copies. (Altium)

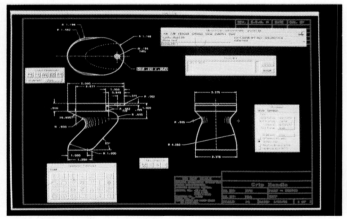

Fig. 2-26. Color can serve a specific function in CADD drawings. (Intergraph Corporation)

controller. This enables the monitor to display the best color and resolution that it is capable of. Typical graphics adapters include: EGA (enhanced graphics adapter), CGA (color graphics adapter), MDA (monochrome display adapter), VGA (video graphics array), and SVGA (super video graphics array).

Input and Pointing Devices

All CADD systems require one or more input devices. Typical input devices include: keyboard, mouse, digitizing tablet, and puck or stylus.

Keyboard

The **keyboard** is the most common and universal input device, Fig. 2-27. It provides an excellent way to input numbers and letters (alphanumeric information). Typing skills are a necessity to use the keyboard efficiently.

Mouse

The **mouse** is a popular input device for CADD programs. Often called a pointing device, it is used to pick objects and identify points on the screen, Fig. 2-28. It is an efficient way to issue commands and draw. Some CADD programs are designed specifically for this type of input device.

Fig. 2-27. A keyboard is used to input numbers, letters, and CADD commands.

Fig. 2-28. A mouse is used for cursor control and picking operations from a screen menu.

Digitizing Tablet

Digitizing tablets consist of a large, smooth surface and a cursor control (puck or stylus). When used with a menu overlay, digitizing tablets can speed the design and drafting process greatly, Fig. 2-29. Many popular CADD packages are developed around menu overlays and require a digitizing tablet for efficient use.

CADD Software

The CADD software provides the operations to be performed on data by the computer. In short, **software** is a computer program designed for a specific application. Each computer program is composed of a series of instructions that are to be executed in a specified sequence. These instructions are written in one of several computer languages.

Many CADD software packages are available for general and specific applications. General drafting packages are used for mechanical and technical drafting. Architectural, engineering, and construction (AEC) packages are specifically made for the design and illustration of structures–houses, commercial buildings, bridges, and roads, Fig. 2-30. Other software programs are even more specific and are used in conjunction with CADD AEC packages.

The "user friendliness" of any CADD system is largely dependent on the design of the software program. In choosing a program, first decide what you want to accomplish. Then, select a program that was designed to perform that task.

Output Devices

Any CADD system must be able to output a paper hardcopy to be useful as a drafting machine.

Fig. 2-29. Digitizing tablets are available in several sizes and resolution. (Summagraphics)

Fig. 2-30. CADD software is available for most any application.

The most commonly used devices are pen plotters, electrostatic plotters, photoplotters, and printers.

The device that most experienced drafters choose is the *pen plotter,* Fig. 2-31. ("Pen plotters" should not be confused with "pin printers," that are also known as "dot matrix" printers.) Pen plotters draw in much the same way that a drafter does. A pen plotter produces a very accurate drawing in ink on a variety of materials–paper, vellum, or polyester film. Pen plotters are classified as either drum or flatbed plotters. Either type will produce a quality drawing.

An *electrostatic plotter* is probably the most versatile type of plotter. It can function as a plotter or printer. This type of machine is much faster than a pen plotter, but does not produce as sharp a line.

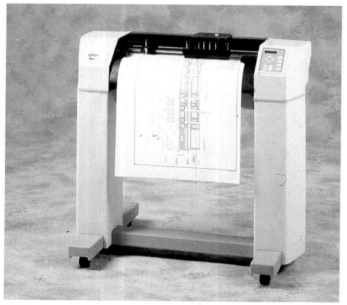

Fig. 2-31. Multi-pen plotters can plot in several colors and line widths to produce very high quality drawings. (Houston Instrument, A Summagraphics Company)

Electrostatic plotters do not draw one line at a time as the pen plotter does. Instead, it produces images in the form of small dots arranged in a matrix. Another consideration is the size of drawing that can be produced on the machine. Most electrostatic plotters are designed for A-size or B-size paper, but some will handle C-size or D-size. Many industrial drawings require larger size drawing sheets.

Photoplotters are very accurate and used for applications such as plotting printed-circuit board masters. These machines use a light beam and special photosensitive paper. They can achieve an accuracy of plus or minus .0005 inch. The main disadvantage is the high cost of the unit.

Printers are designed to produce a hardcopy for alphanumeric data–the typical computer printout. Impact printers, electrostatic printers, and ink jet printers are all used for CADD drawings. *Impact printers* produce images by striking character keys against an inked or carbon ribbon. These printers are also known as *dot matrix printers* or *pin printers. Electrostatic printers* form images by charging a special electrographic paper with an electron beam in the form of the desired character. A special ink spray is passed over the surface and is attracted to the electrostatically charged areas. *Laser printers* are a form of electrostatic printers. Electrostatic printers are faster and typically produce a higher quality image than impact printers. *Ink jet printers* shoot droplets of ink onto the sheet to form the images. These printers are not used that much in CADD applications because they tend to produce a lower quality image. However, these printers are used for small illustrations that might be found in desktop publishing applications.

Laser printers are very popular where line drawings and text material is integrated together, such as desktop publishing. The quality of line drawings is not as sharp as those made with a pen plotter, but is good enough for many applications. One disadvantage, however, is that laser printers cannot handle large sheets.

ALPHABET OF LINES

The American Society of Mechanical Engineers (ASME) has developed a standard for lines that is accepted throughout industry. Known as the *Alphabet of Lines,* this standard is designed to give universal meaning to the lines of a drawing, Fig. 2-32.

The Alphabet of Lines reveals shape, size, hidden surfaces, interior detail, and alternate positions of parts. The lines shown in Fig. 2-32 should be studied carefully, and each drawing produced should conform to this standard. Note that the lines differ in width (sometimes referred to as thickness

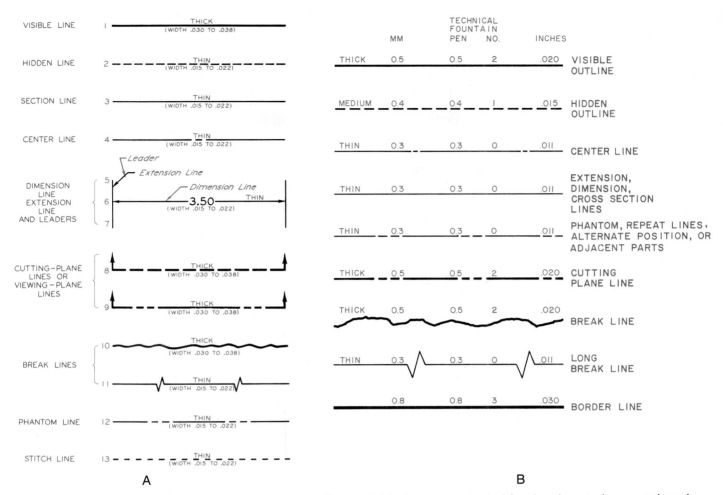

Fig. 2-32. A–The Alphabet of Lines, that uses two line weights, is recommended for drawings to be reproduced by available photographic methods. B–There is also an Alphabet of Lines that uses three line weights. (American National Standards Institute)

or weight). They also differ in character. Each is easily distinguishable from another. Each conveys a particular meaning on the drawing. The use of these lines on a drawing is illustrated in Fig. 2-33.

Different widths of lines are used on drawings. (Refer back to Fig. 2-32.) All lines should be dense black, regardless of width, as thin pencil lines tend to "burn out" during the reproduction process of making prints. Any variation in lines should be in width and character only. Pencil lines are likely to be slightly thinner than corresponding ink lines. However, pencil lines should be as thick as practical to provide acceptable reproductions.

Visible Lines

Visible lines are used to outline the visible edges or contours of the object that can be seen by an observer. Visible object lines should stand out sharply when contrasted with other lines on the drawing. Visilbe lines should be darkest lines in the drawing, with the exception of the border.

Hidden Lines

Hidden lines indicate edges, surfaces, and corners of an object that are concealed from the view of the observer. They are thin or medium lines made up of short dashes, evenly spaced. The dashes are approximately 1/8 inch long and the spaces approximately 1/32 inch long. They may vary slightly with the size of a drawing.

Hidden lines start and end with a dash. They make contact with visible lines or other hidden lines from where they start or end, Fig. 2-34A. If the hidden line is a continuation of a visible line, then a gap is shown, Fig. 2-34B. A gap is also shown when a hidden line crosses but does not intersect another line, Fig. 2-34C. Dashes should join at corners and arcs start with dashes at tangent points, Fig. 2-34D and E.

Hidden lines should be omitted wherever they are not needed for clarity. By omitting hidden lines when not needed, very complex drawings will be easier to read. However, until you gain experience in drafting, you should include all required lines unless otherwise directed by your instructor.

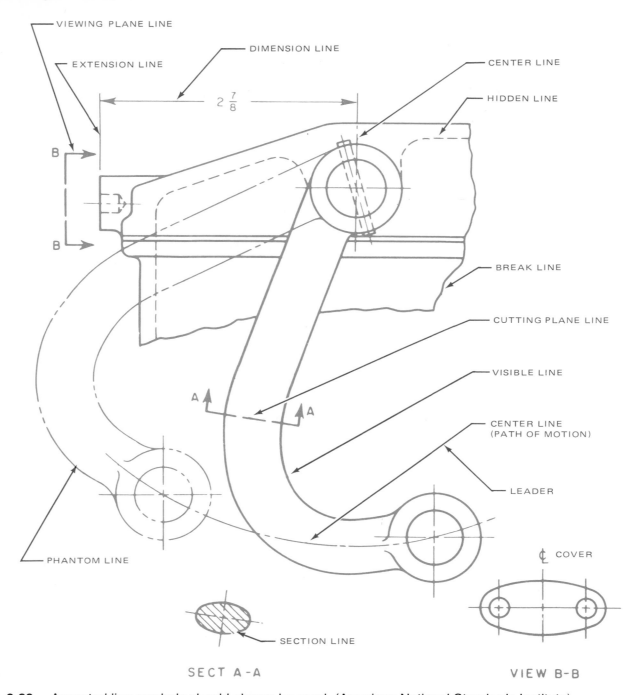

VIEWING PLANE LINE

EXTENSION LINE

DIMENSION LINE

CENTER LINE

HIDDEN LINE

$2\frac{7}{8}$

BREAK LINE

CUTTING PLANE LINE

VISIBLE LINE

CENTER LINE
(PATH OF MOTION)

LEADER

COVER

PHANTOM LINE

SECTION LINE

SECT A-A

VIEW B-B

Fig. 2-33.   Accepted line symbols should always be used. (American National Standards Institute)

Section Lines

Section lines are sometimes called *cross-hatching.* These lines represent surfaces exposed by a *cutting plane* (an invisible plane passing through an object). Section lines are usually drawn at an angle of 45° with a sharp 2H pencil, Fig. 2-35. If more than one object is sectioned on the same drawing, vary the angle and direction of the section lines to demonstrate the different parts. Draw the lines dark and thin to contrast with the heavier object lines. On average size drawings, space the lines by eye about 1/16 inch apart (small drawings, 1/32 inch; large drawings, 1/8 inch). Spacing of section lines should be uniform. The Ames lettering instrument can be used along your triangle as an aid in obtaining uniform section lines.

Section lines can also be used to indicate the type of material that the part is constructed from. The section lining shown in Fig. 2-35 is for cast iron and malleable iron, and it is recommended for general sectioning of all materials on detail and assembly drawings. The section lines for other materials are shown on page 231.

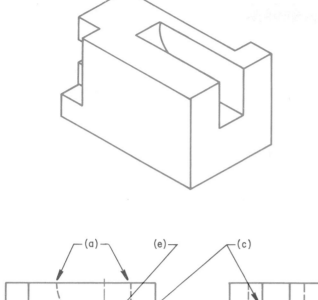

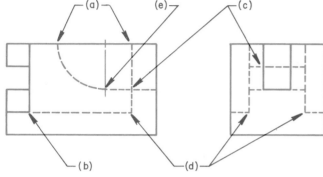

Fig. 2-34. Hidden lines should be represented using accepted standards.

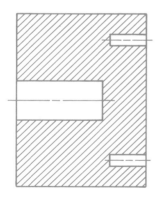

SECTION **A-A**

Fig. 2-35. Section lines are thin, dark lines that are uniformly spaced.

Centerlines

Centerlines are thin lines composed of long and short dashes alternately spaced with a long dash at each end. They indicate axes of symmetrical parts, circles, and paths of motion. (Refer back to Fig. 2-33.) Depending on the size of the drawing, the long dashes vary from approximately 3/4 to 1 1/2 inches (or longer). The short dash is approximately 1/16 to 1/8 inch in length. The space between the dashes should be about 1/16 inch in length. Center lines should intersect at the short dashes if possible. They should extend only a short distance beyond the object lines of the drawing, unless needed for dimensioning or other functions.

Dimension Lines, Extension Lines, and Leaders

Dimension lines indicate the extent and direction of dimensions and are terminated by arrowheads. (Refer back to Fig. 2-33.) **Extension lines** indicate the termination of a dimension. The extension line begins approximately 1/16″ inch from the object and extends to 1/8 inch beyond the last arrowhead. **Leaders lines** are drawn to notes or identification symbols used on the drawing.

Cutting-plane Lines

Cutting-plane lines indicate the location of the edge view of the cutting plane. (Refer back to Fig. 2-33.) Two forms of lines are approved for general use. The first is composed of alternating long dashes (approximately 3/4 to 1 1/2 inches or longer, depending on drawing size) and pairs of short dashes (approximately 1/8 inch with 1/16 inch spaces).

The second form is composed of equal dashes, approximately 1/4 inch (or more) in length. This form contrasts well and is quite effective on complicated drawings. Both forms have ends bent at 90° and terminated by arrowheads to indicate the direction of viewing the section. If the heavyweight line will obscure some of the detail, portions of the line overlaying the object may be omitted.

Break Lines

Break lines are used to limit a partial view of a broken section. For short breaks, a thick line is drawn freehand. (Refer back to Fig. 2-32.) A long break line is drawn with long, thin, ruled dashes joined by freehand "zig-zags." It is used for long breaks, particularly in structural drawing.

Phantom Lines

Phantom lines show alternate positions, repeated details, and paths of motion. (Refer back to Fig. 2-33.) They consist of thin, long dashes. The dashes are approximately 3/4 to 1 1/2 inches in length, alternated with pairs of short dashes, 1/8 inch in length. The space between the dashes is about 1/16 inch in length.

Datum Dimensions

Datums are lines, points, and surfaces that are assumed to be accurate. They are placed on drawings as datum dimensions since they may be used for exact reference and location purposes. The word datum is used in the dimension line for such revealing information. These are lightweight lines with the same characteristics as phantom lines.

Construction Lines

Construction lines are very light, gray lines used to lay out all work. They should be light enough on a drawing so they will not reproduce when making a print. On drawings for display or photoreproduction, they should not be visible beyond an arm's length.

Border Lines

Border lines, while actually not a part of the standard, are used as a "frame" for the drawing. These lines are the heaviest of all lines on a drawing. (Refer back to Fig. 2-32.)

ERASING AND ERASING TOOLS

It would be desirable to make a drawing without erasing. However, mistakes do happen and changes in existing drawings must frequently be made.

Erasing is a technique that the drafter must perfect to do good work. Much erasing time and damage to drawings may be saved by drawing all lines first as construction lines. The lines can be "heavied-in" for final finish.

Two types of *erasers* are useful in drafting: the firm textured rubber eraser for erasing ink lines, and the soft vinyl eraser for erasing pencil lines and cleaning drawings, Fig. 2-36. Erasers containing gritty abrasives (such as ink erasers) should not be used. These will damage the drawing surface and produce *ghosting* on reproductions.

Steel erasing knives should not be used for general line erasing. They are useful in removing ink lines that have overrun, are too wide, or have been made by error, Fig. 2-37. They must be kept sharp. Use them with a light sideways stroke. Special *nonabrasive vinyl erasers* and *erasing fluids* are used on tracing cloth and polyester film.

Erasing Procedure

When making erasures, follow this procedure:
1. Clean eraser by rubbing it on a scrap of paper.
2. With your free hand, hold drawing securely to avoid wrinkling.

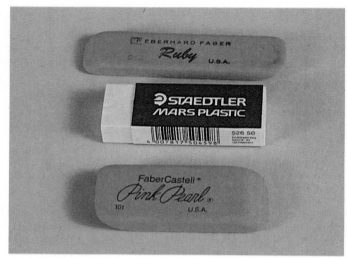

Fig. 2-36.   The red rubber and soft vinyl erasers are recommended for drafting.

Fig. 2-37.   A steel erasing knife is used for removing overruns in inking. (Keuffel & Esser Co.)

3. Rub soft vinyl eraser lightly back and forth to erase detail or line.
4. For erasing deeply grooved pencil or ink lines, place a triangle under the paper for backing.
5. If necessary to protect details close by, use an *erasing shield,* Fig. 2-38. (A piece of stiff paper will serve if an erasing shield is not available.)
6. Clean drawing with vinyl eraser before final finishing of lines.
7. Remove erasure dust with dust brush or soft cloth, Fig. 2-39.
8. After cleaning front of drawing, turn drawing over and inspect back for dirt that may have been transferred from the drawing board to the back of the drawing.

Electric Erasers

Most industries today, because of frequent alterations in product design, find it necessary to make changes on existing drawings. Improvements

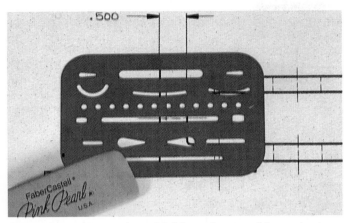

Fig. 2-38. An erasing shield can be used to protect parts of a drawing that are not to be removed.

Fig. 2-39. Frequent use of a dust brush will help keep your drawing clean.

in drafting media, as well as the ***electric erasing machine*** make changes a simple matter, Fig. 2-40. Exercise care in using the electric erasing machine not to press too hard or to remain in one spot too long. This could mar or distort the drawing surface or cause "ghosting" in reproduction of prints.

Only soft-rubber or vinyl erasers should be used in electric erasing machines. A very gentle pressure avoids overheating the drawing surface. Place a piece of thin gage copper, brass, or aluminum sheet under the area to be erased to dissipate the heat and reduce the possibility of damage to the drawing.

NEATNESS IN DRAFTING

The first impression is a lasting one. Remember that the appearance of your drawing is the first

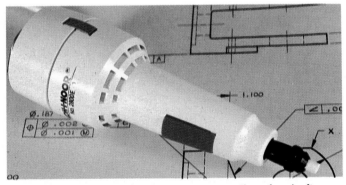

Fig. 2-40. An electric eraser can save time for drafters.

reflection of your ability as a drafter. People have a tendency, and rightly so, to associate *neatness* with *ability* in drafting.

Practice cleanliness from the start. Make it a habit to guard against anything that may make a drawing look dirty. The primary source of dirt on a drawing is pencil graphite (lead). The sliding of the T-square, triangles, shirt sleeves, and hands across the drawing will smear graphite.

Another thing that detracts from the neatness of a drawing is **ghosting.** This occurs when dark, heavy lines are erased. Dark, heavy lines actually create a "groove" in the paper. If the line is erased, this groove will remain and a "ghost image" of the line will be visible.

The following suggestions will help you keep a drawing clean:

1. Wash your hands before starting to draw. Occasionally wash them again while drawing. Since you are continually working over the drawing, clean hands will help to keep the drawing clean.

2. Always wipe the dust and dirt from your instruments with a soft cloth before starting to draw and frequently during use. A thorough cleaning with a soft eraser or erasing solvent keeps instruments in good condition.

3. Lay out all views with light lines using a hard pencil. "Heavy-in" lines only when you are sure all parts are correct.

4. Remove dust as soon as it collects. After each line is drawn, use a soft cloth or dusting brush to remove particles of loose graphite from the sheet. Remove erasure dust immediately with a soft cloth or dusting brush.

5. Do not slide instruments across drawing. Tilt the T-square by pressing down on the head before sliding. Lift the straightedge and triangles to prevent the smear of graphite dust from lines already drawn.

6. Sharpen pencil away from drawing. Enclose the sandpaper pad or file in an envelope before storing in the drawer with drawings or instruments.

7. Keep an orderly drawing area. Have only the tools and equipment needed on top of the desk. This will prevent crowding and avoid the possibility of instruments falling on the floor, Fig. 2-41.

8. Use a paper overlay to cover completed parts of the drawing when lettering or working on other areas of the drawing.

9. Cover drawing at night or store it in a drawer to prevent dust from gathering.

10. Store completed drawings in a portfolio to prevent damage.

Fig. 2-41. An orderly work area will contribute to cleaner drawings and better reproductions.

SCALES

The **scale** is one of the most frequently used drawing instruments. In addition to laying off measurements, it is used to reduce or enlarge the measurements of an object to a suitable size.

A standard scale is made in one of two basic shapes, flat or triangular, Fig. 2-42. The **flat scale** is available in three bevel shapes: regular two-bevel, opposite two-bevel, and four-bevel. The **triangular scale** is available in two styles, regular and concave. The triangular scale has six "faces." This allows multiple scales, typically up to eleven, on one instrument. Many drafters prefer a flat-type scale because it is easier to manipulate.

Scales are made of boxwood, boxwood with plastic faces, all plastic, and aluminum. They may have engine (machine) divided, precision molded, or die-engraved graduations. The better scales are engine divided. Most are made of boxwood with white

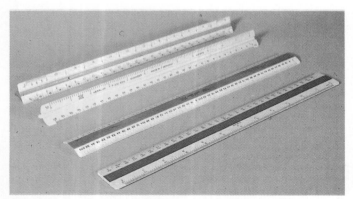

Fig. 2-42. There are several basic shapes of standard drafting scales.

plastic on the faces to make the divisions easy to read. (Refer back to Fig. 2-42.) The divisions on scales are either open divided or fully divided.

Open divided means that the main units are numbered along the entire length of each scale. Also, only the first main unit is subdivided into fractional or decimal segments of the major unit. (Refer back to Fig. 2-42.)

Some open divided scales have two compatible scales on the same face reading from opposite ends. The larger of the two scales is typically twice the size of the smaller. (Refer back to Fig. 2-42.)

The **fully divided** scale has all units along the entire scale subdivided. (Refer back to Fig. 2-42.) This has the advantage of allowing the drafter to lay off several values from the same origin without moving the instrument.

Many types of scales are required to draw objects ranging from small machine parts to large area maps. Therefore, scales are classified according to their use. The following are the most common types of scales.

Architect's Scale

Architect's scales are most commonly used in making drawings for the building and structural industry. These scales are available in all five shapes. They are also used by many mechanical drafters since the major units are divided into feet and inches.

The various scales usually represented on an architect's triangular scale are shown in Fig. 2-43. These scales are arranged in pairs on five faces, plus a full-size scale on the sixth. The full-size scale is marked "16" to indicate that the inch is subdivided into 16ths. This gives a total of eleven different scales on one instrument.

Architect's scales are open divided (except on full-size scale). The end unit beyond the zero is subdivided into 12 parts representing inches of the foot. On the two smaller scales, however, the subdivisions represent 2 inches. On the four larger

1″	=	1″ (full size)
3″	=	1′–0″ (quarter size)
1 1/2″	=	1′–0″ (eighth size)
1″	=	1′–0″ (twelfth size)
3/4″	=	1′–0″ (1/16 size)
1/2″	=	1′–0″ (1/24 size)
3/8″	=	1′–0″ (1/32 size)
1/4″	=	1′–0″ (1/48 size)
3/16″	=	1′–0″ (1/64 size)
1/8″	=	1′–0″ (1/96 size)
3/32″	=	1′–0″ (1/128 size)

Fig. 2-43. The architect's scale typically includes eleven different scales.

scales, the smallest division represents fractional parts of an inch.

Civil Engineer's or Decimal Scale

The *civil engineer's scale* is commonly referred to as the *decimal scale.* This fully divided scale has long been used in civil engineering where large reductions are required for drawings such as maps and charts. However, it is becoming widely used in manufacturing.

Scales on a civil engineer's scale are divided into units representing decimal parts of the inch. Each one inch unit on the 10 scale is subdivided into 10 parts, each 1/10 (.10) inch in size. The 50 scale has 50 parts to the inch, each 1/50 (.02) inch, Fig. 2-44.

The divisions may represent feet, pounds, bushels, time, or any other quantity. The units may be expanded to represent any proportional number. For example, 50 scale could represent 50 feet, 500 feet, or 5000 feet. The subdivisions of the major units would have corresponding values.

Fig. 2-44. Civil engineer's scales are also known as decimal scales. These can be purchased in different styles.

Mechanical Engineer's Scale

The *mechanical engineer's scale* is useful in drawing machine parts where the dimensions are in inches or fractional parts of an inch. Common graduations for mechanical engineer's scales represent one inch and are shown in Fig. 2-45.

These scales are often called the *size* scales. For example, a size scale of "1/8" represents a drawing one eighth the size of the object. The "1" represents the full-size scale.

Combination Scales

A triangular scale combining selected scales from the architect's scale (1/8, 1/4, 1/2, 1 inch to the foot), civil engineer's scale (50 parts to the inch), and mechanical engineer's scale (1/4, 1/2, 3/8, 3/4, and full-size) is available. (Refer back to Fig. 2-42.) These are called *combination scales.* Special scales are also available for such uses as map, aerial photography, and statistical work.

Metric Scales

Metric scales are available for work on metric or dual-dimensioned drawings, Fig. 2-46. Metric scales are "size" scales. For example, 1:20 means the drawing is 1/20th the size of the actual object.

A scale of 1:20 may be referenced on the drawing as 1:20 or 5 cm = 1 m, Fig. 2-47A. A scale of 1:100 means the drawing is 1/100th the size of the actual object and may be referenced as 1:100 or as 1 cm = 1 m, Fig. 2-47B. (Refer back to Fig. 2-46 for an example of the 1:100 scale.) When this scale is used as a reduction scale of 1:100, the 1, 2, and 3 represent 1 meter each (1 cm = 100 cm or 1 meter). The 1:100 scale may also be used as a full-size scale since the smallest division is actually 1 millimeter (mm). The 1, 2, and 3 are actually full-size centimeters (10 millimeters).

Since the metric scales are decimal scales (base-ten number system) their ratios may be changed by multiplying (or dividing) by a multiple of 10. For example, the 1:100 scale may be changed to 1:1000 by multiplying by 10. Each of the numerals, 1, 2, and 3 multiplied by 10 now represent 10 m, 20 m, and 30 m.

The metric system of measurement originated in France in 1793 and has gradually been adopted by most countries throughout the world. In the United States, a sizable number of industries are going to full metric. Most industries will one day go to metric to be competitive in a global market. Some industries will use an interim system called *dual dimensioning* (decimal inch and metric). Dual dimensioning is discussed in Chapter 8.

Fig. 2-46. Metric scales are available for use on metric or dual-dimensioned drawings.

1″ =	1″ (full size)
1/2″ =	1″ (half size)
1/4″ =	1″ (quarter size)
1/8″ =	1″ (eighth size)

Fig. 2-45. There are four common mechanical engineer's scales.

A	SCALE 1:20 (5 cm = 1 m)
B	SCALE 1:100 (1 cm = 1m)
C	SCALE 1:100,000 (1 cm = 1 km)

Fig. 2-47. Metric scales are referenced on a drawing to indicate the units of the ratio.

The metric system was founded on a unit of measure from nature. That unit was one/ten millionth of the distance from the North Pole to the equator. The unit was called **meter.** (In international usage, "meter" is spelled "metre.") More accurate scientific studies in 1961 established the length of the meter as equal to 1,650,763.73 wavelengths of the orange-red light given off by krypton 86, Fig. 2-48. This determination is accurate to 1 part in 100,000,000 and can be reproduced in scientific laboratories throughout the world.   Drafting for Industry

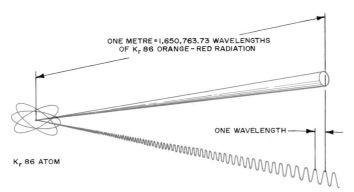

ONE METRE = 1,650,763.73 WAVELENGTHS OF K$_r$ 86 ORANGE - RED RADIATION

ONE WAVELENGTH

K$_r$ 86 ATOM

Fig. 2-48.   The length of the meter is scientifically and accurately established.

International System of Units (SI)

In 1954, and again in 1960, the metric system and its units were redefined in order to correct some deficiencies that had developed over the years. The 1960 General Conference on Weights and Measures formally gave the revised metric system its new title **Systeme International d'Unites** (International System of Units) with the universal abbreviation **SI Units.**

SI units of measure in metric system conform to reason, are consistent, and fit together. The authorized translation from the French definition of the seven basic and two supplementary units of the International System are shown in Fig. 2-49.

The metric system of numbers works in the same way as the decimal-inch system. However, the size of the units and terms are different. Both are on the base-ten number system that makes it easy to shift from one multiple or submultiple to another.

Each of the powers of 10 in the metric system is given a special name. Those multiple and submultiple factors from power 10^3 to 10^6 set off in red in Fig. 2-50 are the most commonly used units.

Any drafter using the metric system of linear measure should learn all common units and their relation to other factors, Fig. 2-51. Actually, changing from kilometers to meters is a simple shift of the decimal point, Fig. 2-52.

BASIC SI UNITS

QUANTITY	NAME OF UNIT	SYMBOL
Length	meter	m
Mass (weight)	kilogram	kg
Time	second	s
Electric current	ampere	A
Temperature	kelvin	K
Luminous intensity	candela	cd
Amount of substance	mole	mol

SUPPLEMENTARY UNITS

UNIT	NAME OF UNIT	SYMBOL
Plane angle	radian	rad
Solid angle	steradian	sr

Fig. 2-49.   There are seven basic and two supplementary units of the International Systems of Units (SI).

MULTIPLICATION FACTORS	NAME	SI SYMBOL
$1\ 000\ 000\ 000\ 000 = 10^{12}$	terameter	Tm
$1\ 000\ 000\ 000 = 10^{9}$	gigameter	Gm
$1\ 000\ 000 = 10^{6}$	megameter	Mm
$1\ 000 = 10^{3}$	kilometer	km
$100 = 10^{2}$	hectometer	hm
$10 = 10^{1}$	dekameter	dam
$1 = 10^{0}$	meter	m
$0.1 = 10^{-1}$	decimeter	dm
$0.01 = 10^{-2}$	centimeter	cm
$0.001 = 10^{-3}$	millimeter	mm
$0.000\ 001 = 10^{-6}$	micrometer	μm
$0.000\ 000\ 001 = 10^{-9}$	nanometer	nm
$0.000\ 000\ 000\ 001 = 10^{-12}$	picometer	pm
$0.000\ 000\ 000\ 000\ 001 = 10^{-15}$	femtometer	fm
$0.000\ 000\ 000\ 000\ 000\ 001 = 10^{-18}$	attometer	am

Fig. 2-50.   The units highlighted are the most commonly used SI units.

Metric Units Used on Industrial Drawings

The basic unit of length in the metric system is the meter. However, the following derived-decimal units are used in various industrial fields as the standard unit of measure on metric drawings, Fig. 2-53.

On drawings in the topographical field where the distance is great, the kilometer is used and a sizable scale reduction results. The meter serves as the official unit of measure in the building and

Kilometer	=	1 000 meters (thousands)
Hectometer	=	100 meters (hundreds)
Dekameter	=	10 meters (tens)
Meter	=	1 meter (units of linear measure)
Decimeter	=	0.1 meter (tenths)
Centimeter	=	0.01 meter (hundredths)
Millimeter	=	0.001 meter (thousandths)
Micrometer	=	0.000 001 meter (millionths)

Fig. 2-51. The relationship between submultiple and multiple units of measure in the metric system are all based on multiples of 10.

construction fields. The centimeter (1/100 m) is the official unit in the lumber and cabinet industries. In the precision manufacturing industries (aerospace, automotive, computer, and machine parts) where close tolerances must be maintained, the official unit is the millimeter (1/1000 m), Fig 2-54.

Drawing to Scale

All drawings should be drawn to scale, except schematics and tables. **Drawing to scale** refers to a drawing that has been reduced proportionally from actual size so that it can be placed on the drawing sheet. A drawing may have to be enlarged proportionally over the actual object size for clarity and detail, Fig. 2-55. This is also called "drawing to scale." Drawing to scale is also used in referring to drawings of objects drawn full size. In this case, the scale would be 1″ = 1″.

The scale selected for a particular drawing depends on the size of the object to be drawn. In gen-

REDUCTIONS		ENLARGEMENTS
1:2	1:50	2:1
1:2.5	1:100	5:1
1:5	1:200	10:1
1:10	1:500	20:1
1:20	1:1000	50:1

Fig. 2-54. There are several typical metric scale designations for drawings that have been reduced or enlarged.

eral, the drawing of the object should be as large as possible while still fitting a standard sheet size.

As indicated in previous sections, there are numerous scale sizes. The first reduction on the architect's scale is to quarter size. (Refer back to Fig. 2-43.) The first reduction on the mechanical engineer's scale is half size. (Refer back to Fig. 2-45.) If these reductions are insufficient, then a smaller scale must be selected.

Reading the Scale

Here is how to read an open divided architect's scale for a mixed number. To read 2 feet and 3 1/2 inches on the eighth-size scale (1 1/2″ = 1′-0″), start with the numeral 2 in the open divided section. Then move to the fully divided unit to locate the 3 1/2 inches, Fig. 2-56.

A reading of 3 5/8 inches on an open divided mechanical engineer's half-size scale would be read in a similar manner, Fig. 2-57. Remember that units on this scale represent inches. Start with the numeral 3 in the open divided section and move to 5/8 inch in the fully divided end unit.

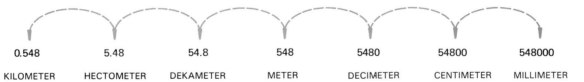

0.548	5.48	54.8	548	5480	54800	548000
KILOMETER	HECTOMETER	DEKAMETER	METER	DECIMETER	CENTIMETER	MILLIMETER

Fig. 2-52. To change a measurement from meters to millimeters, the decimal place is simply moved six places to the right.

INDUSTRY	UNITS OF MEASURE	INTERNATIONAL SYMBOL	MULTIPLE FACTOR
Topographical	kilometer	km	10^3 = 1 000 m
Building, Construction	meter	m	10^0 = 1 m
Lumber, Cabinet	centimeter	cm	10^{-2} = 0.01 m
Mechanical Design, Manufacturing	millimeter	mm	10^{-3} = 0.001 m

Fig. 2-53. Metric units of measure on industrial drawings vary by type of industry.

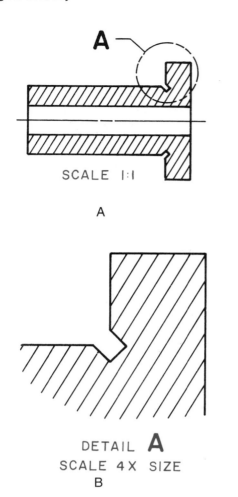

A

SCALE 1:1

A

DETAIL **A**
SCALE 4X SIZE
B

Fig. 2-55. A feature of a part enlarged to show detail should be labeled with the scale.

To read a decimal dimension of 2.125 inches on the civil engineer's fully divided 10-scale, start from the 0 and move past the 2, Fig. 2-58A. Continue past the first tenth (.10) to one-fourth of the next tenth (.025) (judged by eye). This represents the decimal .125. (Refer back to Fig. 2-58A.)

The same reading can be made with greater accuracy on the 50-scale, where each subdivision equals 1/50 or .02 of an inch, Fig. 2-58B. Move from 0 to 10 (5 major divisions to the inch, 10 = 2 inches) and on to 6 subdivisions (6 x .02 = 12). Continue to one-fourth of the next subdivision (1/4 of .02 = .005) for a measurement of 2.125 inches. The 50-scale is the scale most commonly used in the machine-parts manufacturing industries. Decimal dimensioning is standard practice in these industries.

Measuring with the metric scale is shown in Fig. 2-59, using the .01 scale as a full-size scale. The measurement reads 43.5 mm. To use the same scale as a reducing scale of 1:100, let each numbered unit (actually a centimeter) represent a meter—a reduction of 1 to 100. That is, one unit on the drawing represents 100 on the object. The measurement in Fig. 2-59 would then represent 4.35 meters (or 4350 mm).

The difference in the meaning of the words scale and size as used in scale drawings should be noted. The scale 1/4" = 1'-0 is a common scale for drawing house plans (often referred to as "quarter scale"). Refer back to Fig. 2-43 and note that this scale is actually 1/48 size since the quarter inch on the drawing represents one foot on the house plan.

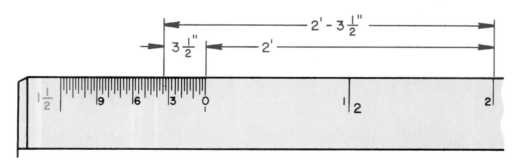

Fig. 2-56. Measuring with the architect's scale.

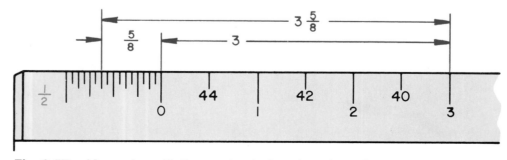

Fig. 2-57. Measuring with the mechanical engineer's scale.

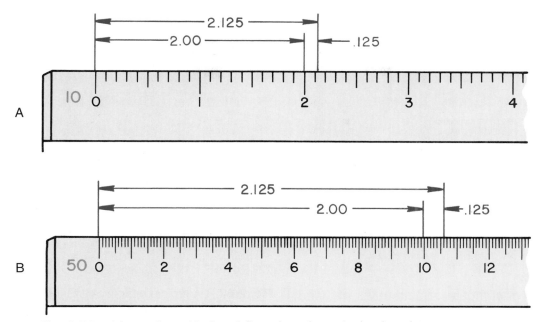

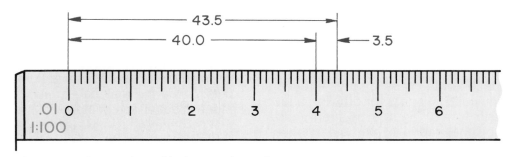

Fig. 2-58. Measuring with the civil engineer's or decimal scale.

Fig. 2-59. Measuring with the metric scale.

The word "scale" refers to the name of a particular scale and not to the size of the drawing.

The word "size" *does* refer to the size of the drawing in relation to the size of the object. Hence, the "quarter size" scale on the mechanical engineer's scale, where 1/4″ = 1″, will produce a drawing one-quarter the size of the object.

The scale of a drawing is usually indicated in the title block of the drawing in a manner similar to that shown below:

- Full size–1/1, or FULL SIZE
- Enlarged–2/1, 4/1, 10/1, 10X, TWICE SIZE
- Reduced–1/4″ = 1′-0″, 1/2, 1/4, 1/10, 1/50, HALF SIZE, and QUARTER SIZE.

Views that have been drawn to a scale other than that indicated in the title block should have the scale noted below the view. (Refer back to Fig. 2-55B.)

Laying Off Measurements

Position the scale on the sheet with the particular scale to be used, face up and away from you,

Fig. 2-60. Eye the scale directly from above. Then, with a sharp conical pencil, mark the desired distance lightly with short dashes at right angles to the scale. Successive distances on the same line should be laid off without shifting the scale, Fig. 2-61.

Fig. 2-60. Eye the scale directly from above when laying off a measurement.

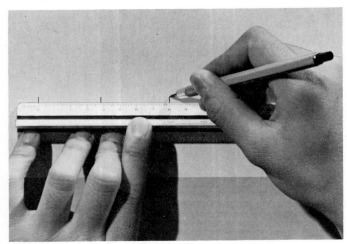

Fig. 2-61. Laying off successive distances.

The dividers or compass should never be used to take distances directly from the scale. This procedure is harmful to the scale. Mark distances on the sheet, then set the dividers or compass to these marks.

SHEET FORMAT

Recommendations are made by the American Society of Mechanical Engineers for drawing sheet borders and basic format data. However, these vary somewhat by the needs of particular industries. Included here are the practices found in most industries.

Sheet Margins

The recommended *margins* for drawing sheets vary from 1/4 inch on A-size and B-size sheets to 1/2 inch for D-size and E-size. Up to one inch may be used on the left edge if the sheet is to be bound. If sheet is to be rolled, 4 to 8 inches should be left beyond the margin for protection.

Title Block

A title block is included on a drawing to provide certain recorded data pertinent to the drawing and to provide supplementary information, Fig. 2-62. The title block is usually located in the lower right hand corner of the drawing just above the border line.

Sometimes a title strip containing the same information is used. The title strip extends partially or completely across the lower portion of the sheet. Suggested title block layouts are given on page 687 for use on your drawings.

INTERNATIONAL HARVESTER COMPANY		
MOTOR TRUCK DIVISION		
FORT WAYNE ENGINEERING		

Fig. 2-62. A typical industry title block will appear in the bottom, right corner of the drawing.

DRAFTING INSTRUMENT PROCEDURES

Basic drafting procedures in instrument usage are presented in this section to assist you in forming good habits. Study the material carefully and refer to it as needed in actual use of the various instruments.

Drawing Horizontal Lines

Horizontal lines are drawn along the upper edge of the drafting machine straightedge (or T-square or parallel), Fig. 2-63A. A T-square should be tight against the working edge of the drawing board. The working edge of the drawing board will be the edge opposite your drawing hand. Lift the drafting machine head to prevent the blade from sliding over the drawing (rubbing graphite across the drawing from existing pencil lines), while bringing it into approximate position for the line to be drawn, Fig. 2-63B.

Let your non-drawing hand slide from the head to the blade with the fingers resting on the blade and the thumb on the drawing board. Your fingers are now in position to bring the blade in perfect alignment where the line is to be drawn. Hold the blade in this position and draw a light line, Fig. 2-63C. Generally, when drawing a line you should move the pencil towards your drawing hand.

Note that the pencil is inclined in the direction the line is being drawn. The pencil is tilted slightly away from the drafter to cause the point to follow accurately along the edge. Let your little finger glide along the blade to help steady your hand. Rotate the pencil between thumb and fingers slowly to

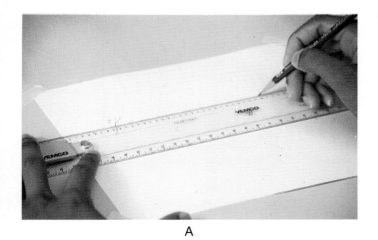

A

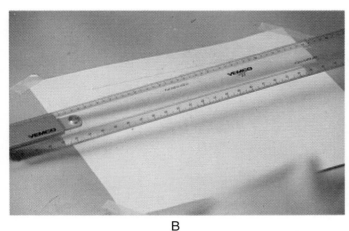

B

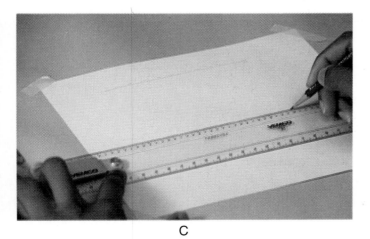

C

Fig. 2-63. Positioning a drafting machine and drawing a horizontal line.

retain a conical point. Draw horizontal lines at the top of the sheet first and work down.

Drawing Vertical Lines

Vertical lines can be drawn with the vertical edge of either the 30-60 degree or 45 degree triangle supported on the upper edge of the T-square or drafting machine blade. (A drafting machine can also be used to draw inclined lines without a triangle since the head rotates.) Position the blade below the starting point of the vertical line and place the triangle on the blade. Hold the triangle and the blade firmly with your palm and fingers, Fig. 2-64.

Draw the lines upward, away from the body. Hold the pencil at a 60° angle to the paper and tilt the pencil away from the triangle so the point will follow accurately along the edge of the triangle.

To maintain accuracy, never draw lines too close to ends of the triangle. Rotate the pencil as you draw to retain a fine point. Draw the vertical lines at the side of the sheet opposite your drawing hand first.

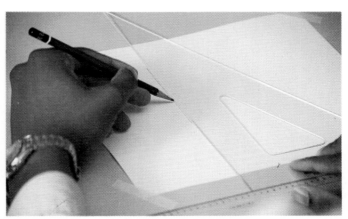

Fig. 2-64. Drawing a vertical line.

Drawing Inclined Lines

Inclined lines at 30°, 45°, and 60° may be drawn by using the appropriate triangle with the T-square, Fig. 2-65. By using a 45 degree triangle and a 30-60 degree triangle in combination, lines at 15° and 75° can be drawn, Fig. 2-66. By using these two triangles individually or in combination, a complete circle may be divided into 24 sectors of 15° each, Fig. 2-67. Inclined lines on degree settings other than these 15° sectors can be drawn with a protractor or a drafting machine. (Refer back to Fig. 2-15 for an example of a drafting machine.)

Drawing Parallel Inclined Lines

Lines parallel to inclined lines can be drawn using a T-square and one triangle, Fig. 2-68A. Adjust the T-square to align the triangle with the given line AB. Hold the T-square firm, place the triangle in the desired location, and draw the parallel line. Lines parallel to inclined lines can be drawn by using two triangles (eliminating the T-square) if the lines are not widely separated, Fig. 2-68B.

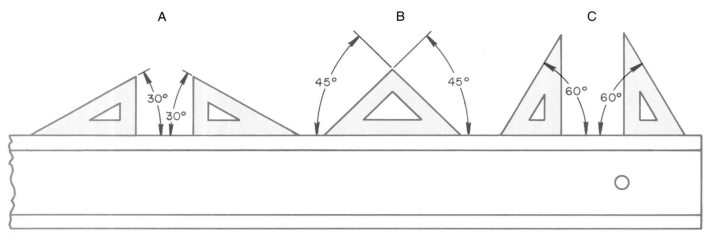

Fig. 2-65. Drawing lines inclined at 30°, 45°, and 60°.

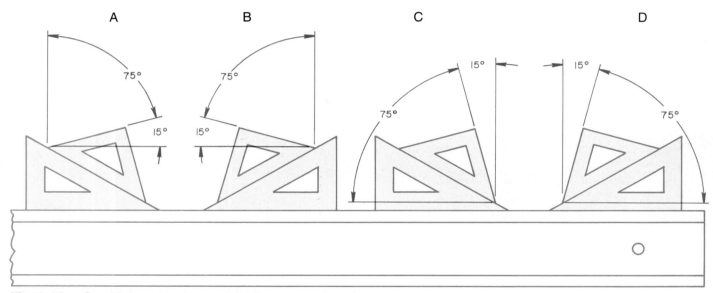

Fig. 2-66. Combining triangles to draw lines at angles of 15° and 75°.

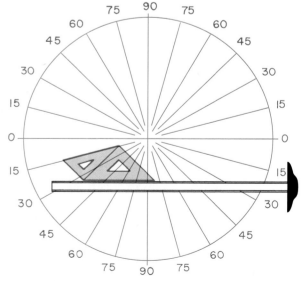

Fig. 2-67. Dividing a circle into 15° sectors with triangles.

Drawing a Perpendicular to an Inclined Line

Place the hypotenuse of a triangle (the line opposite the 90° angle) along the edge of the T-square and adjust until one side of the triangle is aligned with the given line, Fig. 2-69A. Hold the T-square firmly and move the triangle until the second side is in the desired location. Draw the perpendicular line.

When a longer perpendicular line is required, place the shortest side of a triangle against the T-square and adjust until the hypotenuse is aligned with the given line. Hold the T-square firmly and revolve the triangle until the hypotenuse is perpendicular to the line. Then move the triangle to the desired location and draw the line, Fig. 2-69B. This process can be performed with the T-square and either triangle or with the two triangles in combination.

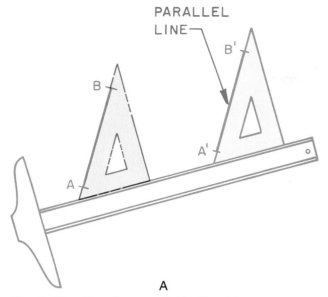

Fig. 2-68. Drawing parallel inclined lines.

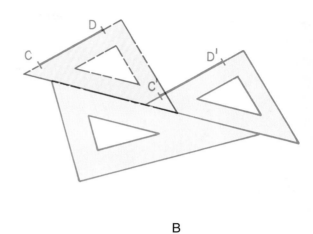

B

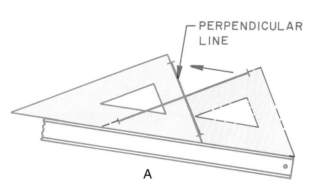

Fig. 2-69. Drawing lines perpendicular to inclined lines.

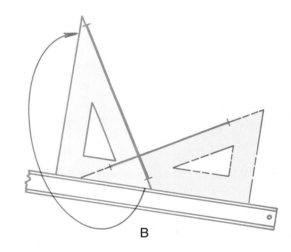

B

The Protractor

The protractor is used to measure and to mark off angles that cannot be done with a T-square and triangles. Protractors are available in several designs, including the simple semicircular type, Fig. 2-70. An adjustable triangle provides more accuracy. The protractor with a vernier attachment is used for extreme accuracy in laying out angles in map work, Fig. 2-71.

To lay out an angle with the semicircular protractor, set the vertex indicator at the vertex of the angle to be drawn, Fig. 2-72A. Mark the desired angle with two fine points, Fig 2-72B. Make sure you use light marks, otherwise they may not erase. Align the triangle with one of the two points by placing your pencil on one point and revolving the triangle in line with the vertex, Fig. 2-72C. Draw a line between the vertex and one of the two points, Fig. 2-72D. Be sure to draw light lines first. If you draw lines too dark to start with, they may "ghost" if

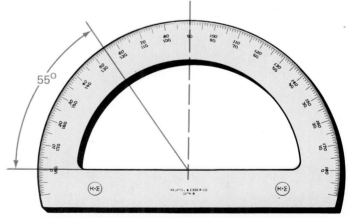

Fig. 2-70. A semicircular protractor can be used to lay out angles. (Keuffel & Esser Co.)

you try to erase them. Repeat this procedure for the other point.

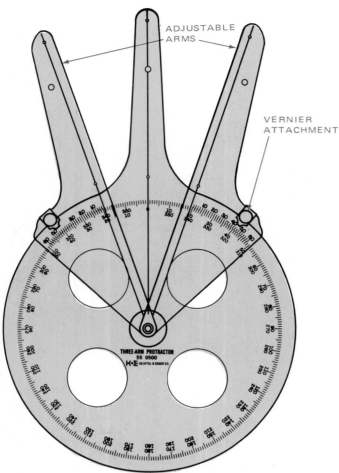

Fig. 2-71. Protractors with vernier attachments are more accurate in laying out angles. (Keuffel & Esser Co.)

Protractors are also available in graduations other than degrees. A percentage protractor is useful in laying out circle graphs, Fig. 2-73.

Pencil Technique with Instruments

To produce accurate and clean drawings, it is important that you develop proper pencil tech-

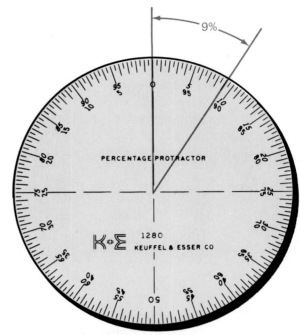

Fig. 2-73. The percentage protractor is useful in constructing circle graphs.

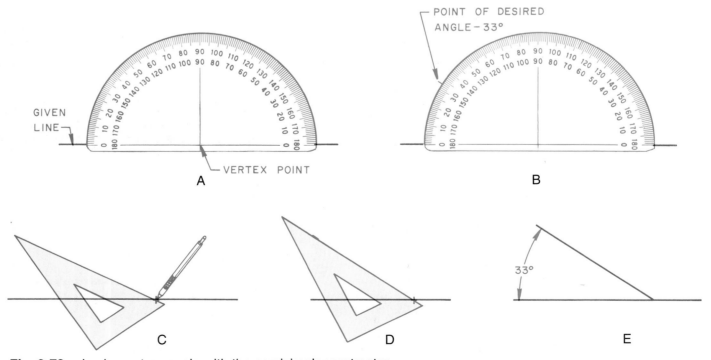

Fig. 2-72. Laying out an angle with the semicircular protractor.

niques when using instruments. Carefully observe, practice, and follow these suggestions as they will aid in developing proper techniques.

1. All layout work on drawings should be done with light construction lines. Use a 4H pencil and rotate it slowly when drawing to help retain the conical point.
2. Before the drawing is "heavied-in," erase all unnecessary lines.
3. "Heavy-in" all lines to their proper line weight. Use the proper grade pencil to give the best results with the paper being used. (Refer back to Fig. 2-32.)
4. For accuracy, and to minimize working over finished lines, pencil in lines in the following order: (1) arcs and circles, (2) horizontal lines starting at the top of the drawing and working down, (3) vertical lines from your non-drawing hand to your drawing hand, (4) inclined lines from top down and from your non-drawing hand to your drawing hand.
5. To prevent smearing of lines, dust loose graphite from the drawing after each line is drawn. Avoid sliding the T-square, triangles, and other instruments across the drawing.

The Compass

Large and small **bow compasses** are used frequently in drafting. The bow compass has a steel ring head and an adjusting screw, Fig. 2-74. The small bow instrument is used for drawing smaller circles with radii of approximately one inch or less. A large bow compass is used for circles with radii up to 5 or 6 inches. A drop bow compass has removable tips that make it a very versatile tool, Fig. 2-75.

When drawing an arc or circle with a large radius, bend the legs of a friction compass at the knees and adjust them to meet the paper in a vertical position, Fig. 2-76. A lengthening bar may be used for drawing circles with larger diameters, Fig. 2-77.

Beam compass

The **beam compass** is used for drawing large arcs and circles. This compass consists of two sets of points, one a needle pivot point and the other a holder for a pen or pencil point, and a beam that the points clamp onto, Fig. 2-78. To draw arcs and circles with the beam compass, hold the pivot point steady with one hand and swing the pen or pencil point with the other.

Sharpening the compass lead

The compass is used with both pencil and pen attachments. Lead used in the compass should be about one grade softer than that used in your pencil. This will allow you to exert less pressure on the compass. The lead should extend approximately 3/8 inch. Sharpen the lead to a chisel point as shown in Fig. 2-79. After sharpening, adjust the lead to a length of 1/32 inch shorter than needle point.

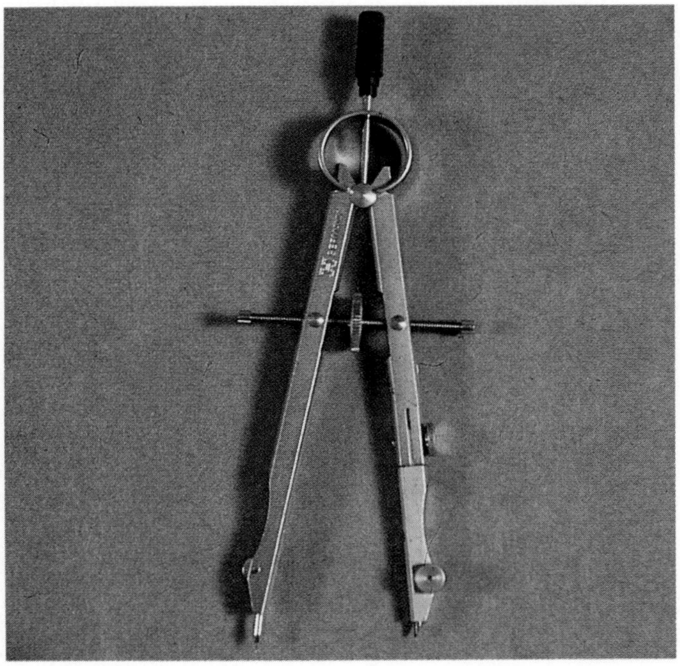

Fig. 2-74. A bow compass can be used to draw arcs and circles.

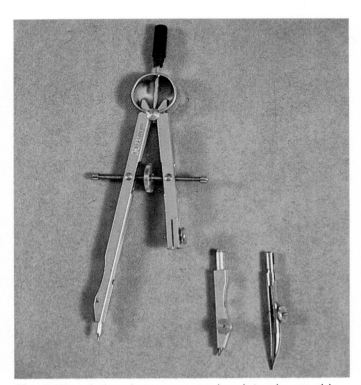

Fig. 2-75. A drop bow compass has interchangeable parts to allow for different applications.

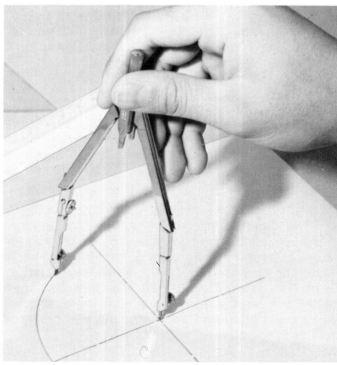

Fig. 2-76. Adjust the legs of a friction compass to meet the drawing surface vertically. (Teledyne Post)

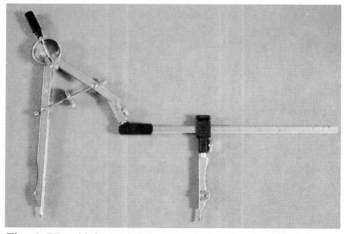

Fig. 2-77. Using a bow compass with a lengthening bar allows large circles to be drawn.

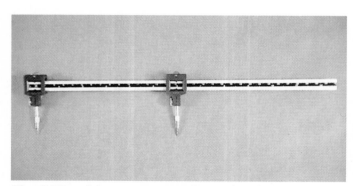

Fig. 2-78. A beam compass is used to draw extremely large circles.

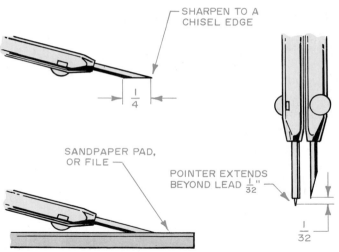

Fig. 2-79. Sharpen the compass lead and adjust so that the pointer extends just beyond the lead.

Drawing arcs and circles

The bow instrument is adjusted to a radius by twisting the adjusting screw between the thumb and forefinger.

To set the compass, measure off the radius on a scrap of paper (or lightly on your drawing) and adjust the compass accordingly, Fig. 2-80. Test the setting by drawing the circle lightly on the drawing or scrap paper, then measure the diameter.

To draw a circle, hold the compass in your drawing hand. Lean the compass slightly forward, and revolve the handle between the thumb and forefinger, Fig. 2-81. Draw the circle or arc lightly at first. When you are ready to "heavy-in," make repeated turns to darken the line.

Arcs and circles to be joined by straight lines should be drawn first. When a number of concentric circles are to be drawn with the compass, draw the smaller circles first since there is a tendency for the needle point hole to become enlarged. A center tack is sometimes used to prevent this, Fig. 2-82.

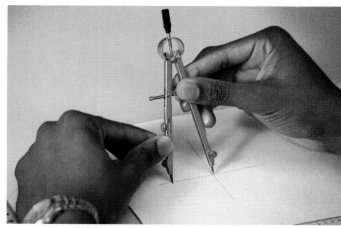

Fig. 2-80. Set the bow compass with the adjusting screw.

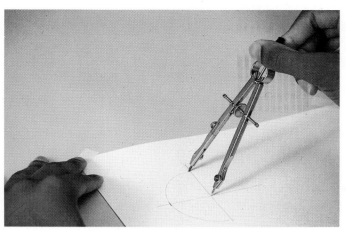

Fig. 2-81. When drawing a circle with a compass, tilt the compass a little so that the lead "follows" the compass.

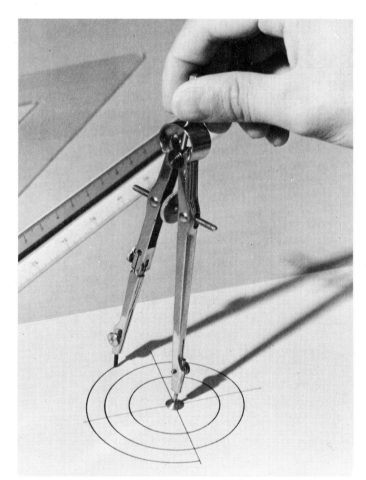

Fig. 2-82. A center tack prevents damage to the sheet from the compass needle.

Drafting Templates and Their Use

Industrial drafting rooms make extensive use of templates, Fig. 2-83. They provide economy and consistency when drawing commonly used characters and symbols. Templates are available for nearly all standard size circles in fractional, deci-

mal, and metric graduations. Templates for ellipses and bolt heads, and symbols for nearly every field of drafting are available to speed the drafter's work.

Templates are also available for use with lettering equipment for drawing many types of symbols, Fig. 2-84. To draw circles, arcs, or ellipses with the aid of a template, first lay out the centerlines on the drawing, Fig. 2-85. Then align the centerlines of the template and draw the figure.

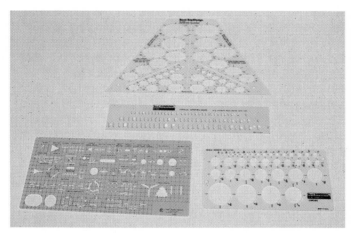

Fig. 2-83. Many templates are used in drafting.

ELECTRICAL SYMBOLS

WELDING SYMBOLS

ALSO CAPS & NUMBERS, SIZE 100

Fig. 2-84. Some symbol templates are made for use with lettering equipment. (Keuffel & Esser Co.)

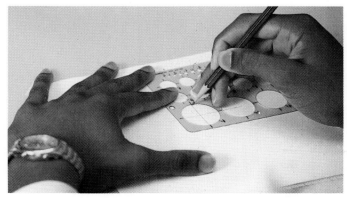

Fig. 2-85. Templates can be used to draw circles. The centerlines should be drawn first. The centerlines on the template should then be aligned with those on the drawing.

Dividers

Two types of dividers are used extensively by drafters: **bow dividers** and **friction joint dividers,** Fig. 2-86. Dividers are used to transfer distances and to divide straight and circular lines into equal parts.

Dividers are adjusted in the same manner as the compass. To transfer or step off distances, hold the knurled handle and place the dividers in position, Fig. 2-87. Mark the distances by making a slight dent in the paper with the divider point. Mark this dent with a light pencil mark or circle the dent.

To divide a line into three equal spaces, for example, set the dividers for an estimated 1/3 of the distance and step off, Fig. 2-88. Correct any error in estimation by decreasing or increasing the divider setting by 1/3 of the error and making another trial. Careful estimation of the distance and adjustment after the first trial should enable you to complete the division in 2 or 3 trials. Avoid puncturing the paper with the divider points.

Proportional Dividers

A **proportional divider** is a special instrument used for dividing linear and circular measurements into equal parts. It is also used to lay off measurements in a given proportion.

The instrument consists of two legs held together by a sliding pivot. It can be adjusted to obtain various ratios between the two sets of points on the ends of the legs, Fig. 2-89.

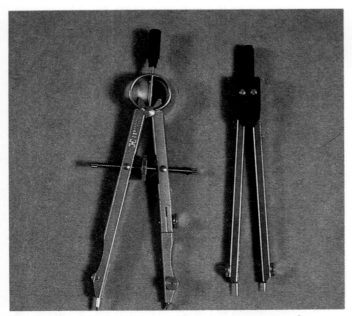

Fig. 2-86. Friction joint dividers are one type of dividers. Many bow compasses can also be used as dividers by placing an additional point where the lead normally goes.

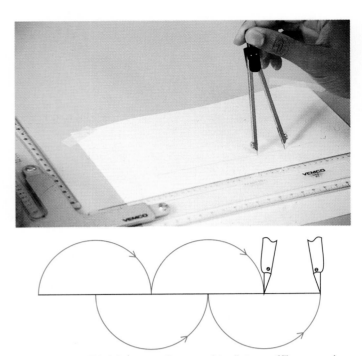

Fig. 2-87. Dividers can be used to "step-off" several lines of the same length.

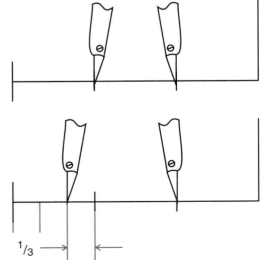

1/3

Fig. 2-88. Dividing a line into three equal parts can be done with dividers.

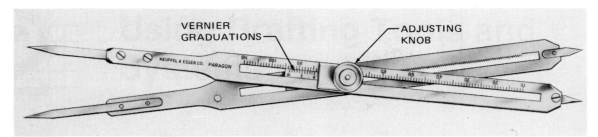

Fig. 2-89. Proportional dividers with vernier can be used to change the scale of a measurement. (Keuffel & Esser Co.)

Graduations on the dividers vary from less expensive proportional dividers (with ratios for division of lines) to dividers with vernier graduations. Settings for any desired ratio between 1:1 and 1:10 and ratios for circles, squares, and area may be made.

Examples of the use of proportional dividers include:

- Dividing straight lines into any number of equal parts.
- Lengthening straight lines to any given proportion.
- Dividing the circumference of a circle into any number of equal parts.
- Laying off the circumference of a circle from the diameter of that circle.
- Laying out a square equal in area to a circle based on the diameter of that circle.

With the more expensive instruments, a table of settings is provided for use in setting the dividers for various proportions.

Irregular Curves

All curves that do not follow a circular arc are known as ***irregular curves.*** These curves are common in sheet metal developments, cam diagrams, aerospace drawings, and various charts.

The instrument used for drawing the final smooth curve through plotted points is called ***irregular*** or ***French curves.*** (These instruments should not be confused with the "irregular curves" that may appear on a drawing.) These instruments are available in many shapes, Fig. 2-90. The instruments are made up of a series of geometric curves in various combinations.

To draw irregular curves, plot a series of points to accurately establish the curve, Fig. 2-9A. Then, lightly sketch a freehand line to join the points in a smooth curve, Fig. 2-91B. Draw in the final smooth line with the irregular curve. Match the irregular curve with three or more points on the sketched curve. Draw a segment at a time until the line is complete, Fig. 2-91C.

Check to see that the general curvature of the irregular curve is placed in the same direction as the curve of the line to be drawn. Do not draw the

Fig. 2-90. Irregular, or French, curves are available in many shapes and sizes.

full distance matched by the irregular curve. Stop short and make the next setting of the irregular curve flow out of the previous one, Fig. 2-91D. When the plotted points of the curve reverse direction, watch for the point of tangency where the irregular curve (instrument) should be reversed and overlap the previous setting with a smooth flow, free of "humps," Fig. 2-91E.

If the curve is symmetrical in repeated phases (the development of a cam diagram, for example), the same segment of the irregular curve should be used, Fig. 2-91F. Marking the curve with a pencil when the first symmetrical segment is drawn will aid in locating that segment for successive phases of the curve.

Flexible curve rules and splines with lead weights are useful in ruling a smooth curve through a number of points, Fig. 2-92.

INKING

Inking in industrial drafting for the production of working drawings is used very little today, but inking drawings for technical publications is quite common. The difficulty of inking has been greatly reduced with the introduction of improved instruments and drafting media (papers and polyester films).

Inked drawings provide a sharper line definition and make cleaning of the finished drawing much easier. It is possible to erase right over inked lines with a soft eraser and remove penciled construction lines.

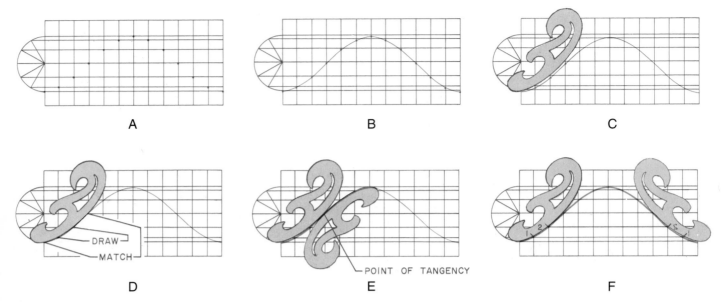

A   B   C

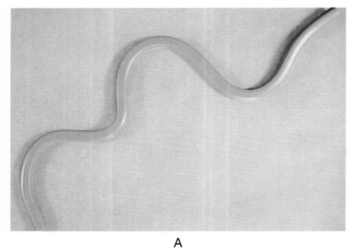

DRAW
MATCH

D   E   F

POINT OF TANGENCY

Fig. 2-91.   Draw smooth irregular curves with the aid of an irregular (French) curve instrument.

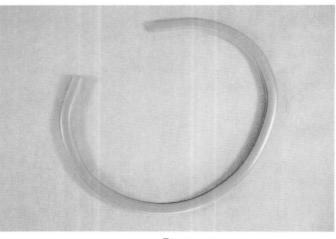

A

B

Fig. 2-92.   Flexible rules can be used to draw many different irregular curves.

Instruments for Inking

The ***ruling pen*** is included in most drafting sets for inking lines. It has two adjustable, sharp nibs that permit lines of varying widths to be drawn, Fig. 2-93. The ink is added to the pen between the nibs from a bottle designed for this purpose. The pen should not be filled too full as the weight of the ink may cause a line wider at the start than at the finish. Usually a level of 1/4 inch is sufficient. Practice will help in arriving at the right amount.

The ruling pen is used to draw a line along the straightedge, triangle, or irregular curve. A ruling pen attachment for the compass is usually included in a set of instruments. Clean ruling pens frequently during use and before storing. Use a damp tissue, wiping both the inside and outside of the nibs.

The ***technical pen*** has largely replaced the use of ruling pens in industrial and technical illustration, Fig. 2-94. Pens are available in a series of

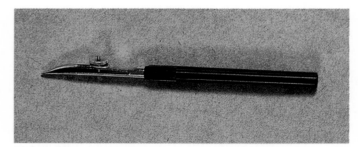

Fig. 2-93.   A ruling pen can be used for inking a drawing. However, technical pens have largely replaced ruling pens.

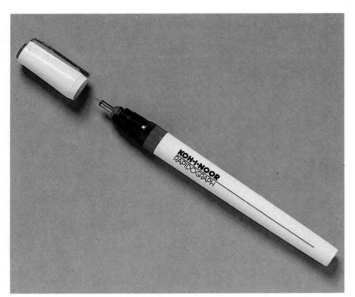

Fig. 2-94. Technical pens are versatile instruments for inking drawings. Technical pens have largely replaced ruling pens. (Koh-I-Noor Rapidograph, Inc.)

point sizes, assuring uniformity of line weight throughout a single drawing or several drawings.

The technical pen has a supply reservoir of ink and does not require filling for each use. Today's technical pens offer instant start-up in inking, even after weeks of storage. The technical pen is most versatile and can be used with straightedges, irregular curves, compasses, templates of all sorts, and in instrument lettering devices. There is very little additional skill required to use a technical pen over a pencil or leadholder.

Procedure for Inking

The following procedures tend to produce neat inked drawings.

1. Lay out the drawing, using light construction lines with a 2H or harder pencil.
2. Ink arcs from tangent point to tangent point (points where an arc is joined by a straight line). These points are located by drawing a light circle and drawing the tangent line lightly so that it just touches the circle. This point is the point of tangency.
3. Ink full circles and ellipses.
4. Ink irregular curves.
5. Ink all straight lines of one line weight. Move pen in one direction. Continue inking remaining straight lines of different weight.
6. Ink notes, dimensions, arrowheads, and title block.

PROBLEMS AND ACTIVITIES

The problems in Fig. 2-95, Fig. 2-96, and Fig. 2-97 are one-view drawing problems to provide you an opportunity to become familiar with the use of basic drafting instruments. Use A-size drawing sheets and draw the objects. Your instructor will tell you whether to use the inch or metric scale. Select a scale size to make good use of available drawing space without crowding. Do not dimension these drawings.

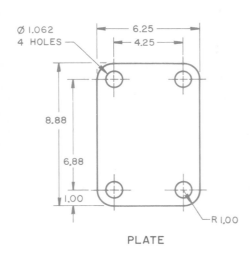

PLATE

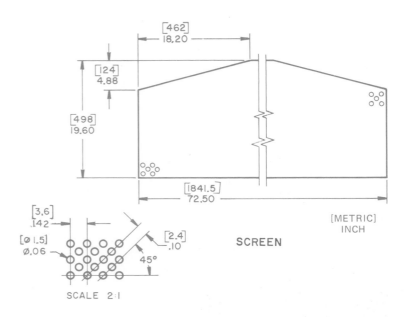

SCREEN

[METRIC]
INCH

SCALE 2:1

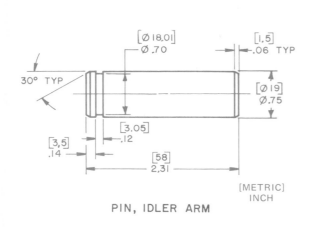

PIN, IDLER ARM

[METRIC]
INCH

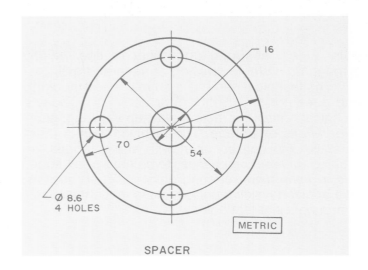

SPACER

METRIC

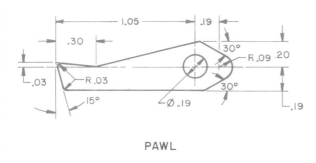

PAWL

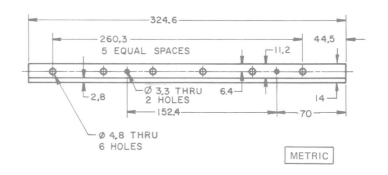

BACK WEAR PLATE

METRIC

Fig. 2-95.   Beginning one-view drawing problems.

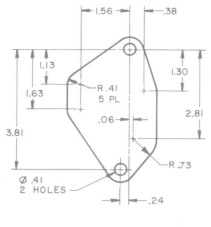

PLATE, COVER

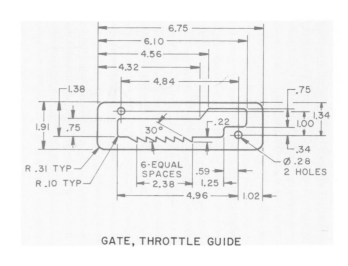

GATE, THROTTLE GUIDE

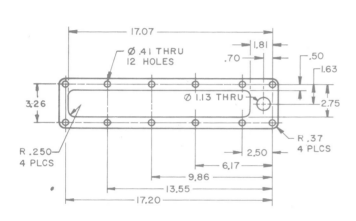

GASKET, WATER INLET CONN

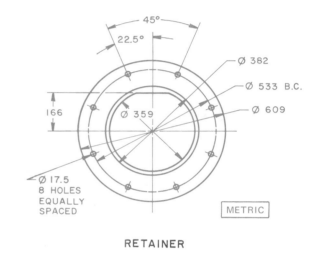

RETAINER

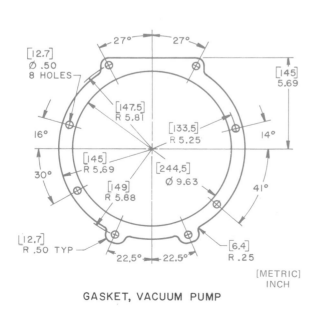

GASKET, VACUUM PUMP

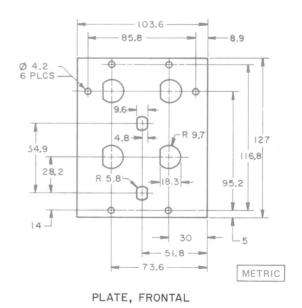

PLATE, FRONTAL

Fig. 2-96. Intermediate one-view drawing problems.

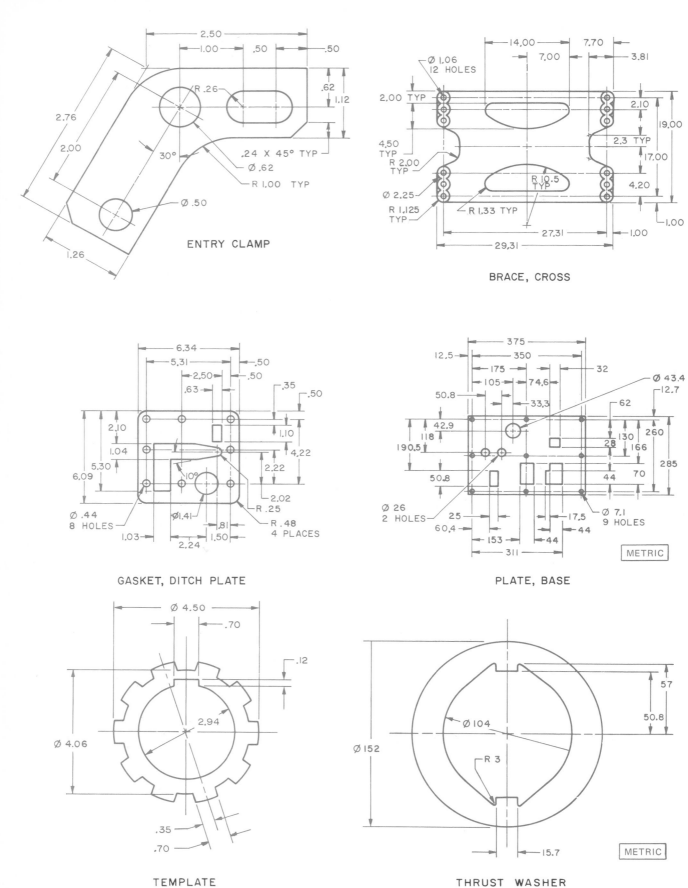

Fig. 2-97. Advanced one-view drawing problems.

Technical Sketching for Communication

KEY CONCEPTS

- [ ] Sketching is an important part of technical communication.
- [ ] Sketching serves a specific purpose in technical communication.
- [ ] There is an accepted technique for technical sketching.

Freehand technical sketching is a method of making a drawing without the use of instruments. This is a technique essential to all who work in a technical field. A good description of the process is "thinking and drafting." When a person sketches, they can concentrate on the solution to a problem without being encumbered with manipulation of instruments.

Most drafters and engineers use freehand sketching to "think through" solutions to drafting problems before starting an instrument drawing. Sketching also permits ideas to be quickly conveyed to others. This is especially important in the area of design improvement. Once the design or problem solution has been sketched, it is given to a drafter to prepare an accurate instrument drawing.

This chapter introduces you to the skills and procedures necessary to do freehand technical sketching. The techniques you learn will be particularly useful when combined with material presented in the chapters on instrument drafting, working drawings, pictorial drawings, and dimensioning.

SKETCHING EQUIPMENT

Freehand sketching requires very little equipment or material. Sketching readily lends itself to use by a drafter in the field or shop who is away from the drafting room. A pencil (typically F or HB grade), soft eraser, and some paper are all that is needed.

Paper

Several types of papers are suitable for sketching, depending on the nature of the job. You can use plain, cross-section, or isometric grid paper.

("Isometric" is a type of pictorial drawing discussed in Chapter 11.) Also available are bond typing paper, drafting paper, and tracing paper.

Cross-section paper is available in varying grid sizes. This paper is helpful in line work and proportions, Fig. 3-1. Isometric grid paper also aids in these areas and in obtaining the proper position of the axes, Fig. 3-2A. Cross-section and isometric grid papers may be purchased with fade-out grid lines. When the sketch is completed and prints are made, the grid lines do not reproduce, Fig. 3-2B.

You may want to start with cross-section paper, but it is best for you to learn to sketch on plain paper as soon as possible. This will help develop your skill and accuracy in freehand sketching without the use of aids.

SKETCHING TECHNIQUE

When sketching, hold the pencil with a grip firm enough to control the strokes. However, do not hold the pencil so tight that you stiffen your strokes or cramp your hand. Your arm and hand should have a free and easy movement. The point of the pencil should extend approximately 1 1/2 inches beyond your fingertips, a little farther than in normal

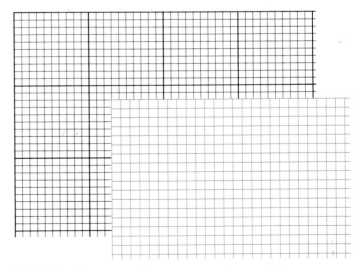

Fig. 3-1. Cross-section paper can be used for freehand sketching.

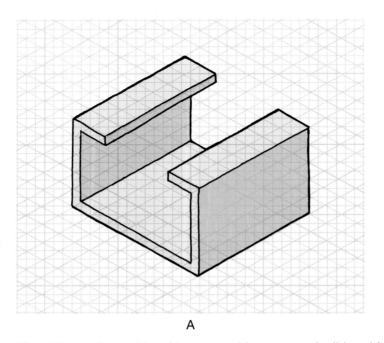

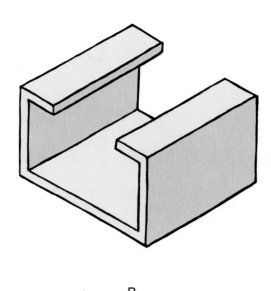

A   B

Fig. 3-2.   A–Isometric grid papers with nonreproducible grid can be used for sketching. B–When the sketch is reproduced, the grid will not reproduce.

drawing or lettering, Fig. 3-3. This will permit better observation of your work and provide a more relaxed position. In addition, your third and fourth fingers can be rested lightly on the drawing surface to steady your hand.

Rotate the pencil slightly between strokes to retain the point longer and produce sharper lines. Initial lines should be firm and light, but not fuzzy. Avoid making grooves in your paper by applying too much pressure. These grooves are difficult to remove. When sketching straight lines, your eye should be on the point where the line will end. Use a series of short strokes to reach that point. When all lines are sketched, go back and darken in the

lines. When darkening in lines, your eye should be on the tip of the lead.

While you should strive for neatness and good technique in freehand sketching, you should expect that freehand lines will look different than those drawn with instruments. Good freehand sketches have character all their own, Fig. 3-4.

SKETCHED LINE

INSTRUMENT LINE

Fig. 3-4.   Your drawings should be neat even when you sketch. However, a sketched line will look different from a line drawn with instruments.

Sketching Horizontal Lines

Horizontal lines are sketched with a movement that keeps the forearm approximately perpendicular to the line being sketched, Fig. 3-5. You will find that four steps are essential in sketching horizontal lines. First, locate the end points of the line, Fig. 3-6A. Next, position your arm for a trial movement, Fig. 3-6B. Then, sketch a series of short, light lines, Fig. 3-6C. Finally, darken the line in one continuous motion, Fig. 3-6D.

Sketching Vertical Lines

Vertical lines are sketched from top to bottom, using the same short strokes in series as for horizontal lines. When making the strokes, position your

Fig. 3-3.   When sketching, hold your pencil farther back than you normally would.

Fig. 3-5. When sketching horizontal lines, you should keep your arm approximately perpendicular to the line being sketched.

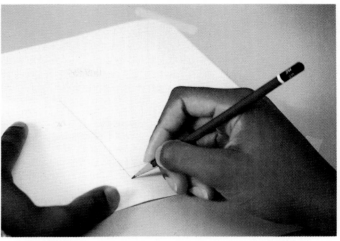

Fig. 3-7. When sketching vertical lines, you should have your arm in a comfortable position with the pencil at about 15° to the line being drawn.

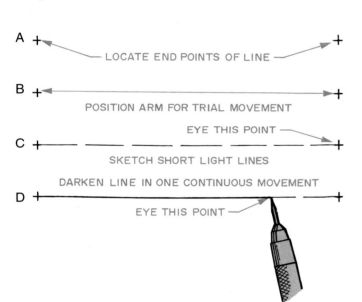

Fig. 3-6. There are four basic steps to sketching a horizontal line.

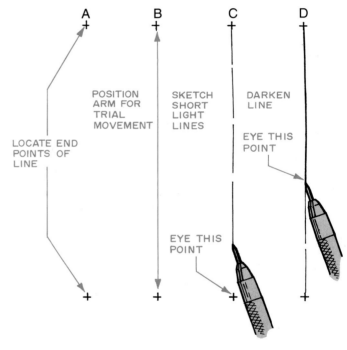

Fig. 3-8. There are four basic steps in sketching a vertical line. These steps are very similar to those used to sketch a horizontal line. (Refer back to Fig. 3-6).

arm comfortably at about 15° to the vertical line, Fig. 3-7. A finger and wrist movement, or pulling arm movement, are best for sketching vertical lines. First, locate the end points of the line, Fig. 3-8A. Next, position your arm for a trial movement in drawing the line, Fig. 3-8B. Then, sketch several short, light lines. When sketching these lines, you should focus on the end point of the line, Fig. 3-8C. Finally, darken the line. When darkening the line, you should focus on the point of the lead, Fig. 3-8D.

You may find it easier to sketch vertical or horizontal lines if the paper is rotated to form a slight angle. (Refer back to Fig. 3-7.) Straight lines that are parallel to the edge of the drafting board, such as border lines, may be drawn by letting the third and fourth fingers slide along the edge of the board as a guide, Fig. 3-9.

Sketching Inclined Lines and Angles

All straight lines that are not horizontal or vertical are called ***inclined lines.*** To sketch inclined lines, sketch between two points or at a designated angle. Use the same strokes and techniques as for sketching horizontal and vertical lines, Fig. 3-10A. If you prefer, rotate the paper to sketch these lines as if they are horizontal or vertical lines, Fig 3-10B.

Angles can be estimated quite accurately by first sketching a right angle (90°), then subdividing

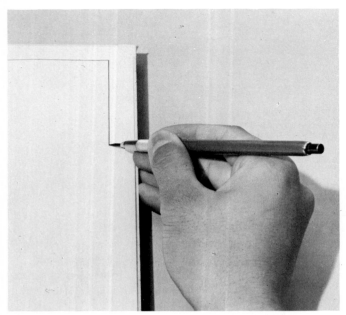

Fig. 3-9. Use your fingers along edge of drawing board as a guide for sketching straight lines.

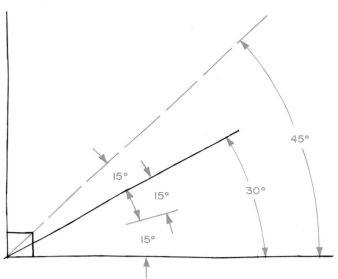

Fig. 3-11. Estimating an angle by freehand sketching is first done by sketching a 90° angle. The 90° angle is then divided (estimation only, no instruments) into the appropriate angle. Remember that 1/2 of a 90° angle is 45°.

it to get the desired angle. Refer to Fig. 3-11 for an illustration of how to obtain an angle of 30°.

Sketching Circles and Arcs

There are several methods of sketching *circles* and *arcs.* All are sufficiently accurate. Familiarize yourself with various techniques to use method best suited to a particular problem.

Center-line method

Six steps are used in the *center-line method* of freehand sketching circles. These steps are: locate centerlines, use a scrap piece of paper with radius marked to locate several points on the circle, position arm for trial movement, sketch the circle, and darken the circle, Fig. 3-12. Be sure to use light lines first, then darken the final shape later.

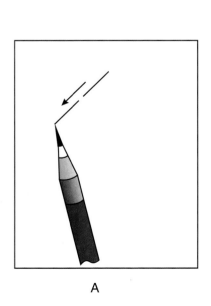

A

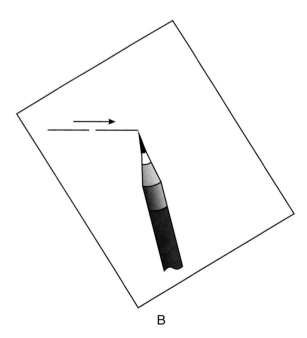

B

Fig. 3-10.
Use the same techniques to sketch inclined lines as for horizontal and vertical lines. A–Inclined lines can be drawn with the paper "square" with the drawing surface. B–Or they can be drawn by rotating the paper so that they "become" horizontal or vertical lines.

Enclosing square method

The following steps are necessary when sketching a circle by the **enclosing square method.** First, locate the centerlines of the circle. Second, sketch a box with sides the same length as the diameter of the circle, Fig. 3-13A. Next, sketch arcs where the centerlines meet the box, Fig. 3-13B. Finally, sketch the circle, Fig. 3-13C.

Hand-pivot method

The **hand-pivot method** is a quick and easy method of sketching circles. There are two ways of using this method. For both methods, first locate the centerlines of the circle. Next, you can use your small finger as a pivot point while holding the pencil, Fig. 3-14A. Rotate the paper while holding your finger on the center of the circle, Fig. 3-14B. A second way to use this method is by holding two pencils in your drawing hand and using one pencil as a pivot point instead of your small finger, Fig. 3-14C.

Free-circle method

The **free-circle method** of freehand sketching circles involves more skill in performance but can be developed with practice. With this method, you do not use any "guides" to help you sketch. You

sketch the circle using only your hand-to-eye coordination and your judgment. First, lightly sketch two or three circles in the approximate shape of the desired circle. Then, darken in the shape that is most accurate. Practice this method until you can draw a nearly perfect circle as shown in Fig. 3-15.

Sketching Ellipses

Occasionally it is necessary to sketch an **ellipse.** Three methods are presented here to aid you in producing a good sketch.

Rectangular method

The **rectangular method** is similar to sketching a circle with the enclosing square method. First, locate the centerlines of the ellipse. Then, draw a box with side lengths equal to the major (longest) and minor (shortest) diameters of the ellipse, Fig. 3-16A. Next, sketch arcs where the centerlines meet the box, Fig. 3-16B. Finally, sketch the ellipse, Fig. 3-16C.

Trammel method

There are four steps used to do freehand sketches by the **trammel method.** First, sketch the

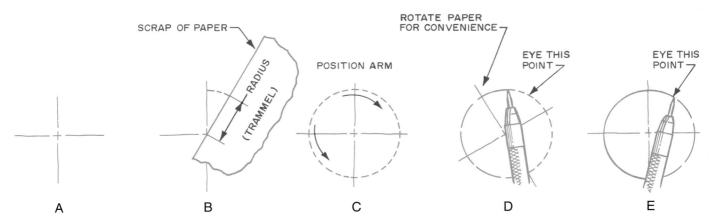

Fig. 3-12. The center-line method is one method of sketching a circle.

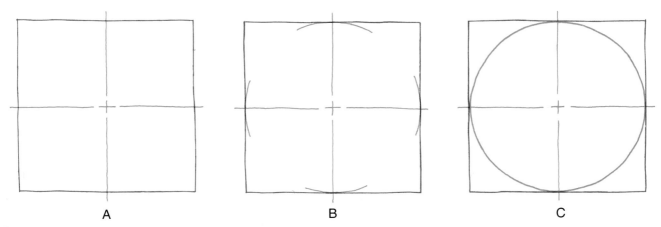

Fig. 3-13. The enclosing square method is a method of sketching a circle.

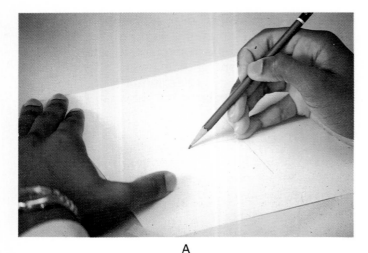

A

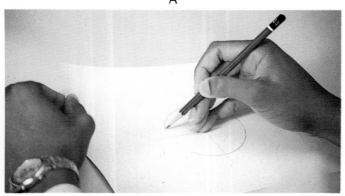

B

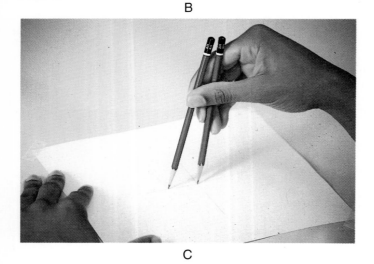

C

Fig. 3-14.   There are two ways to use the hand-pivot method of sketching a circle.

major and minor axes. Second, on a scrap piece of paper (the trammel), mark off three points A, B, and C. The distance from point A to point B should be 1/2 the minor axis and the distance from point A to point C should be 1/2 the major axis. Next, move the trammel around the ellipse, always keeping point B on the major axis and point C on the minor axis. Mark several spots at

Fig. 3-15.   When using the free-circle method of sketching a circle, first sketch two or three light circles that are approximately the correct size. Then, darken in the most accurate shape.

point A. Finally, sketch the ellipse using the marks that you made at point A, Fig. 3-17.

Free-ellipse method

The *free-ellipse method* is the same technique as used for sketching by the free-circle method. In this method, you use only your hand-to-eye coordination and your judgment. First, lightly sketch two or three ellipses in the approximate shape of the desired ellipse. Then, darken in the shape that is most accurate, Fig. 3-18. This method will require some practice to produce an acceptable ellipse.

Sketching Irregular Curves

An *irregular curve* may be sketched freehand by connecting a series of points at intervals of 1/4 to 1/2 inch along its path, Fig. 3-19. Include at least three points in each stroke. "Lead out" of the previous curve into the next.

Proportion in Sketching

There is more to sketching than making straight or curved lines. Sketches must contain correct proportions. *Proportion* is the relation of one

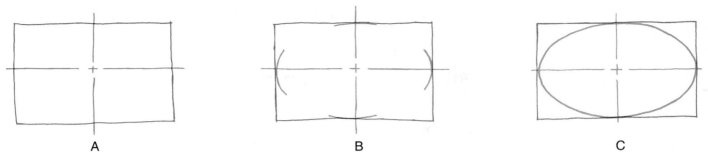

A   B   C

Fig. 3-16. The rectangular method of sketching an ellipse is very similar to the "enclosing square method" of sketching a circle. (Refer back to Fig. 3-13).

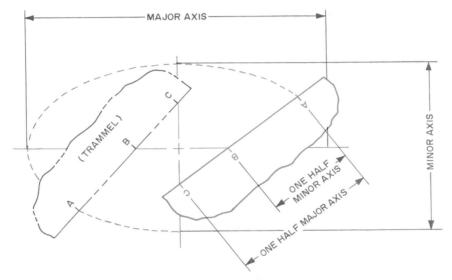

Fig. 3-17. When using the trammel method of sketching an ellipse, first locate the major and minor axes. Next, create a trammel with the distances of 1/2 both the major and minor axes indicated. Use this tool to make an approximation of the ellipse. Finally, darken the ellipse.

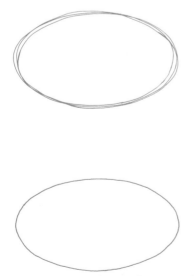

Fig. 3-18. When using the free-ellipse method of sketching an ellipse, first sketch two or three light ellipses in the approximate size and shape of the desired ellipse. Then, darken in the correct shape.

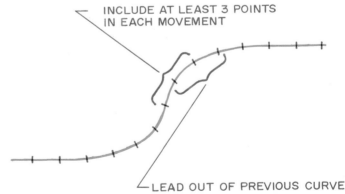

Fig. 3-19. When sketching an irregular curve, include at least three points in each segment. Also, overlap each section as you sketch.

part to another, or to the whole object. You must keep the width, height, and depth of the object in your sketch in the same proportion to that of the object itself. If not, the sketch may convey an *inaccurate* description. There are several techniques which may be used by the drafter to obtain good proportions.

Pencil-sight Method

One of the methods for estimating proportion is the ***pencil-sight method.*** With pencil in hand, extend your arm forward in a stiff arm position and use your thumb on the pencil to gauge the proportions of an object, Fig. 3-20. These distances may be laid off directly on your sketch. Vary the size by moving closer or further from the object. The pencil-sight technique is particularly useful in making sketches of an actual object rather than from a picture of the object.

Unit Method

Another useful technique in estimating proportions is the ***unit method.*** This method involves establishing a relationship between distances on an object by breaking each distance into units. Compare the width to the height and select a unit that will fit each distance, Fig. 3-21. Distances laid off on your sketch should be the same proportion although the units may vary in size. This method is useful when making a sketch from a picture of the object.

Other Methods

You may find it helpful in sketching to first enclose the object in a rectangle, square, or other ap-

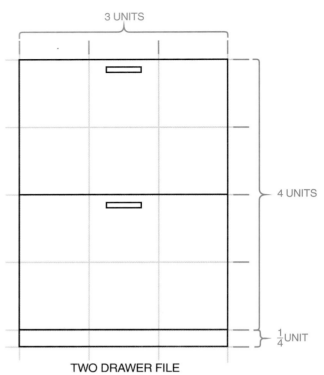

TWO DRAWER FILE

Fig. 3-21. To use the unit method of gauging proportions, divide the object into equal-sized units. The proportion can be changed by either increasing or decreasing the size of the units on your sketch.

propriate geometric form of the correct proportion, Fig. 3-22. Then subdivide this form to obtain the parts of the object. Once the outside proportions are established, the smaller parts are easily divided.

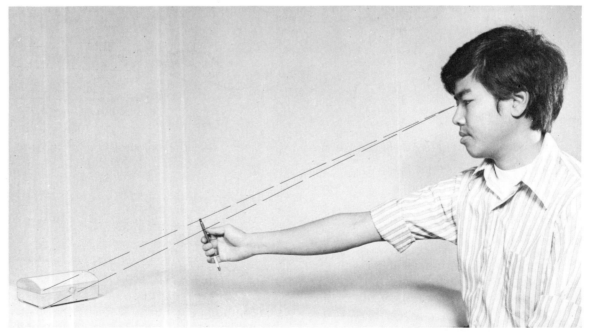

Fig. 3-20. To gauge proportions by the pencil-sight method, hold a pencil at arm's length and place your thumb to indicate measurements. Then, transfer these measurements to your sketch. The proportion can be changed by moving closer or farther away from the object.

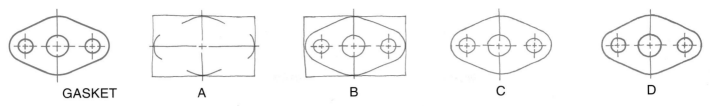

GASKET   A   B   C   D

Fig. 3-22. A—To sketch the gasket shown at the left, first layout the centerlines and sketch a box of the proper size. B—Then, lightly sketch in all of the features of the gasket. C—Next, erase the box that you drew as a guide. D—Finally, darken in the lines.

PROBLEMS AND ACTIVITIES

The problems presented here have been carefully selected to provide meaningful practice in freehand technical sketching. Use A-size (8 1/2 x 11 inch) plain paper–unless directed otherwise by your instructor–plus a pencil and an eraser. **Do not use scales and straightedges.**

Practice sketching strokes on scrap paper as you review each section prior to doing the assigned problems. Do those problems assigned by your instructor.

Sketching Straight Lines and Angles

Sketch a border, then divide your sheet into four rectangles (estimate the dividing point, do not measure), Fig. 3-23. Proceed as follows.

1. Sketch horizontal lines in rectangle No. 1. Allow 1/2 inch space between lines and strive for straight sharp lines.

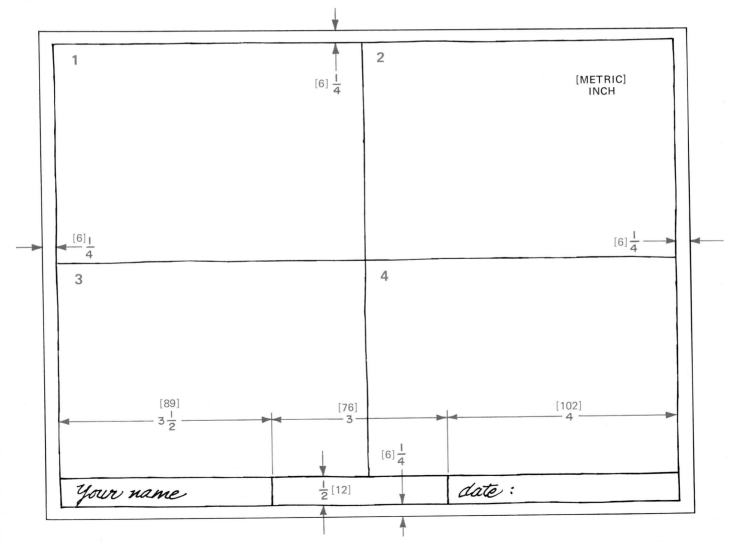

Fig. 3-23. This is one possibility for laying out a sheet for sketching problems. The numbers in parentheses are millimeters.

2. Sketch a series of vertical lines in rectangle No. 2. Allow 1/2 inch between lines and work to achieve true vertical lines.

3. Sketch inclined lines in rectangle No. 3. Sketch the first line as a diagonal between opposite corners of the rectangle. Space additional parallel lines 1/2 inch from this line. Your lines should be straight, sharp, parallel, and uniformly spaced.

4. Use the bottom line of rectangle No. 4 as a reference line. Starting 1/4 inch from the left end of this line, sketch the following angles at 1/4 inch intervals (about 2 inches in length) slanting upward and to the right: 75°, 45°, and 20°. Sketch a second series of angles in the same manner from the right end of the reference line. Slant these lines upward and to the left at 60°, 35°, and 15°. Sign your name and date the sketch.

Sketching Circles and Arcs

Divide your sheet into four rectangles as shown in Fig. 3-23 and sketch the following circles and arcs.

5. Sketch, by the center-line method, a 2 1/2 inch circle in rectangle No. 1. Center the circle in the rectangle. Your finished circle should appear as one sharp, freehand line.

6. Sketch two circles in rectangle No. 2, using the enclosing square and hand-pivot methods. Select the size of the circles so that space is well used but not crowded. Darken the finished circles, but retain the light construction lines for review by your instructor.

7. In rectangle No. 3, sketch a 2 inch diameter circle, using the free circle method. Locate the circle in the center of the rectangle. Erase your light "trial" circle and darken the finished circle.

8. Lightly sketch a rectangle inside rectangle No. 4 at a distance of 1/2 inch inside the border lines. Sketch an arc in each of the corners starting at the lower left-hand corner and working clockwise around the rectangle. Draw the arcs so that they have radii of: 1/2, 3/4, 1, and 1 1/2 inch. Darken the finished rectangle and arcs. Sign your name and date the sketch.

Sketching Ellipses and Irregular Curves

Divide your sheet into four rectangles as shown in Fig. 3-23. Sketch the following ellipses and irregular curve.

9. Sketch an ellipse in rectangle No. 1 with a major axis of 4 inches and a minor axis of 2 1/2 inches, using the rectangular method. Darken the finished ellipse but do not erase your construction lines.

10. In rectangle No. 2, sketch an ellipse with a major axis of 3 inches and a minor axis of 2 inches, using the trammel method. Darken the finished ellipse.

11. Sketch an ellipse in rectangle No. 3 with a major axis of 2 1/2 inches and a minor axis of 1 1/2 inches, using the free ellipse method. Your ellipse should be uniform on both ends. Darken the finished ellipse.

12. On scrap paper, draw an irregular curve similar to the one in Fig. 3-19. Make the curve a suitable size to fit rectangle No. 4 and position it over the rectangle. With a sharp pencil, press lightly to locate points along the curve on the drawing sheet approximately 1/2 inch apart. Sketch the irregular curve through these points. To obtain a smooth curve, make sure your strokes "lead out" of the previous curve into the next set of points. Sign your name and date the sketch.

Sketching Objects on Plain Paper

13. Sketch on A-size plain paper (bond, drawing, or tracing) those problems in Fig. 3-24 assigned by your instructor. Place only one problem on a sheet and do not dimension the object. Select a size for the object that presents a pleasing appearance on the sheet. Strive for good line quality and proportion. Sign your name and date the sketch.

Sketching Objects on Cross-section Paper

14. Sketch on 8 1/2 x 11 cross-section paper (4, 5, 8, or 10 squares per inch) those problems assigned by your instructor from Fig. 3-25. Place only one problem on a sheet and do not dimension the object. Select a size for the object that provides a suitable appearance. Strive for good line quality and proportion. Sign your name and date the sketch.

Sketching Real Objects

15. Sketch one view of objects in the drafting room assigned by your instructor. Use plain or cross-section paper and center the view on the sheet. Review the steps in sketching an object presented in this chapter. Your sketch should reflect your best sketching technique.

Out-of-Class Activity

16. Select some object at home, work, or in the community and sketch one view that best describes the object. Use plain or cross-section paper and present your sketch in class.

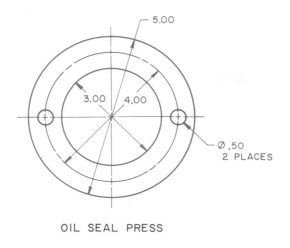

OIL SEAL PRESS

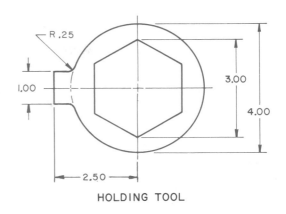

HOLDING TOOL

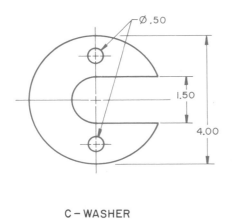

C – WASHER

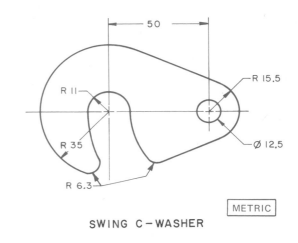

METRIC

SWING C – WASHER

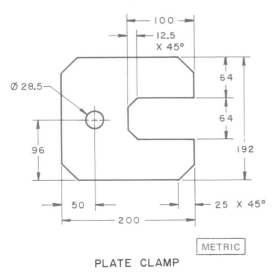

METRIC

PLATE CLAMP

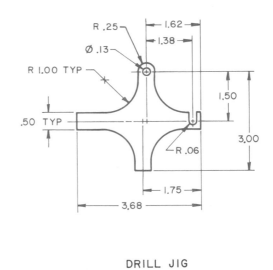

DRILL JIG

Fig. 3-24. Sketch the problems assigned by your instructor on plain paper.

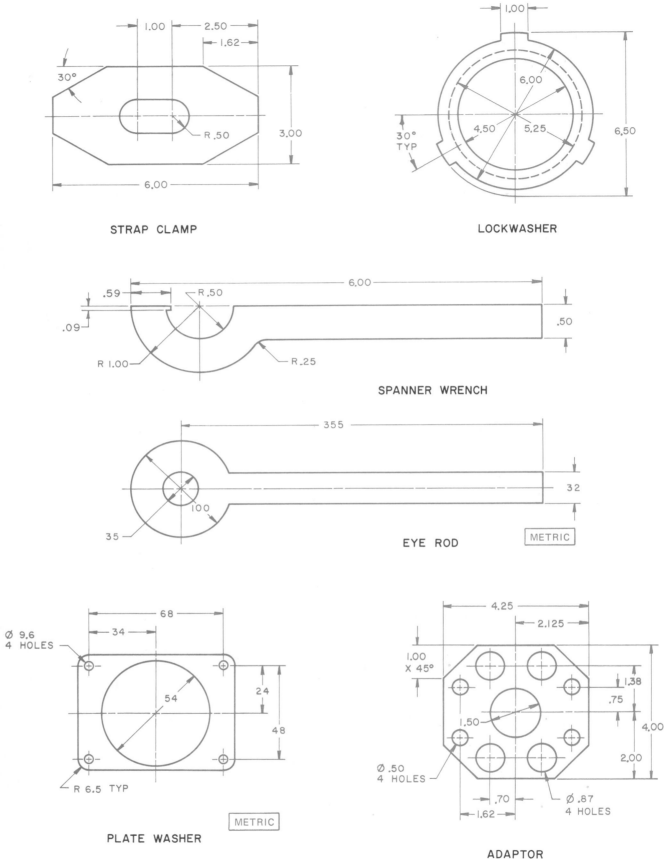

STRAP CLAMP

LOCKWASHER

SPANNER WRENCH

EYE ROD   METRIC

PLATE WASHER   METRIC

ADAPTOR

Fig. 3-25.   Sketch the problems assigned by your instructor on cross-section paper.

Technical Lettering

KEY CONCEPTS

☐ There are different styles of lettering and industry standards.

☐ Guidelines are used for height and slope of letters.

☐ Single-stroke Gothic letters are commonly used in industry.

The purpose of lettering on a drawing is to further clarify the projections (views). Notes may specify materials and processes. Although lettering typewriters, transfer letters, and other devices are being used more, most drafters are still required to letter some drawings freehand.

It is estimated that lettering consumes 20 percent of a manual drafter's time. It is generally agreed that lettering affects the appearance of a drawing more than any other single factor. Lettering that is difficult to read could contribute to costly errors in the manufacturing and servicing of parts and machines. The mastery of lettering techniques is essential.

A careful study of the techniques presented in this chapter and concentrated practice sessions will enable you to develop skill in freehand technical lettering.

PENCIL TECHNIQUE

When freehand lettering, use less pressure on your pencil than when using a drawing instrument. A softer lead (having less clay) is used to maintain equal density (darkness) with lines on the drawing. An HB, F, or H pencil sharpened to a conical point will produce lettering of sufficient quality to reproduce good prints.

When lettering, your forearm should be fully supported on the table with your hand resting on its side. Your third and fourth fingers should rest on the board and your forefinger should be on top of, and in line with, the pencil, Fig. 4-1.

Hold the pencil firmly, but not too tight. Should your fingers tire, pause and flex them a few times. Resting your fingers may help improve your work. Rotate the pencil frequently to maintain a conical point and produce letters of uniform width.

When lettering, as with drawing, you should move your pencil from your non-drawing hand to-

Fig. 4-1. When lettering, your hand and forearm should be fully supported on the drawing surface. Your index finger should be on top of, and in line with, the pencil.

wards your drawing hand. This will "pull" the pencil across the page. If you "push" the pencil, the point will tend to "dig into" the paper. When drawing vertical components of letters, it is also important to "pull" the point across the page.

STYLES OF LETTERING

Hand letter styles can be classified in four groups: Roman, Italic, Text, and Gothic. **Roman** is characterized by thick and thin lines with "accented" strokes, Fig. 4-2A. (The "accented" strokes are called "serifs.") **Italic** letters are similar to Roman, but inclined, Fig. 4-2B. **Text** includes all styles of Old English Cloister, Church, Black, and German Text, Fig. 4-2C. **Gothic** has been used in drafting for many years, Fig. 4-2D. Gothic letters are made up of lines that are all the same "thickness" with no "accented" strokes (serifs). (Letters without serifs are called "sans serif.") Italic and text lettering are useful in printing, various sign applications, and on specialty drawings such as map work and technical illustrations.

Gothic letters have become the standard style used in industry. Gothic is known as **single-stroke Gothic.** "Single-stroke" refers to the width of various parts of letters being formed by a single stroke rather than a number of strokes.

A. ROMAN   **BRAZIL, CHILE AND OTHER**
Advanced ancient cultures flourished in

B. ITALIC   *BRAZIL, CHILE AND OTHER*
Advanced ancient cultures flourished in

C. TEXT   Announcement Big Social Gathering

D. GOTHIC   BRAZIL AND SOUTH AMERICAN
Advanced ancient cultures

Fig. 4-2.   The four main styles of hand lettering are Roman, Italic, Text, and Gothic. Gothic, also called single-stroke Gothic, is the most common font in industry.

Uppercase (capital) letters are recommended for use on machine drawings. They may be either vertical or inclined, but never mixed on the same drawing. Lowercase letters are used for notes on maps and other topographical drawings.

Architectural lettering is similar in style to vertical uppercase Gothic lettering, but less "mechanical" in appearance, Fig. 4-3. Lowercase letters are sometimes used in architectural work.

GUIDELINES

Two types of guidelines are used in freehand technical lettering to maintain uniformity in height and slope: horizontal and vertical, Fig. 4-4. Guide-

ARCHITECTURAL PLANS
NO. 12738

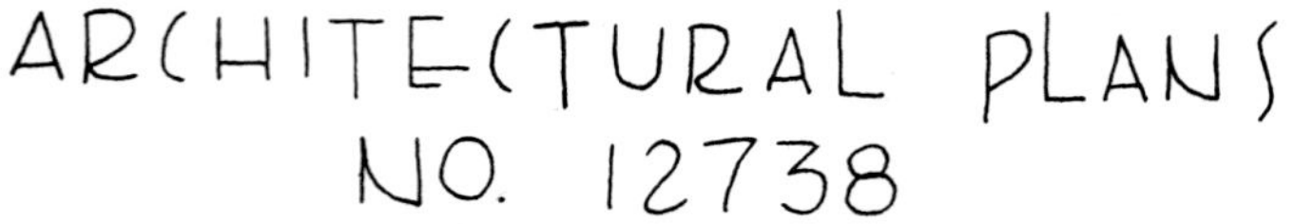

Fig. 4-3.   Lettering used in architectural work is similar to Gothic, but does not appear as "mechanical."

lines are very light lines drawn with a sharp pencil. A 4H or harder pencil is preferred, but your regular drawing pencil may be used. However, take care to draw lines that are light enough not to be seen at arm's length from the drawing. This makes sure that the guidelines will not reproduce.

Horizontal guidelines may be spaced with dividers or with a scale, Fig. 4-5. Lowercase letters are two-thirds the height of uppercase letters. Small uppercase letters, when used with large uppercase letters, are two-thirds to four-fifths the height of large uppercase letters.

The spacing between lines of text appears best when the distance is from one-half to one letter in height. (Refer back to Fig. 4-5.) Vertical or inclined guidelines are drawn at random with a triangle against the T-square or straightedge.

Several useful devices are available for drawing horizontal and vertical guidelines. The Ames Lettering Guide may be used for drawing guidelines for letters 1/16 to 2 inches in height, Fig. 4-6.

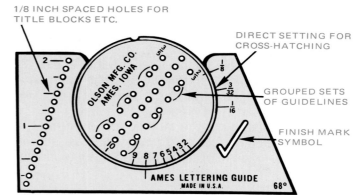

Fig. 4-6.   The Ames Lettering Guide has a variety of uses.

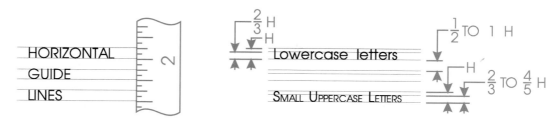

Fig. 4-4.   Guidelines assist in keeping letters straight and at the correct angle.

Fig. 4-5.   Lowercase letters should be 2/3 of the full uppercase letter height. Small uppercase letters should be 2/3 to 4/5 of the full uppercase (large) letter height. The spacing between lines of text should be a minimum of 1/2 of the full uppercase letter height.

To use the guide, place it in position along the T-square or straightedge. Next, insert the point of a sharp pencil in the holes at the desired spacing. Then, slide the guide along the straightedge to draw the horizontal guidelines. For an example of vertical or inclined guidelines, refer back to Fig. 4-4.

The Braddock-Rowe Triangle is also useful for drawing horizontal and slope guidelines, Fig. 4-7. Numbers on the triangle indicate the height for capital letters in thirty-seconds of an inch. For example, No. 8 is 8/32 or 1/4 inch from top to bottom lines between the group of three holes.

A third device for drawing guidelines is the Parallelograph, Fig. 4-8. This instrument provides horizontal guideline spacing in 32nds of an inch and in millimeters. Slopes for 68° and 75° inclined letters are also provided.

SINGLE-STROKE GOTHIC LETTERS

Lettering strokes are the "guide posts" to forming good letters. These strokes consist of straight-line stems, cross bars, and well-proportioned ovals, carefully combined to produce a well balanced letter form.

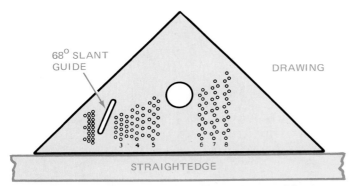

Fig. 4-7. The Braddock-Rowe Triangle simplifies drawing lettering guidelines. (Teledyne Post)

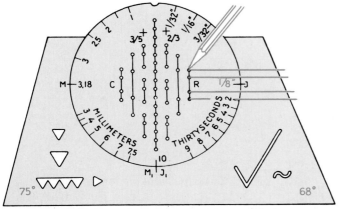

Fig. 4-8. The Parallelograph lettering guide has a variety of uses. (Gramercy)

The vertical Gothic alphabet and numerals shown in Fig. 4-9 are broken into groups of letters and numerals of similar strokes. Close study of this illustration will assist you in learning the order and direction of strokes used in forming each letter. The width of letters will vary from one space for the letter "I" to six spaces for the "W."

Draw vertical and inclined strokes with a movement of the fingers, Fig. 4-10. Form horizontal strokes by pivoting the hand at the wrist, with a

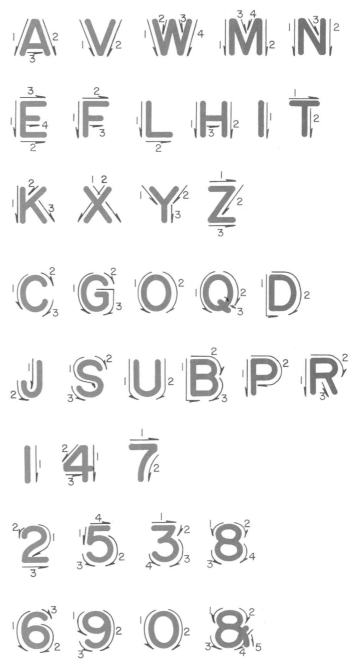

Fig. 4-9. Suggested pencil strokes for making vertical uppercase letters and numerals. Remember, it is important to make pencil strokes in such a way that you are "pulling" the lead across the sheet, not "pushing" it.

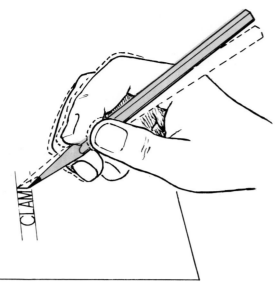

Fig. 4-10. Vertical and inclined strokes are made by finger movements only.

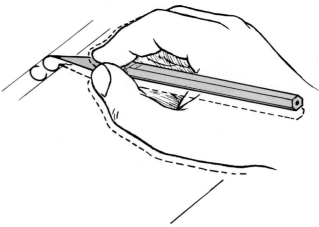

Fig. 4-12. Ovals are formed by movement of the hand and fingers in combination.

slight finger movement as needed to maintain a straight line, Fig. 4-11. Ovals are formed with a combination of hand and finger movement, Fig. 4-12. Ovals are perfect ellipses with a major and a minor axis.

Inclined Gothic uppercase letters and numerals are drawn in a manner similar to vertical letters and numerals. However, the vertical axis is at an angle between 68°and 75°, Fig. 4-13.

Lowercase vertical Gothic letters are formed as shown in Fig. 4-14. The body of the lowercase letters is two-thirds the height of uppercase letters. Ascending or descending stems are equal in length to the height of the uppercase letters. Lowercase inclined Gothic letters are formed as shown in Fig. 4-15. These letters should be inclined at 68° to 75°.

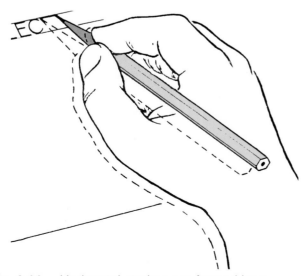

Fig. 4-11. Horizontal strokes are formed by a movement of the hand at the wrist, along with a slight finger movement.

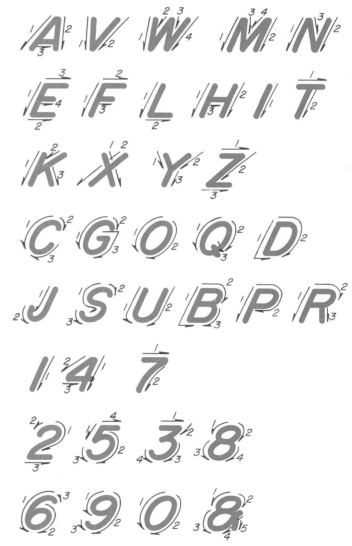

Fig. 4-13. Suggested pencil strokes for making inclined Gothic uppercase letters and numerals. Remember, it is important to make your pencil strokes in such a way that you are "pulling" the lead across the sheet.

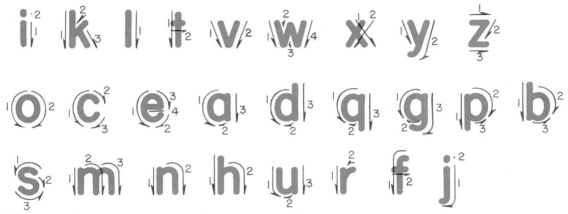

Fig. 4-14. Suggested pencil strokes for vertical Gothic lowercase letters. Remember to "pull" the pencil lead across the sheet.

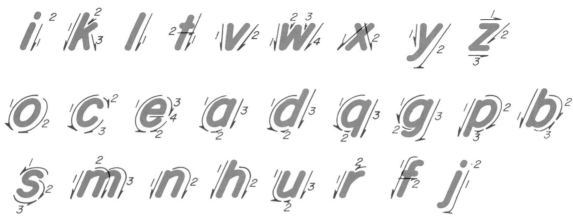

Fig. 4-15. Suggested pencil strokes for making inclined Gothic lowercase letters. Remember to select pencil strokes that will "pull" the lead across the sheet.

Combining Large and Small Uppercase Letters

Some drafters use large and small uppercase (capital) letters in combination for titles or notes on drawings. This combination is more easily read than all uppercase letters of the same height, Fig. 4-16. Height of the small letters should be two-thirds to four-fifths that of the large letters.

Skill in lettering comes with careful study of the form of the letters–and by diligent practice. With practice, any student with a talent for drafting can learn to produce good letters.

WOOD PATTERN TO BE
ENGINEERING APPROVED

Fig. 4-16. Large and small uppercase letters can be used in combination. This combination is easier to read than all large uppercase letters.

Proportion in Letters and Numerals

Once the technique of forming letters and numerals is understood, care must be given to proportioning each element. Proportion is necessary to present a neat appearance. This is especially important where letters are formed into words and sentences, Fig. 4-17.

However, there may be times when it is necessary, or desirable, to compress or expand letters or words. This occurs when space is limited or if you want to attract attention, Fig. 4-18. Keep in mind that you should maintain good proportion when possible.

GOOD
PROPORTION

B A D
PROPORTION

Fig. 4-17. Proportion in forming individual letters is important. Incorrect proportions may result in words that look "pulled apart" (top right) or "smashed together" (bottom right).

NEW HOMES
PROJECT

Fig. 4-18. Good proportion must be maintained even when it is necessary to expand or compress words.

Stability in Letters and Numerals

Optical illusion is also a factor in lettering. For this reason, it is necessary to place the horizontal bar on letters such as "B," "E," and "H" slightly above center. If not, they will appear top-heavy and unstable, Fig. 4-19. It is also necessary to draw the upper portion of letters such as "B," "K," "R," "S," "X," and "Z" and the numerals "2," "3," "5," and "8" slightly smaller than the lower portion to show stability, Fig. 4-20.

Quality of Lettering

The appearance of lettering on a drawing is enhanced when style, height, slope, spacing, and line weight are uniform. The appearance of the drawing and skill of the drafter are reflected in the quality of lettering.

Care should be taken to form each letter correctly. Letters should be formed within the lightly drawn guidelines. Letters should also be of the proper line density. Lettering is a special technique that differs from writing, Fig. 4-21.

BEH   BEH

STABLE   UNSTABLE

Fig. 4-19. Some letters such as "B," "E," and "H" must actually be drawn with the horizontal "bar" just above the true center to make the letter appear correct. This makes lettering stable. If the "bar" is drawn at or below the true center of the letter, then the letter will be unstable.

KRSX   KRSX
2358   2358

STABLE   UNSTABLE

Fig. 4-20. Stable letters and numerals are more pleasing in appearance.

BRACKET   BRACKET
A-125407   A-125407
LETTERING   WRITING NOT
APPROVED   APPROVED

Fig. 4-21. Lettering requires a special technique that reflects the skill of the drafter. Lettering is different from writing.

Composition of Words and Lines

The way that letters are combined into words and sentences tends to reveal the drafter's lettering skill. Many beginning drafters tend to space the letters of words too widely and crowd the spacing between words. Lettering of this nature is difficult to read, Fig. 4-22.

Correct spacing of letters within words is based upon the total *area* between two letters, not just the *distance* between letters. When letters are spaced based only on *distance,* poor composition results, Fig. 4-23.

The shape of the letters themselves determines how much spacing should occur between any two letters. For example, when "A" and "M" are next to each other, less distance is required than for "I" and "M," Fig. 4-24. When "T" and "C" are next to each other the letters nearly overlap. This is also true when "L" and "T" are next to each other.

THESE LETTERS AND
WORDS REPRESENT
GOOD COMPOSITION

THESELETTERS AND
WORDS   RE PRESENT
PO OR CO MPOSITION

Fig. 4-22. Composition of words and lines is very important when lettering a drawing. If the words are improperly spaced, the result is a drawing that appears to be of poor quality.

BORE TO OBTAIN
.0002 CLEARANCE

Fig. 4-23. Equal spacing of letters results in poor composition and is hard to read.

AM IM TC BOLT

Fig. 4-24. The total *area* between letters determines the spacing, not the total *distance.*

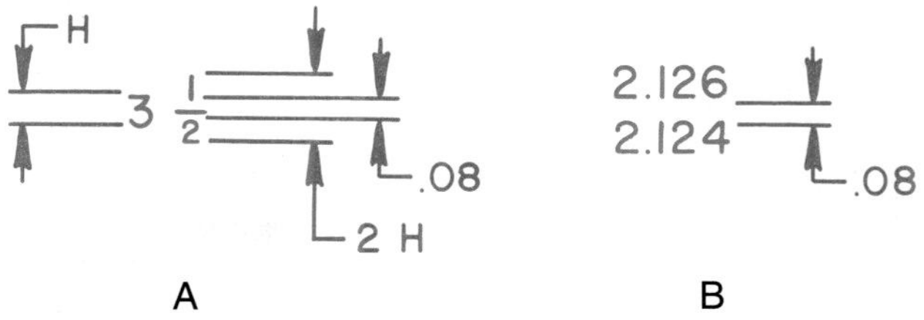

Words should be separated by a space equivalent to the letter space "O," Fig. 4-25. Two letter spaces of "O" are allowed between the end of one sentence and the start of the next.

The space between lines of letters should equal two-thirds of the letter height. This makes text easy to read and also gives a good appearance, Fig. 4-26. Spacing may vary from one-half to one full letter height, depending on the space available, but it should be consistent on a single drawing. Where fractions are involved, spacing between lines should be a full letter height.

Size of Letters and Numerals

Uppercase letters and whole numerals are usually a minimum of 1/8 inch in height for notes. Height for the drawing title and number is generally 1/4 inch.

The overall height of fractions is twice the height of whole numbers, Fig. 4-27A. Note that the numerator and denominator are smaller than the whole number, and they are separated by .08 inch so that they **do not** touch the fraction bar. This avoids confusion in reading fractions on a drawing and provides legible copy from microfilm enlargements. Limit dimensions are shown with a space of .08 inch minimum between the upper and lower dimensions, Fig. 4-27B.

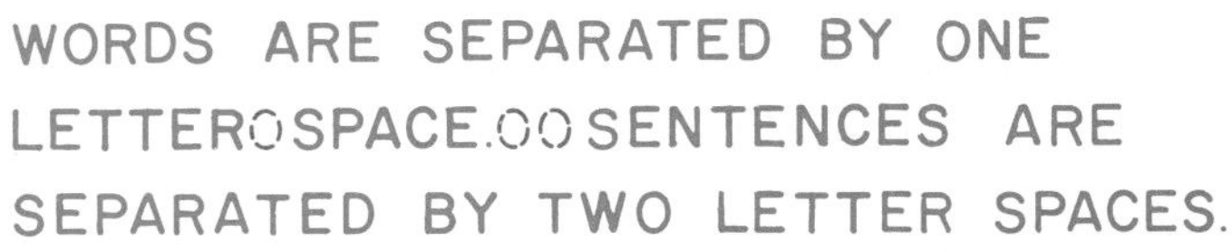

WORDS ARE SEPARATED BY ONE
LETTER○SPACE.○○SENTENCES ARE
SEPARATED BY TWO LETTER SPACES.

Fig. 4-25. The letter "O" is used to determine spacing between words and sentences.

Fig. 4-27. Fractions should be a *total* of twice the full letter height, including the fraction bar. The denominator and numerator should be .08 inch from the line. When upper and lower tolerances are given, they should be separated by .08 inch.

However, some companies have created their own specifications for the sizes of letters and numbers on drawings, Fig. 4-28. These specifications have been created to meet certain needs and to make sure that every drawing produced by, or for, the company is consistent.

Notes on Drawings

Notes on drawings supplement the graphic presentations. Notes are lettered on the drawing to be read horizontally. Uppercase letters at least 1/8 inch high are preferred.

Minimum spacing between lines within a note is 2/3 letter height. (Refer back to Fig. 4-26.) A full letter height should be used on drawings to be microfilmed. Spacing between two separate notes should be at least two letter heights, Fig. 4-29.

Balancing Words in a Space

It is necessary at times to **balance** (evenly space) words within a limited space, such as a title block, Fig. 4-30. This may be done by first lettering the title or note on scrap paper using appropriate guidelines. Then, slip the copy under your tracing paper or vellum and adjust it to suit. If you are working on an opaque surface, measure the scrap copy and transfer the starting point to the drawing.

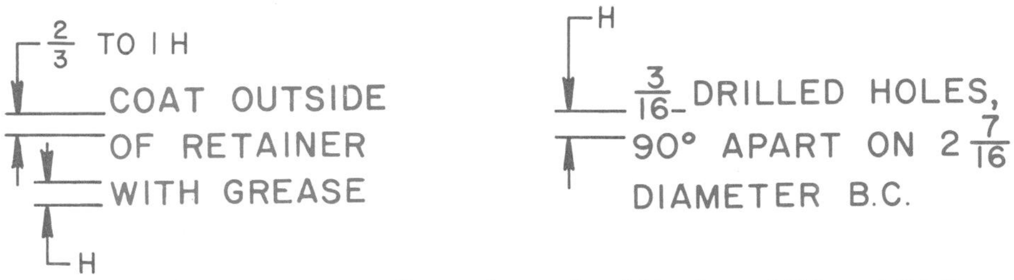

$\frac{2}{3}$ TO 1 H
COAT OUTSIDE
OF RETAINER
WITH GREASE

$\frac{3}{16}$ DRILLED HOLES,
90° APART ON 2$\frac{7}{16}$
DIAMETER B.C.

Fig. 4-26. Spacing between lines of text should be from 2/3 to a full letter height in distance. When fractions appear in a line, the lines should be separated by a full letter height in distance.

USE	SIZE (MIN.)
DRAWING NUMBER	.38
DRAWING TITLE, CODE IDENTIFICATION NO.	.20
LETTERS AND NUMERALS	.156*
TABULATED AND SECTION LETTERS	.40
THE WORDS "SECTION" AND "VIEW"	.20
* THIS APPLIES TO LETTERS AND NUMERALS ON FIELD OF DRAWING, IN GENERAL NOTES, REVISION BLOCK AND PARTS LIST. ALL TOLERANCES ARE THE SAME CHARACTER SIZE AS THE DIMENSION.	

MINIMUM LETTER SIZES								
ITEM	SIZES							
	A		B		C		D AND LARGER	
	INCHES	MM	INCHES	MM	INCHES	MM	INCHES	MM
DRAWING AND PART NUMBER	.250	6	.250	6	.312	8	.312	8
TITLE	.125	3	.156	4	.156	4	.188	5
LETTERS AND FIGURES FOR BODY OF DRAWING (INCLUDING DIMENSIONS AND NOTES)	.125	3	.156	4	.156	4	.188	5
TOLERANCES	.100	2.5	.125	3	.125	3	.156	4
DESIGNATION OF VIEWS, SECTIONS, DETAILS AND DATUM: "VIEWS," "SECTION," AND "DETAIL".	.156	4	.188	5	.188	5	.250	6
"A–A," "B," "X–X," etc.	.188	5	.250	6	.250	6	.312	8
SECURITY CLASSIFICATION, DRAWING STATUS, REFERENCE DRAWING, etc.	.250	6	.250	6	.250	6	.250	6
TYPED SCHEMATICS, DIMENSIONS, AND NOTES ON BODY OF DRAWING	.100	2.5	.125	3	.142	4	.142	4

Fig. 4-28. Some organizations have developed their own specifications for the sizes of letters and numbers. These specifications serve the company's own needs and makes certain that all drawing for the company are consistent. (Globe Engineering Documentation Services, IBM)

Fig. 4-29. The minimum spacing between separate notes is two full letter heights.

Fig. 4-30. When words must be contained within a given space, like a title block, the words should be "balanced" in that space.

Lettering to be Microfilmed

Drawings to be *microfilmed* require special lettering. The letters must be large enough and dense enough to reproduce clearly. Generally, the lettering must be readable on a print one-half the size of the original.

Some companies have modified certain letters and numerals used on drawings to be microfilmed. These modifications help improve clarity on blowbacks (enlargements), Fig. 4-31.

LETTERING WITH PEN AND INK

Most drawings to be used for photographic reproductions in technical publications are inked. These drawings are usually presentation drawings. Inked drawings are also made if a number of copies of prints are to be made over a long period of time.

Inking of freehand letters on drawings requires a special technique and proper equipment. Several styles of pens are available. Standard lettering pen points come in a range of sizes from very fine to heavy.

Technical Pen

Freehand inked letters should be done with a *technical pen,* Fig. 4-32. Technical pens have the

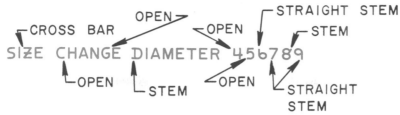

Fig. 4-31. When drawings will be transferred to microfilm, some letters and numbers may need to be modified. This will make certain that the letters and numbers will be accurately reproduced to, and later from, the microfilm.

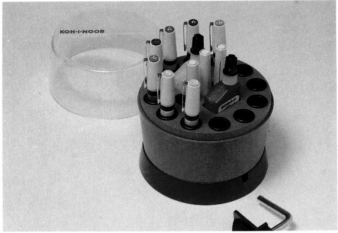

Fig. 4-32. Technical pens are used to ink drawings. (Koh-I-Noor Rapidograph, Inc.)

advantage of enabling uniform stroke widths and do not vary, as the more flexible pen points may. Technical pens consist of a pen point, a needle running through the point to maintain the ink flow, and an ink reservoir, Fig. 4-33. Technical pens are made for use with specially formulated inks.

Most technical pens are shaped to permit their use with Leroy lettering devices, Fig. 4-34. Technical pens also come in a range of sizes suitable for use with various lettering templates. Technical pens also may be used for inking lines on a drawing.

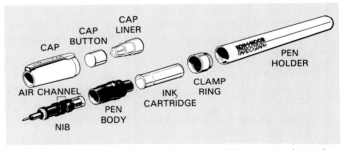

Fig. 4-33. Technical pens have an ink reservoir and a pen point (nib). Some pens have reservoirs that are actually cartridges that can be quickly removed and changed. (Koh-I-Noor Rapidograph, Inc.)

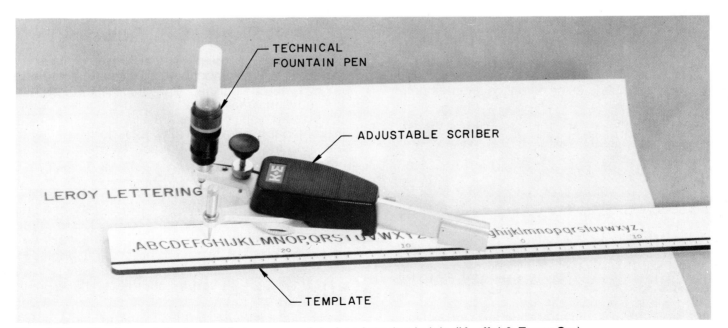

Fig. 4-34. The Leroy lettering guide is very useful when lettering in ink. (Keuffel & Esser Co.)

Mechanical Lettering Devices

Since inking letters and numerals requires considerable skill to achieve satisfactory results, many inked drawings are lettered with mechanical devices. (Refer back to Fig. 4-34.)

The lettering guide consists of a scriber, pen, and template. The scriber is adjustable for making inclined as well as vertical lettering, and templates are available in a variety of letter styles and sizes. Templates are also available for various graphic symbols, such as electronic and architectural symbols, Fig. 4-35.

The procedure for using the Leroy lettering guide is as follows.

1. Select the template size and style desired. Place it along the straightedge.

2. Place the tail pin of scriber in the straight groove of the template and the tracer pin in the letter to be traced.

3. Adjust the straightedge and the template to correct position and trace the letter.

4. Letters are spaced by "eying" the space between the letter being positioned and the previous letter.

5. Words are spaced by positioning the letter on the template that just precedes the first letter of the next word. For example, in the note "SEAM WELD", the "V" is positioned as if it were the next letter after the "M" in "SEAM," Fig. 4-36. The "W" is then correctly spaced to start the next word. If the letter preceding the next word is narrower ("I" for example) or wider ("W" for example) than a normal letter, make the adjustment by eye.

6. Sentences are spaced similar to words, except two letter spaces precede the first letter of the next sentence, Fig. 4-37.

7. To estimate the space required for a word or group of words, count the number of letters and spaces. Then use the scale on the lower edge of each template to lay off space requirements.

Fig. 4-36. When spacing words with the Leroy lettering guide, use the letter that comes immediately before the first letter of the next word to indicate the space (in this case, the letter "V" is used as the space).

Fig. 4-37. When spacing between sentences with the Leroy lettering guide, use two letters to indicate the space (in this case, the letters "Q" and "R" are used to indicate the space).

Another type of lettering device is a lettering guide, Fig. 4-38. A third type of mechanical lettering device is the height and slant control scriber. This

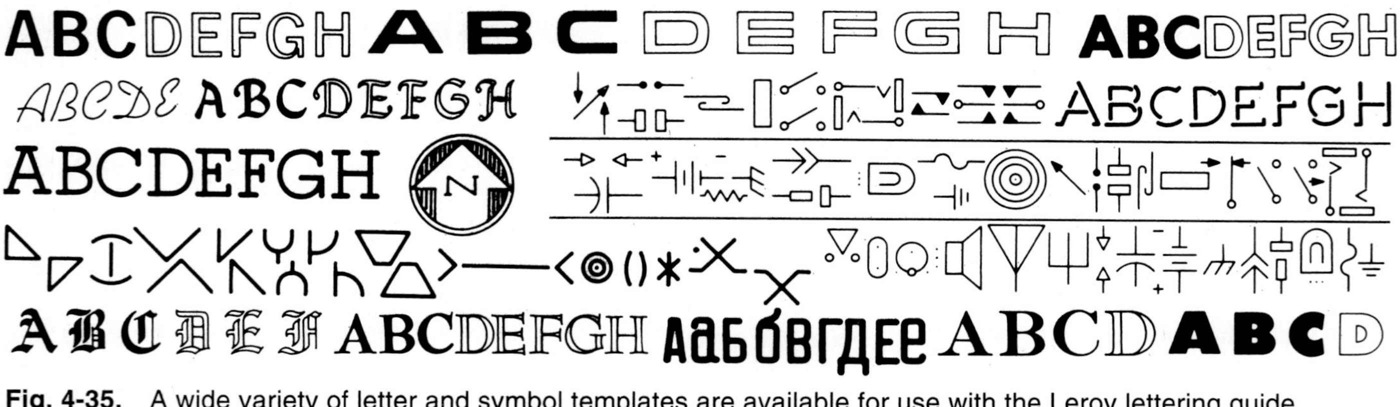

Fig. 4-35. A wide variety of letter and symbol templates are available for use with the Leroy lettering guide. (Keuffel & Esser Co.)

Fig. 4-38. Lettering templates are a useful tool for lettering drawings.

device operates from a template like the Leroy, but it is constructed in a way that permits the expansion or compression of letters. This device allows the ratio of the height of the letter to the width to be varied.

Notes and Symbols in Computer Systems

The computer is used by many industries today to prepare notes and symbols for drawings, Fig. 4-39. Computer-aided design and drafting speeds the lettering process and provides uniform, legible notes and symbols.

The notes and symbols are entered into the computer. These can then be output on a drawing by a plotter or printer. When the entire drawing is developed by computer graphics, the notes may be input through the keyboard and the symbols selected from the menu tablet, Fig. 4-40. The notes and symbols appear on the drawing in the proper position and are printed along with the graphical solution.

Overlays and Transfer Type

Special printed sheets and tapes of pressure sensitive letters and symbols are available. This is called **transfer type.** Transfer type can be used for lettering, dimensioning, or applying certain symbols to drawings, Fig. 4-41.

When using transfer type, the letter, number, or symbol that you will use is aligned in the correct position. Then pressure is applied with a special tool and the type is "transferred."

Overlays are adhesive backed sheets that "lay over" the drawing. These sheets then adhere to the drawing sheet. Many title blocks and borders are

Fig. 4-39. Notes and text can be directly entered into a computer and then reproduced on a hard-copy drawing.

Fig. 4-40. Digitizer tablets allow the drafter to select the desired symbol and input it to the computer. (Summagraphics)

overlays. These materials save industry considerable time in drafting. They are particularly useful where standard notes or symbols are frequently used.

Fig. 4-41. Transfer type can be used to quickly place standard symbols and text onto a drawing.

PROBLEMS AND ACTIVITIES

The following problems and activities are designed to help you practice lettering. Follow the procedures outlined and the directions of your instructor.

Forming Single-Stroke Gothic Letters

Prepare a lettering layout sheet. Use one of the guideline devices previously described to lay out horizontal and slope guidelines as shown in Fig. 4-42, or use a commercial lettering sheet as directed by your instructor. When all materials are ready, do the following Gothic lettering activities.

1. Draw vertical uppercase letters as shown in Fig. 4-9. Letter the alphabet and numerals as many times as space permits. Spacing for letter height is 3/8 inch (10 mm) and spacing between lines of letters is 3/16 inch (5 mm). Use vertical guidelines to keep letters uniform.

2. Draw inclined uppercase letters as shown in Fig. 4-13. Letter the alphabet and numerals as many times as space permits. Spacing for letter height is 1/4 inch (6 mm) with 1/8 inch (3 mm) spacing between lines. Use 68° inclined guidelines.

3. Draw vertical lowercase letters as shown in Fig. 4-14. Repeat the alphabet as many times as space permits. Spacing for letter height is 3/8 inch (10 mm) with 3/16 inch (5 mm) spacing between lines. Use vertical guidelines.

4. Draw inclined lowercase letters as shown in Fig. 4-15. Repeat the alphabet as many times as space permits. Spacing for letter height is 1/4 inch (6 mm) with 1/8 inch (3 mm) spacing between lines. Use inclined guidelines.

Lettering Drawing Notes

5. Letter the following drawing note in 1/8 inch (3 mm) vertical uppercase letters on a drawing

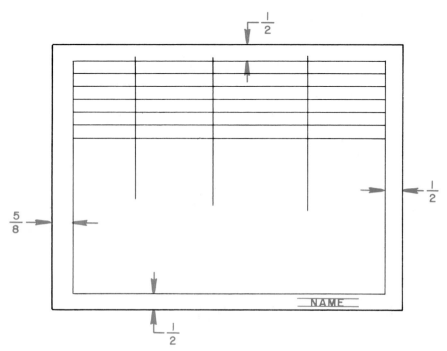

Fig. 4-42. Use this layout to complete the problems at the end of the chapter. The dimensions are given in inches.

sheet. Decide on a pleasing spacing between lines and a line length not to exceed four inches. Center the note in the upper left-hand quadrant (quarter) of the sheet, starting one inch from the top border. Do not include the quotation marks in your note.

"DIMENSIONS THROUGHOUT ARE TO 90° BEND AS ASSEMBLED ON PRODUCT. PART SHOULD BE OVERBENT TO 91° FOR ADDITIONAL TENSION."

6. Use 3/16 inch (5 mm) inclined uppercase letters to letter the following notes in the upper right-hand quadrant of the same sheet.

"SLOT MUST HAVE NO EXTERNAL BURRS"
".203 (5.2) DIA HOLE–PIERCE FROM BOTH SIDES–FOUR PLACES."

7. Use 1/8 inch (3 mm) vertical uppercase and lowercase letters in lettering the following note in the lower left-hand quadrant of the sheet above.

"Note: An Easement Of Four Feet (1.3 m) On Either Side Of This Line Is Reserved For Utility Line Through Properties."

8. Use 3/16 inch (5 mm) initial large uppercase letters and small uppercase letters to letter the following note in the lower right-hand quadrant of the sheet used for problems 5, 6, and 7.

"EMBOSS 5/16 (7.9) DIA x 1/16 (1.5) DEEP–INSIDE ONLY–FOUR PLACES."

9. Review several prints from industry and evaluate the lettering and dimension numerals as to style, size, uniformity, and quality of reproduction. Be prepared to discuss your findings in class.

10. Select a note from an industrial print and letter the note in the same size and style lettering. Compare your work with that done by the industrial drafter.

11. Select a title block from an industrial print. Lay this out on a sheet and letter the content in the same size and style. Compare your work with that on the print.

12. Lay out a lettering sheet as shown in Fig. 4-42 for 3/16 inch (5 mm) letters. Letter the alphabet and numerals lightly in pencil on the upper half of this sheet and then ink the copy freehand.

13. With one of the mechanical lettering devices, letter in ink the following note in .175 inch (4.5 mm) high uppercase letters on the lower half of the sheet used in No. 12. Space allowed for the note is 2 1/2 inches (64 mm) in length and depth as required.

"RIVETS MUST NOT BE CURLED TOO TIGHTLY SINCE THEY ARE AT MOVING JOINTS. RIVET HEADS SHOULD BE OUTSIDE."

14. Using transfer type, prepare the following title.
"FORK BRACKET, SPINDLE RAM ASS'Y."

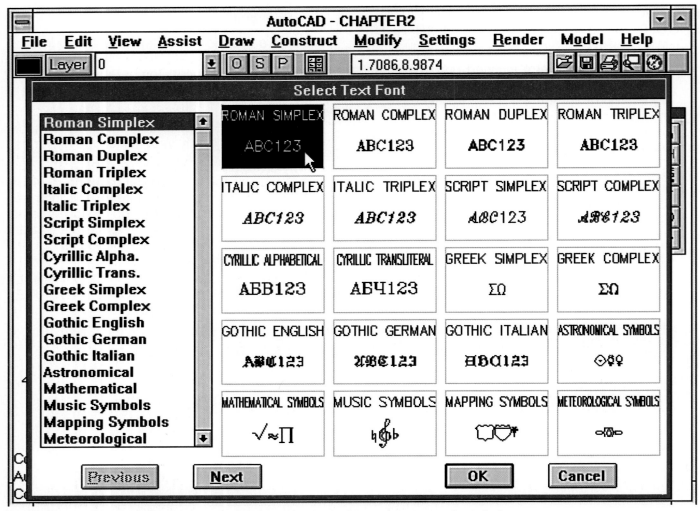

CADD allows many different text styles to be used, including many types of symbols.

5 Basic Geometry Used In Drafting

KEY CONCEPTS

☐ Straight lines can be used to perform basic geometric functions.

☐ Straight lines can be used to construct, bisect, and transfer angles.

☐ Polygons are geometric objects enclosed with straight lines.

☐ Squares are a special type of polygon called a regular polygon.

☐ Regular polygons have all sides of equal length and all angles equal.

☐ Circles and arcs are geometric objects constructed with curved lines.

Engineers, architects, designers, and drafters regularly apply the principles of geometry to the solution of technical problems. This is clearly evident in highway interchanges, architectural structures, and complex machine parts, Fig. 5-1.

Much of the work done in the drafting room is geometric constructions. Every person engaged in technical work should be familiar with solutions to common problems in this area. Methods of problem solving presented in this chapter are based on the principles of *plane geometry,* Fig. 5-2. However, constructions are modified and many short cuts are introduced to take advantage of the instruments available in modern drafting rooms.

Accuracy is very important in drawing geometric constructions. A slight error in laying out a problem

Fig. 5-1. Geometric construction is used in building highways and architectural structures and in manufacturing machine parts. (U.S. Department of Transportation, The Gillette Company)

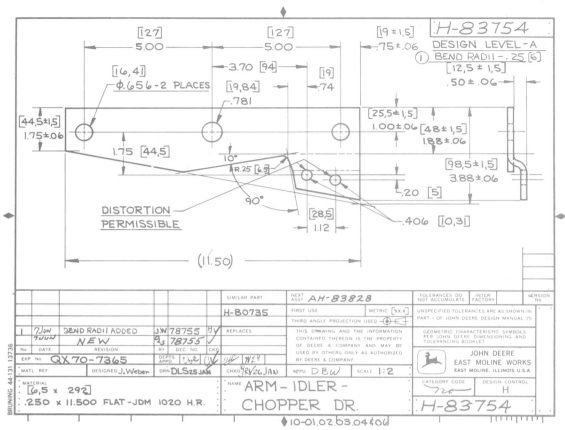

Fig. 5-2. Drawing a machine part may require the application of different geometric concepts.

could result in costly errors in the final solution. Use a sharp 2H to 4H lead in your pencil and compass. Make construction lines sharp and very light. They should not be noticeable when the drawing is viewed from arm's length.

STRAIGHT LINES

There are a number of geometric constructions the drafter needs to perform using straight lines. Some examples are presented in the following paragraphs.

Bisecting Lines

Bisect a line by the triangle and T-square method

A line can be **bisected** (divided into two equal lengths) by using a triangle and a T-square. Use the following procedure for this method.

1. Line AB is given.
2. Draw 45° or 60° lines AC and BC, Fig. 5-3A. The intersection of these two lines is point C.
3. Line CD, drawn perpendicular to line AB, bisects line AB.

Bisect a line by the compass method

A compass and a straightedge can be used to draw a perpendicular bisector to any given line. This method is quick and very useful for lines that are not vertical or horizontal. The following procedure describes this method.

1. Line EF is given.
2. Using E and F as centers, strike arcs with equal radii greater than one-half the length of EF, scribing points G and H, Fig. 5-3B.

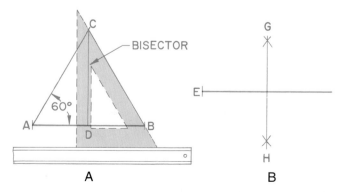

Fig. 5-3. A line can be bisected using a triangle and a T-square.

3. Use a straightedge to draw line GH. This line is the perpendicular bisector of the given line EF.

Line Drawing Technique

Draw a line through a given point and parallel to given line using a triangle and a T-square

1. Line AB and point P are given. A line parallel to AB is to be drawn through point P.
2. Position a T-square or straightedge and a triangle so the edge of the triangle lines up with line AB, Fig. 5-4A.
3. Without moving the T-square, slide the triangle until its edge is in line with point P.
4. Without moving the T-square or the triangle, draw line CD through point P. This line is parallel to line AB.

Draw a line through a point and parallel to another line using a compass

1. The line EF and the point P are given. A line parallel to EF and passing through point P is to be drawn.
2. With point P as the center, strike arc GK that intersects line EF. This radius is chosen arbitrarily. The intersection is point K. With the same radius and point K as the center, strike arc PJ, Fig. 5-4B.
3. With the distance from point P to point J as the radius and point K as the center, strike an arc to locate point G. Point G is the point where this arc and the first arc drawn with point P as the center intersect.
4. Draw line GP. This line is parallel to EF.

Draw a line through a point and perpendicular to a given line using a triangle and a T-square

1. The line AB and the point P are given. A line perpendicular to AB and passing through point P is to be drawn.

2. Place a triangle so that its hypotenuse is on the T-square. Position the T-square and the triangle so an edge of the triangle lines up with the line AB, Fig. 5-5A.
3. Without moving the T-square, slide the triangle so its other edge is in line with point P.
4. Draw the line PC. This line is perpendicular to line AB.

Draw a line through a point and perpendicular to a given line using a compass

1. The line DE and the point P are given. A line perpendicular to DE and passing through point P is to be drawn.
2. With point P as the center and using any convenient radius, strike arc FG, Fig. 5-5B. This arc should intersect line DE at two places. These intersections become points F and G.
3. Using points F and G as centers and the same radius as arc FG, strike intersecting arcs at point H.
4. Draw line PH. This line is perpendicular to line DE.

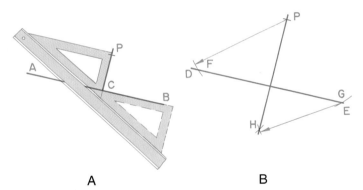

Fig. 5-5. A–Drawing a line through a given point and perpendicular to a given line with a T-square and a triangle. B–A compass can also be used.

Dividing a Line

Divide a line into a given number of equal parts by the vertical line method

1. Line AB is given and to be divided into eleven equal parts.
2. With a T-square and a triangle, draw a vertical line at point B, Fig. 5-6A.
3. Locate the scale with one point at point A and adjust so that eleven equal divisions are between point A and the vertical line BC.
4. Mark vertical points at each of the eleven divisions. Project a vertical line parallel to line BC

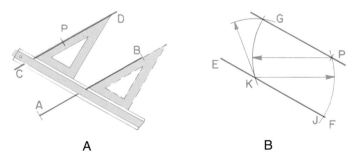

Fig. 5-4. A–A line through a given point and parallel to a given line can be drawn with a T-square and a triangle. B–A compass can also be used for this construction.

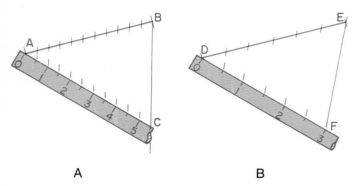

Fig. 5-6. A–A line can be divided into any number of equal parts using the vertical line method. B–The inclined line method can also be used for this construction.

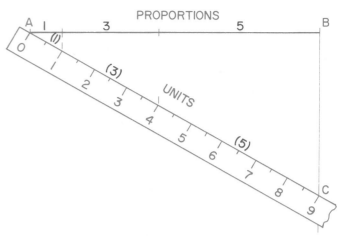

Fig. 5-7. A line can be proportionally divided using the vertical line method.

from these divisions to line AB. These verticals divide line AB into eleven equal parts.

Divide a line into a given number of equal parts by the inclined line method

1. Line DE is given and to be divided into six equal parts.
2. Draw a line from point D at any convenient angle. With a scale or dividers, lay off six equal divisions, Fig. 5-6B. The length of these divisions is chosen arbitrarily, but all should be equal.
3. Draw a line between the last division, point F, and point E.
4. With lines parallel to line FE, project the divisions to line DE. This will divide line DE into six equal parts.

Divide a line into proportional parts by the vertical line method

1. Line AB is given and to be divided into proportional parts of 1, 3, and 5 "units."
2. With a T-square and a triangle, draw a vertical line at point B, Fig. 5-7.
3. Locate the scale with one point on point A. Adjust so that nine equal units (1 + 3 + 5 = 9 units) are between point A and the vertical line BC.
4. Mark each of the proportions. Project a vertical line from these divisions to line AB. These verticals divide line AB into proportional parts of 1, 3, and 5 units.

ANGLES

Another geometric element in drafting is angle construction. The common terms applied to angles in drafting are illustrated in Fig. 5-8. The following symbols are used to represent angles and angular constructions: ∠ = angle; Δ = triangle; ⊥ = ⌐ =

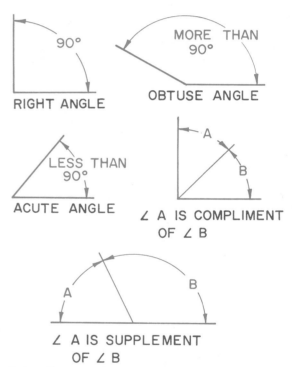

Fig. 5-8. Several different terms are used to describe the properties of angles in drafting.

perpendicular. Techniques for drawing angular constructions are discussed in this section.

Angular Constructions

Bisect an angle

1. Angle BAC is given and to be bisected, Fig. 5-9A.
2. Strike arc D at any convenient radius, Fig. 5-9B.
3. Strike arcs with equal radii. The radii should be slightly greater than one-half the distance from

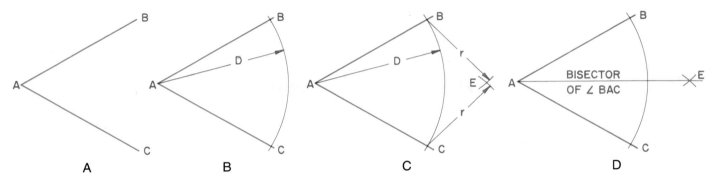

Fig. 5-9. An angle can be bisected using a compass.

point B to point C. The intersection of these two arcs is point E, Fig. 5-9C.

4. Draw line AE. This line bisects angle BAC, Fig. 5-9D.

Transfer an angle

1. Angle BAC is given and to be transferred to a new position at A′B′, Fig. 5-10A.

2. Strike arcs D and D′ at any convenient radius at centers A and A′, Fig. 5-10B. (Note that D and D′ are the same radius.)

3. Adjust compass for arc between points E and F. Strike an arc of the same radius at center E′, Fig. 5-10C. The intersection of these two arcs is point F′.

4. Draw line A′C′ through point F′. Angle BAC has been transferred to angle B′A′C′.

Draw a perpendicular (90°) to a line by the 3, 4, 5 method

1. Line AB is given. Draw line BC perpendicular to line AB.

2. Select a unit of any convenient length and lay off 3 units on line AB, Fig. 5-11A. Note that these units do not have to cover the entire length of the line.

3. With a radius of 4 units, strike arc C with center at point B.

4. With a radius of 5 units, strike arc D with center at point A. The intersection of arc C and arc D is point F, Fig. 5-11B.

5. Draw line BF. This line is perpendicular to line AB.

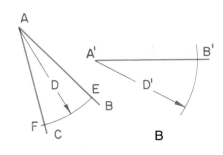

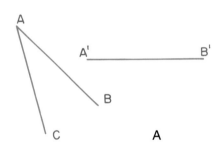

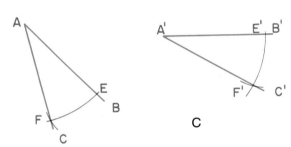

Fig. 5-10. An angle can be transferred using a compass.

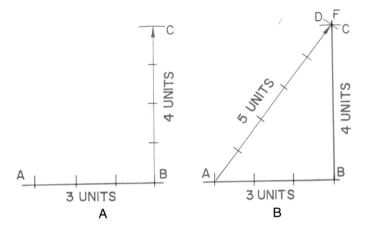

Fig. 5-11. A line perpendicular to a given line can be constructed using the 3, 4, 5 method.

Lay out any given angle with a protractor

1. Line AB is given with an angle of 35° counter-clockwise to be drawn at point C.
2. Locate the protractor accurately along line AB with the vertex indicator at point C, Fig. 5-12.
3. Place a mark at 35°. This mark is point D.
4. Draw a line from point C to point D. The angle DCB equals 35°.

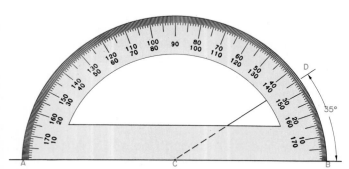

Fig. 5-12.   A protractor is very useful in laying out angles.

POLYGONS

A geometric figure enclosed with straight lines is called a **polygon.** Such figures are called **regular polygons** when all sides are equal and all interior angles are equal. Fig. 5-13 illustrates some common polygons.

Constructing Triangles

Construct a triangle with three side lengths given

1. The side lengths A, B, C are given, Fig. 5-14A.
2. Draw a base line and lay off one side (side C in this case), Fig. 5-14B.
3. With a radius equal to one of the remaining side lengths (side A in this case), lay off an arc (arc A in this case). (Refer back to Fig. 5-14B.)
4. With a radius equal to the remaining side length (side B in this case), lay off an arc (arc B in this case) that intersects with the first arc, Fig. 5-14C.

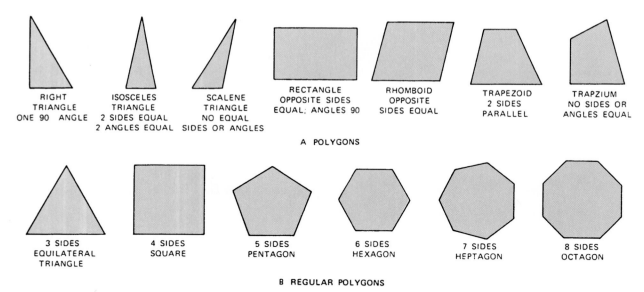

Fig. 5-13.   A–A polygon is a geometric figure enclosed with straight lines. B–A regular polygon has all sides of equal length and all angles equal.

Fig. 5-14.   If three side lengths are given, a triangle can be constructed using a compass.

5. Draw the sides to form the required triangle, Fig. 5-14D.

Construct a triangle with two side lengths and included angle given

1. The sides A and B, and the included angle are given, Fig. 5-15A.
2. Draw a base line and lay off one side (B), Fig. 5-15B.
3. Construct the given angle at one end of the line and lay off the other side (A) at this angle.
4. Join the end points of the two given lines to form the required triangle, Fig. 5-15C.

Construct a triangle with two angles and included side length given

1. Angles A and B, and the included side length AB are given, Fig. 5-16A.
2. Draw a base line and lay off side AB, Fig. 5-16B.
3. Construct angles A and B at opposite ends of line AB, Fig. 5-16B.
4. Extend the sides of angles A and B until they intersect at point C, Fig. 5-16C. Triangle ABC is the required triangle.

Construct an equilateral triangle

An **equilateral triangle** has three equal sides and three equal (60°) angles.

1. The side length AB is given, Fig. 5-17A.
2. Draw a base line and lay off side AB, Fig. 5-17B.

3. There are two different ways of completing this triangle. The "two included angles and one side length given" method described above can be used with a 30-60 degree triangle, Fig 5-17C. Also, the "three given side lengths" method described above can be used with a compass, Fig. 5-17D. With either method, the resulting triangle ABC is an equilateral triangle.

Construct an isosceles triangle

An **isosceles triangle** has two equal sides and two equal angles, Fig. 5-18. In order to construct an isosceles triangle, one of two conditions must be given. One set of conditions is all of the side lengths given (remember that two of the sides are the same length). The other condition is if two angles and one side length are given. Remember that two of the angles are equal, so if only one angle is given, you will actually know two of the angles.

If the length of the two equal sides and the base are given, construct the triangle using the "three sides lengths given" method described above. Refer back to Fig. 5-14 for an example of this construction.

If the two equal angles and one side length are given, construct using the "one side length and two included angles given" method described above. Refer back to Fig. 5-16 for an example of this construction.

Construct a right triangle

A **right triangle** has one 90° angle, Fig. 5-19A. The side directly opposite the 90° angle is called the **hypotenuse.**

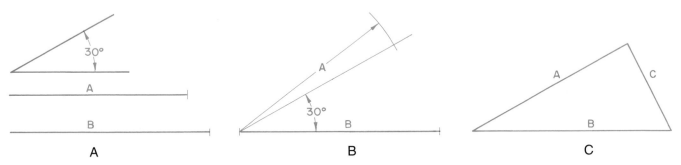

Fig. 5-15. A compass can be used to construct a triangle if two side lengths and the included angle are known.

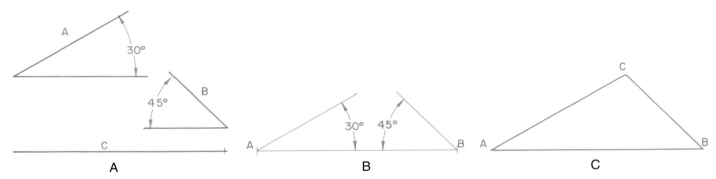

Fig. 5-16. If two angles and the included side length are given, a triangle can be constructed using a protractor.

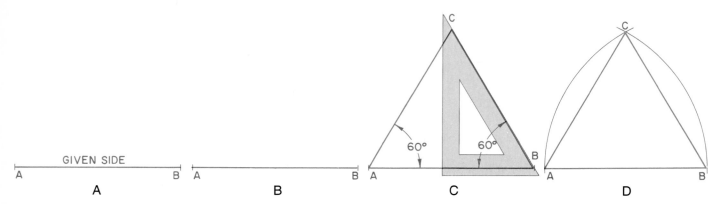

GIVEN SIDE

A   B   A   B   A   60°   60°   B   A   B

A   B   C   D

Fig. 5-17. An equilateral triangle has all sides of the same length and all angles equal. Since the interior angles must all add up to 180°, each of the three equal angles is 60°. If one side length is known, an equilateral triangle can be constructed using the "two given angles and included side length given" method. Another method that can be used is the "three side lengths given" method, since all side lengths of an equilateral triangle have the same length.

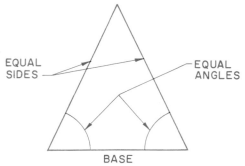

Fig. 5-18. An isosceles triangle has two equal sides and two equal angles.

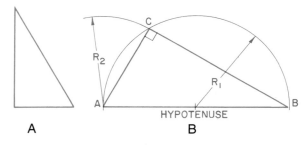

Fig. 5-19. A—A right triangle has one 90° angle. The side opposite the 90° angle is called the hypotenuse. B—If the length of the hypotenuse and one other side is known, a right triangle can be constructed.

If the lengths of the two sides are known, construct a perpendicular. (Refer back to Fig. 5-11 for an example of this construction.) Lay off the lengths of the sides and join the ends to complete the triangle.

If the length of the hypotenuse and the length of one other side are known, use the following procedure to construct the triangle.

1. Lay off the hypotenuse with endpoints A and B.
2. Draw a semicircle with a radius (R_1) equal to one-half the length of the hypotenuse AB, Fig. 5-19B.

3. With a compass, scribe arc AC equal to the length of the other given side (R_2). The intersection of this arc with the semicircle is point C.
4. Draw lines AC and BC. The triangle ACB is the required right triangle. The 90° angle has a vertex at point C.

Constructing Squares

A *square* is a polygon with four equal sides and four right angles.

Construct a square with the length of a side given using the triangle and T-square method

1. Given the length of one side, lay off line AB as the base line, Fig. 5-20A.
2. With a triangle and a T-square, project line BC at a right angle to line AB. Measure this line to the correct length.
3. Draw line CD at a right angle to line BC. Measure this line to the correct length.
4. Draw line AD. This line should be at right angles to both line AB and line CD. This line completes the required square ABCD.

Construct a square with the length of the side given using the compass method

1. Given the length of one side, lay off line EF as the base line, Fig. 5-20B.
2. Construct a perpendicular to line EF at point E by extending base line EF to the left and striking equal arcs on the base line from point E. (The arcs can be any convenient length.) These points are point E_1 and point E_2.
3. From these two points, strike equal arcs to intersect at point J.
4. Draw a line from point E through point J. Set the compass for the distance between point E

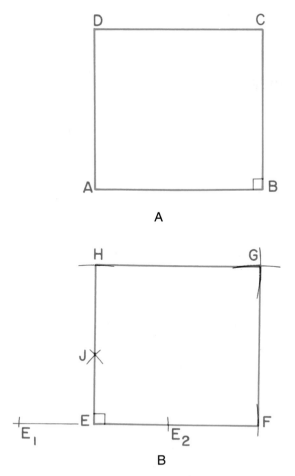

Fig. 5-20. A–If a side length is known, a square can be constructed using a straightedge and a triangle. B– A compass can also be used for this construction.

and point F. Mark off this distance on the line running through point J. This point becomes point H.

5. From points F and H, using the same compass setting, lay off intersecting arcs FG and HG.

6. Lines drawn to this intersection complete the required square EFGH.

Construct a square with the length of the diagonals given

1. The length of the diagonal AC is given, Fig. 5-21A.
2. With radius R_1 equal to one-half the diagonal, construct a circle, Fig. 5-21B.
3. Draw 45° diagonals AC and DB through the center of the circle. The points where the diagonals intersect the circle become points A, B, C, and D, Fig. 5-21B.
4. Draw lines joining points A and B, B and C, C and D, and D and A. These lines form the required square ABCD, Fig. 5-21C.

Constructing Other Polygons

Construct a regular pentagon with the side length given using the protractor method

1. The length of side AB is given, Fig. 5-22A.
2. With protractor, lay off side BC equal in length to side AB and at an angle of 108°. The center of the angle should be at point B, Fig. 5-22B.
3. Lay off side AE in a similar manner. Continue until the required regular pentagon ABCDE is formed, Fig. 5-22C.

Construct a regular pentagon with the circumscribed circle given using the compass method

1. The radius FG of the circumscribed circle is given.
2. Draw a circle with a radius of FG with a center at point F.

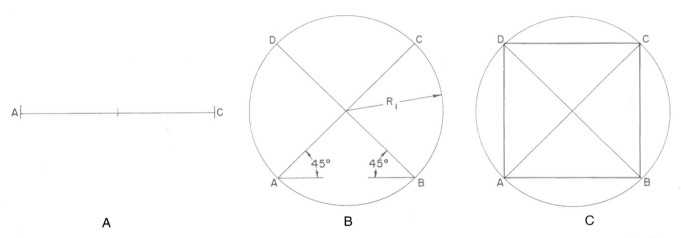

Fig. 5-21. If the diagonal length of a square is known, a square can be constructed using a compass and a 45 degree triangle.

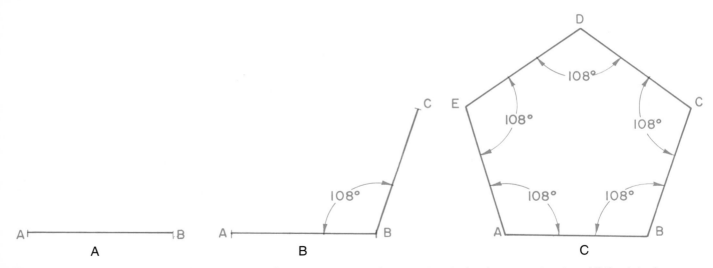

Fig. 5-22. If a side length is known, a regular pentagon can be constructed using a protractor. All the interior angles of a regular pentagon will always be 108° each.

3. Bisect the radius (centerline) FG, Fig. 5-23A. The point of bisection is point H.

4. With point H as the center, scribe an arc with radius equal to HD. This arc intersects the diameter (centerline) of the circle at point J, Fig. 5-23B.

5. With the same radius and point D as the center, scribe arc DJ. This will intersect with the circle at point E.

6. From the same center at point D, scribe an equal arc to intersect with the circle at point C.

7. With the same radius and with points E and C as centers, scribe arcs EA and CB. Draw lines between the points of intersection to form the required regular pentagon ABCDE, Fig. 5-23C.

Construct a hexagon with the distance across the flats given

1. Draw a circle equal in diameter to the distance across the flats, Fig. 5-24A.

2. With T-square and 30-60 degree triangle, draw lines tangent to the circle, Fig. 5-24B.

3. Draw the remaining sides to form the required hexagon, Fig. 5-24C.

4. An alternate position for the hexagon with distance across the flats is shown in Fig. 5-24D.

Construct a hexagon with the distance across corners given

1. Draw a circle equal in diameter to the distance across the corners, Fig. 5-25A.

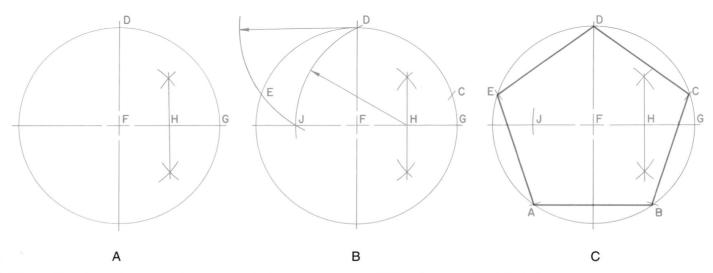

Fig. 5-23. A pentagon can be constructed using a compass if the circumscribed circle is given.

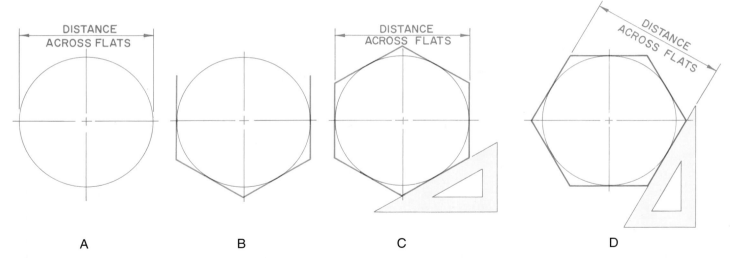

Fig. 5-24. If the distance across the flats is given, a hexagon can be constructed using a compass, a straightedge, and a 30-60 degree triangle. The hexagon can be drawn with two parallel flats in a vertical position or in a horizontal position.

2. With T-square and 30-60 degree triangle, draw 60° diagonal lines across the center of the circle, Fig. 5-25B.

3. With the T-square and triangle, join the points of intersection of the diagonal lines with the circle. This forms required hexagon, Fig. 5-25C. An alternate position for the hexagon is shown in Fig. 5-25D.

Construct an octagon with the distance across the flats given using the circle method

1. Draw a circle equal in diameter to the distance across the flats, Fig. 5-24A.

2. With a T-square and a 45 degree triangle, draw the eight sides tangent to the circle, Fig. 5-26B.

Construct an octagon with the distance across the corners given using the circle method

1. Draw a circle equal in diameter to the distance across the corners, Fig. 5-26C.

2. With a T-square and a 45 degree triangle, lay off 45° diagonals with the horizontal and vertical diagonals (diameters or centerlines).

3. With the triangle, draw the eight sides between the points where the four diagonals intersect the circle, Fig. 5-26D.

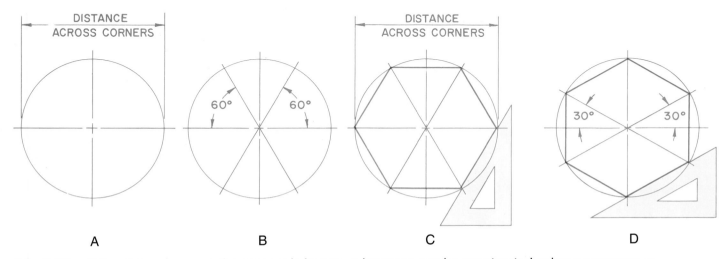

Fig. 5-25. If the distance across the corners is known, a hexagon can be constructed using a compass, a straightedge, and a 30-60 degree triangle. The hexagon can be drawn with two parallel flats in a horizontal position or in a vertical position.

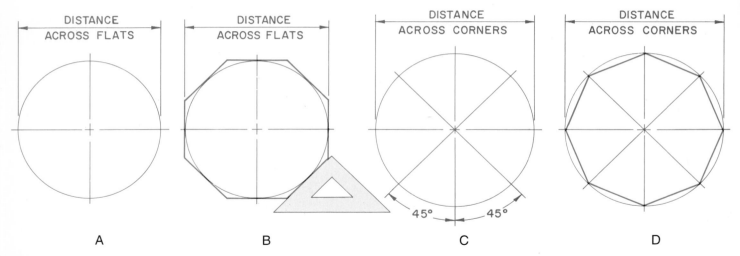

Fig. 5-26.   If the distance across the flats is known, an octagon can be constructed using a compass and a 45 degree triangle. If the distance across the corners is known, an octagon can be constructed by drawing a circle and 45° diagonals.

Construct an octagon with the distance across the flats given using the square method

1. Draw a square with the sides equal to the given distance across the flats, Fig. 5-27A.

2. Draw diagonals at 45°. With a radius equal to one-half the diagonal and using the corners of the square as vertices, scribe arcs, Fig. 5-27B.

3. With a 45 degree triangle and T-square, draw the eight sides by connecting the points of intersection. This completes the required octagon, Fig. 5-27C.

Construct a regular polygon having any number of sides and the diameter of a circumscribed circle given

1. Draw a circle with the given diameter. Divide its diameter into the required number of equal parts (seven in the example), Fig. 5-28A. Use the Inclined Line Method illustrated in Fig. 5-6.

2. With a radius equal to the diameter and with centers at the diameter ends A and B, draw arcs intersecting at point P, Fig. 5-28B.

3. Draw a line from point P through the second division point of diameter AB until it intersects with the circle at point C, Fig. 5-28C. The second point, point P, will always be the point used. The chord AC is one side of the polygon.

4. Lay off the distance AC around the circle using dividers. This will complete the regular polygon with the required number of sides, Fig. 5-28C.

Construct a regular polygon having any number of sides and the side length given

1. Draw side AB equal to the given side, Fig. 5-29A.

2. Extend line AB and draw a semicircle with the center at point A and a radius equal to line AB.

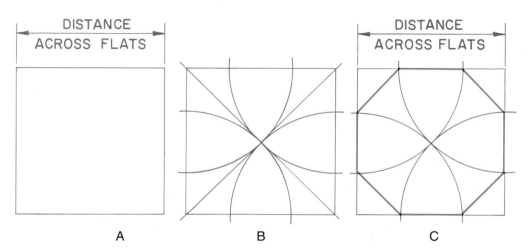

Fig. 5-27.   If the distance across the flats is known, an octagon can be constructed using the square method. A compass, a straightedge, and a 45 degree triangle are needed.

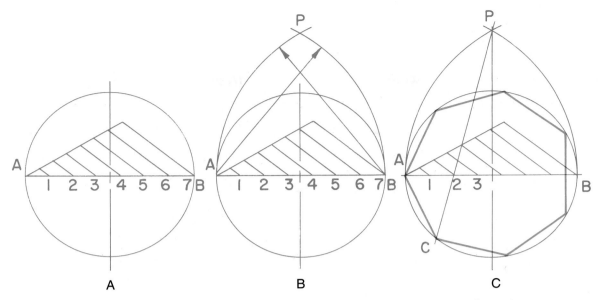

Fig. 5-28. A regular polygon having any number of sides (seven in this case) can be constructed if the diameter of a circumscribed circle is given.

3. With dividers, divide the semicircle into a number of equal parts equal to the number of sides (nine in this example).

4. From point A to the second division point, draw line AC. (The second point will always be used for this construction.)

5. Construct perpendicular bisectors of lines AB and AC. Extend the bisectors to meet at point D, Fig. 5-29B.

6. Using point D as the center and a radius equal to DB, construct a full circle to pass through points B, A, and C.

7. From point A, draw lines through remaining division points on the semicircle to intersect with the larger circle, Fig. 5-29C.

8. Join the points of intersection on the larger circle to form the regular polygon. This polygon

will have the correct number of sides and each side will be of the correct length, Fig. 5-29D.

Transferring Plane Figures

Transfer a plane figure by the triangle method

Plane figures with straight lines may be transferred or duplicated by the triangle method.

1. Given the polygon ABCDE, rotate the figure, Fig. 5-30A.

2. At corner A, draw straight lines AC and AD to form triangles, Fig. 5-30B.

3. Using point A as the center, draw an arc outside the polygon cutting across the extended lines.

4. Draw the line A'B' in the desired position that will have the proper rotation, Fig. 5-30C.

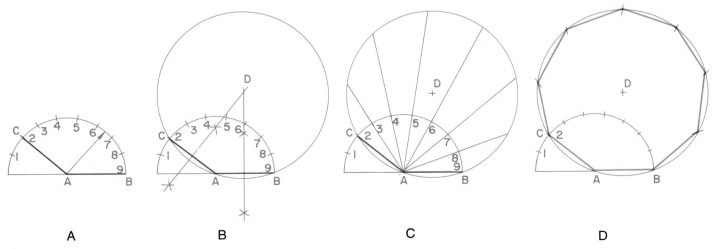

Fig. 5-29. A regular polygon having any number of sides (nine in this case) can be constructed if the side length is given.

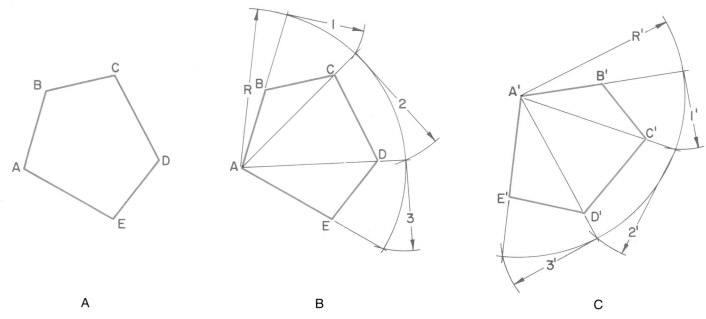

Fig. 5-30. A compass can be used to transfer a given plane figure to a new location. The transferred figure can also be rotated using this method.

5. Draw an arc with radius R′ and a vertex at point A′. (Note that R = R′.)

6. Draw arcs 1′, 2′, and 3′ to locate triangle lines. (Note that 1 = 1′, 2 = 2′, and 3 = 3′.)

7. Set off distances A′C′, A′D′, and A′E′ equal to AC, AD, and AE.

8. Draw straight lines to form transferred polygon A′B′C′D′E′.

Transfer a plane figure involving irregular curves

Plane figures involving irregular curves may be transferred or duplicated by the squares method.

1. Given a plane figure, duplicate that figure, Fig. 5-31A.

2. Draw a rectangle to enclose the plane figure.

3. Select points on the irregular curve that locate strategic points or change of direction and draw coordinates, Fig. 5-31B.

4. Reproduce the rectangle and coordinate points in the new position. Draw the straight lines and irregular curves between the reference points, Fig. 5-31C.

Plane figures may also be duplicated, enlarged, or reduced in size by the squares method shown in Fig. 5-32. This method is similar to the coordinate lines method except squares are drawn, and the irregular curve plotted on the squares.

CIRCLES AND ARCS

A *circle* is a closed plane curve. All points of a circle are equally distant from a point within the circle called the *center.* An *arc* is any part of a circle or other curved line, Fig. 5-33. The *diameter* is the distance across a circle passing through the center. The *radius* is 1/2 the diameter.

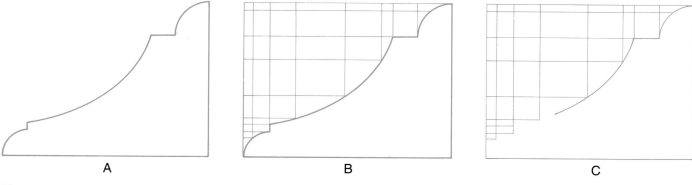

Fig. 5-31. The coordinate lines method can be used to duplicate an irregular curve.

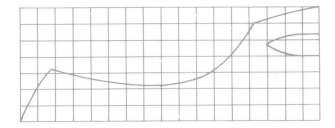

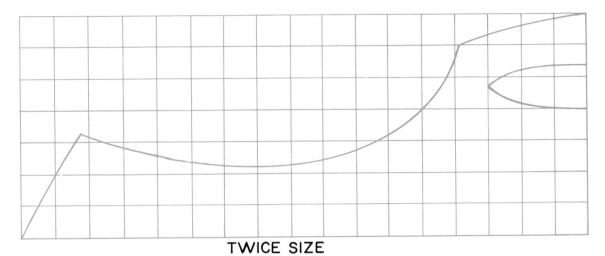

TWICE SIZE

Fig. 5-32. An irregular curve can be duplicated by the squares method. This method can also be used to enlarge or reduce an object (enlarged to twice size in this case).

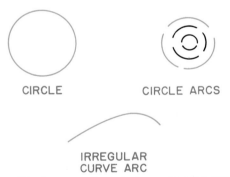

CIRCLE   CIRCLE ARCS

IRREGULAR
CURVE ARC

Fig. 5-33. Circles are closed plane regular curves. Circle arcs are any part of a circle. Irregular curves or arcs are any part of a curved surface that is not regular.

Constructing Circles

Construct a circle through three given points

1. Given the three points A, B, and C, draw a circle through these points, Fig. 5-34A.
2. Draw lines between points A and B, and points B and C.
3. Construct the perpendicular bisector of each line, Fig. 5-34B.
4. The point of intersection of the bisectors is the center of the circle which passes through all three points, Fig. 5-34C.

5. The radius of the circle is equal to the distance from the center to any of the three given points.

Locate center of a circle or an arc

1. Draw two nonparallel chords, AB and BC. (This is similar to the lines drawn in Fig. 5-34C.)
2. Construct the perpendicular bisector of each chord.
3. The point of intersection of the bisectors is the center of the circle or arc.

Construct a circle within a square

1. Given the square ABCD, draw diagonals AC and BD to locate center of the circle, Fig. 5-35A.
2. Locate the center of one side of the square, Fig. 5-35B.
3. The distance from the center of the circle to the center of a side is the radius. This radius will inscribe a circle within the square, Fig. 5-35C.

Drawing Tangents

A *tangent* is a line or curve that touches the surface of an arc or circle at only one point. A tangent can be a straight line. A tangent can also be a circle or an arc.

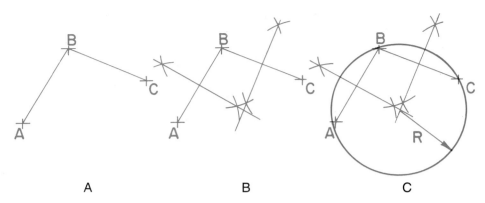

Fig. 5-34. A circle that passes through three given points can be constructed with a straightedge and a compass.

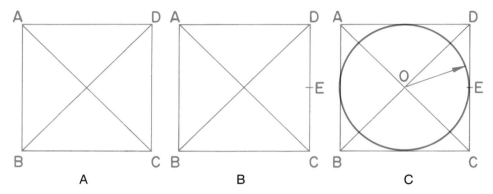

Fig. 5-35. If a square is given, a circle can be constructed within that square using a 45 degree triangle and a compass.

Construct a line tangent to a circle or arc by the triangle method

1. Place a triangle on a straightedge so that the hypotenuse is against the straight edge. Adjust so that one edge of the triangle is in line with the circle center and a point of tangency P, Fig. 5-36A.

2. Slide the triangle along the straightedge until the other edge passes through P. This will form the tangent line.

Construct a line tangent to a circle or arc using the compass method

1. With a point of tangency P as the center and the circle radius PC, construct an arc through point C. This arc should pass through the center C and another point on the circle, point A, Fig. 5-36B.

2. With A as the center and the same radius PC, construct a semicircular arc to pass through P. This semicircle should also extend to the opposite side of point A as point P.

3. Extend line CA until it intersects with semicircle at B.

4. Line BP is perpendicular to line PC. This line is also the required tangent line.

Construct a circle or arc of a given radius tangent to a straight line at a given point

1. Draw a perpendicular to line AB at the given point P, Fig. 5-37A.

2. Lay off the radius CP on the perpendicular, Fig. 5-37B.

3. Draw the circle with the center at point C, Fig. 5-37C. The circle is tangent to line AB at point P.

An application of this construction to a drawing problem is shown in Fig. 5-37D.

Construct a circle or arc with a given radius through a given point and tangent to a given straight line

1. Point P, line AB, and radius R are given, Fig. 5-38A.

2. Strike arc C with radius R using point P as the center, Fig. 5-38B.

3. At distance equal to R, draw line DE parallel to line AB.

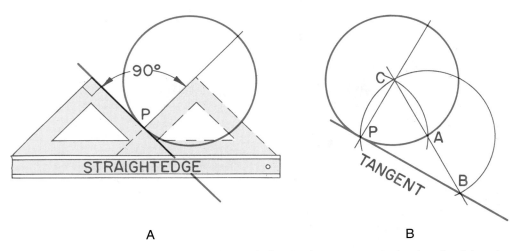

A

B

Fig. 5-36. A–A line tangent to a given circle can be constructed using the triangle method. B–The compass method can also be used for this construction.

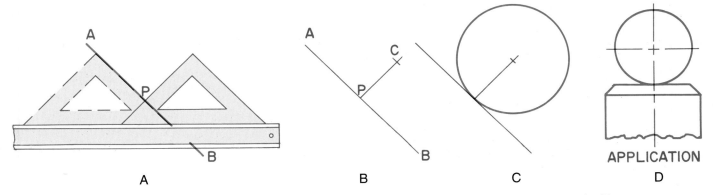

A

B

C

D

Fig. 5-37. A circle of a given radius that is tangent to a line at a given point can be constructed with a straightedge, a triangle, and a compass. There are many applications for this construction in drafting.

4. The intersection of arc C and line DE is the center of the tangent circle or arc, Fig. 5-38C.

An application is shown in Fig. 5-38D.

Construct a circle or arc with a given radius through a given point and tangent to a given circle

1. Point P, radius R, and circle O are given, Fig. 5-39A.
2. Strike arc A with radius R using point P as the center, Fig. 5-39B.
3. Strike arc B using the center of the circle as the arc center and a radius of R + r (r is the radius of the given circle).
4. The intersection of arcs A and B is the center of the required arc. The arc will pass through the given point and be tangent to the circle, Fig. 5-39C.

An application of this construction is shown in Fig. 5-39D.

Construct a circle or arc of a given radius tangent to two given circles

1. Circles A and B, and radius R are given, Fig. 5-40A.
2. Using point A as a center and a radius of R + ra (ra is the radius of circle A), strike arc C, Fig. 5-40B.
3. Using B as a center and a radius of R + rb (where rb is the radius of circle B), strike arc D.
4. The intersection of arcs C and D is the center of the required arc. This arc will be tangent to the two given circles A and B, Fig. 5-40C.

An application of this construction is shown in Fig. 5-40D.

Note: In order to locate the center of an arc tangent to two enclosed circles (circles that are *inside* of the arc), subtract the circle radii from the arc radius (R - ra), Fig. 5-40D.

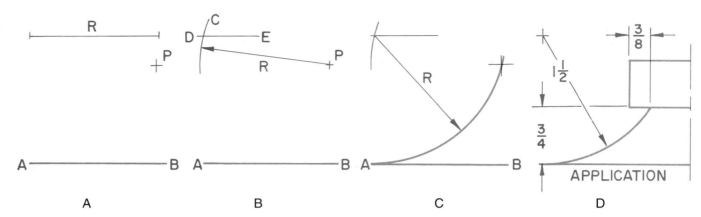

Fig. 5-38.   An arc of a given radius that passes through a given point and is tangent to a given line can be constructed using a straightedge and a compass. A drafter may encounter many cases where this construction must be used.

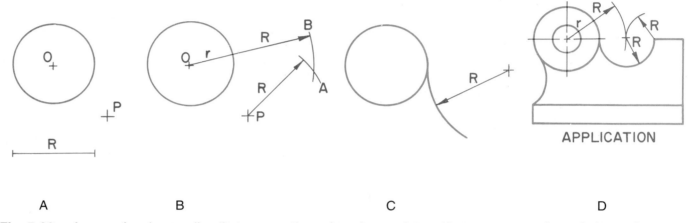

Fig. 5-39.   An arc of a given radius that passes through a given point and is tangent to a given circle can be constructed using a compass. This construction can be used for many different applications in drafting.

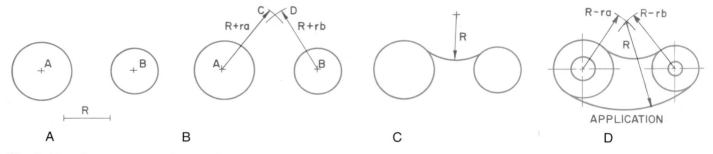

Fig. 5-40.   A compass can be used to construct an arc of a given radius and tangent to two given circles. Note that if the circles are to be inside of the arc that the radii of the two given circles are subtracted from the given radius.

Construct a circle or arc of a given radius tangent to a given straight line and a given circle

1. Line AB, circle O, and radius R are given, Fig. 5-41A.
2. At a distance equal to R, draw line CD parallel to line AB, Fig. 5-41B. Line CD should be located between line AB and circle O.

3. Strike arc E, to intersect line CD, using point O as the center and a radius of R + ro (ro is the radius of circle O).
4. The intersection of arc E and line CD is the center of the required arc. This arc will be tangent to line AB and circle O, Fig. 5-41C.

An application of this construction to a drawing problem is shown in Fig. 5-41D.

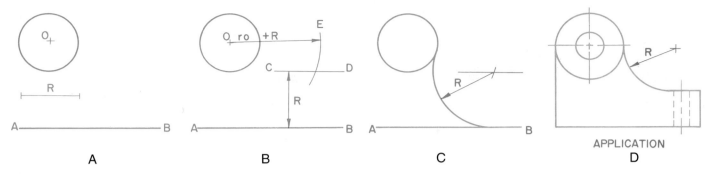

Fig. 5-41. An arc of a given radius and tangent to a straight line and a circle can be constructed using a straightedge and a compass. There are many applications for this construction in drafting.

Construct a circle or arc tangent to two given parallel lines

1. Parallel lines AB and CD are given, Fig. 5-42A.
2. Construct parallel line EF equidistant between lines AB and CD, Fig. 5-42B.
3. The distance from line AB (or line CD) to line EF is the radius of the tangent arc. Set the compass to that radius. Place the compass on line EF and strike required arc, Fig. 5-42C.

An application of this construction to a drawing problem is shown in Fig. 5-42D.

Construct a circle or arc of a given radius and tangent to two given nonparallel lines

1. Nonparallel lines AB and CD, and the radius R are given, Fig. 5-43A.
2. At a distance of R, construct lines EF and GH parallel to lines AB and CD," Fig. 5-43B.

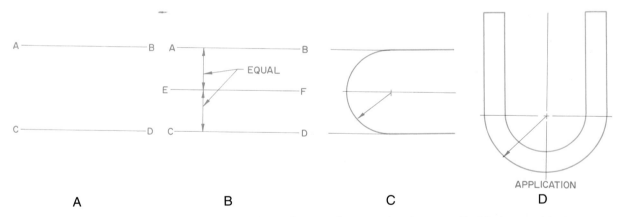

Fig. 5-42. A straightedge and a compass can be used to construct an arc that is tangent to two given parallel lines. A drafter will find this construction useful in many situations.

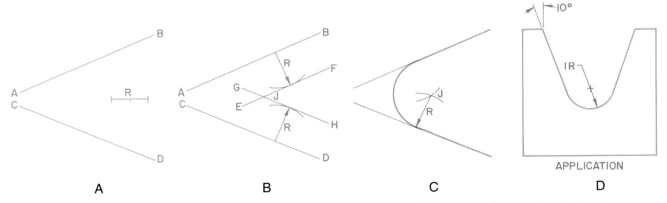

Fig. 5-43. An arc of a given radius that is tangent to two nonparallel lines can be constructed using a straightedge and a compass. There are many applications for this construction in drafting.

3. The intersection of the lines at point J is the center of the required arc. The arc will be tangent to lines AB and CD, Fig. 5-43C.

An application of this construction to a drawing problem is shown in Fig. 5-43D.

Connect two parallel lines with reversing arcs of equal radii

1. Lines AB and CD are given, Fig. 5-44A.
2. Draw a line between points B and C. Divide line BC into two equal parts BE and EC, Fig. 5-44B.
3. Construct perpendicular bisectors of lines BE and EC, Fig. 5-44C.
4. Draw perpendiculars to lines AB and CD at points B and C.
5. The points of intersection of the bisectors and perpendiculars at points F and G are the centers for drawing the equal arcs FB and GC. These form the required arcs. These arcs are tangent to each other and to the given parallel lines.

An application of this construction in a highway drawing is shown in Fig. 5-44D.

Connect two given parallel lines with reversing arcs of unequal radii with one of the radii given

1. Lines AB and CD, and radius R_1 are given, Fig. 5-45A
2. Draw the line BC, Fig. 5-45B.
3. Draw a line perpendicular to line AB at point B. Lay off radius R1 on this line. The endpoint becomes point E. With point E as the

center, strike arc R1 from point B and intersecting line BC. This intersection is point F, Fig. 5-45C.

4. Draw a line perpendicular to line CD at point C. Extend line EF to intersect this perpendicular. This intersection is point G.
5. Using point G as the center and the distance from point G to point C as the radius, draw arc GC from point C to point F. These form the required arcs. These arcs are tangent to each other and to the given parallel lines.

Connect two given nonparallel lines with reversing arcs with one of the radii given

1. Nonparallel lines AB and CD, and the radius R_1 are given, Fig. 5-46A.
2. Draw a line perpendicular to line CD at point C. Lay off radius R_1 on this perpendicular line. The endpoint is point E. With point E as the center, strike arc R_1 from point C, Fig. 5-46B.
3. Draw a line perpendicular to line AB at point B. (Note that this line is drawn on the opposite side of line AB as line CD.) Lay off line BF equal to line CE.
4. Draw line FE and bisect with perpendicular bisector GH.
5. Extend lines FB and GH until they intersect at point J. Draw line JE, Fig. 5-46C. Where line JE intersects arc R_1 is point K.
6. Using J as a center and JB as the radius, draw an arc to connect line AB to arc R_1 at point K. These are the required arcs. These arcs are tangent to each other and to the given lines.

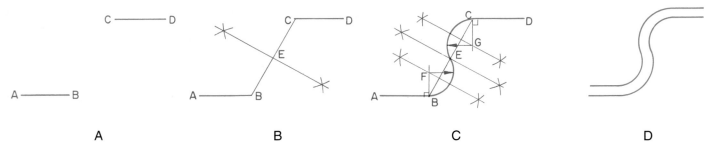

Fig. 5-44.   A straightedge, a triangle, and a compass can be used to connect two parallel lines with reversing arcs of equal radii. A civil engineer may use this construction when drawing a highway "S" curve.

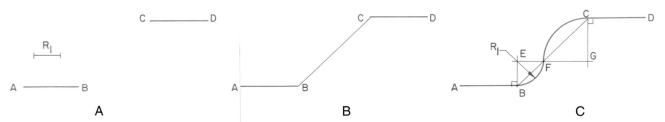

Fig. 5-45.   Two parallel lines can be connected with reversing arcs of unequal radii if one of the radii is given.

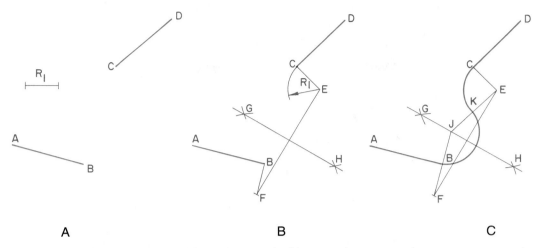

Fig. 5-46. Two nonparallel lines can be connected with reversing arcs using a compass and a straightedge.

Lay off the length of the circumference of a circle using the construction method

This procedure is also referred to as locating the **rectified length.** This means that a curved surface (such as a circle) has its true length laid out on a straight line.

1. The circle O is given.
2. Draw line AB tangent to the point where the vertical centerline crosses the circle. The length of the line should be equal to three times the diameter of the circle, Fig. 5-47A.

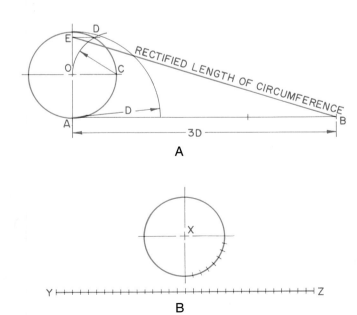

Fig. 5-47. A–The rectified length of a circle can be found using the construction method. This method is accurate to .005%. B–The rectified length can also be found using the equal chord method. This method is not as accurate as the construction method.

3. Locate point C where the horizontal centerline crosses the circle. With the center at point C, strike an arc of a radius equal to that of the circle. This arc will intersect the circle at point D. (Note that point D is on the opposite side of the circle center as point A.)
4. From point D, draw line ED perpendicular to the vertical centerline.
5. Line EB is the approximate length of the circumference of circle O. The error of this line is equal to less than one inch in 20,000 inches or .005%. This error is well within the accuracy range of mechanical drafting instruments.

Lay off the length of the circumference of a circle using the equal chord method

1. The circle X is given.
2. Draw line YZ to an approximate length, Fig. 5-47B.
3. Using dividers, divide one quarter of the circle into an equal number of chord lengths. The accuracy of the projection is increased when a greater number of chords is used.
4. Lay off on line YZ four times the number of chord lengths in the quarter circle. The length YZ is the required approximate length of circle X. This method is not as accurate as the "construction method."

Lay off the length of the circumference of a circle using the mathematical method

The circumference of a circle may also be calculated very accurately by multiplying the diameter by π. The mathematical symbol π represents the approximate number 3.1416. Use the calculated number and lay off the length on a straight line. Refer to Fig. 5-48 for an example of this calculation.

Example: D = 3 inches

$$\begin{array}{r} 3.1416 \\ \underline{3} \\ 9.4248 \end{array}$$

Fig. 5-48.   The rectified length of a circle can be calculated using π. This mathematical symbol represents the approximate number 3.1416. To find the rectified length, multiply the diameter by π. The result is a very accurate calculation of the rectified length.

Lay off the length of a circle arc using the construction method

1. The circle arc AB and the center of the arc, point E, are given, Fig. 5-49A.
2. Draw chord AB. Extend the line and locate point C so that CA is equal to one-half AB.
3. Draw line AD tangent to arc AB at point A and perpendicular to line EA.
4. Using C as the center, strike an arc with a radius of BE to intersect line AD. The length AD is the approximate length of arc AB. The error for this projection is less than one inch in 1000 inches, or .1%, for angles up to 60°.

Lay off the length of a circle arc using the equal chord method

1. The circle arc FG is given, Fig. 5-49B.
2. Draw line FH tangent to the arc. The length of this line should be an estimation of the rectified length.

3. Using dividers, divide arc FG into an equal number of chord lengths. The accuracy of this projection is increased when a greater number of chords is used.
4. Lay off the same number of chord lengths along line FH. The length FH is the approximate length of arc FG.

An application of this construction would be in ascertaining the "stretchout" length of a sheet metal part, Fig. 5-49C.

Lay off a specific length of a given circle arc using the construction method

Note that to use this method the specified length must be given as a line tangent to the given arc.

1. Circle arc AD and the length of the tangent line AB are given, Fig. 5-50A.
2. Divide line AB into four equal parts.
3. Using the first division point (point C) as the center, strike an arc with a radius equal to CB. This arc will intersect the circle arc at point D.
4. Arc AD is equal in length to line AB. This method is accurate to six parts in a thousand, or .6%, for angles less than 90°.

Lay off a given length on a circle arc using the equal chord method

Note that to use this method, the specified length must be given as a line tangent to the given arc.

1. The circle arc EG and the length of the tangent line EF are given, Fig. 5-50B.
2. Using dividers, divide tangent line EF into an equal number of parts. The accuracy of this projection is increased when a greater number of divisions is used.

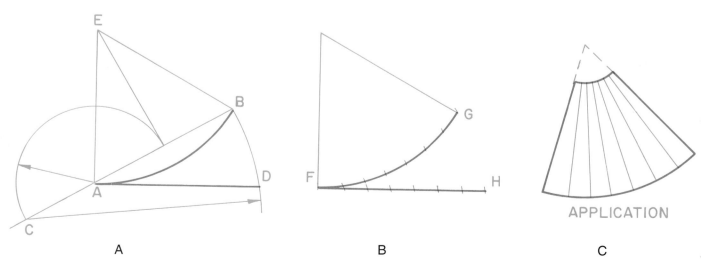

| A | B | C |

Fig. 5-49.   A–The rectified length of a circle arc can be found using the construction method. This method is accurate to .1%. B–The rectified length can also be found using the equal chord method. This method is not as accurate as the construction method. C–There are many applications for this construction in drafting.

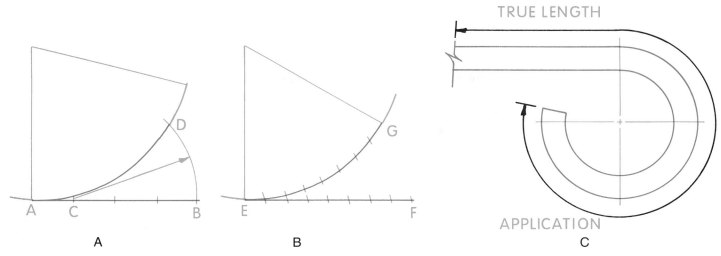

Fig. 5-50. A–A given length can be laid out on a circle arc using the construction method. This method is accurate to .6% for angles off less than 90°. B–The equal chord method can also be used for this construction, but it is not as accurate. Note that the specific length must be given as a line tangent to the given arc. C–Drafters may use this construction in many different applications.

3. Lay off the same number of chord lengths along circle arc. The length of circle arc EG is the approximate length of line EF.

This type of construction is used in figuring out the true length of a formed metal part, Fig. 5-50C.

PROBLEMS AND ACTIVITIES

The following problems have been designed to give you experience in performing simple geometric constructions used in drafting. Practical applications of geometry applied to drafting are included to acquaint you with typical problems the drafter, designer, or engineer must solve.

All problems are to be drawn using Layout I in the Reference Section. Place drawing paper horizontally on the drawing board or table. Use the title block shown in Layout I and lay out your problems carefully, making the best use of space available. A freehand sketch of the problem and solution will help.

Accuracy is extremely important in drawing geometrical constructions. Use a hard lead (2H to 4H) sharpened to a fine conical point and draw light lines. When the construction is complete, darken the required lines and leave all construction lines as drawn to show your work.

Problems which call for "any convenient length" or "any convenient angle" should not be measured with the scale or laid out with the T-square and triangles.

Do the problems assigned by your instructor. A suggested layout for the first four problems is shown in Fig. 5-51.

1. Draw a horizontal line three inches long. Mark the ends with a vertical mark. Bisect the line by the triangular method.

2. Draw a horizontal line of any convenient length. Bisect the line by the compass method.

3. Draw a line AB at any convenient angle. Construct another line one inch from it and parallel to it. Use the triangle-straightedge method.

4. Using the triangle-straightedge method, draw a line at any convenient angle. Construct a perpendicular to it from a point not on the line.

5. Draw a line of any convenient length and angle. Divide the line geometrically into seven equal parts. Use the vertical line method.

6. Lay out an angle of any convenient size. Bisect the angle.

7. Transfer the angle in Problem 6 to the next section on your sheet. Revolve the angle approximately 90° in the new location.

8. Draw a horizontal line. Construct a perpendicular to the line by the 3, 4, 5 method.

9. Construct a triangle given the following: side AB = 3 1/4 inches, ∠ A = 37°, ∠ B = 70°.

10. Construct a triangle given the following: sides AB = 3, BC = 1 1/4, and CA = 2 1/8 inches. Measure each angle with the protractor and add the three together. If your answer is 180°, you have measured accurately. If not, check your measurements again.

11. Construct a triangle given the following: sides AB = 1 1/2, AC = 2 inches, ∠ A = 30°.

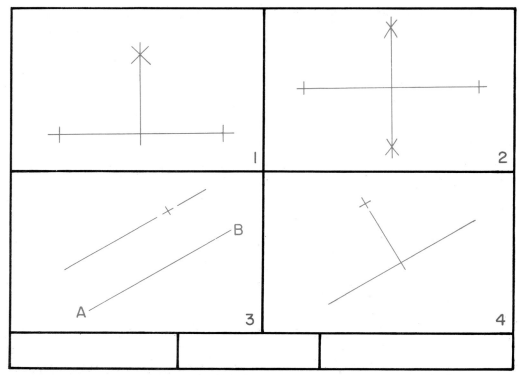

Fig. 5-51. Use Layout I from the reference section as a guide when drawing your border and title block. For the first four problems in the Problems and Activities section, divide your sheet as shown here.

12. Construct an equilateral triangle given side AB = 2 3/4 inches.

13. Draw a 1 1/2 inch line inclined slightly (approximately 10°, but do not measure) from the horizontal. Using this line as the first side, construct a square by any method described in this chapter.

14. Draw a circle 2 1/2 inches in diameter in the center of one of the four sections of the sheet. Inscribe (the square is contained *inside* the circle) the largest square possible.

15. Construct a pentagon within a 2 1/2 inch diameter circle using the compass method.

16. Construct a hexagon measuring three inches across the corners.

17. Construct a hexagon with a distance of 2 1/2 inches across the flats when measured horizontally.

18. Construct an octagon with a distance across the flats of 2 1/2 inches using the square method.

19. Construct a seven sided regular polygon given the length of one side as 1 1/4 inches. Use the method shown in Fig. 5-29.

20. Draw an irregular shaped pentagon approximately two inches across corners in the upper left-hand portion of one section of a sheet. Transfer the polygon to the lower right-hand corner in a 180° revolved position. Use the triangle method.

21. Without measuring, place three points approximately 1 1/2 inches from the approximate center of the sheet section. Place the points at approximately the three, six, and ten o'clock positions. Construct a circle to pass through all three points.

22. Using a circle template, draw a semicircular arc 1 1/2 inches in radius. Locate the center of the arc.

23. Draw a circle 2 1/2 inches in diameter. Construct a tangent at the approximate two o'clock position.

24. Draw a two inch diameter circle in the lower left-hand portion of a sheet section. Draw a line approximately 1 inch above it and inclined toward the upper right corner. Construct a circle arc with a radius of 1 inch tangent to the circle and straight line.

25. Draw two nonparallel lines. Construct a circular arc of any convenient radius tangent to the two lines.

26. Draw a circle whose diameter is 1 1/2 inches. Obtain the length of the circumference of the circle by the construction method, the equal

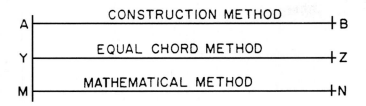

A |———————— CONSTRUCTION METHOD ————————| B

Y |———————— EQUAL CHORD METHOD ————————| Z

M |———————— MATHEMATICAL METHOD ————————| N

chord method, and the mathematical method. Lay off each distance as shown below. Compare the results.

27. Draw a 2 1/4 inch diameter circle. Find the length of the arc from the six o'clock to the approximate four o'clock position using the construction method.

28. Draw a 2 1/2 inch diameter circle. Start at the six o'clock position and lay off 1 inch along the arc. Use the construction method.

Conic sections have many uses in construction. The half circle of this railroad tunnel is a special type of parabola.

Advanced Geometry Used in Drafting

KEY CONCEPTS

☐ An ellipse is formed when a plane is passed through a right circular cone. The angle of the plane to the axis is less than the angle formed by the elements to the axis.

☐ A parabola is formed when a plane cuts a right circular cone at the same angle with the axis as the elements.

☐ A hyperbola is formed when a plane cuts two right circular cones that are connected at their vertices.

☐ Other geometric curves used in drafting include the Spiral of Archimedes, the helix, the cycloids, and the involute.

Complex and advanced geometric constructions are difficult to solve as drafting problems. However, each problem at hand can be worked out by applying the principles of plane geometry and utilizing your drafting instruments.

The preceding chapter covered basic geometric problems. This chapter is designed to provide specific definitions and details on more advanced forms of geometric construction. Some of the constructions you will look at include conic sections, the Spiral of Archimedes, the helix, cycloids, and the involute.

CONIC SECTIONS

Conic sections are curved shapes produced by passing a plane through a right circular cone. A right circular cone has a circular base and an axis perpendicular to the base at its center, Fig. 6-1. An **element** is a straight line drawn from any point on the circumference of the base to the peak of the cone.

Four types of curves result from cutting planes at different angles. These curves are a circle, an ellipse, a parabola, and a hyperbola, Fig. 6-2.

The Ellipse

An **ellipse** is formed when a plane is passed through a right circular cone, making an angle with

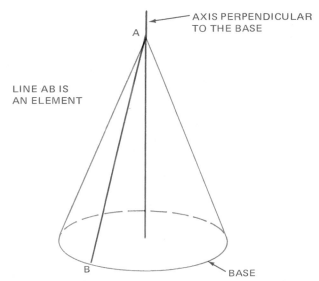

Fig. 6-1. A right circular cone has a circular base and an axis perpendicular to the base at its center.

the axis greater than the elements, Fig. 6-2C. An ellipse also results when a circle is viewed at an angle.

An ellipse can be defined as a curve formed by a point moving in a plane so that the sum of the distances from two fixed points is constant and equal to the major axis. The **major axis** is the largest diameter and the **minor axis** is the smallest diameter. The two fixed points, called **foci** (foci is the plural of **focus**), are often used in constructing an ellipse.

Construct an ellipse using the foci method

1. The major axis AB and the minor axis CD of the ellipse are given, Fig. 6-3A.

2. Locate the foci E and F on the major axis by striking arcs CE and CF with radii equal to one-half of the major axis.

3. On the major axis between point E and point 0, mark points at random, Fig. 6-3B. These points will be used to locate the ellipse. To insure a smooth curve, space the points closely near point E.

4. Begin construction with a point in the upper left-hand quadrant of the ellipse. Using points

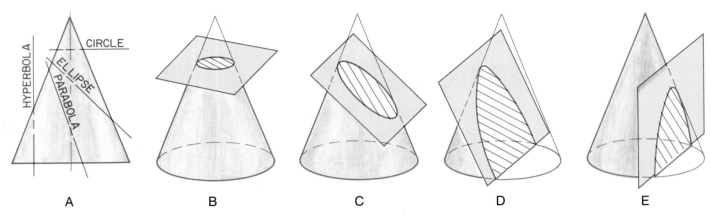

Fig. 6-2. Conic sections are formed when a plane passes through a right circular cone. Conic sections include: B–circle, C–an ellipse, D–a parabola, E–and a hyperbola.

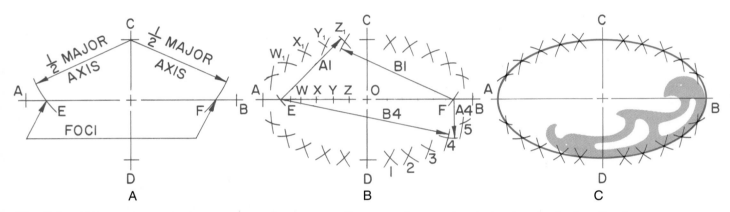

Fig. 6-3. If the major and minor axes are known, an ellipse can be constructed using the foci method.

E and F as centers, and radii equal to the distance from point A to point Z, and from point B to point Z, strike intersecting arcs at point Z1.

5. Use the distance from point A to point Y and from point B to point Y as radii for intersecting arcs at point Y_1. Continue in a similar manner until all points are plotted.

6. The lower left quadrant may be plotted using the same compass settings. Reverse the centers for the radii to plot the points in the two right-hand quadrants.

7. Sketch a light line through the points. Then, with the aid of an irregular curve, darken the final ellipse, Fig 6-3C.

Construct an ellipse using the concentric circle method

1. The major axis AB and the minor axis CD are given.

2. Draw circles of diameters equal to the major and minor axes, Fig. 6-4A.

3. Draw diagonal EE at any point.

4. At points where the diagonal intersects with the major axis circle, draw lines EF parallel to the minor axis. There will be two points of intersection, one in the upper left-hand quadrant and one in the lower right-hand quadrant. (Refer back to Fig. 6-4A.)

5. At points where the diagonal intersects with the minor axis circle, draw lines FG parallel to the major axis. The intersections of the lines at point F are points on the ellipse curve.

6. Two additional points, points H and J, may be located in the other two quadrants by extending lines EF and FG. (Refer back to Fig. 6-4A.)

7. Draw as many additional diagonals as needed to produce a smooth ellipse curve and project their points of intersection, Fig 6-4B.

8. Sketch a light line through the points. Then use an irregular curve to darken the final ellipse, Fig. 6-4C.

Construct an ellipse using the parallelogram method

1. The major axis AB and the minor axis CD are given, Fig. 6-5A. Alternatively, the conjugate diameters can be given, Fig. 6-5B. (Two diameters are conjugate when each is

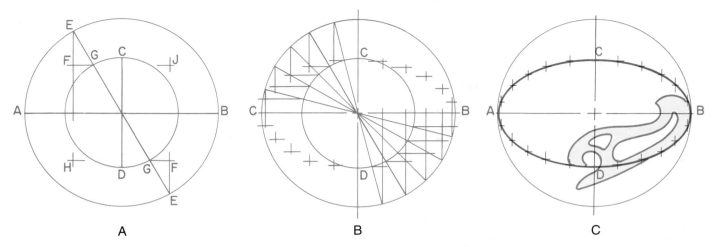

Fig. 6-4.   The concentric circle method can be used to construct an ellipse if the major and minor axes are known.

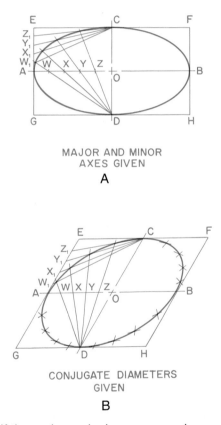

MAJOR AND MINOR
AXES GIVEN

A

CONJUGATE DIAMETERS
GIVEN

B

Fig. 6-5.   If the major and minor axes are known, or if the conjugate diameters are known, the parallelogram method can be used to construct an ellipse.

parallel to the tangents at the extremities of each other.)

2. Construct a circumscribing parallelogram using the given axes as centerlines.

3. Divide lines A0 and AE into the same number of units. All of the units on the line should be of equal length, but the units will not necessarily

be the same length as the units on the other line.

4. Draw line DW to intersect with line CW$_1$, line DX to intersect with line CX$_1$, and so on. These points of intersection are plotting points for the ellipse.

5. Locate points in the remaining quadrants in a similar manner.

6. Sketch a light line through the points. Use an irregular curve to darken the final ellipse.

Find the major and minor axes when the conjugate diameters are given

1. An ellipse and its conjugate diameters are given.

2. Draw a semicircle using point 0 as the center and a diameter of CD, Fig. 6-6. This semicircle will intersect the ellipse at point E.

3. Draw line ED. The minor axis FG is parallel to line ED and passes through the center 0.

4. Draw line EC. The major axis HJ is parallel to line EC and passes through center 0.

Construct an ellipse using the four-center approximate method

1. The major axis AB and the minor axis CD are given, Fig. 6-7A.

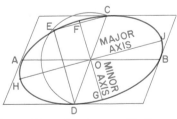

Fig. 6-6.   If an ellipse is given with the conjugate diameters known, the major and minor axes can be found.

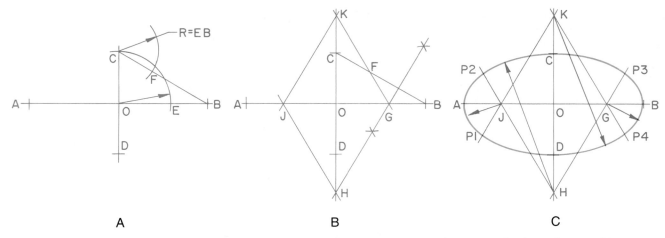

Fig. 6-7. If the major and minor axes are known, the four-center approximate method can be used to construct an ellipse.

2. Draw the line CB. Using the center 0 and the radius 0C, strike an arc intersecting 0B at point E.

3. With the radius EB and using point C as the center, strike an arc intersecting CB at point F.

4. Construct a perpendicular bisector of line FB. Extend this bisector to intersect with the major and minor axes at points G and H, Fig. 6-7B.

5. Points G and H are the centers for two of the four arcs of the ellipse. With a compass and using point 0 as the center, locate points J and K symmetrically with points G and H.

6. Draw a line from point H extending through point J. Draw lines from point K through points J and G.

7. Using centers J and G, strike the arcs JA and GB from point P1 to point P2 and from point P3 to point P4.

8. With points H and K as centers, strike arcs HC and KD from point P2 to point P3 and from point P4 to point P1.

These four arcs will be tangent to each other, forming the four-center approximate ellipse.

Construct an ellipse using the trammel method

A *trammel* is an instrument used for constructing ellipses. Commercially produced trammels are available. You can make a trammel from anything that has a straight edge, such as a piece of paper.

1. The major axis AB and the minor axis CD are given, Fig. 6-8A.

2. On a straight edge, lay off points E, F, and G so that EF is equal to one-half the minor diameter 0C and EG is equal to one-half the major diameter 0A. This marked straightedge serves as a trammel.

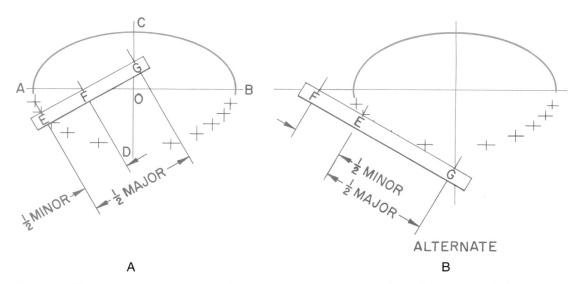

Fig. 6-8. The trammel method can be used to construct an ellipse if the major and minor axes are known.

3. Place the trammel so that point G is on the minor axis and point F is on the major axis.

4. As the trammel is moved, keep point G on the minor axis and point F on the major axis. Point E will mark points on the ellipse. Mark at least five points in each quadrant.

5. Sketch a light line through the points then, with the aid of an irregular curve, darken the final ellipse.

An alternate method of marking and using the trammel is shown in Fig. 6-8B. This trammel method is one of the most accurate means of constructing an ellipse.

Draw an ellipse using the template method

Considerable time can be saved in ellipse construction by using an ellipse template, Fig. 6-9.

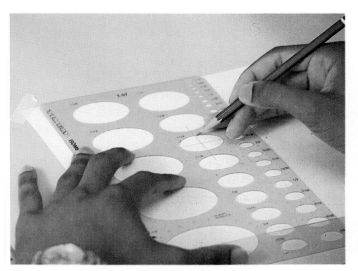

Fig. 6-9. Templates are a quick and easy way to construct ellipses.

Templates come in a variety of sizes. They are usually designated by the angle at which a circle of that size is viewed, Fig. 6-10. The selection of proper ellipse templates is discussed in Chapter 11. To use an ellipse template, line up the centerline marks on the template with the major and minor axes. This properly aligns the ellipse.

The Parabola

The *parabola* is formed when a plane cuts a right circular cone at the same angle with the axis as the elements. (Refer back to Fig. 6-2D.) The parabolic curve is used in engineering and construction for vertical curves (overpasses) on highways and dams, Fig. 6-11. Parabolas are also used in designing bridge arches and in forming the shape of reflectors for sound and light. Parabolas are frequently used in industrial and product design because of their pleasing appearance.

Fig. 6-11. Parabolic curves are used in the design of bridges and dams.

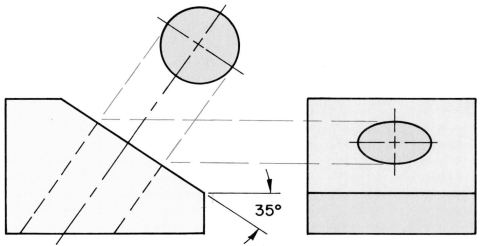

Fig. 6-10. Ellipse templates are designated by degrees. The degree number refers to the angle at which a circle is viewed to produce an ellipse. In this example, a 35 degree ellipse template is required to make the ellipse in the right-hand view.

The parabola may be defined mathematically as a curve generated by moving a point so that its distance from a fixed point (the **focus**) is always equal to its distance from a fixed line (the ***directrix***), Fig. 6-12.

Construct a parabola using the focus method

1. The focus F and the directrix AB are given, Fig. 6-13A.
2. Draw line CD parallel to the directrix at any distance. Draw a line perpendicular to the directrix and passing through the focus. The point at which this line and line CD intersect is

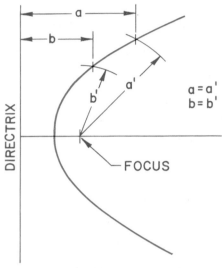

Fig. 6-12. A parabola has a directrix and a focus. Any point on the curve is the same distance from the directrix as it is from the focus.

point G. The point on the directrix at which the perpendicular line starts is point E.

3. With a radius of EG and using point F as the center, strike an arc to intersect line CD at points H and J. These points of intersection are points on the parabola.
4. In a similar manner, locate as many points as necessary to draw the parabola.
5. Sketch a light line through the points and use the irregular curve to darken the line.
6. The vertex V of the parabola is located half way between the origin E and the focus F.

Construct a tangent to a parabola

1. The parabola AB, its axis CD, the focus F, and the point of tangency P are all given, Fig. 6-13B.
2. Draw line P0 parallel to the axes and line PF through the focus. Bisect the angle 0PF. The bisector PQ is tangent to the parabola at point P.

Construct a parabola using the tangent method

1. The points A and B, and the distance from line AB to the vertex D are given, Fig. 6-14.
2. Extend line CD to point E so that line DE is equal to line CD.
3. Draw lines AE and BE. These lines are tangent to the parabola at points A and B.
4. Divide lines AE and BE into the same number of equal parts. The accuracy of this construction increases with the number of divisions. Number the points from opposite ends.

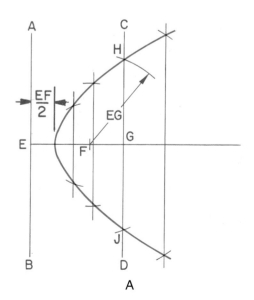

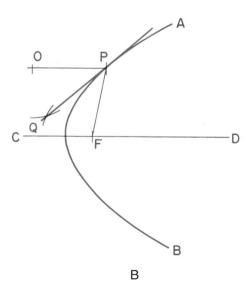

Fig. 6-13. A–If the directrix and the focus are known, the focus method can be used to construct a parabola. B–A tangent to a given parabola can easily be constructed at any given point.

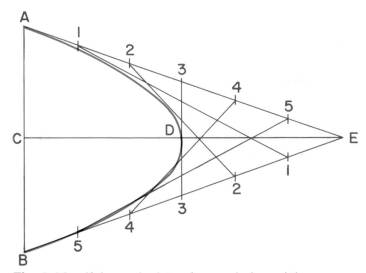

Fig. 6-14. If the endpoints of a parabola and the distance from a line drawn between them and the vertex of the parabola are known, the tangent method can be used to construct a parabola.

5. Draw lines between the corresponding numbered points.
6. These lines are tangent to the required parabola.
7. Sketch a light line tangent to these lines. Use an irregular curve to darken the lines.

Construct a parabolic curve through two given points

1. Points A and B are given, Fig. 6-15.
2. Choose any point as point C and draw tangents CA and CB.
3. Construct the parabolic curve using the tangent method. (Refer back to Fig. 6-14 for an example of this construction.)

Note that the distances A0 and B0 are not necessarily equal. (Refer back to Fig. 6-15B and Fig. 6-15C.) When these two distances are equal, the bisector of angle A0B is also the axis of the parabola. (Refer back to Fig. 6-15A.)

The Hyperbola

A *hyperbola* is formed when a plane cuts two right circular cones that are joined at their vertices, Fig. 6-16. Mathematically, a hyperbola is defined as a plane curve traced by a point moving so that the difference of its distance from two fixed points (the foci) is a constant equal to the transverse axis. The *transverse axis* is the distance between the vertices of the two curves. *Asymptotes* are lines that intersect at the midpoint of the transverse axis. The lines of the hyperbola will approach, but not intersect, the asymptotes if extended to infinity.

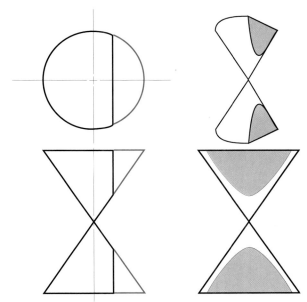

Fig. 6-16. A hyperbola is formed when a plane passes through two right circular cones that are joined at their vertices.

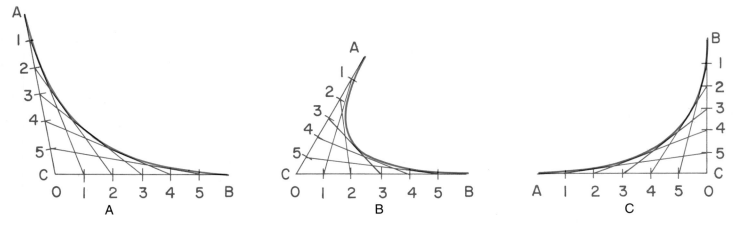

Fig. 6-15. If two points are given, the tangent method can be used to construct a parabola, using an assumed third point.

Hyperbolic curves are used in space probes. The equilateral hyperbola can be used to indicate varying pressure of gas as the volume varies. Gas pressure varies inversely as the volume changes.

Construct a hyperbola using the foci method

1. The foci F_1 and F_2, and the transverse axis AB are given, Fig. 6-17A.
2. Lay off a convenient number of points to the right of point F_2.
3. With points F1 and F2 as centers and A4 (in the example) as radius, draw arcs C, D, E, and G.
4. With points F_1 and F_2 as centers and the radius of B4, draw arcs that intersect the existing arcs at points C, D, E, and G. These points of intersection are points on the hyperbola.
5. Continue to lay off intersecting arcs, using radii of A1 and B1, A2 and B2, and so on.
6. Sketch a light line through the points. Use an irregular curve to darken the final curve.

Locate the asymptotes of a hyperbola

1. The transverse axis, the foci, and the hyperbola are given.
2. Draw a circle which has a center 0 that is the midpoint of the transverse axis and passes through the foci, Fig. 6-17B.
3. Construct perpendiculars to the transverse axis at points A and B. These points are the vertices of the hyperbola.

4. The asymptotes extend through the points where the perpendiculars intersect the circle.

Construct a tangent to a hyperbola

1. The hyperbola LBK and the point of tangency P are given. (Refer back to Fig. 6-17B.)
2. Draw lines from point P to the foci F1 and F2.
3. Bisect the angle F_1PF_2. The bisector HP is the required tangent.

Construct a hyperbola with the asymptotes and one point on the curve given

1. Asymptotes 0A and 0B and point P on the curve are given, Fig. 6-18 A.
2. Through point P, draw lines CD and EG parallel to the asymptotes.
3. From the origin 0, draw a number of radial lines intersecting line CD at points 1, 2, 3, 4, 5, and line EG at points 1′, 2′, 3′, 4′, 5′.
4. Draw lines parallel to the asymptotes at points 1 and 1′, 2 and 2′.
5. The intersections of these lines are points on the hyperbola. Continue until a sufficient number of points have been located to produce a smooth and accurate curve.
6. Sketch a light line through the points. When you are satisfied with the shape use an irregular curve and darken the curve.

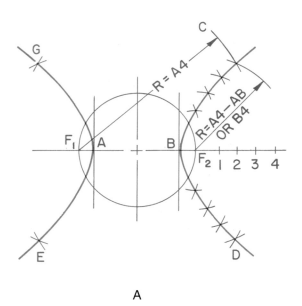

A

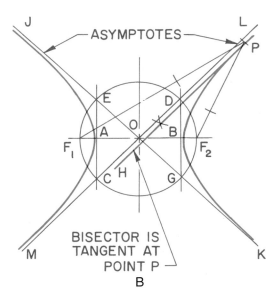

B

Fig. 6-17.   A–If the foci and the transverse axis are given, a hyperbola can be constructed using the foci method. B–If a hyperbola with known foci is given, a tangent to any point on that hyperbola can be easily constructed.

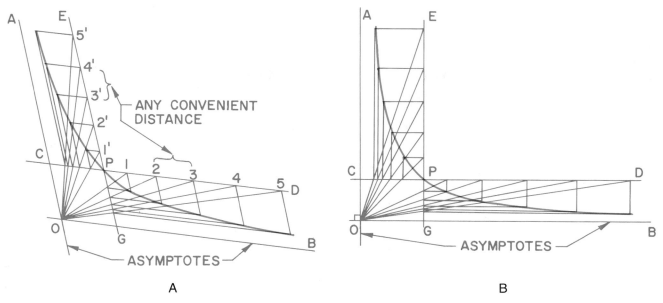

Fig. 6-18. A–A hyperbola can be constructed if the asymptotes and a point on the curve are given. B–If the asymptotes form a right angle, the hyperbola is called a rectangular or equilateral hyperbola.

Asymptotes which are at right angles to each other produce a hyperbola that is called a ***rectangular hyperbola*** or an ***equilateral hyperbola,*** Fig. 6-18B.

OTHER CURVES

Other curves commonly used in engineering, design, and drafting are the Spiral of Archimedes, the helix, the cycloids, and the involute. The construction and application of these curves are discussed in this section.

The Spiral of Archimedes

The ***Spiral of Archimedes*** is formed by a point moving uniformly around and away from a fixed point, Fig. 6-19. The Spiral of Archimedes curve is used in design of cams to change uniform rotary motion into uniform reciprocal (straight line) motion.

Construct a Spiral of Archimedes

1. The rise of one revolution, OB, is given, Fig. 6-19.
2. Draw horizontal line AB through point O. Lay off a convenient number of equal parts totaling 1 1/2 inches (for example, 12 parts of 1/8 inch each) on line OB (line OB is 1 1/2 inches long).
3. Using point O as the center and O12 as the radius, draw a circle.
4. Divide the circle into the same number of equal parts as line OB (in this example, 12 equal

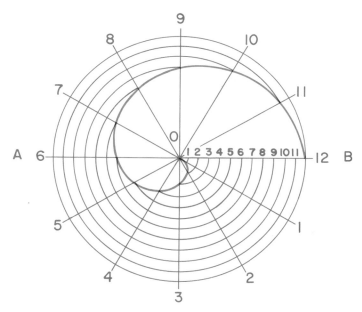

Fig. 6-19. A Spiral of Archimedes is formed by a point moving uniformly around and away from a fixed point.

parts of 30° each). Number each line, starting with first line after line OB.

5. Using the center O, draw an arc with a radius equal to the distance from point O to equal part 1. The arc should start on line OB and end on line 1.
6. Continue with concentric arcs for each of the equal parts. Start the arc on line OB and end the arc on the corresponding numbered line (the arc starting on the second equal part

would end on line 2, the arc starting on the seventh equal part would end on line 7).

7. The points of intersection of the concentric arcs and radial lines are points on the spiral curve.

8. Sketch a light line through these points. Finish with an irregular curve.

The Helix

The **helix** is a space, or three-dimensional, curve rather than a plane curve as those previously discussed. A helix can be described as a point moving around the circumference of a cylinder at a uniform rate while moving parallel to the axis at a uniform rate. The **pitch** or **lead** of a helix is the distance, parallel to the cylinder's axis, that it takes for the curve to make one complete revolution on the circumference of the cylinder. All basic threads are based on a helix.

Construct a helix

1. The diameter of the cylinder and the pitch or lead are given, Fig. 6-20A.

2. Draw the top view as a circle equal to the diameter of the cylinder and divide into any number of equal parts. For example, 12 parts of 30° each. Number the divisions.

3. Draw the front view of the cylinder with a length equal to the pitch or lead. The centerline of the cylinder must be the centerline of the circle.

4. Divide the front view along the axis into the same number of equal parts as for the top view. Number the divisions.

5. Project the points of intersection of the radial lines with the circumference of the circle to the corresponding numbered division on the cylinder.

6. These points of intersection in the front view are points on the helix curve.

7. Sketch a light line through the points. Finish with the aid of an irregular curve.

A stretchout of the development of the helix angle is shown in Fig. 6-20B.

Typical uses of the helix are bolt and screw threads, auger bits used in boring wood, flutes on a drill, and helical gears. The helix shown in Fig. 6-20 is a right-hand helix and advances into the work or mating part when turned clockwise. On a left-hand helix, the path moves from right to left and advances into the work or mating part when turned counterclockwise.

Cycloids

Cycloids are formed by the path of a fixed point on the circumference of a rolling circle. Cy-

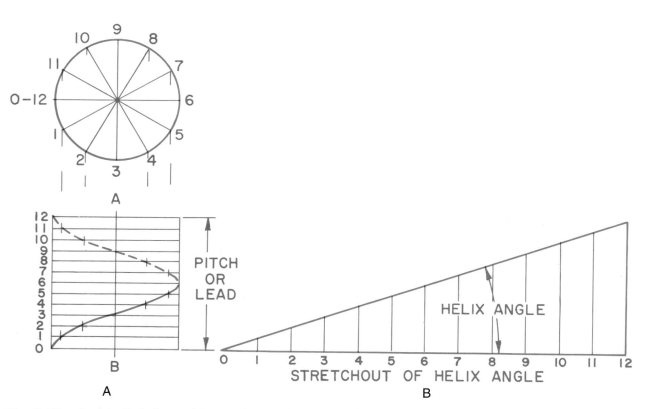

Fig. 6-20.   A–A helix is formed by a point moving at a uniform rate around the diameter of a cylinder and parallel to the axis of the cylinder. B–It is a three-dimensional curve. The stretchout of a helix will appear as a right triangle. Threads are based on a helix.

cloids are useful in the design of cycloidal gear teeth. When the circle rolls along a straight line, the path of the fixed point forms a cycloid.

Construct a cycloid curve

1. The generating circle and a tangent line AB equal to the rectified length of the generating circle are given.
2. Divide the circle and line AB into the same number of equal parts, Fig. 6-21.
3. Draw lines C, D, E, F, and G through the division points on the circle and parallel to line AB.
4. Project the division points on line AB to line E (the line that passes through the center of the given circle) by drawing perpendiculars.
5. Using the intersections of the perpendiculars with line E as centers, draw circle arcs of radius OP representing the various positions of the rolling circle as it moves to the left.
6. Assume the fixed point P is at its highest point on the curve at division line 6 (the middle division). When the center of the circle moves to the next division (the intersection of line 5 and line E), point P will have moved to the next division as well (line C). Therefore, the intersection of circle arc 5 and line C is the next point on the cycloid curve.
7. Similarly, the next point on the curve will be at the next division, or at the intersection of circle

arc 4 and line D. Locate the remaining points on the curve by intersecting circle arcs.
8. Sketch a light line through the points of intersection. Finish the cycloid curve using an irregular curve.

Construct an epicycloid

An epicycloid is similar to a cycloid. However, an *epicycloid* is formed by the generating circle rolling on the outside of another circle as opposed to a straight line. The construction is similar to that of the cycloid except that line AB and the other horizontal lines are concentric circle arcs, Fig. 6-22.

Construct a hypocycloid

An *hypocycloid* is formed by a fixed point on the generating circle rolling on the inside of another circle. (An epicycloid is formed by rolling on the *outside* of a circle.) The construction of this curve is similar to that of the epicycloid. The hypocycloid is shown in Fig. 6-23.

Involutes

An *involute* is the curve formed when a tightly drawn chord unwinds from around a circle or a polygon. Involute curves may start on the surface of the circle or polygon, or they may begin a distance away from the geometric form.

Construct an involute of an equilateral triangle

1. The triangle ABC is given.

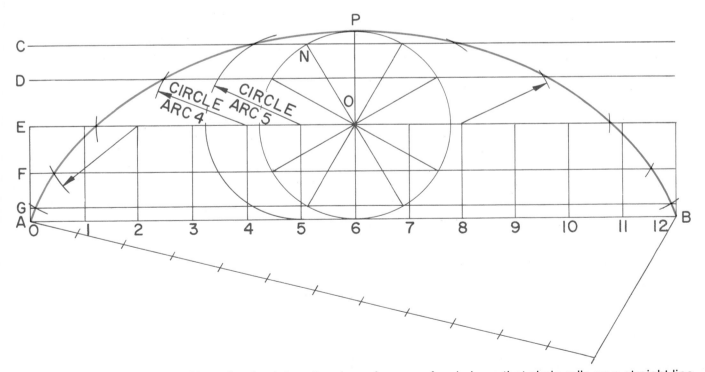

Fig. 6-21. A cycloid is formed by a fixed point on the circumference of a circle as that circle rolls on a straight line.

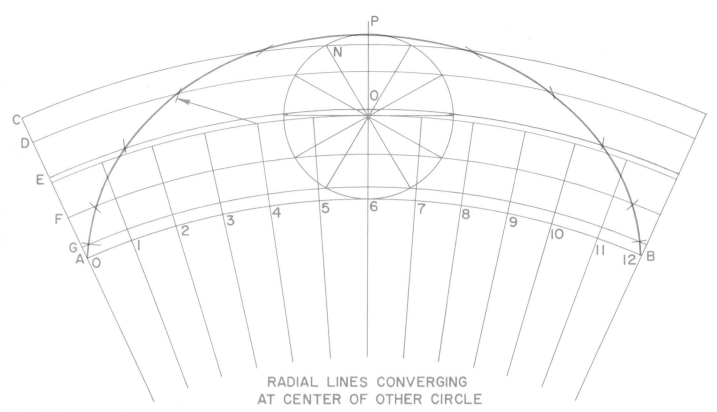

Fig. 6-22. An epicycloid is formed similar to a cycloid, except that the generating circle rolls on a curved line as opposed to a straight line.

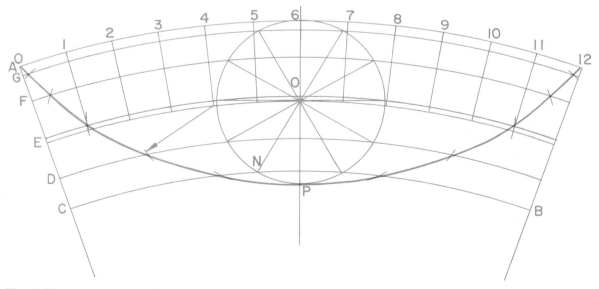

Fig. 6-23. A hypocycloid is formed just as an epicycloid except that the generating circle rolls on the *inside* of a curved line. The generating circle of an epicycloid rolls on the *outside* of a curved line.

2. Extend side CA to point D, Fig. 6-24A.

3. With point A as the center and AB as the radius, strike arc BD.

4. Extend side BC to the approximate location of point E. Using point C as the center and CD as the radius, strike arc DE.

5. In a similar manner, strike arc EF.

6. Continue process until a curve of the desired size is completed.

Construct an involute of a square

1. The square ABCD and a starting point P are given, Fig. 6-24B.

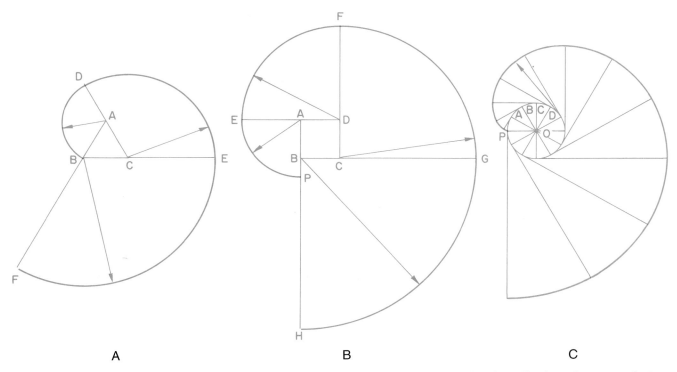

A   B   C

Fig. 6-24. Involutes can be constructed with polygons or circles. They can be described as the curve that results when a tightly wound chord unwinds from around the object.

2. Extend side AB to point P and side DA to the approximate location of point E. Using point A as the center and a radius of AP, strike arc PE.

3. Extend side CD to the approximate location of point F. Using point D as the center and a radius of DE, strike arc EF.

4. Continue until an involute of the desired size is achieved.

Construct an involute of a circle

1. The circle O and the starting point P are given, Fig. 6-24C.

2. Divide the circle into a number of equal parts. Draw tangents at the division points.

3. Beginning at first division point after the starting point P, lay off on tangent A a distance equal to length of arc AP. (Point A is one division away from point P.)

4. On tangent B, lay off a distance equal to the length of arc AP + BA (the length of two circle arcs).

5. Continue with tangent C with a distance equal to three circle arcs, and so on until the distance on the final tangent has been set off.

6. Sketch a light line through these points. Finish the involute of a circle with an irregular curve.

The involute of the circle is the curve form used in the design of involute gear teeth, Fig. 6-25.

Fig. 6-25. Some gears are designed with involute curve teeth. (Cincinnati Gear Co.)

PROBLEMS AND ACTIVITIES

The following problems involve complicated geometric constructions. Practical applications are included to acquaint you with typical geometric problems the drafter, designer, or engineer must solve.

All problems are to be drawn on drawing sheets of suitable size and in accordance with the layouts shown in the Reference Section. Place drawing sheets horizontally on the drawing board or table. Use the title block shown in Layout I in the Reference Section. Lay out your problems carefully to make the best use of space available. A freehand sketch will help.

Accuracy is extremely important. Use a hard lead (2H to 4H) sharpened to a fine, conical point. Draw light lines to start. Darken the required lines when the construction is complete. Leave all construction lines to show your work.

1. Using an appropriate scale, draw the outline of an elliptical swimming pool with a major diameter of 20 feet and a minor diameter of 12 feet. Use the foci method.

2. The six spokes in a gear wheel have an elliptical cross section with a major diameter of 2.50 inches and a minor diameter of 1.50 inches. Construct the ellipse twice size, using the concentric circle method.

3. The design for a bridge support arch is elliptical in shape. It has a span of 36 feet (the major diameter) and the rise at the center of the ellipse is 12 feet above the major diameter. Construct the half ellipse representing the arch. Use the trammel method.

4. Using an ellipse template, draw an ellipse. Label the size and degree of the ellipse.

5. A parabolic reflector for a flood light has a focus five inches from the directrix and the rise is four inches. Draw the parabolic form of the reflector.

6. A highway overpass has a horizontal span of 200 feet and a rise of 25 feet. The curve form is parabolic. Draw the form of this curve. Hint: The apex of the two tangent lines from the end points of the span must be 50 feet above the end points.

7. Construct a hyperbola and its asymptotes, using the foci method and given a horizontal transverse axis of 3/4 inch. This axis and the foci are 1 1/4 inches apart.

8. Construct a equilateral hyperbola with asymptotes located near, and parallel to, the lower and left-hand borders of your sheet. A point 1/4 inch to the right of the vertical asymptote and 3/4 inch above the horizontal asymptote is located on the hyperbola.

9. Starting in the center of a sheet section, draw one revolution of a Spiral of Archimedes with the generating point moving uniformly in a counterclockwise direction and away from the center at the rate of 1 1/4 inches per revolution.

10. Construct a section of a horizontal right-hand helix with a diameter of 2 inches, a length of 1 1/2 inches, and a pitch of 1 inch.

11. Construct a cycloid generated by a one inch diameter circle rolling along a horizontal line.

12. Construct the involute of a point starting at the apex of a 1/2 inch equilateral triangle for one clockwise revolution.

Part II
Drafting Techniques and Skills

Drafting techniques and skills include the skills and knowledge that basically define the graphic language and the process of making, reproducing, and storing drawings. The theories of orthographic projection and pictorial drawing are presented in this section. Basic and precision dimensioning are also presented. In short, Part II covers the most fundamental aspects of technical drawing. A thorough understanding and mastery of this material is necessary before proceeding to related or advanced topics.

Part II includes the following chapters: *Multiview Drawings, Dimensioning Fundamentals, Geometric Dimensioning and Tolerancing, Section Views, Pictorial Drawings,* and *Reproducing and Storing Drawings.*

Sometimes it is necessary to see the internal features of an assembly. Section view drawings are very helpful in these instances. This gear pump has been "cut open" to reveal the internal features. The cutting planes of a section view also "cut open" an object or assembly. The red surfaces in this photo represent the cutting planes. In a section view, these surfaces will be "cross hatched."

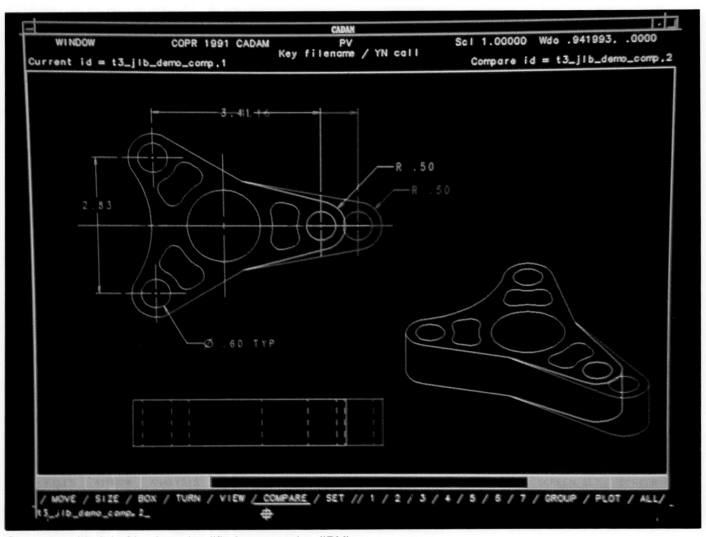

Computer-aided drafting has simplified many tasks. (IBM)

Multiview Drawing

KEY CONCEPTS

☐ Orthographic projection is the process of explaining three-dimensional objects in two-dimensions (a flat sheet of paper, for example).

☐ Orthographic projections are called orthographic projection drawings or multiview drawings.

☐ Multiview drawings typically have three views, but may have as few as one or two for simple objects or as many as four or more for complex objects.

☐ First-angle projections are used mainly in Europe. Third-angle projections are used mainly in the United States.

The multiview drawing is the major type of drawing used in industry. It is a projection drawing that incorporates several views of a part or assembly on one drawing, Fig. 7-1.

The various views of an object are carefully selected to show every detail of size and shape. Information about the processes to be performed on the part is also included. Usually, three views are drawn. However, drawings may vary from one or two views for a simple part to four or more views for a complicated part or assembly.

In addition, the views are arranged in a manner that is standard throughout industry. The top view always appears above the front view. The right-side view normally appears to the right of the front view.

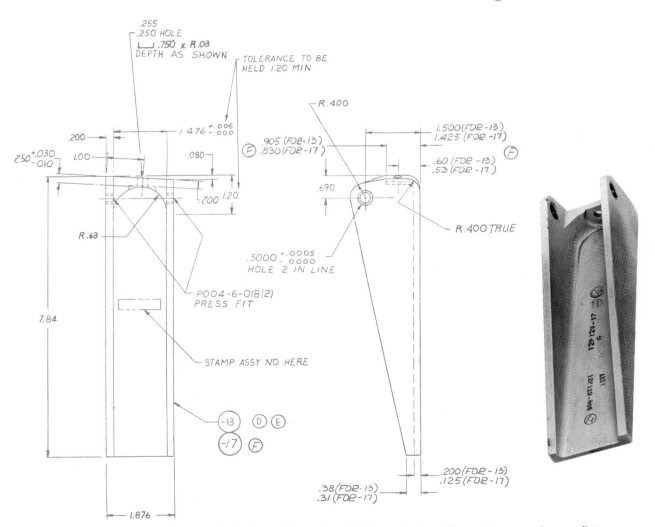

Fig. 7-1. Multiview drawings explain three-dimensional objects in two dimensions, such as a flat sheet of paper or a computer screen.

When used, left-side view usually is placed directly to the left of the front view.

This system of drawing is known as **orthographic projection.** The terms **multiview drawings** and **orthographic projection drawings** are used to describe orthographic projections.

ORTHOGRAPHIC PROJECTION

An **orthographic projection** drawing is a representation of the separate views of an object on a two-dimensional surface. It shows the width, depth, and height of the object, Fig. 7-2A.

The projection is achieved by viewing the object from a point assumed to be at infinity (an indefi-

nitely great distance away). The **lines of sight,** or **projectors,** are parallel to each other and perpendicular to the plane of projection, Fig. 7-2B.

The Projection Technique

Persons experienced in drafting are readily able to "picture" different views of an object in their minds. This mental process is known as **visualizing** the views. Visualizing is done by looking at the actual object or a three-dimensional picture of the object. This is one of the most important aspects of drafting technique that you can learn.

The **glass box** is helpful in developing skill in visualizing a view, Fig. 7-3. Each face of an object is viewed from a position that is 90°, or perpendicular, to the projection plane for that view. (Refer to

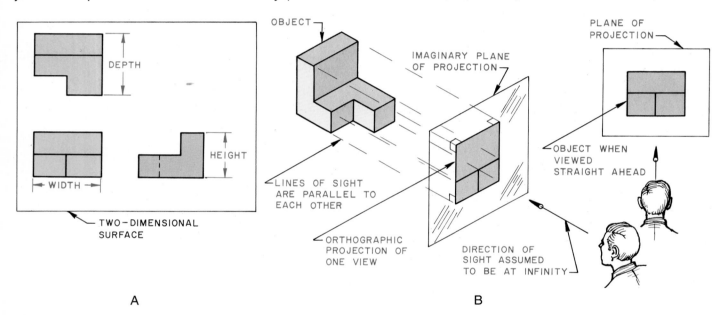

A   B

Fig. 7-2. In orthographic projection, all the features of an object are projected onto a perpendicular viewing plane. The viewing point is assumed to be at a distance of infinity from the object.

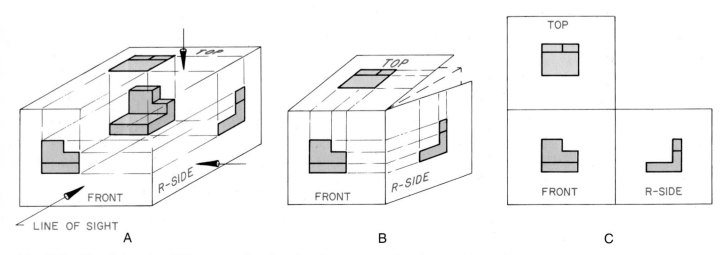

A   B   C

Fig. 7-3. The "glass box" illustrates the visualization system of orthographic projection.

Fig. 7-3A.) The views are obtained by projecting the lines of sight to each plane of the glass box.

The projection shown in the **frontal plane** is called the **front view** or **front elevation,** Fig. 7-4. On the **horizontal plane**, the projection is called the **top view** or **plan view.** If the projection is on the **profile plane**, it is called the **side view** or **end view,** or **side elevation** or **end elevation.**

These three planes are at right angles to each other when in their natural position (with the glass box closed), Fig. 7-5. The frontal plane is considered to be lying in the plane of the drawing paper. The horizontal and profile planes are revolved into position on the drawing, so that they are in the same plane as the drawing paper. These three planes are referred to as **principal planes** because they are the views shown on most industrial drawings.

These three planes are also called **coordinate planes** because of their right-angle relationship in the folded box. When they are unfolded, they establish a definite coordinate relationship between all views of an orthographic projection.

However, six views are possible from the six sides or planes of the glass box, Fig. 7-6. Note the

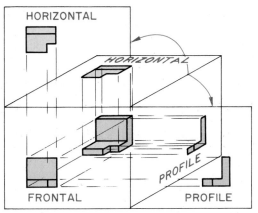

Fig. 7-4. There are three principal or coordinate planes of projection that are typically used in orthographic projection.

manner in which the box is unfolded as shown in Fig. 7-6B and Fig. 7-6C. This establishes the coordinate relationship of the three additional views. Also note the lines of projection from one view to the next. This makes certain that the height dimension will be the same for the rear, left-side, front,

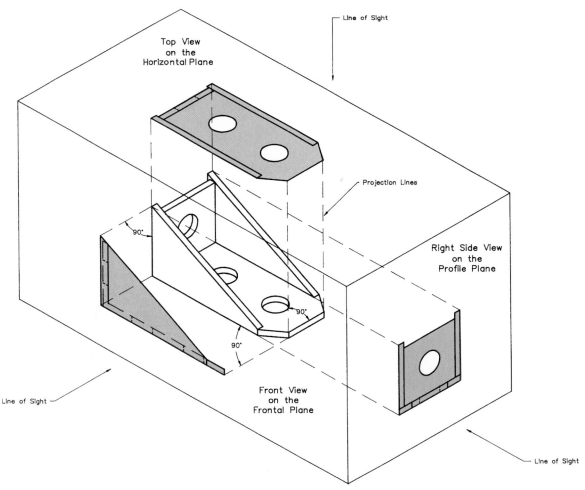

Fig. 7-5. The "glass box" is used to project the features of an object onto a two-dimensional surface.

and right-side views, and that they are all aligned with coordinates. (Refer back to Fig. 7-6C.) The top, front, and bottom views are all aligned and have the common dimension of width, as does the rear view. The top, right-side, bottom, and left-side views have depth dimensions that are common.

Alternate Location of Views

There are occasions when the preferred location of views shown in Fig. 7-7A is not feasible due to space limitations. For example, an expanded "List of Materials" on a drawing, which usually appears above the title block, may crowd the usual location of the right side view.

To compensate for this lack of space, the right side view is projected directly across from the top view. This view is located as if the profile plane were hinged to the horizontal plane, Fig. 7-7B. The projection would then appear on the drawing as shown in Fig. 7-7C.

Although the usual practice is to locate required views in normal projected positions, the side view or profile planes may be projected off the top or bottom views. Likewise, the rear view may be projected upward from the top view, or downward from the bottom view. Each is revolved into position in the same plane as the front view when alternate locations are necessary.

Projection of Elements

Drawing the several views of an object is done by making measurements of certain points, lines,

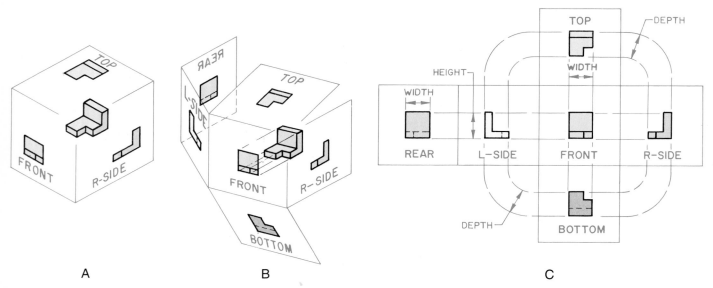

Fig. 7-6. When the "glass box" is "unfolded," there are six coordinate planes that may be used in orthographic projection.

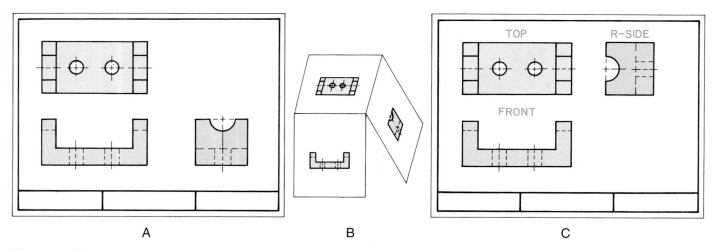

Fig. 7-7. Sometimes it is necessary to use an alternate location for the right-side view. This may be because of crowding by an expanded "List of Materials."

and surfaces in one view and projecting these to the other views. Features such as circular holes or arcs should be located initially in the view where they appear as circles or arcs, then projected to the remaining views, Fig. 7-8.

Projection of the elements provides for greater accuracy in the alignment of views. Projection is also faster than measuring each view separately with the scale or dividers. A single, 45° miter line is drawn to project point, line, and surface measurements from one view to another. (See line PC in Fig. 7-8.) This miter line meets the projection planes AP and PB at point P equidistant between the views.

Projection of Points

A *point* is defined as something having position, but not extension. It has location in space, but has no length, depth, or height. A point on a drawing should be indicated by a small cross (+) mark and never by a dot. It may be the intersection of two lines, the end of a line, or the corner of an object.

Point P in space is located by measuring three directions (length, depth, and height) from the planes of projection. These measurements are made from the frontal, the horizontal, and the profile planes. This is represented by point P_F, point P_H, and point P_P in Fig. 7-9A. Point P in the orthographic projection is located by making each measurement once in the appropriate view and projecting measurements with a triangle and straightedge to the other views, Fig. 7-9B.

Measurements need not be made from planes of projection on a drawing. Usually, the first point representing a corner, centerline, or other feature of the object is properly located. Then space between views is allotted to produce a balanced drawing. Other points, lines, or surfaces are located from this first point, Fig. 7-10.

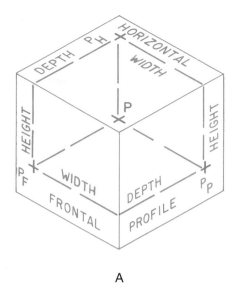

A

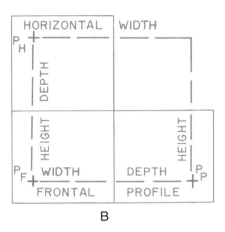

B

Fig. 7-9. A point in space is located by measuring in three directions: length, depth, and height.

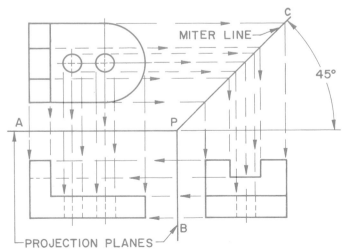

Fig. 7-8. A 45° miter line is used to help project elements between views.

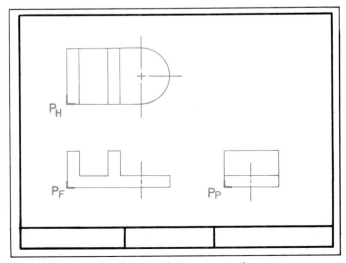

Fig. 7-10. Typically, a point representing a corner or other convenient feature of an object is first located on a drawing. All other points, lines, and surfaces are then located from this first point.

Projection of Lines

A *straight line* is defined as the shortest distance between two points. A *curved line* is a line following any of a variety of arcs or curved forms.

In a drawing, lines may represent the intersection of two surfaces, Fig. 7-11A. Lines may also represent the edge view of a surface or the limits of a surface, Fig. 7-11B.

There are four kinds of straight lines found on objects in drawings. These lines are horizontal, vertical, inclined, and oblique. Each line is projected by first locating its endpoint.

Horizontal lines are parallel to the horizontal plane of projection and one of the other planes, while being perpendicular to the third plane. Refer to line AB in Fig. 7-12. A horizontal line appears *true length* (viewed perpendicular to the line) in two of the planes and as a point in the third.

Vertical lines are parallel to both the frontal and profile planes, and perpendicular to the horizontal plane. (Refer to line EG in Fig. 7-12.) A vertical line appears true length in the frontal and profile planes and as a point in the horizontal plane.

Inclined lines are parallel to one plane of projection and inclined in the other two planes. (Refer to line CE in Fig. 7-12.) An inclined line appears true length in one of the planes and *foreshortened* (not as long) in the other two.

Oblique lines are neither parallel nor perpendicular to any of the planes of projection. (Refer to line DF in Fig. 7-12.) An oblique line appears foreshortened in all three planes of projection.

Curved lines may appear as a circle, an ellipse, a parabola, a hyperbola, or some other geometric curve form. They may also be irregular curves. For curves other than circles and arcs of circles, a number of points must be located on the curve and projected to the view concerned.

Projection of Surfaces

Plane and curved surfaces represent most of the surface features found on machine parts. Examples of plane surfaces are found on cubes and pyramids. Examples of circular curved surfaces are found on cylinders and cones.

Surfaces may be horizontal, vertical, inclined, oblique, or curved. These surfaces are drawn by locating the end points of the lines that outline their shapes.

Horizontal surfaces are parallel to the horizontal projection plane and appear in their true size and shape (the view is perpendicular to the surface) in the top view. Refer to surfaces A and B in Fig. 7-13. Horizontal surfaces appear as lines in the frontal and profile planes of projection. (Refer to surface A, surface B, and lines 4 through 13 in Fig. 7-13.)

Vertical surfaces are parallel to one or the other of the frontal or profile planes. They appear in their true size and shape in this plane. (Refer to surfaces G and F in Fig. 7-13.) They are perpendicular to the other two planes and appear as lines in these planes. (Refer to lines 8 through 9 in the frontal plane and lines 6 through 8 in the profile plane in Fig. 7-13.)

Inclined surfaces are neither horizontal nor vertical. (Refer to surface D in Fig. 7-13.) They are perpendicular to one of the projection planes and

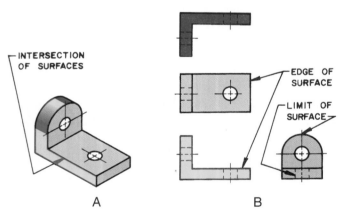

Fig. 7-11. Lines can represent the intersection of surfaces, the edge of surfaces, or the limits of surfaces.

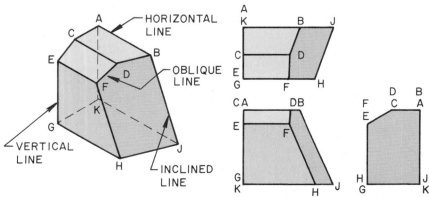

Fig. 7-12. There are four basic types of straight lines used on drawings.

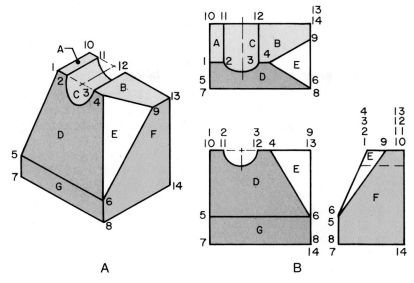

Fig. 7-13. Surfaces will appear differently in each projection of a drawing.

appear as a true length line in this view. (Refer to lines 1 through 5 in the side view in Fig. 7-13.) In the other two planes or views, inclined surfaces appear foreshortened. (Refer to the top and front views in Fig.7-13B.)

Oblique surfaces are neither parallel nor perpendicular to any of the planes of projection. (Refer to surface E in Fig. 7-13.) They appear as a surface in all views but not in their true size and shape.

Curved surfaces may be a single curved surface (cone or cylinder), double curved surface (sphere, spheroid, or torus) or warped surface (machine screw thread or helix), Fig. 7-14.

Curved surfaces of the circular curve forms (cylinder) appear as circles in one view and as rectangles in the other views. (Refer back to Fig. 7-14.) Three views of curved surface objects are shown here, but two are usually sufficient.

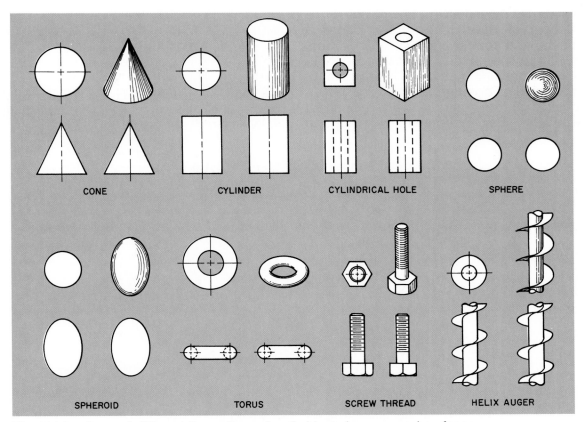

Fig. 7-14. Several different three-dimensional objects have curved surfaces.

Methods of obtaining the true size and shape of various lines and surfaces is covered in detail in Chapter 13.

Projection of Angles in Multiview Drawings

Angles that lie in a plane parallel to one of the projection planes will project their true size on that plane. Refer to angle A in Fig. 7-15. Angles that lie in a plane inclined to the projection plane will project smaller or larger than true size, depending on their location. (Refer to angle B in Fig. 7-15.)

Also, when a 90° angle on an inclined plane has one of its legs parallel to two projection planes, it will project true size in the other plane. (Refer to angle C in Fig. 7-15.) Angles on an oblique plane will always project smaller or larger than true size depending on their location. (Refer to angle D in Fig. 7-15.)

To project angles lying in an inclined or oblique plane, locate their end points by projection and draw lines between these points.

Selection of Views

First considerations in making multiview drawings are the selection and arrangement of views to be drawn. Select views which clearly describe the details of the part or assembly. The front view should best describe the shape of the object. Typically, the front view also shows the most details of the object.

The number of views to be drawn depends upon the shape and complexity of the part. Often, two views will provide all the details necessary to construct or assemble the object. This is particularly true of cylindrical or round objects, Fig. 7-16A.

Flat objects made from relatively thin sheet stock may be adequately represented with only one view by noting the stock thickness on the drawing, Fig. 7-16B.

The views for the object in Fig. 7-16C were not well chosen. This results in a poor arrangement of the views. However, the views in Fig. 7-16D have been well selected to best describe the part. The number of hidden lines is reduced in all views. This is a balanced arrangement of the three required views.

Summary of Factors Involved

Four factors serve as guidelines in the selection of views. No single factor should determine the selection. Rather, consideration of all factors is most likely to result in the best selection of views.

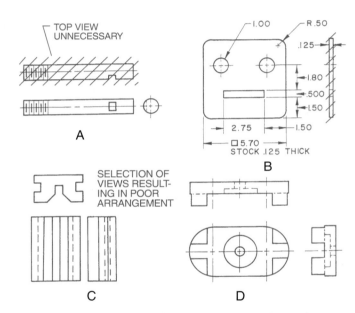

Fig. 7-16. When selecting views, select views that best describe the object. Eliminate views that are redundant or unneeded.

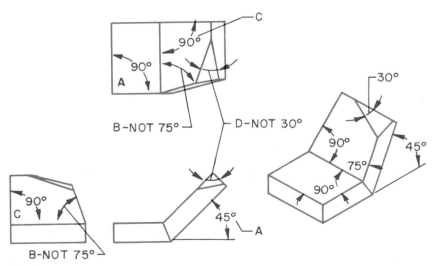

Fig. 7-15. In the different views of an orthographic projection, angles may not appear as the true measurement.

Give prime consideration to selection of front view

1. Select a view that is most representative of contour or shape of object.
2. Consider natural or functioning position of object.
3. Place the principal surface area parallel or perpendicular to one or more planes of projection.
4. Consider an orientation of the view which produces the least hidden lines in all views.

Consider space requirements of entire drawing

1. Long and narrow objects may suggest a top and a front view.
2. Short and broad objects may suggest a front and a side view.
3. When the "Title Block" or "List of Materials" tends to crowd the right side view, consider an alternate position of the view or select another view.

Choose between two equally important views unless space requirements or other factors prohibit

1. The right-side view is preferred over the left-side when a choice is available.
2. The top view is preferred over the bottom view when a choice is available.

Consider the number of views to be drawn

1. Use only the number of views necessary to present a clear understanding of the object.
2. One or two views may be sufficient for a relatively simple object. Three or more views may be required for more complex objects.

Space Allotment for Multiview Drawing

Crowding the views in multiview drawings detracts from the appearance of the drawings. Crowding makes reading and understanding drawings more difficult.

The practice in industrial drafting rooms is to provide ample space between the views of a drawing for dimensions, callouts (specific notes), and general notes. This serves to make the views more distinct and provides a neat appearance. Anticipate the number of dimension lines or notes to be used and allow sufficient space.

Once the scale of the drawing has been determined, figuring space allotment is rather easy. Add the combined width and depth of the front and side views, Fig. 7-17. Lay off this total (6.5 inches, for example) to scale along the lower border. Allow approximately 1 inch between views and lay off this distance beyond the first measurement. Divide the amount remaining between the two end spaces, which will allow about 1.5 inches for each. Vertical spacing is figured similarly by laying off the distances along a vertical border line. (Refer back to Fig. 7-17.)

Projection of Invisible Lines and Surfaces

Surfaces and intersections that are hidden behind a portion of the object in a particular view are usually represented by *hidden* or *invisible lines.* Obviously, these terms are used to refer to the surface or intersection that the invisible line represents, rather than the line itself being invisible.

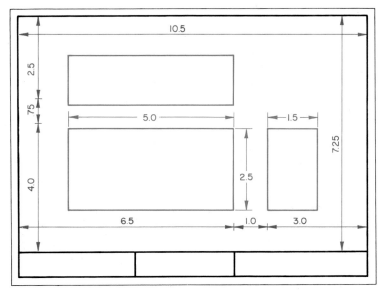

Fig. 7-17. Once the overall dimensions of all of the views are known, the spacing of the views on a drawing sheet is a simple procedure.

Invisible lines and arcs begin with a dash. If the invisible line is a continuation of a line or arc, a space is left to show exactly where the invisible line or arc begins, Fig. 7-18 (1). When invisible surfaces actually intersect on the object itself, the dashes join at the intersection with a "+" or a "T," Fig. 7-18 (2). Invisible lines that cross but do not intersect each other on the object are drawn with a gap at the crossover on the drawing, Fig. 7-18 (3).

Angular lines that come to a point are shown with the dashes joined at the point. An example of this would be the bottom of a drilled hole, Fig. 7-18 (4). Invisible lines meet at the corners with an "L," unless the corner is joined by a visible line. In that case, a gap is used. See Fig. 7-18 (5) and Fig. 7-18 (6). Parallel invisible lines have their dashes staggered, Fig. 7-18 (7).

Invisible lines are usually omitted in sectioned views unless absolutely necessary for clarity, Fig. 7-18 (8). Industrial drafters frequently omit some invisible lines from views when the drawing is clear without them. This avoids a cluttered appearance. However, it is good practice for the beginning drafter to include all invisible lines in regular views until a better understanding of the problem of clarity in a drawing is gained.

Precedence of Lines

Occasionally you will find that certain lines coincide in the projection of views in multiview drawings. Should this occur, visible lines take precedence over all others. The following priority of line importance governs precedence of lines.

1. Visible lines.
2. Invisible lines.
3. Cutting-plane lines.
4. Centerlines.
5. Break lines.
6. Dimension and extension lines.
7. Section lines (crosshatching).

All lines have precedence over any line that appears below it. For example, a centerline (number 4) will have precedence over a break line (number 5), a dimension or extension line (number 6), and a section line (number 7).

Removed Views

Sometimes it is desirable to show a complete or partial view on an enlarged scale to clarify the detail of the part, Fig. 7-19. This particular view is removed to a nearby area of the drawing. However, the same orientation of the view is maintained. The removed view is appropriately identified and referred to on the regular view.

Partial Views

Because of their shape, some objects may not require all views to be full view. Objects which are symmetrical in one view may require only a half view in that plane, Fig. 7-20A. A partial view may be broken on the centerline or at another place with a broken line, Fig. 7-20B.

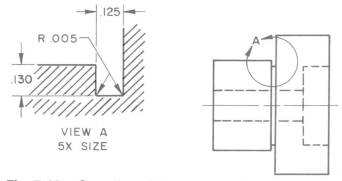

VIEW A
5X SIZE

Fig. 7-19. Sometimes it is necessary to provide a removed partial view that is enlarged to clarify a machining detail.

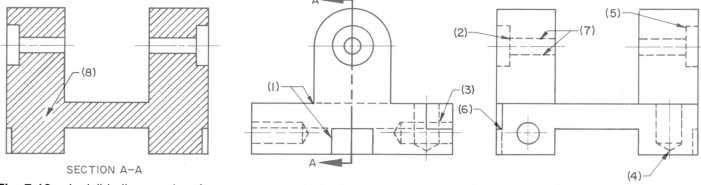

SECTION A–A

Fig. 7-18. Invisible lines and surfaces appear as dashed lines on a drawing. There are certain conventions that need to be followed for drawing these lines.

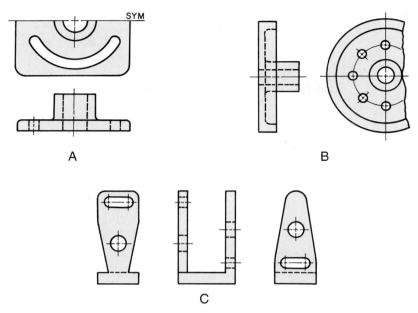

Fig. 7-20. If an object is symmetrical, it is not always necessary to draw the complete object. A–A line of symmetry can be shown. B–Or a break line can be used. C–On the other hand, sometimes it is necessary to show two side views if the object is different on both sides. In this case, show only the features that can be seen on that side. This improves the clarity of the drawing.

Objects with different side views should have two partial side views drawn. Each of these views should include lines for that view only, thus avoiding a confusion of lines from the other view, Fig. 7-20C.

CONVENTIONAL DRAFTING PRACTICES

A number of conventional drafting practices are used in American industry to reduce costs, speed the drafting process, and clarify drawings. Some of these practices are presented here. Other conventions appear in chapters related to the specific practice being described.

Fillets and Rounds

When making metal castings, it is necessary to avoid sharp interior corners. This prevents fractures of the metal. Also, sharp exterior corners are difficult to form in the mold. To eliminate these problems, patterns for the castings are made with rounded corners. A small, rounded, internal corner is known as a *fillet,* Fig. 7-21A.

A small, rounded, external corner is known as a *round.* (Refer back to Fig. 7-21A.) Since there is no sharp line intersecting the two surfaces, an assumed line of intersection of the two surfaces is drawn, Fig. 7-21B. The view in Fig. 7-21C shows

fillets and rounds represented on plane surfaces. Notice that the top view, which has no fillets drawn on it, lacks clarity.

Runouts

A *runout* is the intersection of a fillet or round with another surface. The shape of an arm, spoke, or web also affects the shape of the runout, Fig. 7-22. The arc of the runout should be the same radius as the fillet or round. A runout may be drawn freehand, with an irregular curve, or with a compass.

Right-hand and Left-hand Parts

Whenever possible, industry will make opposite parts identical to reduce the number of different parts required for an assembly. Examples include automobile wheels, tires, and some doors for kitchen cabinets. In some cases, opposite parts are not identical. Examples include automobile door handles, cabinet drawer slides, and many cabinet doors that can only be hung one way, Fig. 7-23.

When opposite parts cannot be made interchangeable, the conventional practice is to draw one part and to note "RH PART SHOWN, LH PART OPPOSITE," Fig. 7-24. This works quite well for most simple parts and saves considerable drafting time. Where there is a chance for confusion in the details of the opposite part, both parts should be drawn.

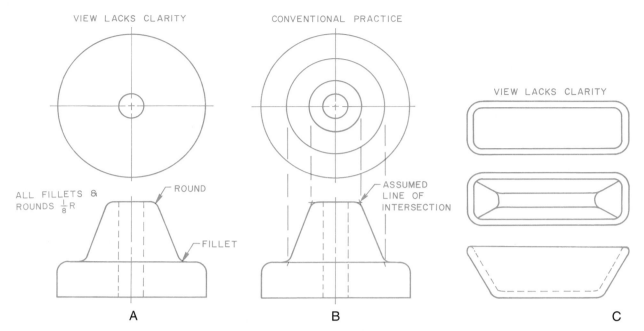

Fig. 7-21. Fillets and rounds are typically shown as lines. This provides much more clarity than if the drawing is shown as it truly appears.

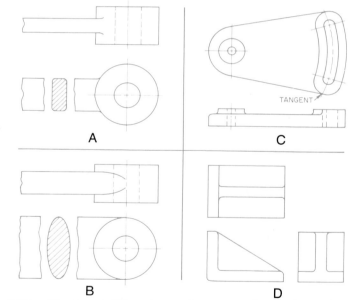

Fig. 7-22. There are different ways to represent runouts on drawings.

Cylinder Intersections

The conventional representation for small cylinder intersections with plane or cylindrical surfaces is shown in Fig. 7-25A. The same is true for cuts in small cylinders, Fig. 7-25B. For keyseats and small drilled holes, the intersection is typically unimportant. Clarity on the drawing is accomplished by treating these intersections conventionally, Fig. 7-25 C. Intersection of larger cylinders should be plotted as discussed in Chapter 15.

Revolving Radial Features

Some objects that have radially arranged features appear confusing when true orthographic projection techniques are followed. For example, the lugs on a flange appear awkward and out of position with the flange in true projection, Fig. 7-26A. When revolved to a position on the centerline, the ribs appear to be symmetrical, Fig. 7-26B. This drawing has more clarity, even though the projection is not a true projection.

Fig. 7-23. Automobile door handles are examples of opposite parts that are not always identical. (Fisher Body Division, GMC)

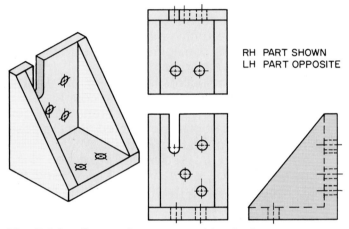

RH PART SHOWN
LH PART OPPOSITE

Fig. 7-24. If opposite parts are identical, a note on the drawing is a way for the drafter to save time.

A number of small holes arranged radially in a plate are also confusing in the side view when true projection techniques are followed, Fig. 7-26D. Conventional practices call for the holes to be revolved to positions of symmetry, Fig. 7-26E. Ribs on a hub are another example of radial features that appear more clearly when revolved to a position of symmetry, Fig. 7-26H.

Parts in Alternate Positions

At times it is necessary to draw an alternate position of parts to show the limits and necessary clearance during operation. The part is drawn in its alternate position by using a phantom line, Fig. 7-27.

Repeated Detail

Drawings of coil springs, radial flutes, and other repeated details would require considerable drafting time if these were drawn in full views. Conventional drafting practices require the drawing of one or two of the individual details, with the remainder represented by phantom lines, Fig. 7-28.

First-angle and Third-angle Projections

In orthographic projection, drawings are referred to as *first-angle* or *third-angle* projections. These two projections are derived from a theoretical

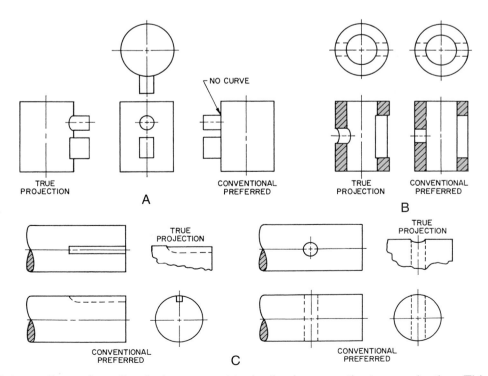

Fig. 7-25. The intersections of small cylinders are not typically shown as the true projection. This is also the case with keyseats and small, drilled holes.

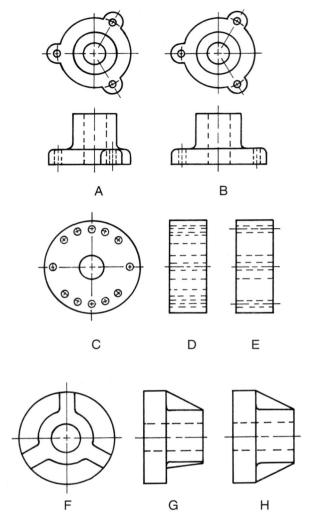

Fig. 7-26. Radial features are typically revolved to achieve symmetry. Although this is not the true projection, it makes the drawing more understandable.

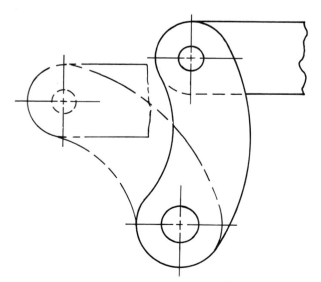

Fig. 7-27. Phantom lines are used to show alternate positions of parts. This is done in many cases where a part has specific movement limits or clearance requirements.

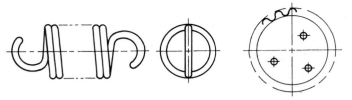

Fig. 7-28. Phantom lines can be used to represent repeated details.

division of all space into four quadrants by a vertical plane and a horizontal plane, Fig. 7-29. The quadrants are numbered 1 through 4, starting in the upper left quadrant and continuing clockwise, when viewed from the right side. The viewer of the four quadrants is considered to be in front of the vertical or frontal plane, and above the horizontal plane. The position of the profile plane is not affected by the quadrants. It is considered to be either to the right or left of the object, as desired.

Third-angle projection is used in the United States and Canada. Most European countries use first-angle projection. The main difference between the two is how the object is projected and the positions of the views on the drawing.

In *third-angle projection,* the projection plane is considered to be between the viewer and the object. The views are projected forward to the pro-

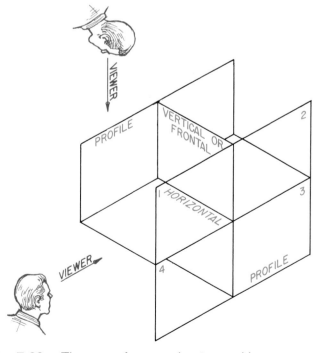

Fig. 7-29. There are four quadrants used in projections. These quadrants are created by the intersection of a horizontal and a vertical, or frontal, plane. The quadrants are numbered clockwise from the upper left quadrant (when viewed from the right side). These quadrants are used for "first-angle" and "third-angle" projections.

jection plane, Fig. 7-30A. The views appear in their natural positions when the views are revolved into the same plane as the frontal plane, Fig. 7-30B. The top view appears above the front view, the right-side view is to the right of the front view, and the left view to the left of the front view.

In *first-angle projection,* the projection plane is on the far side of the object from the viewer, Fig. 7-31. The views of the object are projected to the rear instead of being projected forward.

The individual views are the same as those obtained in third-angle projection, but their arrangement on the drawing is different when revolved into the frontal plane, Fig. 7-32. The "glass box" is still hinged to the frontal plane, but the frontal plane is behind the object. The top view appears below the front view, right side appears to the left of the front view, and left side appears to the right of the front view.

It is possible to place an object in any of the four quadrants. However, the second and fourth quadrants are not practical. The third-angle of projection is followed entirely in this text. However, it is good for you to understand first-angle projection as well. This will allow you to interpret a drawing prepared in another country.

Industries which serve customers in the international market sometimes mark their drawings to indicate first-angle or third-angle projection, Fig. 7-33.

VISUALIZING AN OBJECT FROM A MULTIVIEW DRAWING

Most students in beginning drafting have difficulty visualizing a machine part or assembly from a

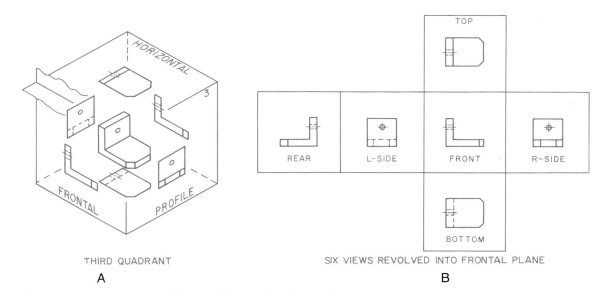

THIRD QUADRANT
A

SIX VIEWS REVOLVED INTO FRONTAL PLANE
B

Fig. 7-30. Views are projected forward in third-angle projection.

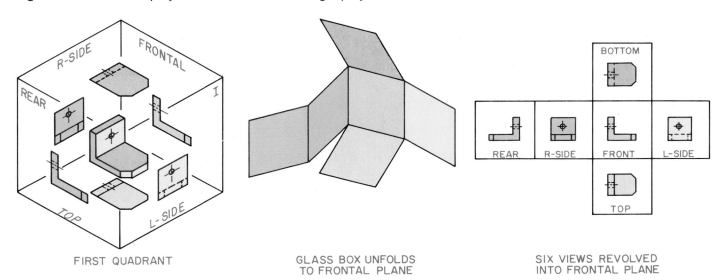

FIRST QUADRANT

GLASS BOX UNFOLDS TO FRONTAL PLANE

SIX VIEWS REVOLVED INTO FRONTAL PLANE

Fig. 7-31. Views are projected towards the rear in first-angle projection.

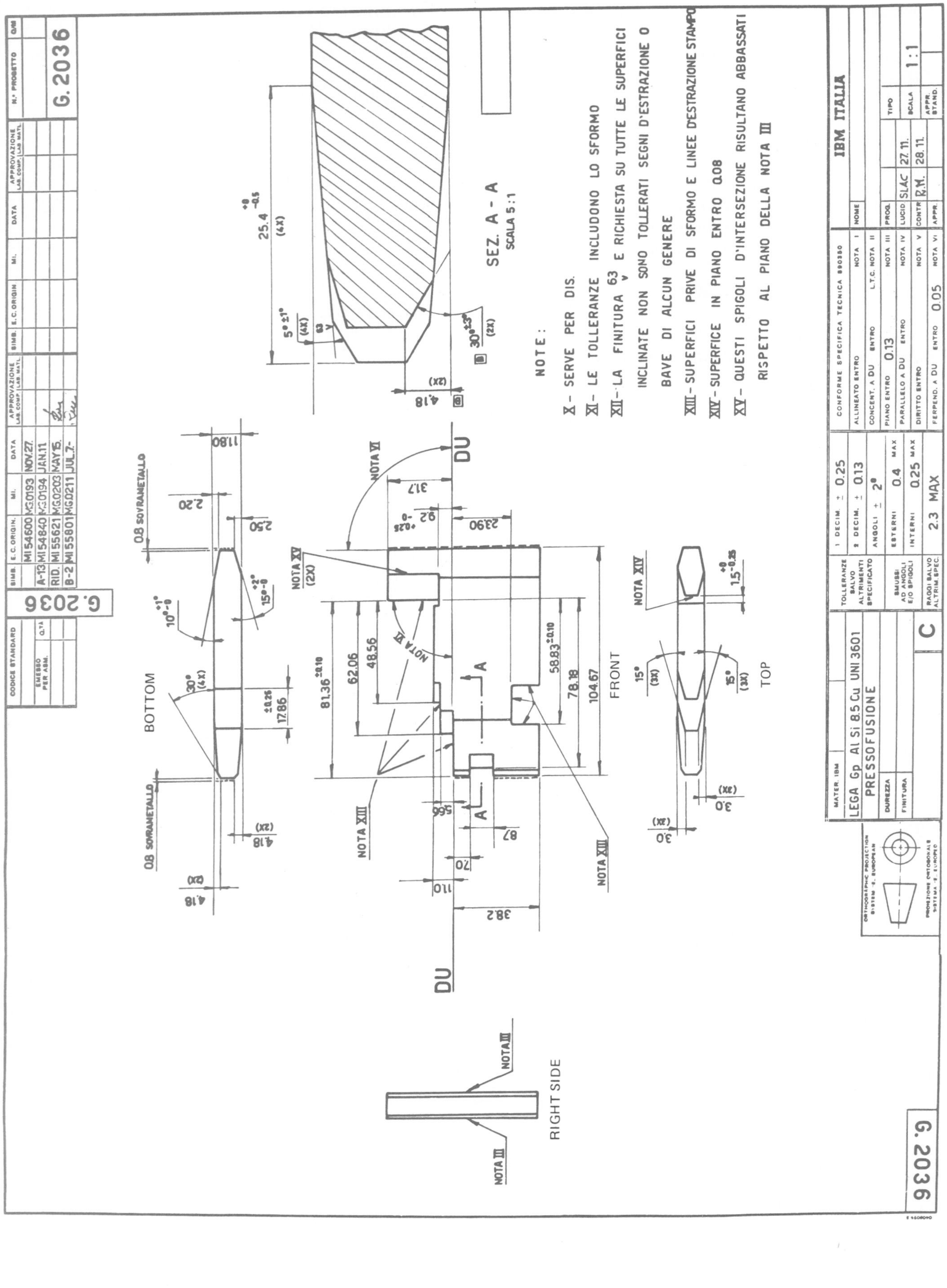

Fig. 7-32. First-angle projection is used in most European countries. (IBM)

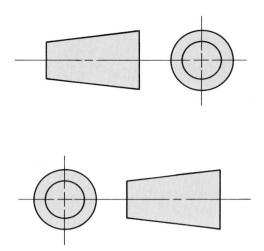

Fig. 7-33. Symbols are used to indicate whether the drawing is a first-angle (top) or a third-angle (bottom) projection.

multiview drawing. Mastery of a few simple techniques will enable you to solve the "mystery." Follow the steps listed below to help you visualize an object, Fig. 7-34.

1. Break each view down to a basic geometric shape (rectangle, circle, cone, or triangle).

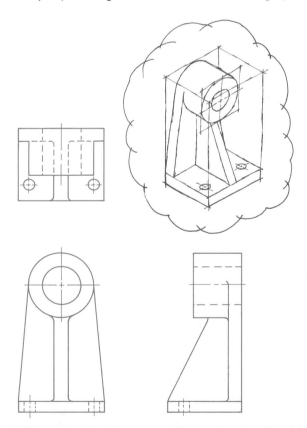

Fig. 7-34. An object from a multiview drawing should be visualized three-dimensionally. If the drawing is broken down into basic geometric shapes, it is much easier to visualize the three-dimensional representation of that object.

2. Consider possibilities for each shape. Is a circle a hole or a protruding shaft? Is a rectangle a base plate or web? Is a triangle a brace or support?

3. Check the basic geometric shape in one view against its shape in another view. What do hidden lines and centerlines represent?

4. Check the length, depth, or height dimensions in two or more views. Compare these dimensions with geometric shapes.

5. Put various geometric components together mentally and you should begin to visualize the object.

6. Make a freehand sketch of the object to clear up any uncertain details.

LAYING OUT AND FINISHING A DRAWING

It is good practice to draw light layout and construction lines until you have solved the essential problems in each view. All lines should then be darkened to complete the drawing. If changes are necessary, they are relatively simple to make when lines are lightly drawn. Observe the following steps until they become an unconscious part of your drafting.

1. Draw light lines at first (heavy lines tend to "ghost" and are difficult to erase).

2. Select a sheet size and drawing scale that will avoid crowding the views, dimensions, and notes.

3. Check your measurements carefully in "blocking out" required views.

4. Locate and lay out arcs and circles first, then straight lines.

5. During layout, do not include hidden lines, centerlines, or dimension lines. A short mark, or a dimension figure lightly noted near its location, will serve as a reminder that it is to be included.

6. Check your layout carefully for missing lines, dimensions, notes, or special features required in problem assignment.

7. Remove unnecessary construction lines. Give the drawing a general cleaning.

8. Darken the lines. Start with arcs and circles, then do the lines from top down. Next, darken the rest of the lines from your non-drawing hand to your drawing hand.

9. Letter the notes and title block.

10. Check finished drawing carefully for spelling, line weight, and general appearance.

PROBLEMS AND ACTIVITIES

Problems with missing lines and missing views, and other problems are given to provide you with experience in multiview projection techniques.

Missing Lines

Study the views in Fig. 7-35. Use Layout II in the Reference Section. Sketch in the views of the problems and add the missing lines. Have your sketch approved by your instructor before proceeding.

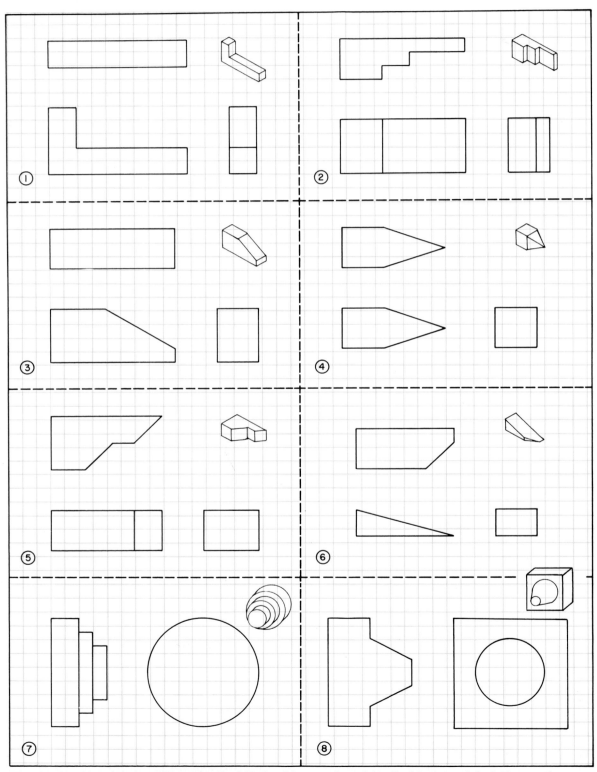

Fig. 7-35.   Use these problems for the Missing Line section of the Problems and Activities section.

Missing Views

Study the views given in Fig. 7-36. Make a sketch of the two views given and the missing view. Have your sketch approved by your instructor. Then make drawings using Layout II in the Reference Section.

Identification of Points, Lines, and Surfaces

This problem calls for identification of points, lines, and surfaces in four multiview drawings to be listed in chart form, Fig. 7-37. A sample entry has

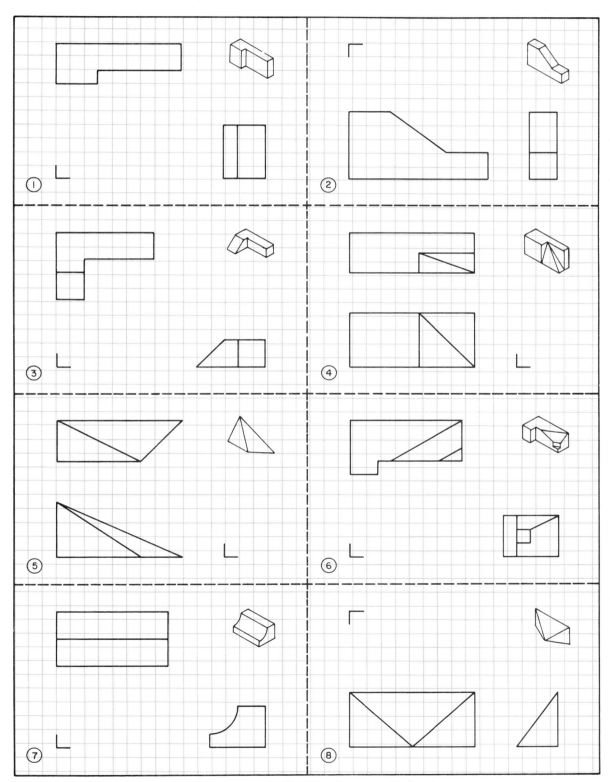

Fig. 7-36. Missing view problems for use in the Problems and Activities section.

PICTORIAL VIEW			FRONT VIEW			SIDE VIEW			TOP VIEW		
PROBLEM A											
POINT	LINE	SURFACE	POINT	LINE	SURFACE	POINT	LINE	SURFACE	POINT	LINE	SURFACE
1.	9-8	D	30				35-36			23-26	V
2.		C									
3.										24-26	
4.										23-26	
5.				27-28							
PROBLEM B											
1.											

Fig. 7-37. Use this chart for recording information about the multiview drawings found in Fig. 7-38. Follow the directions given in the Problems and Activities section.

been completed in the first row for problem "A" shown in Fig. 7-38.

Check the views in problem "A" for the projection of points, lines, and surfaces listed in the first row of the chart in Fig. 7-37. Use the following example as a model.

Given: line 35-36 in the right-side view in Fig. 7-38. Identify its elements in the two remaining views of the multiview drawing and in the pictorial drawing.

A study of the views reveals that line 35-36 appears as line 9-8 in the pictorial view. It also is in the same plane as line 7-6, but only the nearest point or line in the line of projection is listed. Line 35-36 also appears as surface D. Line 35-36 appears as point 30 in the front view, as line 23-26 in the top view, and surface V in the top view. All of this information can be found in the chart in Fig. 7-37 in the row labeled as 1.

Prepare a chart with nineteen rows for problems and identification items using Layout II in the Reference Section. Follow the example illustrated in Fig. 7-37. Letter the headings and the information required. Use 1/8 inch capital letters and numerals.

Continue identification of the remaining items for problem "A." Fill in the missing information for the corresponding given items in row 2 through row 5 in Fig. 7-37. Then use the following information for problems B, C, and D in Fig. 7-38.

Problem B

1. Line 3-6, front view.
2. Surface Z, front view.
3. Line 6-10, front view.
4. Surface B, pictorial view.
5. Line 1-2, top view.

Problem C

1. Line 28-30, front view.
2. Surface V, top view.
3. Line 4-7, pictorial view.
4. Surface C, pictorial view.
5. Line 22-27, top view.

Problem D

1. Line 26-27, front view.
2. Surface Z, right-side view.
3. Line 23-25, top view.
4. Surface B, pictorial view.

Reading a Drawing

Reading an industrial drawing will aid you in visualizing the views. The following questions will guide your reading of the drawing in Fig. 7-39. Answer these questions on a separate sheet of paper.

1. What views are shown?
2. What is the name of the part?
3. What is the number of the part?
4. From what size stock is the part to be made?
5. Starting with the circled letters, match the lines and surfaces in the two views. Note there are two extra numbers for which no letter matches.
6. Why were the two views chosen by the drafter? Would a top view make the drawing any clearer?
7. Is the circle at number 7 a recessed or protruding cylinder? What confirms this?
8. What size hole is drilled through at number 1?

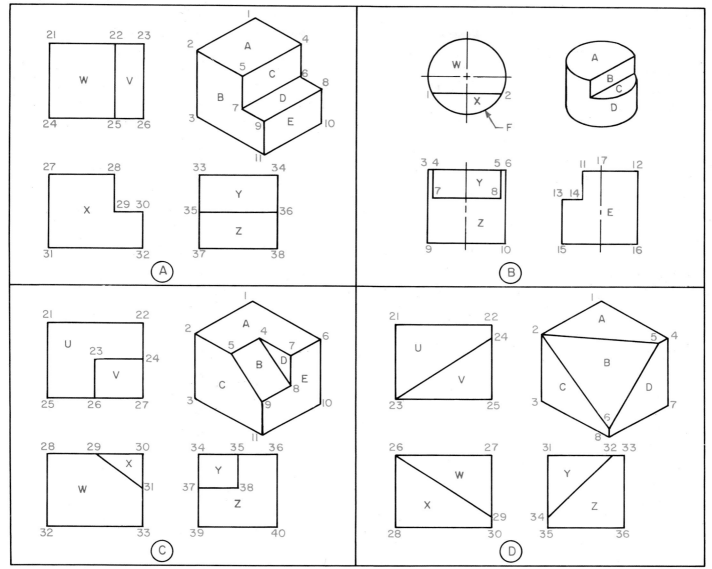

Fig. 7-38. Use these drawings and identify the points, lines, and surfaces indicated in the instructions in the Problems and Activities section.

9. Give the pilot drill size for the threaded hole at number 8.

10. How thick is the piece at S?

Multiview Problems

Study the problems shown in the pictorial views in Fig. 7-40. Select and sketch the necessary views for each problem assigned. Have these checked by your instructor. Prepare a multiview drawing using Layout I in the Reference Section. Do not dimension the drawing.

Multiview Drawings with Removed Views

Study the parts shown in Fig. 7-41. Then select and draw the necessary views, including a removed view of the feature indicated. Unless otherwise indicated, the removed view is to be drawn two times size. A freehand sketch is not required. However, it is a good practice to first sketch all drawn views until you have become proficient in visualizing and spacing multiview drawings on a sheet. Use Layout I and select an appropriate scale for the drawing.

Multiview Drawings with Partial Views

When a machine part is symmetrical in one view, a partial view may be drawn. In others, because of differences in side views, two partial-side views should be drawn. Select views carefully and draw necessary partial views for objects in Fig. 7-42. Use Layout I found in the Reference Section. Select an appropriate scale to use on the drawing.

Multiview Problems Involving Conventional Drafting Practices

Select and draw the necessary views for the parts shown in Fig. 7-43. Use standard conventional drafting practices where applicable. Prepare Layout I and select an appropriate scale for the drawings.

First-angle Projection

Although most of your work in drafting will be done in third-angle projection, making a drawing in first-angle projection is the best way to understand the system used in most European countries. Make the necessary first-angle projection views of the "BEARING SUPPORT" in Fig. 7-40 or the "ACTUATING LEVER" in Fig. 7-43. Use Layout I in the Reference Section.

Problems in Machine Parts

Study the machine parts in Fig. 7-44. Draw the necessary views of each part to adequately describe it. Use Layout I in the Reference Section.

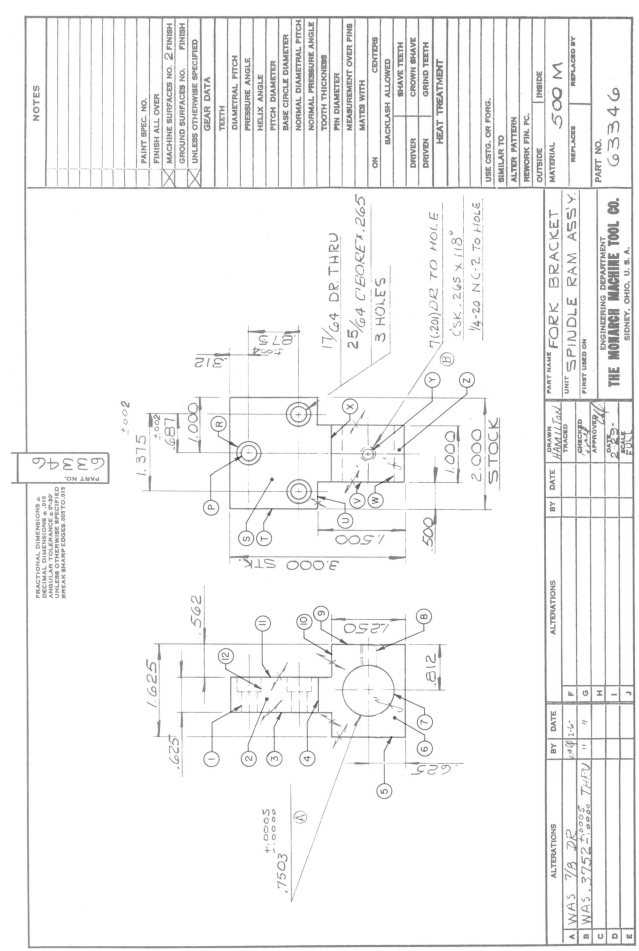

Fig. 7-39. Use this industrial print to help you practice reading a drawing. Answer the questions in the Problems and Activities section.

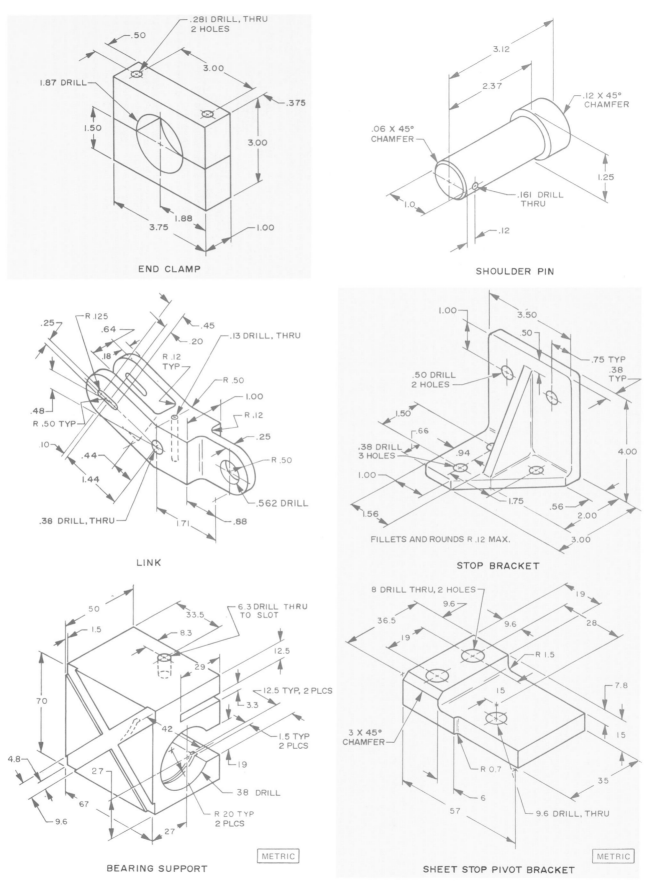

Fig. 7-40. Make a multiview drawing of the problems assigned by your instructor. First make a sketch. When your instructor has approved your sketch, then make the multiview drawing.

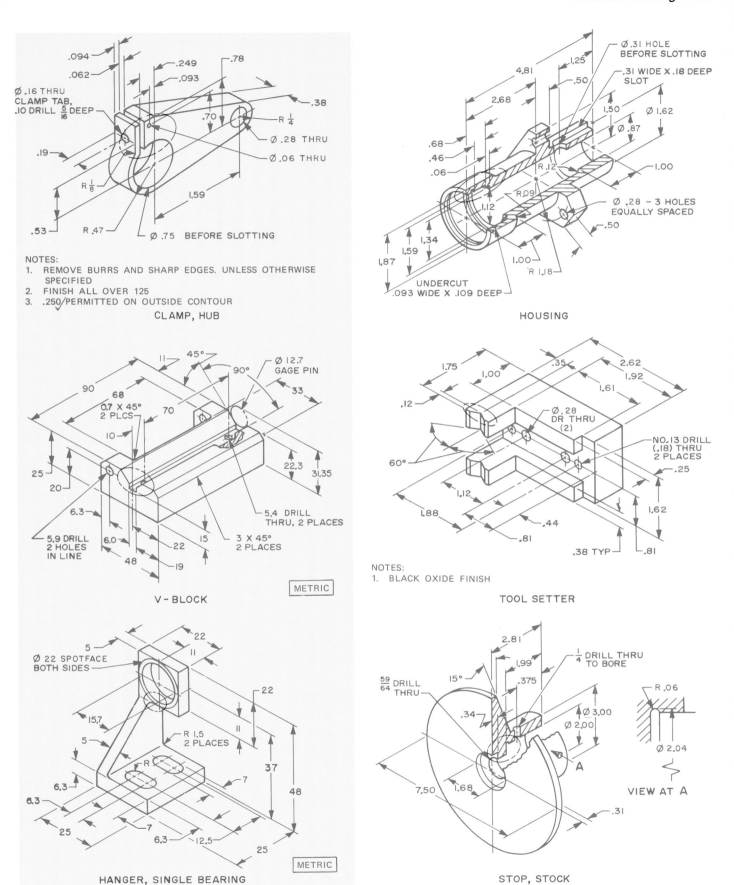

Fig. 7-41. Select and draw all of the required views. The part of the drawing circled in red requires a removed view to be drawn in order to fully communicate all of the details.

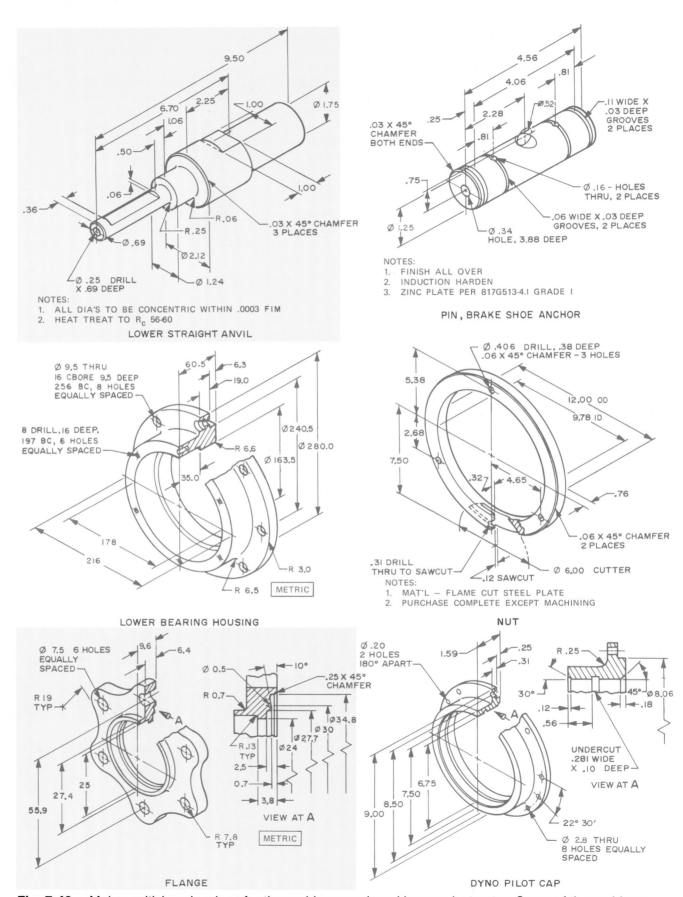

Fig. 7-42. Make multiview drawings for the problems assigned by your instructor. Some of the problems may be symmetrical and require only a partial view drawing.

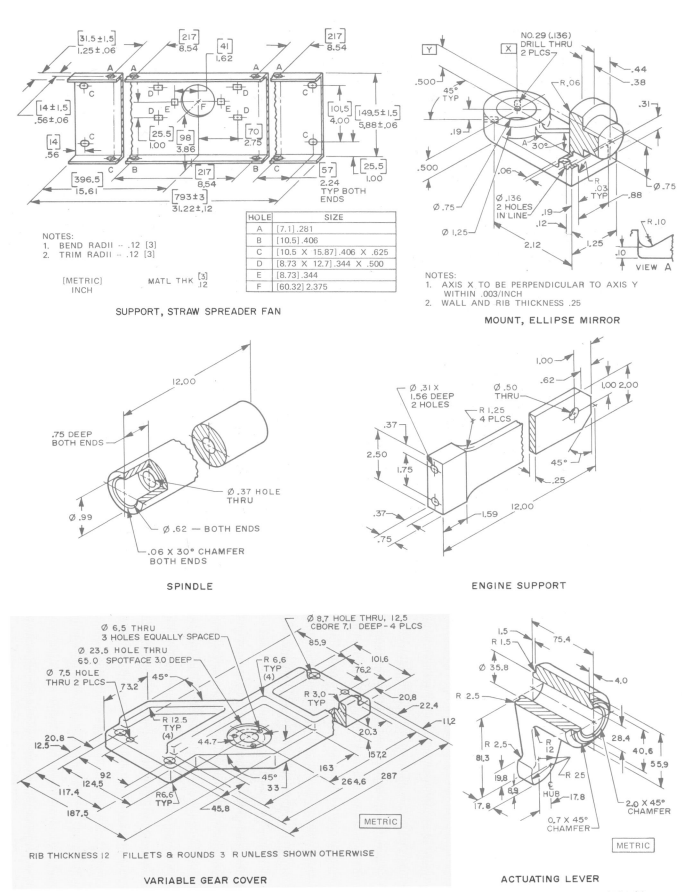

NOTES:
1. BEND RADII -- .12 [3]
2. TRIM RADII -- .12 [3]

[METRIC]
INCH

MATL THK [3]
.12

HOLE	SIZE
A	[7.1] .281
B	[10.5] .406
C	[10.5 X 15.87] .406 X .625
D	[8.73 X 12.7] .344 X .500
E	[8.73] .344
F	[60.32] 2.375

SUPPORT, STRAW SPREADER FAN

NOTES:
1. AXIS X TO BE PERPENDICULAR TO AXIS Y WITHIN .003/INCH
2. WALL AND RIB THICKNESS .25

MOUNT, ELLIPSE MIRROR

SPINDLE

ENGINE SUPPORT

RIB THICKNESS 12   FILLETS & ROUNDS 3 R UNLESS SHOWN OTHERWISE

VARIABLE GEAR COVER

ACTUATING LEVER

Fig. 7-43. Make multiview drawings for the problems assigned by your instructor. Follow accepted drafting conventions when completing these drawings.

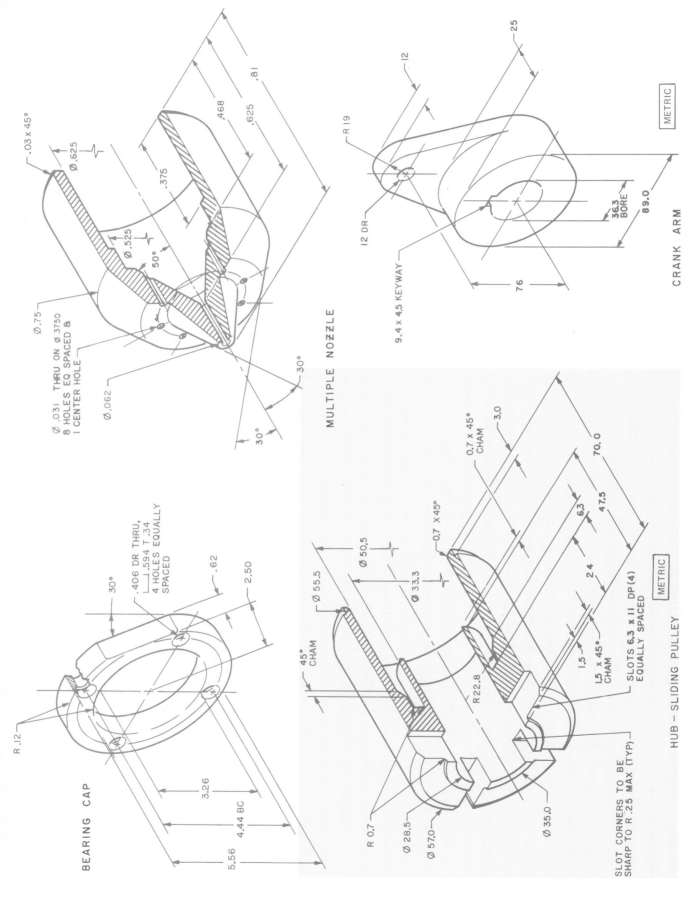

.03 x 45°

Ø .625

Ø .525

50°

Ø .75

Ø .031 THRU ON Ø .3750
8 HOLES EQ SPACED &
1 CENTER HOLE

Ø .062

.468

.375

.625

.81

30°

30°

MULTIPLE NOZZLE

R 19

12 DR

9.4 x 4.5 KEYWAY

12

25

36.3
BORE

89.0

76

METRIC

CRANK ARM

R .12

30°

.406 DR THRU,
.594 T .34
4 HOLES EQUALLY
SPACED

.62

2.50

3.26

4.44 BC

5.56

BEARING CAP

45°
CHAM

Ø 55.5

Ø 50,5

Ø 33.3

R 22.8

R 0.7

Ø 28.5

Ø 57.0

Ø 35.0

SLOT CORNERS TO BE
SHARP TO R .25 MAX (TYP)

0.7 x 45°
CHAM

0.7 X 45°

1.5 x 45°
CHAM

SLOTS 6.3 x 11 DP (4)
EQUALLY SPACED

3,0

6.3

47.5

2.4

1.5

70.0

METRIC

HUB – SLIDING PULLEY

Fig. 7-44. Make multiview drawings of the problems assigned by your instructor. Be sure to include all of the views required to accurately and fully describe the part.

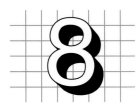

Dimensioning Fundamentals

KEY CONCEPTS

☐ Dimensions are used to define and describe size and locations.

☐ Dimension, extension, and leader lines are used on drawings to assist in the communication of information.

☐ Fractional and decimal are two types of inch dimensioning.

☐ Metric dimensioning is also used on industrial drawings.

☐ General and local notes are used on drawings to further clarify details.

☐ There are general rules for good dimensioning that should always be followed.

Dimensioning is the process of defining the size, form, and location of geometric components on engineering or architectural drawings. It is one of the most important operations in producing a detail drawing and should be given very careful attention.

Two general types of dimensions are used on drawings. These two types are size dimensions and location dimensions, Fig. 8-1. **Size dimensions** are dimensions that define the size of geometric components of a part. The diameter of a cylinder and the width of a slot are examples of size dimensions. **Location dimensions** are dimensions that define the location of these geometric components in rela-

tion to each other. The distance from the edge of a part to the center of a hole is an example of a location dimension.

ELEMENTS IN DIMENSIONING

A standard set of lines and notes are recommended for use on industrial drawings. All lines used in dimensioning are drawn as thin lines, using a 2H or 4H lead, Fig. 8-2.

Dimension Lines

A **dimension line** is a line with termination symbols (generally arrowheads) at each end to indicate the direction and extent of a dimension, Fig. 8-2A. The dimension line may be broken and the dimension numeral inserted, Fig. 8-2B. The dimension line may also be a full, unbroken line with the dimension numeral located above or below it. (Refer back to Fig 8-2A.) However, broken and full dimension lines should not be used on the same drawing.

The first dimension line is spaced 3/8 to 1 inch (or 10 to 25 mm) from the view, depending on space available on the drawing, Fig. 8-2A.

The practice in industry is to keep dimension lines away from the view for greater clarity. When

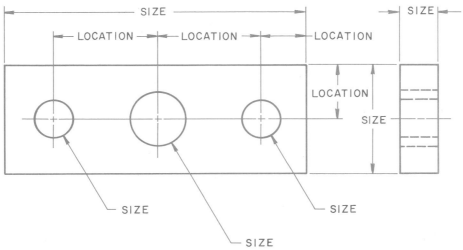

Fig. 8-1. Dimensions provide information concerning size and location on drawings.

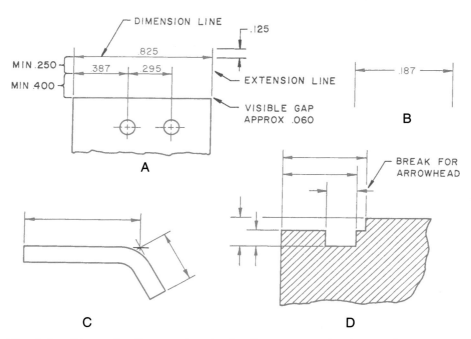

Fig. 8-2.   Dimension lines and extension lines are used to locate dimensions and other information. They should be drawn consistent with accepted industry standards.

the minimum distance of 3/8 inch is used, adjacent dimension lines should be spaced at least 1/4 inch apart. Dimensions spaced one inch or more from the view may have subsequent lines spaced less than one inch, depending on the size of the drawing.

Extension Lines

Extension lines are used to indicate the termination of a dimension. (Refer back to Fig. 8-2A.) They are usually drawn perpendicular to the dimension line with a visible gap of approximately .06 inch (1.5 mm) from the object. Extension lines extend approximately 1/8 inch (or 3 mm) beyond the dimension line. When extension lines are used to locate a point, they must pass through the point as in Fig. 8-2C.

Crossing of dimension or extension lines should be avoided by placing the shortest dimensions nearest the object and progressing outward according to size. Dimension lines should be located so they are not crossed by any line. When it is necessary to cross a dimension line with an extension line, the extension line is broken, Fig. 8-2D.

Leaders

Leaders are thin, straight lines that lead from a note or dimension to a feature on the drawing. Leaders terminate with an arrowhead or dot, Fig. 8-3. Leaders that terminate on an edge or at a specific point should end with an arrowhead. Dots are used with leaders that terminate inside the outline of an object, such as a flat surface.

The dimension or note end of the leader contains a horizontal bar approximately 1/8 inch in length. Preferably, the leader angle (angle that the leader line is projected from the object) should be 45° to 60°. Leaders should never be drawn parallel to extension or dimension lines. Leaders drawn to a circle or circular arc should be in line with center of the particular feature.

Features Indicated by "X"

Some industries use an "X" to indicate the number of repetitive features. For example, "2 X ø.375" indicates two holes each with a diameter of .375, Fig. 8-3. This provides for clarity and speed in drafting.

Dimensional Notes

Dimensional notes are notes used to describe size or form, such as specifying holes, chamfers, and threads, Fig. 8-4. They serve the same purpose as dimensions. Notes are used with dimension and extension lines to provide specific information about details on the drawing. Dimensional notes always appear horizontally and parallel to the bottom of the drawing.

Arrowheads

Arrowheads are drawn at the termination of dimension lines and leaders. They can be drawn freehand, but must be distinct and accurate. Arrow-

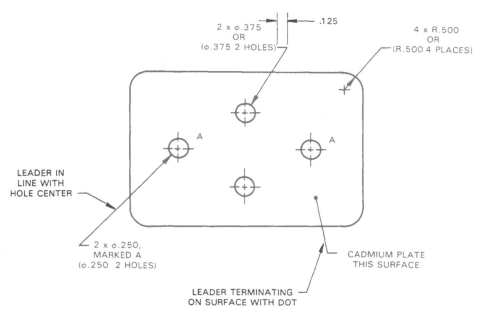

Fig. 8-3. Leaders are used in dimensioning to provide dimensions for holes, radii, and to provide an indication of the feature being described in a note.

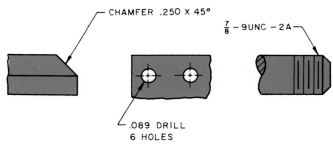

Fig. 8-4. Dimensional notes with leaders are used to describe size and form, or provide additional information not communicated elsewhere.

heads for all dimension lines and leaders on the same drawing should be approximately the same size as the height of whole numerals, usually 1/8 inch, Fig. 8-5A.

The width of the base of the arrowhead should be one-third of its length. It is drawn with a single stroke forming each side, either toward the point or away from it depending on the position of the arrowhead and the preference of the drafter, Fig. 8-5B. A third stroke forms the curved base, Fig. 8-5C. The arrowhead is then filled in for a distinctive appearance, Fig. 8-5D.

Dimension Figures

Dimension figures should be clearly formed to prevent any possibility of being misread. Some industries are using a style of numerals that provides a positive recognition even when a portion of the numeral is lost in reproduction or is not clear on the drawing (see Fig. 4-31, page 93).

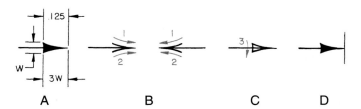

Fig. 8-5. Arrowhead can be drawn freehand, however, they should be neat and consistent. The filled-in arrowhead is widely accepted for use on industrial drawings.

The height of dimension figures is the same as the letter height on the drawing, usually 1/8 inch. Fractions are twice the height of whole numbers.

Placement of Dimension Figures

Common fractions are centered in a break in the dimension line, Fig. 8-6A. Decimal dimensions may be centered in a break in the line, Fig 8-6B. Decimal dimensions may also be placed above the line and the metric dimensions below the line, Fig 8-6C. Some industries practicing dual dimensioning (decimal and metric) show the decimal dimension above the line followed by the metric dimension in parentheses, Fig. 8-6D.

A comma is used in place of a decimal point in metric dimensioning in many European countries. However, it has not yet become standardized for worldwide practice.

Staggered Dimensions

Where a number of dimensions appear in the same area of a drawing, the dimension figures are staggered. This gives the drawing a better appearance and saves space, Fig. 8-7.

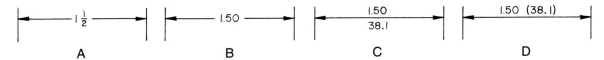

Fig. 8-6. There are several different ways to place dimension figures in fractions, decimals, and metric units on a drawing.

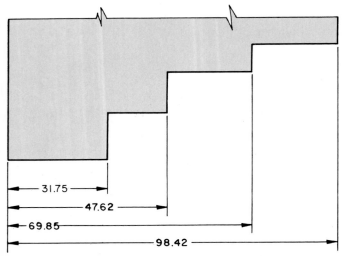

Fig. 8-7. Staggered dimensions are used on drawings to give a better appearance and improve clarity. Dimensions are staggered when a number of dimensions appear in the same area on a drawing.

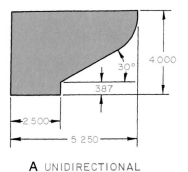

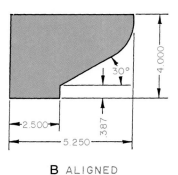

Fig. 8-8. The unidirectional system of dimensioning aligns all of the dimension numbers parallel with the bottom edge of the drawing. The aligned system of dimensioning aligns the dimension numbers parallel with either the bottom or the right side of the drawing, depending on the feature being described.

Unidirectional and Aligned Systems

In the **unidirectional system**, all dimension figures are placed to be read from the bottom of the drawing, Fig. 8-8A. This is the recommended industry standard. In the **aligned system**, all dimensions are placed parallel to their dimension lines and are read from the bottom or right side of the drawing, Fig. 8-8B.

There are zones that should be avoided in the aligned system due to the awkward angle of reading, Fig. 8-9. The unidirectional system of dimensioning has been widely adopted and is gaining favor in industry.

Dimensioning within the Outline of an Object

Dimensions should be kept outside the views of an object whenever possible. Exceptions are permissible where directness of application makes it necessary, Fig. 8-10A. When it is necessary to dimension within the sectioned part of a sectional view, the section lines are omitted from the dimension area, Fig. 8-10B.

MEASURING SYSTEMS

Linear dimensions on a drawing are expressed in decimals or common fractions of an inch in the English system of measurement. In the metric system of measurement, dimensions are given as millimeters. When linear dimensions exceed a certain length (144 to 192 inches, depending on the particular industry), the dimension value is given in feet and inches. In such cases, abbreviations for feet (ft or ′) and inches (in or ″) are usually shown after the values. Linear dimensions in excess of 10,000 millimeters (mm) are expressed in meters (m) or meters and decimal portions of meters.

Decimal Inch Dimensioning

Decimal dimensioning is preferred in most manufacturing industries today because decimals are easier to add, subtract, multiply, and divide. Preferably, decimal dimensioning should use a two-place increment of .02 inch (two hundredths of an inch) such as .04, .06, .08, and .10. However, the particular situation that you are working with may not permit this.

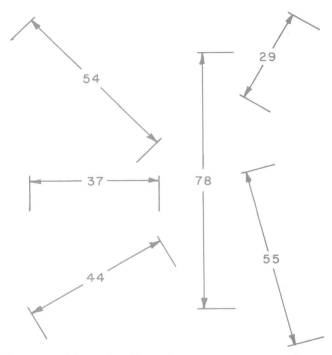

Fig. 8-9. Dimension lines that appear at unusual angles are best dimensioned using the unidirectional dimension system. This makes the dimensions easier to read than aligned dimensions would.

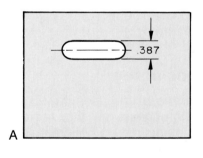

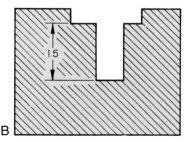

Fig. 8-10. Dimensions should always be placed on the outside of an object when possible. However, in some cases this is not possible. There are a few dimensions that are allowed within the outline of an object.

When decimals of .02 inch increments are divided by two, the quotient is a two-place decimal as well. Decimal dimensions in sizes other than .02 inches should be used where more exacting requirements must be met.

Decimal Dimensioning Rules

1. Omit zeros before the decimal point for values of less than one, Fig. 8-11A.
2. Common fraction decimal equivalents of .250, .500, and .750 may be shown as two-place decimals unless they are used to designate a drilled hole size, a material thickness, or a thread size, Fig. 8-11B. A title block tolerance can be used to specify different tolerances for dimensions having a different number of decimal places. (Refer back to Fig. 8-11B.)
3. Standard nominal sizes of materials, threads, and other features produced by tools that are designated by common fractions may be shown as common fractions, Fig. 8-11C.

   Examples:
   "3/4 - 10 UNC - 2A THD"
   ".250 DIA (1/4)"
   "3/8 HEX"

4. Decimal points must be definite, uniform, and large enough to be visible on reduced size drawings. The decimal point should be in line with the bottom edge of the numerals and letters to which it relates, Fig. 8-11D.

Rules for Rounding Off Decimals

When it is necessary to round off decimals to a lesser number of places, the following rules apply.

1. When the next figure beyond the last digit to be retained is less than 5, use the shortened form unchanged. For example, the number 2.62385 is to be shortened to two decimal places. The third figure beyond the decimal point is three (2.62385), which is less than five. The decimal is rounded off to the number 2.62.
2. When the next figure beyond the last digit to be retained is greater than 5, increase the digit by 1. For example, the number 2.62385 is to be shortened to three decimal places. The fourth figure beyond the decimal point is eight

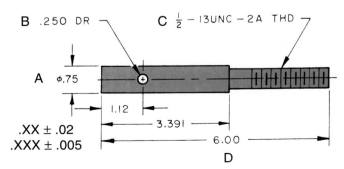

Fig. 8-11. There are rules of good decimal dimensioning that should always be followed when dimensioning a drawing.

(2.62385), which is greater than five. The decimal is rounded off to the number 2.624.

3. When the next figure beyond the last place to be retained is equal to 5 and the last digit of the shortened form is an odd number, increase the last digit by one. For example, the number 2.62375 is to be shortened to four decimal places. The fifth digit beyond the decimal point is 5 (2.62375). The last digit of the shortened form is 7 (2.62375), which is an odd digit. The last digit of shortened form is increased by one and the number becomes 2.6238.

4. When the next digit beyond the last number to be retained is equal to five and the last digit of the shortened form is an even digit, the shortened form remains unchanged. For example, the number 2.62385 is to be shortened to four decimal places. The fifth digit beyond the decimal point is 5 (2.62385). Last digit of the shortened form is an even digit (2.62385). The last digit of shortened form is left unchanged and the number becomes 2.6238.

Fractional Dimensioning

Common fraction dimensioning is used on drawings in architectural and structural fields. Close tolerances are not as important in these fields. Many times the materials involved, such as lumber, are not manufactured to very small tolerances.

A horizontal fraction bar is used with all fractions. It is located at the mid-point of the vertical height of numerals and capital letters. Where older drawings have been dimensioned with common fractions, some manufacturing industries change these fractions to decimals by referring to a conversion chart.

Metric Dimensioning

Metric dimensioning, like decimal inch dimensioning, uses the base-ten number system. This makes it easy to move from one multiple or submultiple to another by shifting the decimal point.

The unit of linear measure in the metric system is the meter. However, the millimeter is used on most drawings dimensioned in the metric system where the linear dimension is less than 10,000 millimeters. The letter abbreviation for millimeters (mm) following the dimension figure is omitted when all dimensions are in millimeters.

Since 1960, the metric system has been referred to internationally as *Système International d' Unités* or the *International System of Units.* The universal abbreviation SI indicates this system.

Metric dimensioning rules

1. A period is used for the decimal point in countries that use the English system of measurement. Most other countries use a comma for the decimal point.
2. Whenever a numerical value is less than one millimeter, a zero should precede the decimal point.
3. Digits in metric dimensions are not to be separated into groups by use of commas or spaces.
4. Use multiple and submultiple prefixes of 1000, such as kilometer (km), meter (m), and millimeter (mm) whenever possible. Avoid the use of centimeter (cm).
5. Do not mix SI units with units from a different system. (An exception is "dual dimensioning" discussed below.)

Dual Dimensioning

Dual dimensioning uses the English inch and metric (SI) dimensions on the same drawing. In dual dimensioning, the English measurement is usually given in decimal inches and the metric measurement in millimeters. If the drawing is intended primarily for use in the United States, the decimal inch dimension will usually appear above the line and millimeters below, Fig. 8-12A. In countries where the metric system is used, the millimeter dimension is shown above the line and the decimal inch dimension below, Fig. 8-12B.

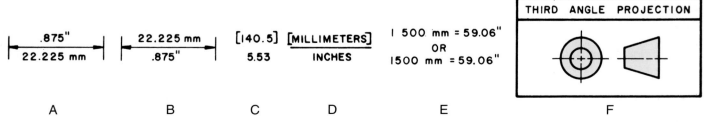

Fig. 8-12. There are several ways that dimensions can appear in the dual dimensioning system. In all cases, both metric and English units of measure will be given. C—Typically, a symbol indicating what type of projection is being used is also given. This eliminates possible confusion between first-angle and third-angle projections.

Some industries using dual dimensioning place the metric dimension in brackets above the decimal dimension, Fig. 8-12C. It is recommended that a note be used adjacent to or within the title to show how the inch and millimeter dimensions are identified, Fig. 8-12D. Whole numbers in the metric system may be separated according to groups by a letter space or simply run together, Fig. 8-12E. It is recommended that all dual dimensioned drawings indicate the angle of projection used, to eliminate any confusion when used in different countries, Fig. 8-12F.

Dual dimensioning is being replaced by drawings that are dimensioned only in metric. Where drawings are dimensioned in a single system, individual identification of linear units is not required. However, a note on the drawing will state "UNLESS OTHERWISE SPECIFIED, ALL DIMENSIONS ARE IN MILLIMETERS" (or "INCHES"). Where some inch dimensions are shown on a millimeter-dimensioned drawing, the abbreviation "IN." should follow the inch value. (Note that in this case, a period follows the abbreviation.)

On decimal-inch drawings, the symbol "mm" should follow the millimeter values.

An example of a drawing using dual dimensioning is shown in Fig. 8-13. Tables are provided in the Reference Section for conversion of common fractions and decimals to millimeters and vice versa.

DIMENSIONING FEATURES FOR SIZE

The features of a part or assembly consists of geometric shapes. These shapes may be cylinders, cones, pyramids, or spheres. Dimensioning consists of describing the size and position of each feature.

Cylinders

Cylinders may be solids (as in a projecting stem) or negative volumes (as in a hole), Fig. 8-14 A.

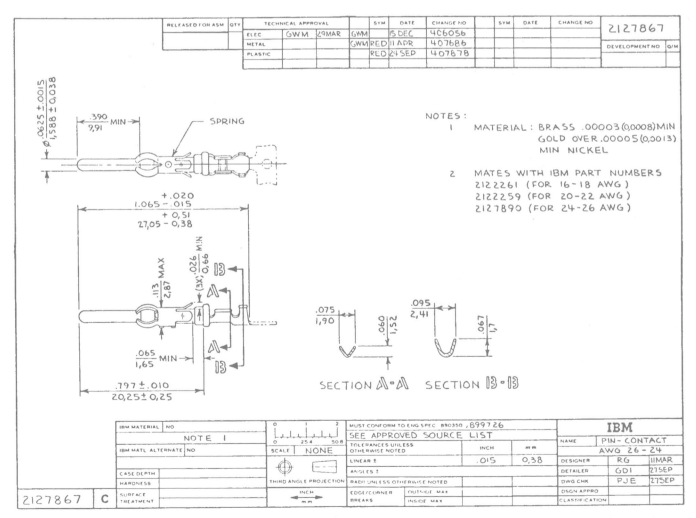

Fig. 8-13. Dual dimensioning is often used in industry, especially by companies that may do business in other countries that use the metric system. (IBM)

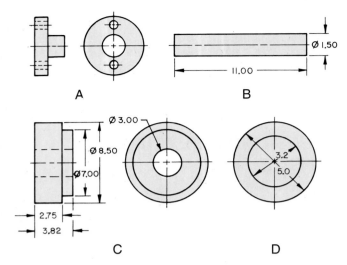

Fig. 8-14. There are certain guidelines that should be followed when dimensioning cylinders.

Cylindrical features are dimensioned for diameter and length. In single view drawings where the feature or part is not shown as a circle, the dimension should be preceded by the symbol for diameter (the international symbol ⌀), Fig. 8-14B.

The diameter symbol is not necessary for a cylinder on the view where it appears as a circle, Fig. 8-14C. The symbol is also unnecessary where the dimension is placed on the circular view itself, Fig. 8-14D. Where a leader is used to indicate the diameter, the value of the diameter should be preceded by the symbol for diameter. (Refer back to Fig. 8-14C.)

Circular Arcs

Circular arcs are dimensioned by indicating their radius with a dimension line, Fig. 8-15. The dimension line is drawn from the radius center and ends with an arrowhead at the arc. The dimension is inserted in the line preceded by "R," Fig. 8-15A. Dimensions of radii, where space is limited, may be indicated outside the arc, Fig. 8-15B. The dimension may also be placed with a leader when space is limited, Fig. 8-15C. A small cross should be used to indicate a radius located by a dimension, Fig. 8-15D.

Foreshortened Radius

Sometimes the center of an arc radius exists outside the drawing itself or interferes with another view. In this case, the radius dimension line should be shown foreshortened and the arc center located with coordinate dimensions, Fig. 8-16. That portion of the dimension line next to the arrowhead and arc is shown radially (in line with the arc center).

True Radius Indication

A true arc on an inclined surface may be misleading to the worker when dimensioned as shown in Fig. 8-17A. This dimension is clarified by the addition of "TRUE R" before the radius dimension which means "true radius on the surface," Fig. 8-17B.

Fillets and Corner Radii

Fillets and corners may be dimensioned by a leader as shown in Fig. 8-18A. Where there are a

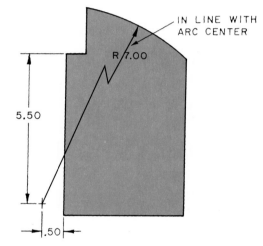

Fig. 8-16. When dimensioning a foreshortened radius, the center point should be located. The part of the leader that is close to the arc itself should be drawn in line with the center of the arc. The rest of the leader should terminate at the center, but not be drawn through the arrowhead.

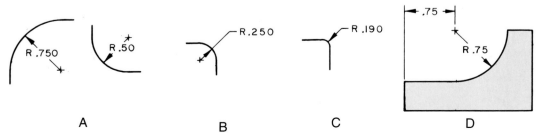

Fig. 8-15. There are several different ways that arcs may be correctly dimensioned.

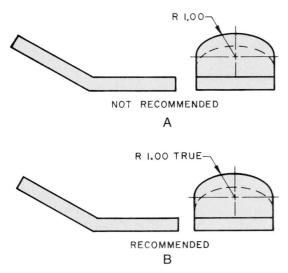

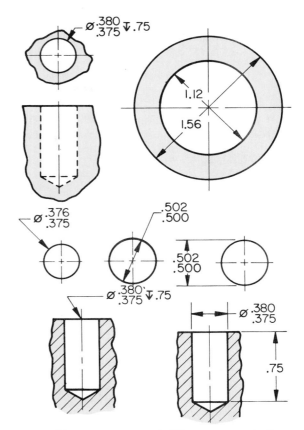

Fig. 8-17. A–When a radius is dimensioned on an inclined surface, the radius will appear out of scale from the dimension. B–This may be confusing. In this case, the term "TRUE" should be added to the dimension for clarity.

Fig. 8-19. There are several different accepted methods for dimensioning round holes. The most appropriate method should be selected for each individual case. (American National Standards Institute)

large number of fillets or rounded edges of the same size on a part, the preferred method is to specify these with a note rather than to show each radius, Fig. 8-18B.

Round Holes

Holes are preferably dimensioned on the view in which they appear as circles. Small holes are dimensioned with a leader and larger holes by a dimension at an angle (usually 30°, 45°, or 60°) across the diameter, Fig. 8-19. Holes that are to be drilled, reamed, or punched are specified by a note. The depth of holes may be specified by a note or dimensioned in a sectioned view. Additional information on the dimensioning of holes is discussed in Chapter 9.

Hidden Features

Dimensions should be drawn to visible lines whenever possible. Where hidden features exist

and are not visible for dimensioning in another view, a section view should be used, Fig. 8-20. An exception to this general rule is a diameter dimension on a partial section where no other view is used to show the circular diameter, Fig. 8-21.

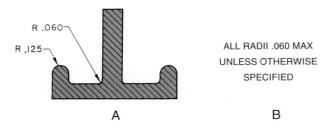

Fig. 8-18. A–Radii for fillets and corners can be indicated with a leader and the radii specified. B–Fillets and corner radii can also be specified with a note.

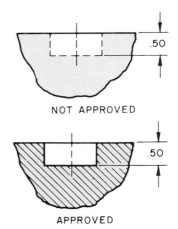

Fig. 8-20. Hidden features should not be dimensioned in a view where the feature appears as hidden lines. Instead, hidden features should be dimensioned in section views whenever possible.

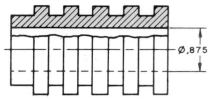

Fig. 8-21.   In a partial section view, a hidden line may be dimensioned to if there is no other way to describe the part. This is an exception to the general rule of avoiding dimensions to hidden lines and should be used infrequently.

Knurls

Knurls are specified by diameter, and type and pitch of the knurl, Fig. 8-22A. When control of the diameter of a knurl is required for an interference fit between parts, this is also specified, Fig. 8-22B. The length along the axis may be specified if required.

Angles

Angular dimensions may be specified in degrees, minutes, and seconds, Fig. 8-23. The sym-

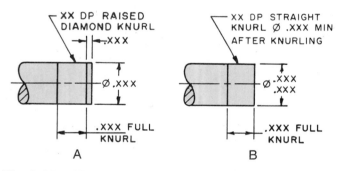

Fig. 8-22.   Knurls are usually explained with a combination of a note and location and size dimensions.

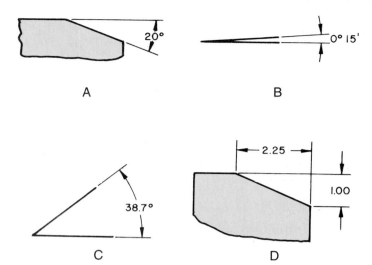

Fig. 8-23.   There are several different ways to dimension angles. Select the most appropriate method for a given instance.

bols for these are: degrees °, minutes ′, and seconds ″. Angles specified in degrees alone have the numerical value followed by the symbol ° or by the abbreviation "DEG," Fig. 8-23A. When an angle is expressed in minutes alone, the number of minutes is preceded by "0°," Fig. 8-23B. Angles may be specified in degrees and decimal parts of a degree, Fig. 8-23C. Angles may also be specified by coordinate dimension, Fig. 8-23D.

Chamfers

Chamfers of 45° may be dimensioned by a note as in Fig. 8-24A. Use of the word "CHAMFER" is optional in the note. All chamfers, other than 45° chamfers, are dimensioned by giving the angle and the measurement along the length of the part, Fig. 8-24B.

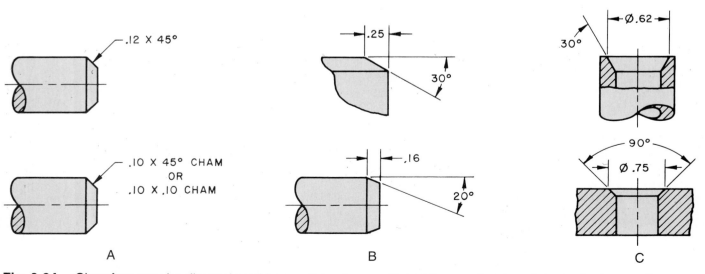

Fig. 8-24.   Chamfers may be dimensioned in a variety of ways. Select the most appropriate method according to the location of the chamfer.

Chamfers should never be dimensioned along their angular surface. Internal chamfers are dimensioned in the same manner except in cases where the diameter requires control, Fig. 8-24C.

Counterbore or Spotface Symbol

A counterbore or spotface is indicated by the symbol shown in Fig. 8-25A. The symbol precedes the dimension.

Countersink Symbol

A countersink is indicated by the symbol shown in Fig. 8-25B. The symbol precedes the dimension.

Counterdrill

Counterdrilled holes are dimensioned by specifying the diameter of the hole and the diameter, depth, and included angle of the counterdrill, Fig. 8-25C.

Depth Symbol

The depth of a feature is indicated by the symbol shown in Fig. 8-25B. The symbol precedes the depth dimension.

Dimension Origin Symbol

The origin of a tolerance dimension between two features is indicated as shown in Fig. 8-25E.

Square Symbol

A square shaped feature is indicated by a single dimension preceded by the symbol shown in Fig. 8-25F.

Arc Length Symbol

Arc length measured on a curved outline is indicated by the symbol shown in Fig. 8-25D. This symbol is placed above the dimension.

Diameter and Radius Symbols

Diameter, spherical diameter, radius, and spherical radius are indicated by the symbols shown in Fig. 8-25G. These symbols precede the value of a dimension or tolerance given for a diameter or radius.

Offsets

Offsets on a part should be dimensioned from the points of intersection of the tangents along one side of the part as shown in Fig. 8-26.

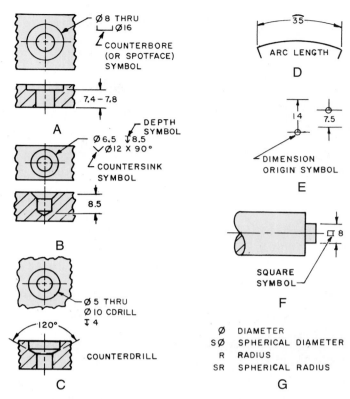

Fig. 8-25. There are standard symbols used for dimensioning. Some of these symbols include symbols for a counterbore, a countersink, depth, a counterdrill, an arc length, a dimension origin, and a square.

Keyseats

Keyseats (a recess in a shaft or hub) are dimensioned as shown in Fig. 8-27A. Woodruff keyseats are dimensioned as shown in Fig. 8-27B. For additional information see the Keys Reference Section.

Narrow Spaces and Undercuts

Dimension figures should not be crowded into narrow spaces. One of the methods shown in Fig. 8-28 will help maintain clarity on the drawing. Note the breaks in extension lines near the arrowheads in the lower part of the illustration.

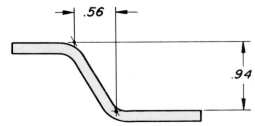

Fig. 8-26. Offsets should be dimensioned from points of intersection. (General Motors Corp.)

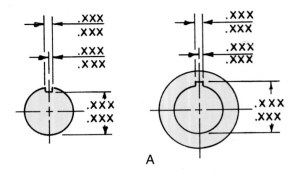

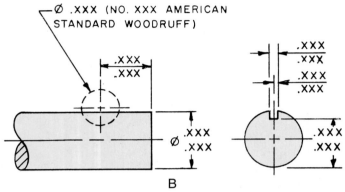

Fig. 8-27. A–Regular keyseats should be dimensioned following the accepted standard. B–A special type of keyseat called a "Woodruff keyseat" has a separate standard that should be followed when dimensioning.

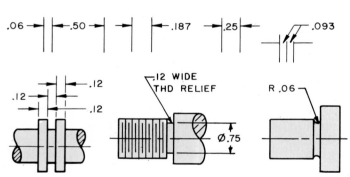

Fig. 8-28. There are several techniques generally used for dimensioning narrow spaces.

Tapers—Conical and Flat

Conical tapers are dimensioned by specifying one of the following:

1. A basic taper and a basic diameter, Fig. 8-29A.
2. A size tolerance combined with a profile of a surface tolerance applied to the taper.
3. A tolerance diameter at both ends of a taper and a tolerance length, Fig. 8-29B.

Flat tapers are dimensioned by specifying a tolerance slope and a tolerance height at one end, Fig. 8-29C. Taper and slope for conical and flat tapers are depicted by the symbols shown in Fig. 8-29. The vertical leg of the symbol is always to the left.

Irregular Curves

Irregular curves may be dimensioned by the coordinate, or offset, method as shown in Fig. 8-30. Each dimension line is extended to a datum line.

Symmetrical Curves

Curves which are symmetrical may be dimensioned on one side of the axis of symmetry only,

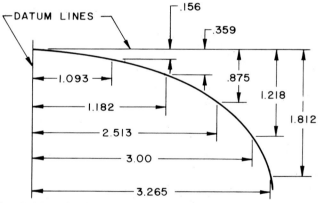

Fig. 8-30. An irregular curve should be dimensioned using datum lines.

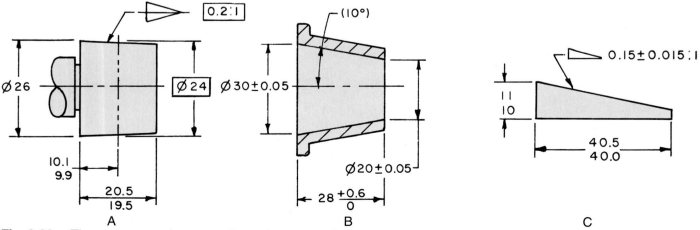

Fig. 8-29. There are several ways to dimension tapers. Select the most appropriate method for a given application.

Fig. 8-31A. When only one-half of a symmetrical part is shown, it is dimensioned as shown in Fig. 8-31B.

Rounded Ends

For parts having fully rounded ends, overall dimensions should be given for the part and the radius of the end indicated but not dimensioned, Fig. 8-32A. Parts having partially rounded ends should have the radii dimensioned, Fig. 8-32B.

Slotted Holes

Slotted holes are treated as two partial holes separated by a space. A slot of regular shape is dimensioned for size by length and width dimensions, Fig. 8-33. The slot is located on the part by a dimension to its longitudinal center and either one end or a center line.

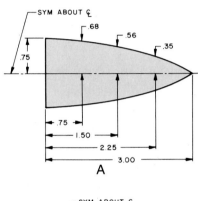

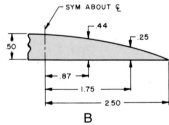

Fig. 8-31. Curves that are symmetrical to a centerline need to be dimensioned on one side of the axis of symmetry only.

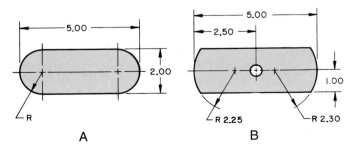

Fig. 8-32. There are two different methods used to dimension rounded ends on a drawing. Select the method that is appropriate for a given application.

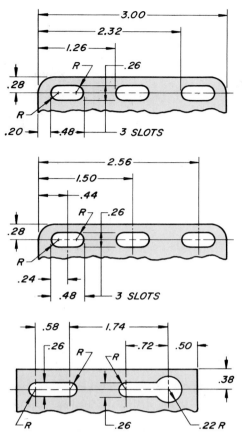

Fig. 8-33. When dimensioning slotted holes, one of the accepted standard practices should be used. (General Motors Corp.)

DIMENSIONING FEATURES FOR POSITION

Position dimensions specify the location or distance relationship of one feature of a part with respect to another feature or datum. Features may be located with respect to one another by either linear or angular expressions.

Point-to-Point Dimensioning

Point-to-point dimensions are usually adequate for simple parts. However, parts which contain features mating with another part should be dimensioned from a datum, Fig. 8-34. (See Chapter 9.) In point-to-point dimensioning, one dimension is omitted to avoid locating a feature from more than one point and possibly causing unsatisfactory mating of parts.

Coordinate Dimensioning Systems

Coordinate dimensioning is a type of dimensioning useful in locating holes and other features

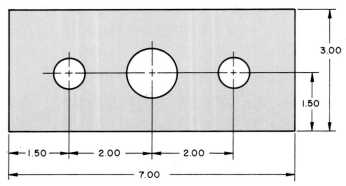

Fig. 8-34. The point-to-point method is widely used for dimensioning locations.

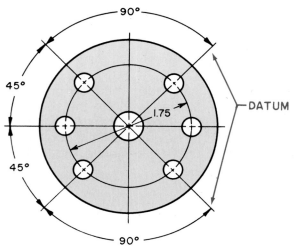

Fig. 8-36. Both radial dimensions and datum planes can be used in polar coordinate dimensioning.

on parts. Basically, two different systems are used: rectangular coordinate and polar coordinate dimensioning.

Rectangular coordinate dimensioning is useful in locating holes and other features that lie in a rectangular or noncircular pattern, Fig. 8-35A. These dimensions are at right angles to each other and from a datum plane.

With the assistance of a computer, holes distributed around a bolt circle can be located accurately for dimensioning, using the rectangular coordinate system as in Fig. 8-35B. Rectangular coordinate dimensioning is used on drawings of parts that are to be numerically machined (CNC machining).

Polar coordinate dimensioning should be used when holes or other features to be located lie in a circular or radial pattern, Fig. 8-36. A radial dimension is given from the center of the pattern and an angular dimension from a datum plane.

When it is necessary to hold closer tolerances for mating features, *true-positioning dimensioning* should be used as discussed in Chapter 9.

Tabular Dimensioning

Tabular dimensioning is a form of rectangular coordinate dimensioning. The location of dimensions for features are given from datum planes and listed in a table on the drawing sheet, Fig. 8-37. Dimensions are not all applied directly to the views. This method of dimensioning is useful where a large number of similar features are to be located.

Ordinate Dimensioning

Ordinate dimensioning is very similar to the rectangular coordinate system. It makes use of *datum dimensioning.* In datum dimensioning, all dimensions are from two or three mutually perpendicular datum planes.

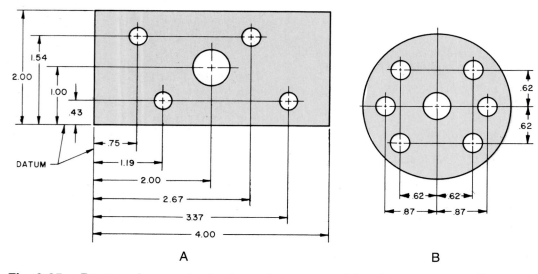

Fig. 8-35. Rectangular coordinate dimensioning uses datum lines and centerlines.

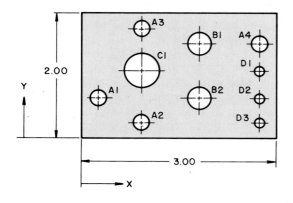

		REQD	4	2	1	3
		HOLE DIA	.250	.312	.500	.125
POSITION			HOLE SYMBOL			
X ⟶	Y ↑		A	B	C	D
.250	.625		A1			
1.000	.250		A2			
1.000	1.750		A3			
2.750	1.500		A4			
1.750	1.500			B1		
1.750	.625			B2		
1.000	1.000				C1	
2.750	1.000					D1
2.750	.625					D2
2.750	.250					D3

Fig. 8-37. Tabular dimensioning includes a drawing and a table listing.

Ordinate dimensioning differs from the coordinate system in that the datum planes are indicated as zero coordinates. Dimensions from these planes are shown on extension lines without the use of dimension lines or arrowheads, Fig. 8-38.

This system of dimensioning is sometimes referred to as **arrowless dimensioning.** It lends itself to work which is to be programmed for numerical machining.

UNNECESSARY DIMENSIONS

Any dimension not needed in the manufacture or assembly of an item is an unnecessary dimension. This includes dimensions repeated on the same or on another view, Fig. 8-39A. Also, it is not necessary to include all "chain" dimensions when the overall dimension is given.

When an entire series of chain dimensions is given, difficulties may arise in manufacturing where there is an accumulation of tolerances (variations permitted in measurements). The exception to providing all of the individual dimensions in a chain is in the architectural and structural industries where the interchangeability of parts and close tolerances are not as important, Fig. 8-39B.

NOTES

Notes are used on drawings to supplement graphic information, dimensions. In some cases, notes are used to eliminate repetitive dimensions. All notes should be brief and clearly stated. Only one interpretation of the note must be possible.

Standard abbreviations and symbols may be used in notes where feasible. Abbreviations do not require periods unless the abbreviated letters spell a word or are subject to misinterpretation.

Size, Spacing, and Alignment of Notes

The lettering in notes is the same size as in the dimensions on a drawing. This size is usually 1/8 inch in height. Spacing between lines within a note should be from one-half to one full letter height. At least two letter heights should be allowed between separate notes.

All notes should be placed on the drawing parallel to the bottom of the drawing. Lines of a single note, and successive notes in a list of general notes, should be aligned on the left side. Notes related to specific features should also be placed parallel to the bottom.

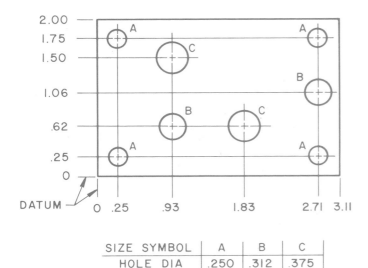

SIZE SYMBOL	A	B	C
HOLE DIA	.250	.312	.375

Fig. 8-38. Ordinate dimensioning uses distances measured from a coordinate of "0,0."

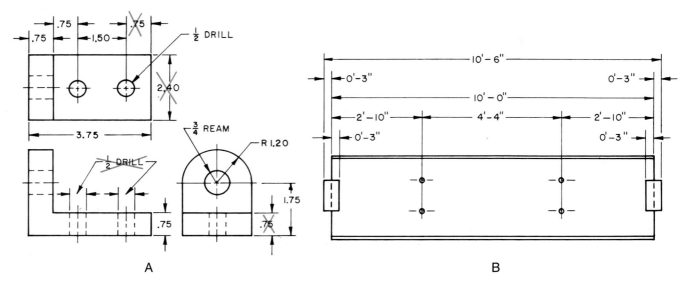

Fig. 8-39.   A–Duplicate dimensions should be eliminated to reduce the "clutter" of a drawing. B–In architectural and structural drawings where tolerances are not as important, the last dimension in a "chain" can be left in.

General Notes

Notes which convey information that apply to the entire drawing are called *general notes.* Some examples are:

1. "CORNER RADII .12 +/-.06"
2. "REMOVE ALL BURRS AND BREAK ALL SHARP EDGES .01 OR MAX"
3. "ALL MARKINGS TO BE PER MIL—STD—130B FOR LIQUID OXYGEN SERVICE"
4. "MAGNETIC INSPECT PER MIL—STD—1-6868

No period is required at the end of the note, unless more than one statement is included within any one note.

General notes are usually placed on the right-hand side of the drawing above the title block or to the left of the title block, Fig. 8-40A. They are numbered from the top down or from the bottom up, depending on the industry policy.

Notes are sometimes included in the title block of the drawing, such as general tolerances, material specification, and heat treatment specification, Fig. 8-40B. The notes shown in Fig. 8-40C are also general notes since they apply to the entire drawing.

Local Notes

Local notes provide specific information, such as designating a machine process, Fig. 8-40D and Fig. 8-40F. Actual dimension may also be designated by a local note, Fig. 8-40E. Designating a standard part is another example of a use for local notes. (For example, a standard size hexagonal cap screw may be designated by the note "No. 6-32UNC-2B HEX CAP SCR.")

Some examples of local notes are:

1. ".344 +/- ø.002, 36 HOLES"
2. "R.06—2 PLACES"
3. "PAINTED AREA 1.75 SQUARE AS INDICATED, ONE SIDE ONLY"

Local notes are usually located close to a specific feature. A leader usually extends from either the first or the final word of the note, Fig. 8-41.

Local notes may be included with the list of general notes. They refer to a specific feature or area on the drawing. In this case, a reference number is enclosed in a square or triangle called a *flag.* The flag is placed on the drawing near the feature being described, Fig. 8-41. A leader is used to indicate the exact feature being referred to.

To avoid a crowded appearance, or the necessity to relocate a note, notes should not be placed on the drawing until after the dimensions have been added.

RULES FOR GOOD DIMENSIONING

The following rules should serve as guides to good dimensioning practices:

1. Take time to plan the location of dimension lines. Avoid crowding by providing adequate

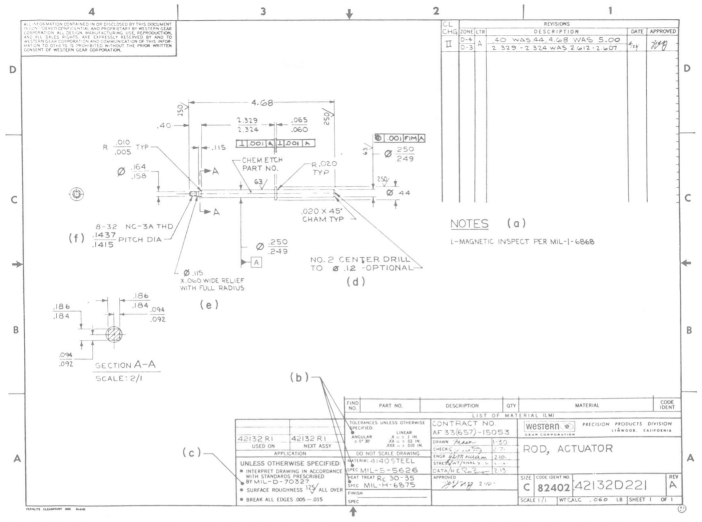

Fig. 8-40. Notes are used on a drawing to help clarify details and to communicate information about the part that cannot otherwise be easily communicated. (Western Gear Corp.)

room for spacing (at least .40 inch (10 mm) for first and .25 inch (6 mm) for successive lines).

2. Dimension lines should be thin and contrast noticeably with visible lines of the drawing.

3. Dimension each feature in the view which most completely shows the characteristic contour of that feature.

4. Dimensions should be placed between the views to which they relate and outside the outline of the part.

5. Extension lines are gapped away from the object approximately .06 inch (1.5 mm) and extend beyond the dimension line approximately .125 inch (3 mm). Extension lines may cross other extension lines or object lines when necessary. Avoid crossing dimension lines with extension lines or leaders whenever possible.

6. Show dimensions between points, lines, or surfaces which have a necessary and specific relation to each other.

7. Dimensions should be placed on visible outlines rather than hidden lines.

8. State each dimension clearly so that it can be interpreted in only one way.

9. Dimensions must be sufficiently complete for size, form, and location of features so that no calculating or assuming of distances or locations is necessary.

10. Avoid duplication of dimensions. Only those dimensions that provide essential information should be shown.

11. One dimension in a chain dimension should be omitted (architectural and structural drawing excepted) to avoid location of a feature from more than one point.

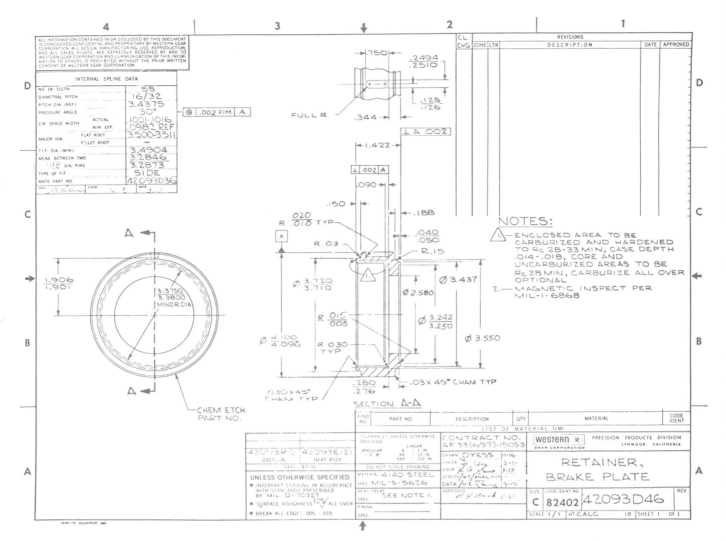

Fig. 8-41. Local notes can be directed to a feature on the drawing by a reference number and "flag." (Western Gear Corp.)

PROBLEMS AND ACTIVITIES

The following dimensioning problems are designed for A-size or B-size drawing sheets. Use a separate sheet for each problem assigned by the instructor. Dimensions should not be crowded. Provide ample space between views. Select the proper scale for the size of drawing sheet being used.

1. Locate and construct by freehand sketching techniques the required extension lines, dimension lines, and leaders to correctly dimension the parts shown in Fig. 8-42. No dimension figures are to be included. Letter the title of each part in the title block of the sketch.

2. Draw and dimension the parts shown in Fig. 8-43. Check each drawing against the rules for good dimensioning.

3. Draw the necessary views and dimension the parts shown in Fig. 8-44. Check each drawing against the rules for good dimensioning.

SELECTED ADDITIONAL READING

Dimensioning and Tolerancing for Engineering Drawings, ASME Y14.5M, American National Standards Institute, 1430 Broadway, New York, NY 10018.

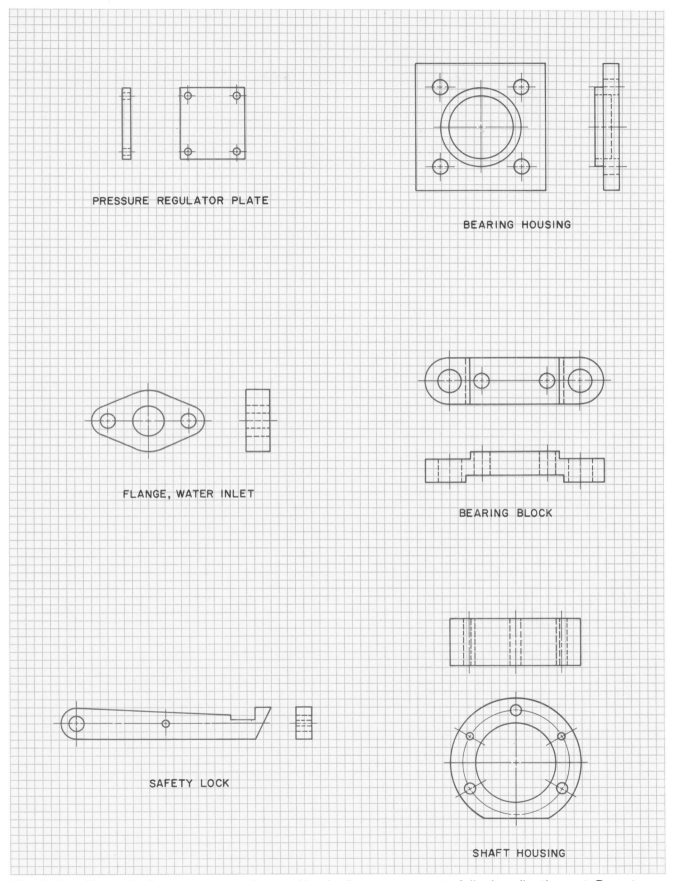

Fig. 8-42. Sketch in dimension, extension, and leader lines necessary to fully describe the part. Do not insert the actual dimensions. Do the items assigned by your instructor.

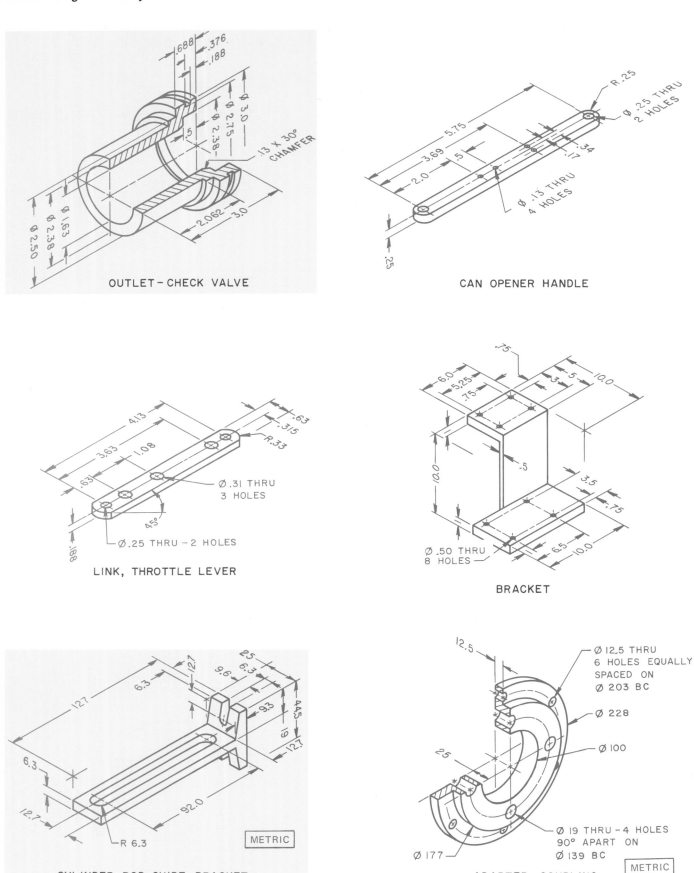

OUTLET – CHECK VALVE

CAN OPENER HANDLE

LINK, THROTTLE LEVER

BRACKET

CYLINDER ROD GUIDE BRACKET

ADAPTER, COUPLING

Fig. 8-43. Evaluate each drawing and determine what, if any, dimensions are missing. Draw the appropriate views of the objects assigned by your instructor and dimension each. Make sure to follow the "rules for good dimensioning."

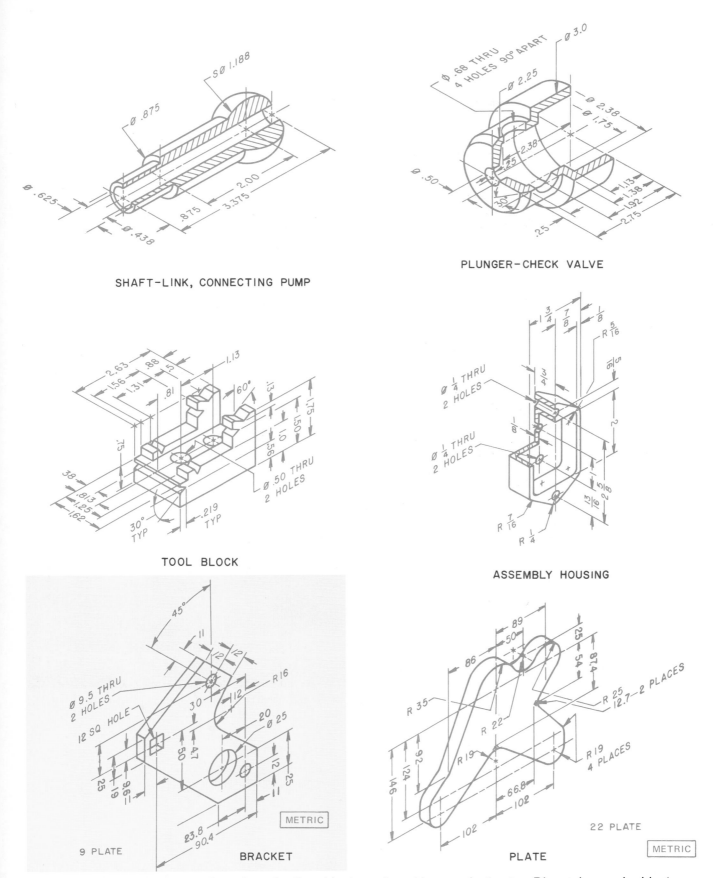

Fig. 8-44. Select the appropriate views for the objects assigned by your instructor. Dimension each object. Make sure that your dimensions fully describe the object. Also be sure to follow the "rules for good dimensioning."

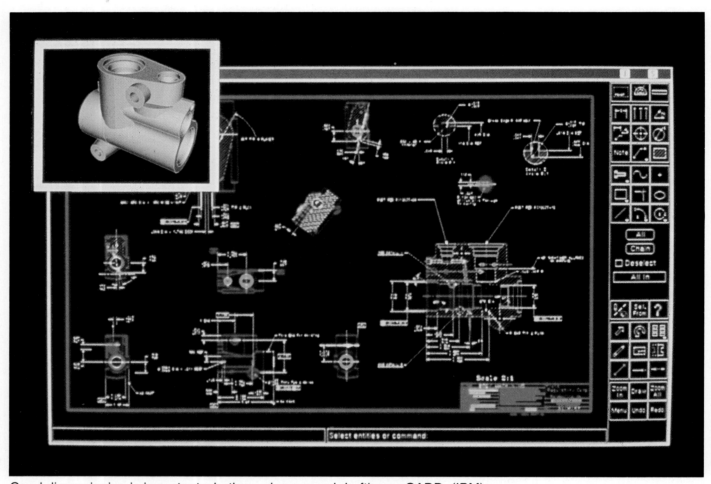

Good dimensioning is important whether using manual drafting or CADD. (IBM)

Geometric Dimensioning and Tolerancing

KEY CONCEPTS

- [ ] There are specific terms used by industry for precision dimensioning and tolerancing.
- [ ] Tolerances describe types of fits for machine parts.
- [ ] There are accepted industry standards for applying toleranced dimensions to drawings.
- [ ] There are specific symbols used for geometric dimensioning and tolerancing.
- [ ] Computer numerical controlled equipment requires specific information on drawings.
- [ ] Precision dimensioning allows the consistent interchangeability of manufactured parts.

The manufacture of a product usually requires the assembly of a number of different components. These components may all be made by the same company in one location. However, in many cases several different industries supply the components for assembly. Therefore, it is necessary to control dimensions very closely so that all of the parts will fit properly. This is called **interchangeable manufacture.** Interchangeability is also essential for replacement parts.

TOLERANCING

The control of dimensions is called **tolerancing.** A toleranced dimension means that the dimension has a range of acceptable sizes that are within a "zone." The size of this zone depends on the function of the part. To achieve an exact size (non-toleranced dimension) is not only very expensive, but virtually impossible under normal conditions. Therefore, tolerances are set as liberal as possible while still being able to achieve the function of the part.

Industrial designers and engineers establish tolerances based on industry standards, practice, and the function of the part. Drafters, too, must understand the application of these tolerances to engineering drawings.

Types of Tolerances

Tolerances are used to control the size of the features of a part. In addition to tolerances on size, there are two general types of tolerances. **Positional tolerances** control the location of features on a part. **Form tolerances** control the form or the geometric shape of features on a part. These basic types and uses of tolerances are presented in this chapter.

Definition of Terms

Standard terms have been adopted by industry to effectively communicate information related to tolerancing. A drafter must understand the meaning of these terms. A drafter must also be able to correctly apply these terms to a drawing.

Basic dimension

A **basic dimension** is an exact, untoleranced value used to describe the size, shape, or location of a feature. Basic dimensions are used as a "base." From this base, tolerances or other associated dimensions are established.

Basic dimensions are not directly toleranced. Any permissible variation is contained in the tolerance on the dimension associated with the basic dimension. An example of a tolerance on a hole diameter is shown in Fig. 9-1A.

Basic dimensions are indicated on the drawing by the word "BASIC." The abbreviation "BSC"

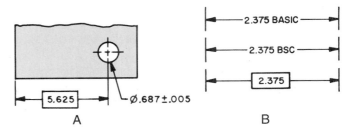

Fig. 9-1. A–Basic dimensions are not toleranced. Instead, the feature that the dimension is referring to is toleranced. B–There are three ways to indicate that a dimension is a basic dimension.

following or immediately below the dimension figure can also be used. The dimension figure can also be enclosed in a rectangular frame to indicate that it is a basic dimension. A general note such as "UNTOLERANCED DIMENSIONS LOCATING TRUE POSITION ARE BASIC" can also be used.

Reference dimension

Reference dimensions are placed on drawings for the convenience of engineering and manufacturing personnel. These dimensions are indicated by enclosing the dimension within parentheses, Fig. 9-2.

Reference dimensions are not required for the manufacturing of a part or in determining the acceptability of the part. Referenced dimensions are untoleranced dimensions with values equal to the **nominal sum.** The nominal sum is the difference of the associated toleranced dimensions. Reference dimensions may be rounded off as desired.

Datums

Datums are exact planes, lines, or points from which other features are located. A datum is usually a plane or point on the part. However, a datum can also be a plane or surface on the machine being used. For example, a datum can be located on the mill table for a part that will be milled.

Care should be exercised in the selection of datums on drawings to make certain they are recognizable, accessible, and useful for measuring. Corresponding features on mating parts should be selected as datums to assure ease of assembly. Datums are indicated by the word "DATUM," Fig. 9-3A. This can also be represented symbolically by a capital letter enclosed in a square frame, Fig. 9-3B.

A machined part may require more than one datum plane in its dimensioning. Letters such as "A," "B," and "C" are assigned to each datum. Datums may be considered to be **primary, secondary,** and **tertiary** depending on the design of the part. (Refer to Fig. 9-18.) Preference for the datums would appear in that order. Dimensions on the part

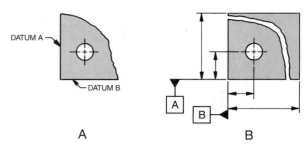

Fig. 9-3. A–Datums can be identified by the word "DATUM" followed by a reference letter and a leader indicating the datum. B–Datums can also be identified by enclosing the reference letter in a square connected to the extension line of the datum.

would be established in that given sequence. In other words, the primary datum would appear first (on the left) in a feature control frame and the tertiary datum would appear last (on the right). If a particular feature can be measured in reference to all three datums, the measurement in reference to the primary datum takes precedence over the other two. So if the measurement is within tolerance when measured from the secondary and tertiary datums, but not from the primary datum, the feature is not within tolerance.

Nominal size

The **nominal size** is a classification size given commercial products such as pipe or lumber. It may or may not express the true numerical size of the part or object. For example, a seamless, wrought-steel pipe of 3/4 (.750) inch nominal size has an actual inside diameter of 0.824 inch and an actual outside diameter of 1.050 inches, Fig. 9-4A. In the case of a cold-finished, low-carbon steel, round rod the nominal 1 inch size is within .002 inch of actual size, Fig. 9-4B.

Basic size

The **basic size** is the size of a part determined by engineering and design requirements. From this

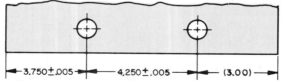

Fig. 9-2. Reference dimensions are not toleranced. They are presented on the drawing for the benefit of engineering and manufacturing personal. Reference dimensions are not used in the manufacture or inspection of the part. Reference dimensions are indicated on a drawing by enclosing the dimension in parentheses.

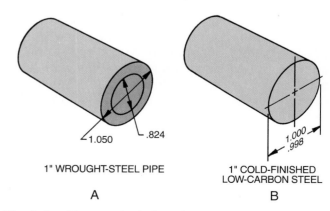

Fig. 9-4. The nominal size of a product does not necessarily reflect the actual size.

size, allowances and tolerances are applied. For example, strength and stiffness of a shaft may require one inch diameter material. This basic one inch size (with tolerance) is usually applied to the hole size and allowance for a shaft, Fig. 9-5.

Actual size

Actual size is the measured size of a part or object. This measurement is taken from the manufactured part.

Allowance

Allowance is the intentional difference in the dimensions of mating parts to provide for different classes of fits. This is not the same as a tolerance. Allowance is the minimum clearance space or maximum interference, whichever is intended, between mating parts. In the example shown in Fig. 9-5B, an allowance of .002 has been made for clearance (1.000 - .002 = .998).

Design size

The **design size** of a part is the size after an allowance for clearance has been applied and tolerances have been assigned. The design size of the

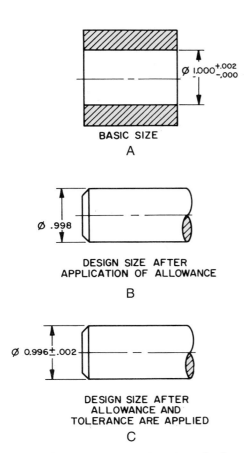

BASIC SIZE

A

DESIGN SIZE AFTER
APPLICATION OF ALLOWANCE

B

DESIGN SIZE AFTER
ALLOWANCE AND
TOLERANCE ARE APPLIED

C

Fig. 9-5. A–The basic size is the size of a feature that a design must follow. B–The allowance is the maximum variance allowed in the design size. C–The final design size has the allowance and the tolerance for the designed part accounted for.

shaft shown in Fig. 9-5B is shown in Fig. 9-5C after tolerances are assigned.

Limits of Size

Limits are the extreme dimensions allowed by the tolerance range. Two dimensions are always involved, a maximum size and a minimum size. For example, the design size of a feature may be 1.625″. If a tolerance of plus or minus two thousandths (±.002) is applied, then the two limit dimensions are maximum limit 1.627″ and minimum limit 1.623″.

Tolerance

Tolerance is the total amount of variation permitted from the design size of a part. This is not the same as allowance. Tolerances should always be as large as possible while still able to produce a usable part to reduce manufacturing costs. Tolerances can be expressed as limits, Fig. 9-6A. Tolerances can also be expressed as the design size followed by a plus and minus tolerance, Fig. 9-6B.

Tolerances can also be given in the title block or in a note, Fig. 9-6C. If a tolerance is given in a note or the title block, it applies to all dimensions on the drawing, unless otherwise noted.

Unilateral tolerance

A **unilateral tolerance** varies in only one direction from the specified dimensions, Fig. 9-7A. This type of tolerance might be "plus the tolerance" or "minus the tolerance," but never both.

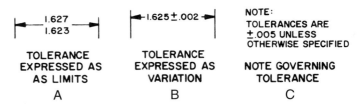

TOLERANCE
EXPRESSED AS
AS LIMITS

A

TOLERANCE
EXPRESSED AS
VARIATION

B

NOTE:
TOLERANCES ARE
±.005 UNLESS
OTHERWISE SPECIFIED

NOTE GOVERNING
TOLERANCE

C

Fig. 9-6. There are three ways of indicating a tolerance on a drawing.

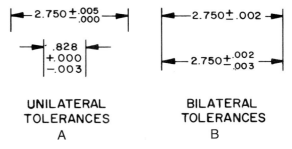

UNILATERAL
TOLERANCES

A

BILATERAL
TOLERANCES

B

Fig. 9-7. A–Unilateral tolerances are tolerances that vary in one direction only from the specified dimension. B–Bilateral tolerances are tolerances that vary in both directions from the specified dimension.

Bilateral tolerance

Bilateral tolerances vary in both directions from the specified dimension, Fig. 9-7B. This type of tolerance might vary by the same amount from the given dimension, or it might vary by a different amount in each direction. However, the dimension will vary in both directions.

Fit

Fit is a general term referring to the range of "tightness" or "looseness." Fit results from the application of a specific combination of allowances and tolerances in the design of mating parts. There are three general types of fits: clearance, interference, and transition.

Clearance fit

A **clearance fit** has a positive allowance, or "air space." The limits of size are defined so that a clearance always results when mating parts are assembled, Fig. 9-8A.

Interference fit

An **interference fit** has a negative allowance, or "interference," Fig. 9-8B. This is often referred to as a **press fit** or a **force fit.**

Transition fit

In a **transition fit,** the limits of size are defined so that the result may be either a clearance fit or an interference fit. For example, the smallest shaft size allowed by the shaft tolerance will fit within the largest hole size allowed by the hole tolerance and a clearance will result. However, the largest shaft size allowed by the shaft tolerance will interfere with the smallest hole size allowed by the hole tolerance. The two mating parts will have to be "pressed" together, Fig. 9-8C.

Basic size systems

In the design of mating cylindrical parts, it is necessary to assume a **basic size** for either the hole or shaft. The design sizes of mating parts are then calculated by applying an allowance to this basic size. Manufacturing costs usually determine which mating part becomes the standard size.

If standard tools can be used to produce the holes, the basic hole size system is used. However, if a machine or an assembly requires several different fits on a cold-finished shaft, the basic shaft size is most economical in manufacturing and is the one used. When standard parts (such as ball bearings) are inserted in castings, the basic shaft size also applies.

Basic hole size

In the **basic hole size system,** the basic size of the hole is the design size, and the allowance is applied to the shaft. **Basic hole size** is the minimum hole size produced by standard tooling, such as reamers and broaches. Allowances and tolerances are specified for this basic or design size to produce the type fit desired.

An example of a basic hole size is shown in Fig. 9-9A. In this example, the basic hole size is the minimum size, .500″. An allowance of .002″ is subtracted from the basic hole size and applied to the shaft for clearance, providing a maximum shaft size of .498″. A tolerance of +.002″ is then applied to the basic hole size and -.003″ to the shaft size to provide a maximum hole size of .502″ and a minimum shaft size of .495″.

The tightest fit (minimum clearance) is .500″ (smallest hole size) - .498″ (largest shaft size) = .002″. The fit giving maximum clearance is .502″ (largest hole size) - .495″ (smallest shaft size) = .007″.

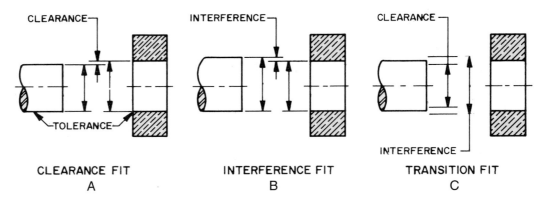

Fig. 9-8. A–A clearance fit is one where a gap, or air space, occurs between two mating parts. B–An interference fit is when the part being inserted is larger than the opening that it is being inserted into. C–A transition fit occurs when the dimensions and tolerances are specified so that the mating parts might fit as an interference fit or a clearance fit, depending on how the two fit in the tolerance range.

These examples have provided a clearance fit in mating parts. To obtain an interference fit in the basic hole size system, add the allowance to the basic hole size and assign this value as the largest shaft size.

Basic shaft size

When the **basic shaft size system** is used, the design size of the shaft is the basic size and the allowance is applied to the hole. The **basic shaft size** is the maximum shaft size. Tolerances and allowances are then specified for this basic or design size to produce the desired fit.

In Fig. 9-9B, the maximum shaft size of .500″ is taken as the basic (design) size. An allowance of .002″ is added to the basic shaft size and applied to the hole size, providing a minimum hole size of .502″. A tolerance of -.001″ is specified for this basic shaft size and +.003″ for the hole size. This provides a minimum shaft size of .499″ and a maximum hole size of .505″.

The minimum clearance provided is .502″ (smallest hole size) - .500″ (largest shaft size) = .002″. The maximum clearance provided is .505″ (largest hole size) - .499″ (smallest shaft size) = .006″.

The example has provided a clearance fit in the mating parts. Interference fits may be obtained in the basic shaft size system by subtracting the allowance from the basic shaft size and assigning this value as the minimum hole size.

Maximum material condition (MMC)

The **maximum material condition (MMC)** is present when the feature contains the maximum amount of material. MMC exists when internal features, such as holes and slots, are at their minimum size, Fig. 9-10A. MMC also occurs when external features, such as shafts and bosses, are at their maximum size, Fig. 9-1B.

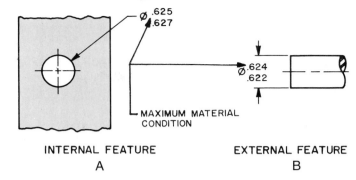

INTERNAL FEATURE
A

EXTERNAL FEATURE
B

Fig. 9-10. A–The maximum material condition (MMC) is when the feature has the most material possible while still staying in the tolerance. For an internal feature, MMC occurs when the feature is at the lower limit. B–For an external feature, MMC occurs when the feature is at the upper limit.

MMC is applied to the individual tolerance, datum reference, or both. The position or form tolerance increases as the feature departs from MMC by the amount of such departure. In other words, if the feature that controls MMC varies (a standard shaft, for example), the position tolerance of the feature will change (the mating hole, for example).

Least material condition (LMC)

Least material condition (LMC) is present when the feature contains the least amount of material within the tolerance range. LMC exists when holes are at maximum size. LMC also occurs when shafts are at minimum size.

Just as with MMC, if the LMC of a controlling feature varies, the position tolerance of the mating part will change as well. The variance of the position tolerance will be equal to the amount of variance in LMC.

Regardless of feature size (RFS)

Regardless of feature size (RFS) means that geometric tolerances or datum references must be met no matter where the feature lies within its size tolerance. Where RFS is applied to a positional or form tolerance, the tolerance must not be exceeded regardless of the actual size of the feature. RFS applies with respect to individual tolerance, datum reference, or both, where no symbol is specified. RFS is assumed on all dimensions unless otherwise indicated.

Dimensions not to scale

All drawings (with the exception of diagrammatic and schematic) should be drawn to scale. However, on a drawing revision, the correction of a dimension in scale may require an excessive amount of drafting. If the drawing remains clear, the dimension can be changed and underlined with a thick, straight line to indicate "not-to-scale," Fig. 9-11.

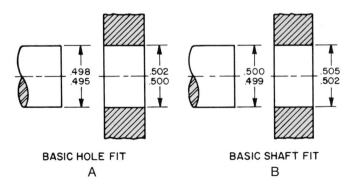

BASIC HOLE FIT
A

BASIC SHAFT FIT
B

Fig. 9-9. A–In the "basic hole size system," the size of the hole is the design size. Allowances are applied to the size of the shaft during the design phase. B–In the "basic shaft size systems," the size of the shaft is the design size and allowances are applied to the hole size during the design phase.

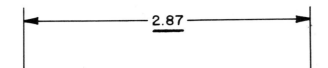

Fig. 9-11. A "not-to-scale dimension" is indicated by a straight, thick line underneath the dimension that is not-to-scale.

Original drawings should not be issued with dimensions "out-of-scale." These dimensions should be kept to an absolute minimum on revisions. Where there is the slightest chance of misinterpretation of an out-of-scale dimension on a revised drawing, the drawing should be redrawn.

Application of tolerances

When manufacturing items requiring interchangeability of parts, tolerancing of all dimensions is required. Exceptions are dimensions labeled "BSC," "REF," "MAX," or "MIN." Tolerances are normally expressed in the same number of decimal places as the dimension. (Refer back to Fig. 9-7.) These tolerances are applied to the dimensions either as limit dimensioning or as plus and minus tolerancing.

Limit dimensioning

With **limit dimensioning,** only the maximum and minimum dimensions are given, Fig. 9-12. Use one of the following methods to arrange the limit numerals.

1. For positional dimensions given directly, the maximum (high) limit is always placed above the minimum (low) limit, Fig. 9-12A. For positional dimensions given in note form, the minimum limit always precedes the maximum limit, Fig. 9-12B.

2. For size dimensions of features given directly, the number representing the maximum material condition (MMC) is placed above the number representing the least material condition, Fig. 9-12C. For size dimensions given in note form, the MMC number precedes the LMC number, Fig. 9-12D.

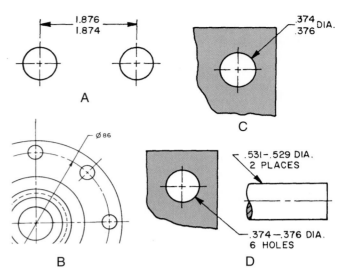

Fig. 9-12. Only the maximum and minimum dimensions are given when using limit dimensioning. A–When the limit dimensions are given directly, the maximum dimension always appears above the minimum dimension. B–When the limit dimensions are given as a note, the minimum dimension always comes before the maximum dimensions. C–When a size dimension of a feature is given directly, the dimension representing MMC is given above the other. D–For size dimensions given in note form, the dimensions representing MMC precedes the other.

Plus and minus tolerancing

In **plus and minus tolerancing,** the tolerances are generally placed to the right of the specific dimension. The tolerances appear as a number with "stacked" plus and minus signs before it. This expression is the allowed variation of the size or location of a feature. (Refer back to Fig. 9-6B.)

Checking tolerances

The tolerances of a drawing are checked to see that all parts will assemble as specified. Two methods of computing the tolerance between two features are shown in Fig. 9-13.

Selective assembly

For the manufacture of mating parts with tolerances of a fairly wide range, interchangeability of

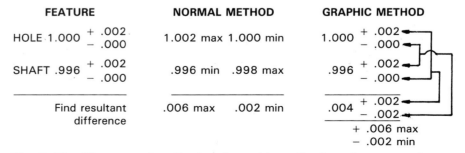

FEATURE		NORMAL METHOD		GRAPHIC METHOD	
HOLE 1.000 $^{+\ .002}_{-\ .000}$		1.002 max	1.000 min	1.000 $^{+\ .002}_{-\ .000}$	
SHAFT .996 $^{+\ .002}_{-\ .000}$		.996 min	.998 max	.996 $^{+\ .002}_{-\ .000}$	
Find resultant difference		.006 max	.002 min	.004 $^{+\ .002}_{-\ .002}$	
				$^{+\ .006\ max}_{-\ .002\ min}$	

Fig. 9-13. The normal method and graphic method are two ways of computing the tolerance between a hole and a shaft. (Sperry Flight Systems Div.)

parts is easily obtained. However, for mating parts with close fits, very small allowances and tolerances are required. The manufacturing costs of producing parts with very small allowances and tolerances is often too expensive.

Selective assembly is a process of selecting mating parts by inspection, and classification into groups according to actual sizes. Small size features (such as cylindrical shafts and holes) are grouped for matching with small size features of mating parts. Medium size features are grouped with medium size mating parts. Large size features are grouped with large size mating parts.

The cost of manufacturing is reduced considerably in selective assembly due to less restricted allowances and tolerances. This method is usually more satisfactory in achieving transition fit mating parts than in interchangeable assembly.

Chain dimensioning

The dimensioning of a series of features, such as holes, from point to point is known as ***chain dimensioning,*** Fig. 9-14A. When these dimensions are toleranced, overall variations in position of the features may occur that exceed the tolerances specified, Fig. 9-14B. The possible variation is equal to the sum of the tolerances on the intermediate dimensions. (Refer back to Fig. 9-14B.) For example, the variation in position between features "A" and "B" in Fig 9-14B range from 1.499" (3.997" - 2.498" = 1.499") to 1.501" (4.003" - 2.502" = 1.501"). This is a difference of .002" instead of the intended ±.001.

Chain dimensioning is used on drawings prepared for incremental positioning CNC operations (see Chapter 22). The tolerances are built into the machine and dimensions are given as basic dimensions. Tolerancing of these dimensions would not change the part being machined.

Datum dimensioning

In ***datum dimensioning,*** sometimes called ***base line dimensioning,*** features are dimensioned individually from a datum, Fig. 9-14C. This system of dimensioning avoids accumulation of tolerances from feature to feature. Where the distance between two features must be closely controlled, without the use of an extremely small tolerance, datum dimensioning should be used. Datum dimensioning is also used for absolute positioning CNC operations.

Angular surface tolerancing

Angular surface tolerancing uses a combination of linear and angular dimensions, or linear dimensions alone, to dimension and tolerance angular surfaces. A dimension and its tolerance

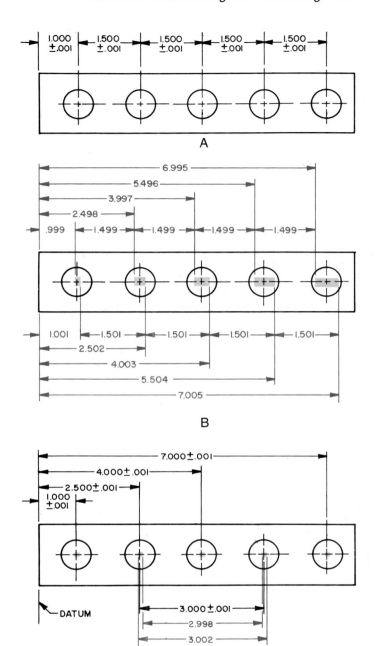

Fig. 9-14. B–In chain dimensioning, error can accumulate even though each dimension stays within tolerance. C–In datum dimensioning, the error is limited by dimensioning every feature of a part from a common point or plane.

specify a tolerance zone that the surface must lie in, Fig. 9-15A. The tolerance zone widens as it moves away from the apex of the angle.

Where a tolerance zone with parallel boundaries is desired, a basic angle may be specified as in Fig. 9-15B. By specifying a basic angle, a tolerance zone with parallel sides is indicated. The surface controlled must lie within the tolerance zone. A tolerance zone with parallel sides is usefull for parts that have long angular surfaces.

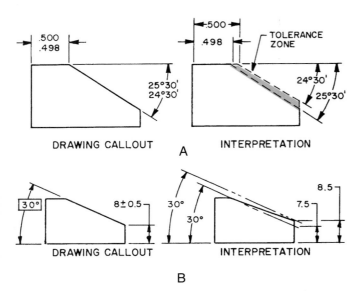

Fig. 9-15. A–An angular surface can be toleranced using a combination of linear and angular dimensions. The tolerance zone in this case will widen as it is moved from the apex of the angle. B–If a tolerance zone with parallel sides is required, a basic angle dimension should be specified.

Selection of Fits

Tables of recommended tolerances have been worked out by many companies, as well as the American National Standards Institute (ANSI) Committee. The designer or drafter should consult these when it is necessary to select tolerances for a specific sized feature and mating part. The use required from a piece of equipment determines the limits of size of mating parts and the selection of type of fit.

Standard fits

A number of types and classes of fits are given in the Reference Section. Any fit of mating parts will usually be required to perform one of three functions: running or sliding fit, locational fit, or force fit. These fits are further divided into classes and assigned letter symbols. Fits are not indicated on a drawing. Fits are specified on a drawing by the tolerances.

Running and sliding fits (RC)

A *running and sliding fit* is a type of fit designed to provide similar running performance, with suitable lubrication, throughout the range of sizes.

- RC1: Close sliding fits that are designed for accurate location of parts which must be assembled without play. (Play is the amount of "movement" or "slippage.")
- RC2: Sliding fits that are intended for accurate location but with greater maximum clearance than RC1. Parts move and turn easily, but are not intended to run freely.

- RC3: Precision running fits that are the closest fits which can be expected to run freely at slow speeds and light journal (shaft) pressure.
- RC4: Close running fits that are designed to run freely on accurate machinery with moderate surface speeds and journal pressures. These are designed for use when accurate location and minimum play is needed.
- RC5 and RC6: Medium fits that are designed for higher running speeds and/or heavy journal pressures.
- RC7: Free running fits that are designed for use where accuracy is not essential or where temperature variations are likely to occur. (A temperature change can change the size of a part.)
- RC8 and RC9: Loose running fits that are designed for use where wide commercial tolerances may be necessary, together with an allowance on the external member.

Locational fits (LC)

Locational fits relate to the location of mating parts. They are subdivided into three classes based on design requirements.

- LC: Locational clearance fits that are designed for parts which can be freely assembled or disassembled. Classes of fits run from snug fits for parts requiring accuracy of location, through medium clearance fits for parts such as ball bearing race and housing, to looser fits for fastener parts requiring considerable freedom of assembly.
- LT: Locational transition fits that are a medium fit, between clearance and interference fits, where accuracy of location is important but some clearance or interference is permissible.
- LN: Locational interference are fits that provide accurate location for parts requiring rigidity and alignment with no special requirements for bore pressure. Such fits are not intended for parts designed to transmit frictional loads from one part to another by virtue of tightness of fit. Those conditions are covered by force fits.

Force fits (FN)

Force fits or *shrink fits* are a type of interference fit normally characterized by constant bore pressures throughout the range of sizes. The interference varies almost directly with the diameter. These are also called *press fits.*

- FN1: Light drive fits that require light assembly pressures and produce more or less permanent assemblies. They are used for thin sections or long fits, or in cast iron external members.
- FN2: Medium drive fits that are designed for ordinary steel parts, or for shrink fits on light sections. They usually are the tightest fits that can be used with high-grade cast iron external members.

- FN3: Heavy drive fits that are suitable for heavier steel parts or for shrink fits in medium sections.
- FN4 and FN5: Force fits that are designed for parts which can be highly stressed, or for shrink fits where heavy pressing forces required are impractical.

GEOMETRIC DIMENSIONING AND TOLERANCING

Geometric dimensioning and tolerancing is a system of dimensioning and tolerancing drawings with emphasis on the actual function and relationship of part features. This system is used where interchangeability is critical. This system does not replace the coordinate dimensioning system. Geometric dimensioning and tolerancing is used in conjunction with coordinate dimensioning. Position and form tolerancing (geometric dimensioning and tolerancing) does not imply tighter tolerance. Rather, geometric dimensioning and tolerancing permits the use of maximum tolerances while maintaining 100% interchangeability.

Geometric dimensioning and tolerancing has become the system used by most industries because of the clarity and preciseness in communicating specifications. Every drafter, designer, and engineer should understand its use.

Symbols for Positional and Form Tolerances

Geometric characteristic symbols for tolerances reduce the number of notes required on a drawing. These symbols are compact, recognized internationally, and reduce misinterpretation.

The standard symbols used for geometric characteristics of part features are shown in Fig. 9-16. Modifying symbols are shown in Fig. 9-16. The meaning and application of these symbols are discussed later in this chapter.

Templates are available for use in drawing the symbols for geometric dimensioning and tolerancing. The template shown in Fig. 9-17 features symbols, specifications for surface characteristics, and a complete alphabet.

Datum Identifying Symbol

Datums are identified on the drawing by a reference letter (any letter except "I," "O," and "Q") which is preceded and followed by a dash, and enclosed in a frame, Fig. 9-18A. Where more than one datum is used on a drawing, the desired order or precedence of datums is shown from left to right in the feature control symbol, Fig. 9-18B.

SYMBOL FOR:	ASME Y14.5
STRAIGHTNESS	⎯
FLATNESS	▱
CIRCULARITY	○
CYLINDRICITY	⌭
PROFILE OF A LINE	⌒
PROFILE OF A SURFACE	⌓
ALL-AROUND PROFILE	⟲
ANGULARITY	∠
PERPENDICULARITY	⊥
PARALLELISM	//
POSITION	⊕
CONCENTRICITY/COAXIALITY	◎
SYMMETRY	⌯
CIRCULAR RUNOUT	* ↗
TOTAL RUNOUT	* ↗↗
AT MAXIMUM MATERIAL CONDITION	Ⓜ
AT LEAST MATERIAL CONDITION	Ⓛ
REGARDLESS OF FEATURE SIZE	NONE
PROJECTED TOLERANCE ZONE	Ⓟ
DIAMETER	⌀
BASIC DIMENSION	30
REFERENCE DIMENSION	(30)
DATUM FEATURE	A◄
DATUM TARGET	Ø6/A1
TARGET POINT	✕
DIMENSION ORIGIN	⊕►
FEATURE CONTROL FRAME	⊕ ⌀0.5Ⓜ A B C
CONICAL TAPER	▷
SLOPE	◿
COUNTERBORE/SPOTFACE	⌴
COUNTERSINK	⌵
DEPTH/DEEP	⤓
SQUARE (SHAPE)	□
DIMENSION NOT TO SCALE	15
NUMBER OF TIMES/PLACES	8X
ARC LENGTH	⌒105
RADIUS	R
SPHERICAL RADIUS	SR
SPHERICAL DIAMETER	S⌀

* MAY BE FILLED IN

Fig. 9-16. Symbols specify positional and form tolerances in geometric dimensioning. (American National Standards Institute)

Fig. 9-17. Templates are available for use with geometric dimensioning and tolerancing. These templates have the standard symbols on them and greatly reduce the time required for geometric dimensioning and tolerancing. (RapiDesign)

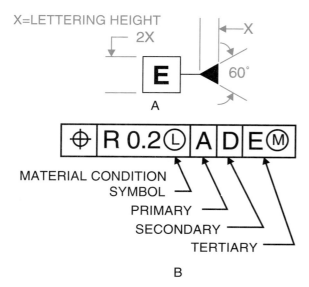

Fig. 9-18. A–A datum identifying symbol is any letter (except "I," "O," or "Q") and enclosed in a frame. (The previous symbol had dashes). B–Datum reference letters are given in the desired order of precedence. (American National Standards Institute)

Datum Targets

It is not always practical to identify an entire feature as a datum feature. For example, a very large feature might not be practical to use as a datum feature. **Datum targets** are used when the whole feature is not to be used as a datum feature. The types of targets that may be used are points, lines, and areas.

Application of material condition symbols for tolerances of position and form

The application of material condition symbols to a drawing is limited to features subject to variations in size. These may be any feature or datum feature whose axis or center is controlled by geometric tolerances. Three different material condi-

tions are used: Maximum Material Condition (MMC), Least Material Condition (LMC), and Regardless of Feature Size (RFS). For individual tolerances, MMC or LMC must be specified on the drawing datum reference, as applicable. RFS is always assumed unless MMC or LMC is specified. The only time that RFS is specified is when a previous standard (earlier than ASME Y14.5M-1994) is specified for use.

The symbol Ⓜ is used to designate **Maximum Material Condition.** The symbol Ⓛ specifies the **Least Material Condition. Regardless of Feature Size** is assumed when no symbol is given. This means that "regardless of feature size" the tolerance must not be exceeded. These symbols may be used only as modifiers in feature control frames. (Refer back to Fig. 9-18B.) The abbreviations MMC, LMC, and RFS should be used in notes and dimensions, not the symbols. Where MMC, RFS, or LMC is specified, the appropriate symbol follows the specified tolerance and applicable datum reference in the feature control frame. (Refer back to Fig. 9-18B.)

Feature Control Frame

The **feature control frame** is the means by which a geometric tolerance is specified for an individual feature, Fig. 9-19. The frame is divided into compartments containing, in order from the left, the geometric characteristic symbol followed by the tolerance. Where applicable, the tolerance is preceded by the diameter or radius symbol and followed by an appropriate material condition symbol, Ⓜ or Ⓛ.

If the tolerance is related to a datum (or multiple datums), the datum reference letter (or letters in the case of multiple datum references) follows in the next compartment. Where applicable, the datum reference letter is followed by a material condition symbol. Datum reference letters are entered in the desired order of precedence, from left to right, and need not be in alphabetical order. (Refer back to Fig. 9-18B.)

The feature control frame is associated with a feature or features by one of the following methods.

1. Attaching a side or end of the frame to an extension line from the feature, provided it is a plane surface. (Refer back to Fig. 9-19.)

2. Attaching a side or end of the frame to an extension of the dimension line pertaining to a feature of size.

3. Running a leader from the frame to the feature.

4. Placing the frame below or attached to a leader-directed callout or dimension that controls the feature.

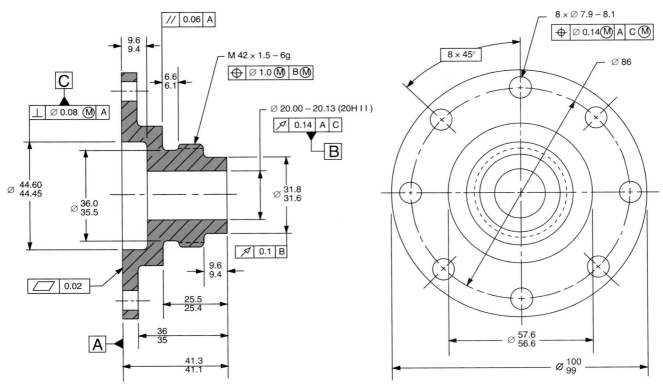

Fig. 9-19. The feature control frame should be connected to an extension line of the feature, to a leader running to the feature, or below a leader-directed callout of the feature.

Tolerances of Location

Tolerances assigned to dimensions locating one or more features in relation to other features or datums are known as *positional tolerances.* There are three basic types of positional tolerances: true position, concentricity, and symmetry.

True positional tolerance

The term *true position* has a meaning similar to basic dimension. It describes the exact (true) location of a point, line, or plane (usually the center plane) of a feature in relation to another feature or datum. A true position is located by basic dimensions, Fig. 9-20. A feature control frame is added to the note used to specify the size and number of the feature, Fig. 9-20C.

A *true positional tolerance* is the total amount a feature may vary from its true position. Note in Fig. 9-20B that the coordinate system of tolerancing results in a square tolerance zone, and the actual variation from true position may exceed the specified variation. The actual variation along the diagonal is .014 or 1.4 times the tolerance specified. By utilizing a true positional tolerance (circular zone), the larger tolerance can be specified and interchangeability of parts still maintained, Fig 9-20C.

The tolerance zone is represented as a circle in one view and is assumed to be a cylindrical zone for the full depth of the hole, Fig. 9-21. The axis of the feature hole must be within the tolerance zone.

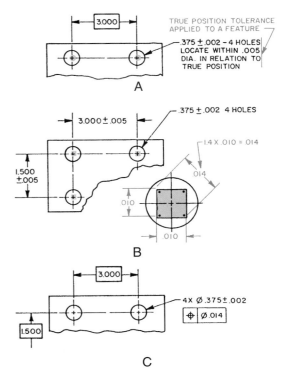

Fig. 9-20. A–A true positional tolerance can be given as a note in a callout. B–A coordinate tolerance of a circle will result in a square tolerance zone. This means that the variation from true position might be outside of the specified tolerance (the detail in B illustrates this). C–A true positional tolerance can also be given as a callout and a feature control frame.

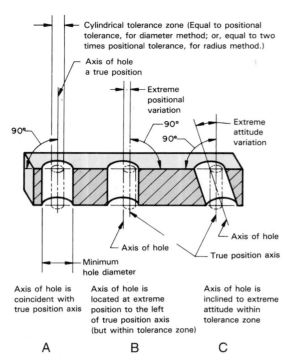

Note that the length of the tolerance zone is equal to the length of the feature, unless otherwise specified on the drawing.

Fig. 9-21. The tolerance zone for a hole is assumed to be a cylinder for the full depth of a feature. (American National Standards Institute)

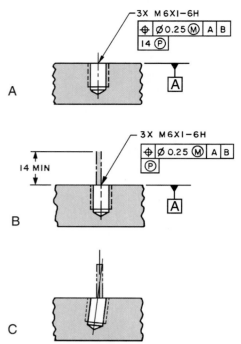

Fig. 9-22. A projected tolerance zone indicates the distance above the surface of the part that is critical to a mating feature. A–The projected tolerance zone can be indicated using a feature control frame. B–The projected tolerance zone can also be indicated using a combination of a feature control frame and a chain dimension. C–The interpretation of the tolerances specified in A and B. (American National Standards Institute)

Projected tolerance zone

A *projected tolerance zone* is specified where the variation in perpendicularity of threaded or press-fit holes could cause fasteners such as screws, studs, or pins to interfere with mating parts. The application of a projected tolerance zone to a positional tolerance is shown in Fig. 9-22. The extent of the projected tolerance zone may be indicated in the feature control frame, Fig. 9-22A The extent can also be shown as a dimensioned heavy chain line drawn closely to the centerline of the hole, Fig. 9-22B.

True position for noncircular features

Noncircular features such as slots, tabs, and elongated holes may be toleranced for position by using the same basic principles used for circular features. Positional tolerances usually apply only to surfaces related to the center plane of the feature, Fig. 9-23.

Where the feature is at MMC, its center plane must fall within a tolerance zone having a width equal to the true position tolerance for the diameter method (or twice the tolerance for the radius method). Note that the tolerance zone also defines the variation limits of "squareness" of the feature.

Concentricity tolerance

Concentricity is the condition of two or more surfaces of revolution having a common axis. A

concentricity tolerance callout and interpretation are shown in Fig. 9-24. In cases where it is difficult to find the axis of a feature (and where control of the axis is not necessary to part's function), it is recommended that the control be specified as a runout tolerance or a true position tolerance.

Symmetry

A *symmetrical feature* or part has the same contour and size on opposite sides of a central plane or datum feature. A symmetry tolerance may be specified by using a positional tolerance at MMC, Fig. 9-25. The true position symbol is recommended where a feature is to be located symmetrically about a datum plane and the tolerance is expressed on an MMC or RFS basis, depending upon the design requirements.

Tolerances of Form, Profile, Orientation, and Runout

Tolerances can also be used to specify the form, profile, orientation, and runout of features on a part. These tolerances, along with the positional tolerances, can be used to fully define every feature of the part, and the parts as a whole.

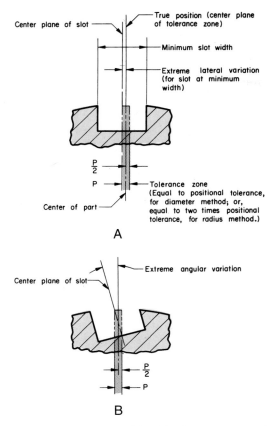

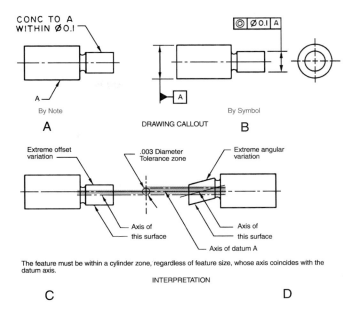

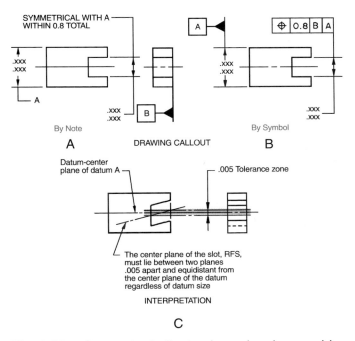

Fig. 9-23. A—The true positional tolerance zone for a noncircular feature (such as a slot) applies to the center plane of the feature. B—At MMC, the center plane of the feature must fall in the tolerance zone. (American National Standards Institute)

Fig. 9-24. Concentricity describes how closely two or more surfaces revolve about a common axis. A—Concentricity can be indicated by a note. B—A feature frame and the concentricity symbol can also be used. C—The maximum offset that the tolerance allows for the examples given in A and B. D—The maximum angular variation for the examples shown in A and B. (American National Standards Institute)

Fig. 9-25. Symmetry indicates how closely one side of a feature matches the other side of a feature (about a centerline or a center plane). A—Symmetry can be indicated with a note. B—Symmetry can also be indicated with a feature control frame and the symmetry symbol. C—The interpretation of the tolerances given in A and B. (American National Standards Institute)

Form tolerance

Tolerances which control the form of the various geometrical shapes and free-state variations of features are called *form tolerances.* Form tolerances are used to control the conditions of straightness, flatness, roundness, and cylindricity. A form tolerance specifies a tolerance zone that the particular feature must lie in.

Straightness tolerance. *Straightness* describes how close all elements of an axis or surface are to being a straight line. Straightness is specified by a tolerance zone of uniform width along a straight line within which all elements of the line must lie. The drawing callout can be indicated by geometric symbol, by note, and interpretation, Fig. 9-26. All elements of the feature in Fig. 9-26 must lie within a tolerance zone of .010 inch for the total length of the part. Straightness may be applied to control line elements in a single direction or in two directions.

Flatness tolerance. *Flatness* describes how close all elements of a surface are to being in one plane. Flatness is specified by a tolerance zone between two parallel planes that the surface must lie in. The specification for a flatness tolerance is

DRAWING CALLOUT   INTERPRETATION

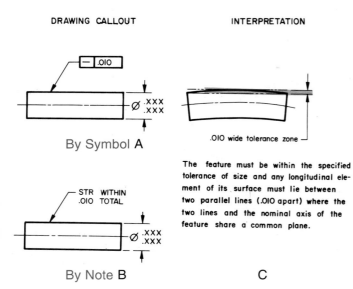

By Symbol A

By Note B   C

The feature must be within the specified tolerance of size and any longitudinal element of its surface must lie between two parallel lines (.010 apart) where the two lines and the nominal axis of the feature share a common plane.

Fig. 9-26. Straightness indicates how closely a surface of an object is to a straight line. A–Straightness can be indicated by a feature control frame and the straightness symbol . B–A note can also be used to indicate straightness . C–The interpretation of the examples given in A and B .
(American National Standards Institute)

shown by symbol and by note, Fig. 9-27. All elements of the surface in Fig. 9-27 must lie within a tolerance zone of .010 inch for the total length and width of the part. The feature control frame is placed in a view where the surface to be controlled is represented by a line.

Circularity (roundness) tolerance. *Circularity* describes the distance from the axis of all the elements of a revolved surface that is intersected

DRAWING CALLOUT   INTERPRETATION

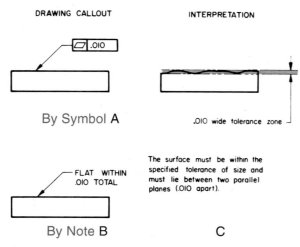

By Symbol A

By Note B   C

The surface must be within the specified tolerance of size and must lie between two parallel planes (.010 apart).

Fig. 9-27. Flatness indicates how closely all elements of a surface are to lying in a single plane. A–Flatness can be indicated by a feature control frame and the flatness symbol. B–A note can also be used to indicate flatness. C–The interpretation of the examples shown in A and B.
(American National Standards Institute)

by any plane. For a sphere, the intersecting plane passes through a common center. The intersection plane is perpendicular to a common axis for a cylinder or cone. A circularity tolerance is specified by two concentric circles confining a zone that the circumference must lie in. The permissible circularity tolerance is specified for a cylinder in Fig. 9-28. An example of a circularity tolerance for a cone is shown in Fig. 9-29. An example of a circularity tolerance for a sphere is shown in Fig. 9-30. Note

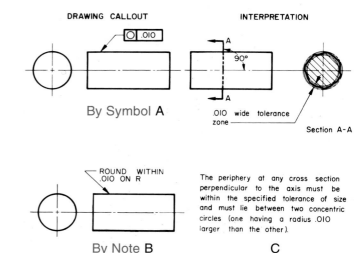

DRAWING CALLOUT   INTERPRETATION

By Symbol A   .010 wide tolerance zone

Section A-A

ROUND WITHIN .010 ON R

By Note B   C

The periphery at any cross section perpendicular to the axis must be within the specified tolerance of size and must lie between two concentric circles (one having a radius .010 larger than the other).

Fig. 9-28. Circularity is indicated by a plane cutting a revolved surface perpendicular to the axis of that surface. How closely all points of the resulting circle are to being equidistant from the center indicates the circularity of the revolved surface. A–The circularity of a cylinder can be indicated by a feature control frame and the circularity symbol. B– A note can also be used to indicate circularity. C–The interpretation of the examples in A and B.
(American National Standards Institute)

DRAWING CALLOUT   INTERPRETATION

By Symbol A   .010 wide tolerance zone

Section A-A

ROUND WITHIN .010 ON R

By Note B   C

The periphery at any cross section perpendicular to the axis must be within the specified tolerance of size and must lie between two concentric circles (one having a radius .010 larger than the other).

Fig. 9-29. A–The circularity of a cone can be indicated by a feature control frame and the circularity symbol. B–A note can also be used to indicate circularity. C–The interpretation of the examples in A and B.
(American National Standards Institute)

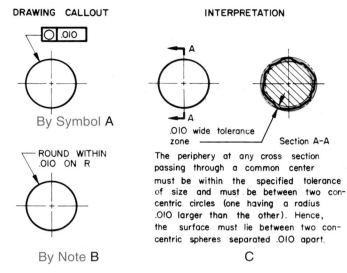

DRAWING CALLOUT   INTERPRETATION

By Symbol **A**

ROUND WITHIN
.OIO ON R

By Note **B**

.OIO wide tolerance zone   Section A-A

The periphery at any cross section passing through a common center must be within the specified tolerance of size and must be between two concentric circles (one having a radius .OIO larger than the other). Hence, the surface must lie between two concentric spheres separated .OIO apart.

C

Fig. 9-30. A–The circularity of a sphere can be indicated by a feature control frame and the circularity symbol. B–A note can also be used to indicate circularity. C–The interpretation of the examples in A and B. (American National Standards Institute)

that the tolerance zone for each of these is established by a radius.

Cylindricity tolerance. *Cylindricity* describes the distance of all elements of a surface of revolution from the axis of revolution. If there is perfect cylindricity, all elements of the surface are equidistant from the common axis. A cylindricity tolerance zone is defined as the space between two concentric cylinders that the specified surface must lie in. This space is called the *annular* space. Note in Fig. 9-31 that the tolerance zone is established by a radius. The cylindricity tolerance also controls

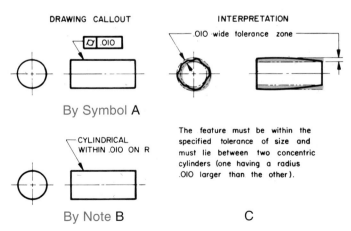

DRAWING CALLOUT   INTERPRETATION

.OIO wide tolerance zone

By Symbol **A**

CYLINDRICAL
WITHIN .OIO ON R

The feature must be within the specified tolerance of size and must lie between two concentric cylinders (one having a radius .OIO larger than the other).

By Note **B**   **C**

Fig. 9-31. Cylindricity indicates how closely all elements of a revolved surface are to being equidistant from a common axis. A–Cylindricity can be indicated by a feature control frame and the cylindricity symbol . B–A note can also be used to indicate cylindricity. C–The interpretation of the examples in A and B. (American National Standards Institute)

roundness, straightness, and parallelism of the surface elements.

Profile tolerancing

The elements of profiles consist of straight lines and curved lines. The curved lines may be either arcs or irregular curves. The elements in a profile line or surface are located with basic dimensions. The **profile tolerance zone** is established by applying a specified amount of permissible variation to these dimensions, Fig. 9-32.

The profile tolerance zone may be specified as **bilateral** (to both sides of the true profile) or **unilateral** (to either side of the true profile). The tolerance zone is shown along the profile in a conspicuous place by one or two phantom lines at a distance greater than the actual tolerance for drawing clarity.

A dimensional part is shown with a profile tolerance specification for a surface in Fig. 9-33. A profile tolerance for a line is specified in the same manner except the symbol is different. (Refer back to Fig. 9-16.)

When a profile tolerance applies to surfaces all around the part, a circle is located at the junction of the feature control frame leader, Fig. 9-33C.

Orientation tolerances

Orientation describes how a feature of an object "sits" on the object. Angularity tolerances, parallelism tolerances, and perpendicularity tolerances all describe the orientation of a feature.

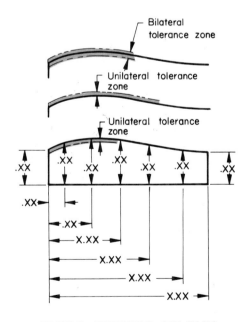

PROFILE DIMENSIONS ARE BASIC

Fig. 9-32. Profile tolerance zones indicate how far from the basic dimension a surface can vary. Profile tolerances can be bilateral or unilateral. The zone is represented by a line (or two lines in the case of bilateral) parallel to the feature. (American National Standards Institute)

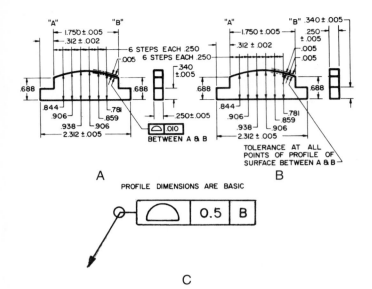

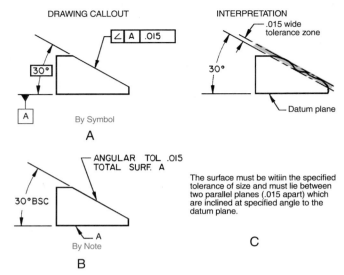

Fig. 9-33. A–A profile tolerance can be specified by a feature control frame and the profile symbol . B–A note can also be used to indicate a profile tolerance. C–An example of a feature control frame with the profile symbol. (American National Standards Institute)

Fig. 9-34. Angularity indicates how closely all elements of a surface are to being in a tolerance zone that is inclined at the specified angle. A–Angularity can be indicated by a feature control frame and a symbol. B–A note can also be used to indicate angularity. C–The interpretation of the examples in A and B. (American National Standards Institute)

Angularity tolerance. The means of controlling the specific angle of a surface or axis (other than 90°) with respect to a datum plane or axis is known as *angularity tolerance.* This is achieved by specifying a tolerance zone confined by two parallel planes, inclined at the required angle to a datum plane or axis. The parallel planes establish the zone that the toleranced surface or axis must lie in, Fig. 9-34.

Parallelism tolerance. *Parallelism* describes how close a line or surface is to being parallel to a datum plane or axis. Parallelism is specified for a plane surface by a tolerance zone confined by two planes parallel to a datum plane that the specified feature (axis or surface) must lie in, Fig. 9-35. Parallelism is specified for a cylindrical surface by a tolerance zone parallel to a datum feature axis. The axis of a feature must lie in this zone, Fig. 9-36.

Perpendicularity tolerance. *Perpendicularity* describes how close a line or a surface is to being at a right angle to a datum plane or axis. Perpendicularity tolerance for a surface is specified by a zone confined by two parallel planes (or cylindrical tolerance zone) perpendicular to a datum plane or axis. The controlled surface of the feature must lie in this zone, Fig. 9-37. A method of tolerancing a cylindrical feature for perpendicularity is shown in Fig. 9-38.

Runout tolerance

Runout is a composite tolerance used to control those surfaces constructed around a datum axis and those at right angles to a datum axis. There are two types of runout control: circular and total.

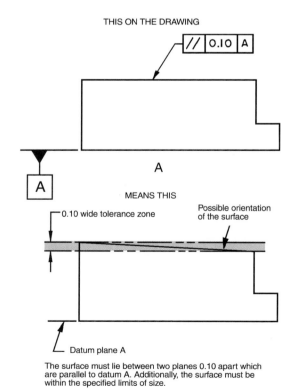

Fig. 9-35. Parallelism indicates how closely all elements of a surface are to being equidistant to a given datum surface. A–Parallelism can be indicated using a feature control frame and the parallelism symbol. A note can also be used to indicate parallelism. B–The interpretation of the example in A. (American National Standards Institute)

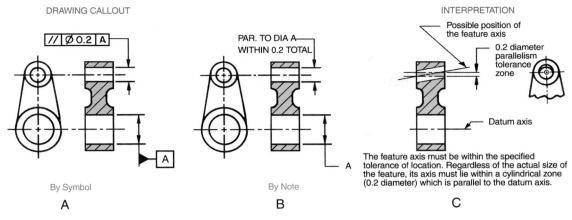

Fig. 9-36. A–Parallelism can be indicated for a cylindrical surface by a feature control frame and a symbol . B–A note can also be used to indicate parallelism. C–The interpretation of the examples in A and B. (American National Standards Institute)

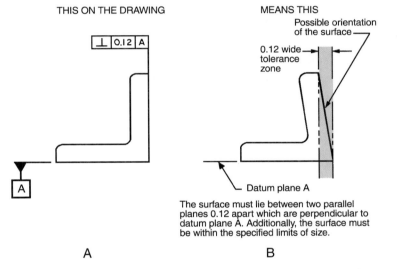

Fig. 9-37. Perpendicularity indicates how closely a surface is to being at a right angle to a given datum surface. A–A feature control frame and the perpendicularity symbol can be used. A note can also be used to indicate perpendicularity. B–The interpretation of the example in A. (American National Standards Institute)

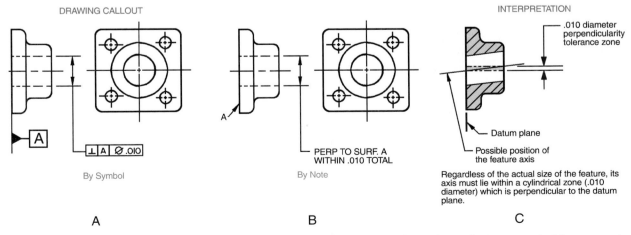

Fig. 9-38. A–Perpendicularity can be indicated for a cylindrical feature by using a feature control frame and a symbol. B–A note can also be used to indicate perpendicularity of a cylinder. C–The interpretation of the examples in A and B. (American National Standards Institute)

Circular runout tolerance. A *circular runout tolerance* is applied to features independently. Circular runout controls the elements of circularity and cylindricity of a surface. The measurement is taken as the part is rotated through 360°, Fig. 9-39B. Where applied to surfaces constructed at right angles to the datum axis, circular elements of a plane surface (wobble) are controlled. Where the runout tolerance applies to a specific portion of a surface, the extent is shown by a chain line adjacent to the surface profile.

Total runout tolerance. A *total runout tolerance* is applied to all (total) circular and profile surface elements. The part is rotated through 360° for the measurement of total runout. Total runout controls circularity and cylindricity for the part as a whole.

Round-off Rules for English and Metric System Conversions

The general rules for rounding off the converted millimeter and inch values in geometric tolerancing are shown in Fig. 9-40. Two methods of rounding tolerances may be used, depending on the degree of accuracy which must be maintained in the part.

Method A
Method A rounding involves rounding to the nearest rounded value of the limit so that, on the average, the converted tolerances are not significantly different from the original tolerances, Fig. 9-41. The limits converted by this method may, in some instances, be outside the original tolerance, Fig 9-41C. Where this variance is acceptable for interchangeability of parts, this method can be used.

Method B
Method B rounding is done systematically toward the interior of the tolerance zone. This means that the converted tolerances are never outside of the original tolerances, Fig. 9-42.

Where tolerance limits must be respected absolutely, "Method B" should be used. Parts manufactured to converted limits, but are to be inspected by means of original gauges, should have "Method B" rounding used.

Limit Dimension for Dual Dimensioned Drawings

Where tolerances are expressed on dual dimensioned drawings as limit dimensions, the dimension is expressed as a limit dimension for the decimal-inch system as well as the metric system, Fig 9-43. If a conversion is required, the methods described above should be used.

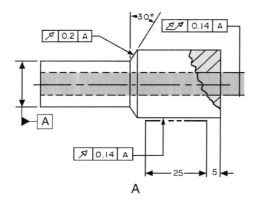

A

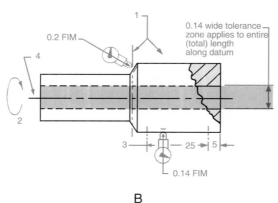

B

Fig. 9-39. A–Runout tolerance zones can be specified with a feature control frame and the runout symbol . B–The interpretation of the example in A . When a runout tolerance of 0.2 is given, for example, the full indicator movement (FIM) must not exceed 0.2 as the object is rotated through 360°.
(American National Standards Institute)

ROUND-OFF RULES		
Total Tolerance in Inches		**Millimeter Conversion Rounded to**
At least	Less than	
.00001	.0001	5 Decimal Places
.0001	.001	4 Decimal Places
.001	.01	3 Decimal Places
.01	.1	2 Decimal Places
.1	1.	1 Decimal Place
Total Tolerance in Millimeters		**Inch Conversion Rounded to**
.005	.05	5 Decimal Places
.05	.5	4 Decimal Places
.5	5.0	3 Decimal Places
5.0 and Over		2 Decimal Places

Fig. 9-40. There are general rules for rounding off converted millimeter and inch values.
(General Motors Drafting Standards)

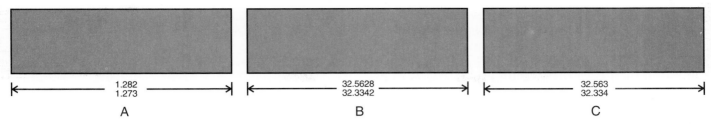

Fig. 9-41. When using "Method A" for rounding converted measurements, conventional rounding practices are used. However, in some cases this method can produce upper and lower dimensions that are outside of the original tolerance range. A–The original decimal-inch measurements. B–The converted millimeters. C–When the millimeters are rounded using "Method A," both the upper and lower limits are outside of the tolerance range. The total range in B is 0.2286 millimeters. In C the total range is 0.229, or 0.0004 millimeters larger than the original range.

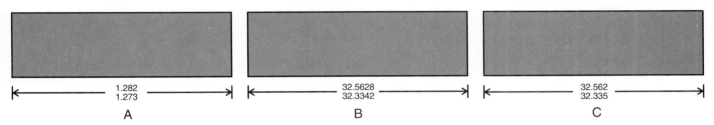

Fig. 9-42. When rounding converted measurements using "Method B," the numbers are always rounded to the inside of the tolerance range. A–The original decimal-inch measurements are given in. B–The converted measurements. C–When the millimeters are rounded using "Method B," both the upper and lower limits remain within the original tolerance range. The total range in B is 0.2286 millimeters. The total range in C is 0.227, or 0.0016 millimeters less than the original tolerance range.

Plus and Minus Tolerancing

The round-off practice described earlier cannot be applied directly to plus and minus tolerances. The inch dimensions and tolerance should first be changed into a limit dimension, then converted to millimeters as illustrated in Fig. 9-44.

$$\begin{bmatrix} 1.135 & 28.829 \\ 1.125 & 28.575 \end{bmatrix}$$

Fig. 9-43. In dual dimensioning, tolerances are expressed in both the decimal-inch and metric systems.

$$1.130 \pm .005 = \begin{bmatrix} 1.135 & 28.829 \\ 1.125 & 28.575 \end{bmatrix}$$

Fig. 9-44. When rounding a "plus and minus" dimension, the dimensions must first be converted to limit dimensions. After this, either "Method A" or "Method B" can be used to round the converted numbers.

DIMENSIONING FOR COMPUTER NUMERICAL CONTROLLED EQUIPMENT

Computer Numerical Control (CNC) is a system of controlling the movements of machines by using microprocessors (computers). Drills, mills, welding, flame cutting, filament winding, paint spraying, and drafting machines can all be controlled by computers, as well as many other machines.

The controlling computer processes numerical information. This information, called a **program,** represents all of the information contained on an engineering drawing. This information can be stored on computer disks, magnetic tapes, or optical disks. Many drawings completed using CADD systems can be "downloaded" directly to the CNC machine. This saves much time in the manufacturing process.

In the past, the information (program) was contained on **punch cards** or **punch tape.** The program information was contained as a series of holes punched in the tape or cards. This information was read through mechanical and electrical means, but no microprocessor (computer) was involved. These types of older machines are called simply **Numerical Controlled (NC)** machines. There was no microprocessor involved in the controlling of these machines.

CNC machines have rapidly replaced older NC machines. A CNC program is a translation of information shown on the drawing to numerical control, or "G code," language Fig. 9-45. The program is prepared by a **part programmer** who knows material characteristics, how to read drawings, machine processes, tool requirements, and the particular CNC machine operation and capabilities. Many drawings created on a CADD system can be downloaded directly to the CNC machine. This eliminates the need for a person to "translate" a drawing into "G code."

Dimensioning Requirements for CNC Operations

Dimensioning requirements for CNC machining are not significantly different from those required for manual machine operation. The drawing must be fully descriptive of the part and its features. Dimensional requirements must be in terms of distance from a point of origin (datum). The dimensions can also be expressed from point to point along two or three mutually perpendicular axes. The rectangular coordinate system of dimensioning should be used to facilitate programming time and to reduce the possibility of error.

The CNC machine receives its numerical instruction based upon the rectangular coordinates shown in Fig. 9-46. The point of origin is the **zero point** (reference plane) on the machine. This point is also called **machine zero.** This point should be identified on the CNC machine drawing. Occasionally, machine zero is located off of the left, front corner of the part. This will make all "X" and "Y" dimensions positive. However, any position can be defined as machine zero. Any position on the part can be described in terms of location on the "X," "Y," and "Z" axes, as defined from machine zero.

Within the rectangular coordinate system, two methods of indicating movement to points or positions are used: absolute (address) and incremental:

1. In the **Absolute Method,** machine positions are given in terms of distance from the origin of the coordinate axes (machine zero). This is similar to datum dimensioning.

Example: Refer to Fig. 9-47. With the center of the machine spindle or cutter positioned at point 1 and ready for movement to point 2, the absolute system command would be "move (2,2)." This is in terms of machine zero.

2. In the **Incremental Method,** machine positions are given in terms of distance from the present machine location. This is similar to chain dimensioning.

```
%
N001    T1M6
N002    G00X12.000Y8.000Z20.000M00
N003    G00X12.000Y12.000Z20.000S1200M03
N004    Z8.000F80
N005    Z8.500
N006    G00X14.000Y11.464
N007    Z8.000F80
N008    Z8.500
N009    G00X15.464Y10.000
N010    Z8.000F80
N011    Z8.500
N012    G00X12.000Y8.000Z20.000M00
N013    M02
```

Fig. 9-45. CNC machines use "G code" programming to define a tool path. In the program shown here, the numbers that follow the "X," "Y," and "Z" define the tool path in terms of an "absolute zero" position. This program will drill holes in three separate locations. Can you make a sketch locating the holes on a block that is 18″ x 14″ and 8 1/4″ thick? Do the holes go all the way through the part? (Hint: Line N002 is a tool change operation, not drilling.)

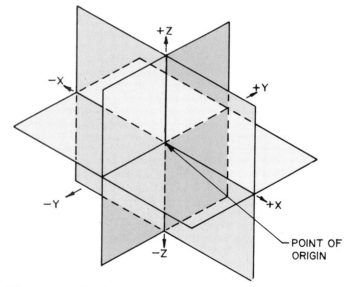

Fig. 9-46. CNC machine rectangular coordinates use three axes to fully describe any point. The intersection of these axes is the point of origin, also known as "machine zero."

Example: Refer back to Fig. 9-47. With center of machine spindle or cutter positioned at point 1 and ready for movement to point 2, the incremental system command would be "move (-1,-1)."

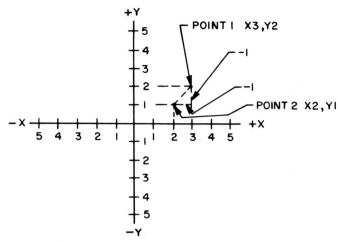

Fig. 9-47. The absolute method of movement defines a given point from "machine zero." The incremental method of movement defines a given point from the machine's current position. In this example, to move from point 1 to point 2 in the incremental method, the command would be "-1,-1."

Tolerancing Requirements for CNC Operations

Drawings for CNC operations should include tolerances based upon the design needs rather than CNC machine capabilities. CNC machines only recognize exact commands (dimensions). These commands give no tolerance. However, tolerances should be assigned.

The first reason tolerances should be assigned is that the accuracy will vary between CNC machines. This cannot be controlled by operator manipulation. Having a toleranced print gives the machine operator a guide for measuring the part. The operator will then have the same "range" for good parts that a manual machine operator would have.

The second reason for tolerancing drawings for CNC use is that the parts may need to be produced on conventional machines with manual operator. Occasionally a CNC machine, like any other machine, will break down. Instead of losing much time as "downtime," the parts might need to be machined by conventional methods. If the print is already toleranced, the time required to produce a toleranced print is saved. Manual machining requires a toleranced drawing or print.

Finally, if the drawing is toleranced, much flexibility in programming is possible. If the programming is done by manual means, the programmer can be flexible, since the part needs to be made to a "range" and not a specific size. Tooling, machine capabilities, or other factors may require that the programmer vary from the "basic" dimensions.

CNC Dimensioning Rules

To facilitate the use and accuracy of drawings in numerical controlled machining, the following rules should be observed:

1. Dimensioning must fully describe the part and all of its features.
2. Sufficient dimensions must be supplied to prevent the part programmer or inspector making assumptions or calculations.
3. All dimensions should be inch-decimal or metric units (or both).
4. Rectangular coordinate dimensions must be used to describe distances from datum planes or surfaces.
5. Angles should be specified by rectangular coordinate position dimensions.
6. Reference planes ("machine zero") may be common with the datum plane of drawing, or they may be assigned outside the part. Reference planes should be clearly marked on the drawing.
7. Tolerances should be based on design requirements rather than CNC machine capability.
8. Geometric position and form tolerances may be used with CNC machining.

CNC Design Considerations

The following design techniques are recommended for use on drawings for parts and assemblies regardless of manufacturing method to be employed (courtesy of Allis-Chalmers, *Engineering Standards*):

1. Draw principally machined surfaces as the primary (front) view. This will keep the drawing and major machining operation in the same relationship.
2. Design for symmetry so that initial programming steps can be reused.
3. As many part features as practical should be located on a common "X" or "Y" axis. Two or more holes, bolt circles, slots, or other features located on a common centerline can reduce total programmed movement with less chance of error than random location.
4. Aim for a minimum number of machine setups. Wherever practical, design for machining from one side only.
5. Drilled or tapped holes, and counterbore depths should always be dimensioned from a finished surface (datum). A finished surface should be provided on rough castings.

6. Avoid blind-tapped holes and back spotfacing operations. This will eliminate "downtime" to clear chips and to add special cutters.

7. A preferred tool list should be followed whenever possible. For example, "Standard 534" lists preferred sizes of tap drills. This recommendation is based on a reduction of tool inventories to essential items compatible with machine capacities and consistent with economical design considerations. It is not practical to call for .250-20, .312-18, and .375-16 tapped holes in the same part when one size could satisfy all three applications.

8. Reduce the number of different hole sizes. A CNC machine can store a certain number of different tools. Operations beyond this number require special handling.

9. Specify the same radius for fillets and rounds where possible. This will reduce the number of different cutting tools required.

NOTE: Consideration should be given to standardized design characteristics that can be programmed on a reusable basis. Specific families of parts that follow an established set of rules are adaptable to computer programming where optimum design can be selected electronically. A typical flow of CNC information is shown in Fig. 9-48.

PROBLEMS AND ACTIVITIES

The following problems provide you with an opportunity to apply geometric dimensioning and tolerancing principles.

Bilateral Tolerancing and Limit Dimensioning

1. Draw the front view and a left-side section of the "DIFFERENTIAL SPIDER" shown in Fig. 9-49. Dimension using bilateral tolerances. Delete the note on tolerances, but include the other notes on the drawing in an appropriate space.

2. Draw two views of the "SHAFT BEARING CARTRIDGE" shown in Fig. 9-49. Include the right-side view as a section. Use bilateral tolerances and identify the surface roughness as indicated.

3. Draw two views of the "TRUNNION IDLER" shown in Fig. 9-49. Make the upper portion of the profile view a broken-out section to show the hole detail. Show bilateral toleranced dimensions as limit dimensions and tolerance of other dimensions as a note. Include all other notes on the drawing as well.

4. Draw the "SLIDE NUT" shown in Fig. 9-50 as a full section. Show threads in simplified form and dimension the drawing using limit dimensions.

Reading Geometric Dimensions and Tolerances

5. Refer to the drawing in Fig. 9-50 and answer the following questions:
   A. What is the name of the part and the drawing number?
   B. What feature of the part is datum "A?"
   C. Interpret the feature control symbol at B .
   D. What type of dimension is at X ? What relationship must exist on the finished part between the surface at Y and datum "A?"
   E. What relationship is specified for feature Z and datum "A"?
   F. Which feature calls for the smoother surface texture, V or W? Explain.

6. Refer to the drawing in Fig. 9-51 and answer the following questions:
   A. Give the name and number of the part.
   B. What general tolerances are specified for the part?
   C. What surface roughness is called for at X ? What surface roughness is called for at Z ?
   D. What specification is given for hole "D?"
   E. Identify datums "A," "B," and "C."
   F. What relationship must exist between hole "B" and datums "A," "B," and "C?"
   G. What relationship is specified between hole "D" and hole "C?"
   H. State the dimension at Y as a limit dimension.

Geometric Dimensioning and Tolerancing

7. Draw the necessary orthographic views of the "BEARING SUPPORT" shown in Fig. 9-50. Include a sectional view. Delete the notes on geometric tolerancing and include these as "feature control symbols" on the drawing.

8. Draw the necessary views of the "SLIDE NUT" shown in Fig. 9-50. Dimension the drawing in metric. Change the general tolerance note to metric.

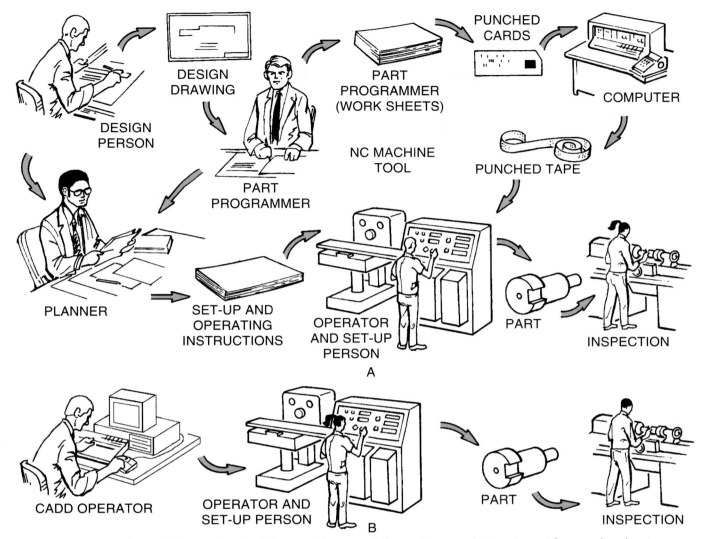

Fig. 9-48. A–The flow of information for NC operations went through several steps to get from a drawing to a manufactured part. B–The flow of information for CNC operations is greatly reduced.

9. Draw the necessary views of the objects shown in Fig. 9-53. Dimension the drawing using geometric "feature control symbols" to replace notes where possible.

Dimensions for CNC operations

10. Draw the necessary views of the "LIMIT SWITCH MOUNTING PLATE" shown in Fig. 9-54. Dimension the drawing for incremental CNC machining.

11. Draw the necessary views of the "MOUNTING PLATE" shown in Fig. 9-54. Dimension the drawing for the absolute method of CNC machining using ordinate dimensioning.

SELECTED ADDITIONAL READINGS

1. *Dimensioning and Tolerancing for Engineering Drawings, ASME Y14.5M,* American National Standards Institute, 1430 Broadway, New York, NY 10018.

2. *Drawing Requirements Manual, Mil-D-1000/MilStd-100,* Global Engineering Documentation Services, Inc., P.O. Box 2060, Newport Beach, CA 92660.

3. *Geometric Dimensioning and Tolerancing,* Madsen, David A., The Goodheart-Willcox Co., Inc., 123 West Taft Drive, South Holand, Il 60473-2089 1-800-323-0440

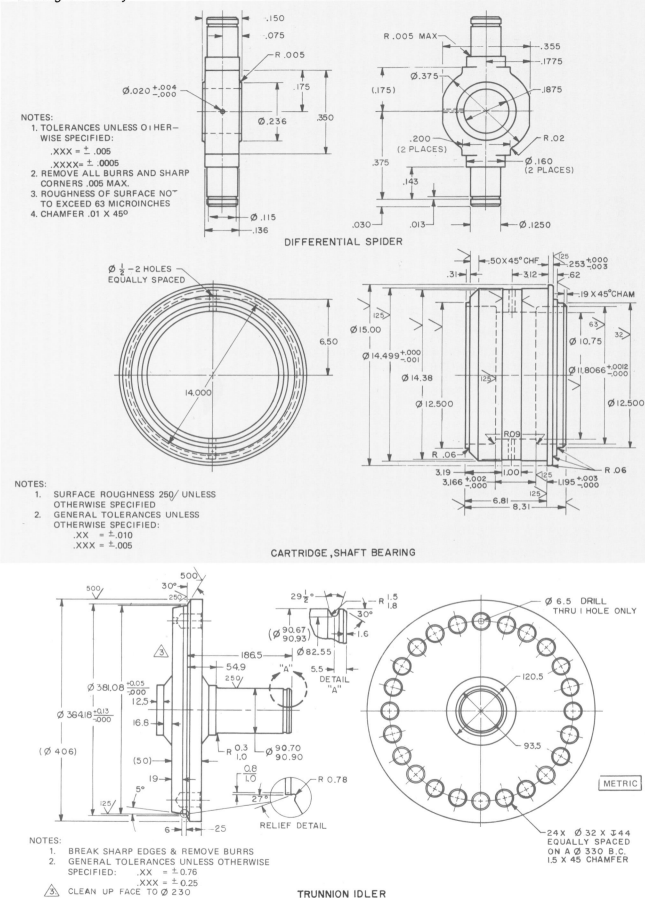

Fig. 9-49. Use these drawings for the problems on bilateral dimensioning. Follow the instructions given in the Problems and Activities section and by your instructor.

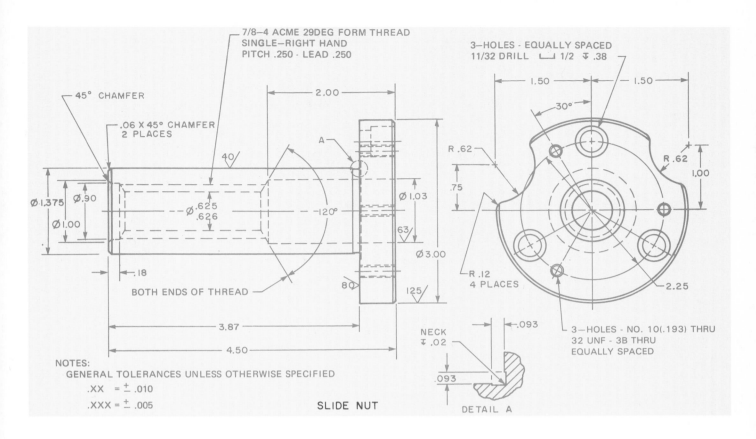

7/8-4 ACME 29DEG FORM THREAD
SINGLE-RIGHT HAND
PITCH .250 - LEAD .250

3-HOLES - EQUALLY SPACED
11/32 DRILL ⌴ 1/2 ⌁ .38

45° CHAMFER

.06 X 45° CHAMFER
2 PLACES

A

2.00

1.50   1.50

30°

R .62   R .62

40/   Ø 1.03   .75   1.00

Ø 1.375   Ø .90   Ø .625
                     .626   -120°   63/   R .12
Ø 1.00   4 PLACES
                                    Ø 3.00   2.25

.18   80/   125/

BOTH ENDS OF THREAD

3.87

4.50

NECK   .093
⌁ .02   3-HOLES - NO. 10(.193) THRU
                           32 UNF - 3B THRU
                           EQUALLY SPACED

NOTES:
GENERAL TOLERANCES UNLESS OTHERWISE SPECIFIED
.XX = ± .010
.XXX = ± .005   SLIDE NUT   .093

                                        DETAIL A

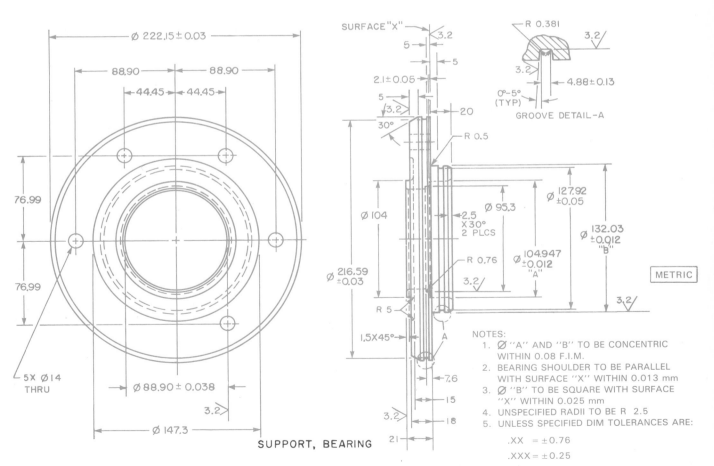

Ø 222.15 ± 0.03

88.90   88.90

44.45   44.45

SURFACE "X"   3.2

5   R 0.381   3.2
       5
2.1±0.05   3.2   4.88±0.13
5   0°-5°
3.2/   20   (TYP)
                                 GROOVE DETAIL-A
30°

76.99   R 0.5

                    Ø 104   Ø 127.92
                              ±0.05
                    2.5
                    X30°   132.03
                    2 PLCS   ±0.012
76.99   "B"
                    R 0.76
         Ø 216.59   Ø 104.947
         ±0.03   ±0.012
                          "A"   3.2/
         R 5

5X Ø14   1.5X45°   A
THRU
         Ø 88.90 ± 0.038   7.6

                    3.2/   15

         Ø 147.3   18

         SUPPORT, BEARING   21

METRIC

NOTES:
1. Ø "A" AND "B" TO BE CONCENTRIC
   WITHIN 0.08 F.I.M.
2. BEARING SHOULDER TO BE PARALLEL
   WITH SURFACE "X" WITHIN 0.013 mm
3. Ø "B" TO BE SQUARE WITH SURFACE
   "X" WITHIN 0.025 mm
4. UNSPECIFIED RADII TO BE R 2.5
5. UNLESS SPECIFIED DIM TOLERANCES ARE:

   .XX = ±0.76
   .XXX = ±0.25

Fig. 9-50. Use these drawings for the problems on limit dimensioning and geometric dimensioning and tolerancing problems. Follow the instructions given in the Problems and Activities section and by your instructor.

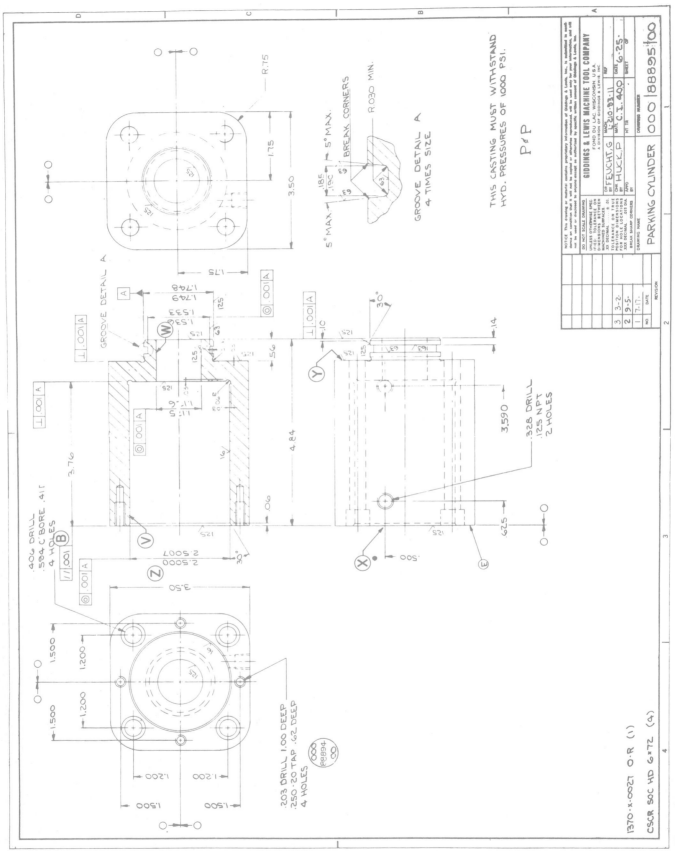

Fig. 9-51. Use this industry print for the problems on reading geometric dimensioned and toleranced drawings. Follow the instructions given in the Problems and Activities section and by your instructor.

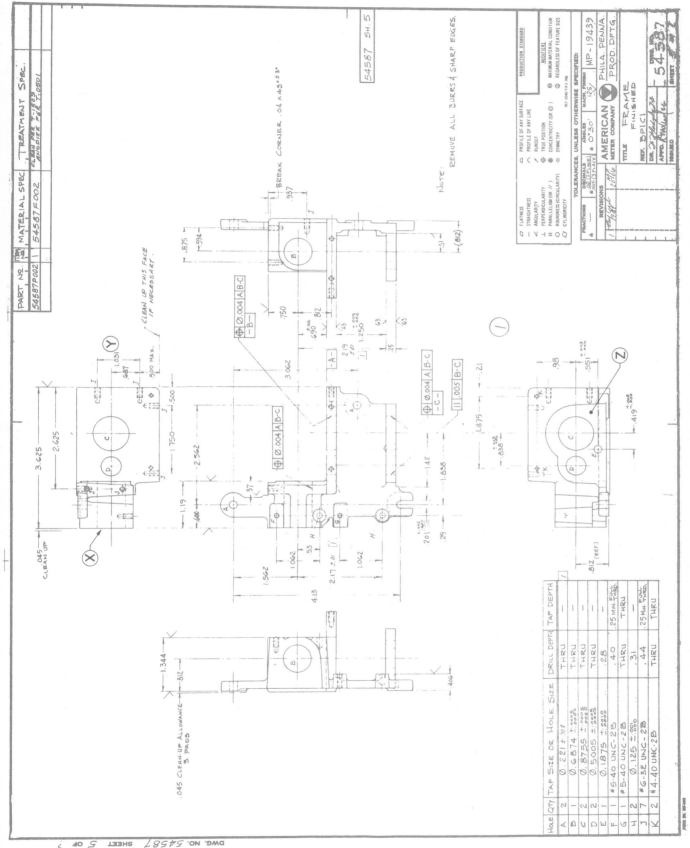

Fig. 9-52. Use this industry print for the problems on reading geometric dimensioned and toleranced drawings. Follow the instructions given in the Problems and Activities section and by your instructor.

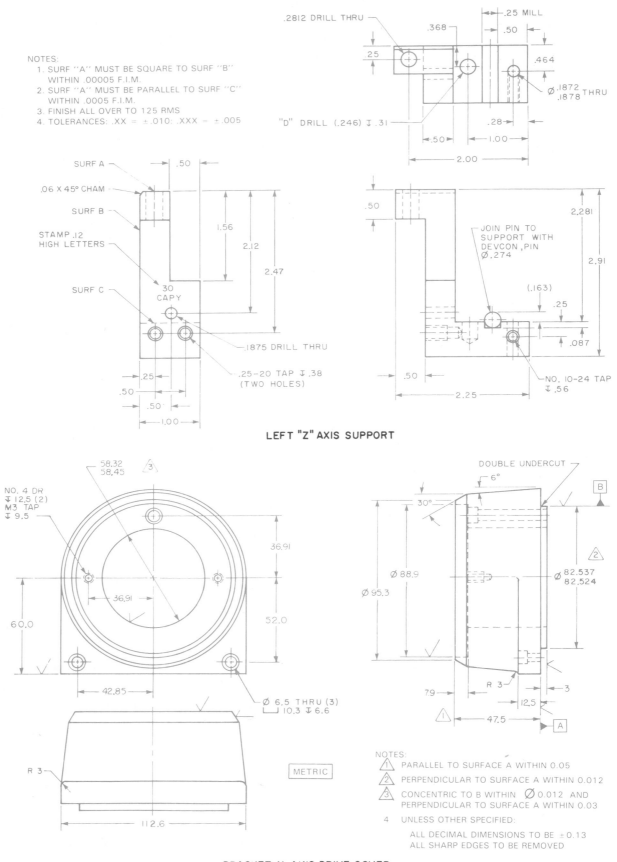

NOTES:
1. SURF "A" MUST BE SQUARE TO SURF "B" WITHIN .00005 F.I.M.
2. SURF "A" MUST BE PARALLEL TO SURF "C" WITHIN .0005 F.I.M.
3. FINISH ALL OVER TO 125 RMS
4. TOLERANCES: .XX = ±.010; .XXX = ±.005

.2812 DRILL THRU

.25 MILL

.50

.368

.25

.464

Ø.1872 / .1878 THRU

"D" DRILL (.246) ⊥ .31

.28

.50

1.00

2.00

SURF A

.50

.06 X 45° CHAM

SURF B

STAMP .12 HIGH LETTERS

SURF C

1.56

2.12

2.47

30 CAP ⦵

.1875 DRILL THRU

.25

.50

.50

1.00

.25-20 TAP ⊥ .38 (TWO HOLES)

.50

2.281

JOIN PIN TO SUPPORT WITH DEVCON, PIN Ø.274

2.91

(.163)

.25

.087

.50

2.25

NO. 10-24 TAP ⊥ .56

LEFT "Z" AXIS SUPPORT

58.32 / 58.45   ⟨3⟩

NO. 4 DR ⊥ 12.5 (2) M3 TAP ⊥ 9.5

36.91

36.91

60.0

52.0

42.85

Ø 6.5 THRU (3) ⊔ 10.3 ⊥ 6.6

R 3

112.6

METRIC

DOUBLE UNDERCUT

6°

B

30°

Ø 88.9

Ø 95.3

Ø 82.537 / 82.524

⟨2⟩

R 3

7.9

12.5

3

47.5

A

⟨1⟩

NOTES:
⟨1⟩ PARALLEL TO SURFACE A WITHIN 0.05
⟨2⟩ PERPENDICULAR TO SURFACE A WITHIN 0.012
⟨3⟩ CONCENTRIC TO B WITHIN Ø 0.012 AND PERPENDICULAR TO SURFACE A WITHIN 0.03
4 UNLESS OTHER SPECIFIED:
   ALL DECIMAL DIMENSIONS TO BE ±0.13
   ALL SHARP EDGES TO BE REMOVED

BRACKET, Y-AXIS DRIVE COVER

Fig. 9-53. Use these drawings for the problems on precision dimensioning. Follow the instructions in the Problems and Activities section and by your instructor.

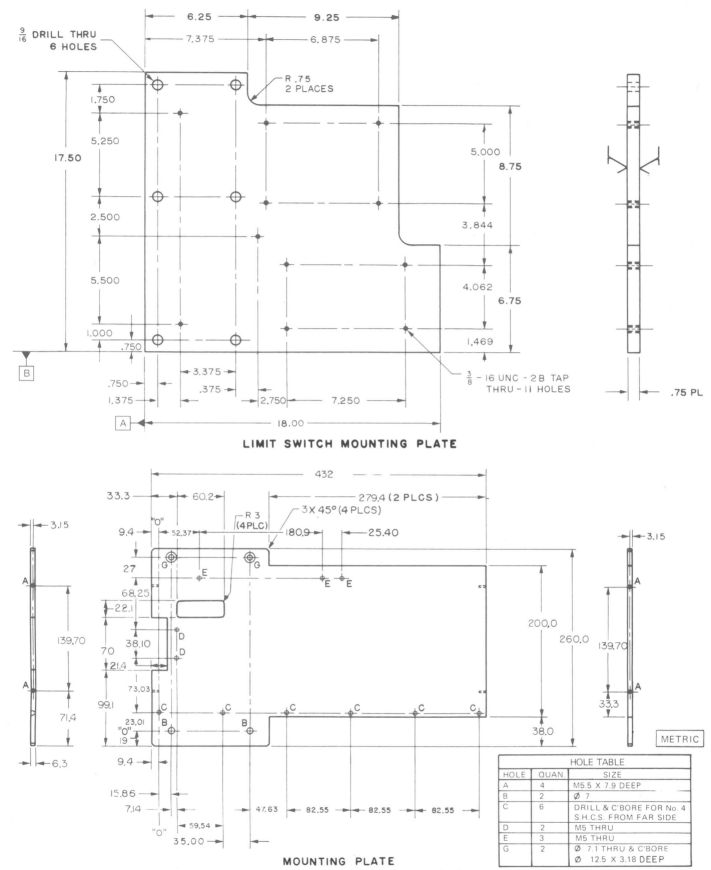

Fig. 9-54. Use these drawings for the problems on dimensioning for CNC operations. Follow the instructions in the Problems and Activities section and by your instructor.

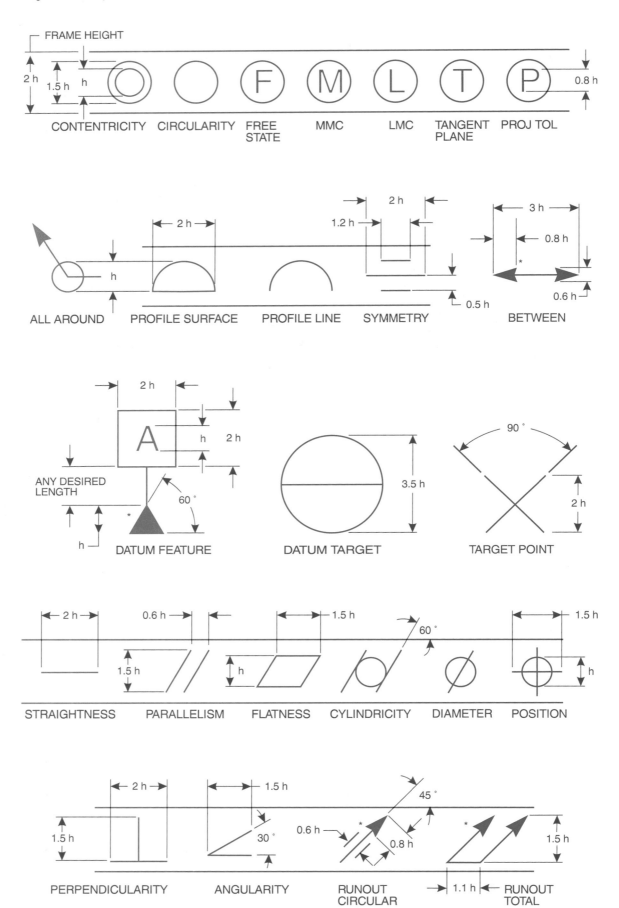

The proper construction of various geometric dimensioning and tolerancing symbols.

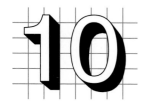

Section Views

KEY CONCEPTS

☐ Visualize a section view along a cutting plane.
☐ Construct section views.
☐ Draw various types of sections.
☐ Apply conventional drafting practices in sectioning.

A **section view** is a view "seen" beyond an imaginary cutting plane. The plane passes through an object at right angles to the direction of sight, Fig. 10-1A. Section views are used to show interior construction, or details of hidden features, that can-

not be shown clearly by outside views and hidden lines, Fig. 10-1B. Section views are also used to show the shape of exterior features. Examples of these types of section views include automobile body components, airplane wings, and fuselage sections.

SECTION CUTTING-PLANE LINES

The cutting plane of a section is indicated by lines called **cutting-planes lines.** (Note that *cutting-plane lines* should not be confused with a *cutting plane.* You can easily tell the difference by the

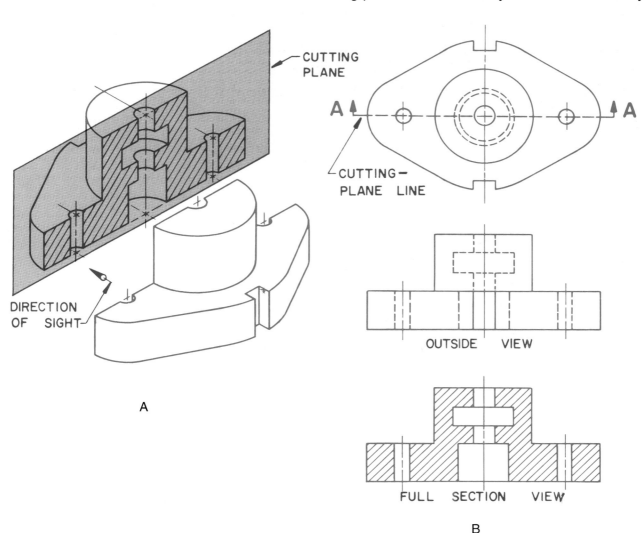

Fig. 10-1. Section views show the construction details and shape of parts.

hyphen in *cutting-plane lines.*) Cutting-plane lines are usually a series of heavy dashes, each about 1/4 inch long, Fig. 10-2A. However, a heavy line of alternating long dashes 3/4 to 1 1/2 inches long with a pair of short dashes 1/8 inch long spaced 1/16 inch apart can also be used. Some industries favor a simplified representation of the cutting-plane line that includes only the ends, Fig. 10-2B.

For objects having one major centerline, the cutting-plane line can be omitted if the section is clearly along the centerline. Arrowheads at the ends of cutting-plane lines are used to indicate the direction in which the sections are viewed, Fig. 10-2B. The cutting plane may be bent or offset to show details of hidden features to better advantage, Fig. 10-2C.

PROJECTION AND PLACEMENT OF SECTION VIEWS

Whenever possible, a section view should be projected from, and perpendicular to, the cutting plane. The section view should be placed behind the arrows, Fig. 10-3. This arrangement should be maintained whether the section is adjacent to, or removed some distance from, the cutting plane.

SECTION LINING

The exposed (cut) surface of the section view is indicated by **section lines.** Section lines are sometimes called **crosshatching.** These lines emphasize the shape of a detail, or differentiate one part from another.

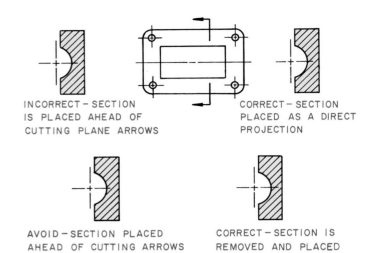

INCORRECT – SECTION IS PLACED AHEAD OF CUTTING PLANE ARROWS

CORRECT – SECTION PLACED AS A DIRECT PROJECTION

AVOID – SECTION PLACED AHEAD OF CUTTING ARROWS

CORRECT – SECTION IS REMOVED AND PLACED BEHIND CUTTING–PLANE ARROWS

Fig. 10-3. The section view should be placed behind the cutting-plane arrows.

Section lines are thin, parallel lines drawn with a sharp pencil or No. 0 (0.3 mm) pen. Section lines should not be drawn around dimensions that are required inside of the section view. This will provide clarity to the drawing.

Direction of Section Lines

Section lines should be drawn at an angle of 45° to the main outline of the view, Fig. 10-4A. On adjacent parts, section lines should be 45° in the opposite direction, Fig. 10-4B. For a third part adjacent part, section lines should be drawn at an angle of 30° or 60°, Fig. 10-4C. Where the angle of section lining is parallel, or nearly parallel, with the outline of the part, a different angle should be chosen, Fig 10-4D. Section lines should not be intentionally drawn to meet at common boundaries.

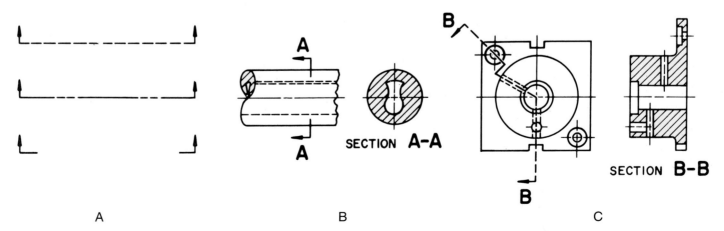

SECTION **A-A**

SECTION **B-B**

A   B   C

Fig. 10-2. A–Section cutting-plane lines have arrowheads on the terminal ends to indicate the direction of viewing. B–Some industries prefer cutting-plane lines that only show the ends and arrowheads. C–Sometimes cutting-plane lines are "bent" through several different planes to show many different features.

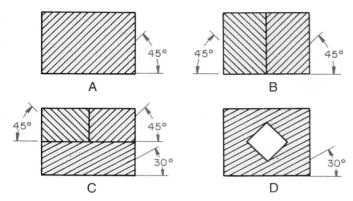

Fig. 10-4. A–Section lines should be drawn at 45° angle to the object lines. B–When two parts are adjacent to each other, the section lines should be drawn at opposite 45° angles. C–If three parts are adjacent to each other, the third part should have section lines at 30°. D–If the object lines are parallel, or nearly parallel, to 45°, a different angle should be used.

Spacing of Section Lines

Section lines should be uniformly spaced throughout the section, Fig. 10-5. However, spacing may be varied according to the size of the drawing. Section lines should be spaced approximately 1/8 to 3/16 inch apart. Refer to Chapter 12

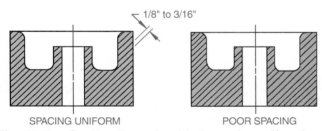

Fig. 10-5. Section lines should always be uniformly spaced.

Reproducing and Storing Drawings for the spacing of section lines on drawings to be microfilmed.

In general, spacing should be as wide as possible to save time. This will also improve the appearance of the drawing. Section lines should end at the visible outline of the part without gaps or overlaps.

Hidden Lines Behind the Cutting Plane

Hidden lines behind the cutting plane should be omitted unless needed for clarity, Fig. 10-6A. In half sections, hidden lines are shown on the unsectioned half *only* if needed for dimensioning or clarity on the drawing, Fig. 10-6B.

Visible Lines Behind the Cutting Plane

In general, visible lines behind the cutting plane are shown. However, they may be omitted where the section view is clear without them, Fig. 10-7. If omitting the lines will not save a large amount of time, they should be left in.

TYPES OF SECTION VIEWS

Many types of sections have been adopted as standard sectioning procedure. Each type has a unique function in drafting. The different types of sections and their uses are discussed in the following sections.

Full Section Views

The cutting plane passes entirely through an object in a *full section view.* The cross section behind the cutting plane is exposed to view, Fig. 10-8. The

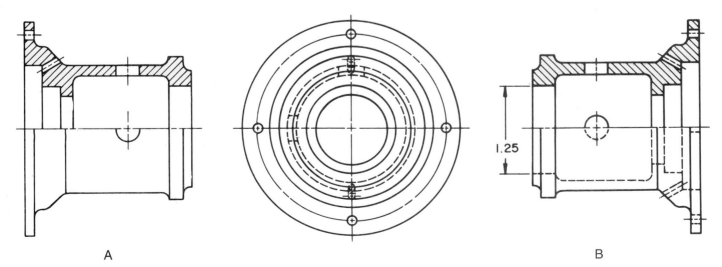

Fig. 10-6. A–Hidden lines behind the cutting plane are typically left out unless needed for clarity. B–In half section views, hidden lines should be included on the unsectioned half if needed for clarity.

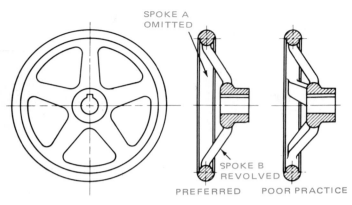

Fig. 10-7. If the view will still be clear, visible lines behind the cutting plane can be omitted. Sometimes it is necessary to rotate parts of the object in the section view so that the clarity of the drawing is improved. (American National Standards Institute)

cutting-plane line and section title may be omitted if the section view is in orthographic projection position. A full section view usually replaces an exterior view in order to show interior features.

Half Section Views

A *half section view* of a symmetrical object shows the internal and external features in the same view, Fig. 10-9. Two cutting planes are passed at right angles to each other along the centerlines or symmetrical axes, 10-9A. One-quarter of the object is "removed" and a half-section view is exposed. The cutting-plane lines and section titles are omitted. A half section is used to show both the interior and exterior features of a symmetrical object in a single view, Fig. 10-9B.

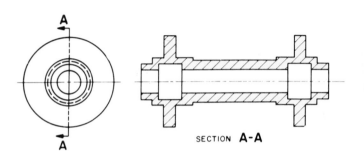

Fig. 10-8. The cutting plane in a full section view passes through the entire object.

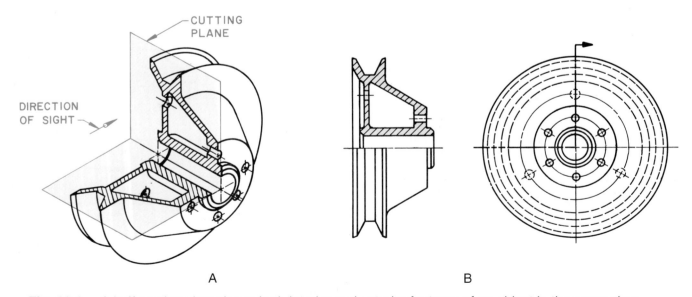

Fig. 10-9. A half section view shows both interior and exterior features of an object in the same view.

Revolved Section Views

A *revolved section view* is obtained by passing a cutting plane perpendicularly through the centerline or axis of the part to be sectioned. The resulting section is "revolved" 90° in place, Fig. 10-10. The cutting-plane line is omitted for symmetrical sections. Visible lines may be removed on each side of the section and break lines used for clarity. A revolved section is used to show the true shape of the cross section of an elongated object such as a bar. Revolved sections are also used to show a feature of a part such as a rib, arm, or linkage.

Removed Section Views

A *removed section view* is one that has been moved out of its normal projected position in the standard arrangement of views, Fig. 10-11. The removed section should be labeled "SECTION A-A" and placed in a convenient location on the same sheet. If the removed section must be located on another sheet of a multiple-sheet drawing, appropriate identification and zoning references should be made. A removed section is similar to a revolved section, except that the cross section is removed from the actual view of the part.

Offset Section Views

The cutting plane for an *offset section view* is not one single plane. The plane is stepped, or offset, to pass through features which lie in more than one plane, Fig. 10-12. The path of the cutting plane is shown on the view to be sectioned. The features are drawn in the section view as if they were in one plane. Offsets in the cutting plane are not shown in

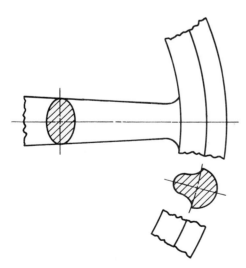

Fig. 10-10. A revolved section view shows the cross section of a feature. The view is rotated in place to show the profile. A revolved view can be shown in a break or inside of the object lines.

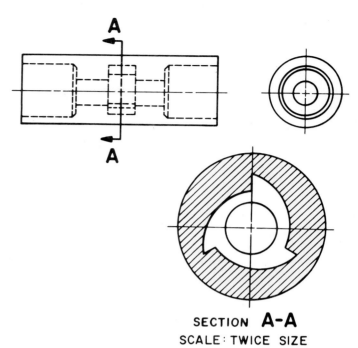

Fig. 10-11. A removed section is not drawn in the normal location. The removed section view can be drawn at a different scale if necessary to show detail. However, if the scale is different, the new scale must be indicated.

the section view. An offset section view is useful when section views of features that lie in more than one plane are needed.

Broken-out Section Views

A *broken-out section view* appears in place on the regular view. The partial section is limited by a break line, Fig. 10-13. Broken-out sections are used to show interior detail of objects where less than a half section is required to convey the necessary information.

Aligned Section Views

Certain objects may be misleading when a true projection is made. In an *aligned section view,* features such as spokes, holes, and ribs are drawn as if rotated into, or out of, the cutting plane, Fig. 10-14. Aligned section views are used when actual or true projection would be confusing. Note in Fig. 10-14 that the offset features have been rotated to align with the centerline and projected to the section view for clarity. If these features were projected directly, their lengths and locations would be distorted. (Also refer back to Fig. 10-2C.)

Thin Section Views

Structural shapes, sheet metal, and packing gaskets are often too thin for section lining. *Thin*

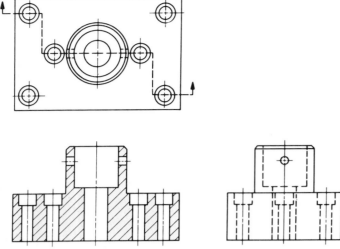

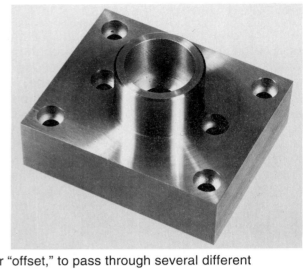

Fig. 10-12. The cutting plane in an offset section view is bent, or "offset," to pass through several different features that do not lie in the same plane.

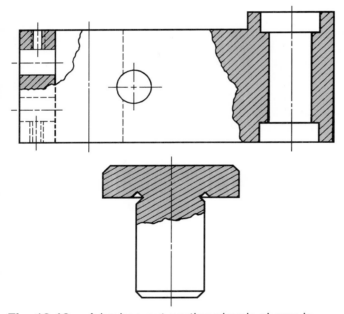

Fig. 10-13. A broken-out section view is shown in place on the regular view. The section view is limited by break lines.

A

A

SECTION **A-A**

Fig. 10-14. An aligned section view has features that are not shown in their true position. Features such as spokes are shown rotated to provide a clear drawing.

section views can be shown solid, Fig. 10-15. Where two or more thicknesses are shown, a space should be left between the thicknesses. The space should be large enough to be acceptable for microfilm reproductions.

Auxiliary Section Views

An *auxiliary section view* is an auxiliary view that has been sectioned, Fig. 10-16. The section should be shown in its normal auxiliary position. If necessary, the auxiliary section should be identified with letters and a cutting-plane line. For information

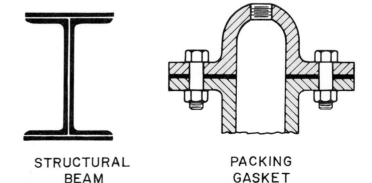

STRUCTURAL
BEAM

PACKING
GASKET

Fig. 10-15. Thin materials that are sectioned are shown solid. Where two or more thin materials are next to each other, a space should be left between them. This space should be large enough to reproduce on microfilm.

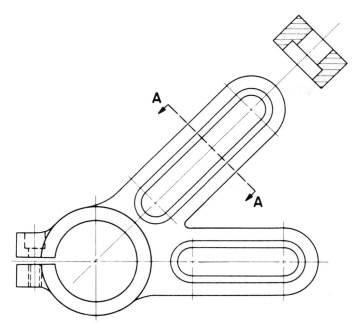

Fig. 10-16. An auxiliary section view is an auxiliary view that is sectioned. The view appears in the normal position for an auxiliary view.

on the construction of auxiliary views, see Chapter 13. The auxiliary section is used to add clarity to critical areas of a drawing.

Partial Section Views

Partial section views are used to show details of objects without drawing complete conventional views, Fig. 10-17.

Phantom Section Views

The *phantom (hidden) section view* is used to show interior construction while retaining the exterior detail of a part. This type of section view is not used much in industry. Two examples of phantom section views are shown in Fig. 10-18. This type of section is also used to show the positional relationship of an adjacent part. The section lines in a

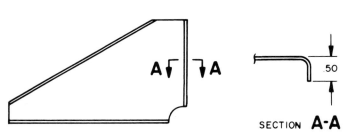

Fig. 10-17. Partial section views save drafting time. Partial section views are used when information can be accurately communicated without drawing a complete view. (General Dynamics, Engineering Dept.)

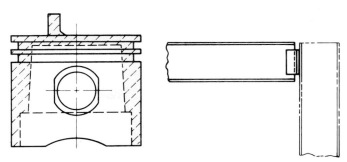

Fig. 10-18. Phantom section views are used to show both the visible and internal features on the same view. This type of section view is not used frequently in industry.

phantom section view are typically a series of short dashes.

Unlined Section Views

Standard parts such as bolts, nuts, rods, shafts, bearings, rivets, keys, pins, and similar objects should not be sectioned if their axis lies in the cutting plane. This is done for clarity on the drawing. (When the axis of these type parts lie at right angles to the cutting plane, the parts should be sectioned.) Refer to Fig. 10-19 and Fig. 10-20.

CONVENTIONAL PRACTICES

Certain practices have become standard procedure for section views. The following practices should be followed, in combination with the practices discussed for specific types of sections.

Rectangular Bar, Cylindrical Bar, and Tubing Breaks

For a long bar, or tubing, with a uniform cross section, it is usually not necessary to draw its full

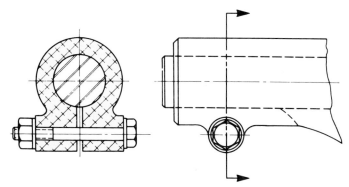

Fig. 10-19. When a shaft is at a right angle to the cutting plane, the shaft is sectioned. If a shaft is parallel to the cutting plane, it is not sectioned.

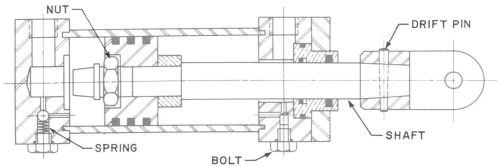

Fig. 10-20.   Shafts, bolts, nuts, and pins all should only be sectioned if they are perpendicular to the cutting plane.

length. Often, the piece is drawn to a larger scale and a break made. The break is sectioned as shown in Fig. 10-21. The true length of the piece is indicated by a dimension, Fig. 10-22.

The conventional breaks for cylindrical bars and tubing are known as "S" breaks. These breaks can be drawn with a template, or constructed as follows.

1. Draw the rectangular view for the bar or tube, Fig. 10-23A.
2. Lay off fractional radius widths on the end to be sectioned.

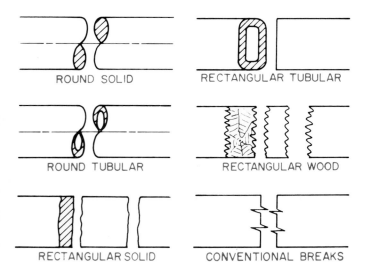

Fig. 10-21.   When sectioning breaks in bar and tubing stock, there are accepted conventions according the stock.

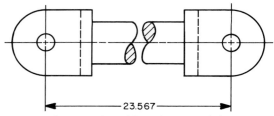

Fig. 10-22.   Conventional breaks permit long, uniformly shaped objects to be drawn to a scale large enough to present details clearly.

3. Scribe 30° construction lines to locate radii centers A, B, C, and D, Fig. 10-23B. (For a wider sectioned face, use a 45° projection on four angles adjacent to the centerline.)
4. Set the compass on the radius center and adjust it so the arc passes through center point P, stopping short of outside edge, Fig. 10-23C. Only three radii are used, depending on placement of sectioned face.
5. Complete the ends of the "S" curve by freehand, Fig. 10-23D.
6. Draw the inside curve by freehand when tubing is being represented.
7. Add section lining to the visible sectioned part, Fig. 10-23E and Fig. 10-23F.

NOTE: When an "S" break is shown with stock continuing on both sides of the break, the sectioned faces are diagonally opposite. (Refer back to Fig. 10-21). The "S" break can also be drawn freehand by estimating the width of the "S" and crossing at the centerline.

Intersections in Section Views

When a section is drawn through an intersection, and the offset or curve of the true projection is small, the intersection can be drawn conventionally without an offset or curve, Fig. 10-24A and Fig. 10-24C. Intersections of a larger configuration can be projected or approximated by circular arcs, Fig. 10-24B and Fig. 10-24D.

Ribs, Webs, Lugs, and Gear Teeth in Section Views

Ribs and webs are used to strengthen machine parts, Fig. 10-25A. When the cutting plane extends along the length of a rib, web, lug, gear tooth, or similar flat element, the element is not sectioned. This will avoid a false impression of thickness or mass.

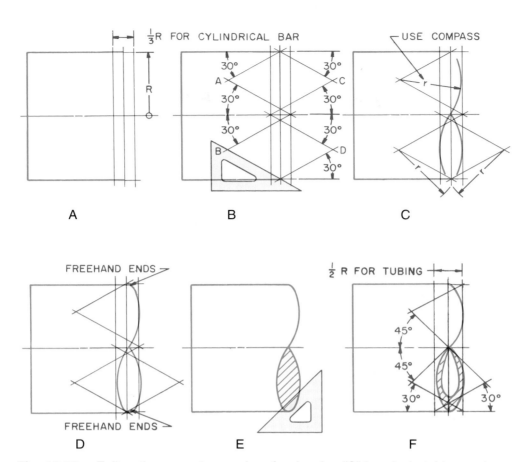

Fig. 10-23. Follow the correct procedure for drawing "S" breaks in tubing and round bar stock.

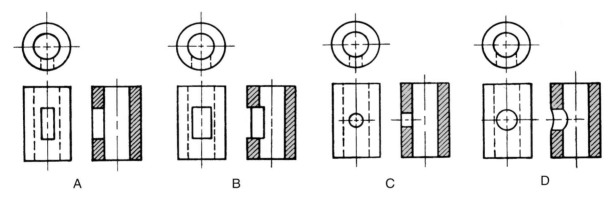

Fig. 10-24. When the offset of the true projection is small, the intersection can be drawn without the offset (A and C). When the offset of the true projection is large, the projection should be approximated with circular arcs (B and D). (American National Standards Institute)

An alternate method of section lining is shown in Fig. 10-25B. The spacing is twice that of the regular section. This spacing is used where the actual presence of the flat element is not sufficiently clear without section lining.

When the cutting plane cuts an element crossways, the element is sectioned in the usual manner, Fig 10-25C. Gear teeth are *not* sectioned when the cutting plane extends through the *length* of the tooth, Fig. 10-25D. Gear teeth *are* sectioned when the cutting plane cuts *across* the teeth. For example, the profile view of a worm gear should be sectioned.

Outline Sectioning

If the drawing remains clear, section lines should be shown only along the borders of the part, Fig. 10-26. This is known as **outline sectioning.**

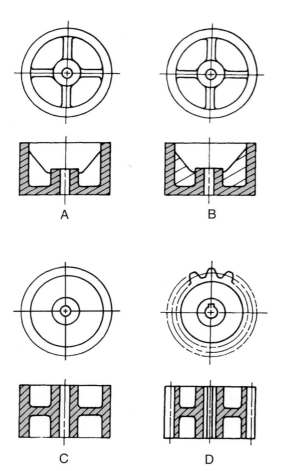

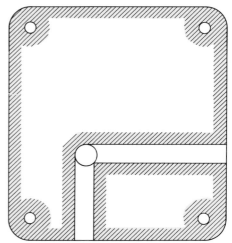

Fig. 10-26. Outline sectioning should be used for large parts where all of the necessary information will still be communicated.

Fig. 10-25. A–Ribs and webs should not be sectioned when they fall on the cutting plane. B–However, an alternative method of sectioning ribs and webs is sometimes used. C–When the cutting plane cuts the element crossways (perpendicular to the plane) the element should be sectioned normally. D–When the cutting plane passes through the end of a gear tooth, the tooth should not be sectioned.

This convention is particularly useful on large parts where considerable time would be required to draw section lines.

Section Titles

Letters used to identify the cutting plane are included in the title of the section, such as "SECTION A-A" or "SECTION B-B." When the single alphabet is exhausted, multiples of letters may be used, such as "SECTION AA-AA" or "SECTION BB-BB." The section title always appears directly under the section view.

When a section view is located on a sheet other than the one containing the cutting plane indication, the sheet number and zone of the cutting plane indication should be referenced for easy location. A similar cross reference should be located on the view containing the cutting-plane line.

Scale of Section Views

Preferably, sections should be drawn to the same scale as the outside view they are taken from. When drawn to a different scale, the scale should be specified directly below the section title, Fig. 10-27.

MATERIAL SYMBOLS IN SECTION VIEWS

The symbolic cross sectioning for various materials is shown in Fig. 10-28. The general purpose (cast-iron) section lining should be used for all materials, except parts made of wood. However, on detail and assembly drawings of multi-material parts, the appropriate material symbols should be used. This calls attention to the different materials that the parts are made of. (Refer back to Fig. 10-19 and Fig. 10-20.)

However, symbolic sectioning serves no practical purpose on many drawings. Therefore, the general purpose symbol should be used whenever possible. The general purpose symbol requires less time to construct. Also, the materials, processes, and protective treatment necessary to meet the

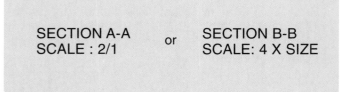

SECTION A-A
SCALE : 2/1   or   SECTION B-B
SCALE: 4 X SIZE

Fig. 10-27. When the scale of a section view is different from the scale of the rest of the drawing, the scale should be indicated directly below the section title.

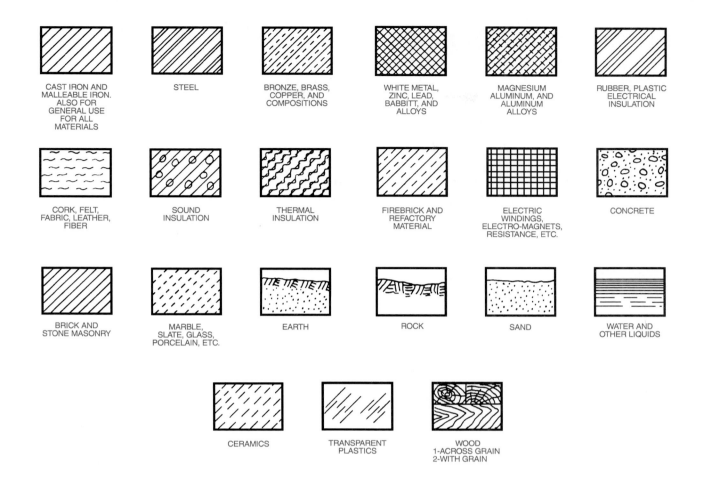

Fig. 10-28. Different symbols are used to indicate different materials in section views. (John Deere & Co.)

design requirements of a part are normally indicated on the drawing or parts list.

PROBLEMS AND ACTIVITIES

The following problems have been selected to provide experience in drawing various types of section views. They are also designed to assist in developing an understanding of their applications. Select an A-size or B-size drawing sheet and an appropriate scale to produce a well balanced drawing.

1. Refer to Fig. 10-29 and draw the necessary views. Include a full section of the parts shown. On those parts where the cutting-plane line has not been given, select the best position of the section to show interior details.

2. Make a half section drawing of the parts shown in Fig. 10-30.

3. Draw the necessary views of the objects shown in Fig. 10-31. Show revolved sections of the appropriate features.

4. Draw the necessary views of the objects shown in Fig. 10-32. Include a removed section view of the features indicated. Use section views where necessary to improve clarity.

5. Draw the necessary views of the objects shown in Fig. 10-33. Include offset sections as indicated.

6. Draw the necessary views of the "CLUTCH PISTON" and "SHEAVE" shown in Fig. 10-34. Include an aligned section.

7. Make a two-view drawing of the "PIVOT PIN" shown in Fig. 10-34. Include a broken-out section.

8. Draw the necessary views to clearly describe the objects shown in Fig. 10-35. Use drafting conventions discussed in this chapter to represent conventional breaks and intersections.

9. Draw the necessary views of the "FAN BRACKET" and "ANGLE BRACKET" shown in Fig. 10-36. Either method discussed in this chapter of showing webs in a section may be used.

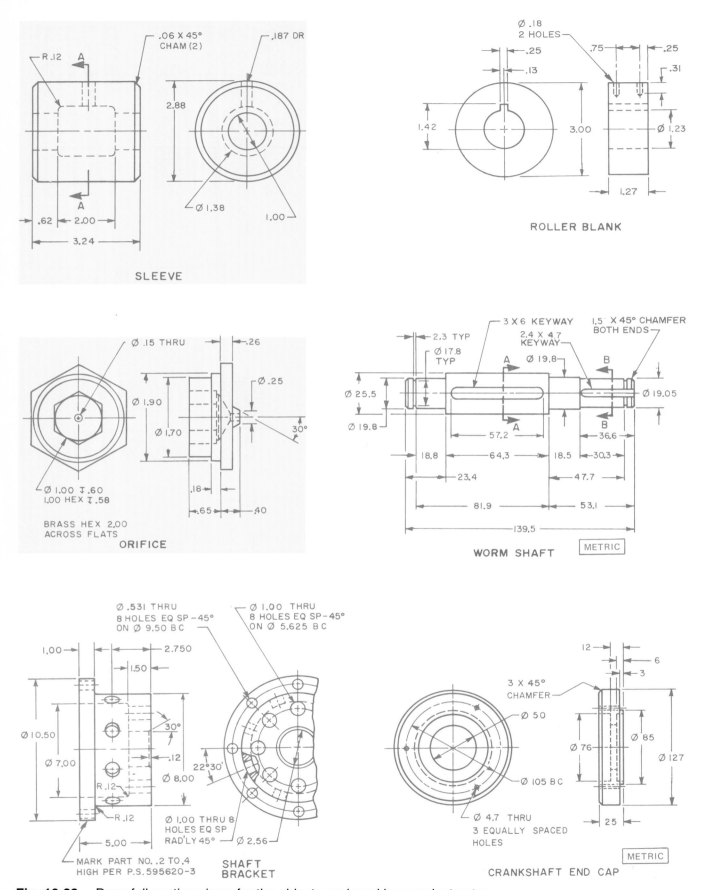

SLEEVE

ROLLER BLANK

ORIFICE

WORM SHAFT   METRIC

SHAFT BRACKET

CRANKSHAFT END CAP   METRIC

Fig. 10-29.   Draw full section views for the objects assigned by your instructor.

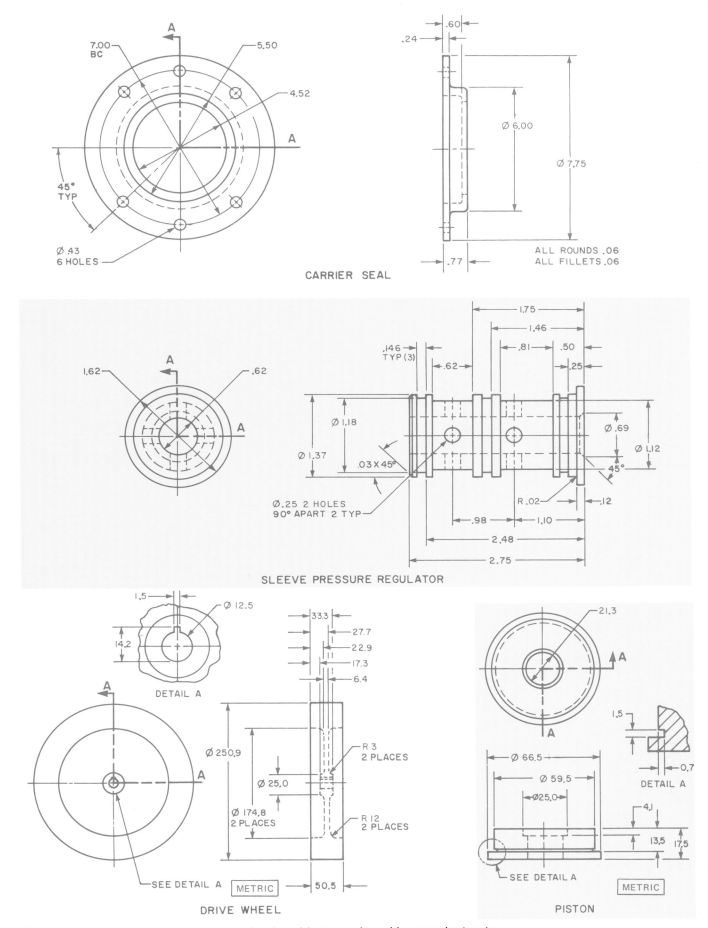

Fig. 10-30. Draw half section views for the objects assigned by your instructor.

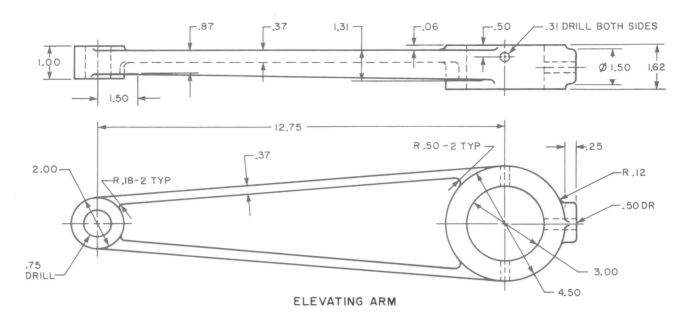

ELEVATING ARM

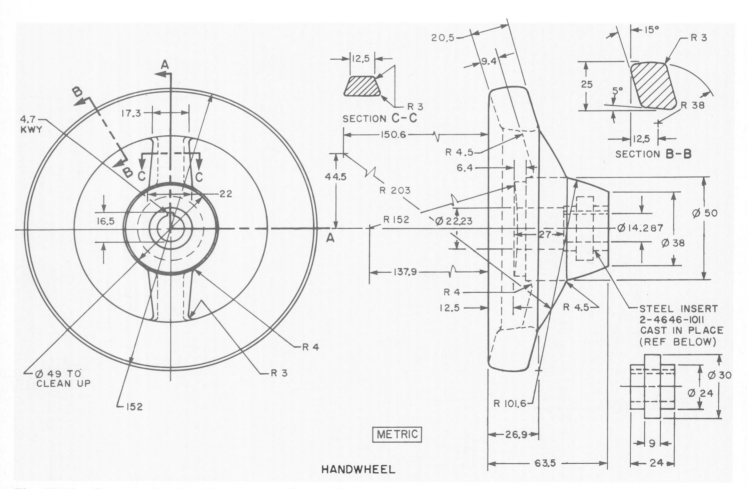

HANDWHEEL

Fig. 10-31.   Draw revolved section views for these objects.

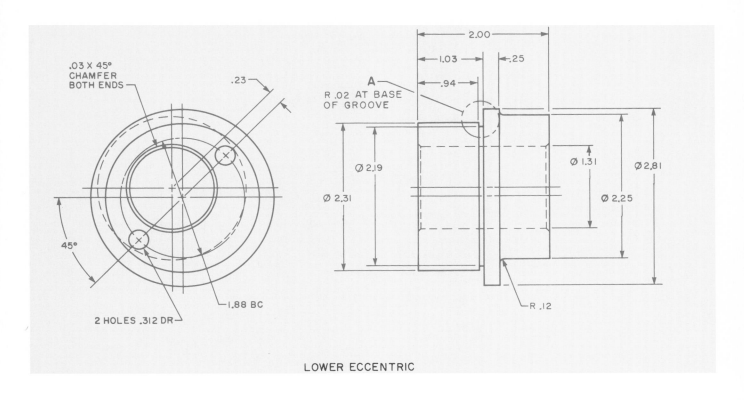

LOWER ECCENTRIC

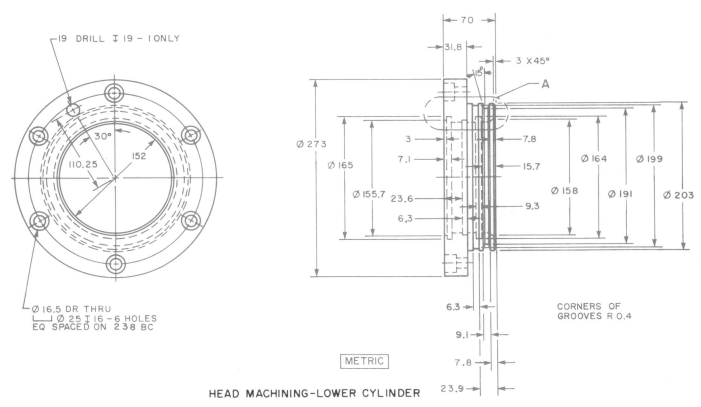

METRIC

HEAD MACHINING-LOWER CYLINDER

Fig. 10-32. Draw removed section views for these objects.

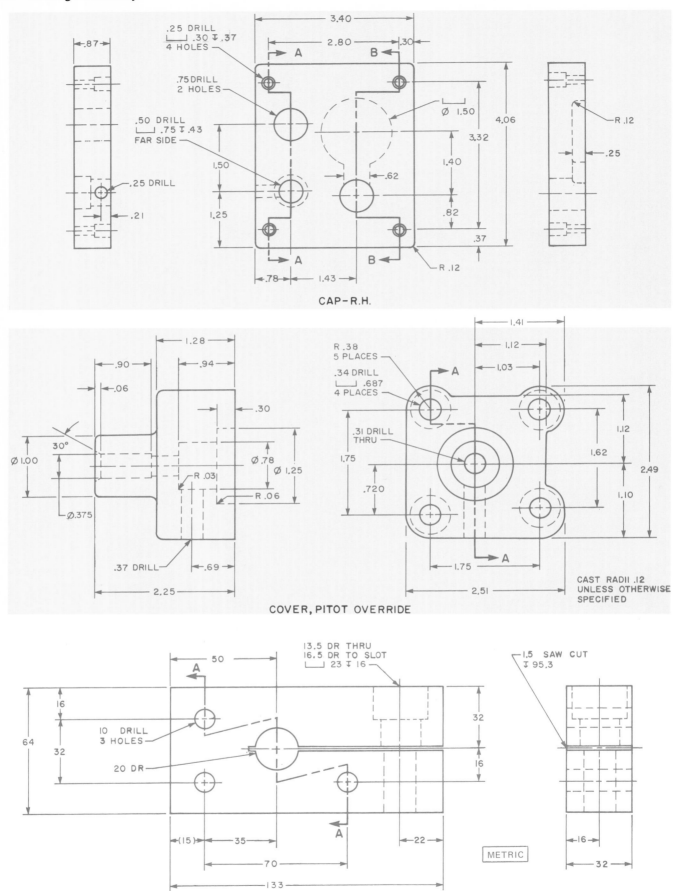

Fig. 10-33.　Draw offset section views for the objects assigned by your instructor.

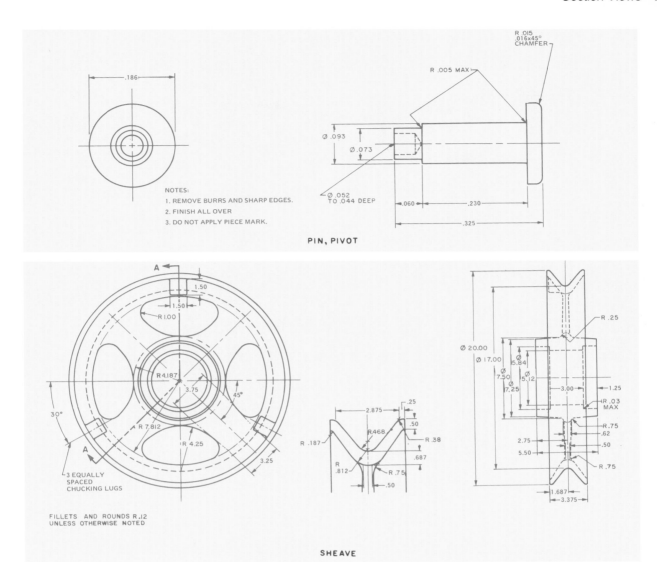

PIN, PIVOT

NOTES:
1. REMOVE BURRS AND SHARP EDGES.
2. FINISH ALL OVER
3. DO NOT APPLY PIECE MARK.

SHEAVE

FILLETS AND ROUNDS R.12
UNLESS OTHERWISE NOTED

3 EQUALLY
SPACED
CHUCKING LUGS

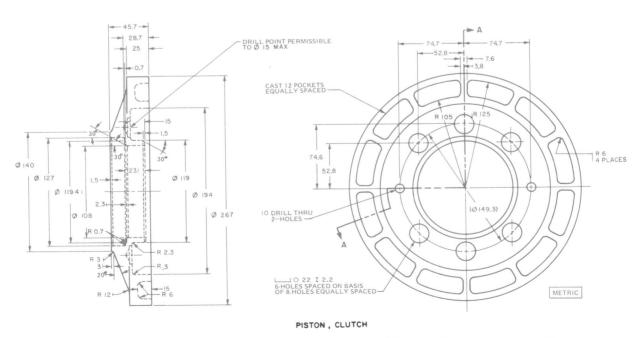

PISTON, CLUTCH

DRILL POINT PERMISSIBLE
TO Ø 15 MAX

CAST 12 POCKETS
EQUALLY SPACED

10 DRILL THRU
2—HOLES

6-HOLES SPACED ON BASIS
OF 8-HOLES EQUALLY SPACED

METRIC

Fig. 10-34. Draw aligned and/or broken-out section views of these objects as assigned by your instructor.

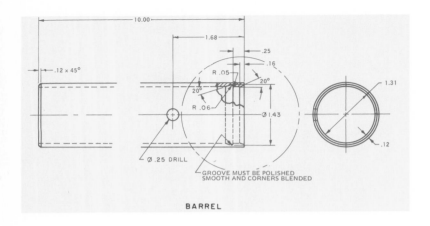

BARREL

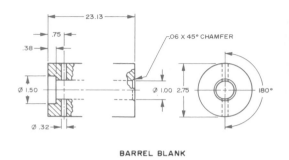

BARREL BLANK

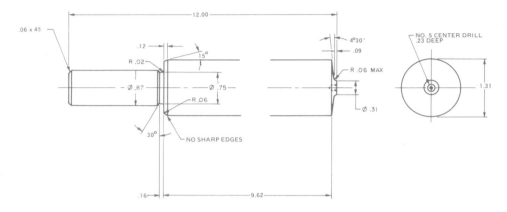

ROD

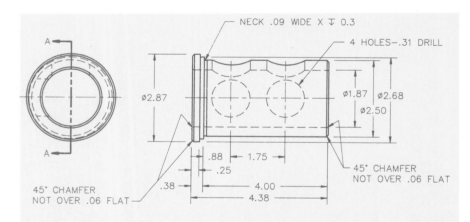

Fig. 10-35.   Draw section views of the problems assigned by your instructor. Be sure to follow drafting conventions for section views.

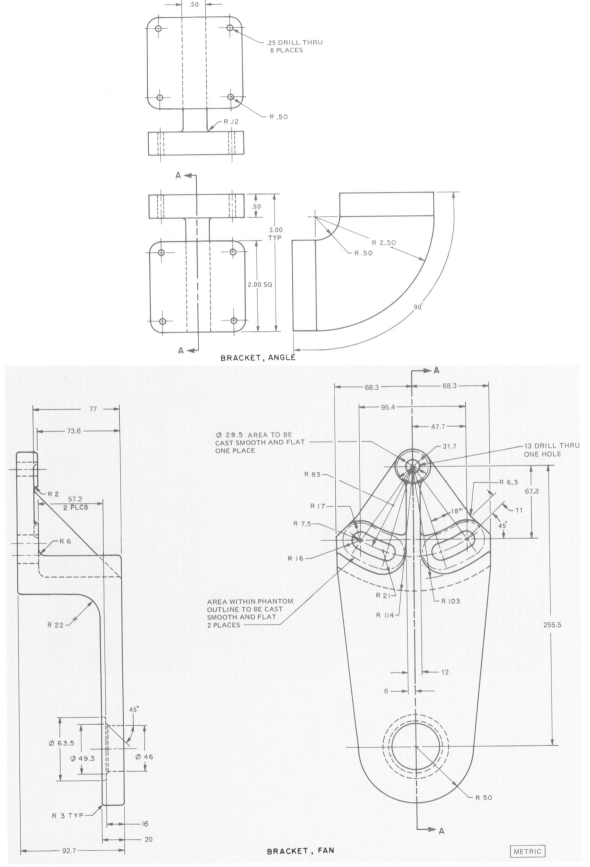

Fig. 10-36. Draw section views of these problems. Note that each of these objects has a web. Use accepted drafting conventions to draw the needed section views.

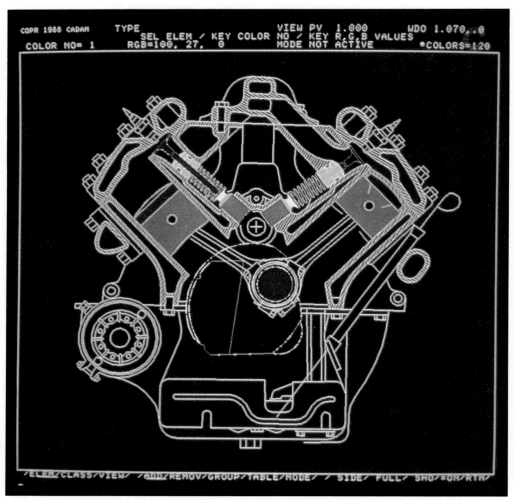

Cadd allows complex parts or assemblies to be "cut open" or sectioned to show internal workings. (IBM)

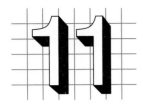

Pictorial Drawings

KEY CONCEPTS

☐ There are three basic types of pictorial drawings: axonometric, oblique, and perspective.

☐ Isometric, dimetric, and trimetric are three subtypes of axonometric views.

☐ One-point and two-point perspectives are subtypes of perspective views.

Pictorial drawings are more "life-like" than multiview-orthographic drawings. These drawings are particularly useful for a "non-technical" person who needs information from a drawing.

Pictorial drawings are used to supplement multiview drawings. A pictorial drawing should clarify information contained in a multiview. Pictorials are sometimes used as a substitute for a multiview drawing. If the task to be performed is not highly complex, a pictorial view might convey all of the required information by itself. For example, a pictorial drawing describes the assembly of machine parts much better than a multiview drawing would, Fig. 11-1.

Pictorial drawings are widely used for assembly drawings, piping diagrams, service and repair manuals, sales catalogs, and technical training manuals. Pictorials are also used by the general

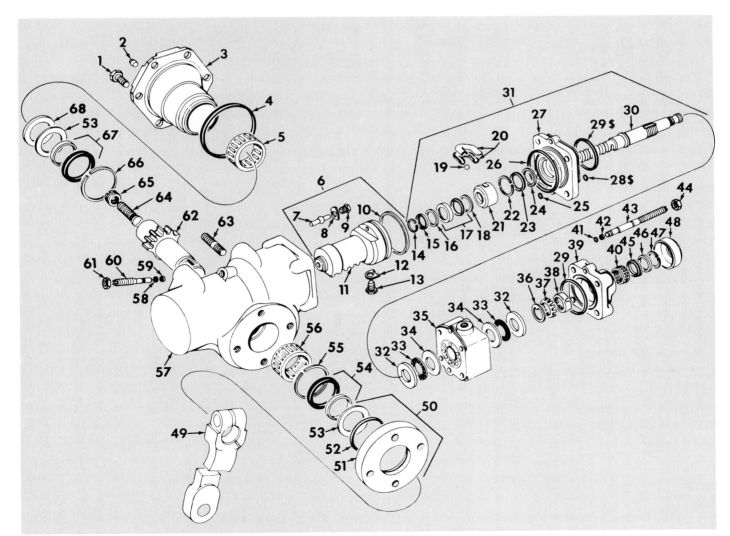

Fig. 11-1. An exploded pictorial drawing clarifies the assembly of parts. (International Harvester Co.)

public in the assembly of prefabricated furniture, swing sets, and "do-it-yourself" kits.

drawing them. Recommended methods of dimensioning and sectioning are also covered.

TYPES OF PICTORIAL PROJECTIONS

There are three basic types of pictorial projections used in industry: axonometric, oblique, and perspective, Fig. 11-2. Under each of these three groupings are several subtypes. This chapter presents the various types of pictorials and the techniques used in

Axonometric Projection

In *axonometric projection,* the lines of sight are perpendicular to the plane of projection. In this sense, axonometric projection is similar to orthographic projection, Fig. 11-3. However, while the lines of sight of the axonometric projection are perpendicular to the plane of projection, the three faces of the object are all inclined to the plane of projection. This

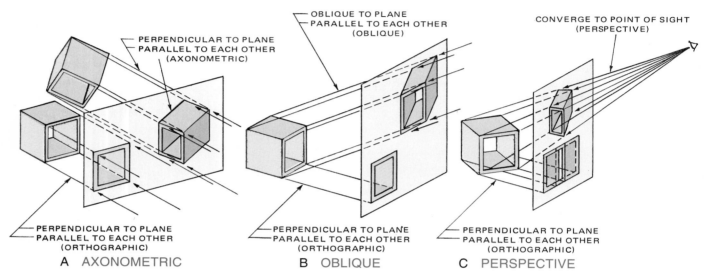

Fig. 11-2.   There are three basic types of pictorial projections: A–axonometric, B–oblique, and C–perspective. (American National Standards Institute)

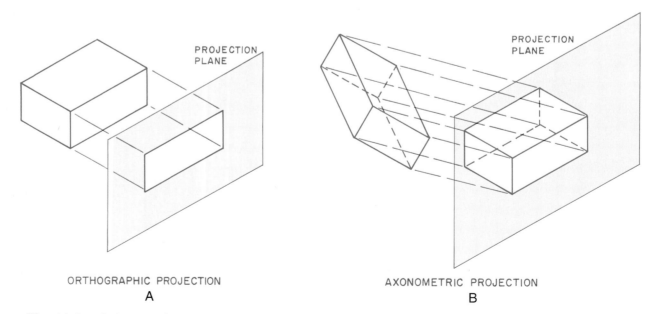

Fig. 11-3.   A–In an orthographic projection, a face of the object is perpendicular to the projection plane. B–In an axonometric projection, while the projection lines are perpendicular to the projection plane, the object faces are all inclined to the projection plane.

gives the projection a three-dimensional pictorial effect. The principal axes of an axonometric projection can be at any angle, except 90°.

There are three types of axonometric projections: isometric, dimetric, and trimetric, Fig. 11-4. There are two differences between the three types of axonometric projections. They differ in the angles made by the faces with the plane of projection. These axonometric projections also differ in the angles made by the principal axes with the plane of projection.

Isometric projection

Isometric means "equal measure." In an *isometric projection,* the three principal faces of a rectangular object are equally inclined to the plane

of projection. The three axes also make equal angles with the plane of projection, Fig. 11-4A.

An isometric projection is a true orthographic projection (parallel projection lines) of an object on the projection plane. It may be produced by revolving the object in the multiview 45° about the vertical axis XC, Fig. 11-5B. The cube is then tilted forward until the body diagonal XY is perpendicular to the plane of projection, Fig. 11-5C.

The vertical axis, XC, is at an angle of 35°16′ with the plane of projection and appears vertical on that plane. Principal axes AX and XB appear on the plane of projection at 30° with the horizontal. The three front edges AX, XB, and XC, called isometric axes, are separated by equal angles of 120° in the isometric projection. (Refer back to Fig. 5-11C.)

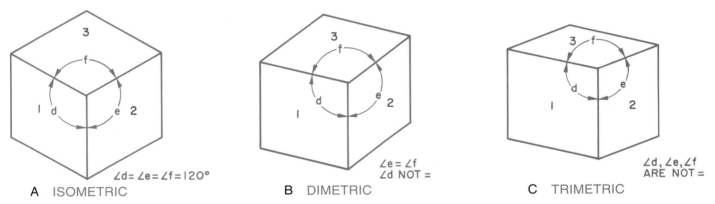

Fig. 11-4. A–In an isometric projection, all three principal faces (1, 2, and 3) of a rectangular object are equally inclined to the projection plane. Also, all three axes make equal angles (d, e, and f) with each other. B–In a dimetric projection, only two faces (1 and 3) make equal angles with the projection plane. Also, only two axes make equal angles (e and f) with each other. C–In a trimetric projection, none of the three faces (1, 2, or 3) make equal angles with the projection plane. In addition, none of the axes make equal angles (d, e, or f) with each other.

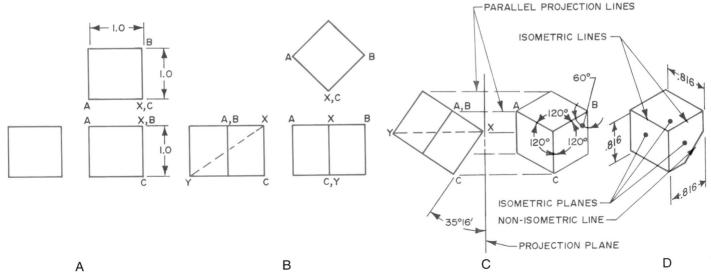

Fig. 11-5. An isometric projection can be constructed by using the revolution method.

Angles of 90° in the orthographic view appear as large as 120° or as small as 60° in the isometric view, depending on the viewing point.

Lines along, or parallel to, the isometric axes are called *isometric lines.* These lines are foreshortened in an isometric projection to approximately 81% of their true lengths. (Refer back to Fig. 11-5.) *Foreshortened* means that the line is shorter than true length.

Lines which are not parallel to the isometric axes are called *non-isometric lines,* Fig. 11-5D. The faces of the cube shown in Fig. 11-5D are called *isometric planes.* Isometric planes include all planes parallel to the "faces." An isometric projection can also be obtained by means of an auxiliary projection, Fig. 11-6.

Isometric projections are true projections. However, for objects more complicated than a cube, the object must first be drawn in orthographic projection. (Refer back to Fig. 11-5.) Then the isometric projection is constructed by revolution or auxiliary projection. A special scale similar to the one shown in Fig. 11-7 can also be used. Since direct measurements can then be made on the isometric axes, it is common practice to make an isometric drawing instead of an isometric projection.

Isometric drawing and projection compared. Basically, an isometric drawing can be constructed without first making a multiview drawing. This simplified process is possible because measurements can be made with a regular scale on the isometric axes of the drawing.

The main difference between isometric projection and isometric drawing is that isometric drawings tend to be larger, Fig. 11-8. Actual measurement of full lengths are used in the drawing, while foreshortened lengths are projected in the isometric projection.

Isometric drawing of an object with normal surfaces. *Normal surfaces* are those surfaces parallel to the principal planes of projection. This does not include inclined or oblique surfaces. It is easier to understand how to construct an isometric drawing if normal surfaces involving only isometric lines are considered at first.

The object shown in Fig. 11-9 is constructed as follows.

1. Draw an isometric block equal to the width, depth, and height of the object shown in the multiview, Fig. 11-9A.

2. Lay off dimensions along the isometric lines for the cut through the top. Draw isometric lines, Fig. 11-9B.

3. Erase unnecessary construction lines and darken the pictorial, Fig. 11-9C.

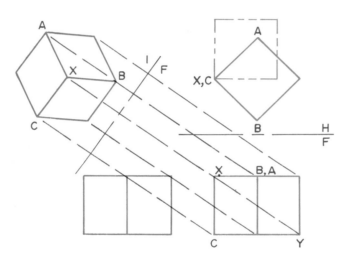

Fig. 11-6. An isometric projection can be constructed by the auxiliary view method.

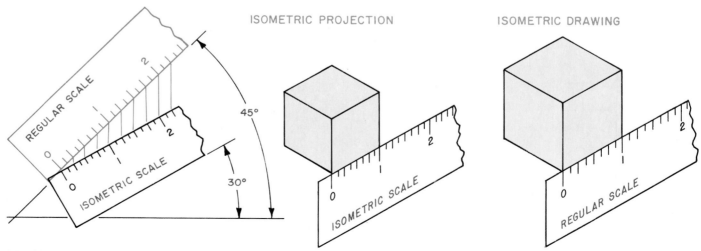

Fig. 11-7. An isometric scale can be constructed for use in isometric projection. An isometric scale is a foreshortened scale of the "regular" scale.

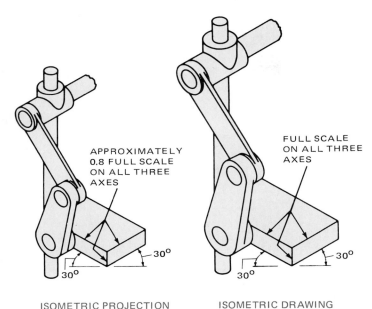

Fig. 11-8. An isometric drawing of an object is similar to an isometric projection of the same object. The main difference is that the drawing is somewhat larger. (American National Standards Institute)

Note: Hidden lines are omitted from isometric drawings unless needed for clarity.

Isometric drawing involving non-isometric lines. Isometric drawings of objects with inclined or oblique surfaces involve non-isometric lines. *Non-isometric* lines cannot be measured directly on the drawing. These lines are drawn by locating the end points on isometric lines and then drawing the lines. All measurements on isometric drawings must be made parallel to the isometric axes. Non-isometric lines are not shown as true length.

To construct an isometric projection of an object with inclined and oblique surfaces, use the following procedure.

1. Draw an isometric block equal to the width, depth, and height of the object shown in the multiview, Fig. 11-10A.
2. Lay off, along isometric lines, the end points of non-isometric lines, Fig. 11-10B.
3. Use a straightedge to join the endpoints, Fig. 11-10C.
4. Erase unnecessary construction lines and darken the pictorial.

Constructing angles in isometric drawings. Angles do not appear as true size in isometric drawings. Therefore, angles cannot be drawn to the true value in an isometric drawing. An angle is drawn by locating the end points of the sides. The end points are then connected to form the required angle.

Use the following procedure to draw an angle in an isometric projection.

1. Draw an isometric block equal to the width, height, and depth of the object in the multiview, Fig. 11-11A.
2. Lay off, along isometric lines, the end points of the lines forming the angle. (Refer back to Fig. 11-11A.)
3. Use a straightedge to join the endpoints. This will form the angle, Fig. 11-11B. Angles may be projected to the opposite side of the block using isometric lines. The angle cut on the opposite end should be drawn parallel to the first cut.
4. Erase unnecessary construction lines and darken the pictorial, Fig. 11-11C.

Coordinate method of laying out an isometric drawing. In the preceding sections, the "block" method of laying out isometric drawings was used. Isometric drawings may also be laid out using the coordinate method shown in Fig. 11-12. For certain

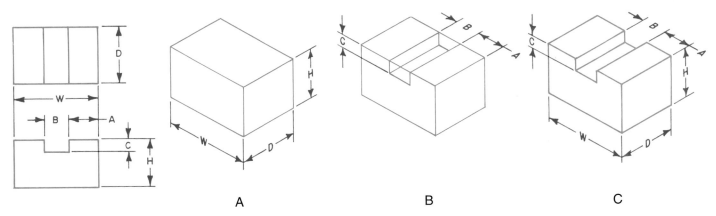

Fig. 11-9. Normal surfaces are those surfaces that are parallel to the planes of projection. Normal surfaces do not include oblique or inclined surfaces. A–Isometric constructions are easier if a "box" of the total outside dimensions is first constructed. B–The details of the object are added. C–Finally, all construction lines are removed and the object lines darkened.

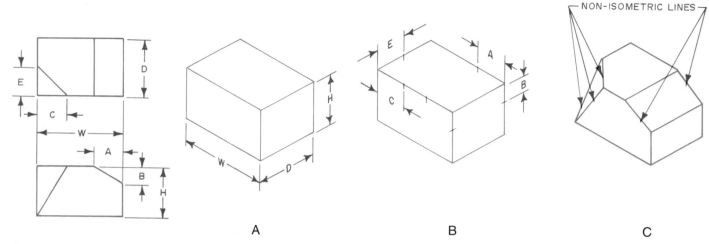

Fig. 11-10. Non-isometric lines and surfaces are those that are not parallel to the projection planes. A–As with constructing normal surfaces, first construct a "box" of the total outside dimensions. B–Then locate the endpoints of the non-isometric lines on the normal lines. C–Finally, use a straightedge to connect the endpoints and create the non-isometric surfaces.

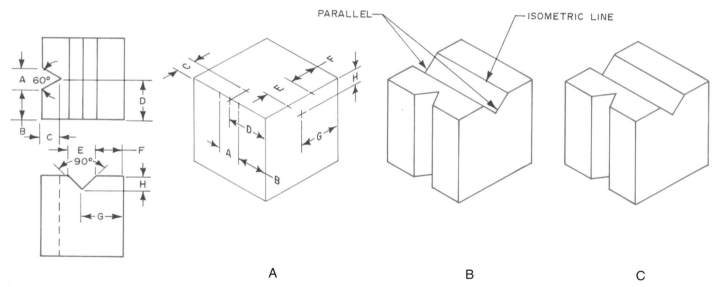

Fig. 11-11. Angles do not appear true size in isometric drawings. A–First, construct a "box" of the total outside dimensions. B–Next, locate the endpoints of the angle by measuring on or parallel to isometric lines . C–Finally, use a straightedge to connect the endpoints and form the angle.

objects which are not basically "cubic" (such as truncated pyramid), the coordinate method is faster.

Use the following procedure to construct an isometric drawing by the coordinate method.

1. Locate a starting point such as the lower front corner. Draw two horizontal axes, Fig. 11-12A. These axes will be inclined upward at 30°. (Note: The technique of centering an isometric drawing in a space is covered later in this chapter.)

2. Lay off the lengths of the two base edges. (Refer back to Fig. 11-12A.)

3. Construct base line FG of the assumed isometric plane passing through point E by measuring the offset of distance X. (Refer back to Fig. 11-12A.)

4. Locate distance Y from corner A of pyramid and project to intersect line FG and Y′. (Refer back to Fig. 11-12A.)

5. Construct a vertical line at Y′ and lay off the height of point E, Fig. 11-12B. Then, use a straightedge and draw line EC.

6. Continue to locate remaining points of truncated cut in a similar manner.

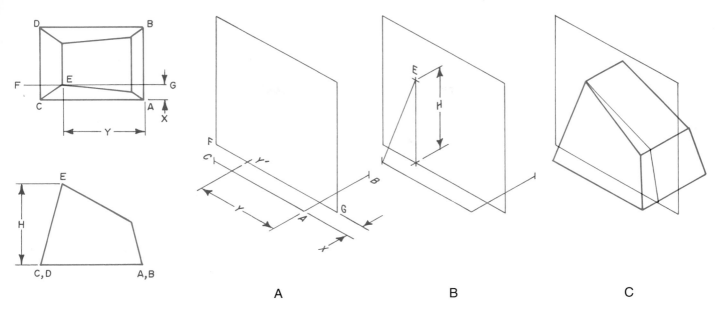

Fig. 11-12. An isometric drawing can be constructed using the coordinate, or offset, method.

7. Erase unnecessary lines and darken finished drawing, Fig. 11-12C.

Circles and arcs in isometric projection— four-center approximate method. Circles and arcs will appear as ellipses or partial ellipses in isometric drawings. One method of constructing these is the four-center approximate method. This method approximates a true ellipse, Fig. 11-13A. True ellipses can also be drawn with an isometric ellipse template, Fig. 11-13B. The coordinate method explained later in this chapter can also be used.

The four-center approximate method is fast and effective for most isometric drawings. If tangent circles are involved, the process is not quite as simple. The procedure for drawing a four-center approximate ellipse is as follows.

1. Locate the center of the circle and draw isometric centerlines, Fig. 11-14A.

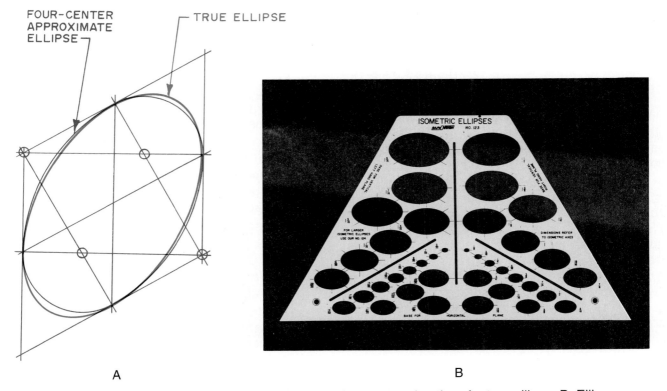

FOUR-CENTER APPROXIMATE ELLIPSE

TRUE ELLIPSE

A

B

Fig. 11-13. A–A four-center approximate ellipse is a close *approximation* of a true ellipse. B–Ellipse templates are available that make drawing ellipses quick and easy.

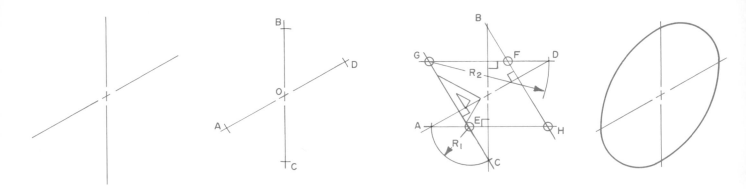

Fig. 11-14. The four-center approximate method of constructing an isometric ellipse is a quick and effective method to use.

2. With O as the center and a radius equal to the radius of actual circle, strike arcs A, B, C, and D to intersect the isometric centerlines, Fig. 11-14B.

3. Through each of the points of intersection, draw a line perpendicular to opposite centerline, Fig. 11-14C.

4. The four intersecting points, E, F, G, and H, of the perpendiculars are the radius centers for the approximate ellipse. The radii of the arcs are the distances from the center along the perpendiculars to the intersection of the isometric centerlines. In Fig. 11-14, the radii are EA, FB, GD, and HA, (as shown in Fig. 11-14C).

5. Draw four arcs CA, BD, AB, and CD to complete the ellipse, Fig. 11-14D.

An isometric arc is constructed using the four-center approximate method by selecting the required portion and drawing that segment, Fig. 11-15.

The construction of isometric circles using the four-center approximate method is shown in each of the three principal faces or planes in Fig. 11-16. The same procedure is used in drawing isometric ellipses by this method, regardless of position.

Coordinate method of drawing isometric circles and arcs. The coordinate method of drawing an isometric circle is a process of plotting coordinate points on a true circle. These points are then transferred to an isometric square that is the same size as the circle to be drawn, Fig. 11-17. This method results in a true projection of the isometric ellipse.

Use the following procedure to construct an isometric circle by the coordinate method.

1. Locate center of isometric circle and draw centerlines, Fig. 11-17A.

2. Strike arcs equal to the radius of the required circle. Construct an isometric square, Fig. 11-17B.

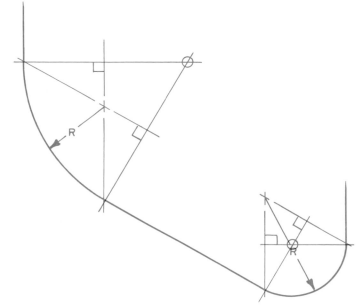

Fig. 11-15. Arcs can be constructed by using the four-center approximate method of constructing an ellipse. Instead of drawing the entire ellipse, select the vertex that will produce an approximation of the required arc.

3. Draw a semicircle adjacent to one side of the isometric square. (Refer back to Fig. 11-17B.)

4. Divide the semicircle into an even number of equal parts. Project these divisions to a side of the square. (Refer back to Fig. 11-17B.)

5. From the intersection points, draw lines across the isometric square parallel to the centerline. (Refer back to Fig. 11-17B.)

6. Transfer points 1, 2, 3, and 4 from the semicircle to the upper left quarter of the isometric square by setting off the appropriate distances, Fig. 11-17C. Repeat this for the upper right quarter. Project these intersections to the two lower quarters.

7. Draw a smooth curve through these points to form an isometric ellipse, Fig. 11-17D.

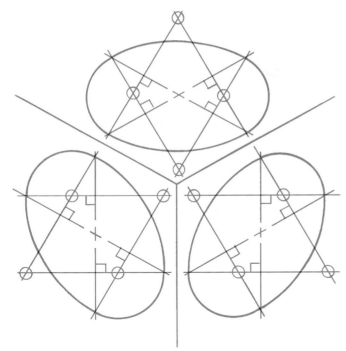

Fig. 11-16. The four-center approximate method can be used in all three principal isometric planes.

Arcs are constructed by the coordinate method in the same manner as circles. The required arc is divided into a number of equal parts and these points are projected to the isometric view. A smooth arc is then drawn through the projected coordinate points.

Constructing irregular curves in isometric drawings. Irregular curves can be constructed in isometric drawings by using the coordinate method shown in Fig. 11-18. The procedure for this construction is as follows.

1. Select a sufficient number of points on the irregular curve in the orthographic view to pro-

duce an accurate representation when transferred to isometric view, Fig. 11-18A. (Be sure to locate a point at each sharp break or turn.)
2. Draw coordinates through each point and parallel to the two principal axes of the orthographic view. (Refer back to Fig. 11-18A.)
3. Draw an isometric rectangle in the corresponding plane of the isometric view. The rectangle should be equal in size to the one containing the irregular curve in the orthographic view, Fig. 11-18B.
4. Draw isometric coordinate lines with spacing equal to the coordinates in the orthographic view. (Refer back to Fig. 11-18B.)
5. Draw a smooth curve through the points of intersecting coordinates. This will form the required irregular curve in the isometric drawing, Fig. 11-18C.
6. Then project coordinate points to form the thickness of the object (if there is a thickness). (Refer back to Fig. 11-18C.)

Section views in isometric. Section views in isometric are an effective means of graphically describing the interior of complex machine parts or assemblies. Half and full sections are frequently used. When an isometric half section is used, the correct view is to position the part where both sides of the removed section are visible, Fig. 11-19A. An isometric full section should be placed so that one of the axes is on the cutting plane, Fig. 11-19C. Occasionally, a broken-out section is useful in showing a particular feature in an isometric view, Fig. 11-20.

Section lines are normally drawn at an angle of 60°, Fig. 11-21A. This angle closely resembles the 45° crosshatching in multiviews. If a 60° angle will cause the section lining to be parallel or

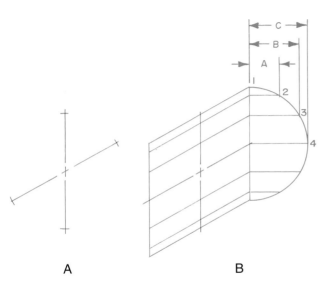
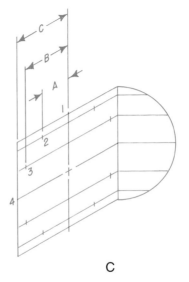
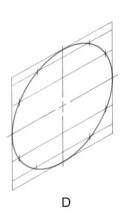

A B C D

Fig. 11-17. An isometric ellipse can be constructed by using the coordinate method.

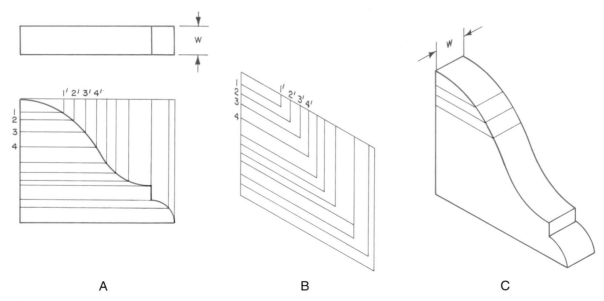

Fig. 11-18. An irregular curve can be transferred to an isometric drawing by using the coordinate method.

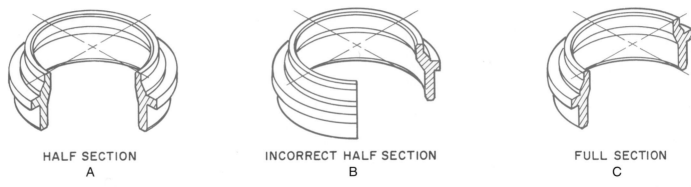

HALF SECTION
A

INCORRECT HALF SECTION
B

FULL SECTION
C

Fig. 11-19. A–The cutting planes for isometric half section views should be parallel to the isometric axes . B–The cutting planes should *not* be parallel to the edges of the drawing sheet. C–For full section views, the cutting plane should be parallel to one of the isometric axes as well.

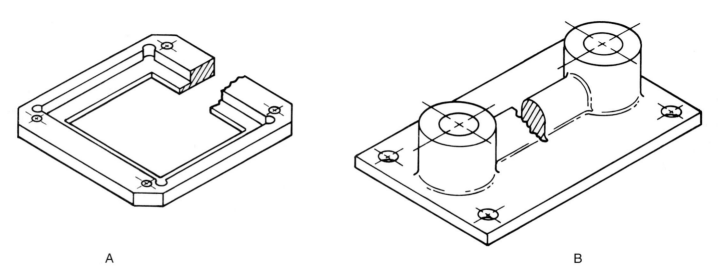

A

B

Fig. 11-20. Broken-out sections can be useful to show a particular detail in an isometric view.

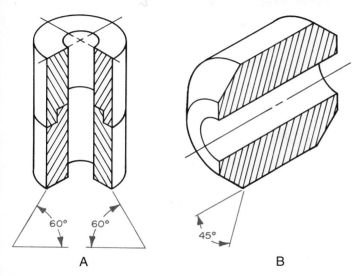

Fig. 11-21. Section lines in isometric are normally drawn at 60°. However, another angle should be chosen if 60° will produce lines parallel, or close to parallel, the object lines. A–In an isometric half section, the lines should be drawn to match if the cutting planes are revolved together. B–However, in a full section the lines remain the same on all parts of the object.

perpendicular to the visible outline of the object, a different angle should be chosen, Fig. 11-21B.

The direction of section lining in a full section remains the same on all portions "cut" by the cutting plane. (Refer back to Fig. 11-21B.) An exception to this is if the view is of an assembly with several parts. Section lining in an isometric half section should be drawn to match if the two planes of the section were revolved together. (Refer back to Fig. 11- 21A.) This will make the lines appear as if they are drawn opposite.

The steps in constructing an isometric section view are as follows.

1. Draw an isometric "box" the size of the three overall dimensions of the object. Lines should be drawn very light, Fig. 11-22A.

2. Draw the outline along the cutting plane. (Refer back to Fig. 11-22A.)
3. Add the remaining details, Fig. 11-22B.
4. Erase the construction lines and add the cross-hatching, Fig. 11-22C.

Isometric Dimensioning. The general rules for dimensioning multiview drawings also apply to isometric drawings. The aligned or isometric plane system is approved, Fig. 11-23A. The unidirectional system is also approved, Fig. 11-23B and Fig. 11-23C. The unidirectional system is gaining favor in industry. All dimensions are easily read from the bottom of the drawing. In either system, the dimension lines, extension lines, and dimension figures should lie in the correct plane for the feature dimensioned. The dimension figures in the unidirectional system may all be shown in one plane for convenience and speed in drawing. (Refer back to Fig. 11-23C.) However, the dimension and extension lines must be properly aligned.

Notes in the aligned system lie in one of the principal planes. (Refer back to Fig. 11-23A.) Notes in the unidirectional system are placed horizontally so that they lie in or are parallel to the picture plane. (Refer back to Fig. 11-23B and Fig. 11-23C.)

Some incorrect practices in isometric dimensioning are shown in Fig. 11-24A. The correct way to indicate these dimensions is shown in Fig. 11-24B.

Representing screw threads in isometric. A great amount of time is required to draw the actual representation of screw threads in isometric projections. The increase in clarity of the drawing is not large enough to justify the time required. As a result, actual representation is seldom used. The practice is to represent crest lines of threads with a series of isometric circles (ellipses) uniformly spaced, Fig. 11-25A. It is not necessary to duplicate the actual pitch of the thread. The dimension

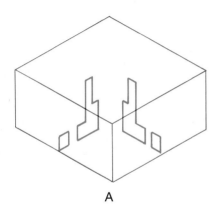

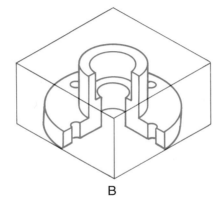

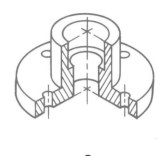

Fig. 11-22. A–To construct an isometric section view, first construct a "box" and locate the sectioned faces inside. B–Next, draw the remaining parts of the object. C–Finally, darken the object lines, add the section lines, and remove the construction lines.

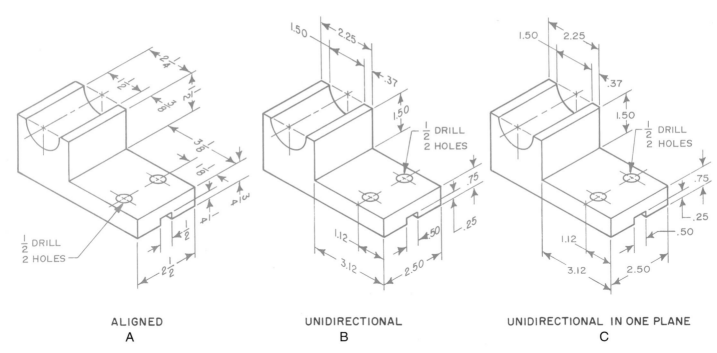

Fig. 11-23. When dimensioning an isometric drawing, the dimension lines should be in the plane corresponding to the feature being dimensioned.

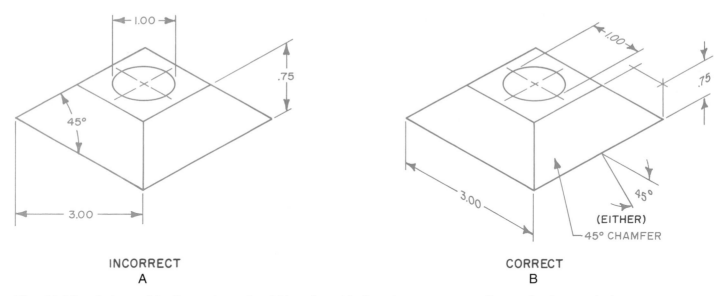

Fig. 11-24. A–Isometric dimensions should be placed in the plane corresponding to the feature being dimensioned. B–If necessary, add additional extension lines to locate a dimension.

will identify thread characteristics. Shading may be used to increase effectiveness of the thread representation, Fig. 11-25B.

Isometric ellipse templates. The four-center approximate and coordinate methods of constructing isometric circles can be very time-consuming. When ellipse templates are available in the correct size, they should be used to speed the process of drafting. The appearance of the finished drawing is also greatly improved by using templates. Isometric

templates are available in a variety of sizes. Their use simply requires alignment of the template along the isometric centerlines of the circular feature. In some cases, vertical and horizontal centerlines can be used with an ellipse template, Fig. 11-26.

Alternate positions of isometric axes. It may be desirable to draw an object in isometric with the axis in a position other than normal. For example, the object could be viewed from below, Fig. 11-27A. This is called **reverse position.** For long

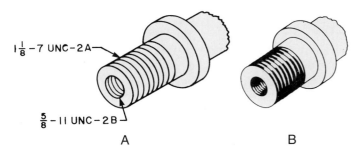

$1\frac{1}{8} - 7$ UNC-2A

$\frac{5}{8}$ - 11 UNC-2B

A   B

Fig. 11-25. A–Screw threads are represented in isometric views by uniformly spaced isometric circles. B–These circles represent the crest of the thread. Shading can be added to better represent the threads. (American National Standards Institute)

Fig. 11-26. Ellipse templates can be aligned using isometric centerlines as guides. Regular horizontal and vertical centerlines can also be used in some cases.

objects, the object could be shown horizontally, Fig. 11-27C. The isometric axis may be located in any of a number of positions as long as equal spacing of 120° is maintained between the three axes.

Centering an isometric drawing. The technique of locating an isometric drawing in the center of a sheet, or at any other location, is a matter of finding the center of the object and positioning this point in the desired location, Fig. 11-28. Match this center point with the desired center location on the sheet. The "starting point" should be located from this point by measuring parallel to the isometric axes. The entire isometric drawing will then be correctly positioned.

Advantages and disadvantages of isometric drawings. The isometric drawing is one of the easi-

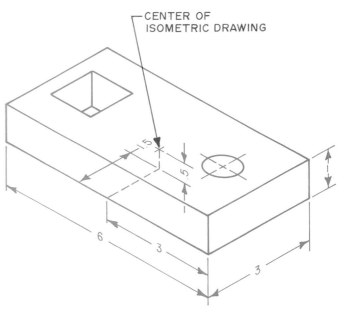

CENTER OF ISOMETRIC DRAWING

Fig. 11-28. To center an isometric drawing, first locate the center of the object. Then place this point in the center of the sheet or workspace. This method can also be used to locate an isometric view at any desired location on the sheet.

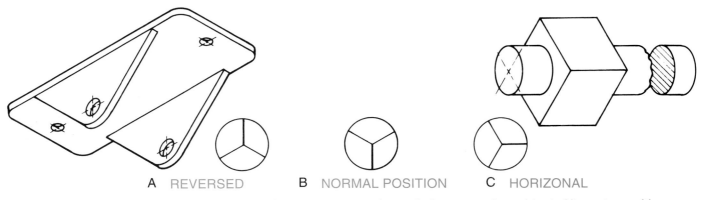

A REVERSED   B NORMAL POSITION   C HORIZONAL

Fig. 11-27. Alternate positioning of isometric axes can reveal certain features of an object. Alternate positions can also be used to view an object in its normal position.

est to construct since the same scale is used on all axes. It has an advantage over orthographic projection in that three sides of the object may be shown in one view. This presents a more realistic representation of the object. Circles are not greatly distorted, as in the receding views of an oblique drawing. (Oblique drawings are discussed later in this chapter.)

There are certain disadvantages inherent in isometric drawings. One is the tendency for long objects to appear distorted. This is so because parallel lines on an object remain parallel, rather than converging toward a distant point. When the actual object is viewed or seen in a perspective drawing, the parallel lines of the object will appear to converge at a point called the **vanishing point.** Also, the symmetry of an isometric drawing causes some lines to meet or overlap. This might confuse the reader of the drawing.

Dimetric projection and dimetric drawing

Dimetric projection is a type of axonometric projection. A **dimetric projection** has two faces equally inclined. (Refer to 1 and 3 in Fig. 11-4B.) Two axes make equal angles with the plane of projection. (Refer to e and f in Fig. 11-4B.) The third face and angle are different.

The two equal angles can be any angle larger than 90° and less than 180° that is not equal to 120°. (If the equal angles are 120°, the third angle must also be 120° and the drawing becomes an isometric projection.) The third axis makes an angle that is either larger or smaller than the two equal angles.

The two axes making equal angles with the projection plane, or lines parallel to these, are foreshortened equally. The third axis and lines parallel to it are foreshortened or enlarged more, depending on how the object is viewed. Like isometric projection, dimetric projection can be constructed by the revolution method or by the auxiliary view method.

Constructing a dimetric projection. The only difference between dimetric and isometric projection is in the angle (line of sight) the object is viewed from. In Fig. 11-29, a cube is revolved in two views to the required line of sight for the two equal angle axes. It is then projected from the orthographic to the dimetric projection. This produces a true measure of the cube, as well as the angle of the dimetric axes.

Careful projection techniques will produce angles and lengths accurate enough for most pictorial drafting requirements. Angles and measurements requiring mathematical accuracy should be calculated by trigonometry.

Given a cube and the angle of rotation of the dimetric axes, use the following procedure to find the angle of the dimetric axes and the true measure on all three axes in the dimetric projection.

1. Draw two views of a cube. The views should be equally inclined to the plane of projection, Fig. 11-29A.

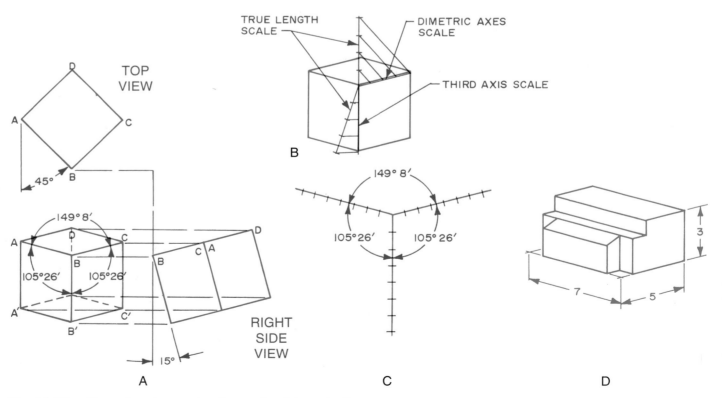

Fig. 11-29. Procedure in constructing a dimetric projection.

2. Project points A, B, C, and D to the front view to form a dimetric projection of the cube.

3. Lay off the actual scale of the orthographic cube on a line at an angle with one of dimetric axes and a second line at an angle along third axis, Fig. 11-29B. Divide each actual measurement line into a number of equal parts. Project these divisions geometrically to divide each axis into proportionally equal parts. The two dimetric axes will be foreshortened, but will have the same measure. The third axis will be foreshortened more due to the line of sight for this particular cube. (In some cases, the third axis may appear longer than dimetric axes, depending on line of sight.)

4. Transfer dimetric units of measure to the dimetric scale, Fig. 11-29C. This scale can be extended with similar units for measuring greater lengths.

The scale can also become a permanent scale for measuring any dimetric projection drawn at this line of sight. An example of a dimetric projection of an object drawn at same angle as the cube and using the dimetric scale just produced is shown in Fig. 11-29D.

The number of variations in the angle of sight for a dimetric projection is infinite. This permits the view of an object that will best portray its features to be drawn. Remember that one of the angles of sight must be 45°. This is the direction of the desired leg for "Y" axis. The other angle of sight may be any angle other than 35°16'. A dimetric projection of a gyro instrument is shown in Fig. 11-30.

A dimetric projection can also be constructed by the successive auxiliary view method shown in Fig. 11-31. This method is discussed in more detail later in this text in Chapter 13.

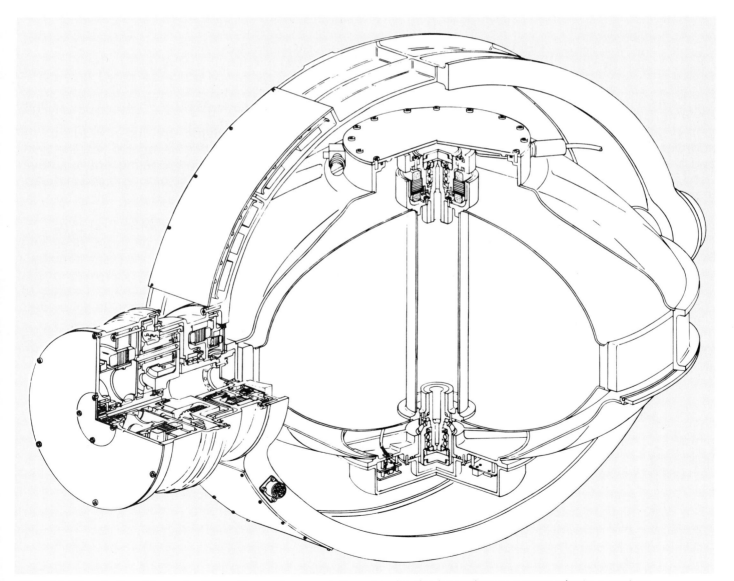

Fig. 11-30. A dimetric projection with a cutaway section may be the best view to communicate a part or assembly. (Sperry Flight Systems Div.)

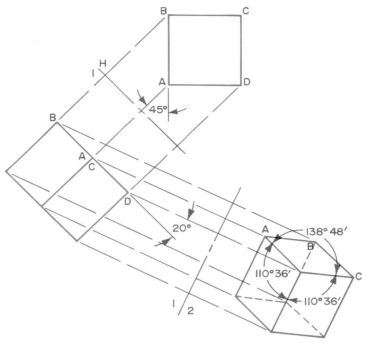

Fig. 11-31. A dimetric projection can be constructed using the successive auxiliary view method.

Constructing an approximate dimetric drawing. Many industries modify their dimetric projection drawings to save time. They construct an approximate dimetric drawing where regular scales and angles common to triangles or a compass are used. The full-size scale, the three-quarter scale, or the half scale is frequently selected for the dimetric axes or the third axis, depending on which is to be reduced.

An *approximate* dimetric drawing can be drawn by substituting angles closely approximating those of a true projection. These angles can easily be achieved with the protractor or adjustable triangle.

The regular full-size scale or a foreshortened scale can be used on the different axes depending on their positions in the dimetric view. Some common axes for approximate dimetric drawings are shown in Fig. 11-32, along with suggested scales for the axes.

Approximate dimetric drawing should not be attempted until the student is familiar with the principles of dimetric projection. One must have a "feel" for the angle of sight and the proportioning of measurements on the axes.

Trimetric projection and trimetric drawings

In a **trimetric projection** all three faces and axes make different angles with the plane of projection. (Refer back to Fig. 11-4C.) The axes of axonometric projections may be placed in a variety of positions, as long as their relationship to each other is maintained.

Angles must be selected that best portray the features of the object in the desired line of sight.

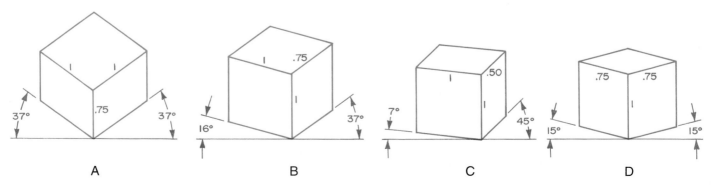

Fig. 11-32. There are common axes and scales used for constructing approximate dimetric drawings. Any of these can be used, depending on the particular application.

Construction of a trimetric projection is essentially the same as a dimetric projection. However, the difference is that no two axes or surfaces are viewed the same on the plane of projection.

Constructing a trimetric projection. The same procedure for dimetric can be used for trimetric. However, the cube is inclined at unequal angles on the viewing plane, Fig. 11-33A. The trimetric scale is constructed in the same manner as the dimetric, except a scale must be developed for each axis, Fig. 11-33C. The object shown in Fig. 11-29D has been constructed as a trimetric projection in Fig. 11-33D.

Constructing an approximate trimetric drawing. The construction of an approximate trimetric drawing is similar to that of an approximate dimetric drawing. However, do not attempt the approximate trimetric projection until thoroughly famil-

iar with true trimetric projection. Some common axes for approximate trimetric drawings are illustrated in Fig. 11-34.

Ellipse guide angles for dimetric and trimetric drawings

Circles will appear as ellipses in dimetric and trimetric drawings. The major axis of the ellipse will appear as a true length line in both dimetric and trimetric drawings.

The four-center approximate method for constructing isometric ellipses is satisfactory for the dimetric axis when the scale is the same on both edges. Ellipses may be drawn by the coordinate plotting method on all axes.

However, the most satisfactory method of drawing ellipses in dimetric and trimetric drawings is to first find the angle that a circle is viewed from in a particular plane. Then select the appropriate

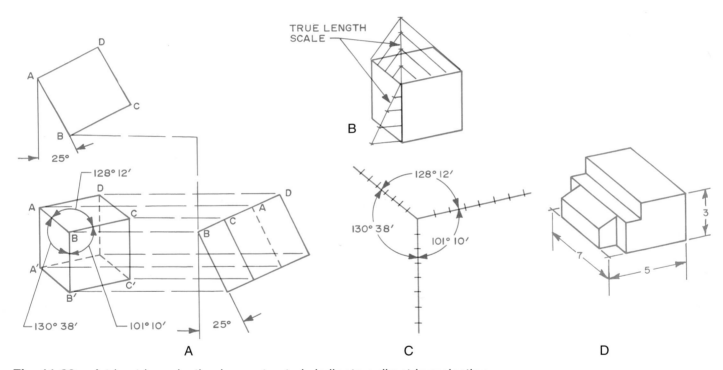

Fig. 11-33. A trimetric projection is constructed similar to a dimetric projection.

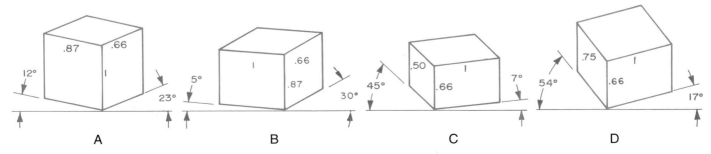

Fig. 11-34. There are common axes and scales used for constructing approximate trimetric drawings. Any of these can be used, depending on the application.

ellipse template and draw the ellipse. This is accomplished by studying auxiliary views (see Chapter 13). Simply find the angle between the edge view of the principal planes in the dimetric or trimetric projection and the line of sight, Fig. 11-35.

The ellipse angle for a dimetric drawing is identical for the two dimetric axes and different for the third. The ellipse angle differs for each of the three trimetric axes.

Given a trimetric projection of the cube in Fig. 11-33, you can find the ellipse guide angle and direction of the major diameter of the ellipse on each principal plane of the trimetric cube by using the following procedure.

1. Line OE is perpendicular to plane ABCO, Fig. 11-35A. Line OC is perpendicular to plane AOEF. Line OA is perpendicular to

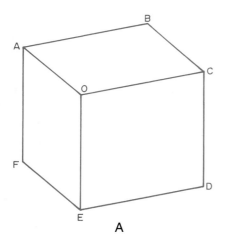

A

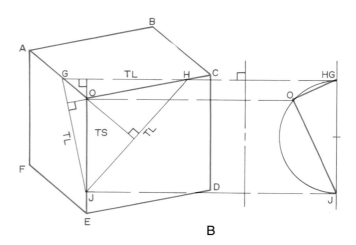

B

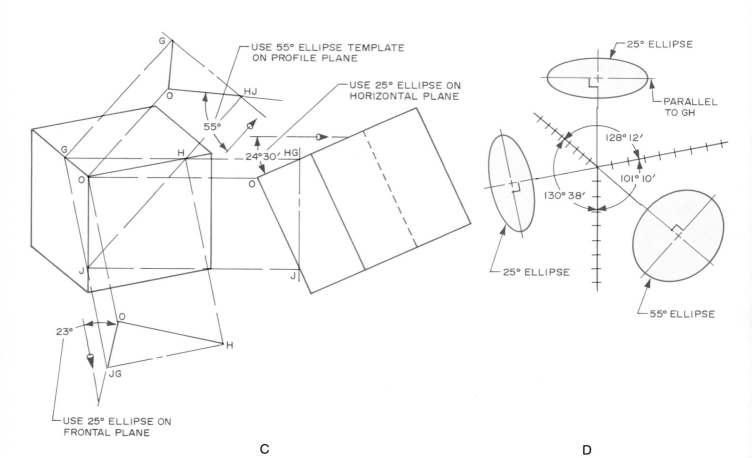

C   D

Fig. 11-35. Determining ellipse guide angles for dimetric and trimetric drawings.

plane OCDE. (This simply means that the planes of the cube are perpendicular to each other.)

2. Construct true length lines GH, HJ, and JG on each plane by drawing them perpendicular to extensions of lines OC, OA, and OE. When two lines that are known to be perpendicular appear on the drawing as perpendicular, one or both of the lines are true length, Fig. 11-35B.

3. The intersection of the true length lines form the true size plane HJ. A plane whose edges appear *true length* will appear *true size*.

4. Plane GHJ is a frontal plane (parallel to the viewing plane). This plane will project as a vertical edge in the side view. (Refer back to Fig. 11-35B.)

5. To find the position of the 90° angle of the cube that is between plane GHJ and the viewing plane, construct a semicircle with a diameter equal in length to the edge view of plane GHJ. Then project point O (the vertex of the 90° angle) to the semicircle. Inscribe the position of the corner of the cube. (Lines extending from the end points of the diameter of a semicircle joined at a point on the semicircle form a right angle.)

6. Draw auxiliary views showing the edge view of the true size plane by drawing "point" views of lines HG, HJ, and JG, Fig. 11-35C.

7. The angle between the edge view and the line of sight for a particular view is the ellipse angle.

8. Position ellipse guide so major diameter is parallel to true length line on that particular plane, Fig. 11-35D.

Advantages and disadvantages of dimetric and trimetric drawings

Dimetric and trimetric drawings permit an infinite number of positions for a machine part or other product to be viewed pictorially. This is very useful for showing a certain feature of a particular object.

The greatest disadvantage of dimetric and trimetric drawings is that they are considerably more difficult to construct than isometric, oblique, or perspective drawings. However, once a desired position has been decided upon and a scale developed, including the angle and location of ellipses, the dimetric and trimetric drawings can be produced quite easily.

Oblique Projections and Drawings

An *oblique projection* or drawing is a type of pictorial drawing similar to axonometric drawings.

Only one plane of projection is used. While the lines of sight are parallel to each other, they meet the plane of projection at an oblique angle, Fig. 11-36.

The object is positioned with its front view parallel to the plane of projection. This view, and surfaces parallel to it, will project true size and shape. The top and side views are viewed at an oblique angle and are distorted along the "Y" axis.

Oblique projection

True oblique projection is obtained by the projection techniques shown in Fig. 11-36. Lines of sight are selected at an angle suitable to show the desired features of the object. The front view should be the one most characteristic of the object.

At least two orthographic views are necessary to construct an oblique projection. The top and right side view have been used in Fig. 11-36. All points are projected to the projection plane by oblique projectors parallel to the line of sight. From the plane of projection, the points are projected horizontally or vertically to intersect with their corresponding point. These points are connected to form the required oblique projection.

Lines along the "X" and "Z" axes will intersect at right angles. The angle formed by the "Y" axis and the horizontal is governed by the lines of sight selected in the two orthographic views. The true size of this angle can be found by methods explained in Chapter 13.

Oblique projection of a plain block is a relatively easy procedure. However, with more complex

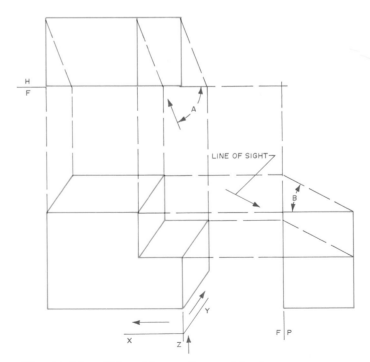

Fig. 11-36. At least two orthographic views are needed to construct an oblique projection.

machine parts or industrial products, the oblique projection is a time-consuming procedure and little use is made of this method in industry.

Oblique drawing

The oblique drawing is based on the principles of oblique projection, although its construction is much faster and the results often more satisfactory than an oblique projection itself.

There are three types of oblique drawings: cavalier, cabinet, and general. The three differ only in the ratio of the scales used on the front axes ("X" and "Z") and the receding axis ("Y").

The **cavalier oblique** is a drawing based on an oblique projection. The lines of sight in a cavalier oblique make an angle of 45° with the plane of projection. In a true cavalier projection, the receding lines along the "Y" axis project in their true length. Therefore, the cavalier oblique drawing is usually drawn with a receding axis of 45° (approximating a line of sight of 45°) and the same scale is used on all three axes, Fig. 11-37.

Using an equal scale on all axes is the principal advantage of the cavalier oblique drawing over other types. However, it presents a distorted appearance for objects when the depth approaches or exceeds the width. Unless otherwise indicated, an oblique drawing refers to a cavalier oblique.

The **cabinet oblique** drawing is also based on an oblique projection. However, the lines of sight make an angle of 63° 26′ with the plane of projection. In this type of projection the receding lines project one-half their true length. Therefore, the scale on the receding axis of the cabinet oblique drawing is one-half that for the other axes, Fig. 11-38.

The **general oblique** drawing is based on a type of oblique projection as well. The line of sight

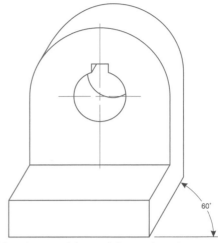

Fig. 11-38. In a cabinet oblique drawing, receding lines typically angle upwards at 60°. These lines appear half scale, while all other lines appear true length.

in a general oblique is not 45° or 63° 26′. The scale on the receding axis is greater than one-half but less than full size. The general oblique drawing may be drawn at any angle between 0° and 90°, Fig. 11-39.

While oblique drawings may be drawn with a receding axis at any angle between 0° and 90°, the most common angles are 45° and 30° with the horizontal. These angles can easily be drawn with triangles. The three types differ mainly in the ratio of the scale used on the receding axis compared to the other axes, Fig. 11-40.

Angles in oblique drawings

Angles that are shown true size in the frontal orthographic view will appear true size in the frontal plane of an oblique. Angles that lie in planes other than the frontal plane must be located by finding the end points of the lines forming the angle, Fig. 11-41.

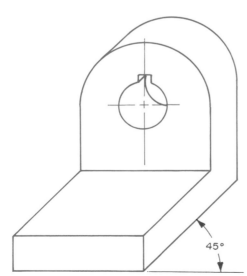

Fig. 11-37. In a cavalier oblique drawing, the receding lines typically angle upwards at 45°. All lines appear true length.

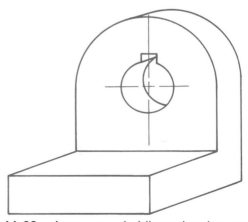

Fig. 11-39. In a general oblique drawing, receding lines are typically drawn upwards at any angle between 45° and 90° other than 45° or 60°. The receding lines also appear somewhere between true length and full scale.

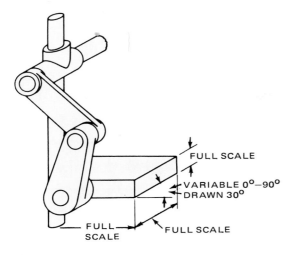

OBLIQUE PROJECTION — CAVALIER

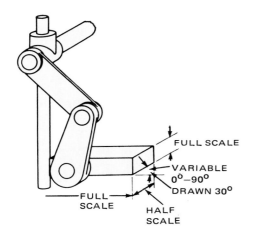

OBLIQUE PROJECTION — CABINET

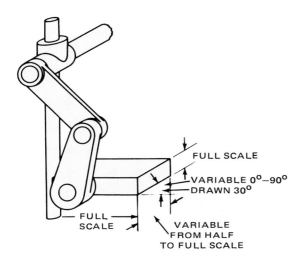

OBLIQUE PROJECTION — GENERAL

Fig. 11-40. The three types of oblique drawings will give different appearances. The appropriate type should be selected to best portray the object. (American National Standards Institute)

The top and front views of the object are enclosed in rectangles. These rectangles outline the overall size of the object and are used in laying out the oblique view.

The angles made by lines AA' and BB' with the back of the object are found by setting off the appropriate measurements from the orthographic views on the oblique cube. All measurements must be made along the oblique axes, or lines parallel to these axes. All measurements are foreshortened on the "Y" axis for the cabinet and general oblique drawings.

Arcs, circles, and irregular curves in oblique drawings

Arcs and circles located in the frontal plane of an oblique drawing will appear in their true shape. When arcs and circles are located in the principal cavalier oblique planes other than the frontal, these may be drawn using the four-center approximate method, Fig. 11-42A. Arcs and circles for cabinet and general oblique drawings are first drawn in their true shape in the frontal plane. They must then be transferred to the other oblique planes using the coordinated method and oblique "squares," due to the foreshortening of the "Y" axis, Fig. 11-42B. Irregular curves are also transferred to oblique drawings by means of the coordinate method, Fig. 11-42C.

Selection of position for oblique drawings

One of the advantages of oblique drawings is that arcs and circles are drawn in their true shape when they are located in the frontal plane. When possible, consideration should be given to selecting the view containing arcs and circles as the front view.

However, there may be other reasons why this choice for the front view may not be the best. For example, elongated objects should be located parallel to the frontal view to minimize distortion, Fig. 11-43A. It may also be desirable to view the object from below. This can be done by drawing the "X" and "Z" axes in their normal positions, but drawing the receding axis "Y" *down* at the oblique angle instead of up, Fig. 11-43B.

Sectional views in oblique

Sectional views may be used in an oblique drawing to provide a better view of interior detail, Fig. 11-44. The half section is used more often than the full section. The full section usually does not show sufficient exterior detail of a part. Correct positioning of the part is as important for oblique sections as it is for exterior oblique drawings.

Oblique dimensioning

Dimensioning of oblique drawings is similar to the dimensioning of isometric drawings. Approved practice calls for the dimension and extension lines

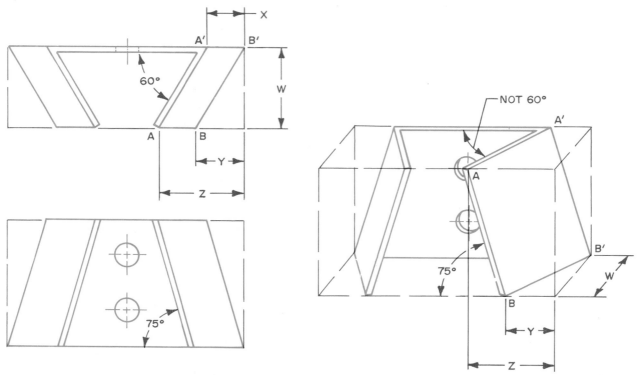

Fig. 11-41. Angles that lie in the frontal plane can be laid off directly. Angles that lie in any other plane must be located by first finding their endpoints. The endpoints must be located by measuring parallel to the isometric planes.

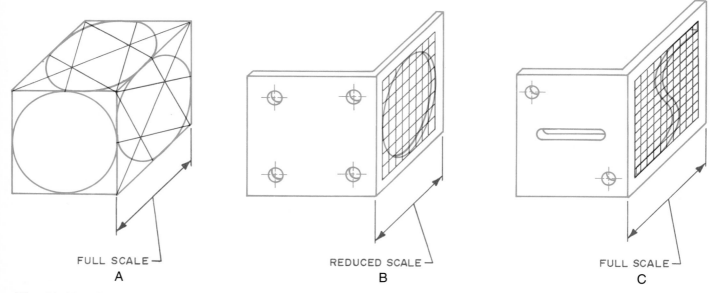

Fig. 11-42. Arcs, circles, and irregular curves that lie in the frontal plane of an oblique drawing are true size. These can be laid off directly. Arcs, circles, and irregular curves that lie in any other plane must be located differently. A–The four-center approximate method can be used to locate arcs, circles, and irregular curves. B–Arcs and circles can also be located by the coordinate method. C–Irregular curves can be located using the coordinate method as well.

to lie in the corresponding plane for the feature dimensioned, Fig. 11-45. When dimensioning oblique drawings, either the aligned or unidirectional system may be used, depending on company guidlines.

Advantages and disadvantages of oblique drawings

The oblique drawing has the advantage of showing an object in its true shape in the frontal

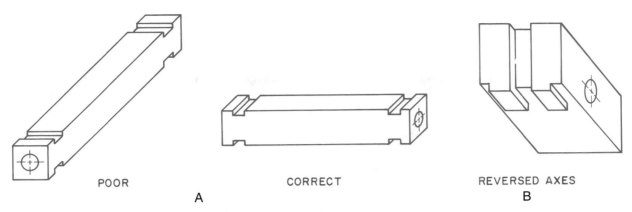

POOR   CORRECT   REVERSED AXES

A   B

Fig. 11-43. A—Elongated objects should be located so that the long side is parallel to the viewing plane. B—This will minimize distortion. Occasionally the bottom of an object needs to be shown. By reversing the direction of the "Y" axis, the bottom of the object can be shown.

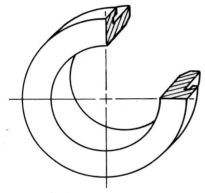

Fig. 11-44. Oblique half section views are sometimes used to show interior details. The half section is most often used because the full section usually does not show enough exterior detail.

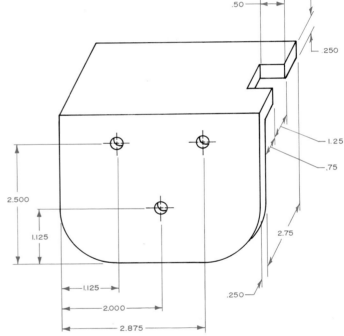

Fig. 11-45. Dimensions on oblique drawings must be in the correct plane for the feature dimensioned.

plane. Therefore, it may be a faster method of pictorial presentation when arcs and circles are located in this plane. The cavalier oblique is somewhat distorted on the receding planes. However, this is compensated for in the cabinet and general oblique drawings. These obliques have reduced scales on the "Y" axis.

A definite limitation of the cabinet and general oblique drawing is that circular and curved features on the receding planes must be drawn by the time-consuming coordinate method. Long objects appear distorted when necessary to draw these features on the "Y" axis, particularly in the cavalier oblique. When cabinet and general oblique drawings are used, a second scale must be used on the "Y" axis.

Perspective

Perspective drawings are the type of pictorial drawings that most nearly represent what is seen by the eye or camera. In the pictorial drawings studied earlier in this chapter, the parallel lines remained parallel. However, in perspective drawings parallel lines tend to converge as they recede from a person's view. This reproduces the effect of looking at a real object or scene, Fig. 11-46.

The three basic types of perspective drawings are the one-point (parallel), two-point (angular), and three-point methods. These methods are named for the number of vanishing points required in their construction, Fig. 11-47.

Terminology in perspective drawings

There are certain terms that must be defined before a discussion of perspective drawings can proceed. Those terms commonly used are pre-

Fig. 11-46. The principles of perspective make train tracks appear to come together in the distance. If a line is drawn along each track, the point of intersection is called a vanishing point.

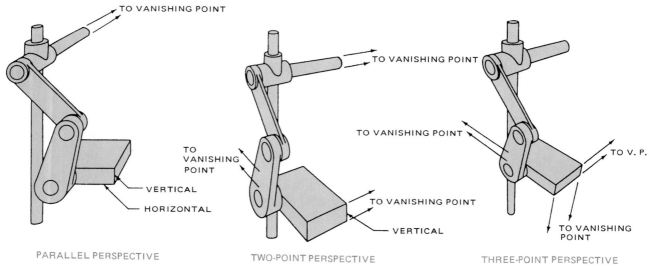

Fig. 11-47. One-point (or parallel), two-point (or angular), and three-point perspectives will produce different appearances. The appropriate type should be selected for the particular application. (American National Standards Institute)

sented here and illustrated for further clarification in Fig. 11-48.

Station point. A *station point (SP)* is an assumed point representing the position of the observer's eye. This is sometimes called the "point of sight." The location of this point greatly affects the perspective produced. To select the best view, lo-cate the station point at a distance from the object to form a viewing angle of the entire object of approximately 30°, Fig. 11-48A. Move the station point to the right or left, depending on the particular view of the object to be emphasized. The elevation of the station point is on the horizon line and determines whether the object is viewed from above, on, or below center.

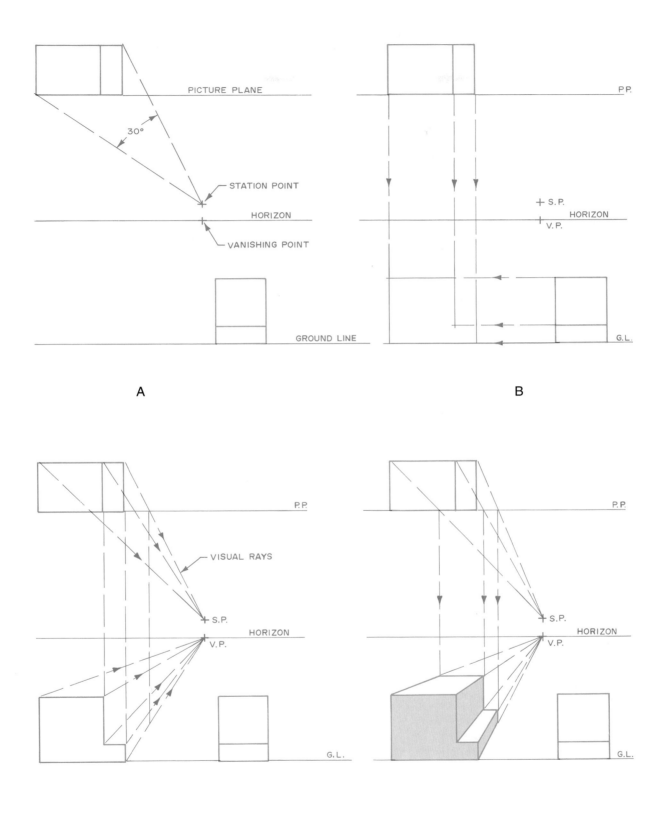

Fig. 11-48. A one-point, or parallel, perspective has one vanishing point located on the horizon line.

Vanishing points. *Vanishing points (VP)* are points in space where all parallel lines meet. (These lines are *not* parallel to the picture plane.) For horizontal parallel lines, the vanishing points are always located on the horizon line, Fig. 11-48C. (Also refer to Fig. 11-48.)

Visual rays. *Visual rays* are lines of sight from the object to the station point. They represent the light rays that produce an image in the viewer's eye. (Refer back to Fig. 11-47C.)

Picture plane. A *picture plane* is the projection plane that the perspective is viewed from. It may be aligned with, in front of, or behind the object. The picture point is a vertical plane for most perspectives. The picture point appears as a line in the top view and as a plane in the front view. However, it could appear in any position. For example, for a "bird's eye" perspective, the plane would be horizontal in the top view and appear as an edge in the front view.

Horizon line. The *horizon line* is a horizontal line that the vanishing points are located on and where receding lines tend to converge. The horizon line can appear at any level. It can be above, behind, or below the object to produce the perspective view desired. The effects of various levels of the horizon line and positions of the station point are illustrated in Fig. 11-49.

Ground line. The *ground line (GL)* is the base line or position of rest for the object.

One-point or parallel perspective

The **one-point perspective** has only one vanishing point. In a one-point perspective, the frontal plane of the object is parallel to the picture plane. A one-point perspective is also known as a parallel perspective.

Given the top view parallel to the picture plane, the side view, the station point, horizon line, and ground line, the construction procedure for one-point perspective is as follows.

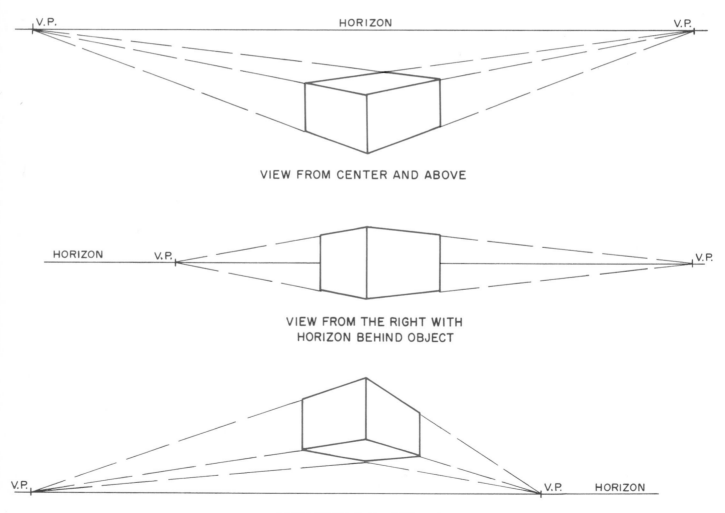

Fig. 11-49. The station point can be placed below, on, or above the horizon line. The station point can also be placed on the center, to the right, or to the left of the object.

1. Locate the vanishing point on the horizon line directly below the station point. (Refer back to Fig. 11-48A.)

2. Construct the frontal plane of the object by projections from the top and side views. This will be a true size (scale) projection since the surface lies in the picture plane. (Refer back to Fig. 11-48B.)

3. Draw projectors from the rear points of the object (top view) to the station point. (Refer back to Fig. 11-48C.)

4. Draw projectors from points on the front view to the vanishing point. (Refer back to Fig. 11-48C.)

5. Construct the depth of the perspective view by drawing vertical projectors from points where the projectors from the top view to the station point cross the picture plane. (Refer back to Fig. 11-48D.) These vertical projectors intersect the projectors to the vanishing points to form a one-point, or parallel, perspective.

The one-point perspective is used in many instances to help in illustrating the interior of a room or structure.

Two-point or angular perspective

In a **two-point perspective,** two sets of principal planes of the object are inclined to the picture plane. This perspective is sometimes called an **angular perspective** because of the angle the object makes with the picture plane. The third set of planes (usually the top view) remains parallel to its plane of projection. Parallel lines of the inclined sets converge at vanishing points on the horizon line, Fig. 11-50. Given the top view, side view, and the station point, the construction procedure for a two-point perspective is as follows.

1. Draw the picture plane, the horizon line, and the ground line, Fig. 11-50A.

2. Draw projectors from the station point parallel to the forward edges of the object to intersect the picture plane, Fig. 11-50B.

3. Project these points vertically down to the horizon line to establish two vanishing points.

4. From where it appears at picture plane, project line VW to the ground line, Fig. 11-50C.

5. Line VW is parallel to, and lies on, the picture plane. Therefore, line VW appears true length in the perspective when projected from the side view.

6. Project the end points of line VW to the left and right vanishing points. This will establish two perspective planes.

7. Draw projectors from the exterior corners of the object in the top view to the station point.

8. Project the intersections of these projectors with the picture plane to the perspective view to establish the limits of the object.

9. Draw projectors from points X and Y to the station point. Where these projectors cross the picture plane, drop vertical projectors to locate these lines in the perspective view, Fig. 11-50D.

10. Project point Z to the perspective true length corner line. From there, project it toward the vanishing points to complete the shoulder cut in the object.

11. Complete the remaining lines of the perspective as shown. Erase all construction lines.

The two-point perspective drawing is very useful in drawing large structures. Buildings are often drawn in two-point perspective for architectural work. Engineering projects such as bridges or petroleum plant piping installations are also commonly drawn in two-point perspective.

Perspective of objects lying behind the picture plane

When objects lie behind, and do not touch, the picture plane, the lines of the object must be extended to the picture plane to establish their piercing points, Fig. 11-51. From these piercing points, verticals are dropped to the ground line. This establishes true length lines. The receding lines are drawn from these true length lines and extend to the vanishing points. On these receding lines, lengths are projected from the intersection of the visual rays with the picture plane.

Measuring-point system in perspectives

The **measuring-point system** is a means of making accurate measurements in perspective drawings without having to retain the top view once the vanishing points and measuring points have been established, Fig. 11-52C. This means that actual measurements can be made on the ground line, rather than projecting these from the top view. The process of locating measurements on the perspective view is more direct. Large top views that require considerable drawing table space (an architectural plan, for example) can be removed.

Given the top view and its position, the station point, the horizon line, and the ground line, follow this procedure for drawing two-point perspective using the measuring-point system as follows.

1. Locate vanishing points as in a regular two-point perspective. (Refer back to Fig. 11-50B.)

2. Draw the front corner line OE true length from the known height measurement in the side view. Project its planes to their respective vanishing points, Fig. 11-52A.

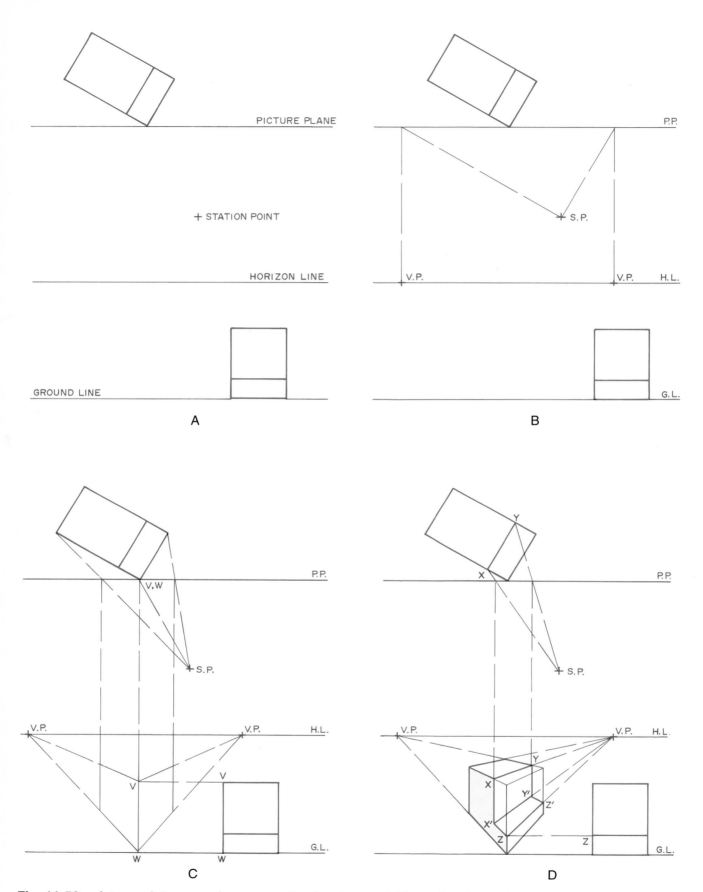

PICTURE PLANE

+ STATION POINT

HORIZON LINE

GROUND LINE

A

P.P.

S.P.

V.P.   V.P.   H.L.

G.L.

B

P.P.

V,W

S.P.

V.P.   V.P.   H.L.

V   V

V

W   W

G.L.

C

P.P.

Y

X

S.P.

V.P.   V.P.   H.L.

Y

X   Y'

Z'

X'

Z   Z   G.L.

D

Fig. 11-50.   A two-point, or angular, perspective has two vanishing points located on the horizon line.

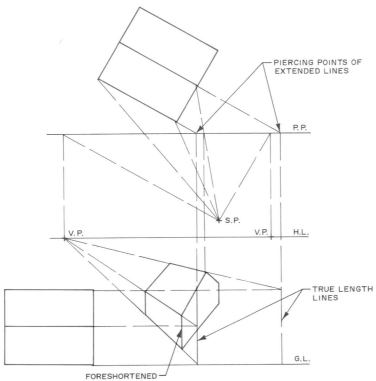

Fig. 11-51. Lines on objects that lie behind the picture plane are foreshortened.

3. Using point O in the top view (the corner of the block on the picture plane) as the center of a radius, revolve edges OA and OB into the picture plane.

4. Draw lines AA′ and BB′.

5. Draw lines parallel to AA′ and BB′ from the picture plane and passing through the station point, Fig. 11-52B.

6. Project the points of intersection with the picture plane in step 5 to the horizon line to establish measuring points.

7. Project lines A′O and OB′ to the ground line to form true length lines AO and OB. (These lines have been revolved into the picture plane where they appear true length. They could have been laid off true length on the ground line by direct transfer of measurements from the top view.)

8. Project the true length height of AD and BC from OE, Fig. 11-52C.

9. Extend planes AD and BC to their respective measuring points. The intersections of these planes with the receding planes from corner OE establish the limits of the receding planes from corner OE.

10. Complete the perspective by drawing the remaining vertical or receding lines.

11. Measurements for other features can be laid off true length on the ground line from known measurements in top. As an example, the shoulder cut X-Y-Z in 11-52D.

Circles in perspective

Circles parallel to the picture plane will appear as circles in perspective drawings, Fig. 11-53. If the circle lays on the picture plane, it will be true size. If the circle is on a plane behind the picture plane, it will appear in a reduced size, but as true circle, Fig. 11-53. These circles may be drawn with a compass or circle template by locating their centers in the perspective view and determining their diameters by projection.

Circles on a plane inclined to the picture plane will appear elliptical in the perspective view. These circles can be located by means of the enclosing square or coordinate method. In Fig. 11-54A, the technique for drawing a perspective representation of a circle in a vertical plane inclined to the picture plane is shown. The procedure for locating this circle is as follows.

1. Divide orthographic front view into a number of parts (30°-60° divisions are suggested).

2. Transfer these division points to their respective locations in the top view.

3. Project points from the front view to the front corner of the block containing the circle in the

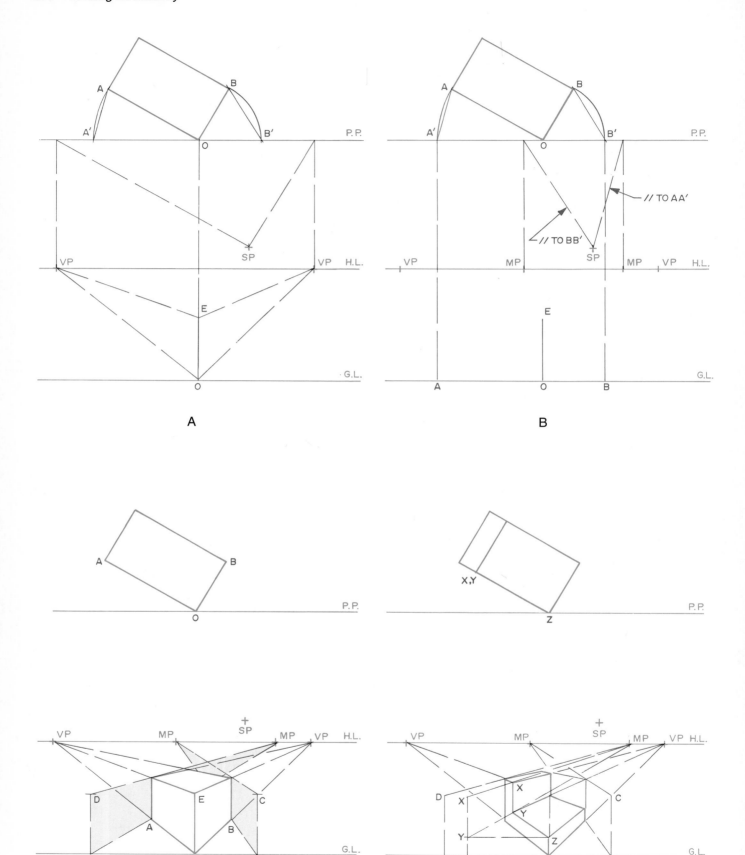

Fig. 11-52. The measuring point system can be used to construct perspective drawings. This system is very useful for drawing large objects, such as buildings.

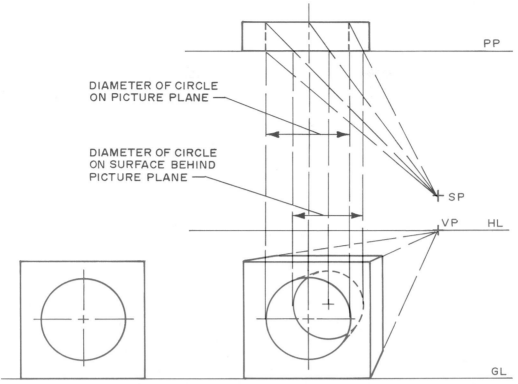

Fig. 11-53. Circles on surfaces parallel to picture plane will appear as circles in perspective drawings. However, only circles that appear on the picture plane will appear true size. All others will be foreshortened.

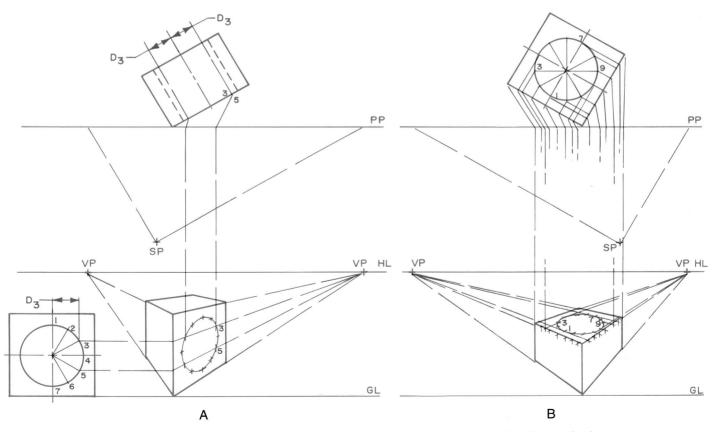

A

B

Fig. 11-54. Circles on surfaces inclined to the picture plane can be constructed using the enclosing square, or coordinate, method.

perspective view. From there project points to the vanishing point.

4. Project division points from the top view to the picture plane by visual rays to the station point. Also, project division points from the picture plane, vertically, to the perspective view to intersect with their corresponding lines.

5. Draw the perspective ellipse with an irregular curve or an ellipse template.

The construction of circles on horizontal surfaces in perspective drawings is shown in Fig. 11-54B. The procedure for this construction is as follows.

1. Divide circle in top view into a number of parts (30°-60° divisions are suggested).

2. Project these division points to the front two edges of the block on lines parallel to adjacent sides.

3. Establish the height of the block on the picture plane corner in the perspective. (This will be true length since the corner is on the picture plane.) Draw receding lines to the vanishing points.

4. Project division points from the top view to the picture plane by visual rays to the station point. Also project from the picture plane, vertically, to the perspective view to intersect with their corresponding front edge lines.

5. From these points of intersection on the frontal edges, draw receding lines to the two vanishing points. The points of intersection of corresponding lines are points on the ellipse.

6. Draw the perspective ellipse using an irregular curve or an ellipse template.

Irregular curves in perspective

Irregular curves may be drawn in perspective using the coordinate method, Fig. 11-55. Points are located along the curve in the orthographic views and projected to the picture plane and vertical true length line in the perspective.

Perspective grid

A **perspective grid** is a graph-oriented method of making accurate perspective drawings without having to establish (and project from) vanishing, measuring, and sighting points, Fig. 11-56. There are many variations, such as the one-point method for interior views, the three-point oblique, the cylindrical grid for representing aircraft fuselages, and others. However, the cube grid is the most widely used for general purpose illustration, Fig. 11-57.

Grids include a scale that measurements can be projected from. This maintains an accurate representation of proportion in the object being drawn. Instead of actually drawing on the grid, an overlay sheet of tracing paper or vellum is used to preserve the grid for further use, as well as to eliminate the grid pattern on the perspective drawing.

Grids save time in the preparation of perspective drawings. Grids also permit several drafters and illustrators to draw related components independent of each other and know the parts will match with respect to size, angle, and perspective.

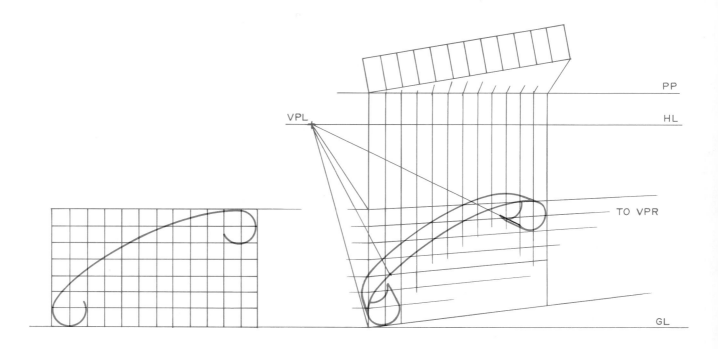

Fig. 11-55.   Irregular curves can be constructed in perspective by using the coordinate method.

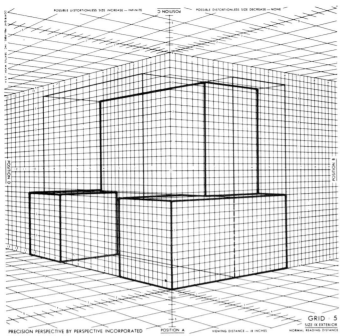

Fig. 11-56. A perspective grid eliminates the need to establish vanishing, measuring, and sighting points. These points have already been established by the grid.

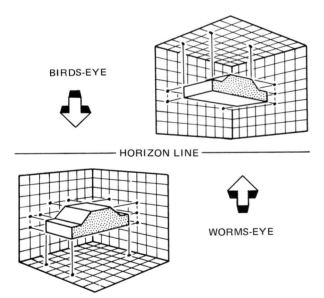

Fig. 11-57. The cube-type grid is the most common type of perspective grid for general purpose illustrations. (General Motors Drafting Standards)

Perspective grids also permit additions or revisions to drawings at a future date with speed and accuracy. Perspective grids may be purchased at local drafting and art supply stores.

Perspective drawing boards

Special perspective drawing boards are available that also speed the process of constructing perspective drawings, Fig, 11-58. These boards are

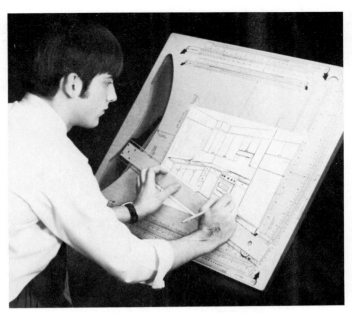

Fig. 11-58. A Klok perspective drawing board simplifies the process of drawing perspectives.

fairly simple to use and come with a variety of scales, permitting direct reading for layout of perspectives. Perspective boards are most valuable to drafters and illustrators who make frequent use of perspective drawings in their work.

Sketching in perspective

The technique of sketching in perspective is based upon an understanding of the principles of sketching discussed in Chapter 3. Also, the mechanical projection of perspective drawings discussed earlier in this chapter is applied to perspective sketches.

Developing a technique for sketching in perspective is dependent on good proportion of the objects being represented. The steps in sketching a two-point, or angular, perspective are as follows.

1. Establish two vanishing points on the horizon line as far apart as desired, Fig. 11-59A.

2. Sketch a true length (scale) vertical line to represent the front corner of the object to be sketched. For Fig. 11-59A, this has been centered between two vanishing points and located below horizon line. This positions front faces of object at 45° with picture plane and a view from above object.

3. Sketch lines from the ends of the front vertical line to the two vanishing points.

4. Sketch vertical lines at a distance of one-half true distance (scale) from the front vertical line to establish the length of the two frontal planes.

5. From the upper-rear corners of these planes, sketch receding lines to the opposite vanishing points to form the top surface of the object.

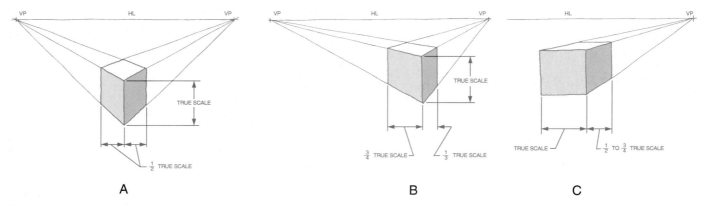

Fig. 11-59. There are several different variations of perspective sketches.

Other features may be added to the sketch by measuring on the front corner and projecting to their proper location of depth on a half-scale basis. The object may be positioned so that its faces make angles of 30° and 60° with the picture plane by locating the front vertical line as shown in Fig. 11-59B. A one-point, or parallel, perspective would be sketched as shown in Fig. 11-59C.

Circles are sketched in perspective by first drawing a square, Fig. 11-60. Then the center points of the sides of the square are located. The enclosed circle (ellipse) is then sketched.

CADD generated pictorials and models

The use of computer-assisted design and drafting brings new methods to create and view three-dimensional drawings. Although you can still create pictorial drawings using the techniques discussed in this chapter, you can also use the 3D drawing functions of CADD to create true 3D models. The computer "thinks" of the design as having material and volume. A 3D model of a milk container is shown in Fig. 11-61. The designer can view the

model from any angle. The designer can also view the object with shading. Two devices that were designed and illustrated using CADD software are shown in Figure 11-62.

PROBLEMS AND ACTIVITIES

The following problems will provide you with the opportunity to apply the techniques of pictorial drawing. The problems may be placed on A-size or B-size sheets.

1. Make isometric drawings of any two of the parts in Fig. 11-63. Dimension the drawings.

2. Prepare a dimetric scale. Use this scale in drawing an approximate dimetric drawing of any of the parts in Fig. 11-63. Dimension the drawing.

3. Make an approximate trimetric drawing of the "BEARING MOUNT" in Fig. 11-64. Select an axis from those shown in Fig. 11-34 which best

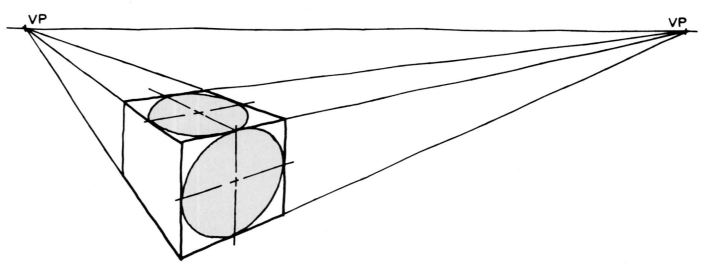

Fig. 11-60. Circles in perspective are sketched by first sketching the circumscribed square and locating perspective centerlines for the circle. Then sketch the circle inside of the square.

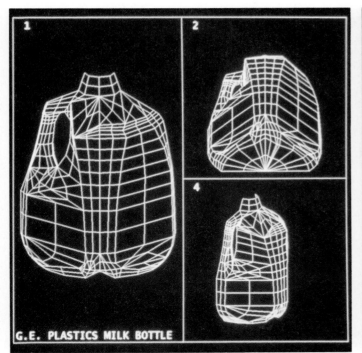
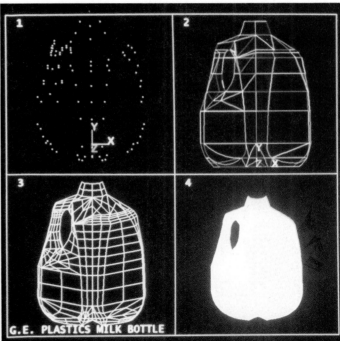

Fig. 11-61. Computer-assisted design and drafting programs allow 3D models of objects to be created. A–The object can be rotated and viewed from various angles . B–A "skin" can then be added to the object to give it a realistic look that is close to the appearance of the final product. (GE Plastics, Inc.)

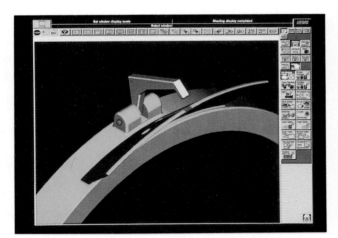
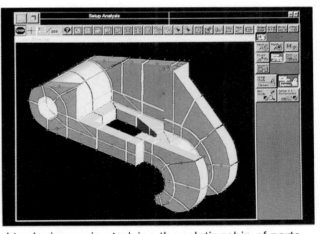

Fig. 11-62. CADD pictorials and models are very useful to designers in studying the relationship of parts and in communicating the overall design to others. (Intergraph Corporation)

displays the features of the object. Find the ellipse guide angle and use an ellipse template to draw the circular features.

4. Make an oblique drawing of the "WEAR PLATE" in Fig. 11-64. Dimension the drawing.

5. Construct a one-point perspective drawing of the "MOUNTING BRACKET" in Fig. 11-64. Do not dimension.

6. Make pictorial drawings of the remaining objects in Figs. 11-64 and 11-65. You may select the type of pictorial drawing to be used, or use the type assigned by your instructor.

7. Select an object at home, school, or work and prepare a pictorial drawing of the object. Use any type of pictorial drawing, or use the type assigned by your instructor.

8. Design some object which you can use around home or work. Discuss your idea with your instructor before starting a drawing. Then, prepare a dimensioned pictorial drawing of the object.

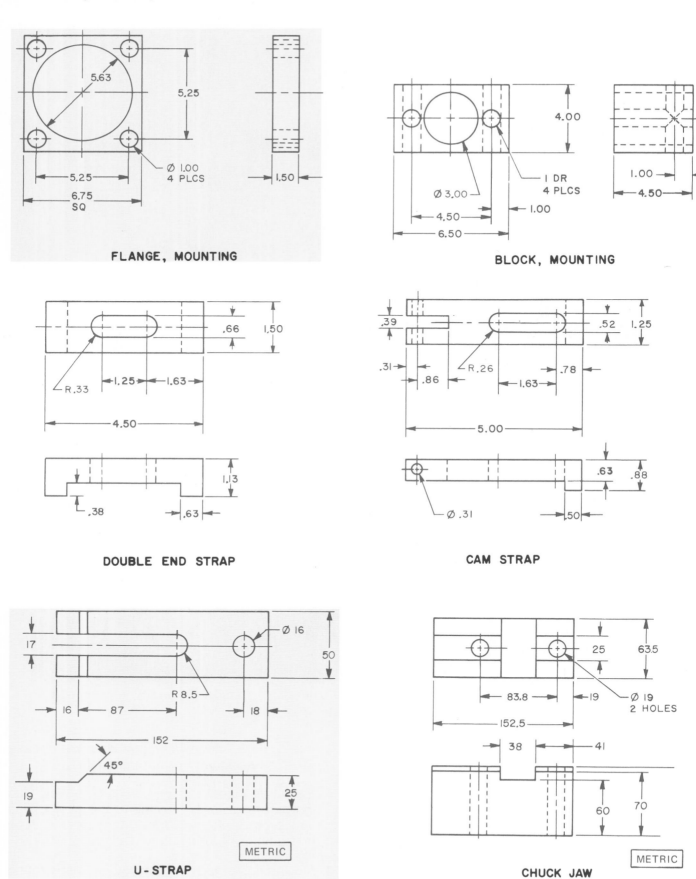

Fig. 11-63.   Use these objects for the problems on isometric and dimetric drawings.

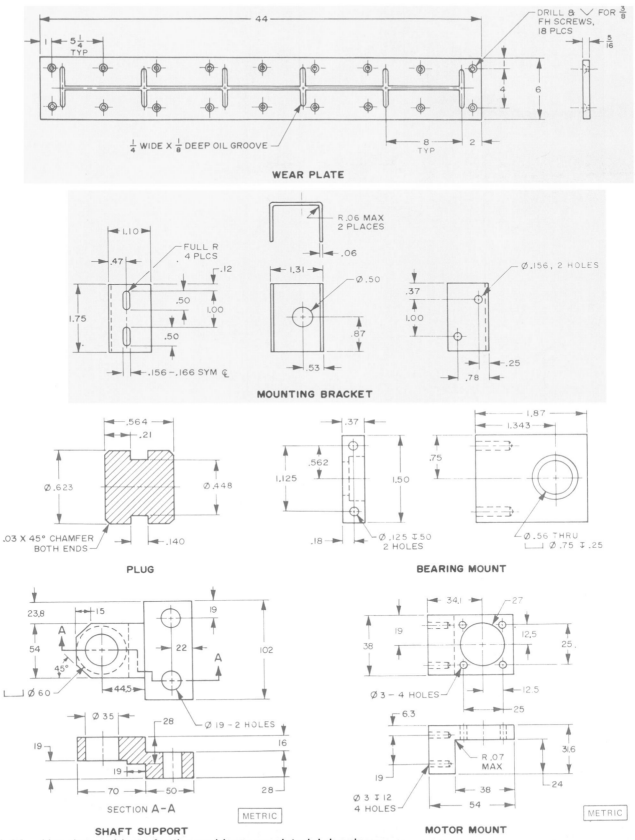

Fig. 11-64.   Use these objects for the problems on pictorial drawings.

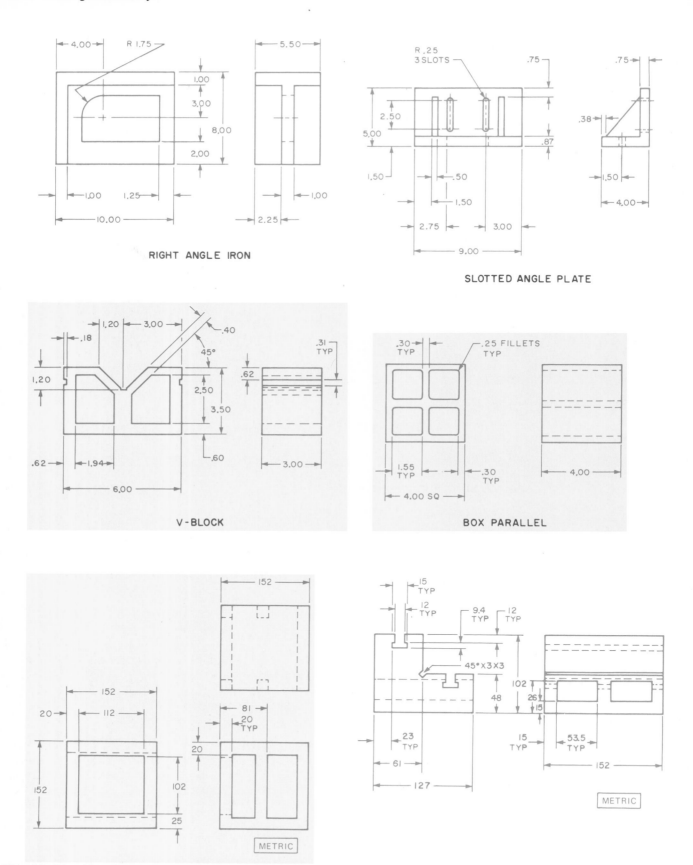

Fig. 11-65.   Use these objects for the problems on pictorial drawings.

12 Reproducing and Storing Drawings

KEY CONCEPTS

☐ There are several ways used by industry to copy, or reproduce, drawings.

☐ Microfilm has certain advantages and is often used to store drawings.

☐ Diazo prints (or whiteprints), blueprints, and electrostatic method prints are three ways to reproduce drawings.

☐ Storing drawings is an important part of the drafting process.

Making prints of original drawings is a major portion of the time that most industrial drafting departments spend on reproductions. However, this is not the only type of reproduction work performed. Other important aspects include microfilming, photodrafting, and scissors drafting, Fig. 12-1. All of these processes have done much to improve the drafting process as a whole.

REPRODUCTION OF DRAWINGS AND THE DRAFTING PROCESS

Elements of reproduction processes are presented and explained in this chapter. This will provide the student with an understanding and appreciation of the importance to drafting of these processes.

Definition of Terms

A knowledge of the following terms will be helpful in understanding the reproduction of drawings.

- **Aperture card:** An electronic accounting machine card with a rectangular hole designed to hold a single frame of microfilm.

- **Autopositive:** A print made on paper or film by means of a positive-to-positive silver type emulsion.

- **Blowback:** An enlarged print made from a micro-image.

- **Generation:** The blowback made from a microfilm of an original drawing is a first generation

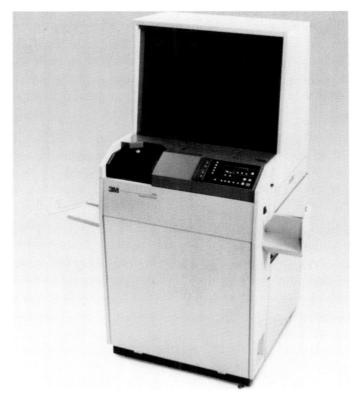

Fig. 12-1. Microfilm readers use aperture cards to view drawings or make hard copies. (3M Co.)

print. A blowback made from a microfilm of this first generation print would be a second generation print. The term generation is used to express the quality required of an original drawing being microfilmed. It is usually one that is capable of being reproduced as a clearly readable fourth generation print. Fourth generation quality is necessary because drawing changes are made on a reproduction copy of the original drawing. This revised copy is reproduced and changes are made on this revision for four or more reproduction sequences. Fourth generation quality must be present in the original drawing so that all lines, notes, and dimensions are clearly readable on the fourth generation print, Fig. 12-2.

- **Hard copy:** An enlarged print, on paper, cloth, or film, made from an original drawing or microfilm image.

- **Intermediate:** A translucent reproduction made on vellum, cloth, or film from an original

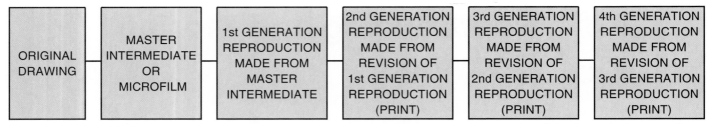

| ORIGINAL DRAWING | MASTER INTERMEDIATE OR MICROFILM | 1st GENERATION REPRODUCTION MADE FROM MASTER INTERMEDIATE | 2nd GENERATION REPRODUCTION MADE FROM REVISION OF 1st GENERATION REPRODUCTION (PRINT) | 3rd GENERATION REPRODUCTION MADE FROM REVISION OF 2nd GENERATION REPRODUCTION (PRINT) | 4th GENERATION REPRODUCTION MADE FROM REVISION OF 3rd GENERATION REPRODUCTION (PRINT) |

Fig. 12-2.   A first generation drawing is a reproduction of a master intermediate or microfilm. A second generation is a reproduction of a first generation print. A third generation is made from a second generation, and a fourth generation is made from a third generation.

drawing to serve in place of the original for making other prints.

- **Negative print:** A print, usually on opaque material, which is opposite to the original drawing. In other words, light lines appear on a dark background.
- **Positive:** A print which is similar to the original drawing. In other words, dark lines appear on a light background.
- **Reproducible:** Capable of being used as a master for making prints by the action of radiant energy (light) on chemically-treated media.
- **Sensitized:** A reproduction material coated with a light-sensitive emulsion.
- **Translucent:** A material that permits the passage of light (partially transparent).

Microfilming

The technique of microfilming has been known since the early 1800s. However, it was not until World War II that microfilming was applied to drafting. Microfilming was used to store **security copies** of original drawings at a separate location in the event of a disaster. Security copies are exact duplications of the original drawings. More recently, microfilming has been developed as an active working technique in creating new drawings, updating old drawings, reducing storage space requirements, and in finding drawings once they have been filed.

The microfilm system has done more to revolutionize drafting operation than any other single factor. The principal advantages are:

- Less storage space is required.
- The retrieval time is nearly instantaneous. This allows for quick viewing, or for quick hard copies.
- The data in drawings is used more because of the ready availability.
- There is reduced handling. This helps to preserve the original drawings.

However, in order to offer these advantages, microfilming places rigid quality control demands on the original drawing. The microfilm process reduces a 34″ x 44″ (E-size) drawing 30 times. This means

there is not much room for error in line weight, lettering, and drawing quality.

Aperture card

The microfilm **aperture card** is an electronic accounting machine (EAM) tabulating card. The card is a punched card with a single frame microfilm insert that contains the image of the original drawing, Fig. 12-3. The card may be retrieved quickly in an electronic card sort, inserted in a reader-printer, and viewed on a screen or used to produce a hard copy.

Drafting techniques for microfilm quality

The quality of any drawing reproduction depends on the quality of line work used, and on clear, legible, well-formed letters and numbers in notes and dimensions. For drawings to be microfilmed, these qualities are especially important.

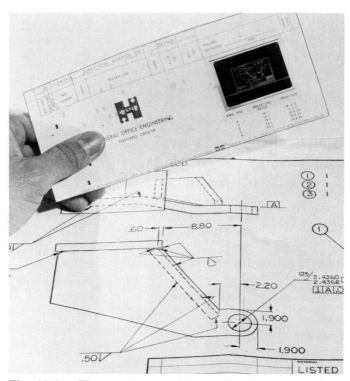

Fig. 12-3.   The aperture card is an effective means of storing, retrieving, and disseminating drawing information. The microfilmed drawing is contained on the small piece of film located in a "window" on the card.

Refer back to Chapter 4 for a discussion of lettering drawings to be microfilmed. In addition to precise lettering, lines should be a minimum of .01″ thick. However, avoid excessively thick lines. Also, the space between lines (or lines of lettering) should be a minimum of .06″, Fig. 12-4. Cleanliness is also very important for drawings that will be microfilmed.

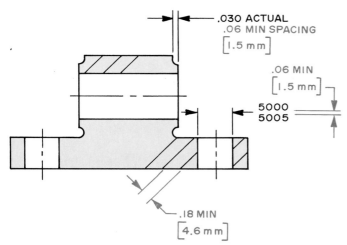

Fig. 12-4. Line and letter quality, cleanliness, and spacing are critical for drawings that will be microfilmed.

Photodrawings

A *photodrawing* is a photograph of either an object or a model of an object that callouts and notes are added to. A photodrawing is made on reproducible drafting film or paper. The necessary lines, dimensions, and notes are added to the paper to create the photodrawing, Fig. 12-5. Photodrawings have been used for some time in aerial mapping of land areas for highway and other construction projects. They have also been widely used in the electronic industry for wire assembly work.

Where a drawing would require a considerable number of hours in layout work, and where a pictorial presentation would aid in interpreting the drawing, a photodrawing might be more descriptive than a conventional drawing. This might mean a considerable time savings as well.

Photodrawings begin with a photograph of a machine part, model, or building, Fig. 12-6. The photo is prepared as a continuous tone or half-tone print. Half-tone photographs produce the best quality diazo prints.

The term *photodrafting* is also used. This term includes photodrawing, but also refers to the process of making a drawing where no photographs are used. In this case, sections of one or more drawings are combined in a new or revised drawing using photographic techniques. The process

of "cutting and pasting" is referred to as "scissors drafting" and is discussed in the next section.

Scissors Drafting

When part (or all) of one drawing is used to create part (or all) of a "second-original" drawing, this process is known as *scissors drafting.* Modern reproduction methods make it possible to merge parts of several drawings onto one drawing media. The necessary lines, notes, and dimensions can then be added to this.

Scissors drafting is similar to photodrawing, except the source of the "add-on" material is from other drawings, prints, parts catalogs, or charts, Fig. 12-7. Unwanted sections on a drawing can be "cut" or "blanked out" during the reproduction process. Changes can be drawn in without erasing.

When the parts have been combined onto one reproducible drawing, an *autopositive print* is made and serves as the "second original" drawing. Callouts, notes, and dimensions are then added to complete the drawing. Considerable drafting time can be saved when the parts added represent complicated and detailed objects.

Reproduction Prints

Today, there are numerous processes used to make prints from original drawings, or from a microfilm copy of an original drawing. The term *print* refers to any hard copy. The prints most commonly used require a translucent drawing (or microfilm copy) and a paper, cloth, film, or other medium that has been coated with a light-sensitive chemical. After passing the drawing and light-sensitive paper past a light source, the materials can be developed chemically to produce legible reproductions or prints.

The lines and lettering on the original drawing, or other master copy, prevent the light from acting on the sensitized coating of the reproduction base material, Fig. 12-8. This unexposed material will react during the chemical treatment process to develop the image of the original drawing. The type of reproduction print made depends upon the type of sensitized material used and its subsequent processing.

Several types of reproductions are made to serve different purposes. One type of reproduction, called an *intermediate,* is developed on a suitable medium such as vellum, film, or photographic paper. This will serve as a drawing medium in preparing "second-original" drawings. Another type intermediate is developed on a translucent material and serves as the "tracing" for use in making additional prints, saving the wear and tear on the

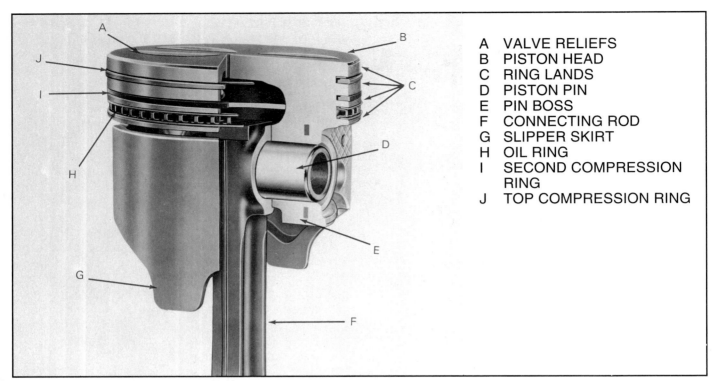

A VALVE RELIEFS
B PISTON HEAD
C RING LANDS
D PISTON PIN
E PIN BOSS
F CONNECTING ROD
G SLIPPER SKIRT
H OIL RING
I SECOND COMPRESSION
   RING
J TOP COMPRESSION RING

Fig. 12-5. A photodrawing is a photograph of an object that is reproduced on drafting film, or paper, with notes and callouts added to complete the drawing. (John Deere & Company)

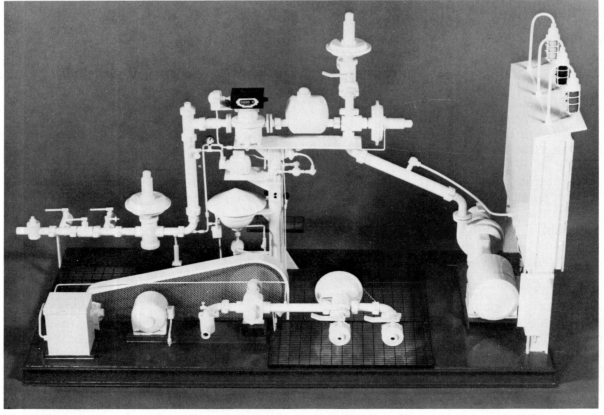

Fig. 12-6. A scale model can be used to make a photodrawing of large objects or buildings. (Cities Service Oil Co.)

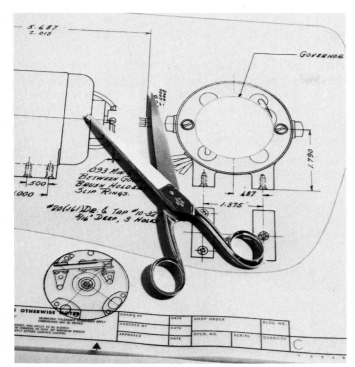

Fig. 12-7. Scissors drafting makes use of photodrafting techniques, taking parts of original drawings and splicing other parts or notes in for a "second original" drawing. (Eastman Kodak Co.)

DRAWING LINES ARE IMAGED BY INHIBITING LIGHT ACTION ON BLUEPRINT, DIAZO OR SILVER EMULSION COATINGS OF REPRODUCTION MATERIAL

LIGHT SOURCE

FEED BELT

REPRODUCTION MATERIAL

SENSITIZED COATING

ORIGINAL DRAWING MATERIAL

LINES ON SURFACE OF DRAWING

GLASS CYLINDER

Fig. 12-8. A typical reproduction machine uses a light source to reproduce the lines on a drawing. The lines on a drawing block light to certain areas of the light-sensitive paper. These "blocked" areas will become the lines on the reproduction. After exposure to the light source, the original returns to the operator and the sensitized material continues through a chemical treatment to develop the print. (General Motors Corp.)

original drawing. The largest number of reproductions used are the opaque type. These are used by the workers who will actually produce the object described on the drawing.

Blueprint

For many years, the **blueprint** was the only type of reproduction made. This print has white lines on a blue background. It is developed from a material coated with light-sensitive iron salts. The blueprint is called a **negative print** because the reproduction is opposite in tone to the original drawing. Dark lines on a light background produce light lines on a dark background.

Diazo process

A more recent and widely used process of making prints is the **diazo process.** This process produces **positive prints** with dark lines on a light background. These prints may have blue, black, brown, red, or other colored lines and are often referred to as **whiteprints.** However, the terms **direct line prints, black line, brown line,** and **red line** are also used. This process utilizes the light sensitivity of certain diazo compounds and is available in dry, moist, and pressurized form.

The original translucent drawing and sensitive diazo paper are inserted into a diazo type print machine, Fig. 12-9. They are then exposed to a light source. The light source destroys the unprotected diazo compound. The original drawing is returned to the operator and the exposed sensitized paper is carried through the developing section of the machine. Here it is exposed to a chemical that develops the print.

The moist form transfers an ammonia solution to the print to cause the development. The print is delivered in a somewhat moist or damp state. Dry form of diazo print making utilizes an ammonia vapor to develop the exposed copy. The copy produced is relatively dry. The pressure form of the

Fig. 12-9. A pressure diazo print machine is a common way to reproduce drawings in industry. (Teledyne Rotolite)

diazo process utilizes a thin film of a special activator delivered under pressure to the exposed copy to complete the development.

Electrostatic process

The **electrostatic process** is a means of producing paper prints and intermediate transparencies from original drawings or microfilm, Fig. 12-10. This process, commonly referred to as **xerography,** will develop a print on unsensitized paper. It will produce same size, enlarged, or reduced copies of the original drawing. The process works by forming an image electrostatically on a selenium coated drum or plate. The image can then be transferred onto most any type of material. The exposed surface is dusted with a dark colored powder, called **toner.** The toner is affixed permanently by heat or solvent action.

The electrostatic reproduction process is increasingly used as a means of producing prints and other documents. The electrostatic process is very convenient and fast. Electrostatic copiers are simple to use as well. Intermediate transparencies can also be produced by this process.

Storing Drawings

The production of engineering drawings represents a sizable investment to a company. Therefore, proper control and storage of the original drawing is important. A typical system provides for

three things. First, the location and status of drawings is known at all times. Second, damage to original drawings through improper handling is minimized. Finally, distribution of prints to appropriate individuals is provided for. Companies with CADD systems store their drawings in computer files. These files have some advantages for production, storage, control, and reproduction.

Companies using traditional drafting procedures store their drawings in flat-drawer files, hung vertically in cabinets, or in tubes, Fig. 12-11. Stor-

A

B

Fig. 12-11. A–Vertical storage cabinets provide easy access to large drawings. B–Flat storage cabinets contain individual compartments, or drawers, for related drawings. (Diazit)

Fig. 12-10. The Xerox™ machine is an electrostatic print making machine. Some machines can handle sheets as large as E-size. Some machines can automatically cut and roll the reproduced prints. (Xerox)

ing drawings in a safe place is important, but rapid location is also necessary. Prints are often folded and stored in standard office file cases. If properly organized, this method does provide rapid retrieval and security.

PROBLEMS
AND ACTIVITIES

The following problems and activities will provide an opportunity for you to gain further knowledge and understanding of the reproduction processes.

1. Select drawing problems from this text, or as assigned by your instructor. Prepare working drawings to microfilm quality standards.

2. Prepare a photodrawing utilizing a suitable photograph. Dimension the drawing and add necessary notes.

3. Prepare a "second-original" drawing using the scissors drafting technique. Select an earlier drawing to be modified or combined with another. Estimate the time saved by this technique over preparing an entirely new drawing.

4. Make a print on a translucent medium of one of your drawings. Use either the diazo process or the blueprint process. Study the reproduction qualities of the drawing as revealed in the print.

5. Make an electrostatic copy of one of your drawings prepared earlier. Compare the quality of this reproduction with the one in Problem 4.

6. Visit a local print service company and find out the types of reproduction processes they perform. Which type of reproduction is in greatest demand? What are the costs for the various types of reproduction prints? Write a short paper summarizing your findings.

7. Visit a drafting equipment supply company and inquire about the type of reproduction equipment they sell. Find out about new processes or trends developing in this field. Write a short paper on the new processes and trends.

Plotters are a way of reproducing CADD drawings. Copies can be created quickly and inexpensively. (Xerox)

Part III
Descriptive Geometry

Descriptive geometry is that phase of drafting dealing with spatial relationships and graphical analysis of problems. An introduction was presented in Chapter 7, Multiview Drawings, on the fundamentals of projection drawing.

In this part of the text, the principles and applications of descriptive geometry useful in problem solving are presented in chapters dealing with *AUXILIARY VIEWS, REVOLUTIONS, INTERSECTIONS,* and *DEVELOPMENTS.*

The manufacture of high performance aircraft requires many technical drawings to detail and specify the many different processes involved. (McDonnell-Douglas-St. Louis)

To create all of the drawings for this building, many auxiliary views, intersections, and developments may have been involved.

Auxiliary Views

KEY CONCEPTS

☐ An auxiliary view can provide necessary information about features that do not appear on the principal surfaces.

☐ Primary auxiliary views are projected from orthographic views.

☐ Secondary auxiliary views are projected from a primary auxiliary view and a principal view.

☐ Successive auxiliary views are projected from secondary (or later successive) auxiliary views.

The principal views of multiview drawings (top, front, and side views) are normally adequate for describing the shape and size of most objects. However, objects having features inclined or oblique to the principal projection planes require special projection procedures, Fig. 13-1. These projection procedures are possible through auxiliary views.

Auxiliary views are also often used in the sheet metal and packaging industries. In many cases, patterns must be developed for various-shaped surfaces at an angle with other surfaces, Fig. 13-2. The projection of these features in normal views always results in foreshortened and distorted views. These surfaces are not true to shape and size.

Fig. 13-2. Auxiliary views were used to develop the inclined surfaces of these satellite dishes. (Winnebago)

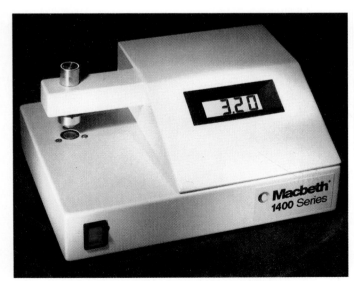

Fig. 13-1. Devices with inclined surfaces require auxiliary views. How many inclined surfaces can you identify on the densitometer? (Macbeth)

Auxiliary views provide the following basic information for features appearing on surfaces other than the principal surfaces.

• The true length of a line.
• Identifies the point view of a line.
• Locates the edge view of a plane.
• Determines the true size of a plane.
• Locates the true angle between planes.

Auxiliary (supplementary) projections are necessary to sufficiently analyze and describe these features for production purposes. Two kinds of auxiliary views are used in describing and refining the design of industrial products: **primary auxiliary views** and **secondary auxiliary views.**

Primary auxiliary views are projected from orthographic views—horizontal, frontal, or profile. Secondary auxiliary views are projected from a primary auxiliary and a principal view. Views projected after a secondary auxiliary are known as successive auxiliary views.

PRIMARY AUXILIARY VIEWS ┼┼┼┼

A primary auxiliary view is a first, or direct, projection of an inclined surface. It is projected perpendicular to one of the principal orthographic views: horizontal, frontal, or profile, Fig. 13-3. A primary auxiliary view is perpendicular to only one of the principal views. (A view that is perpendicular to two of the principal views is a normal view (discussed in Chapter 7) and not an auxiliary view.)

True Size and Shape of Inclined Surfaces

The primary auxiliary view is useful in determining the true size and shape of a surface that is inclined. This view is not useful in finding true size in any of the principal views of projection. (Refer back to Fig. 13-3.) The auxiliary view is projected from the principal view where the inclined surface appears as an "edge." The example in Fig. 13-4 shows the auxiliary view as an edge in the frontal plane. Therefore, the auxiliary projection is called a frontal projection.

Primary auxiliary planes of projection are always numbered "1" and preceded by the letter indicating a frontal, horizontal, or profile plane. In Fig. 13-4B, the reference plane F-1 is drawn at any convenient location and parallel to the line in the front view that represents the edge of the inclined surface.

Note in Fig. 13-4A that the primary auxiliary plane F-1 is perpendicular to the frontal plane. Also, the line of sight is perpendicular to the auxiliary plane. This is how any of the normal projection planes are viewed as well. Distances along the plane of projection are projected perpendicularly in their true length, Fig. 13-4B. True length measurements of depth are transferred with a scale or dividers from the top or side view to the auxiliary view.

This same object could have been arranged so that the inclined surface appears as an edge in the top view, Fig. 13-5. The inclined surface is projected in the same manner as for a frontal auxiliary projection. However, the plane of projection H-1 is a horizontal (top view) auxiliary, perpendicular to the horizontal plane. The line of sight is perpendicular to the inclined surface in the horizontal plane. The result is a horizontal auxiliary.

A profile auxiliary view would be projected from the profile or side view, when the inclined surface appears as an edge in this view, Fig. 13-6.

Location of a Point in an Auxiliary View

The location of points in an auxiliary view is the first important step in understanding the projection

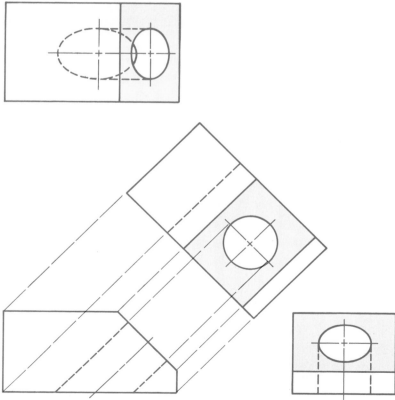

Fig. 13-3.   A primary auxiliary view is a first, or direct, projection of an inclined surface.

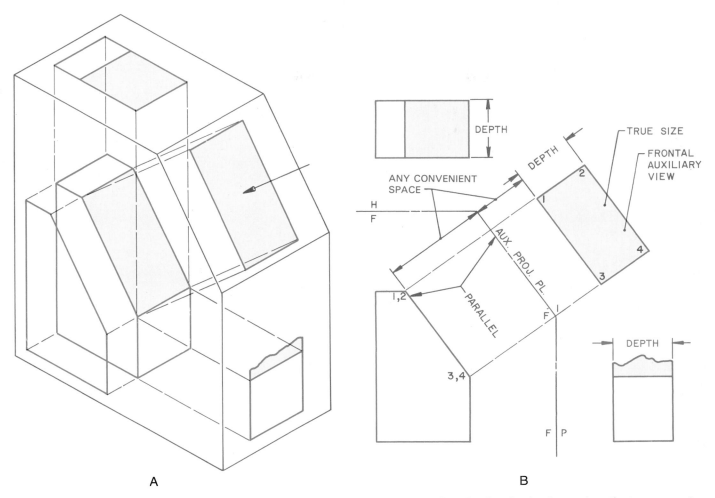

Fig. 13-4. For a frontal projection, the reference plane is drawn parallel to the line in the front view that represents the inclined surface.

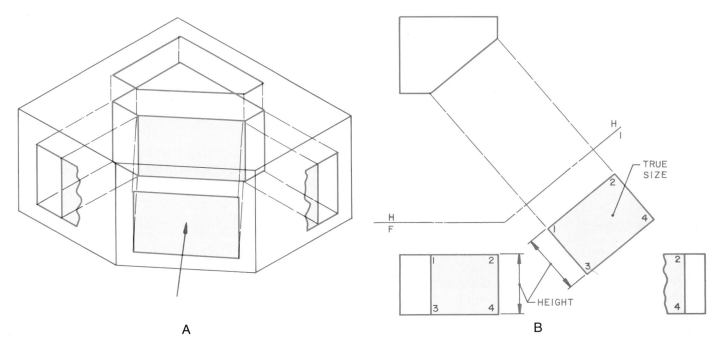

Fig. 13-5. The reference plane in a horizontal, or top, projection is located parallel to the line in the top view that represents the edge of the inclined surface.

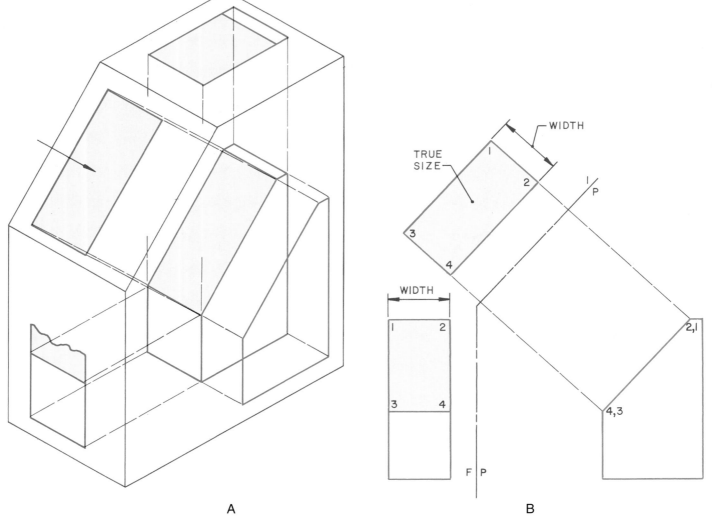

Fig. 13-6. The reference plane in a profile section is parallel to the line in the side view that represents the inclined surface.

and development of auxiliary views. In Fig. 13-7A, point A is shown in the "glass box." This point has been projected to the frontal, horizontal, and auxiliary planes. Since the inclined surface is perpendicular to the frontal plane, a frontal auxiliary is projected. Point A is located in the auxiliary view by projecting it perpendicularly to the plane of projection and setting off the depth from the top view, Fig. 13-7B.

Location of a True Length Line in an Auxiliary View

The *auxiliary view method* is useful in determining the location and true length of a line that is inclined to the principal views in orthographic projection. Location and determination of the line's true length consists of locating its two end points in the auxiliary view and connecting these points to form the required line.

Given the top and front views, the procedure is as follows.

1. Draw an auxiliary viewing plane F-1 parallel to line AB in the front view, Fig. 13-8B. (The viewing plane can also be established for the top view.)

2. The line of sight is perpendicular to viewing plane F-1 and line AB. Any line viewed perpendicularly will be seen in its true length.

3. Measure the distance that line AB lies away from the frontal plane, as viewed in the horizontal plane. Plot these distances perpendicular and away from auxiliary viewing plane F-1 to locate points A and B, Fig. 13-8C.

4. Connect the endpoints for the true length line AB, Fig. 13-8D. Any given line whose location is plotted on a viewing plane parallel to the given line will appear in its true length when viewed perpendicularly to the viewing plane.

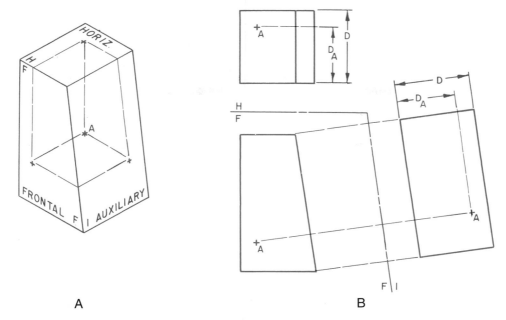

A   B

Fig. 13-7. The location of a point in an auxiliary view is the first important step in understanding auxiliary projection.

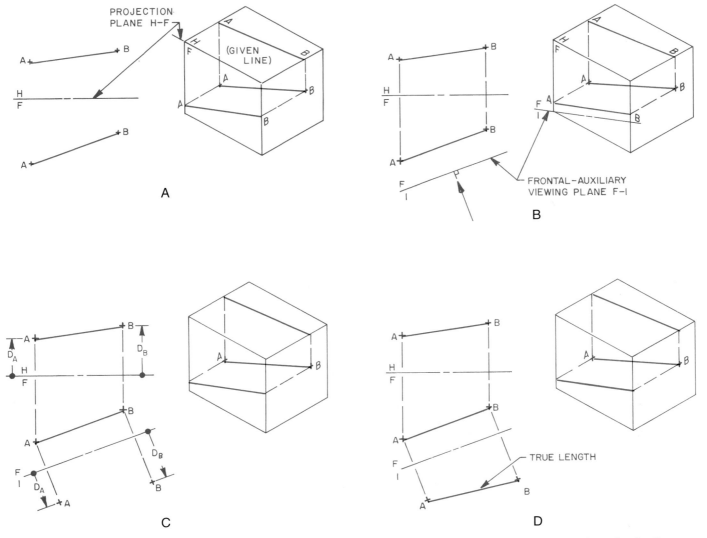

A   B

C   D

Fig. 13-8. The auxiliary view projection method can be used to determine the location and true length of a line.

Slope of a Line

The **slope** of a line is the angle that the line makes with the horizontal plane. The extent of slope may be expressed in percent or degrees of grade, Fig. 13-9A. When the line of slope is oblique to the principal planes, an auxiliary view is required to find the true length of the line. Since slope is the angle the line makes with the horizontal plane, a horizontal auxiliary view is used, Fig. 13-9B. The slope line AB will appear true length. The angle it makes with the horizontal plane is the required angle. The slope of a line may be designated as positive or negative. If positive, the slope is upward from the designated point. If negative, the slope is downward.

Slope has many applications in industry. Gravity-feed delivery systems require the slope of a line to be determined. Ramps designed for wheelchairs have very specific slope requirements, Fig. 13-10. The roads that you drive on every day make use of slope as well.

Point Method Applied to the Development of an Inclined Surface

The application of the point method to the development of an inclined surface in a primary auxiliary view is shown in Fig. 13-11.

True Angle Between a Line and a Principal Plane

The true angle (θ) formed between a line in a primary auxiliary view and a principal plane may be determined in a view where the principal plane ap-

Fig. 13-10. There are specific requirements that must be met when designing vehicles for handicapped individuals. The slope of an access ramp is a very critical part of the design. Automotive engineers and designers must be aware of the slope requirements when completing the design of a vehicle that includes an access ramp. (Homecare Products, Inc.)

pears as an edge and the line is true length (TL). In Fig. 13-12, the angle between the horizontal plane and the true length line AB is a true angle. Note that the reference plane has also been drawn through point A to show the true angle.

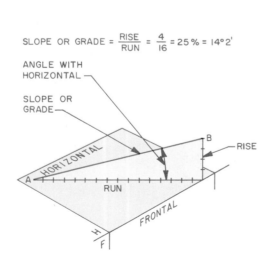

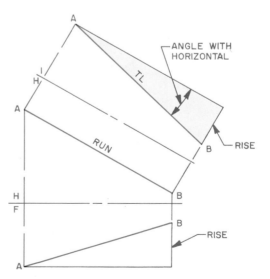

Fig. 13-9. When a line is oblique to the principal planes, an auxiliary view is needed to find the true length.

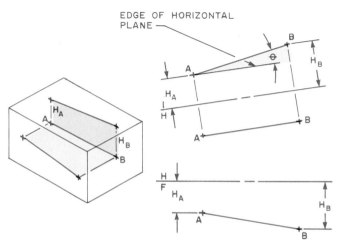

Fig. 13-11. The point method can be used to develop an inclined surface in a primary auxiliary view.

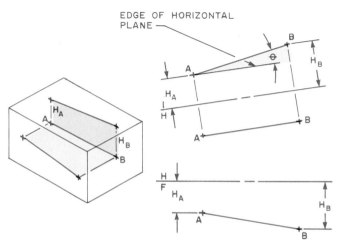

Fig. 13-12. A true angle is the angle that is formed between a line in a primary auxiliary view and a principal plane. This angle can be found in a view where the principal plane appears as an edge and the line is true length.

The true angle could also be developed for the frontal auxiliary plane, Fig. 13-13. The line of sight is parallel to the frontal plane and perpendicular to line AB. A reference plane has been drawn through point A and the true angle projected away from the frontal plane. Refer to the pictorial view in Fig. 13-13.

Determination of True Angle Between Two Planes

Frequently in the design and manufacture of parts, it is necessary to determine the angle between two planes in order to correctly specify the design of the parts, Fig. 13-14. The true angle between two planes, called a *dihedral angle,* can be found in a primary auxiliary view. This occurs when the line of intersection is true length in one of the principal views and the line of intersection appears as a point in the auxiliary view.

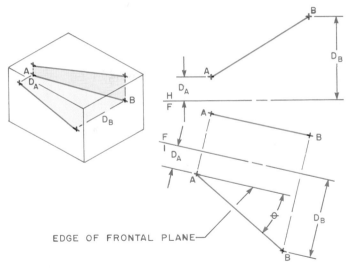

Fig. 13-13. If the true angle is developed from the frontal auxiliary plane, the line of sight is parallel to the frontal plane and perpendicular to line AB.

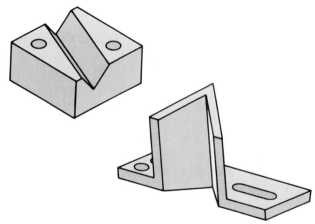

Fig. 13-14. Some manufactured parts require that angles between different planes be defined before the parts can actually be manufactured. Auxiliary views are required to define these angles.

In Fig. 13-15, the line of intersection AB is seen in its true length in the top view since it is parallel to the horizontal plane. A horizontal auxiliary is drawn perpendicular to line AB, providing a point view of line AB. The angle between the two planes is a true angle.

Projecting Circles and Irregular Curves in Auxiliary Views

Circles and irregular curved lines may be projected in auxiliary views by identifying a sufficient number of points in the primary views. These points are projected to the auxiliary view and are plotted to provide a smooth curve, Fig. 13-16A. The steps involved in the projection of circles and irregular curved lines in a primary auxiliary view are as follows.

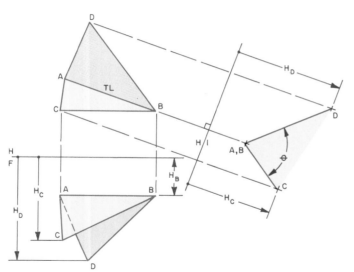

Fig. 13-15. Line AB is parallel to the horizontal plane, and therefore seen as true length in the top view. A horizontal auxiliary drawn perpendicular to AB will provide a point view of AB. The angle formed between these two planes is a true angle.

1. Draw two principal views showing the circle or irregular curve and the angle of the inclined surface, Fig. 13-16B.
2. Locate points on the curve in the principal view. Any desired number and location of points may be selected, as long as enough are used to project the character of the curve.
3. Project points on the curve to the line in the principal view representing the edge of the inclined surface.

4. Draw reference plane F-1 parallel to the edge view of the inclined surface, Fig. 13-16B. In the irregular curve example, the reference plane has been drawn in line with the far side of the object. It could have been drawn to the left of the auxiliary view, or in the center and measured accordingly in the side view. The results would be the same.
5. Project the points marked on the inclined surface perpendicularly to the auxiliary view. Measure off the location of the points as shown.
6. Sketch a light line through these points. Finish with an irregular curve.

Drawing Circles and Circle Arcs in Auxiliary Views by Ellipse Template Method

Circular shapes may be drawn in auxiliary views by the *projection method.* However, ellipse templates speed up the process and produce much better results. Take the following steps in drawing regular circular features appearing as ellipses in auxiliary views.

1. Locate the reference plane for the auxiliary view (in center of view), Fig. 13-17.
2. Project to the auxiliary view points and lines that locate the centerline, the major axis, and the minor axes of the circular feature.
3. Identify the angle that the inclined surface makes with the principal plane.

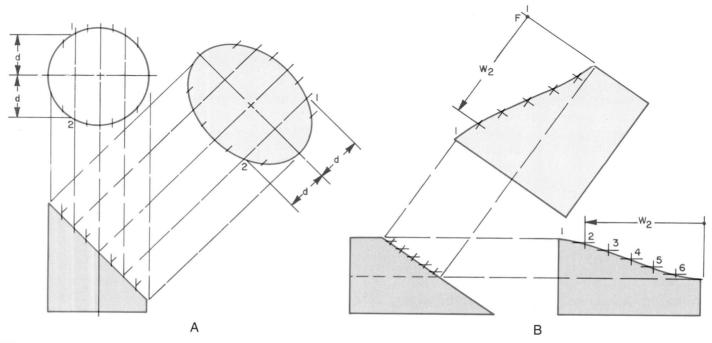

A   B

Fig. 13-16. Circles and irregular curves can be projected in auxiliary views by identifying points and projecting the points.

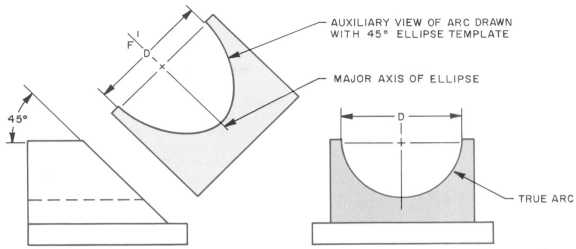

Fig. 13-17. Ellipse templates speed the drafting process. In order to use an ellipse template, the major and minor axes must first be located.

4. Select an ellipse template of the correct angle and major diameter. Draw the required ellipse.

Construction of a Principal View with the Aid of an Auxiliary View

Sometimes the true size and shape of a feature exists on the inclined surface of a part, such as a circular curve or hole. When this occurs, it is necessary to construct the auxiliary view, then reverse the projection procedure in constructing the principal view where the feature is shown foreshortened, Fig. 13-18. Take the following steps to complete this procedure.

1. Construct the view where the inclined surface appears as an edge (profile view in Fig. 13-18). It may be helpful to partially complete the view until the features involved have been drawn in the auxiliary view.

2. Construct the auxiliary view perpendicular to a reference plane that is parallel to the inclined surface. (Line P-1 in Fig. 13-18. This line is located in the center of the feature.)

3. Locate a sufficient number of points on the outline of the feature to assure an accurate representation.

4. Project the points from the auxiliary view to its principal view. Then project these points to the

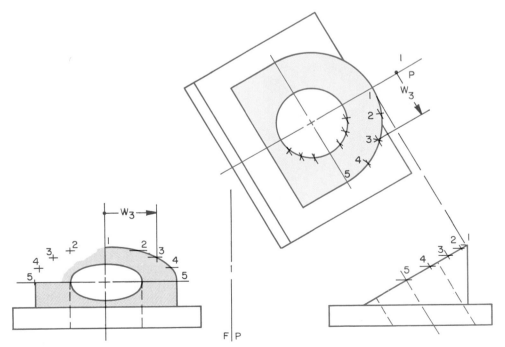

Fig. 13-18. Sometimes an auxiliary view is needed to construct a principal view.

other principal view where the feature appears in a foreshortened plane. Measurements are taken from reference plane P-1 in the auxiliary view and transferred to the principal view.

5. Complete the view by joining the points.

Note: The auxiliary view in Fig. 13-18 is a complete, rather than a partial auxiliary, since the entire part is shown, not just the inclined surface. Hidden lines are normally omitted in auxiliary views unless required for clarity.

SECONDARY AUXILIARY VIEWS

Oblique lines and surfaces are not parallel or perpendicular to any of the principal planes of projection. Therefore, a secondary auxiliary view is required to describe their true size and shape.

In structures having many oblique surfaces, a number of auxiliary views are required, Fig. 13-19. All auxiliary views beyond the secondary auxiliary view are called **successive auxiliary views.** Successive auxiliary views are projected from secondary or prior successive auxiliary views.

SECONDARY AUXILIARY VIEW OF AN OBJECT

Primary auxiliary views require two principal views for their projection. Secondary auxiliary views are projected from a primary auxiliary view and one of the principal views, Fig. 13-20.

Fig. 13-19. Imagine the number of auxiliary views required to describe complex equipment such as the space shuttle Endeavor's cargo bay. (NASA)

The line of sight that the object is viewed from is indicated in the top and front views, Fig. 13-20A. The line of sight is shown in its true length in the primary auxiliary view, Fig. 13-20B.

An auxiliary must be viewed in the direction of its line of sight. To do this, the secondary auxiliary viewing plane 1-2 is constructed at right angles to the true length line of sight. This gives a point view of the line of sight, Fig. 13-20C.

Two plane surfaces are projected to the secondary and given letters to aid in identifying the formation of the object. In the section, the object is completed, showing visible surfaces and lines, plus those edges that are hidden from view, Fig. 13-20D.

The steps in the construction of a secondary auxiliary view are shown in Fig. 13-20. No oblique surfaces were involved in the projection of this object. (How to obtain point-views of lines, true sizes, and true shapes of oblique surfaces will be covered later in this chapter.)

Given the top and front views, follow this procedure for drawing a secondary auxiliary view of an object to find the view of the object when viewed in the line of sight.

1. Project a primary auxiliary view of the object off of a principal view so that it is perpendicular to the chosen line of sight. (Refer back to Fig. 13-20B.)
2. Construct the true-length line of sight.
3. Draw a secondary auxiliary projection plane 1-2 perpendicular to the true-length line of sight in the primary auxiliary. (Refer back to Fig. 13-20C.) Projection of this line will give a point view in the secondary auxiliary.
4. Project two planes by locating their points of intersections and dimensions from plane H-1. (Refer back to Fig. 13-20C.)
5. Locate the remaining points and draw connecting lines to complete object. (Refer back to Fig. 13-20D.) The line of sight will indicate the surfaces and edges that are visible and the ones that are hidden.

Point View of a Line in a Secondary Auxiliary

Finding the point view of a line is basic to the projection of a secondary auxiliary view. Given the top and front view of a line proceed as follows.

1. Construct a primary auxiliary view of line AB by drawing reference plane F-1 parallel to the front view of the line, Fig. 13-21B. (The top view can also be chosen.)
2. Find the true length of line AB by projecting it perpendicular to the reference plane. Determine

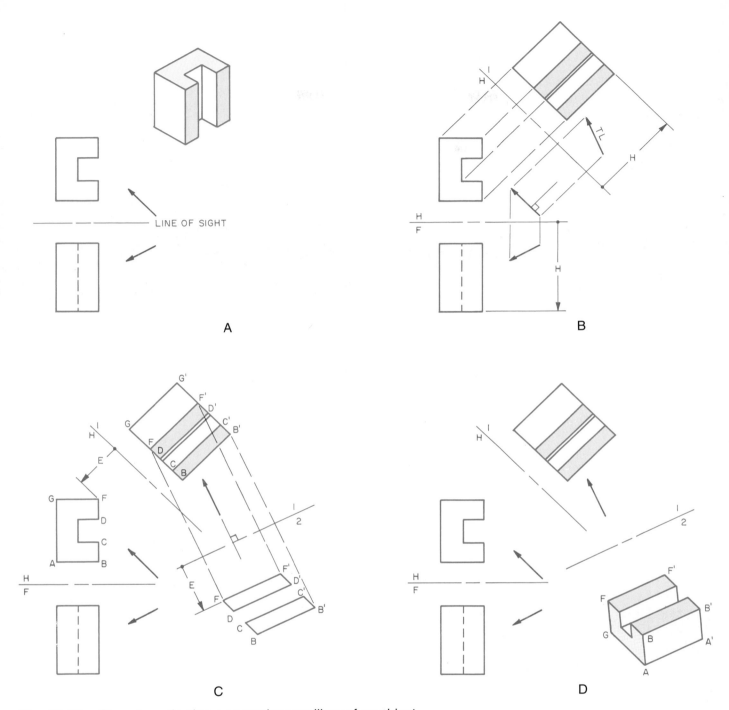

Fig. 13-20. Steps in projecting a secondary auxiliary of an object.

the distance that the endpoints are located from the reference plane H-F in the top view, Fig. 13-21C.

3. Draw reference plane 1-2 for secondary auxiliary perpendicular to line AB. Measure off distance M to locate view point for line AB, Fig. 13-21D.

4. Distance M is perpendicular to reference planes F-1 and 1-2. Both of these planes appear as an edge in the front and secondary auxiliary views.

Determination of True Angle Between Two Planes in a Secondary Auxiliary

The true angle between two planes can be determined when the line of intersection is viewed as a point, and the planes appear as edges. However, when the line of intersection of an angle between two planes is an oblique line, a secondary auxiliary is required to determine the true angle.

The two planes shown in Fig. 13-22 intersect in an oblique line to form an angle. Given the top and

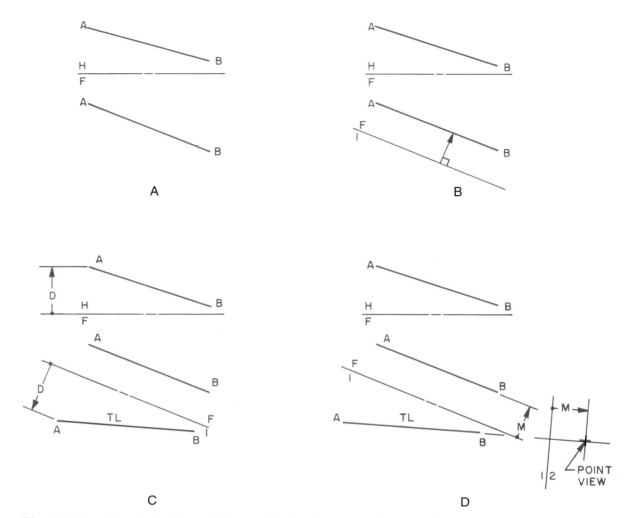

Fig. 13-21. Steps in finding point view of a line in a secondary auxiliary.

front views of the two planes, proceed as follows to find the true size of an angle between two planes.

1. Construct a primary auxiliary view to develop true length of line of intersection AC of angle, Fig. 13-22B.

2. Construct a secondary auxiliary view to develop point view of line of intersection, Fig. 13-22C. (Refer back to Fig. 13-21 as well.) The plane of the angle is perpendicular to the true length line of intersection in the primary auxiliary.

3. True angle is formed by locating points B and D. These points complete the edge view of the two planes enclosing the angle, Fig. 13-22D.

4. When the line of intersection of two planes appears as a point and the planes appear as edges, the true size of the angle can be measured.

True Size of an Oblique Plane

Frequently it is necessary for the drafter or engineer to specify the exact size and shape of an oblique surface of an industrial product. Since some of these

surfaces must function within a very close tolerance range, they must be located and specified on the drawing in a precise manner, Fig. 13-23.

The true size and shape of inclined surfaces may be determined by the use of a primary auxiliary. Except, however, for oblique surfaces which lie in a plane that is not perpendicular to any of the principal planes of orthographic projection. These surfaces require the use of a secondary auxiliary to identify their true size and shape.

Given the top and front views of an oblique plane, proceed to develop its true size as follows.

1. Draw horizontal line AA' in top view, Fig. 13-24B. (Front view could also have been used.)

2. Project the line to the front view where it appears true length. (A line which is parallel to one of principal planes will appear in its true length in an adjacent view.)

3. Draw primary auxiliary plane F-1 perpendicular to line AA' and the line of sight. Project a point view of line AA' in primary auxiliary view, Fig. 13-24C.

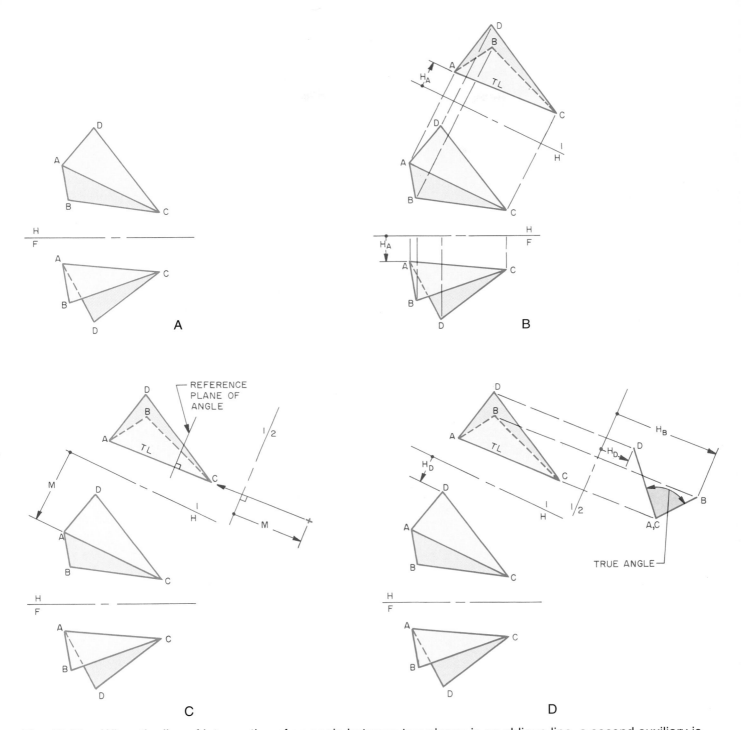

Fig. 13-22. When the line of intersection of an angle between two planes is an oblique line, a second auxiliary is required to find the true angle.

4. Project points B and C to this view where the plane will appear as an edge.

5. Draw secondary auxiliary view plane 1-2 parallel to edge view of plane ABC, Fig. 13-24D.

6. Construct projectors for all three points of plane ABC. Locate points on these projectors by taking measurements perpendicular from the primary auxiliary plane edge view F-1. This produces plane ABC in its true size.

Application of Secondary Auxiliary Method in Developing True Size of an Oblique Plane

Assume that the object shown in Fig. 13-25 is a sheet metal cover for a special machine. One problem that must be resolved in making a drawing of the object is to find the true size and shape for the surface on the oblique plane.

Fig. 13-23.   Auxiliary projection was used to identify the true size of oblique planes on the members of these structures. (DuPont)

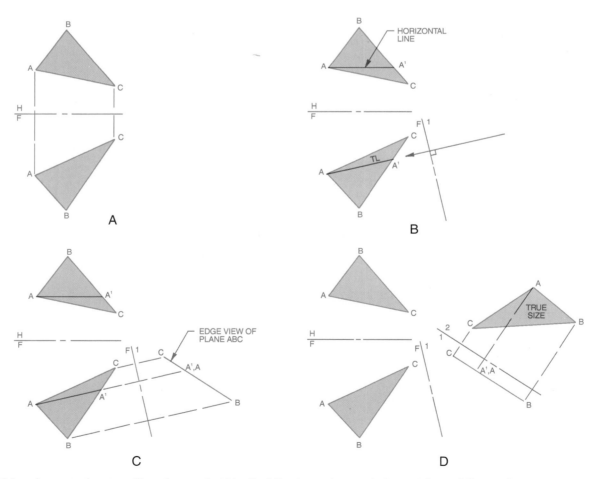

Fig. 13-24.   A secondary auxiliary is required to find the true size and shape of an oblique plane.

Given the front and top views of the sheet metal cover with an oblique surface, take the following steps to define the surface ABCD.

1. Draw reference plane H-F between the top and front view at any desired location, Fig. 13-25.
2. Draw the horizontal line CC′ in the front view. Project it perpendicular to the top view to establish a true length line on surface ABCD.
3. Draw reference plane H-1 perpendicular to the line of sight. This will produce a point view of line CC′ in the primary auxiliary view.
4. Project points A, B, and D and measure their location from reference plane H-F in the front view to form an edge view of ABCD. (All should lie in a straight line.)
5. Draw reference plane 1-2 perpendicular to the line of sight for the edge view of plane ABCD.
6. Project points A, B, C, and D and measure their location from reference plane H-1 in the top view.
7. Connect the points to complete the true size of plane ABCD.

Note: The secondary auxiliary view shown in Fig. 13-25 is a partial auxiliary view. It only shows oblique surface. The remainder of the object can be projected, but it is not needed in this view and would be out of true size and shape.

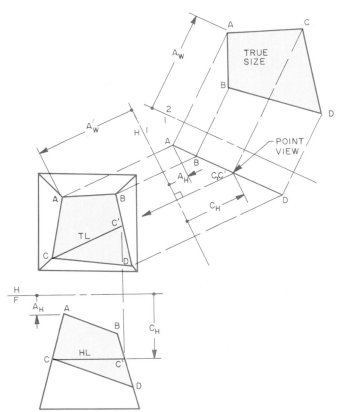

Fig. 13-25. The secondary auxiliary method is used to develop the true size and shape of an oblique plane.

PROBLEMS IN AUXILIARY VIEWS

The following have been planned to provide experience in primary and secondary auxiliary projection in the solution of problems in drafting and design.

Use A-size drafting sheets for these problems. The problems are shown on 1/4″ section paper to help locate the objects on the drawing sheet. Label the points, lines, and planes and show your construction procedure.

Primary Auxiliary Projection

1. Refer to Fig. 13-26A and Fig. 13-26B. Construct an auxiliary view of the inclined surface X of each object. Use a frontal auxiliary for Fig. 13-26A and a horizontal auxiliary for Fig. 13-26B.
2. Refer to Fig. 13-26 and solve each section as follows. In Fig. 13-26C, find the true length of the line utilizing a frontal auxiliary. In Fig. 13-26D, find the slope of the line.

3. Refer to Fig. 13-27A and Fig. 13-27B. Develop the true angle between the planes for both.
4. Refer to Fig. 13-27C and Fig. 13-27D. Develop the true size and shape of the features in the auxiliary view for both.
5. Refer to Fig. 13-28A and Fig. 13-28B. Draw the elliptical features in the auxiliary view, using ellipse templates.

Secondary Auxiliary Projection

6. Refer to Fig. 13-28C and Fig. 13-28D. Draw a secondary auxiliary view of the object as indicated by the lines of sight.
7. Refer to Fig. 13-29A and find the point view of the line in the secondary auxiliary view.
8. Refer to Fig. 13-29B and find the true angle between the oblique planes.
9. Refer to Fig. 13-29C and Fig. 13-29D. Construct the true size and shape of the oblique surfaces by means of secondary auxiliary views.
10. Refer to Fig. 13-30 and draw the necessary views, including auxiliary views, of the parts shown. Dimension your drawings.

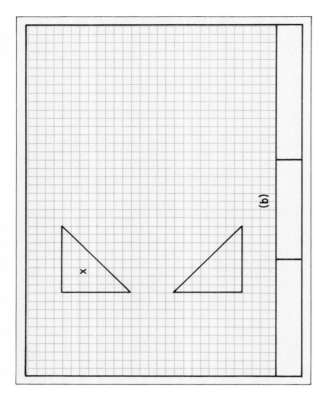

(a)

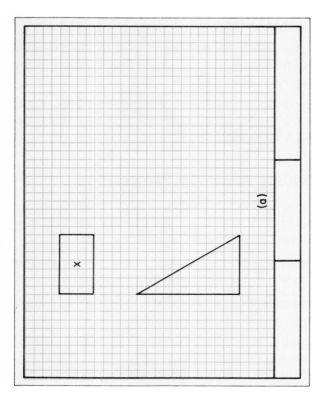

(b)

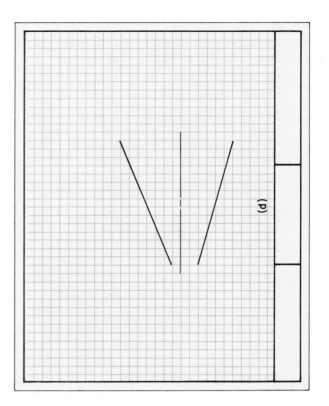

(d)

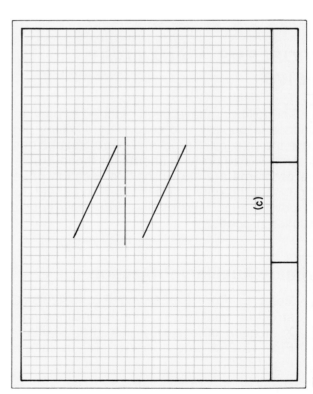

(c)

Fig. 13-26.   Problems in primary auxiliary projection.

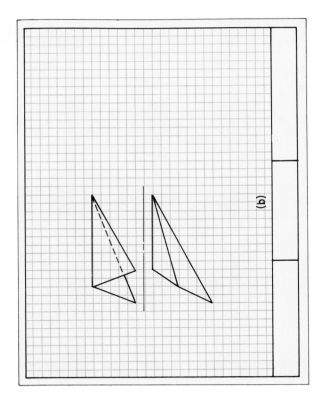

(a)

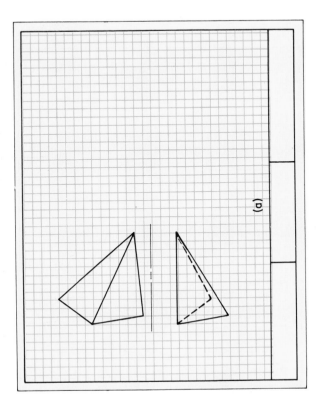

(b)

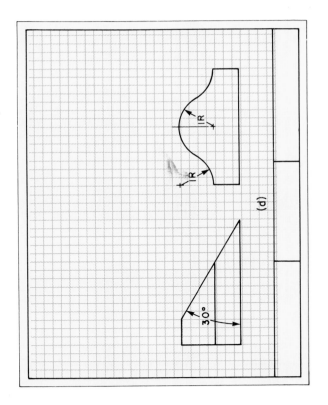

(d)

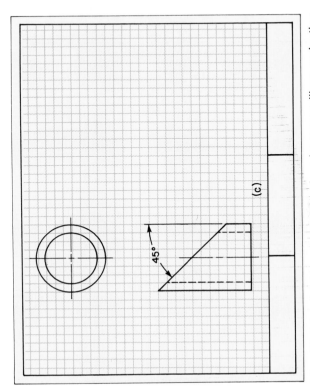

(c)

Fig. 13-27. Additional problems in primary auxiliary projection.

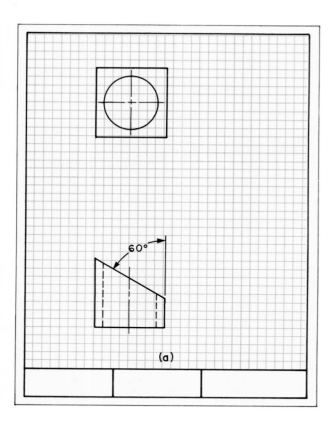

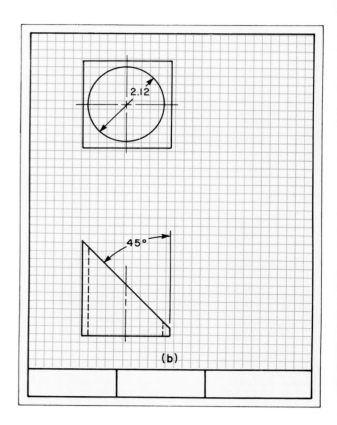

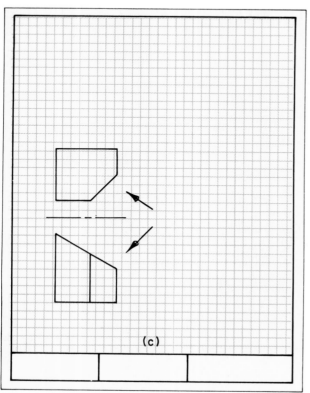

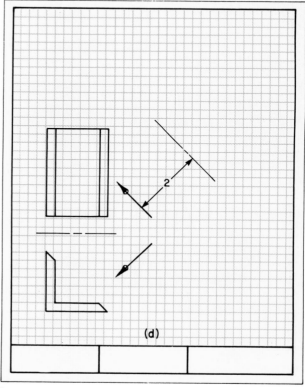

Fig. 13-28. Problems in primary and secondary auxiliary projections.

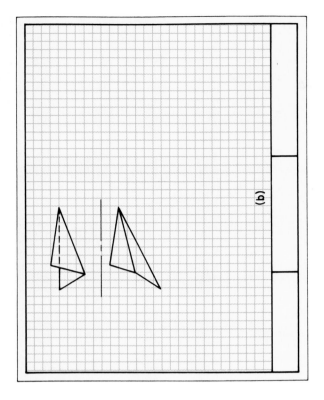

(b)

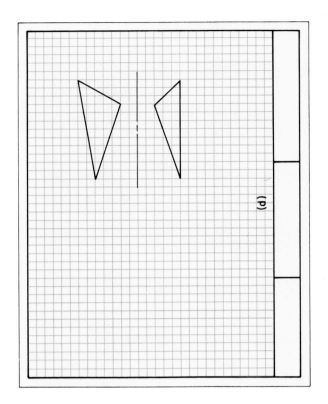

(d)

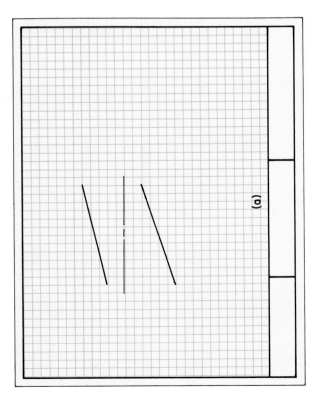

(a)

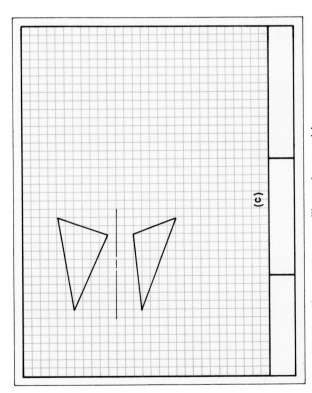

(c)

Fig. 13-29. Secondary auxiliary view problems.

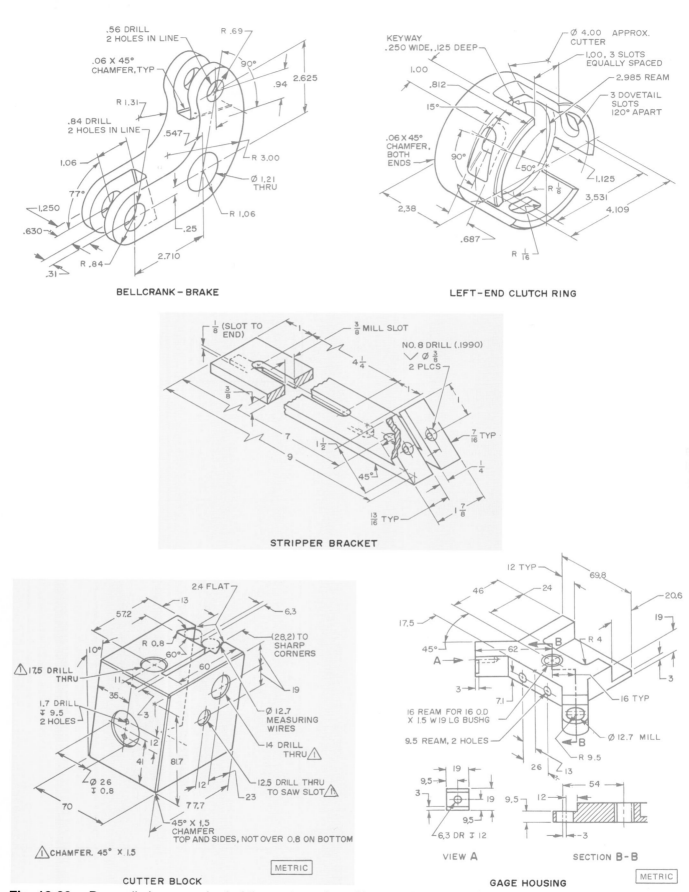

BELLCRANK – BRAKE

LEFT-END CLUTCH RING

STRIPPER BRACKET

CUTTER BLOCK

METRIC

GAGE HOUSING

VIEW A

SECTION B-B

METRIC

Fig. 13-30. Draw all views required of the parts assigned by your instructor. These machine parts require auxiliary views in their description.

Revolutions

KEY CONCEPTS

☐ In a revolution, the object is "moved" and the viewer "remains" in one place.

☐ The true length of a line in space can be found using the revolution method.

☐ The revolution method can be used to find the true size of an angle between a line and a principal plane.

☐ A plane can be revolved to find its true size.

☐ The edge view of a plane can be located using the revolution method.

☐ The revolution method can be used to find the true size of an angle in space.

Revolution is a method available to the drafter/designer in defining spatial relationships of rotating or revolving parts. As the principles of revolution are learned and applied, the similarities between this method and the auxiliary view method will be apparent. Each method tends to enhance one's understanding of the other.

To obtain an auxiliary view of a surface or object, the observer moves to a point where the line of sight is perpendicular to the inclined surface, Fig. 14-1A. In a revolution of an object, the observer is assumed to remain in the original position while the object is revolved, Fig. 14-1B. The auxiliary view of a surface appears exactly the same as a revolution of that surface.

The method of revolution may be further shown as an imaginary plane section of a cone. When the section is inclined to the observer, it appears foreshortened and not in its true size, Fig. 14-2A. If the cone is imagined as being "revolved" so that the plane AB'C' is perpendicular to the line of sight, the plane appears true size, Fig. 14-2B. Line AC is not shown true length because it is not on a plane perpendicular to the line of sight. However, line AC' *is* a true length line since it lies on a plane perpendicular to the line of sight.

SPATIAL RELATIONSHIPS IN REVOLUTION

The revolution method can be used to define several spatial relationships. Nine spatial relation-

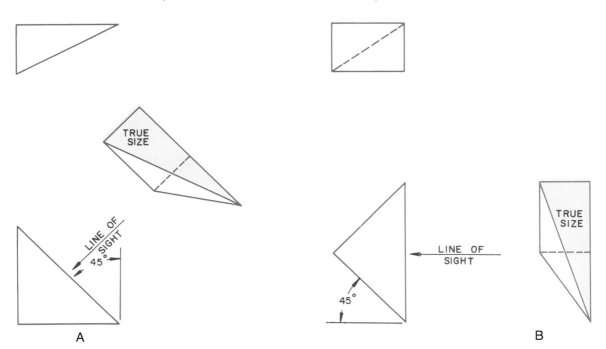

Fig. 14-1. A–In an auxiliary view, the observer "moves" so that the line of sight is perpendicular to the inclined surface. B–In a revolution, the observer "remains still" and the object is "moved" so that the line of sight is perpendicular to the inclined surface.

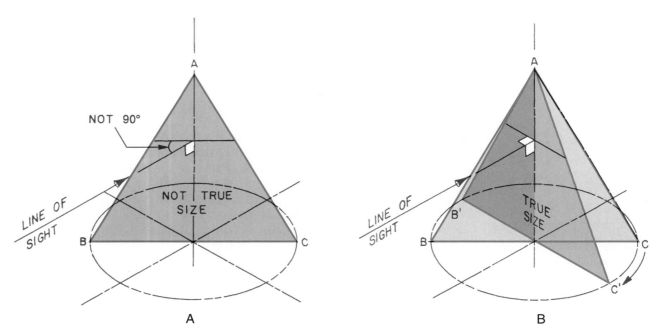

Fig. 14-2. A–When a cross section is viewed at an angle other than 90°, the surface is not seen true size or shape. B–However, if the surface is "revolved" so that the line of sight is 90°, the surface will be seen true size and shape.

ships that can be obtained by using the revolution method are discussed in the following sections.

Revolution of a Line to Find its True Length

The revolution method may be used to find the true length of a line in space. A frontal line that is revolved to find its true length is shown in Fig. 14-3A. When the line is thought of as an element of a cone, the revolution process is easier to visualize.

Given the top and front views of an oblique line AB in space, proceed as follows to obtain its true length.

1. A line is drawn conventionally in the horizontal and frontal views, Fig. 14-3B.

2. Assume the line appears as an element of a cone section in the top and front views, Fig. 14-3C.

3. Revolve the line in the top view to a position parallel to the horizontal plane, Fig. 14-3D. Use point A as the center of the radius.

4. Any line parallel to the horizontal plane in the top view will be seen true length in the adjacent (front) view. Therefore, line AB′ is shown true length in the front view, Fig. 14-3D.

True Size of an Angle Between a Line and a Principal Plane by Revolution

The true size of an angle formed between a line and a principal plane may be found by using the

revolution method. To use the revolution method, the line is revolved to a position parallel to the principal plane.

The true size of the angle between the front view and the horizontal plane in Fig. 14-4A is required. Note in Fig. 14-4A and Fig. 14-4B that the line AB is shown in both the front (A_FB_F) and top (A_HB_H) views. (The front view is on the frontal plane and the top view is on the horizontal plane.)

The line is first revolved in the top view to a position parallel to the frontal plane, Fig. 14-4B. This line, line FL, is then projected to the front view to produce a true length line. The angle between a true length line, line TL, and the principal plane is a true size angle. This angle is the required angle.

Revolution of a Plane to Find the Edge View

The edge view of a plane can be found using the revolution method. A true length line on the plane in an adjacent view must first be located. Then a point view of that line is projected to the view where the edge view is desired, Fig. 14-5.

Given the top and front views of the plane ABC, proceed as follows, Fig. 14-5.

1. Draw the frontal line AD (FL) in the top view parallel to plane H-F, the plane of projection, Fig. 14-5A. Project line AD (FL) to the front view for a true length line AD (TL) on plane ABC.

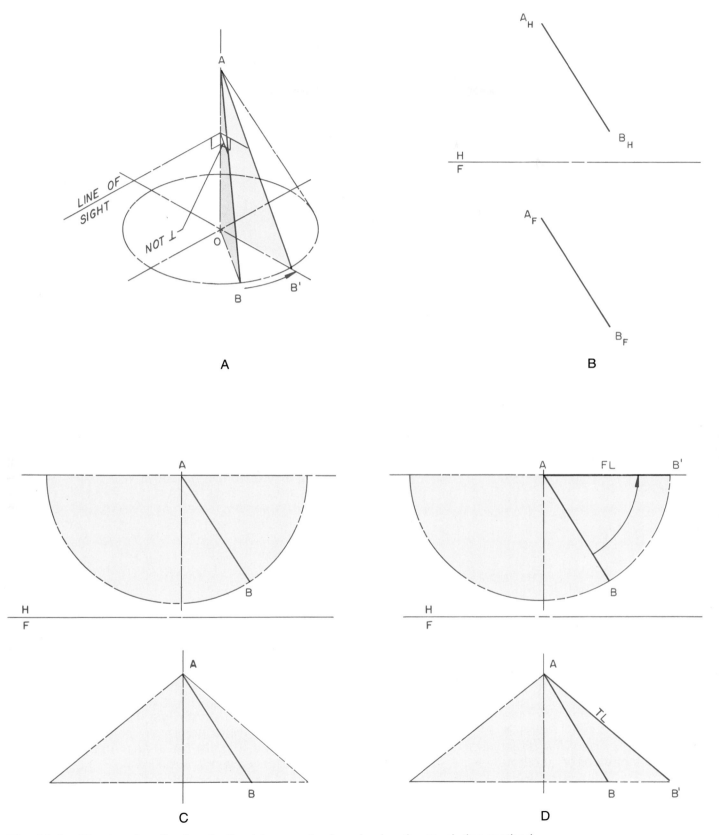

Fig. 14-3. The true length of an inclined line can be found using the revolution method.

2. Revolve true length line AD' (TL) to the vertical position AD', using A as the radius center. Then, using a compass and dividers, transfer plane ABC to its corresponding position. This position will be around line AD', Fig. 14-5B.

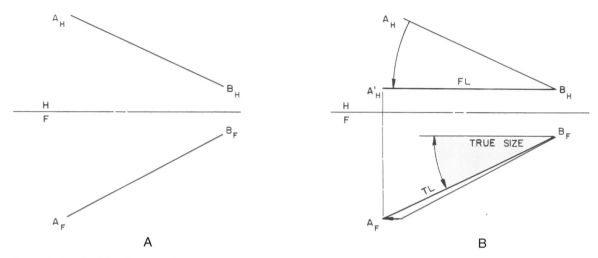

Fig. 14-4. A–The line AB is shown in the front view (A$_F$B$_F$) and the top view (A$_H$B$_H$). B–The line is first revolved in the top view (A$_H$B$_H$) to a position parallel to the frontal plane. The true-length line FL is then projected to the front view. The angle between the projected true-length line and a principlal plane is true size.

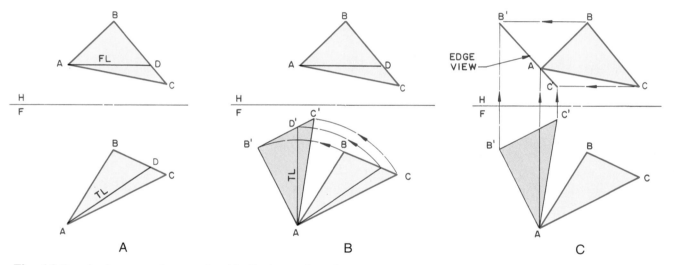

Fig. 14-5. A plane can be revolved to find an edge view.

3. Project points A, B′, and C′ to the top view, intersecting with their corresponding projectors, Fig. 14-5C. These projectors are perpendicular to plane H-F.

4. Join these points to form an edge view of plane ABC, Fig. 14-5C. Points A, B′, and C′ will lie in a straight line if the projections and measurements have been accurate.

Revolution of a Plane to Find its True Size

To find the true size of a plane, an edge view of the plane must first be constructed in a primary auxiliary view. This construction is done by revolving the edge, then projecting it back to the principal view.

Given the top and front views of plane ABC, proceed as follows, Fig. 14-6.

1. Find a point view of the plane by first drawing the horizontal line HL in the front view. This will give the true-length line TL in the top view. From line TL, a point view primary auxiliary is projected, Fig. 14-6A. Plane ABC appears as an edge view in this auxiliary.

2. Revolve the edge view of plane ABC to a position parallel to plane H-I, using point B as the radius center, Fig 14-6B. (Plane H-I is perpendicular to plane H-F.)

3. Project points B, C′, and A′ to the top view. These points will intersect corresponding projectors that are drawn parallel to plane H-I, Fig. 14-6C.

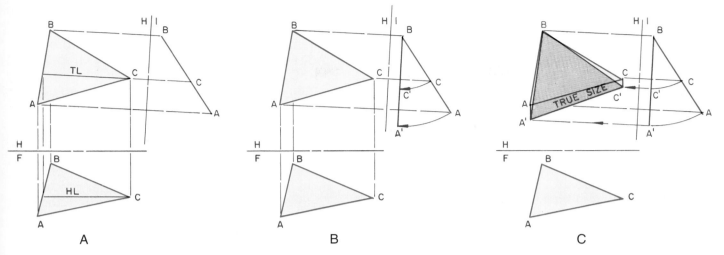

Fig. 14-6. The revolution method can be used to find the true size of a plane.

4. Connect these points of intersection to produce the true size plane A′BC′.

A true size of the plane could have been found in the front view in a similar manner.

True Size of an Angle Between Two Intersecting Planes by Revolution

If the line of intersection between two planes appears in true length in a principal view, the angle between the two planes can be found using the revolution method. A right section is first drawn through the two planes. This section is then revolved to the horizontal view and then projected to the adjacent view.

Given the top and front views of two intersecting planes, proceed as follows to find the true size of the angle between the planes, Fig. 14-7.

1. Draw a right section through the view where the line of intersection between the two planes is true length, Fig. 14-7A. (In this example, the line appears true length in the front view.)

2. Using D as the center, revolve the section line CD to a position (C′D) parallel to plane H-F, Fig. 14-7B. Revolve the point where the line of intersection crosses the right section line, point E.

3. Project points C′, E′, and D to their corresponding points of intersection in the top view. Note that the horizontal line C′D in the front view projects true length in the top view.

4. Connect these points of intersection in the top view to form true size angle C′E′D, Fig. 14-7C. This angle is the required angle between the two intersecting planes.

True Size of an Angle Between Two Intersecting Oblique Planes

The true size of angles between intersecting planes oblique to the principal planes of projection can be found by revolution. First construct a primary auxiliary view where the line of intersection

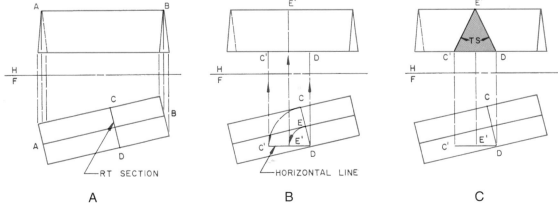

Fig. 14-7. The revolution method can be used to find the true size of an angle between two intersecting planes.

between the two planes appears in its true length. Then follow the preceding four-step procedure.

Revolution of a Point About an Oblique Axis to Find its Path

Machine designs occasionally include parts that meet the machine surface at an oblique axis. Hand cranks frequently meet the machine surface at an oblique angle. The path of rotation of such a machine part may be traced by revolution with **successive auxiliary views.**

Given the top and front views of an oblique axis and a point of rotation, proceed as follows, Fig. 14-8.

1. Develop the true length of line AB in a primary auxiliary view. Project a point view of line AB in a secondary auxiliary view, Fig. 14-8A.

2. Project point P to the primary and secondary auxiliaries. Draw the path of rotation in the secondary auxiliary. As a radius, use the distance from point P to the axis of line AB. The path of revolution appears as an edge in the primary auxiliary. This "edge" is perpendicular to line AB and passes through point P.

3. Next, locate point C so that line AC is true length in the front view. To locate the highest point on the path of point P, draw vertical line AC to plane HF. Then project line AC as a point in the top view, Fig. 14-8B. Project this line back through the primary and secondary auxiliary views. Where the directional arrow crosses the circular path in the secondary auxiliary, point P′, is highest point on path of point P.

4. Project point P′ back through successive views to the top view where it lies on line AB, verify-

ing it as the highest point, Fig. 14-8C. The position of point P′ in the views should be established by careful measurements from appropriate reference planes, as shown in Fig. 14-8C. Any other position could have been established such as the lowest or forward position on the path of rotation. This would be done by drawing a line in the required direction in the appropriate principal view and then projecting it into all views.

5. To draw the elliptical path of revolution of the point in the principal views, use an ellipse template of appropriate major diameter and angle. The major diameter is the diameter of the circular path. The ellipse angle is the angle formed by the line of sight from the front view with the edge view of the circular path in the primary auxiliary. A horizontal auxiliary view would need to be constructed from the top view to find the ellipse angle to use in the top view.

Primary Revolutions of Objects About Axes Perpendicular to Principal Planes

Revolutions are made to obtain one or more of three advantages: a clear view of an object, the true length of a line, or the true size of a surface. **Primary revolutions** are drawn perpendicular to one of the principal planes of projection. An object is shown in its normal position in Fig. 14-9A. It is revolved in the horizontal plane in Fig. 14-9B. The object is revolved in the frontal plane in Fig. 14-9C. It is revolved in the profile plane in Fig. 14-9D. Regular orthographic principles are used in projecting revolved views.

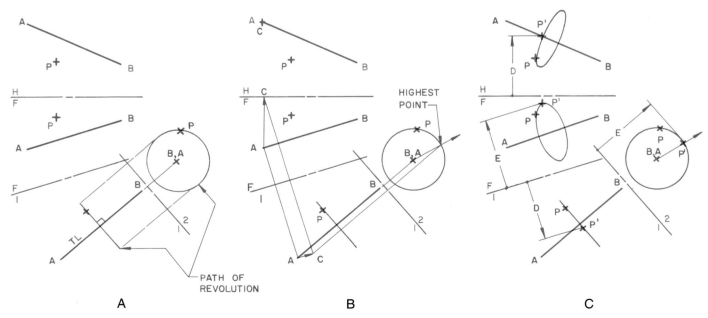

Fig. 14-8.   A point can be revolved about an oblique axis to find its path and highest point.

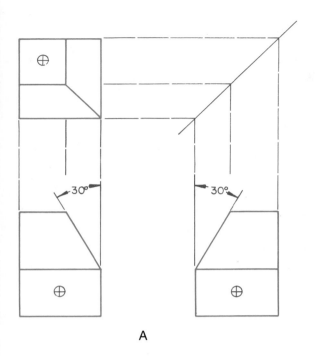

A

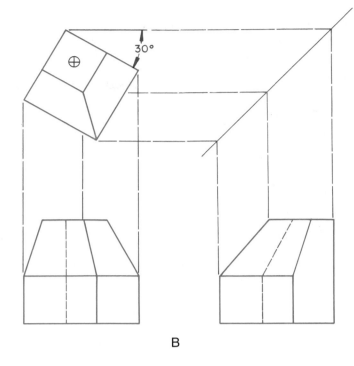

B

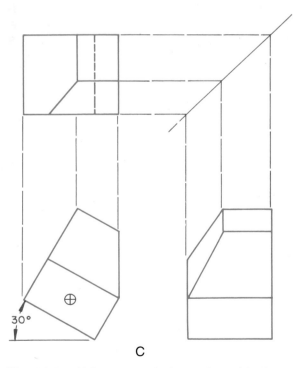

C

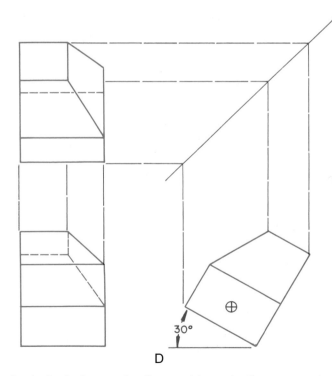

D

Fig. 14-9. Primary revolutions of an object around the axes of principal planes of orthographic projection.

Successive Revolutions of an Object With an Oblique Surface

The true size of an oblique surface may be obtained by *successive revolutions,* Fig. 14-10.

This method is similar to finding the true size of an oblique surface through successive auxiliary views.

A pictorial view of an object with an oblique surface is shown in Fig. 14-10A. The object is shown in its normal orthographic position in Fig. 14-10B,

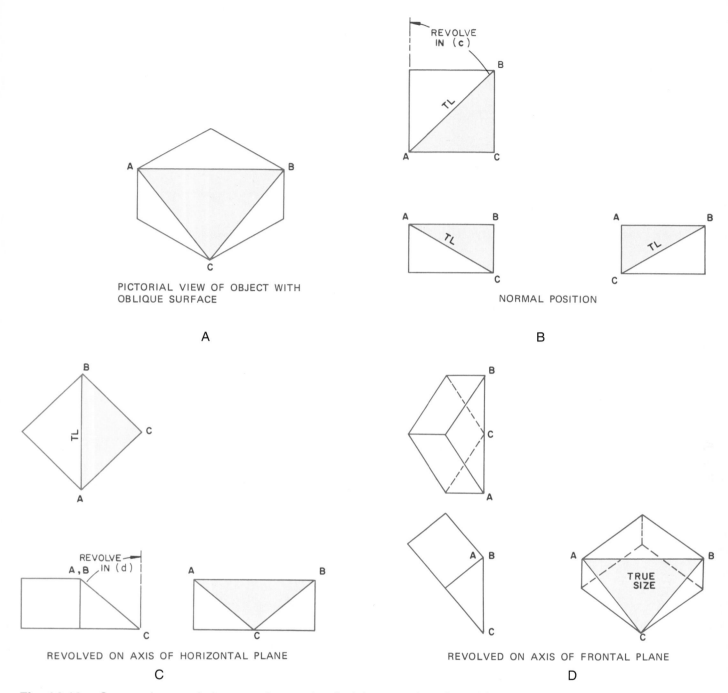

REVOLVE IN (c)

PICTORIAL VIEW OF OBJECT WITH
OBLIQUE SURFACE

A

NORMAL POSITION

B

REVOLVE IN (d)

REVOLVED ON AXIS OF HORIZONTAL PLANE

C

REVOLVED ON AXIS OF FRONTAL PLANE

D

Fig. 14-10. Successive revolutions can be used to find the true size of an oblique surface.

with an indication of the revolution to be made in the first of two successive revolutions. The first revolution of the object is performed in Fig. 14-10C, while the second revolution (in Fig. 14-10D) produces the true size view of the oblique surface.

PROBLEMS AND ACTIVITIES

The following problems give you practice in using the revolution method of problem solving.

Use A-size drafting sheets. Label all reference planes, points, and lines. Some of the problems are shown on 1/4″ section paper to help locate the drawing on the sheet. Use revolution method in solving problems.

1. Refer to Fig. 14-11, then lay out and solve each section as follows. In Fig. 14-11A, find the true length of the line in the horizontal view. In Fig. 14-11B, find the true length of the line in the front view. Also, indicate the angle that it makes with the horizontal plane. In Fig.

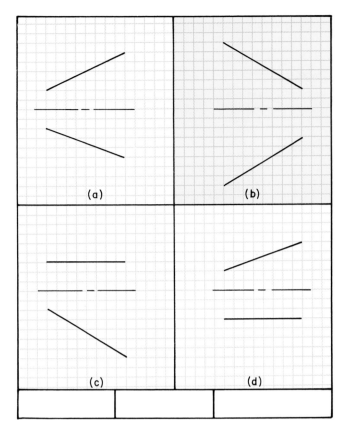

Fig. 14-11. Use these lines for the true length line problems. These lines are drawn on 1/4″ section paper to help you locate them on your drawing sheet.

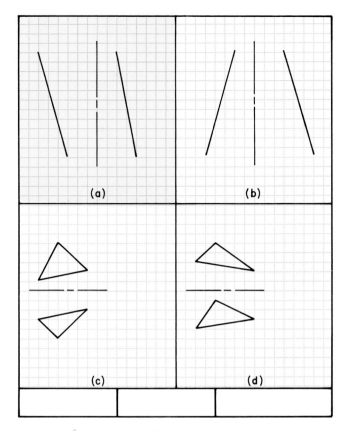

Fig. 14-12. Use these lines and objects for the true size angles and true size plane problems. These lines and objects are drawn on 1/4″ section paper to help you locate them on your drawing sheet.

14-11C, find the true length of the line in the horizontal view. Indicate the angle that it makes with the frontal plane. In Fig. 14-11D, find the true length of the line in the horizontal plane. Indicate the angle that it makes with the frontal plane.

2. Refer to Fig. 14-12, then lay out and solve each section as follows. In Fig. 14-12A, find the true size of the angle that the line in the profile view makes with the frontal plane. In Fig. 14-12B, find the true size of the angle that the line in the frontal view makes with the profile plane. In Fig. 14-12C, find the true size of the plane by projecting an edge view in a horizontal auxiliary and by revolution. In Fig. 14-12D, find the true size of the plane by a primary auxiliary view and by revolution.

3. Refer to Fig. 14-13, then lay out and solve each section as follows. In Fig. 14-13A, find the true size of the plane by projecting an edge view in a horizontal auxiliary view and by revolution. In Fig. 14-13B, find the true size of the plane by a primary auxiliary view and by revolution. In Fig. 14-13C, find the true size of the angle between the intersecting planes. In Fig.

14-13D, find the true size of the angle between the intersecting planes.

4. Refer to Fig. 14-14A and Fig. 14-14B. Find the path of revolution of the points about the lines. Lay out measurements and projections accurately. Allow approximately 1″ space between the frontal view of the line and the primary auxiliary projection plane, and the same amount in the secondary auxiliary. Locate the highest point and project this to all views. Measure and dimension the diameter of the path in the view where the path appears as a circle.

5. Refer to Fig. 14-15 and revolve the objects about their axes, perpendicular to the principal planes as follows. In Fig. 14-15A, revolve the objects about their horizontal axes. In Fig. 14-15B, revolve the objects about their frontal axes. In Fig. 14-15C, revolve the objects about their profile axes. Draw three views, starting with the view perpendicular to the axis of revolution. Place each drawing on a separate A-size sheet. Estimate dimensions to retain proportions of the object shown in the pictorial

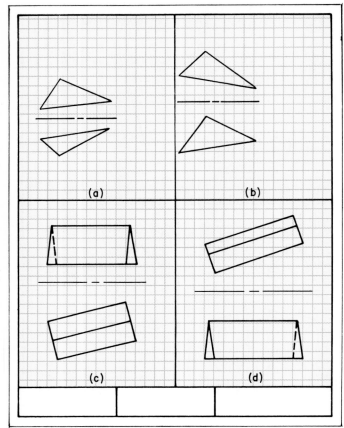

Fig. 14-13. Use these objects for the problems on finding edge views of planes and true size of angle between intersecting planes. These objects are drawn on 1/4″ section paper to help you locate them on your drawing sheet.

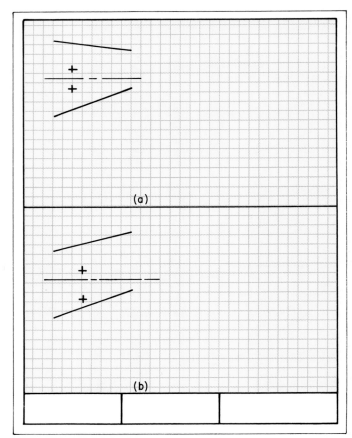

Fig. 14-14. Use these lines for the problems on revolving a point about a line. These lines are drawn on 1/4″ section paper to help you locate them on your drawing sheet.

view. Indicate the angle of rotation, but do not dimension further.

6. Refer to Fig. 14-15D and prepare a successive revolution of the object to produce a true size view of the oblique surface. Use a B-size sheet and divide the drawing space into four sections. Then prepare similar to Fig. 14-10, starting with a pictorial view. Select an appropriate scale and estimate the dimensions to retain the proportions shown. Indicate the true size

surface in the view where it appears, but do not dimension the drawing.

7. Find the true length of the centerline for all of the legs of the "SUPPORT BRACKET" shown in Fig. 14-16.

8. Find the true length of all three antenna guy wires, Fig. 14-17.

9. Find the true distance between the transmission towers A, B, and C shown in Fig. 14-18. Each tower is 40′ tall.

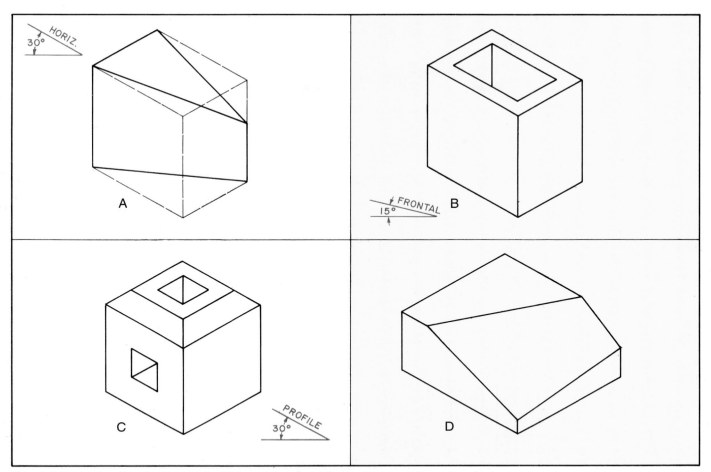

Fig. 14-15. Use these objects for the problems dealing with revolution of objects.

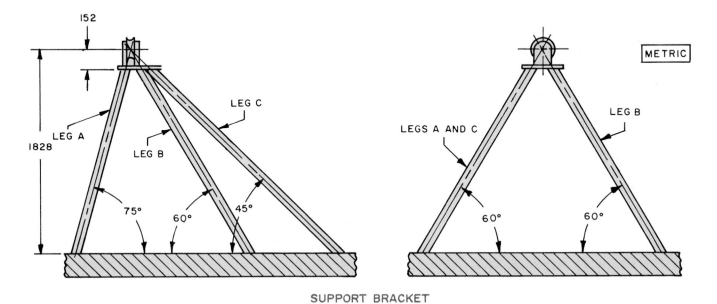

SUPPORT BRACKET

Fig. 14-16. Use this "SUPPORT BRACKET" to find the true lengths of the lines assigned by your instructor.

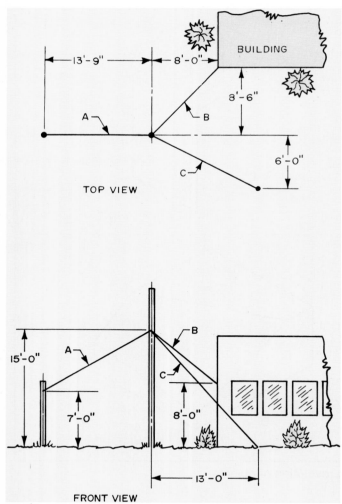

Fig. 14-17. Find the true length of the guy wires for this antenna.

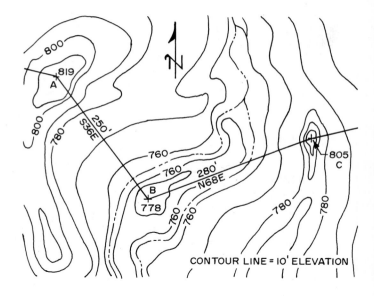

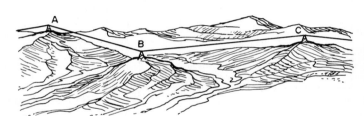

Fig. 14-18. Use this topographic map to find the true distance between each transmission tower. Notice that this map gives both horizontal distances and elevations.

Intersections

KEY CONCEPTS

- [ ] An intersection is where two or more objects join or pass through each other.
- [ ] There are two broad classifications for geometric surfaces: ruled geometric surfaces and double-curved geometric surfaces.
- [ ] Ruled geometric surfaces include plane surfaces, single-curve surfaces, and warped surfaces.
- [ ] A surface that can be "developed" can be "unfolded" into a single flat surface.
- [ ] Plane surfaces and single-curve surfaces can be developed.
- [ ] Warped surfaces cannot be developed, however, a development can be approximated.
- [ ] Double-curved surfaces cannot be developed, but a development can be approximated.
- [ ] Many different aspects of construction and manufacturing require locating the intersections of different surfaces.

When two or more objects (such as two planes or a cylinder and a square prism) join or pass through each other, the lines formed at the junction of their surfaces are known as *intersections*. Numerous examples of intersections can be found in industry. Frequently, the design and specification of buildings require architects and engineers to define the intersection of surfaces. The aerospace and automotive industries also work with intersections of various shapes in the manufacture of instrument panels, body sections, window openings, wings, and fuselages, Fig. 15-1.

This chapter covers the basic principles and geometric forms of intersecting objects. Once these principles and the techniques of their application are understood, most intersection problems can be solved.

TYPES OF INTERSECTIONS

Intersections and their solutions are classified on the basis of the types of geometrical surfaces involved. Two broad classifications of geometrical surfaces are: ruled geometrical surfaces and double-curved geometrical surfaces.

A

B

Fig. 15-1. A–Architects and engineers must resolve intersections created in the design of buildings and clearly specify them on the drawings. B–Defining the intersection of surfaces, such as the wings of the space shuttle, requires study and planning for proper assembly. (NASA)

Ruled Geometrical Surfaces

Ruled geometrical surfaces are surfaces generated by moving a straight line. They may be subdivided into *plane surfaces, single-curved surfaces,* and *warped surfaces,* Fig. 15-2.

Fig. 15-2. There are three types of ruled geometrical surfaces: plane, single-curved, and warped. The plane and single-curve surfaces can be developed, but the warped surfaces cannot.

Plane surfaces and single-curved surfaces can be developed. If a surface can be **developed,** it can be "unfolded" or "unrolled" into a single plane.

Warped surfaces cannot be developed into a single plane. Usually, these surfaces are formed to true shape by peening, stamping, spinning, or by a vacuum or explosive process, Fig. 15-3. Warped surfaces can be divided into sections and developed. However, this sectioning produces only an approximation of the true warped surface.

Double-curved Geometrical Surfaces

Double-curved geometrical surfaces are surfaces generated by a curved line revolving around a straight line in the plane of the curve, Fig. 15-4. Double-curved surfaces, like warped surfaces, cannot be developed into single plane surfaces. However, the **gore method** can be used to approximate the development of the surface, Fig. 15-5.

SPATIAL RELATIONSHIPS

In Chapter 7, an introduction was given to normal points, normal lines, and normal surfaces in space. Chapter 13 presented the techniques of

DOUBLE-CURVED SURFACES

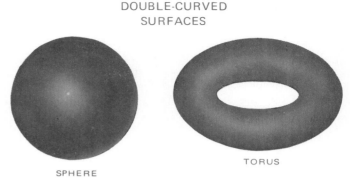

SPHERE

TORUS

PARABOLOID

OBLATE SPHEROID

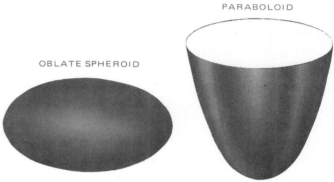

Fig. 15-4. Double-curved geometrical surfaces are formed by a curved line revolving around a straight line that is in the plane of the curve.

Fig. 15-3. How many warped surfaces can you identify on this "advanced-concept" tanker truck? How many plane and single-curve surfaces can you identify? (Freightliner/Heil)

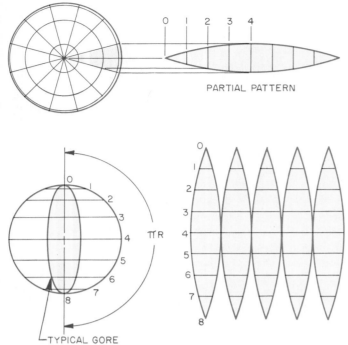

PARTIAL PATTERN

TYPICAL GORE

Fig. 15-5. The "gore" method can be used to develop an approximate flat representation of the surface of a sphere.

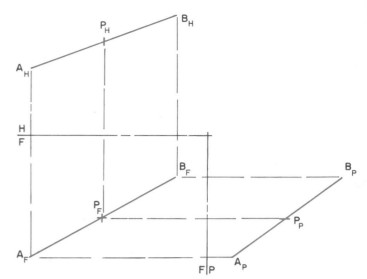

Fig. 15-6. Lines consist of an infinite number of points. Any point on a line can be located in the frontal (F), horizontal (H), of profile (P) planes by using perpendicular projectors.

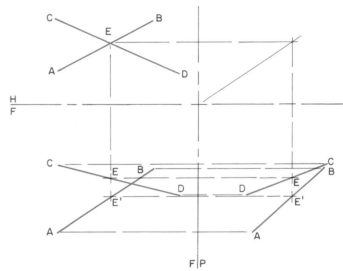

Fig. 15-7. Lines AB and CD appear to cross in the horizontal (H) plane. However, when point E is projected to the frontal (F) and profile (P) planes, it is clear that point E is not common to both lines. Therefore, the lines do not cross.

locating inclined and oblique lines and surfaces in space. Still remaining are a few basic spatial relationships that must be understood before you can solve problems of intersections and development.

Point Location on a Line

Lines are composed of an infinite number of points. To solve problems in space, specific points on lines and surfaces must be located. Line AB is shown in the horizontal, frontal, and profile views in Fig. 15-6. Endpoints A and B are located in the three views by projectors. Any point, point P for example, can be located in a similar manner.

Intersecting and Nonintersecting Lines in Space

Lines that cross in space are not necessarily intersecting lines. **Intersecting lines** have a common point that lies at the exact point of intersection.

If crossing lines AB and CD are intersecting lines, they will have a point common to both lines, Fig. 15-7. Point E is located in the horizontal view and projected to intersect lines AB and CD in the frontal view. It is apparent in the front view that point E is not a single point common to both lines. Therefore, the lines do not intersect. This is further shown in the right side, or profile, view.

In Fig. 15-8, lines GK and LM *do* intersect. They have a common point that lies on both lines. This is revealed by orthographic projection. Two views are sufficient to determine whether crossing lines are intersecting lines. Note that point O is common to both lines in any two views, verifying the intersection of the lines.

Visibility of Crossing Lines in Space

The visibility of crossing lines is established by projecting the crossing point of the lines from an

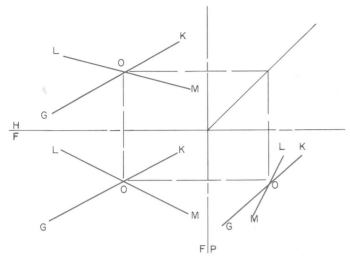

Fig. 15-8. Lines GK and LM appear to cross in the horizontal (H) plane. When point O is projected to both the frontal (F) and profile (P) planes, it is clear that point O is common to both lines. Therefore, the lines do cross.

Visibility of a Line and a Plane in Space—Orthographic Projection Method

Determining the visibility of a line and a plane that cross in space is similar to the procedure for two crossing lines. Given line AB and the plane CDE that cross in the horizontal and frontal views, the line AB crosses two "edges" of the plane, Fig. 15-10A. The lines CE and ED represent two edges of the plane, and line AB crosses these two lines.

The visibility in the horizontal view is determined by projecting the crossing points to the front view, Fig. 15-10B. The projectors "touch" line AB before they "touch" the plane. This indicates that the line is higher than the plane at these points and, therefore, visible in the horizontal view.

Since the projectors "touch" the plane before the line, line AB is invisible in the frontal view. This indicates that the plane is nearer the frontal view and that line AB crosses behind plane CDE.

Triangular planes and prisms have been used here and in other sections to illustrate the intersection of lines and surfaces. Triangular planes and prisms make understanding the principles and procedures less complex. The process described for the intersection of planes is the same regardless of the number of edges a plane has.

adjacent view, Fig. 15-9A. Lines AB and CD are crossing lines. To determine the visible line at the crossing point in the top view, project the crossing point from the top view, Fig. 15-9B. The projection line "touches" line CD first in the front view. This indicates that line CD is higher. Therefore, line CD is visible in the top view at the point of crossing.

To check the visibility of the crossing point in the front view, project the point of crossing from the front view, Fig. 15-9C. The projector "touches" line AB first. This indicates that line AB is nearer the frontal viewing plane. Therefore, line AB is visible at the crossing point of the two lines.

Location of Piercing Point of a Line with a Plane—Orthographic Projection Method

The point of intersection between a plane and a line inclined to that plane is called the *piercing*

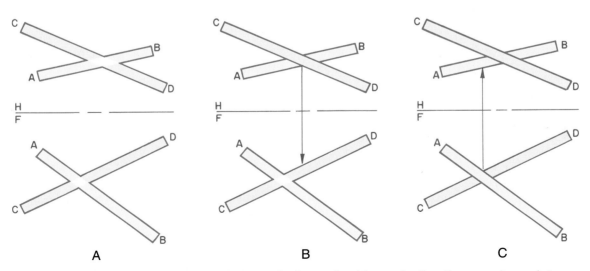

     A   B   C

Fig. 15-9. A–The visibility of crossing lines is determined by projecting the crossing point from an adjacent view. When the crossing point is projected from the top view, line CD is "touched" first. B–Therefore, line CD is visible in the top view at the crossing point. C–When the crossing point is projected from the front view, line AB is "touched" first. Therefore, in the front view line AB is visible at the crossing point.

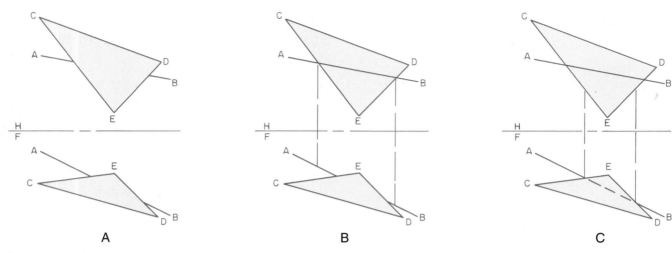

Fig. 15-10. A–The visibility of a line crossing a plane is determined by projecting the crossing points of the line with the "edge" of the plane from an adjacent view. B–If the crossing points are projected from the top view, the line is "touched" first . C–Therefore, the line is *below* the plane in the front view. (Note that this is different from crossing lines.) If the crossing points are projected from the front view, the plane is "touched" first. Therefore, the plane is *below* the line in the top view.

point. The location of the piercing point is essential to the solution of many technical problems. The intersection of pipes and tubing with valves and cylinders requires the location of piercing points, Fig. 15-11.

Given the top and front views of a line and a plane, proceed as follows to locate the piercing point of a line and a plane, Fig. 15-12A.

1. Pass a vertical cutting plane through the top view of line AB (see the pictorial view in Fig. 15-12) that intersects plane DCE at G and K, Fig. 15-12B.

2. Project points G and K to the front view.

3. The intersection of the imaginary cutting plane and plane CDE is represented by the trace line GK in the front view, Fig. 15-12C.

4. Intersecting line AB lies in an imaginary cutting plane and will intersect plane CDE along line GK at point O (front view). Point O is the piercing point.

5. Project point O to establish the piercing point in the top view.

6. The visibility of line segment OB in the front view is determined by projecting crossing point L to the top view. The line segment OB is found to be behind line DE.

7. The visibility of line segment OB in the top view is found by projecting crossing point K to the front view. The line segment OB is found to be higher than line DE.

8. The line segment AO is found to be visible in front view and invisible in top view by the projecting crossing points M and G.

Location of Piercing Point of a Line with a Plane—Auxiliary View Method

The piercing point of a line and a plane may also be located by the auxiliary view method, Fig. 15-13. An edge view of the plane ABC is projected in a horizontal auxiliary view. Then the point where the line intersects the edge view of the plane in the auxiliary is projected to the horizontal view. The piercing point is where this projection intersects line DE. The piercing point is then projected to the other view.

Accuracy of the projection may be checked in the front view by measuring the distance the piercing point lies below the H-F and H-1 projection planes. (Refer to Fig. 15-13.) The visibility of the line and plane is found by determining whether the line or the plane lies nearer the projection plane in the adjacent view at the crossing point. The closest one is visible.

Given the top and front views of plane ABC and line DE that intersects the plane, proceed as follows to locate the piercing point and the visibility of the line and plane, Fig. 15-13A.

1. Draw horizontal line AF in the front view. Project this line to the top view to produce a true length line, Fig. 15-13B.

2. Project this true length line as the line of sight. Construct auxiliary projection plane H-1 perpendicular to line AF.

3. Project the edge view of plane ABC by finding the point view of line AF, Fig. 15-13C. Project line DE. Where this line intersects plane ABC is the piercing point O.

Fig. 15-11.   The locations of piercing points of lines and planes was necessary in the design of these valves. How many piercing points can you identify for each valve? (Sporlan Valve Co.)

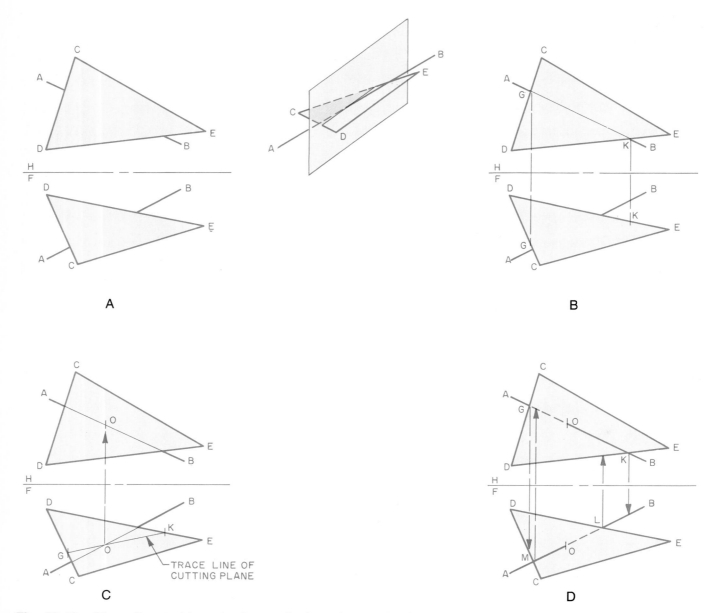

Fig. 15-12.   The orthographic projection method can be used to find the piercing point of a line with a plane.

4. Project point O to the top and front views to intersect with line DE. This will locate the piercing point O in these views, Fig. 15-13D. Point O may be checked for accuracy in the front view by measuring for H.

5. The visibility is determined by the method previously described for orthographic projection.

A Line Through a Point and Perpendicular to an Oblique Plane

It is sometimes necessary to locate the shortest distance from a point to an oblique plane, or to construct a perpendicular to the plane. By drawing an auxiliary view, an edgewise view can be shown

and a perpendicular erected through the point to the plane.

Given the top and front views of a plane and a point in space, proceed as follows, Fig. 15-14A.

1. Draw a horizontal line in the front view. Project this line to the top view where it will appear true length, Fig. 15-14B.

2. Construct a primary auxiliary view of plane ABC that will appear as an edge.

3. Project point P to the auxiliary.

4. Draw line PO perpendicular to the edge view of plane ABC in the auxiliary, Fig. 15-14C.

5. Project the piercing point O to the top view, where it will join line PO. Line PO is parallel to the projection plane H-1 since it is a true

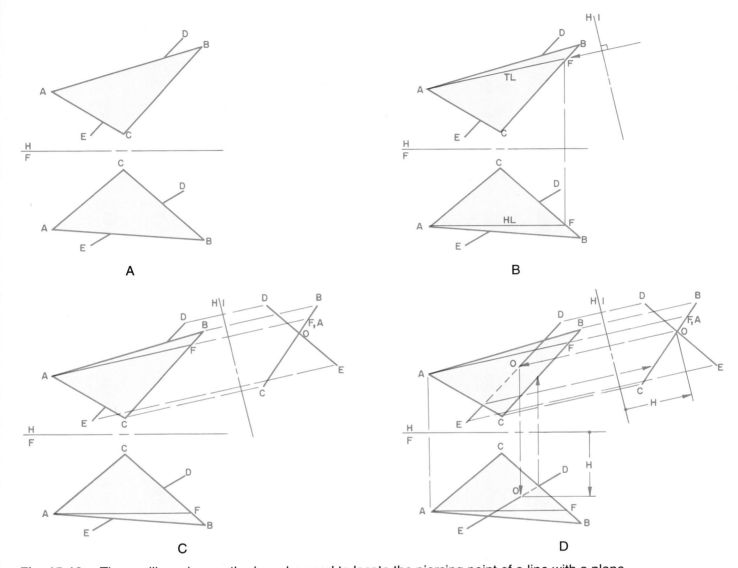

Fig. 15-13. The auxiliary view method can be used to locate the piercing point of a line with a plane.

length line in the auxiliary view. Line PO will also be perpendicular to the line of projection that projects true length in the top view.

6. Project point O to the front view where it will intersect line EF. Line EF is projected from a frontal line through point O in the top view. Point O will be distance H from planes H-1 and H-F.

Intersection of Two Planes—Orthographic Projection Method

The intersection of two planes is a straight line and can be found by locating, on one plane, the piercing points of the lines representing the edges of the second plane. The two points are then connected for the line of intersection, Fig. 15-15.

Given the top and front views of two intersecting planes, proceed as follows to locate their line of intersection, Fig. 15-15A.

1. Pass an imaginary cutting plane vertically through line AB in the top view to establish points 1 and 2, Fig. 15-15B.

2. Project points 1 and 2 to the front view where they will cross lines DE and FE.

3. Line AB pierces plane DEF at G where it crosses line 1-2. Project point G to the top view.

4. In a similar fashion, pass a cutting plane through line AC to establish points that are projected to the front view, crossing lines DE and FE, Fig. 15-15C.

5. Line AC pierces plane DEF at H where it crosses line 3-4. Project point H to the top view.

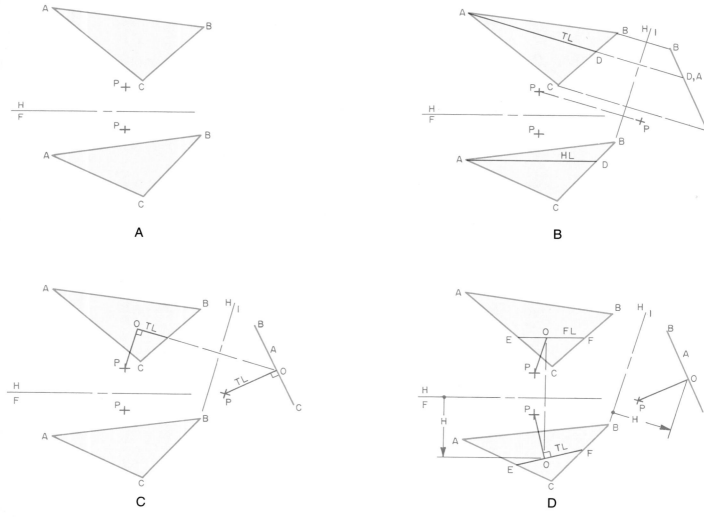

Fig. 15-14.   By drawing an auxiliary view, a line through a point and perpendicular to an oblique plane can be drawn.

6. Analyze the crossing of lines AH and DE in the top view for visibility. Line AH is found to be higher and, therefore, visible in the top view, Fig. 15-15D.

7. Analyze the crossing of AG and DE in the top view for visibility. Line AG is found to be higher and visible.

8. The segment GAH of plane ABC is visible in the top view.

9. The visibility in the front view is determined in a similar way. The segment HGBC of plane ABC is found to be visible.

Intersection of Two Planes—Auxiliary View Method

In some cases, it is easier to find the intersection between two planes by using the auxiliary view method. One plane is viewed as an edge by constructing a reference plane (H-1) perpendicular to the point view of a true length line on that plane, Fig. 15-16.

Given the top and front views of two intersecting planes, proceed as follows to locate the line of intersection, Fig. 15-16A.

1. Draw a horizontal line in the front view and project it to the top view. This will produce a true length line, Fig. 15-16B.

2. Project the point view of this horizontal line to the horizontal auxiliary. Construct an edge view of plane ABC.

3. Project plane DEF to the auxiliary view.

4. Label the points of intersection G and H between the two planes in the auxiliary view, Fig. 15-16C.

5. Project G to the top view where it will intersect line ED.

6. Project H to its line of intersection, EF. Line GH is the line of intersection in the top view.

7. Project points G and H to their intersecting lines in the front view. This will establish the line of intersection there.

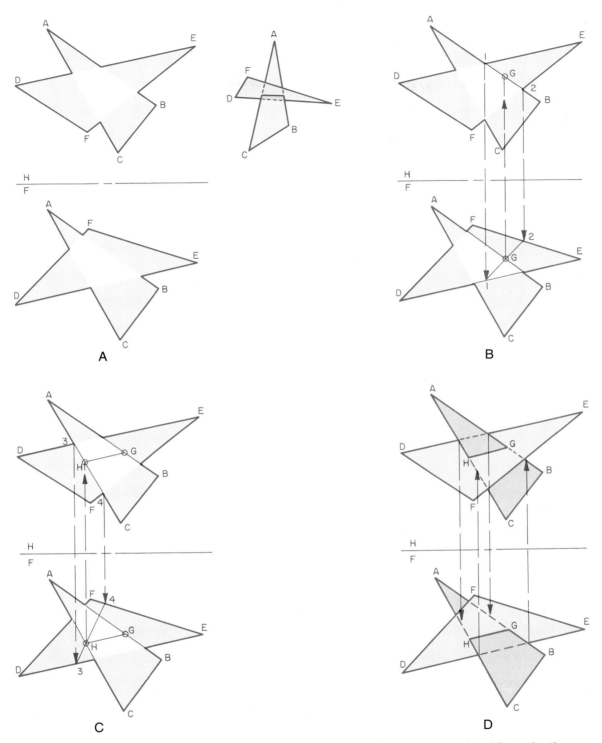

Fig. 15-15. The intersection of two planes can be found by using the orthographic projection

8. Analyze the front view by looking at the top view from the crossing in the front view of lines AB and DE. Line DE is found to be nearer and therefore visible in the front view, Fig. 15-16D.

9. Analyze the top view for the visibility by viewing the front view or auxiliary view as shown by the crossing lines AC and DE. AC is closest in both views and ,therefore, visible in the top view.

Intersection of an Inclined Plane and a Prism—Orthographic Projection Method

When a plane cuts a prism and appears as an edge in one of the principal views, the lines of intersection are found by locating the piercing points of the lines of intersection. The following procedure explains this process.

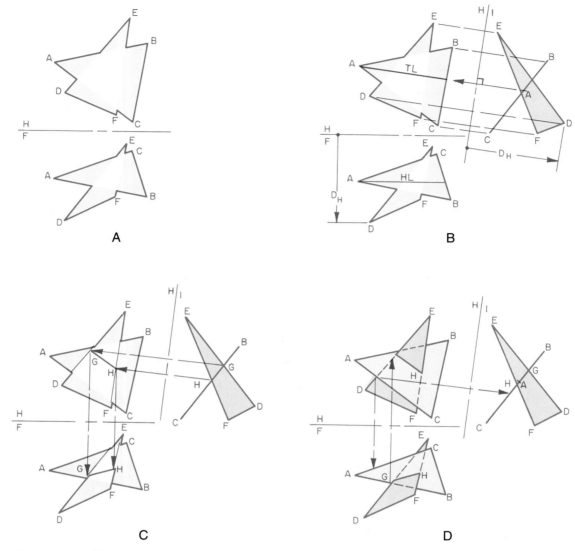

Fig. 15-16. The intersection of two planes can be located by using the auxiliary view method.

Given at least two principal views, including one with an edgewise view of the cutting plane, proceed as follows.

1. Label the intersections of the plane and the lateral corners of the prism in the top and side views, Fig. 15-17A.
2. Project these points to their corresponding lines in the front view, Fig. 15-17B.
3. Join points A, B, and C in the front view to form the line of intersection.
4. Determine the visibility of the prism and the line of intersection by viewing the top and side views from the front view. The vertical corners of the prism that appear in front of and below the cutting plane are visible in the front view. The lateral corners that fall behind the plane are invisible in the front view. The line of intersection AB in the front view is invisible, as is a portion of the upper edge of the plane. This is because their locations are farther away from the frontal profile projection plane than the prism.

Intersection of an Oblique Plane and a Prism—Cutting Plane Method

When a cutting plane is oblique to the principal planes of projection (not perpendicular to any one of them), the cutting plane method is used to find the intersection of a plane and a prism.

Given two principal views of a plane and a prism that intersect proceed as follows, Fig. 15-18A.

1. Label the intersections of the plane and the lateral corners of the prism in the top view, Fig. 15-18B.

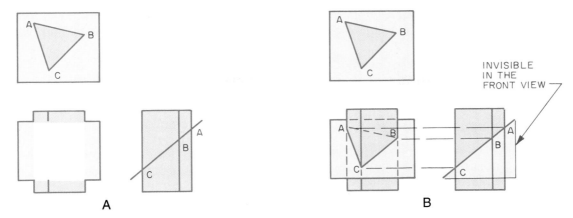

Fig. 15-17. The orthographic projection method can be used to locate the intersection of an inclined plane and a prism.

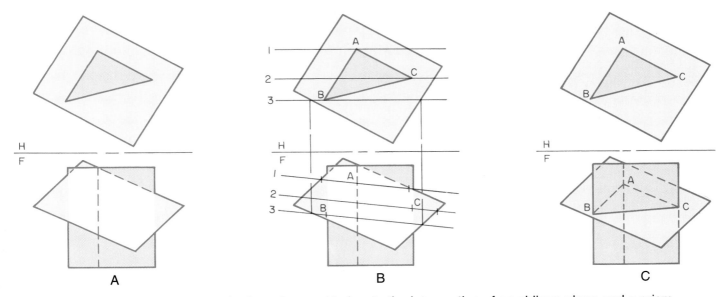

Fig. 15-18. The cutting plane method can be used to locate the intersection of an oblique plane and a prism.

2. Pass cutting planes 1, 2, and 3 through the corners of the prism in the top view. Extend these through the edges of the oblique plane. These planes have been drawn horizontally, but can be in any direction *except* perpendicular to the projection plane.

3. Project the intersections of cutting planes 1, 2, and 3 with the oblique plane edges to the front view, as shown with line 3 in Fig. 15-18B. This line in the front view is the line of intersection of cutting plane 3 with the oblique plane.

4. Project points in top view, where the three cutting planes intersect the corners of the prism, to the front view. (That is, point B of line 3 in the top view to point B, line 3 in the front view.)

5. These points in the front view represent the piercing points of the lateral corners of the prism. Since the prism and the oblique plane are intersecting plane figures, their lines of intersection will be straight lines. Join the piercing points with light construction lines and a straightedge.

6. The visibility is established in each view by viewing the adjacent view for nearness of lines and surfaces, Fig. 15-18C. Once the visibility has been determined, darken the appropriate construction lines.

Intersection of an Oblique Plane and an Oblique Prism—Auxiliary View Method

When neither the cutting plane nor the prism appear as an edge in any of the principal views, construct an auxiliary view to provide an edgewise view of the cutting plane. Once this view is constructed, the solution is similar to that for the inclined plane.

Given two principal views where neither the cutting plane nor prism appear as an edge, proceed as follows, Fig. 15-19A.

1. Construct a primary auxiliary view by taking the point view of a line on the cutting plane to produce an edgewise auxiliary view of cutting plane 1-2-3-4, Fig. 15-19B. Note: Either a horizontal or frontal auxiliary can be used.

2. Project the prism to this auxiliary view.

3. Label the intersections of the plane and the lateral corners of the prism in the auxiliary view, Fig. 15-19C.

4. Project these intersections to their corresponding lines in the top view. These points are the piercing points of the lateral corners of the prism and the cutting plane. Join these points to form the line of intersection in the top view.

5. Project the piercing points from the top view to their corresponding lines in the front view. Draw the line of intersection.

6. The visibility is established by checking adjacent views for nearness of surfaces.

Intersection of Two Prisms—Cutting Plane Method

The surfaces of prisms consist of a number of single planes. Therefore, the intersection of two prisms may be thought of as a prism intersecting with one or more single planes. The solution should be approached as outlined in the preceding sections, working with one plane at a time. Two applications are given. The first is an illustration of the cutting plane method. The second is an application of the auxiliary projection method.

In Fig. 15-20, two intersecting prisms are shown in top and front views. Two of the lateral edges of the triangular prism intersect the square

prism on the front plane and one intersects a rear plane, as seen in the top view. The piercing points of all three lateral edges of the triangular prism are shown in the orthographic projection in the top and front views.

The points at which lateral edge CC′ (Fig. 15-20) intersects planes E and F cannot be obtained directly by projection in the principal views. However, a vertical cutting plane, parallel to the lateral edges of the triangular prism, can be passed through lateral edge CC′ (where the undetermined points lie) and projected to the front view.

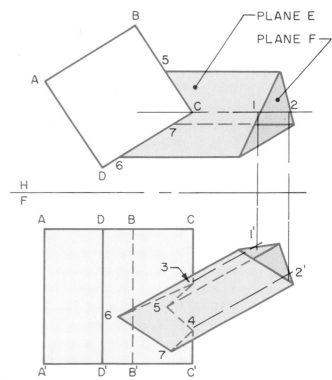

Fig. 15-20.   The cutting plane method can be used to find the intersection of two prisms.

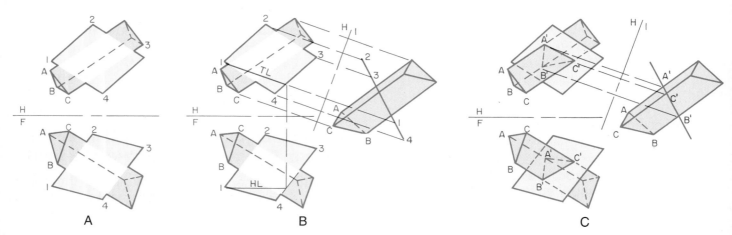

Fig. 15-19.   The intersection of an oblique plane and oblique prism can be located by using the auxiliary view method.

Given the top and front views, proceed as follows.

1. Draw a line in the top view parallel to the edges of the triangular prism and through corner C, Fig. 15-20. This line represents the vertical cutting plane.
2. Project points 1 and 2, where the cutting plane intersects the edges of planes E and F, to the front view to intersect the edges of the same planes at 1′ and 2′.
3. Project lines parallel to the triangular prism through points 1′ and 2′ to intersect with edge CC′ at points 3 and 4, piercing points of edge CC′ with planes E and F.
4. Join piercing points 5, 6, and 7 of the lateral edges of the triangular prism and piercing points 3 and 4 to complete the intersection between the two prisms.

Intersection of Two Prisms—Auxiliary View Method

A second method of finding the intersection between two prisms is the auxiliary view method. The same two prisms used in the cutting plane method are used in the auxiliary view method, Fig. 15-21.

The piercing points of the lateral edges of the triangular prism are found by regular orthographic projection. The intersections of lateral edge CC′ with planes E and F are found by drawing a frontal-auxiliary view where a point view is shown of the end of the triangular prism. Lateral edge CC′ is shown as a line in this view as well.

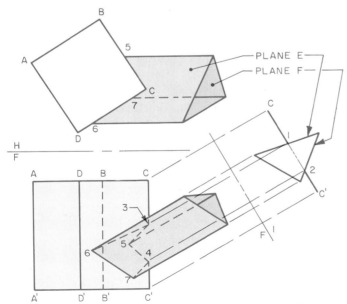

Fig. 15-21. The intersection of two prisms can be found by using the auxiliary view method.

Given the top and front views, proceed as follows, Fig. 15-21.

1. Construct a frontal-auxiliary view that shows a point view of the triangular prism. Use a reference plane that is perpendicular to the lateral edges of the prism.
2. Project lateral edge CC′ as a line. This is only part of the square prism that is necessary to find unknown points of intersection of the two prisms (points 1 and 2). The other piercing points can be located in primary orthographic views.
3. Project points 1 and 2 in the auxiliary view to intersect with lateral edge CC′ at points 3 and 4, piercing points of edge CC′ with planes E and F.
4. Join piercing points 5, 6, and 7 of the lateral edges of the triangular prism and piercing points 3 and 4 to complete the intersection between the two prisms.

Intersection of a Plane and a Cylinder—Orthographic Projection Method

When the intersecting plane appears as an edge view in one of the views, the line of intersection can be found by projection between the principal views, Fig. 15-22.

Given the three principal views, proceed as follows.

1. Project the points of intersection of several randomly spaced parallel lines in the circular (top) view of the cylinder to the front and side views, Fig. 15-22.
2. Project points where these lines intersect with the edge view of the plane. Extend them to the front view to intersect with their corresponding lines of projection from the top view.
3. Connect these points of intersection to form the line of intersection between the plane and the cylinder.
4. The visibility is determined by checking adjacent views.

Intersection of an Oblique Plane and a Cylinder—Cutting Plane Method

When the plane is oblique to the principal views, the cutting plane method can be used to find the line of intersection, Fig. 15-23.

Given the top and front views, proceed as follows:.

1. Pass randomly spaced vertical cutting planes through the top view of the cylinder and plane.

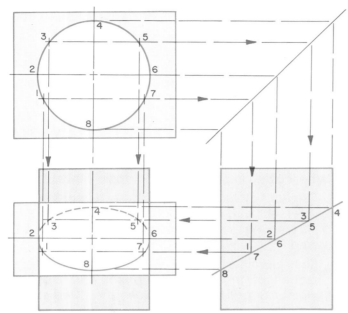

Fig. 15-22. The intersection of a plane and a cylinder can be found by using the orthographic projection method.

These cutting planes should run parallel to edges of the plane, Fig. 15-23. Select any number of planes at random locations. A greater number will produce a more accurate line of intersection.

2. Project the intersections of the cutting plane lines with the edges of the plane to their corresponding location in the front view. Connect two intersections of the edges of the plane with line C (as an example) in the front view. Complete the location of cutting plane lines in the front view.

3. Project the points of intersection of the cylinder and the cutting plane lines in the top view to their corresponding line in the front view. For example, points 6 and 10. Complete the projection of intersecting points. These points are the piercing points of the cylinder and the oblique plane in the front view.

4. Sketch a light line between these points in the front view. Finish with an irregular curve.

5. Visibility is determined by checking the two views.

Intersection of an Oblique Plane with an Oblique Cylinder—Auxiliary View Method

When a plane and a cylinder oblique to the principal views intersect, the line of intersection must be found through an auxiliary view, Fig. 15-24.

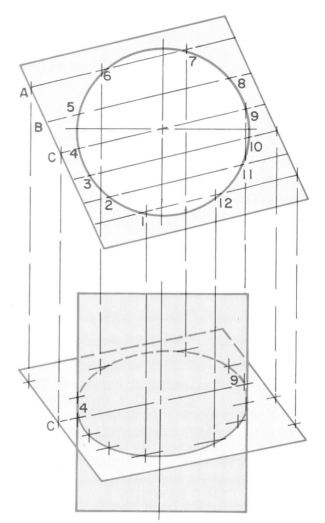

Fig. 15-23. The cutting plane method can be used to find the intersection of a plane and a cylinder.

Given the frontal and profile views, proceed as follows to find the line of intersection between the oblique plane and the oblique cylinder.

1. Construct an edge view of the plane in a primary auxiliary, Fig. 15-24. The cylinder will appear foreshortened. Elliptical ends are constructed as previously described for orthographic projection.

2. Pass randomly spaced cutting planes through the elliptical end of the auxiliary view of the cylinder to extend to the edge view of the plane.

3. Project points of intersection of the cutting planes from the elliptical end and the edge view of the oblique plane in the auxiliary to the profile view. This forms the line of intersection in this view.

4. Project points of intersection on the ellipse in the profile view to the frontal view. Transfer measurements of the points from the auxiliary

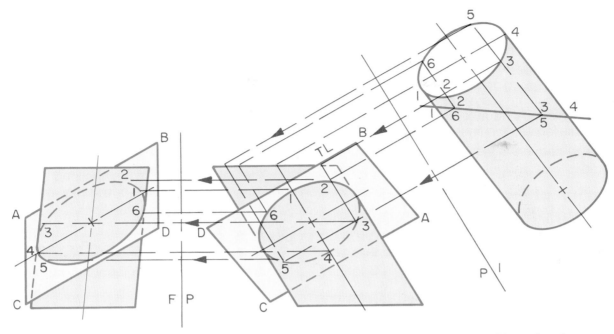

Fig. 15-24. Intersection of an oblique plane and an oblique cylinder can be found by using the auxiliary view method.

view to the frontal view. This forms the line of intersection in this view.

5. The visibility is determined by checking adjacent views.

Intersection of a Cylinder and a Prism

The intersection of a cylinder and a prism can be found by approaching the solution as a series of single planes intersecting a cylinder, Fig. 15-25. Each plane is treated one at a time. Since one or more of the planes of the prism are oblique in the principal views, an auxiliary view is needed to project the true shape of the prism, Fig. 15-26.

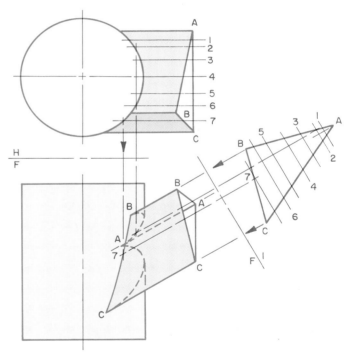

Fig. 15-26. When determining the intersection of a cylinder and a prism, an auxiliary view may be necessary to show the true size of the prism.

Given the top and front views of the cylinder and prism, the procedure is as follows.

1. Construct a primary auxiliary view showing the true size and shape of the prism, Fig. 15-26.
2. Pass randomly spaced cutting plane lines through the top view, intersecting the cylinder and prism.

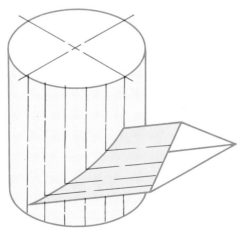

Fig. 15-25. A cylinder can be thought of as being composed of a number of single planes.

3. Transfer the cutting plane lines to the auxiliary view so they are perpendicular to the line of sight. The spacing of the cutting plane lines must equal that in the top view.

4. Project corresponding points of intersection between the cutting planes and the cylinder and prism in the top and auxiliary views to the front view. This locates the lines of intersection.

5. Check the adjacent views for visibility.

Intersection of Two Cylinders

The line of intersection of two cylinders can be found by the cutting plane method, Fig. 15-27. An auxiliary view is constructed to show the true size and shape of the inclined cylinder. Cutting planes are passed through the intersection of the two cylinders to identify piercing points used to plot the curved line of intersection.

Given the top and front views of the two intersecting cylinders, proceed as follows, Fig. 15-27.

1. Construct an auxiliary view of the inclined cylinder to show its true size and shape.

2. Pass randomly spaced cutting plane lines through the intersection of the two cylinders in the top view.

3. Transfer the cutting plane lines to the auxiliary view. These lines must be perpendicular to the line of sight. The spacing of the cutting plane lines must equal that in the top view.

4. Project the corresponding points of intersection between the cutting planes and the two cylinders to the front view from the top and auxiliary views. This locates the line of intersection.

5. Check adjacent views for the visibility.

Intersection of an Inclined Plane and a Cone

The intersection of an inclined plane and a cone may be found in the principal views where one view of the plane is an edge view, Fig. 15-28.

Given the top and front views, proceed as follows.

1. In the top view where the base of the cone appears as a circle, draw a number of randomly spaced diameters intersecting this circle, Fig. 15-28.

2. Project these points of intersection to the base of the cone in the front view. Connect them with the apex of the cone.

3. The lines from the base to the apex locate the piercing points in the inclined plane. Project these points to the top view to their corresponding diametric lines.

4. Connect these points of intersection in the top view to form the line of intersection. The line of

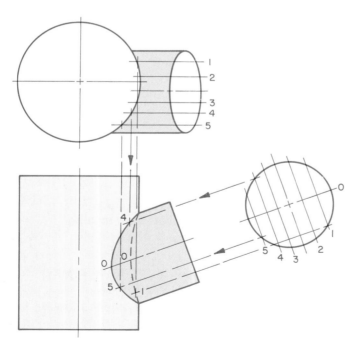

Fig. 15-27. The line of intersection between two cylinders can be found by using the cutting plane method.

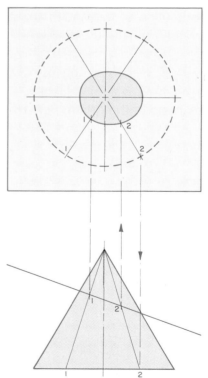

Fig. 15-28. The intersection of a plane and cone can be found in the principal view where the plane appears as an edge.

intersection in the front view coincides with the inclined plane.

Intersection of a Cylinder and a Cone

The line of intersection between a cylinder and a cone may be found in principal views when the intersecting cylinder is parallel to the principal views of orthographic projection, Fig. 15-29.

Given the three principal views, proceed as follows.

1. In the top view, where the base of the cone appears as a circle, draw a number of randomly spaced diametric lines intersecting the circle. (Refer to Fig. 15-29.) Note: If a pattern of the surface is to be developed later for the piece, 12 to 16 diametric lines should be equally spaced. (If this object is to be made out of sheet metal, for example, a pattern will need to be made later.)

2. Project the points of intersection to the front and profile views, intersecting first the base line and then the apex of the cone. If the lines in the profile view do not coincide with the outer surface of the cylinder, draw lines that do coincide. Project back to the top view where the diametric lines should be drawn.

3. The lines in the profile view from the base to the apex locate the piercing points of the cylinder. Project these points to the front view to intersect their corresponding lines. Then pro-

ject these points to the top view to form the piercing points in the line of intersection.

4. Connect these points to form the line of intersection in the top and front views.

SUMMARY OF PROJECTION METHODS APPLICATION IN LOCATING LINES OF INTERSECTION

Three methods of locating lines of intersection between geometric forms have been presented. These three methods are the orthographic method, the cutting plane method, and the auxiliary view method. When selecting the appropriate method to use, look at the following considerations.

- The *orthographic projection method* should be used when the object is parallel to one or more principal planes of projection. Also, the cutting plane representing the line of intersection must be shown as an edge in one of the principal planes of projection.

- The *cutting plane method* should be used when the plane representing the line of intersection is oblique to all of the principal planes of projection.

- The *auxiliary view method* should be used when *both* the intersecting plane *and* the object are oblique to the principal planes of projection.

PROBLEMS AND ACTIVITIES

Accuracy is extremely important in the graphic solution of intersection problems. Use a sharp pencil. Guidelines and construction lines should be drawn very lightly. This will help to increase your accuracy in locating points and intersections, as well as minimizing need for erasures.

Intersecting and Nonintersecting Lines in Space

1. For the problems in Fig. 15-30, find out if the lines intersect or merely cross in space. The lines are shown located on 1/4″ section paper. Use sheet layout A-4 with only a vertical division. Place two problems on a sheet. Use orthographic projection to solve these problems. Label your solution. After having solved the problems, consider the following question: Is there a way of studying views to determine whether lines intersect? Explain your answer.

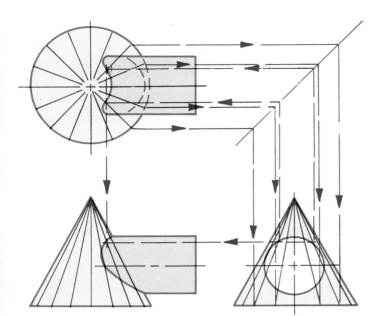

Fig. 15-29. The line of intersection between a cylinder and cone can be found in any principal view where the intersecting cylinder is parallel to the principal view.

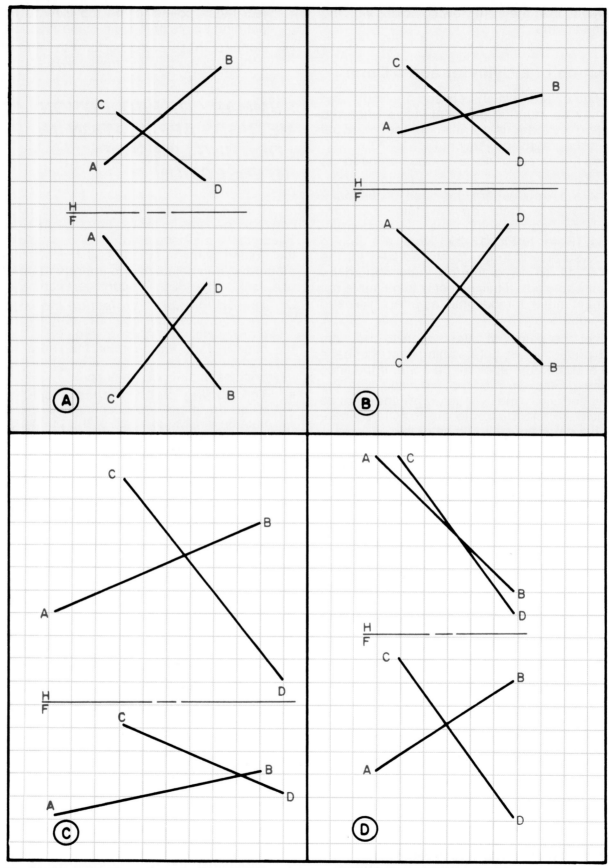

Fig. 15-30.   Use these lines for the problems on intersecting and nonintersecting lines.

Drawing Lines that Intersect in Space

2. Draw the top and front views of two randomly located crossing lines. Determine by orthographic projection whether these lines intersect. If they do not, try to lay out two that do. Check with your instructor if you need assistance.

Visibility of Lines

3. For the problems in Fig. 15-31, find out if the line or the plane is visible in each view. Complete the line as a visible or hidden line, as you have determined.

Piercing Points of Lines with Planes

4. For the problems in Fig. 15-32, locate the piercing point of each line. Also indicate the visibility of the line in each view. Use the orthographic projection method for the first two problems. Use the auxiliary view method for the second two problems.

A Line Perpendicular to an Oblique Plane

5. Construct lines perpendicular to the oblique planes through the points shown in the first two problems in Fig. 15-33. Locate the line in the top and front views.

Intersection of Planes

6. For the problems in Fig. 15-34, locate the lines of intersection of the planes. Indicate the visibility for each plane in the two views. Use orthographic projection for the first problem. For the last three problems, use auxillary views to determine the lines of intersection.

Intersection of a Plane and a Prism

7. Find the line of intersection between the oblique plane and the prism shown in the third problem in Fig. 15-34. Use the cutting plane method to determine the intersection between the plane and the prism. Lay out the problem a second time and find the line of intersection using the auxiliary view method.

Intersection of Two Prisms

8. Find the line of intersection between the square and the triangular prisms shown in the fourth problem in Fig. 15-34. Use the auxiliary view method. Indicate the entire line of intersection using visible and hidden lines.

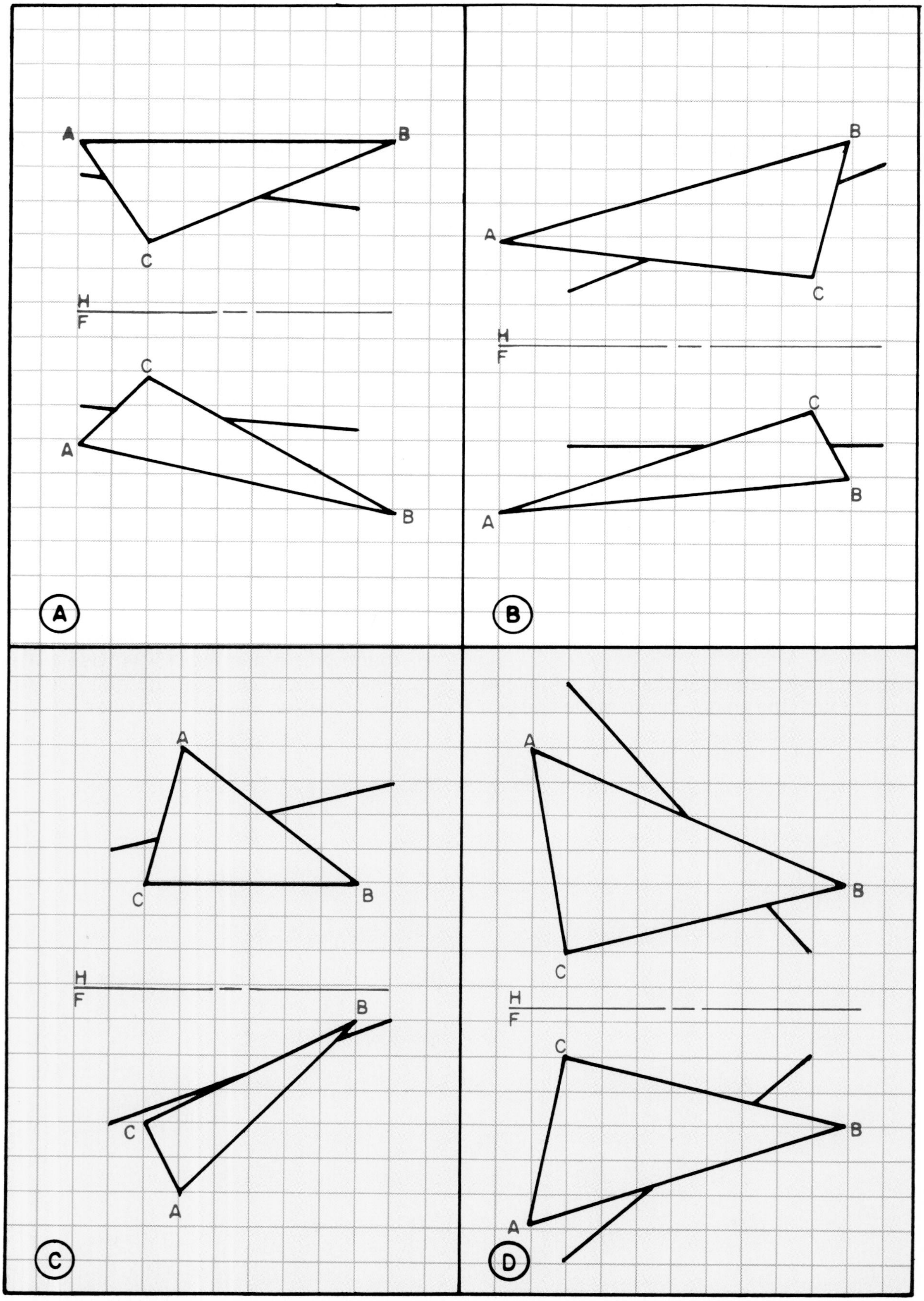

Fig. 15-31.   Use the lines and planes for the problems on identifying the visibility of lines.

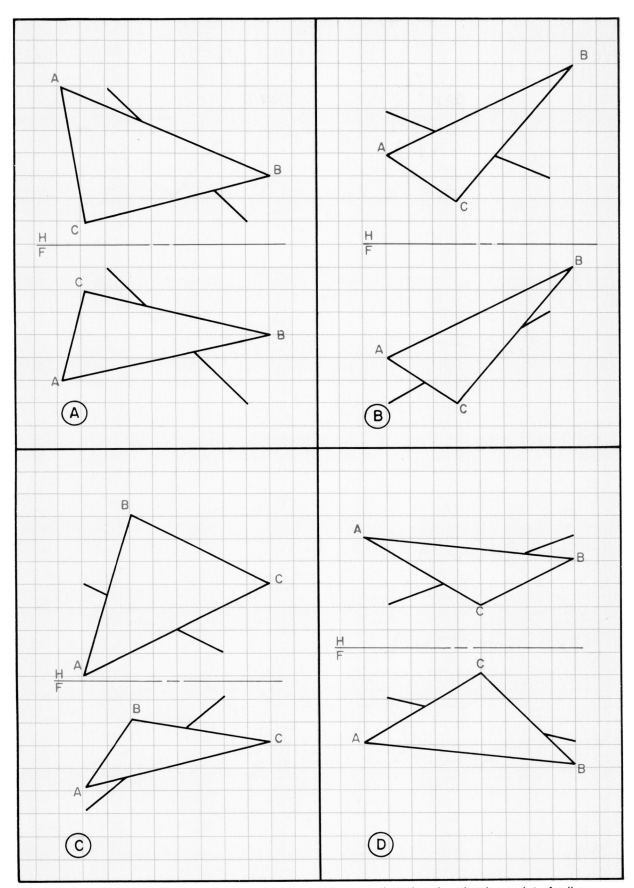

Fig. 15-32. Use these lines and planes for the problems on locating the piercing point of a line.

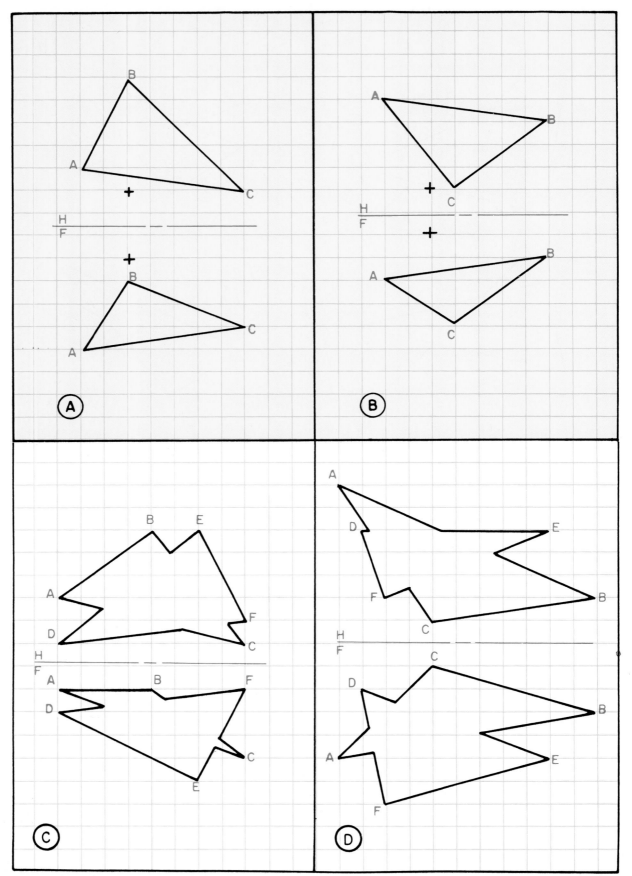

Fig. 15-33. Use these planes for the problems on constructing a line perpendicular to an oblique plane and locating the intersection of two prisms.

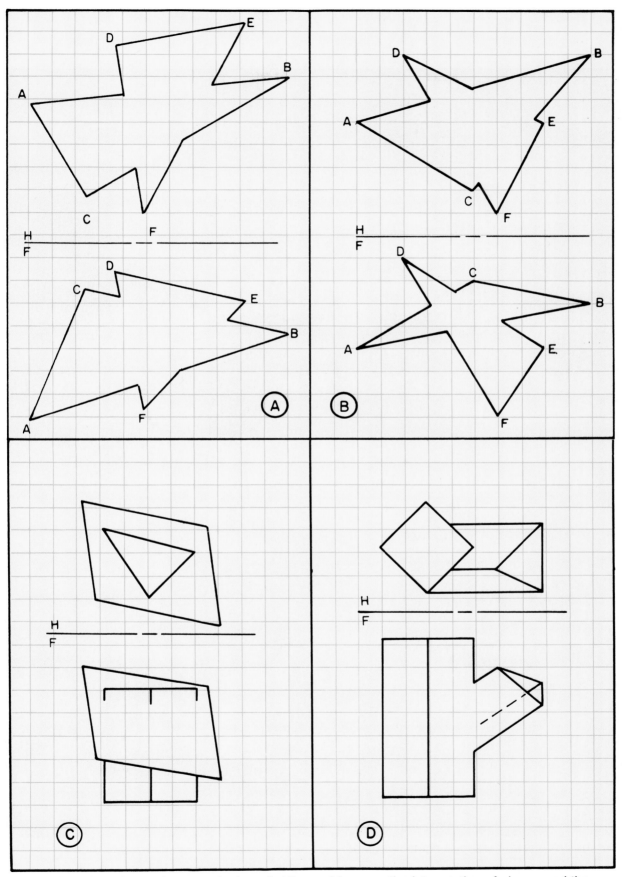

Fig. 15-34. Use these planes and prisms for the problems on the intersection of planes and the intersection of two prisms.

Boxes, like the box shown here for a computer game, require a flat pattern layout. (NovaLogic, Inc.)

Developments

KEY CONCEPTS

- ☐ A development is the layout of a pattern on flat sheet stock.
- ☐ Right angle objects are developed by laying out the true length lines on a stretch-out line.
- ☐ The development of oblique objects require auxiliary projection to find the true length lines.
- ☐ The development of truncated objects is the same as if the object was not truncated, only the truncated portion must be located as well.
- ☐ The development of a cone inclined to its base is different from the development of a right cone.
- ☐ Transition pieces are used in duct work and piping to join different sizes or shapes of pipes or ducts.
- ☐ A warped surface cannot be developed, however a development can be approximated.

A **development** in drafting refers to the layout of a pattern on flat sheet stock. The development might be a pattern for a carton, pan, heating or air conditioning duct, hopper, or any other manufactured product that requires folding or rolling of sheet materials. A flat pattern development for a package design is shown in Fig. 16-1. The heating, ventilation, and air conditioning (HVAC) and the petroleum industries depend heavily upon developments in the design and construction of systems, Fig. 16-2.

Developments are closely related to the material on intersections presented in the previous chapter. In many instances, intersections have to be identified before a development can be completed. This chapter presents basic principles, types of developments, and their industrial applications.

TYPES OF DEVELOPMENTS

Surfaces may be classified into ruled surfaces and double curved surfaces (see pages 321-323). Ruled surfaces are subdivided into planes, single-curved surfaces, and warped surfaces. Plane surfaces and single-curve surfaces can be developed. Warped surfaces can only be approximated by flat pattern development. Double-curved surfaces cannot be developed into single plane surfaces.

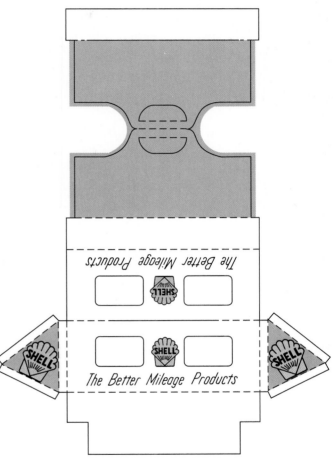

Fig. 16-1. Many containers require a flat pattern to be developed. (Chet Johnson, Industrial Designer)

Development of a Rectangular Prism

The development for a rectangular prism with an inclined bevel is shown in Fig. 16-3. It is laid out along a **stretch-out line.** This stretch-out line represents, and is parallel to, the right section of the prism.

Given top and front views of prism, proceed as follows.

1. Draw an edge view of the right section in the front view, below the bevel line, Fig. 16-3A. Lay off a line to one side of the right section. Make this line parallel to the right section. This will serve as a stretch-out line, Fig. 16-3B.

2. Identify the corners of the right section A, B, C, and D in the top and front views. Move in a clockwise order, since the stretch-out will be

Fig. 16-2.   Flat patterns are needed to lay out patterns for pipes, ducts, and water tower sections. (The Gleason Works; Central Foundry Division, GMC)

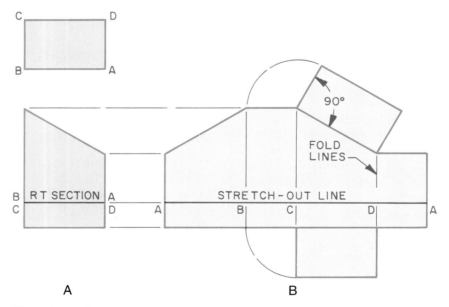

Fig. 16-3.   A rectangular prism can be developed by the orthographic view method.

from an inside view. (Most developments are made from an inside view. This makes the work piece easier to handle in folding and bending machines. An outside view could be laid out by working in a counterclockwise direction.)

3. The true lengths of the prism along the right section line are shown in the top view. Transfer these measurements to the stretch-out line, starting with A-B, B-C, C-D, and D-A. Draw vertical fold lines through these points.

4. The heights of the lateral edges of the prism are projected from the front view. These lines appear true length in this view.

5. Join the points to form the development or pattern for the prism.

6. Project lines at 90° to form the bevel surface and bottom, if required.

7. Lay off their widths with a compass, using an adjacent side as a radius. Join the points to complete these surfaces.

8. Add material for a seam if required.

Development of an Oblique Prism

The development of a prism that is oblique to all principal planes requires an auxiliary projection. The true size of the right section and the true length of the lateral lines are found in the auxiliary projection.

Given a prism that is oblique to all principal planes, proceed as follows.

1. Construct a primary auxiliary view to find the true length of the lateral edges, 16-4.

2. Construct a secondary auxiliary to find the true size and shape of the right section of the prism.

3. Draw a right section line perpendicular to the lateral edges in the primary auxiliary. Extend the right section line for a stretch-out line.

4. Transfer the true size measurements from the secondary auxiliary to the stretch-out line. Start with the lateral corner A.

5. Draw perpendiculars through these points. Project the length of each side directly from the primary auxiliary.

6. Join these points to form the development for the oblique prism.

7. If end pieces are required, construct these as indicated in Fig. 16-3.

8. Allow material for a seam if required.

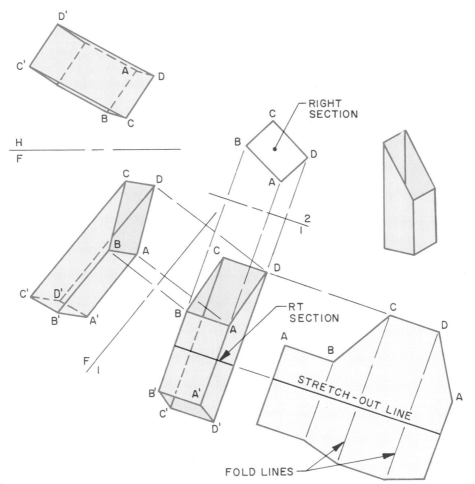

Fig. 16-4. An oblique prism can be developed by the auxiliary view method.

Development of a Cylinder with Inclined Bevel

The development of a cylinder with an inclined bevel is laid out along a stretch-out line. The stretch-out line represents, and is parallel to, the right section of the cylinder, Fig. 16-5.

Given the top and front views with the inclined bevel appearing as an edge, proceed as follows.

1. Draw an edge view of the right section in the front view below the bevel cut. Extend the edge view to the side for the stretch-out line. Note: The stretch-out line can also coincide with the base line, so long as the stretch-out represents a right section.

2. Divide the true size circular view into a number of equal parts (12 in the example). Project these to the front view for the true length lines along the height of the inclined bevel.

3. Transfer these chord measurements to the stretch-out line. Note: Figured mathematically, the length of the stretch-out line is: $\pi \times D$ = circumference, where D = the diameter. The distance is divided geometrically for greater accuracy, Fig. 16-5.

4. Draw a perpendicular through these points on the stretch-out line.

5. Project the base line across the full length of the stretch-out. Project the height of each true length line in the front view to its corresponding line in the stretch-out.

6. Sketch a smooth freehand curve between these points. Finish the curve with an irregular curve.

7. A base cover is the same size as the right section. The cover for inclined bevel is constructed as an ellipse. Refer to Fig. 13-16A.

8. Allow material for a seam if required.

Development of Two-piece and Four-piece Elbow Pipe

The procedure for the development of a two-piece or four-piece right elbow pipe is the same as for the bevel cut. Mating pieces are laid out as shown in Fig. 16-6. The stretch-out for the two-piece elbow was done by figuring the circumference mathematically and dividing the distance geometrically, Fig. 16-6B.

Development of an Oblique Cylinder

The procedure for laying out an oblique cylinder is quite similar to that for the inclined bevel-cut cylinder. The chief difference is the oblique cylinder's true lengths must be found in an auxiliary view, Fig. 16-7. The end covers, if required, are developed from a secondary auxiliary.

Development of a Pyramid

The development of a right pyramid is a type of radial line development. This development involves finding the true length of the chords. These chords must then be laid-out around a radius point, Fig. 16-8.

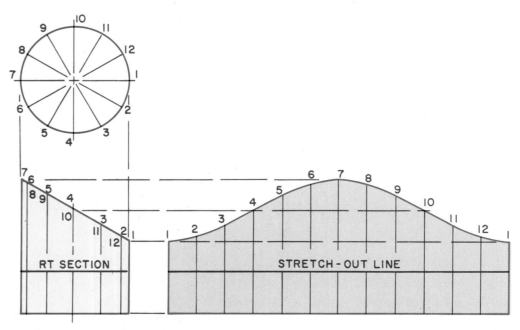

Fig. 16-5. The development for a cylinder with a bevel cut is laid out along a stretch-out line.

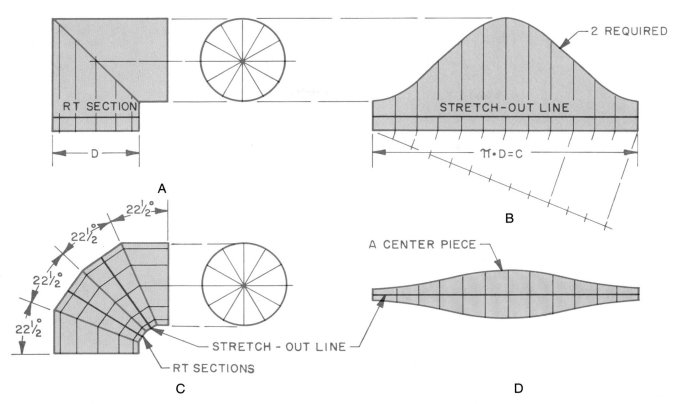

Fig. 16-6. Two-piece and four-piece right elbow pipes are developed in the same way that a cylinder with a bevel is developed.

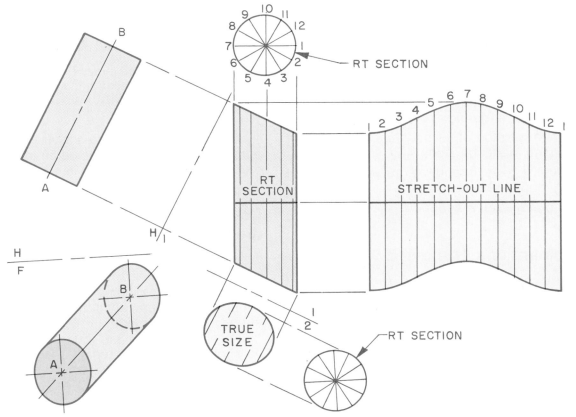

Fig. 16-7. In the development of an oblique cylinder, the true lengths must be found in an auxiliary view. The rest of the development procedures is similar to the development of an inclined bevel-cut cylinder.

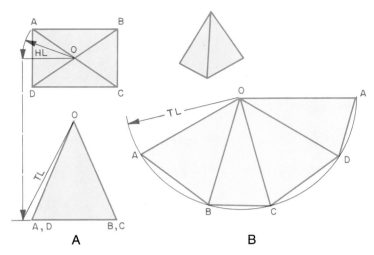

Fig. 16-8. To develop a pyramid, the true length lines are found and then laid out around a radius.

Given the top and front views of a right pyramid, proceed as follows.

1. Find the true length of the corner lines of the pyramid by revolving line OA to a horizontal position in the top view. Then project OA to the front view where it appears true length, Fig. 16-8A. This is the true length for all the corner lines, since the pyramid is a right pyramid.

2. With true length line OA as the radius, strike arc OA for the stretch-out line of the pyramid, Fig. 16-8B.

3. Lines AB, BC, CD, and DA appear in their true length in the top view since the base is in a horizontal plane in the front view. Lay off line AB as a chord on arc OA. Join the end points with O to form triangular side OAB.

4. Continue with the other base lines to form the remaining triangular sides.

5. Lay out the base, if required, adjacent to one of the triangular sides.

6. Allow material for a seam if required.

Development of a Truncated Pyramid

The truncated right pyramid shown in Fig. 16-9 is developed in the same manner as any other right pyramid. However, the truncated portion must also be located. ***Truncated*** means that the apex of the pyramid is "cut off." The following additional steps are necessary to locate the truncated portion.

1. Project the true lengths of the lines from the truncated plane horizontally to the true length line. Transfer these to the stretch-out.

2. Join the lines along the truncated cut.

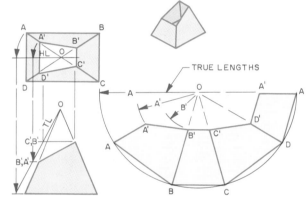

Fig. 16-9. A truncated pyramid is developed as a regular pyramid, however the "cut off" portion must also be located.

If a top cover is required for the truncated cut, a primary auxiliary view projected off the front view will produce the desired cover in its true size and shape.

Development of a Pyramid Inclined to the Base

The development of a pyramid inclined to its base is shown in Fig. 16-10. The procedure is very similar to the development of a right pyramid, except that the sides vary in their true lengths due to the offset of the apex.

The true lengths are found by rotating the corners of the lateral sides in the top view into a horizontal line. These points are then projected to the front view, Fig. 16-10A. Each surface is laid out in the development as a triangle with three sides given, starting with a side involving the shortest seam (seam OA'A), Fig. 16-10B.

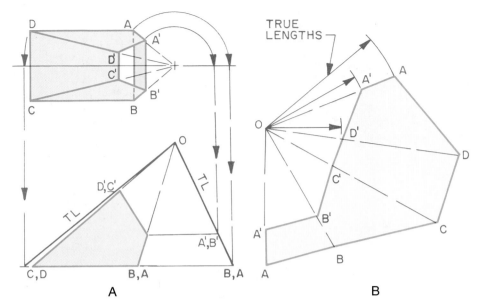

Fig. 16-10. The development for a pyramid inclined to its base is the same as for a right pyramid, however the true length lines will be of different lengths. Note that the object shown here is a truncated pyramid inclined to its base. The development is the same, with the exception of locating the "cut off" portion.

Development of a Cone

The stretch-out of cones, like pyramids, involves radial line development. These stretch-outs can be thought of as the development of a series of triangles around a common radius point, Fig. 16-11. The true length line of a side element of a right cone is shown in the frontal view as line O-1, Fig. 16-11A. The base circle is divided into a number of equal parts and transferred to the radial arc in the stretch-out. This will give the circular length of the development. The base line of the development is drawn as an arc rather than as chords, since all elements of the lateral surface of the cone are the same length.

Given the top and front views of a right cone, proceed as follows to lay out development for a cone.

1. Divide the base circle into number of equal parts, Fig. 16-11A.
2. With true length line O-1 as the radius, draw an arc as a stretch-out line, Fig. 16-11B.
3. Transfer the chord lengths of the base circle from the top view to the stretch-out line to determine the circular length of development.
4. Add material for a seam if required.

If a base is desired, the true size is shown in the top view.

Development of a Truncated Right Cone

The development for a truncated right cone is similar to that of the right cone, with an additional layout for the truncated part, Fig. 16-12.

Given the top and front views of a truncated right cone, proceed as follows.

1. Lay out the development for the cone as described earlier.
2. Determine the true length of the line elements of the cone intersecting the inclined surface, Fig. 16-12A.
3. Transfer the lengths to the appropriate line in the stretch-out, Fig. 16-12B.
4. Sketch a light line through these points. Finish the line with an irregular curve.

If a cap is required for the inclined surface, the development is achieved through a primary auxiliary as shown.

Development of a Cone Inclined to the Base

A cone that is inclined to its base may be developed as shown in Fig. 16-13. The development of this cone is significantly different from other cones.

Observe that the cone in Fig. 16-13 is not a right circular cone. The base is a partial true circle. The intersection of an inclined plane with a right cone would appear as an ellipse. The cone shown in Fig. 16-13 is actually an approximate cone.

The development of a cone inclined to its base involves the radial line method. This includes finding the true lengths of lines and the division of the surface development into triangles, as in the development of a right cone. It differs from the develop-

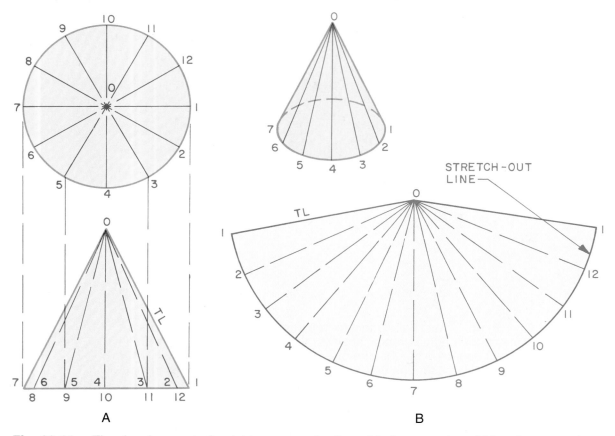

Fig. 16-11. The development of a right cone can be thought of as a series of triangles around a common radius.

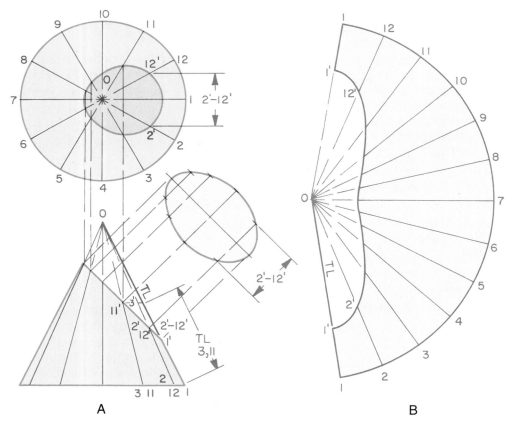

Fig. 16-12. The development of a truncated right cone is the same as a right cone, however the truncated portion must also be located.

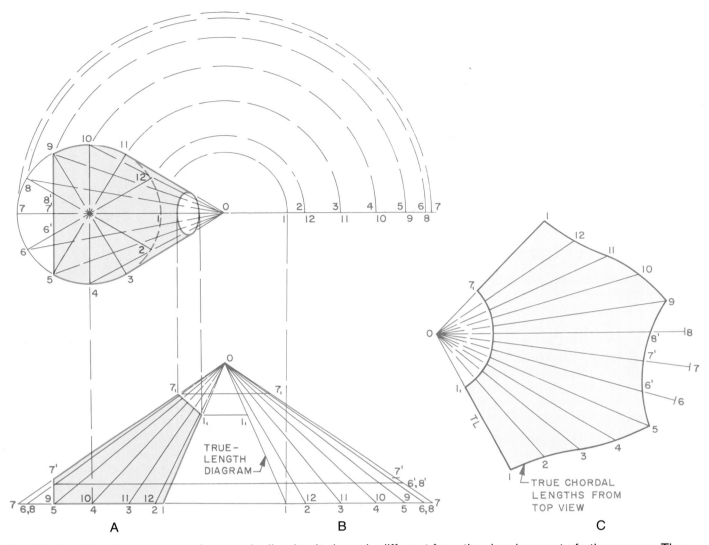

Fig. 16-13.   The development of a cone inclined to its base is different from the development of other cones. The development of a cone inclined to its base requires the radial line method. Note that the cone shown here is a truncated cone inclined to its base. The procedure is the same as for a cone inclined to its base, with the exception of locating the truncated portion.

ment of a right cone in that the base line of the cone is not laid out along a circular arc. Here, the lateral element lines differ in true lengths. In the development of a right circular cone, all of the lateral elements are the same length. Each triangle on the surface must be constructed by laying off the true lengths of its three sides: two element lines and a chord length.

Given the top and front views, proceed as follows to develop a cone inclined to its base.

1. Divide the circular base (top view) into a number of equal parts. Project these points to the base line in the front view, Fig. 16-13A.

2. Also project these points to the apex of the cone in the top and front views.

3. Construct a true length diagram to find the true lengths of the lateral element lines, Fig. 16-13B.

4. Start the stretch-out of the development by laying off the true length of O-1, Fig. 16-13C.

5. Lay off chord length arc 1-2 from point 1 in the stretch-out. This is obtained from the base circle in the top view.

6. Lay off an arc equal to true length line O-2 to intersect with chord arc 1-2 at point 2 on the stretch-out. (Refer back to Fig. 16-13C.)

7. Continue laying out intersecting arcs of true length lateral element lines and chord lengths for the remaining points on the base circle. Disregard, at this time, the vertical cut shown in the front view.

8. Project points 6′, 7′, and 8′ from the front view to the true length diagram. These points represent the vertical cut.

9. Transfer true lengths 6', 7', and 8' to the appropriate lateral in the stretch-out.

10. Project points of intersection of the cut at the upper part of the cone and lateral lines to the corresponding line in the true length diagram. (Refer back to Fig. 16-13B.)

11. Transfer these true lengths to the corresponding lines in the stretch-out. (Refer back to Fig. 16-13C.)

12. Join the points of intersection to form the required shape of the development.

13. Add material for a seam if required.

Development of Transition Pieces

Transition pieces are used to join pipes or ducts of different cross-section shapes. Some examples of transition pieces are shown in Fig. 16-14. A transition piece is needed to join a square duct to a round one, Fig. 16-15. Transition pieces are developed by dividing the surface into triangles. The true lengths of the lateral elements are then found. The true lengths are then transferred to a stretch-out.

Given the top and front views for the transition piece shown in Fig. 16-15, proceed as follows.

1. Divide the circular opening in the top view into a number of equal parts, Fig. 16-15A.

2. Project these division points to the edge view of the circular opening in the front view.

Fig. 16-14. Transition pieces in duct work systems come in a variety of shapes. (The Gleason Works)

3. Connect the points of the four quadrants of the circle to the adjacent corners of the base in the top and front views. These lines represent bend lines in the transition piece.

4. Determine the true lengths of the bend lines by using corners A and B in the top view as centers. Rotate the lengths of the lines into a plane parallel to the frontal plane (plane A-B). Then project these lines perpendicularly to the height line in the front view. Join these points

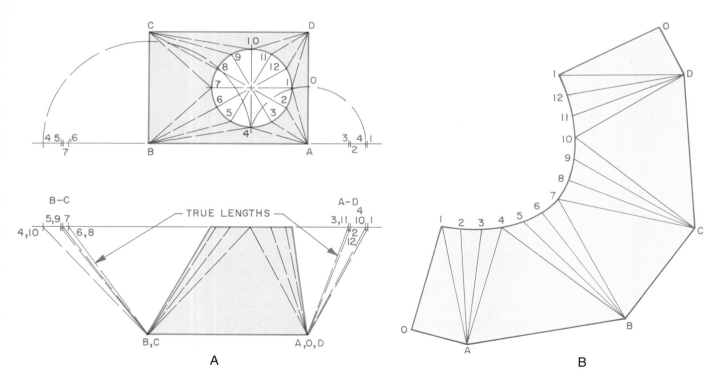

Fig. 16-15. The development of a square to round transition piece involves dividing the surface into triangles with the apex at the appropriate corner of the square.

with corners A and B where the lines appear true length.

5. Make a seam in a flat section of the development by starting the stretch-out with the lay out of true length line 1-O, Fig. 16-15B.

6. Strike true length intersecting arcs 1-A and O-A (O-A is shown true length in top view) to form triangle 1AO.

7. Strike true length intersecting arcs 1-2 (true length in top view) and A-2 to form adjacent triangle 1A2.

8. Continue with the layout of successive adjacent triangles to complete the development.

9. Allow material for a seam if required.

Development of a Warped Surface

A warped surface is a ruled surface that cannot be developed into a single plane. However, an approximation can be developed into a single plane by triangulation.

The piece shown in Fig. 16-16 is a transition from a right circular cylinder on an incline to an elliptical opening at the top. The surface is divided into a number of triangles whose sides appear as straight lines in the views. However, these triangles

will actually be slightly curved when the development is fabricated.

Given the top and front views, proceed as follows to approximate the development of this warped surface.

1. Find the true size of the elliptical base by rotating major diameter AG (from the front view in Fig. 16-16) to the horizontal. Project this true length to the top view and construct a half ellipse. (Minor diameter OD appears true length in top view.)

2. Using dividers, divide this half ellipse into a number of equal parts. Project these horizontally to intersect with the base circle in the top view. Then project these to the other half of the top view and to the front view. The base appears as an edge in the front view.

3. Divide the top elliptical opening into the same number of equal parts. Connect these points with the points on the base circle, forming triangles as shown.

4. Find the true length of the lateral lines by constructing two true length diagrams to keep lines separate and identifiable. The true length of the base line segments (on half of the ellipse) and the top opening segments are shown in the top view.

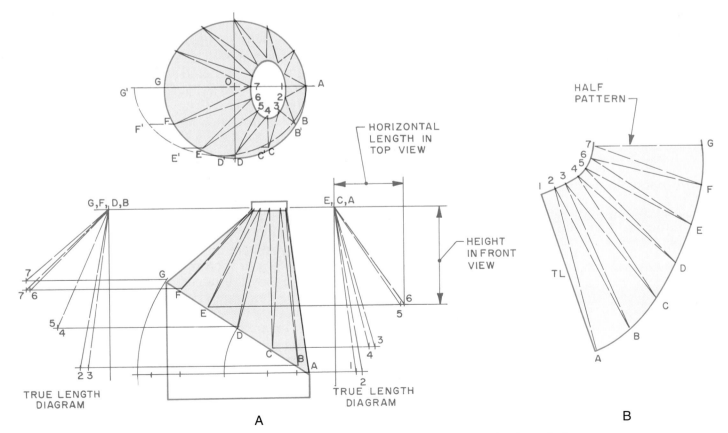

Fig. 16-16. The development of a warped surface can be approximated through triangulation.

5. Construct the stretch-out by starting with line A1 (shown in its true length in front view). Lay off triangle 1A2, using the "three-sides-given method," Fig. 16-16B.

6. Continue to lay off adjacent triangles in their true length until all are complete.

7. Draw a line forming irregular curved lines by connecting points 1 through 7 and A through G.

8. Allow material for a seam if required.

PROBLEMS AND ACTIVITIES

The following problems provide you with an opportunity to practice the skills and knowledge involved in laying out and developing patterns for various geometrical forms.

Patterns

The first nine problems are concerned with making patterns. For these problems, use a B-size sheet. Transfer your pattern layout to a stiff paper, cut it out, fold or roll it into shape, then test the accuracy of your layout.

Rectangular prism

1. Lay out the inside patterns for the development of the first four objects in Fig. 16-17.

Oblique Prism

2. Develop the inside patterns for objects 5 and 6 in Fig. 16-17.

Cylinders with an inclined bevel

3. Develop the inside patterns for objects 7 and 8 in Fig. 16-17.

Multi-piece elbow pipe

4. Lay out the inside patterns for objects 9 through 11 in Fig. 16-18. These objects are pipe elbows.

Oblique cylinders

5. Develop the inside patterns for objects 12 and 13 in Fig. 16-18.

Pyramids

6. Develop the inside patterns for objects 14 through 16 in Fig. 16-18.

Cones

7. Lay out the developments for objects 17 through 20 in Fig. 16-19.

Transition pieces

8. Lay out the developments for the objects 21 through 24 in Fig. 16-19. These objects are transition pieces.

Applied problems

9. Lay out the developments for the "ORCHARD HEATER HOUSING" and the "DUST COLLECTOR HOUSING" shown in Fig. 16-20.

Design Problems

Make use of the skills and knowledge you have learned in the last four chapters on descriptive geometry, and utilize the techniques of design problem solving studied in Chapter 1, to solve the following problems.

10. Design a horizontal cylinder to hold 1000 gallons of water. The ends of the tank are to be flat. The tank is filled through a 4″ pipe that enters the center top of the tank at an angle that extends directly back. The angle meets the plane in line with the back side of the tank at an elevation of 5′ above the tank. Make a drawing of the necessary views and prepare a scaled model of the tank and supply pipe. (Note: For this problem you will need to know how to figure out the volume of a cylinder. The formula for this is $V = \pi \times D \times H$, where V = the volume, D = the diameter, and H = the height.)

11. Design a storage device for a liquid, powder, or granular material. Use two or more of the geometrical shapes studied in these chapters on descriptive geometry. Prepare a scaled model of the design.

12. Design a transition piece to solve a problem or need that you have identified in your home, at school, or somewhere in your community. Draw the necessary orthographic views, develop a full size inside pattern, and form the piece out of the required material. Write a short paper explaining the need and how you have solved the problem.

13. Select an item that is commonly sold in stores (hair care or food related products, for example). Study the design of the package. Applying the design method studied in Chapter 1, try to improve the design of the package. Develop a prototype of your new design.

14. Select some problem that needs to be solved at school, at home, or in the community involving the application of descriptive geometry. Make a drawing and scale model of the solution that best meets the problem needs. (Select something other than a transitional piece, as you have already done this.)

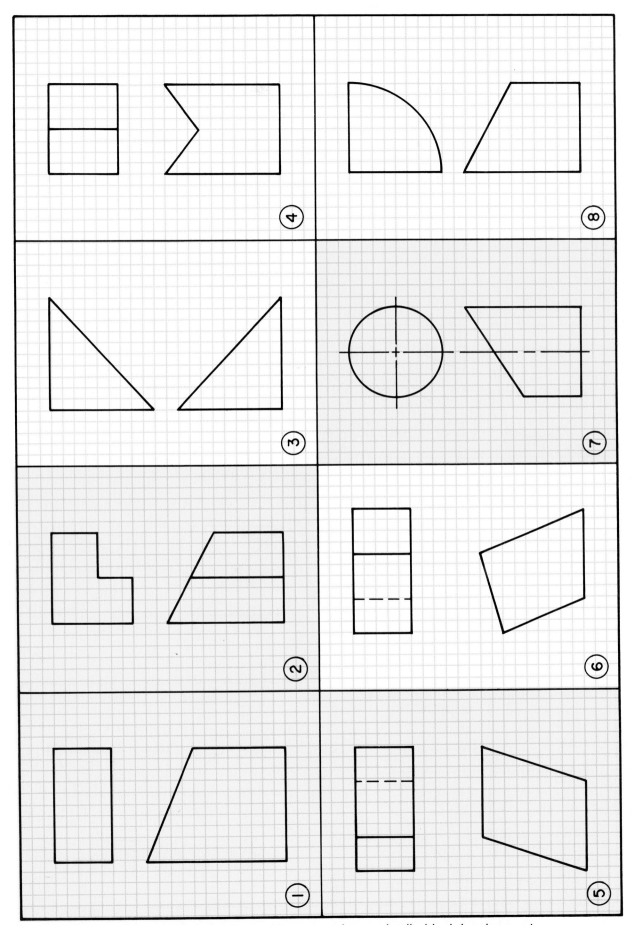

Fig. 16-17.   Use these objects for the problems on prism and cylindrical development.

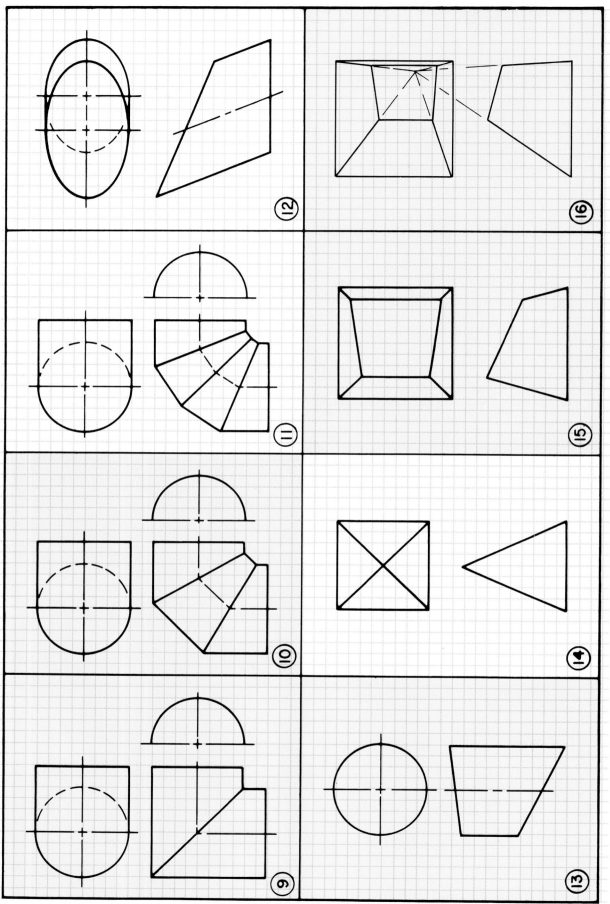

Fig. 16-18. Use these objects for the problems involving the development of cylinders, pyramids, and cones.

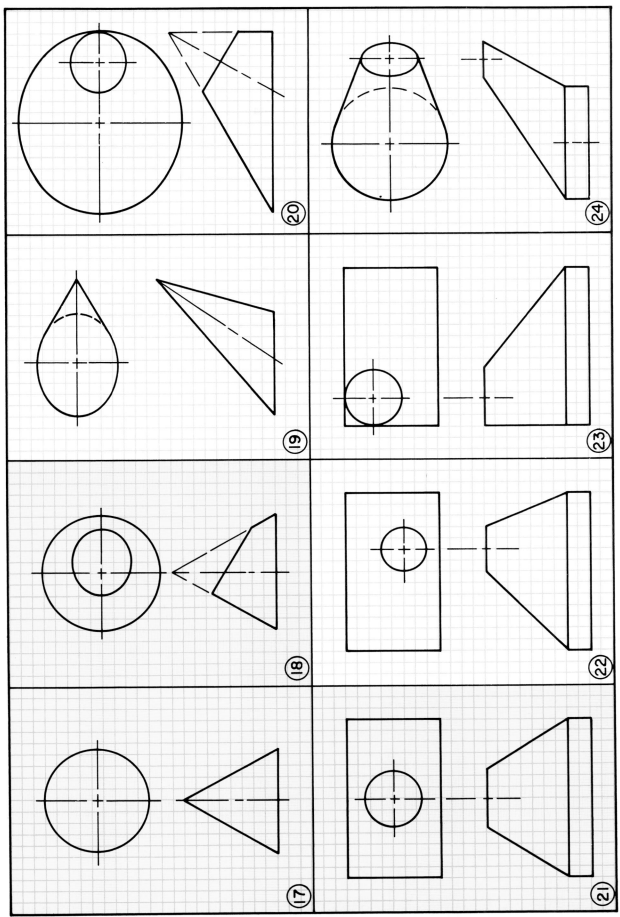

Fig. 16-19. Use these objects for the problems on the development of cones, transition pieces, and warped surfaces.

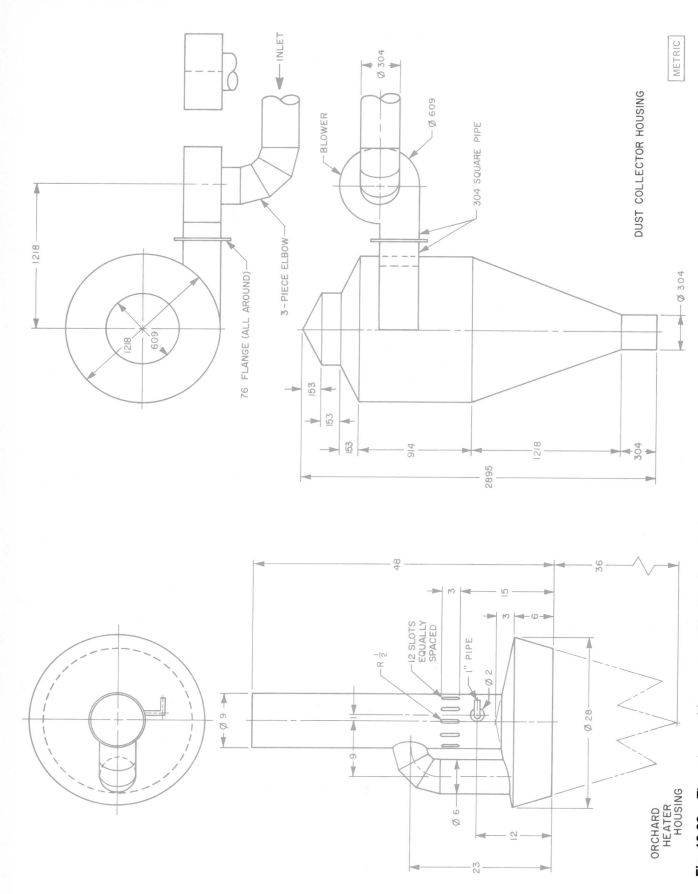

Fig. 16-20. These two problems are applications from industry of developments.

Part IV
Computer-assisted Design and Drafting

The role of the drafter has changed dramatically with the utilization of CADD in the workplace. Drafters and designers are now able to design, create, and evaluate a product before it is produced. Industry is also able to produce the parts with limited human intervention.

This part of the text covers *Introduction to CADD, The CADD Workstation and Software, Constructing and Editing Drawings,* and *Detailing and Advanced CADD Functions.*

The integration of computer-assisted design and drafting (CADD) with computer-assisted manufacturing (CAM) has enabled industry to become more efficient and productive. (Caterpillar, Inc.)

CADD has greatly improved the efficiency of the drafting process. CADD has also expanded the capabilities of drafting as a design and analysis tool. (Emerson Electronics)

Introduction to CADD

☐ Computer-assisted design and drafting (CADD) is an important part of today's industry.

☐ Many components comprise a CADD workstation. These include a computer, software, display screen, input device, and hardcopy device.

☐ CADD has evolved a great deal since the 1950s.

☐ CADD has brought about significant changes and benefits.

☐ Many opportunities exist for properly qualified CADD design-drafters.

☐ CADD plays an important role in a wide variety of industries.

The growing use of computer graphics has affected all areas of our technological society. Medical technicians use computer-generated images to track biological functions. Cartographers use computer graphics to map out land contours and terrain. Graphic artists use computers to lay out designs for printed materials. The manufacturing industry is also affected by computers. Computer-controlled systems are revolutionizing the methods used to design and manufacture a product. At the front end of the manufacturing process, designers use computer graphics to create and document product designs, Fig. 17-1. During the production process, computers may control machine tool movement and transport products. This chapter will focus on a specific subset of computer graphics, otherwise known as CADD.

The term "CADD" commonly refers to computer-assisted design and drafting. CADD means the use of a computer not only to draw, but also to analyze and test product prototypes. Each of these acronyms refer to one concept–the use of computers to assist the design-drafter in developing an idea into a product. The key word is "assist." The computer cannot create a drawing without human skill. The process of drafting still requires the knowledge and talents of the drafter. However, with CADD, the speed of a computer is teamed up with the skills of the drafter. The result is an increase in productivity, both in the quality and number of drawings completed as compared to traditional drafting.

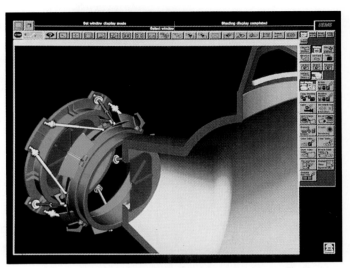

Fig. 17-1. Design engineering problem on a display screen. The growing application of computer-assisted design and drafting/computer-assisted manufacturing (CADD/CAM) technology is used by engineers and drafters. (Intergraph Corporation)

TOOLS OF CADD: THE CADD WORKSTATION

Like a pencil, scale, and T-square, a CADD system is a "tool" of the drafter. The equipment included in a CADD system is referred to as the **workstation.** A typical workstation includes a computer, CADD software, display screen, input device, and hardcopy device, Fig. 17-2.

The **computer** is the heart of any CADD workstation. Equally important is CADD **software,** which is the program used to instruct the computer to perform CADD functions. Advancements in both computer technology and software now allow personal computers to perform functions which only a few years ago were limited to larger mainframe systems.

Input and output devices allow the drafter to interact with the computer system. **Input devices** are used to enter or "input" commands. They are also used to locate positions on the drawing when constructing and editing lines, circles, and other entities. **Output devices,** such as the display screen or hardcopy device, allow you to view the created drawing. The display screen, much like a television, shows a temporary image of the drawing. When the

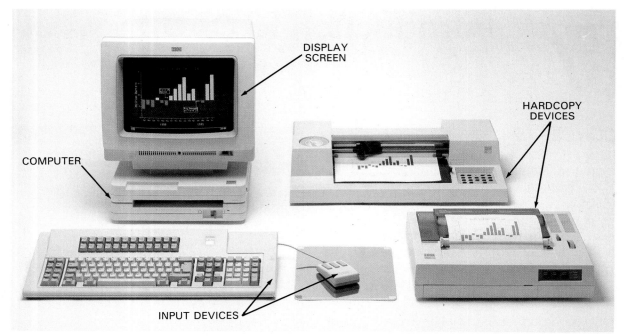

Fig. 17-2.   The CADD workstation includes computer, display screen, input device, and output device. (IBM)

computer is shut off, the image disappears. Hardcopy devices prepare a paper or film copy of the drawing. Various equipment comprising the CADD workstation is discussed in greater detail in Chapter 18.

HISTORY OF COMPUTER GRAPHICS

As discussed earlier, computer-assisted design and drafting is a part of the broader area of computer graphics. The development of computer graphics has progressed greatly since its beginning at the Massachusetts Institute of Technology.

In the 1950s, MIT developed the Automatically Programmed Tools (APT) program. The APT program, still used today, is a method of generating computer code to define the geometry of simple products. The code is then used by numerical control (NC) equipment to control machine movement while producing a part. Although this program isn't a truly interactive graphics program, it raised the idea of using computer code to define a product, rather than using pencil lead on a piece of paper.

Later that decade, the Air Force developed the SemiAutomatic Ground Environment (SAGE) project, which was a simulation program that displayed radar images showing enemy attacks. Military personnel would view the screen, then point to a sector of the screen to indicate interceptor aircraft. This was the first limited use of *interactive graphics,* or the ability to communicate with the computer.

The 1960s saw the first true application of interactive graphics as we know it today. In 1962, a doctoral student named Ivan Sutherland developed software which allowed a light pen to draw on a display screen. The light pen could not only point (as it did in the SAGE project), it could also draw images. The project was refined the next year by T.E. Johnson to draw multiview and perspective views.

Throughout the 1960s, automotive, aerospace, defense, and computer industries began further research and development of computer design and engineering systems. Most of these systems were proprietary systems (specific to the company); they were not developed to be sold to other companies. Later, many of these companies marketed their systems such as *Unigraphics* by McDonnell-Douglas and *CADAM* by Lockheed. Shortly thereafter, companies specializing in CADD systems appeared. They developed and sold **turnkey systems,** workstations which include all of the hardware (computer and related equipment) and software needed to run the system.

These graphics systems introduced some of the first highly sophisticated examples of CADD. They were capable of advanced work, such as solids modeling. Solid models are representations of an object in three dimensions, Fig. 17-3. The computer "thinks" of the object as a solid; it recognizes that the object has not only a surface, but material on the inside as well. Using the three-dimensional concept, a product design can be translated into code for use with CNC equipment.

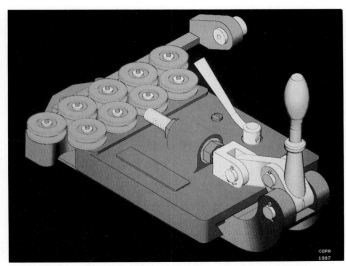

Fig. 17-3. Solid models represent an object in three dimensions. (IBM)

The 1970s saw advances in hardware technology and the introduction of more CADD systems. The primary developments in hardware were the progressive use of miniature circuits and displays which were less expensive and produced clearer images. The 1970s also saw the growth of CADD software developers. These companies specialized in CADD programs, while computer companies specialized in developing computers.

The 1980s were most noted for the development of microcomputer CADD systems, Fig. 17-4. Although companies continue to market mainframe and minicomputer systems, the personal computer (PC) has revolutionized CADD. The low cost of CADD on the PC has allowed most industries and schools to run full-fledged computer-assisted design and drafting programs. Due to the advances in computer hardware, it is now possible for microcomputers to perform most of the CADD functions of larger systems.

The 1990s have provided greater PC computing power, larger and more powerful software programs, and greater integration of computer graphics tools. There is little doubt that the future of CADD is bright. As new technology upgrades hardware and software, industry will combine CADD with all aspects of manufacturing, construction, and electronics design. The use of computer graphics will continue to grow as industrial and graphic engineers realize the benefits of CADD.

BENEFITS OF CADD

The productivity gained by using a CADD system lies in many areas. Five of the primary areas are:

- Drawing speed.
- Drawing quality.
- Quick modifications.
- Better communication.
- Analysis tools.

Drawing Speed

Research indicates that skilled CADD users are up to three times faster than drafters using traditional tools. This can mostly be attributed to the automation of tedious work. A traditional drafter spends a large amount of time "laying lead" to create each image on paper. Using a CADD system, most of your input is points. Instead of drawing a line, you select endpoints and the computer completes the line between the endpoints. Another example is a circle. The CADD user simply specifies the center and radius, Fig. 17-5. The computer then generates the circle.

Fig. 17-4. Microcomputer CADD systems perform many of the tasks of larger mainframe systems. (IBM)

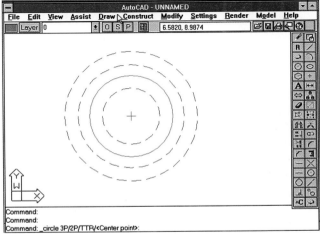

Fig. 17-5. Drawing speed is increased with a CADD system. Here the center point and radius are specified and the computer generates the circle.

Another factor in drawing speed is repetition. A rule of thumb with CADD is never to draw the same item twice. Instead, draw the object once and copy, then copy it to the desired location. The object could be as simple as a rectangle or as complex as a subassembly. In addition, commonly used symbols and parts can be held in a symbol library. Basically, a *symbol* is a miniature drawing, which can be inserted into the current drawing in any position. Symbol libraries may be included with the CADD software or they might be marketed separately. For example, an architectural symbol library might contain all the symbols for doors, windows, plumbing, electrical, and heating equipment.

Quality

Quality gained using CADD includes line quality and accuracy. When using the computer, all additions and modifications can be saved and viewed on the display screen. The drawing is sent to a hardcopy device for reproduction only when it is complete. This eliminates the constant drawing and erasing commonly done on a traditional drawing.

Another key to quality is the precision of the drawing. Lines, circles, and other entities created on a CADD system can be positioned with accuracy greater than one-thousandth of an inch. This ensures accurate tolerances when manufacturing the product.

Modifications

Modifying a drawing is one of the most time-consuming tasks of the drafter. There may be one way to draw a line, but a hundred ways to change it. A full range of CADD editing functions make modifications quick and easy. A revised drawing is simply replotted using a hardcopy device.

Communications

Application of computers in the production process has increased communications among departments of a company. First, information about the drawing (including the drawing itself) can be stored in a main computer, or *server.* Each department can access this information with their terminal to verify the status of a product, Fig. 17-6. Communications are also improved because guidelines can be standardized. These guidelines might include linetypes, lettering style, and layout of the working drawings. In addition, drafters can use the same database for symbols and subassemblies previously drawn. Mistakes in documentation are greatly reduced and confusion is kept at a minimum.

Analysis Tools

Sophisticated CADD systems are capable of a wide variety of analysis and testing functions. Some

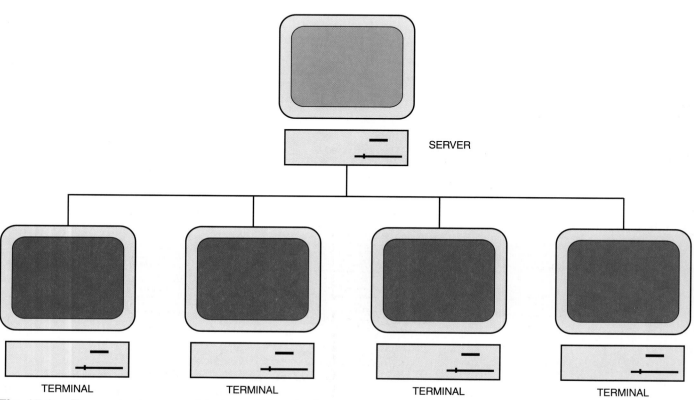

Fig. 17-6. The server is accessible from all terminals.

application programs can use the drawing data created by the CADD system to study many aspects including durability, strength, and potential failures of the product. See Fig. 17-7. The end result is a better quality product without having to build expensive prototype models for destructive testing.

QUALIFICATIONS OF THE CADD DRAFTER

Computer-assisted design and drafting has changed several of the traditionally required skills of the drafter. Tedious tasks, such as refining linework and lettering, are becoming less important. More emphasis is placed on analysis and problem-solving abilities. Even though a CADD system is simply a drafting tool, the task of actually drawing objects will decline as systems become more advanced. More significance will be placed on the drafter's ability to specify design features.

The qualifications of the CADD operator include a background in drafting, software proficiency, and hardware knowledge. A CADD operator must thoroughly understand drafting standards and conventions to be effective in the workplace. An operator with poor knowledge of drafting standards will create poor drawings, regardless of the type of computer hardware and software being used.

Knowing the capabilities of the software is an important step to becoming proficient with the system. You should be able to interact freely with the computer. Know the command you want, and know how to find it in the most efficient manner. Your drafting speed will increase as you become familiar with the system.

Most employers will not require that you know a high-level computer programming language. However, having some programming experience is an advantage. In time you will encounter *parametric programming*, which is listing a series of CADD commands used to create a certain object. Parametric programs automate the drafting process. You only need to supply certain values to the computer, and it will do the rest. See Fig. 17-8. For example, to draw a screw, you may only need to provide the diameter, length, and threads per inch. These values are the *parameters.* Using these values, the CADD system draws the necessary entities to describe the screw. Understanding these parametric programs by using them will provide insight into how they are actually developed.

As a CADD operator, you will be expected to understand the function of different hardware. Although you may use a specific workstation, take time to study the different types of computers, input devices, display screens, and output devices, as well as the proper care of this equipment. At some time, you may be required to recommend equipment purchases. Know what aspects of the equipment make one piece of equipment better than another.

CADD DRAFTING POSITIONS

Drafting positions have remained relatively stable with the inclusion of CADD in the workplace.

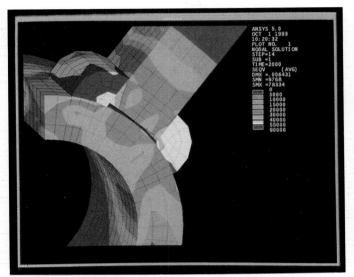

Fig. 17-7. Analysis software can be used to detect any problems with a product before actual production begins. Here, welding stresses are calculated by applying a simulated heat to the weld areas. (Swanson Analysis Systems, Inc.)

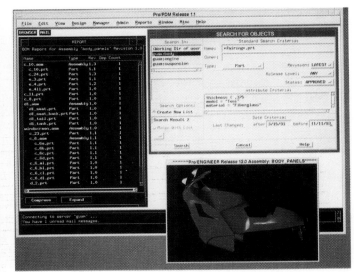

Fig. 17-8. Parametric programming allows an operator to input certain values to the computer (parameters), and have the design adjust automatically. This parametric software allows simultaneous modifications to be made to a design and CNC information. (Parametric Technology Corporation)

The position of layout and detail drafters has generally changed job title to **CADD operator.** Design drafters are now referred to as *graphic designers.* The biggest growth has been in positions related only to operation of the CADD system. **Systems managers** load software, start the system, and notify operators when the system is ready for use. During operation of the system, managers report hardware problems and software bugs. They might also schedule operation of the system. CADD **programmers** develop and maintain the functions of the CADD software. They load new software updates as they become available and correct "bugs" (errors in the software). They may also research new applications of CADD software. The programmer may design interface software which allows increased communication between the CADD system and production equipment.

THE DRAFTING ENVIRONMENT

The environment of drafting departments has changed rather drastically in companies using CADD. Modern facilities include comfortable settings. Tables and stools are replaced by ergonomically designed workstations with ample room for equipment. See Fig. 17-9. The atmosphere (temperature and humidity) is carefully controlled to ensure reliable operation of the computer.

Due to the high initial cost of large CADD systems, most companies have drafters work in shifts.

Fig. 17-9. Computers have added a new look to the drafting department. (IBM)

The number of drafters generally does not decline. However, most companies invest in fewer workstations, and to reduce payback time, run the system from 14 to 24 hours a day. For companies just switching to computers, old drawings are digitized into CADD drawing files almost constantly. This is in addition to creating, modifying, and revising existing drawings.

JOB OUTLOOK

Traditionally, the drafting department was recognized as the bottleneck of workflow. The time required to create and revise drawings often exceeded that for design and production. Computer-assisted design and drafting has lessened much of the bottleneck. The need for new CADD operators presents a great opportunity for all persons.

APPLICATIONS OF COMPUTER-ASSISTED DESIGN AND DRAFTING

Computer-assisted design and drafting is a vital part of various drafting disciplines. In addition to a basic CADD software, specialized software is available for certain disciplines. Customized CADD packages are generally available for three areas:

- **Mechanical.** Mechanical CADD includes machine and product design related to manufacturing, Fig. 17-10A. Mechanical designers might have special software to analyze a product design and generate code for numerical control equipment.

- **Architectural Engineering and Construction (AEC).** AEC applications include architecture, facilities planning, landscape architecture, and structural design. See Fig. 17-10B. AEC CADD systems speed the layout of space and materials in residential and commercial structures.

- **Electronics.** Electronic CADD involves design and layout of components for circuit boards and electronic products, Fig. 17-10C. Electrical engineers use programs which automatically place electrical components and their connections.

These three areas make up more than 90 percent of the CADD applications. Other areas include civil drafting, business graphics, and publishing. This section will discuss those CADD functions that enhance a certain drafting discipline.

Mechanical Applications

Mechanical design and drafting remains the most prominent use of CADD systems. The accu-

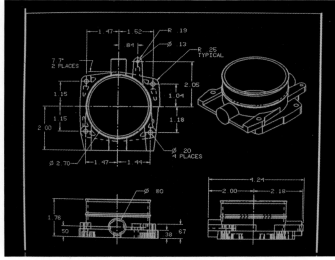

A

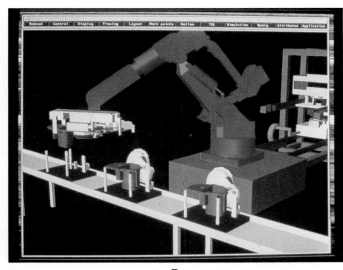

B

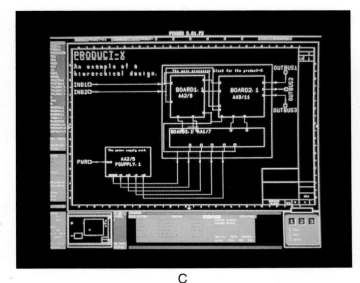

C

Fig. 17-10. Customized CADD packages are generally available for A—mechanical, B—architectural engineering and construction, and C—electronics. (IBM, Tecnomatix)

racy of CADD increases the precision and dimensioning of mechanical designs since measurements are taken from the drawing data, not measured by the drafter. This precision is necessary if the drawing data will be sent directly to equipment for production. The drafter can also magnify the connection of two parts to inspect how they match. Motion may also be analyzed to make sure movable parts do not interfere with each other.

Modeling

One of the largest mechanical CADD applications is modeling. *Modeling* is the process of creating a three-dimensional view of an object. The drafter identifies faces, or surfaces, of an object from which the computer creates the model.

Three-dimensional models help drafters visualize a product. Models can be shaded for appearance and rotated to help the designer see the object from different angles. Certain types of models are used to test products for strength, durability, etc. There are three types of models:

- Wireframe.
- Surface.
- Solid.

Wireframe models are created by connecting "points" of an object. "Points" refers to intersections of lines of the object. Connecting these points makes the object appear three dimensional, Fig. 17-11.

Surface models are constructed by connecting edges. Basically, a surface model begins with a wireframe model. Then each plane is converted to a surface as if a sheet of plastic covered the wireframe. Many CADD systems allow you to assign color to surfaces to enhance the appearance, Fig. 17-12. Since the model has solid surfaces, a drafter can produce shading by directing a light source at the model at a given angle.

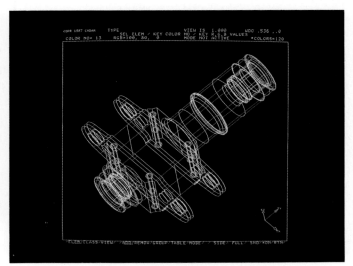

Fig. 17-11. Wireframe models are transparent views. (IBM)

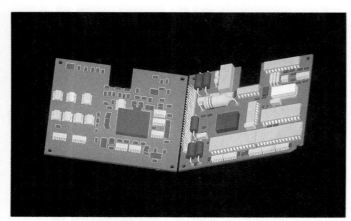

Fig. 17-12. Electronics design utilizing surface modeling techniques. (Intergraph Corporation)

A special application of surface models is sheet metal design. After making the surface model of the product, it can be "unfolded" to view the pattern required to make the sheet metal product. This ability saves much time and effort for patternmakers.

Solids modeling is the most refined type of modeling. Instead of recognizing only surfaces, the computer "thinks" of the object as solid material. In wireframe modeling, the computer recognizes only exterior surfaces. However, in solids modeling the computer recognizes both exterior and interior features of the object, Fig. 17-13. Solid models are created using very small elements, called primitives. ***Primitives*** (including cubes, spheres, cones, pyramids, and other shapes) form the building blocks of the model.

Since solid models are made of small elements, like a real object is made of atoms, a solid model can be cut, pressed, pulled, and twisted much like a real object. This type of testing is re-

ferred to as ***finite element analysis***. If you put force on one area of the model, you can analyze the effect of that force in all areas of the model.

The use of solid models is on the increase for industrial applications. Designers are now creating solid models first, then generating detailed drawings from the solid model. Section views, as well as other standard views, are created by rotating the solid model and projecting the necessary features.

Computer-aided manufacturing

The increased use of computers is seen throughout the manufacturing process. The area of producing drawings using a CADD system is part of a much broader concept known as ***computer-assisted manufacturing (CAM).*** At first, computer-assisted design and drafting systems are used to define the geometry of the object. This geometry is then converted to numerical code (NC) by an NC processor. The code is read by a controller, which is connected to a milling machine, lathe or other processing equipment, to direct the speed and direction of the machine tool, Fig. 17-14. The code may be sent directly to the controller or stored in a data storage device for later use. CAM speeds up the manufacturing process since the same information used to create the design is used by machinery to machine the part.

Before being sent to a machine, the numerical code can be used for machine tool simulation. Using this type of program, the computer displays the tool path on the screen, Fig. 17-15. Errors in tool movement can be detected before a part is ma-

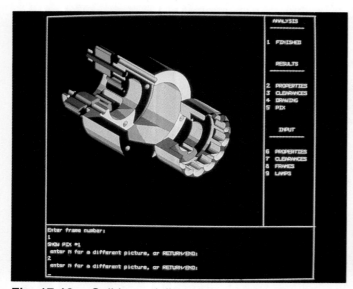

Fig. 17-13. Solids modeling allows designers to view interior and exterior features of a product. (IBM)

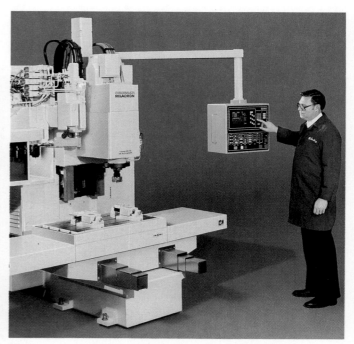

Fig. 17-14. The machine tool is numerically controlled by the computer. (Cincinnati Milacron)

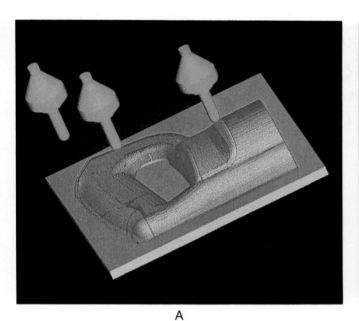

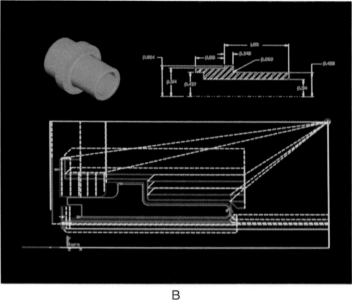

A   B

Fig. 17-15. Machine tool simulation. A–Verifying milling toolpaths. B–Verifying lathe toolpaths. (Intergraph Corporation)

chined. If the tool moves too fast or too deep, the designer can edit the tool path. Without CADD/CAM interface, most errors are caught when the tool breaks because of a wrong move.

Computer-integrated manufacturing

CAM systems generally control one or several machines, called ***machining cells.*** A ***computer-integrated manufacturing (CIM)*** system controls the entire product assembly line, Fig. 17-16. CIM interfaces machining cells with robots and transport systems which load, process, remove, and transport parts as they pass through the production line. A CIM system reduces the amount of tedious human labor required in the total production process.

Fig. 17-16. CIM systems involve the use of machining cells, robots, and transport systems. (Cincinnati Milacron)

Architectural Engineering and Construction Applications

Architectural engineering firms are increasing their use of CADD in the design and layout of residential and commercial buildings. *AEC applications* include facilities design, plans, elevations, details, structural design, and landscape architecture.

Facilities design

A growing use of computers in architecture is *facilities design*. These programs help determine space and room arrangements required by the client. Using facilities design software, an engineer identifies the type of activities in certain departments. In addition, the relationship of these departments is indicated, whether it be high or low. The computer then calculates a proposed arrangement of rooms and space for optimum communication and traffic.

Plans

Plans include layouts for foundation, floor, electrical, plumbing, and HVAC systems. When drawing floor plans, the CADD functions allow a designer to move, copy, and mirror rooms. If one room shape is commonly used, the designer may choose to make a copy of the room for other parts of the building.

Special inquiry commands are used to calculate properties of space allocations. Area can be found to report the square footage of each room, or of the total

building. The volume of rooms can be calculated to determine heating and cooling requirements.

An important feature of CADD when producing plans is symbol libraries. The most commonly used symbols for construction, electrical, and HVAC systems are stored in a library. When inserting a door, window, outlet, vent, or other commonly used item, the designer needs only to insert the appropriate symbol. Many CADD systems also keep track of the number of items inserted into the drawing. When the plan is complete, a bill of materials can be listed automatically based on the number of components inserted into the plan.

Elevations

The presentation of a proposed building is critical to contract sales. Certain CADD capabilities, such as color and three-dimensional imaging have greatly enhanced presentation graphics. Instead of a two color, two-dimensional drawing of the building, the architect can create a realistic perspective view complete with color, Fig. 17-17. In addition, the architect can rotate the image, showing clients views of the proposed structure from different angles.

Section views

Section views in both AEC and mechanical applications are quickened by the use of hatching. Instead of drawing single section lines, the CADD operator simply defines a border around the area to be hatched. When the type of hatch is selected,

Fig. 17-17. Computer-generated model of a commercial structure. (Landcadd International, Inc.)

section lines are automatically drawn at the correct angle and spacing within the selected border.

Structural design

A CADD system capable of analysis aids the work of structural designers. Members, such as girders, columns, and beams can be placed to provide maximum strength for the structure. With three-dimensional commands, the structure can be viewed from various angles. Clearance between members can be analyzed. This information is then used by other AEC engineers when adding other systems, such as heating and cooling ducts.

Piping plans

Another growing use of CADD in AEC applications is the design and layout of piping systems consisting of motors, pumps, and valves to transport water or other fluids. The initial layout is made in two dimensions. Symbols for pipes, pumps, valves, regulators, and other components, are arranged to provide proper flows. Once the entire system is laid out, the designer makes a three-dimensional model of the design, Fig. 17-18. The three-dimensional model helps the designer check whether any pipes or components interfere with each other. Advanced piping systems can even indicate errors in flow direction and help in valve or pump selection.

Landscape architecture

Applications programs in *landscape architecture* allow the drafter to create and easily change the layout of trees, shrubs, and other groundcover surrounding a building, Fig. 17-19. Symbols are used to represent each of these items. The features of each tree or other foliage are included when symbols are entered into the drawing. Cost estimates can be automatically produced from the quantity of each item in the landscape design. The ease of modification allows landscape architects to edit layouts according to client's needs and wants.

Electronic Applications

The evolution of many modern electronic components can be attributed to use of CADD electronic application programs. This software allows designers to more easily lay out and test the logic of circuits.

The initial layout is done with use of a symbol library. The library contains a list of prepared symbols representing standard electronic components. Once all components are identified, pin connections can be identified for automatic *tracing,* which is the task of connecting the "wires" from one component to another, Fig. 17-20. The computer finds the shortest route which doesn't interfere with other traces. This job typically took weeks manually; using the computer the same practice can be done in minutes. Routes for different circuits are given color to clarify the design.

Once tracing is complete, logic testing is performed. Historically, *logic testing* was done by breadboarding the components manually. Automatic logic testing simulates the operation of the circuit board. Tests are done for open circuits, shorts, or other improper connections. Errors can be easily edited on the display screen.

Other Applications

Any industry which has reason to draw is a candidate for CADD. CADD software is used for such applications as stage set design and book

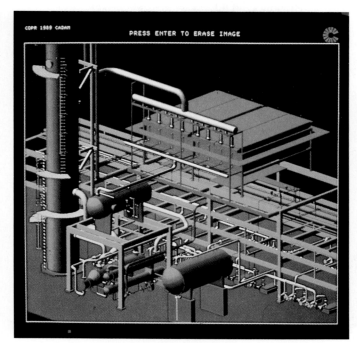

Fig. 17-18. Three-dimensional piping layout can disclose errors in the design. (IBM)

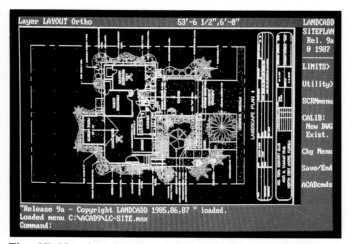

Fig. 17-19. Landscape architecture benefits from the use of applications programs. (Landcadd International, Inc.)

Fig. 17-20. Integrated circuits designed using CADD. (IBM)

illustration. Each year, a wider variety of CADD applications are found to replace traditional drafting techniques. Applications discussed in this chapter cover only a few of the areas where computer graphics might be used.

Civil drafting

Civil drafting involves making surveys, charts, and maps that describe land terrain, road systems, utility systems, and other topographic features using CADD, Fig. 17-21. Changes due to construction can be easily modified on the existing map. Data concerning mineral deposits can be analyzed on the screen to determine mining and drilling operations.

Business graphics

In addition to product drawings, companies use CADD to create charts and graphs for sales and market data to production flow and process se-

quences. Various CADD features, such as color and hatching, allow graphic artists to create appealing presentations.

Publishing

The benefits of computer-assisted design and drafting make it cost-effective to replace T-squares and technical pens for layout of artwork. More recently, CADD drawings are being transferred directly into electronic page layouts. Standard (paper-based) and electronic publications (on-line procedures) benefit from this technology, Fig. 17-22. When corrections are necessary, the illustration can easily be modified using CADD editing features. Artwork is then produced using a hardcopy device, such as a pen plotter. Also, entire pages, including copy and imported CADD artwork can be reproduced using laser printers or other high-resolution output devices.

QUESTIONS FOR DISCUSSION

1. Present several benefits of computer-assisted design and drafting over manual drafting.
2. List the components of a CADD workstation.
3. Briefly summarize the history of computer graphics.
4. Compare the qualifications of a CADD drafter with those of a manual drafter.
5. Describe the changes that occur in the environment of a drafting department that has converted to CADD.

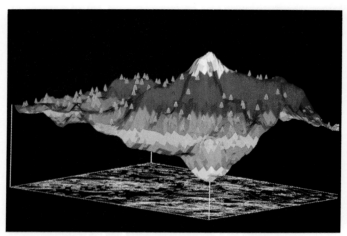

Fig. 17-21. The CADD system permits vivid environmental studies associated with specific land masses. (Landcadd International, Inc.)

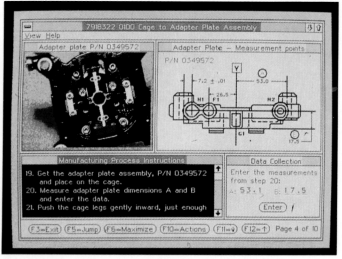

Fig. 17-22. Electronic publications, such as this on-line procedures manual, benefit from the capability of incorporating CADD-generated drawings with the text. (IBM)

6. Describe the impact computers have made on mechanical design and manufacturing.

7. Describe the impact computers have made on the architectural engineering and construction field.

8. What do you believe the future holds in store for computer graphics applications?

PROBLEMS AND ACTIVITIES

1. Review current literature on the applications of computer graphics for drafting. Check your school or community library for magazines containing current articles. Noted magazines are *Computer Graphics World*, the *S. Klein Computer Graphics Review*, *Design Graphics World,* and *Plan and Print*. Note the trends in the field and prepare a report for presentation to your class.

2. Plan a visit to the CADD/Drafting department of a local industry. Make a list of the equipment they are using. Ask whether the department has experienced any increase in productivity since converting to computers.

3. Interview an engineer, industrial designer, architect, or drafter. Ask about the uses and advantages of CADD in his/her field of work and prepare a report for your class.

CADD systems are very flexible and can be used for many applications. (IBM)

The CADD Workstation

KEY CONCEPTS

☐ Each component of a CADD system provides a specific function.

☐ The two primary elements of a computer are the central processing unit and memory.

☐ Data storage devices are required for a CADD system.

☐ Display screens allow the CADD design-drafter to view the drawing.

☐ Input devices are used to "communicate" with the computer.

☐ CADD hard copy devices are an important part of a CADD system.

The equipment used by the CADD drafter is very different than that of the "traditional" drafter. The combination of the software, the computer, and the peripheral devices is referred to as the **workstation,** Fig. 18-1. A workstation consists basically of six components:

- **Computer.** The computer is the heart of any workstation. Its speed and memory largely determine the power of the workstation.
- **Software.** The CADD program determines the available functions of the workstation. Programs range from simple drawing software to sophisticated design and analysis programs.

Fig. 18-1. A CADD workstation is composed of six essential components (four are shown here–computer, data storage device, input device, and display screen). (Intergraph Corporation)

- **Data storage device.** A data storage device holds the CADD program and drawing data.
- **Display screen.** The display screen allows the drafter to view the drawing being created.
- **Input device.** The input device is used to enter commands and digitize point positions.
- **Hardcopy device.** The hardcopy device makes a paper or film copy of the drawing held in computer memory or stored on a data storage device.

COMPUTERS

The progress of computer-assisted design and drafting (CADD) is largely attributed to the advancement of computer technology. Early computers were large and cumbersome, and most did not have the power of today's microcomputers. The development of integrated circuits and microprocessors have led to powerful CADD systems capable of complex designs and product analysis.

A computer, by definition, simply adds, subtracts, compares, and stores data. A human is capable of the same activities. However, the computer is able to process this data at high speeds with consistent results. The integral elements of a computer include the central processing unit and memory.

Central Processing Unit

The **central processing unit (CPU)** consists of the control unit and arithmetic logic unit (ALU). The **control unit** directs the flow of data. It accepts data from an input device or from memory. It then sends the data to be added, subtracted, or otherwise compared to the ALU. The control unit also governs the sequence of instructions included in the CADD software. When the processing is complete, the control unit sends the result either back to memory or to an output device such as a display screen or hardcopy device.

The speed of the CPU is rated in cycles and number of bits processed. A **cycle** is the time it takes for the computer to process an instruction. Most modern computers are rated in megahertz (MHz). One megahertz is equal to one million cycles completed in one second. A typical microcomputer has a cycle rating of 25 to 66 MHz.

The amount of information that a computer can work with at one time is rated in bits. A *bit* is the basic unit of all digital computer operation. Eight bits make up a byte. One byte represents one character such as a letter, number, or symbol. An 8-bit computer can process one character, or 8 bits, at a time. A 16-bit computer can process two characters at a time. A 32-bit computer can process four characters at one time. A 64-bit computer can process eight characters at one time. Thus, the more bits the computer can process at one time, the faster it is.

Memory

The computer's memory stores information to be processed. Data is held in one or more microchips. All computers use at least two types of memory: Read Only Memory (ROM) and Random Access Memory (RAM). (Memory is not the same as storage, or disk, space.)

Read only memory

Read only memory (ROM) contains data and instructions that cannot be changed by the user. This data is typically used when the computer is first turned on. The computer may also look to ROM for specific functions, such as outputting data to the display screen. The data in ROM is not lost when the computer is turned off; it is permanently held on one or two microchips in the computer.

Random access memory

Random access memory (RAM) holds information while you are using the computer. When the power is turned off, all information stored in RAM is lost. When working with a CADD program, RAM retains instructions used by software. It may also hold your drawing while you are creating or editing entities.

RAM is rated in kilobytes (KB) or megabytes (MB). A kilobyte is 1024 bytes, or characters, of information. A megabyte is 1,024,000 bytes. The amount of memory your computer has is very important for efficient operation. Most CADD programs depend on large amounts of memory for storage of a large number of instructions from the CADD software. In addition, a very large drawing might require a considerable amount of memory. Most CADD programs require a minimum of 4MB of RAM to maintain effective productivity.

Categories of Computers

Computers are classified by their speed, memory, and size of programs they can run. The three categories of computers are mainframe, minicomputer, and microcomputers. The distinction between these computers has blurred because of advances in computer technology. Modern microcomputers can perform functions once limited to mainframe and minicomputers.

Mainframe computers

A mainframe computer is capable of great speed and memory, Fig. 18-2. Mainframe computers typically have 32-bit or 64-bit CPUs and may include hundreds or thousands of integrated circuits.

Mainframe systems are networked computers. CADD operators work at terminals, that are connected to the mainframe system. Each terminal includes an input device and display screen. Many drafters can access the features of the CADD system simultaneously. Completed drawings are stored in a data storage device connected to the mainframe. Since the storage can be accessed by all drafters, it is called a **common database.** Engineering drawings from the database may be accessed by the manufacturing department for machine tool setup, or by the accounting department for inventory of materials.

Mainframe computers are capable of **multitasking.** Multitasking is the computer's ability to run more than one program at the same time. Mainframe computers typically contain more than one CPU to handle increased demand for processing power.

Minicomputers

A minicomputer is less powerful than a mainframe computer, Fig. 18-3. Although they are typically 32-bit machines, minicomputers are generally smaller, slower, have less memory, and are less capable of multitasking. However, they do allow networking of several terminals.

A minicomputer is powerful enough to run even the most complex CADD program. This, compared with its low price relative to mainframe computers,

Fig. 18-2. Mainframe systems are capable of great speed and memory. (IBM)

Fig. 18-3. A minicomputer has similar capabilities to a mainframe, but requires less space than one. (Intergraph Corporation)

makes it a desirable alternative to mainframes. Many industries use CADD systems that run on minicomputers.

Microcomputers

The category of microcomputer has emerged with the development of the microprocessor. Microcomputers are smaller, slower, and less powerful than mainframe and minicomputers. Their primary advantage is their low cost compared to other computers. Another advantage is size; most are small enough to fit on a desktop.

The category of microcomputers is divided into two areas: microsystems and personal computers (PCs), Fig. 18-4. Microsystems are typically 32-bit machines. Many microsystems are based on the UNIX operating system. Two examples of microsystems are SUN and APOLLO workstations. Personal computers are mostly 32-bit machines as well. However, 64-bit processors are being used more in personal computers.

Microcomputers may be stand-alone systems or networked to a host (server) computer. With a stand-alone system, each drafter works on a separate computer. Networked PCs are connected to a host PC that contains the CADD program and drawing files. The terminal is a separate PC. Once the program and drawing is loaded into the terminal PC, it uses its own processing capability, not that of the host PC.

DATA STORAGE DEVICES

Data storage devices permanently save information. (Storage space, or disk space, is not the

A

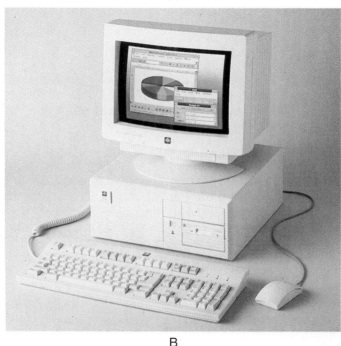

B

Fig. 18-4. A–Microsystems are generally 32-bit machines. (Intergraph Corporation) B–PCs are typically 32-bit computers, however, 64-bit processors are now being used as well. (Zenith Data Systems)

same as computer memory.) They hold both programs and data created by those programs. During a drawing session, the drafter loads the CADD program from data storage into computer memory. When finished with a drawing, the drawing data is sent to a storage device for use at a later time.

Data storage devices are classified in two ways. First, they may store data on magnetic or optical medium. Magnetic storage devices hold data by charging magnetic particles on the surface of the tape or disk. The combination of charged and uncharged particles can be translated into data. A read/write head can sense (read) the charge of particles and can also change (write) the charge when saving new information. Optical storage devices use a laser to read the presence of indentations (pits) on a disk. The number and sequence of pits is translated by the computer into information. Several types of data storage devices are shown in Fig. 18-5.

The second way to classify data storage devices is as disk storage or as tape storage. The biggest difference between tape and disk storage is the speed of accessing and writing information. With tape storage, a magnetic tape is passed along a read/write head, much like a cassette audio tape is played. To access information on the tape, you must wind the tape to the proper point. With disk storage, a disk is rotated and a read/write head (or laser) moves to the proper position above the rotating disk. This allows the computer to access and store data much faster.

Disk Storage Devices

Disk storage includes diskettes, hard disks, disk cartridges, and optical disks. Each of these devices differs in the amount of data storage and their appearance.

Diskettes

Diskettes are circular pieces of film coated with magnetic particles, and enclosed in a cardboard or plastic jacket. They are available in 3.5″, 5.25″, and 8″ sizes. Diskettes measuring 5.25″ and 8″ are generally referred to as "floppy disks," while the 3.5″ version is called a "microdisk." The 3.5″, 5.25″, and 8″ diskettes are shown in the bottom row of Fig. 18-5.

Fig. 18-5. A wide variety of data storage devices are available. Applications and storage requirements should govern the selection of a device. Storage devices shown here (starting at the top and moving clockwise) are: reel-to-reel tape, compact disk, 5.25″ floppy disk, 3.5″ microdisk, tape cartridge, and disk cartridge. (IBM)

Floppy disks. The floppy disk is held in a floppy disk drive that spins it at about 300 rpm. Holes cut in the jacket allow the read/write head to lower onto the disk surface to read and write information to and from the disk. The write-protect notch on the jacket can be covered with an adhesive tab to prevent data from being written to the disk.

The capacity of a floppy disk is determined by the number of sides used and the density of the magnetic coating. Most disk drives use both sides of the disk. Disks for these drives are labeled as DS, meaning double-sided. Both sides of the disk are guaranteed to store data. Single-sided (SS) disks are only guaranteed to hold data on one side.

The density of magnetic particles on the disk may be DD (double-density) or HD (high-density). The DS/DD (double-sided, double-density) disks hold 720,000 bytes (characters) of data. DS/HD (double-sided, high-density) disks will hold 1.2 MB of data. Floppy drives designed for DD disks will not read to or write from an HD disk.

Floppy disks (as well as any storage media) must be handled very carefully. Bending the disk or writing on the jacket will damage the magnetic coating. Exposing the disk to magnetic devices (TV, stereo speaker), heat, or moisture may erase the data.

Microdisks. Microdisks are 3.5″ rigid magnetic diskettes sealed in a plastic package. Since the disk is rigid, more magnetic particles can be applied to the disk. Microdisks will commonly hold 720 K, 1.4 MB, or 2.8 MB of data. The disk is inserted into a disk drive that spins the disk at about 600 rpm.

Hard disk drives

Hard disk drives consist of magnetic-coated aluminum "platters" sealed in an airtight enclosure, Fig. 18-6. Due to the number and rigidity of the platters, hard drives can store from 40 MB to over 1 gigabyte of data. They are used for personal computers and minicomputer CADD systems. The drive

Fig. 18-6. Hard disk drives are more durable than floppy or microdisk drives. (IBM)

may be internal or external. However, the platters of an external hard drive are not removable. A light on the front of the enclosure indicates when the drive is in use.

Disk cartridges

Disk cartridges are 8″ flexible disks held in a plastic housing, Fig. 18-7. Most cartridges hold in excess of 20 MB of data, while it is common for many to hold 40 to 1000 MB. This storage capacity, and the fact that they are removable, make them a desirable alternative to hard disk drives.

Disk packs

Disk packs are used on mainframe and mini-computers. They consist of several rigid, magnetic coated disks that are stacked together as a unit or pack, Fig. 18-8. The pack is capable of tremendous amounts of data storage.

Optical disk storage

Optical disks are plastic disks with an aluminum coating, Fig. 18-9. Pits that are chemically etched in the coating are read by a laser. The number and sequence of the pits is interpreted by the computer as data. Most people are familiar with audio optical disks, commonly referred to as compact disks.

Optical disks offer two major advantages. Most optical disks can hold over 200 MB of data, and they are removable. Optical disks are "read/write" meaning that data can be written to and read from the disks. However, due to their expense per mega-byte of storage capacity, optical disks are primarily used for archival purposes. Drawings that will never have to be modified are kept for historical and reference use.

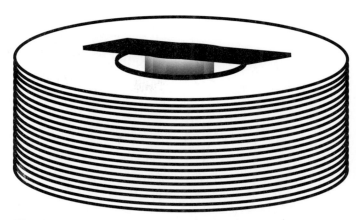

Fig. 18-8. Disk packs are composed of several rigid disks stacked together as a unit.

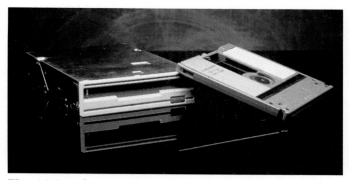

Fig. 18-9. Optical disk storage device and drive. Nonmagnetic medium, such as this optical disk, are more reliable than magnetic storage devices. (IBM)

Tape Storage Devices

Tape storage devices use magnetic-coated plastic tape, that is wound against the read/write head, to transfer data to and from the tape. Tape systems, as a rule, are slow. The read/write head cannot instantly locate any position on the tape to access data.

Tape devices are primarily used for backup purposes. They are not quick enough for data accesses while using a computer. Some tapes can hold over one gigabyte (billion bytes) of data. There are two types of tape storage devices: reel-to-reel and tape cartridges.

Reel-to-reel devices

Reel-to-reel tape drives consist of two removable reels (a data reel and take-up reel) that are installed on a tape transport system to wind the tape past the read/write head, Fig. 18-10. The reels are available in standard sizes of 7″, 8.5″, and 10.5″. The 8.5″ and 10.5″ reels with 1″ wide tape are used on main-frame and minicomputer systems. Reel-to-reel drives for microcomputers use 7″ and 8.5″ diameter reels and smaller width tape. One type of reel-to-reel tape is shown in the back row of Fig. 18-5.

Fig. 18-7. These disk cartridges have 150 MB of data storage capacity. (Iomega)

Fig. 18-10. Reel-to-reel magnetic tape transport system. (IBM)

Tape cartridges

Tape cartridges contain both data and take-up reels in a plastic enclosure, Fig. 18-11. Many are smaller than a cassette tape. New 4 mm and 8 mm data cartridges are commonly being used for data storage. The 4 mm cartridges are capable of holding up to 4 gigabytes of information, while 8 mm cartridges can hold up to 5 gigabytes of data. An 8 mm data cartridge is shown in the middle of Fig. 18-5.

DISPLAY SCREENS

In traditional drafting, you create and view the drawing on paper. With CADD, you enter drawing commands, and entities appear on the display screen, Fig. 18-12. The produced images can be modified many ways because the screen simply displays the drawing data held in memory. For example, the image can be magnified to make detail work much easier. You could also rotate an object. Instead of erasing and redrawing, as would be done with traditional tools, you simply access a rotate

Fig. 18-11. Tape cartridges are popular for backup and archival purposes.
(Tallgrass Technologies Corporation)

Fig. 18-12. The display screen is used to view drawing data. (Intergraph Corporation)

command. The drawing data is changed in memory and the rotated entity is displayed.

The following section will discuss the many types of display devices. Their function is the same, yet the methods used to create the image are different. The most important difference between display screens is resolution. *Resolution* refers to how clear the image on the screen is. Clearer images using "high-resolution" monitors make drawings easier to see.

Cathode Ray Tube

Displays are commonly referred to as monitors. The most common display technology for monitors is the *cathode ray tube (CRT).* It makes an image by propelling an electron beam against a phosphor-coated screen surface. When the beam hits the surface, individual dots of phosphor (called *pixels*) glow. From the drafter's point-of-view, the glowing pixels form images. The path of the electron beam is controlled by drawing data sent from the computer to form entities on the screen. Color is created activating three separate color pixels on the screen. There are two technologies used to control the CRT's electron beam: raster scan and stroke writing.

Raster scan displays

In raster scan displays, the path of the CRT's electron beam is controlled to scan the phosphor-coated surface. It scans from left to right and top to bottom, much like you are reading this text, Fig. 18-13. The beam scans continuously–over 60 times a second–since the phosphor pixels glow only for a short time and must be reactivated regularly. Raster scan displays are the most common type of display used for CADD workstations.

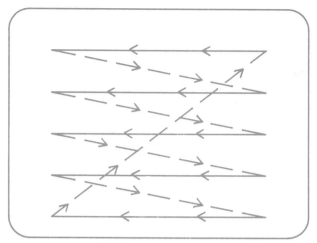

Fig. 18-13. A raster scan display scans the screen in a regular pattern.

To form an image, the beam selectively activates phosphor pixels as it scans across the screen. The more pixels that can be used to draw the image, the clearer the image will be. This is referred to as **resolution.** Low-resolution screens have about 640 pixels horizontally and 480 pixels vertically. Medium-resolution screens have 1024 pixels horizontally and 768 vertically. High-resolution screens are 1600 x 1280. To illustrate this example, a 1″ horizontal line on a low-resolution screen might consist of 24 pixels. A high-resolution screen might use 60. There are two different types of raster scan displays: interlaced and non-interlaced.

Interlaced monitors. An **interlaced monitor** uses the raster scan method. It will scan every line, from left to right and top to bottom. However, two passes are required for every line. On the first pass, the monitor will scan every other pixel. On the second scan, the remaining pixels are scanned. In other words, the second pass "interlaces" the line. This type of monitor can produce a "flicker." When doing drafting or other graphics work, this flicker can be a problem.

Non-interlaced monitors. **Non-interlaced monitors** also use the raster scan method. The difference between interlaced and non-interlaced is that non-interlaced monitors scan every pixel in one pass. This means that each line is scanned only once. Since only one pass is made, the "flicker" is eliminated. Non-interlaced monitors are preferred for drafting and graphics work.

Stroke writing displays

Stroke writing displays use a different method of controlling the electron beam. The beam does not scan the screen in a regular pattern. Instead, it creates images by actually "drawing" on the screen. The beam performs much like you use a pencil to draw. For a step block, the beam would be control-led to draw six straight lines, Fig. 18-14. Since the phosphor will glow for only a short time, two methods are used to keep the image lit: refresh displays and direct-view storage tubes.

Refresh displays constantly "redraw" the image to prevent it from becoming dim. This is similar to scanning, yet the electron beam refreshes the image by drawing the same line pattern of the drawing. However, there is a disadvantage to this method. If there are many entities in the drawing, it takes the beam longer to refresh the image. In this case, the screen may flicker as the beam redraws a complex drawing.

To eliminate screen flicker, direct-view storage tube technology was developed. It also has the electron beam "draw" the image. However, once the beam draws the entities on the phosphor, the image is saved by flooding the screen with electrons. These electrons do not create images. They only keep pixels lit that were previously activated by the beam. The image remains constant until the screen is "redrawn." You can add entities, but none are removed until the image is regenerated.

Flat Displays

Flat displays are thin display screens commonly seen on laptop computers. They do not use cathode ray technology because the CRT requires a certain distance to fire electrons at the screen. Instead, they produce the image near the surface of the screen. This is done using three different technologies: liquid crystal, gas plasma, and electroluminescent.

Liquid crystal displays

Liquid crystal displays (LCDs) are common on laptop computers requiring very small, flat screens, Fig. 18-15. Liquid crystal displays require very little

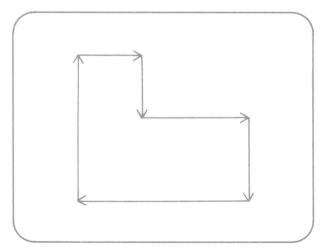

Fig. 18-14. Scanning pattern of a stroke writing display.

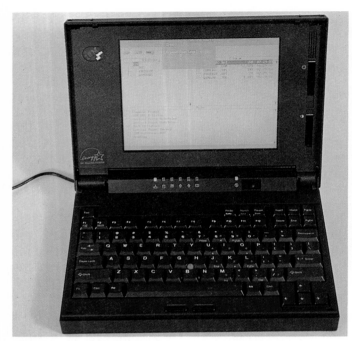

Fig. 18-15. Liquid crystal displays (LCDs) are commonly used on laptop computers. Many of these displays are color.

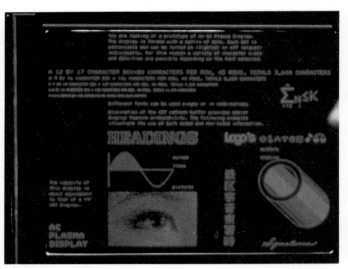

Fig. 18-16. Gas-plasma displays are characterized by their orange-red color. (IBM)

power and can be battery-driven. You likely have seen liquid crystal displays on digital watches and hand-held TVs.

The method of producing an image with an LCD is done with liquid crystals and a grid of wires, called electrodes. When the wires are not energized, the crystals lay flat. To the viewer, no image is seen. However, when two intersecting wires are energized, the crystals stand up. To the viewer, this is seen as a black dot. By selectively energizing wires, the combinations of dots form the image. LCD devices are available in monochrome and color units (unlike gas-plasma and electroluminescent displays). The LCD image is filtered through color lenses to give the illusion of color.

Gas-plasma displays

Gas-plasma displays are made of a thin glass enclosure containing a low-pressure neon or neon/argon gas. A grid of wires runs through the enclosure. When two crossing wires are energized, the gas at their intersection glows. The glowing gas appears as a single dot, comparable to the pixel on a CRT. By energizing different wires, an image is created by a number of glowing gas pixels, Fig. 18-16.

Gas-plasma displays are monochrome (one color) that limits their use with CADD. They are a medium-resolution display, with a typical 17″ diagonal x 3″ thick display. This makes the laptop unit very cumbersome and inconvenient to use when compared to other portable units.

Electroluminescent displays

Electroluminescent displays work much like gas-plasma displays, except that instead of gas, they use a luminescent chemical. Pixels are made by the intersection of two energized wires that pass along the chemical. These displays are also monochrome.

Dual Screens

Dual screens are commonly used for CADD systems needing one display for text and one for graphics. When analysis is done, the results are displayed on the text display, while the design is shown on the graphics display. Another use for dual screens is separating communications text from graphics. Communications text refers to commands used to create entities. The sequence of commands used can be viewed and recorded while also viewing the design image on the graphics display.

INPUT DEVICES

Unlike drawings produced using traditional drafting, you cannot "touch" a CADD drawing. Instead, you must use an input device to select commands and pick point positions on the display screen. Actually, the input devices let you interact with the computer by allowing you to enter instructions, and then receive some type of response or feedback.

When drawing, a cursor is displayed on the screen to indicate the position of the input. For example, when drawing a line you would use the input device to move the cursor to the position of line's first endpoint. Pressing a specified button on the input device selects that location.

There are three general categories of input devices: keyboards, digitizing tablets, and cursor control devices. **Keyboards** are typically used to type in text and enter commands. **Digitizing tablets** are used to move the cursor across the screen, enter commands, and trace existing drawings. **Cursor control devices** also move the cursor across the screen and are used to enter commands. However, you cannot trace existing drawings with cursor control devices.

Keyboards

There are two types of keyboards used with CADD. The most common type is an alphanumeric keyboard. This device, similar to the layout of a typewriter, is simply called a keyboard by most drafters. The second type of keyboard is the function keyboard.

An alphanumeric keyboard is always used with a CADD system, even when another input device is attached, Fig. 18-17. A typical keyboard has characters for letters, numbers, and symbols. It also has special keys called **function keys.** When pressed, function keys perform a series of commands that otherwise would have to be entered separately. On a microcomputer, function keys are labeled F1, F2, up to F12. Alphanumeric keyboards are used to type in text and enter commands by typing a letter or function key.

A function keyboard has a number of keys that perform specific commands of the CADD system, Fig. 18-18. They are not used to enter text. The function keys are used in conjunction with an alphanumeric keyboard. The keys of a function keyboard are typically assigned to perform certain functions that may otherwise take several minutes to perform.

Digitizing Tablet

A digitizing tablet is a flat, rectangular plastic pad with a tracking device (usually a puck or a stylus) attached by a cable, Fig. 18-19. A matrix of hundreds of evenly spaced wires is sandwiched within the plastic pad. A specific location is determined by each intersection of wires. To sense the intersection of

Fig. 18-17. An alphanumeric keyboard is used to input data as well as control the screen cursor.

Fig. 18-18. The function keyboard, left, may be used instead of function keys. (CADAM)

Fig. 18-19. Digitizing tablets are available in sizes ranging from 9″ x 9″ to 36″ x 48″. (Kurta Corporation)

wires, the tracking device is moved across the pad. As the tracking device is moved across the surface of the pad, the screen cursor also moves.

To select a point position using a digitizing tablet, first move the tracking device across the pad until the cursor is properly positioned on the display screen. Then, press down on the stylus or press the appropriate button on the puck to select that location.

Digitizing tablets are often used to convert existing paper drawings into CADD drawing files for easy modification at a later date. The paper drawing is taped to the surface of the tablet. Then by selecting CADD commands and picking point positions of the existing drawing, you create a CADD drawing.

The digitizing process is now being automated using scanners. The paper drawing is placed on a scanner drum or under a camera. The camera detects the lines on the drawing and converts this information into computer data. The data can then be loaded in as a CADD drawing file and edited as you would a drawing originally made using CADD.

Cursor Control Devices

Cursor control devices allow you to move the cursor across the screen and select point locations. However, unlike a digitizing tablet, they are not designed to trace an existing drawing. However, the primary advantage of cursor control devices over digitizing tablets is reduced cost.

Mouse

A mouse is a hand-held device rolled on top of a flat surface, Fig. 18-20. A ball and wheels, or a light sensor on the bottom of the mouse detect direction and distance of movement. When moving the mouse, the cursor on the display screen also moves. To select a point position, such as the endpoint of a line, press the appropriate button on the mouse. There may be more than one button, but most programs use the left button as the select button.

Mice are very common input devices because they are easy to use and inexpensive. However, you cannot trace an existing drawing with a mouse because there is no way to locate the mouse precisely over a point on a paper drawing.

Joystick

Although not as popular as mice, joysticks are still used on many CADD systems. The joystick consists of a small box with a movable shaft, Fig. 18-21. Pushing the shaft in a direction causes the screen cursor to also move in that direction. Buttons on the joystick allow you to select point positions.

Trackball

A trackball is basically an upside-down mouse. It consists of a rectangular base with a control ball set in the top, Fig. 18-22. Moving the ball with the palm of your hand causes the screen cursor to move. Buttons on the device are used to pick point locations and to select commands from a screen menu.

Fig. 18-21.   Joysticks can move the cursor in any direction. (CH Products)

Fig. 18-22.   A trackball can be used to control the position of the cursor and pick points.

Light pen

The light pen is different from other cursor control devices because you hold it against the display screen, Fig. 18-23. The screen senses the position of the pen and places the cursor directly under the pen tip. Light pens are primarily used with proprietary CADD software used on mainframe systems.

A light pen is efficient because you can point exactly to the position or entity on the screen. There is a pressure-sensitive tip on the pen that, when pressed, enters a point location. One drawback to the light pen is it is very tiring to use. You must keep your hand lifted to the screen.

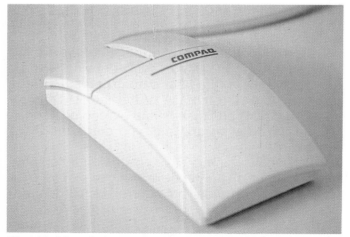

Fig. 18-20.   A mouse is a hand-held cursor control device.

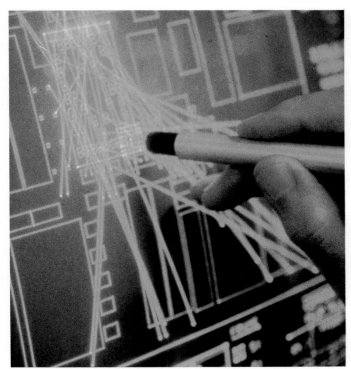

Fig. 18-23. A light pen is used to draw directly on special CRT screens. (IBM)

HARDCOPY DEVICES

A CADD drawing is held in computer memory or stored on a data storage device until you are ready for a paper or film hardcopy. Then you use commands of the CADD software to send the drawing data to a hardcopy device. The device duplicates the drawing held in memory by creating an image on paper or on photographic film.

There are many types of hardcopy devices. They vary in the quality of generated drawing and the speed that the hardcopy is completed at. Offices using CADD will generally have a hardcopy device specifically designated for final output, and others used for preliminary design work.

Pen Plotters

A pen plotter most closely resembles how a drafter draws by hand, Fig. 18-24. The plotter moves a pen over paper to create the image. Depending on the type of plotter, the paper may also move. After drawing an entity, the pen carriage picks up the pen, relocates it, and sets it down to draw another entity.

Pen plotters may use a single pen or may be capable of holding four to twenty pens. Multipen plotters automatically select one assigned pen to draw a different lineweight or color. Pen plotters are classified as flatbed, drum, or microgrip.

A

B

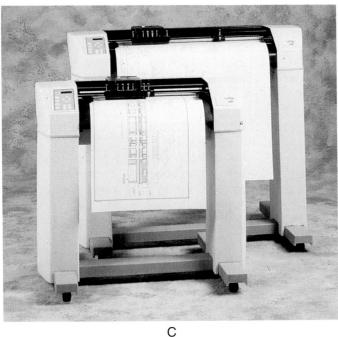

C

Fig. 18-24. Pen plotters. A–Flatbed plotter. (Houston Instrument, A Summagraphics Company) B–Drum plotter. (Gerber Scientific, Inc.) C–Microgrip plotter. (Hewlett-Packard Co.)

Flatbed plotters

With flatbed plotters, the paper is held on a flat surface by tape, vacuum, or electrostatic charge. A pen carriage moves the pen over the paper in two directions. The dimensions of a flatbed plotter refer to the maximum size paper it can hold. Flatbed plotters are typically utilized for A-size and B-size drawings.

Drum plotters

Drum plotters have the paper attached to a drum. The pen moves across the drum to produce vertical ("Y" axis) lines on the drawing. The drum rotates to produce horizontal ("X" axis) lines on the drawing. Simultaneous movement of both the pen and the drum permits curves and angled lines. The paper size that can be used on a drum plotter is limited by the width and circumference of the drum. Many drum plotters have continuous feed options that allow a roll of paper to be attached to the plotter. This allows the plotter to plot drawings while unattended.

Microgrip pen plotters

Microgrip plotters resemble drum plotters in that "X" axis direction is produced by moving the paper and "Y" axis direction is produced by moving the pen. However, small rubber rollers grip the paper at the edges rather than having the paper attached to the circumference of the drum.

Electrostatic Plotters

Electrostatic plotters are much like photocopying machines, Fig. 18-25. They selectively place electrostatic charges on paper that then attract ink. The charges are made by a writing head that passes back and forth across the paper. As many as 400 electrostatic "dots" are placed per inch. The paper is then fed through toner, consisting of ink suspended in a liquid, where the ink is attracted to the paper to form the image.

The position of the charges is determined by a process called *rasterization.* A processor in the plotter converts the drawing data into series of dots that the plotter can reproduce.

There are three types of electrostatic plotters: multipass web-fed, single-pass web-fed, and drum. Each is capable of multicolor plots. A multipass web-fed plotter feeds paper from a roll, against the writing head, through the toner, and draws the paper up through a take-up reel. To produce different colors, the paper must be rewound, the color of toner changed, and the paper fed through the plotter again. A single-pass web-fed plotter feeds the paper against four writing heads and four different color toners all in one pass. In drum plotters, the paper is attached to a cylinder that rotates the

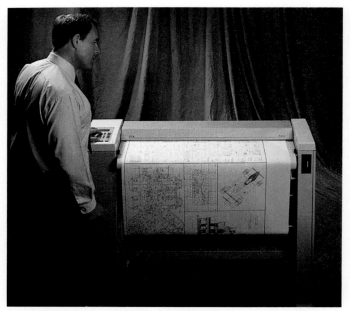

Fig. 18-25. Electrostatic plotters use a rasterization process to convert the lines of a drawing into a series of dots that are printed. (Calcomp, Inc.)

plotter against the writing head and through the toner. A different color toner is used for each rotation to produce multicolor plots.

Ink Jet Plotters

Ink jet plotters form images by propelling individual droplets of ink onto paper, Fig. 18-26. The droplets are guided in flight to hit the paper in the proper position. Ink jet plotters are capable of producing multicolor images by using different colors

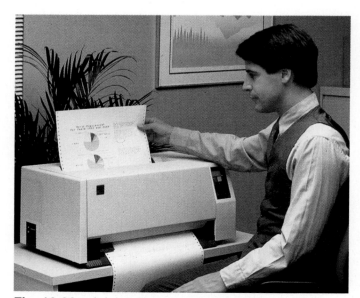

Fig. 18-26. Ink jet plotters propel ink droplets onto the plotting medium to form images. (IBM)

of ink. Ink jet plotters are useful in developing presentation drawings for clients.

Laser Plotters

Laser plotters use a beam of light to create an image, Fig. 18-27. The beam scans across the surface of a photosensitive belt mounted on a drum. An electrical charge is applied to the belt by the beam. The drum then rolls the belt through a toner bath. Toner is attracted to the belt and then transferred to a sheet of paper.

Thermal Plotters

Thermal plotters melt ink on a separate ribbon or transfer sheet and transfer the hot ink to the paper to produce the image, Fig. 18-28. Ribbons or

sheets that carry several colors are deposited in several layers to produce a multicolor image.

Dot Matrix Printing

A dot matrix printer is often used for printing letters and other business documents. In CADD, the dot matrix printer is used to produce rough check plots, Fig. 18-29. During the design, a check print is made to verify the status of the project. A dot matrix plot is lower in quality, but much less expensive and time-consuming than other hardcopy devices.

Photo Plotter

Photo plotters produce drawings by exposing photographic paper against a CRT. The dry silver-coated paper darkens in those areas where lines are present on the CRT. It is a poor reproduction and used primarily as a rough plot.

Fig. 18-27. High-resolution laser plotters provide good-quality results. (Gerber Scientific, Inc.)

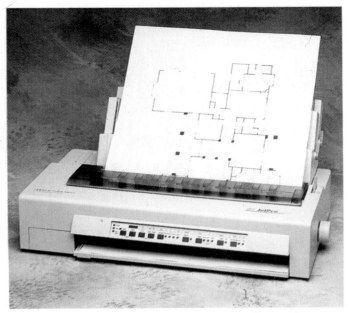

Fig. 18-28. Thermal plotters are primarily used to produce presentation materials and business graphics. (Calcomp, Inc.)

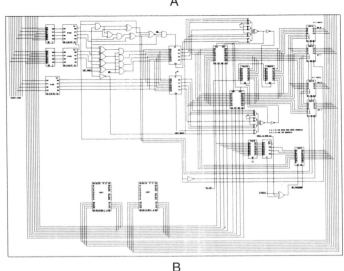

A

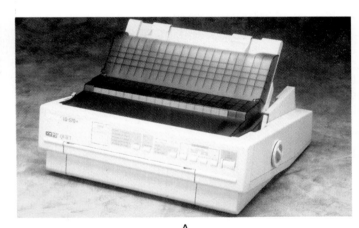

B

Fig. 18-29. A—A dot matrix printer can be used as hardcopy device. (Epson American, Inc.) B—The hardcopy produced by a dot matrix printer is usually not a very high-resolution image.

CRT Image Recorder

Cathode ray tube (CRT) image recorders are an improvement over low-quality photo plotters. CRT image recorders expose photographic film against a color CRT unit. The image can be made either on slides or into photographs, Fig. 18-30.

Recently, there has been a movement to capture screen images electronically using various types of utility software on the market. These utility programs allow a drafter to display the drawing as desired, and then press a "hot key" to capture the image in an electronic format. This method produces high-quality images, that can be used for presentation purposes. The screen image shown in Fig. 18-31 was captured using a screen capture utility program. The captured images can be made into slides, photographs, or transparencies. They can also be "imported" directly into page layout programs for technical documents.

COM Devices

Computer output-to-film (COM) devices produce hardcopies by directing a laser onto microfilm.

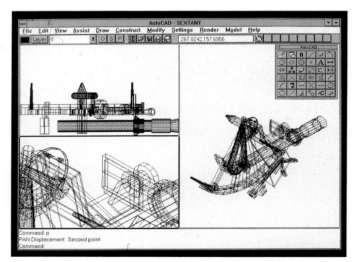

Fig. 18-31. Image saved using a screen capture utility.

Fig. 18-32. COM devices utilize a laser to form images on microfilm. (3M Co.)

Fig. 18-30. CRT image recorders expose photographic film against a high-quality display. (Polaroid)

The laser is controlled to "draw" on the film, much like a pen plotter. The film image is inserted into an aperture card that is coded with data related to the drawing, Fig. 18-32. Most large engineering firms use COM devices because it is easier to store aperture cards than to store full-size paper prints.

QUESTIONS FOR DISCUSSION

1. Describe the difference between read only memory and random access memory.
2. Summarize the differences between mainframe computers, minicomputers, and microcomputers.
3. Explain why disk storage devices access data much faster than tape storage devices.
4. Define "resolution" and the impact it has on the displayed image.
5. Classify the three types of input devices.
6. Name the input device that can be used to input existing drawings by tracing.
7. Explain the difference between an alphanumeric keyboard and a function keyboard.
8. Identify the three models of pen plotters.
9. Why do most large engineering firms use COM devices?

PROBLEMS AND ACTIVITIES

Review current literature or visit a local computer graphics vendor to find answers to the following questions.

1. What is the highest display resolution currently available?
2. What type of hardcopy device is advertised most often?
3. How do features of personal computer (microcomputer) CADD programs differ from larger systems?
4. What is the range of storage capacities for different data storage devices?

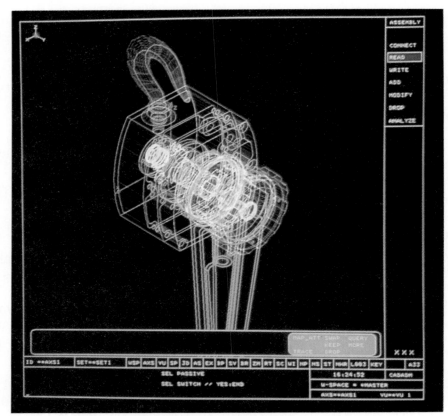

CADD programs allow complex parts to be easily constructed and changed. (IBM)

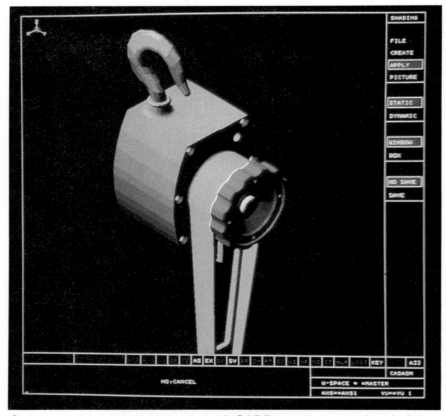

Once complex parts are constructed, CADD programs can "render" the objects to provide a life-like representation. (IBM)

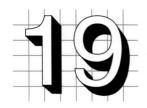

Constructing and Editing Drawings

19

KEY CONCEPTS

- [ ] Most CADD programs require the design-drafter to log on.
- [ ] Commands can be selected and entered into a computer in several ways.
- [ ] Menus provide access to commands.
- [ ] File management is an important aspect of a design-drafter's responsibility.
- [ ] The Cartesian coordinate system is used by most CADD systems.
- [ ] Drawing setup procedures should be performed prior to beginning a drawing.
- [ ] Graphic entities can be constructed in a variety of ways.
- [ ] Drawing aids are used to maintain accuracy when constructing CADD drawings.
- [ ] Editing commands allow a design-drafter to modify a drawing.
- [ ] Display controls provide a means for the design-drafter to view desired parts of a drawing.

Computer-assisted design and drafting systems offer a wide variety of functions and features. Although no two systems offer the exact features and capabilities, the basic structure and command sequences among them are similar. The main difference between CADD software is the names used for drawing and editing commands. This chapter covers drafting and detailing functions that are common to a wide variety of programs.

LOGGING ON TO THE SYSTEM

When starting up a CADD system, a drafter must first progress through a series of startup steps, called a *logon procedure.* See Fig. 19-1. For a minicomputer system, this may require turning on the terminal and entering a password. With a microcomputer, the drafter usually types in the program name and enters directly into the system. If several microcomputers are networked to a com-

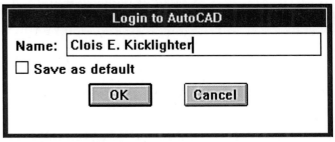

Fig. 19-1. The log on procedure for CADD programs may include password or user name.

mon server, the drafter may have to proceed through additional steps to access the network. The exact procedure is different for every drafting department. Once you have successfully logged onto the system, the CADD program's main screen will appear, providing you with access to the features of the CADD software.

CADD COMMANDS

When using a computer, each instruction you give is considered a command. Likewise, when using a CADD program, each direction you enter to create or edit a line, circle, or other entity is a command. In its simplest form, a *command* is a short word which, when entered, achieves some function.

To create or edit a drawing, you select functions, usually by entering a series of commands. This series of commands is commonly referred to as *command syntax.* For example, to draw a horizontal line, you might have to select the Draw menu, followed by picking Line, and then adding a Horizontal modifier, *in that order.* The command structure described in this chapter is treated in a general manner and is intended to cover a number of systems.

MENUS

In most cases, you must know which command to select to perform drawing or editing functions. The computer cannot anticipate your next move.

Remember that the computer is only a machine, not a drafter. Therefore, it is your responsibility to select commands to create drawings that adhere to drafting standards. To help you identify available drawing functions, the computer supplies menus. A *menu* is a list of commands displayed on the screen or printed on a digitizing tablet overlay, Fig. 19-2. Menus allow you to "pick" the commands, rather than having to type in an entire command.

Commands can be selected in one of three ways: typing on the keyboard, picking from the screen, or picking from a digitizing tablet. On most systems, you can use the keyboard along with the menus to select commands. However, it is often easier to "pick" the command you need from a screen menu or tablet menu.

Some CADD systems interface with voice-recognition systems. A CADD operator simply speaks directly into a microphone headset using a predetermined set of key words or phrases. The computer responds to the command names.

Screen Menus

A *screen menu* shows available commands on the display screen. This area, known as the *menu pick area,* is usually found along the top or side of the screen. To select a command from the screen menu, move the screen cursor (seen as crosshairs) to the menu pick area by moving your mouse, puck, or stylus. Locating the cursor in the menu pick area usually highlights single commands. Highlight the command you want to use or the menu you want to access and press the pick button on the input device.

Since there are often more than 100 commands in a CADD program, only a certain number can be displayed at one time. Thus, screen menus are divided into levels. The menu commands and menu names are initially shown on the screen. By selecting one of these commands, a series of options appear. Using this technique of *nesting,* all commands can be accessed. The primary commands might appear in a bar or as a list alongside the screen. After selecting one of these commands, another menu replaces the main menu. The second menu may also appear to pull down from the command or even pop up in the middle of the screen.

In addition to command names, some screen menus show functions as *icons,* or small pictures that represent the command, Fig. 19-3. When you pick an icon, a command is issued, allowing you to pick a point on the screen.

Tablet Menus

A tablet menu is a plastic overlay placed on a digitizing tablet. The overlay is divided into com-

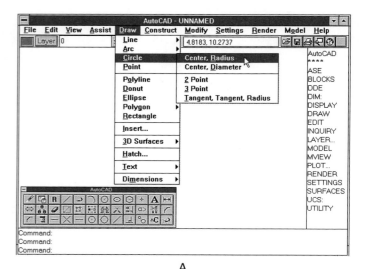

A

B

Fig. 19-2. Types of menus. A–Screen menus for a popular CADD software. The drafter can select commands from the pull-down menus at the top of the screen or from the list along the right side of the screen. B–Digitizing tablet overlay. Picking commands from the overlay generally reduces the number of picks you must make to access a command.

Fig. 19-3. An icon menu allows you to pick commands using graphical icons.

mand areas and an area in which moving your puck or stylus controls the screen cursor. The tablet menu provides instant access to commands without having to pick a series of "nested" commands as you might on a screen menu. On the other hand, a tablet menu cannot contain all commands. Therefore, you will still rely on the screen menu or typed command entry to access all functions. In addition, when using a tablet menu, you must remove your eyes from the screen and then reorient them when returning to the screen. With on-screen menus, you do not need to look away from the screen.

COMMON CADD FUNCTIONS

Standard functions found in most CADD programs can be grouped into several categories. These categories are defined as follows.

- *File Management Commands.* Allow you to begin, save, and load drawings from a data storage device.
- *Drawing Setup Commands.* Allow you to determine your drawing area and units of measurement (U.S. Customary, SI Metric, and decimal degrees).
- *Drawing Commands.* Add graphic entities—lines, circles, arcs, curves, polygons, and ellipses—to the drawing.
- *Drawing Aids.* Help you precisely locate position on a drawing. In most applications, the screen cursor moves freely about the drawing as you move the input device. To select exact location, select a drawing aid which "snaps" the cursor to a specific location on an object or grid dot.
- *Editing Commands.* Allow you to change previously drawn entities. Moving, copying, and changing color or linetype are typical editing tasks. Editing commands also allow the drafter to construct a complex drawing by copying and manipulating just a few simple objects.
- *Display Controls.* Determine what part and how much of the drawing is shown on the display screen.
- *Dimensioning Commands.* Let you automatically place linear and radial dimensions. You simply have to choose the points to dimension.
- *Text Commands.* Place notes and lettering on the drawing.
- *Hatch and Fill Commands.* Place section lines and graphic patterns on drawings.
- *Layer Commands.* Allow you to separate different parts of the drawing into levels, much like using overlays.
- *Inquiry Commands.* Let you ask questions about your drawing, such as distance, perimeter, area, and drawing status.
- *Symbol Library Commands.* Allow you to store and insert standard drafting symbols in drawings as needed. This prevents having to draw any symbol twice.
- *Plotting Commands.* Allow you to make a hardcopy of the drawing stored on disk or in memory.
- *Help Commands.* Offer on-screen help if you forget how to use a function.

File Management

Each time you use a computer, you are working with files. A *file* is a group of related data, which in the case of a CADD drawing file, describes a drawing. When loading and using a CADD program you are using files. When working with a CADD system, you create, save, load, and otherwise manipulate drawing files. This process of manipulating drawing files is called *file management* and includes the following functions.

Starting a new drawing
After loading the CADD software, you typically have a choice between beginning a new drawing or editing an existing drawing. Some systems require that you select a specific command from a main menu to begin a new drawing. Other CADD programs allow you to begin a drawing immediately.

Saving the current drawing
At some point in time, you will want to save information added to a drawing. Save your drawings frequently. Every 10 to 15 minutes is recommended. This ensures that your drawing remains intact in the event of a power failure or system crash.

Loading a drawing
The process of *loading* a drawing recalls a drawing file for continued work. Remember to save the drawing again when finished to update any changes made during the drawing session.

Listing drawing files
A *listing* shows the drawings stored on the current data storage device. It may also show the number of entities and time spent on the drawing, Fig. 19-4. The information is useful when billing clients for services performed, such as developing detailed drawings for a project.

Deleting drawing files
Delete drawings with caution! Delete a drawing from the data storage device only with the supervisor's or instructor's permission. While newer versions of operating systems (such as MS-DOS) allow you to undelete files once they have been deleted, some corruption may still occur.

Coordinate Systems

Each line, circle, arc, or other entity you add to a drawing is located by certain points. A line is

```
Drawings pathname: \acad\drawings
```

drawing name	date last modified	time last modified	cumulative drawing time	object total	symbol total
Mi04001	10-July-1994	9:04 am	13 hrs 6 mins	635	3
Mi04007	10-June-1994	9:16 am	11 hrs 9 mins	592	59
Mi00006	24-May -1994	2:17 pm	8 hrs 10 mins	293	18
Mi03221	2-June-1994	5:14 pm	5 hrs 53 mins	1094	27
Mi04412	2-May -1994	3:15 pm	13 hrs 5 mins	687	0
Pd17AB24	22-July-1994	4:33 pm	7 hrs 2 mins	2948	0
Pd17AA43	3-July-1994	10:42 pm	21 hrs 7 mins	3301	32

```
There is a total of 7 files.
Press [Enter] to continue.
```

Fig. 19-4.   A directory of files may help you determine the most recent version of a drawing. Cumulative drawing time may be useful for billing purposes in a professional environment.

defined by its two endpoints. A circle is defined by its center point and a point along the circumference. A square is located by its four corner points. To precisely locate entities, all CADD programs use a standard point location system, called the *Cartesian coordinate system.*

Cartesian coordinate system

The Cartesian coordinate system consists of two axes, Fig. 19-5. The horizontal axis is the *"X" axis.* The vertical axis is the *"Y" axis.* The intersection of these two axes is the *origin.* The location of any point can be determined by measuring the distance, in units of measurement, from the origin along both the "X" and "Y" axes. These two values are called the "X" and "Y" coordinates, or the *coor-*

dinate pair. Each axis is divided into equal units. Each unit may refer to inches, feet, or metric measurements, such as meters or millimeters. The unit of measurement along the "X" and "Y" axes is the same.

Coordinates for absolute point locations can be positive or negative. This depends on their location in reference to the origin. A point located to the left of the origin has a negative "X" value. A point located below the origin has a negative "Y" coordinate value. The origin for most CADD programs is the lower-left corner of the display screen. When entering point locations to draw an entity, you can use either the absolute, relative, or polar coordinate entry.

Absolute coordinates.   *Absolute coordinates* refer to exact point locations measured from the origin. The line and rectangle shown in Fig. 19-6 are defined by absolute coordinates.

Relative coordinates. *Relative coordinates* define distance from a previous point. For example, suppose you have entered the first endpoint of the line at "1,1" as shown in Fig. 19-7. Typing the relative coordinate "@6,8" places the second endpoint six units to the right and eight units above the first endpoint. (The "@" symbol is used by one CADD system to designate a relative coordinate entry.)

Polar coordinates. When using the *polar coordinate entry* method, points are located by an angle and a distance from a previous point. For example, suppose you have entered the first point of a line as absolute "1,1." You could enter the coordinate "@14" for the second endpoint to draw a line 14 units long at a 30° angle from the previous point, Fig. 19-8. (The "@" and "..." symbols are used by one CADD system to designate distance and angle.) Polar coor-

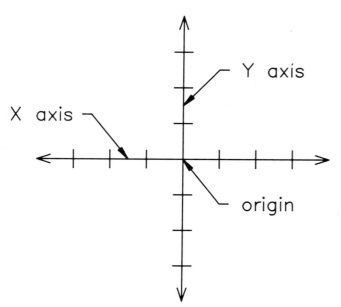

Fig. 19-5.   The Cartesian coordinate system.

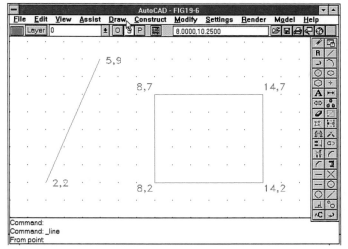

Fig. 19-6. Absolute coordinates are measured from the origin.

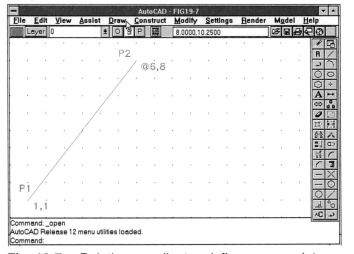

Fig. 19-7. Relative coordinates define a new point from a previous point.

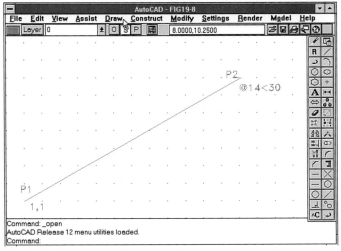

Fig. 19-8. An angle and a distance are used to define polar coordinates.

dinates are typically relative, meaning that the distance is measured from the previous point.

Coordinate display. The *coordinate display* is a readout of the current position of the cursor on the drawing, Fig. 19-9. The coordinate display is usually found in the status line. It may show absolute position on the drawing, relative position from the previous point, or polar position from the previous point.

Drawing Setup

When beginning a new drawing, you must specify certain parameters, such as units of measurement, drawing area, and drawing scale. This process is called *drawing setup.* The CADD system will assign default values to each parameter. However, to operate a CADD system efficiently in industry, values can be assigned to these parameters.

Actually, there are a number of parameters which can be set before or during a drawing session. These might include linetype, text height, and drawing color. This section focuses on only those that should be considered before adding the first entity.

Units of measurement

The units of measurement are the values assigned to linear and angular increments. Linear units may be inches, feet, millimeters, meters, or user-defined values. Angular units include methods of measuring angles, such as in degrees or in radians. Once you set the units of measurement, it is not a good idea to switch to the other. However, it is acceptable to change the notation, such as from inches in fractional format to inches in decimal format.

Most CADD software allows you to set the degree of accuracy desired in your drawing. The *unit resolution* is the number of decimal places used to

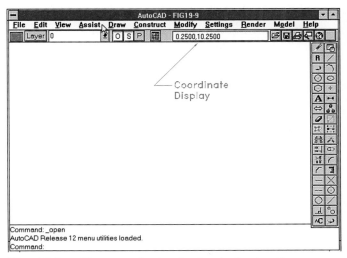

Fig. 19-9. The coordinate display indicates the position of the cursor on the display screen.

calculate dimensions you enter. Having more digits behind the decimal point results in more accurate coordinate entry. However, for some drawings two decimal places may be sufficient.

Drawing area

In traditional drafting, your drawing area is limited by the size of the sheet of paper. With CADD, your drawing area may not be limited. Most systems allow you to set the size of the drawing area as large as necessary. Since you are using "real world" measurements, the drawing area for a house plan might be 80′ wide by 40′ high. However, you will need to set the drawing scale to make the drawing fit on a standard-size sheet.

The drawing area is defined by specifying the lower-left and upper-right corners. The lower-left corner is typically 0,0 (the origin). For the house plan just mentioned, you would set the upper-right corner of the drawing area at "100,60." (This would provide for a 80′ by 40′ drawing area, plus 20′ horizontally and vertically for dimensions.)

Some systems base the drawing area on standard drafting paper sizes. In this case, you must select a sheet size large enough for your drawing. The drawing area is then confined to the sheet's dimensions. Sheet sizes are limiting. You obviously cannot place a 40′ line on a D-size (36″ wide) sheet. Therefore, you must select an appropriate drawing scale.

Drawing scale

Drawing scale describes the relationship of drawing measurements to "real life" measurements. For example, a 1″ line on the drawing may only plot out to be 1/4″ long. Scale is needed especially for CADD systems which offer a limited drawing area. Drawing scale will reduce the size of an entered line by some factor. However, the coordi-

nate display and coordinates you enter should still be full-size.

Constructing Graphic Entities

The "substance" of any CADD drawing is the geometry. The drawing you create consists of graphic entities, including points, lines, polygons, circles, and arcs.

Drawing points

Points, sometimes called **nodes,** mark an exact coordinate position. They are helpful as a reference for placing other entities. After entering the appropriate command, enter coordinates or pick the location on screen with your input device. A dot or plus mark (+) should appear, Fig. 19-10.

Drawing lines

Lines are geometric entities defined by their two endpoints. The points may be entered as absolute or relative to another point. You can enter coordinates using the keyboard or digitizing tablet or pick a location on screen.

Several different methods can be used to draw lines. Each of these methods allows you to draw a line using different ways of selecting position. Some lines are placed in relation to an existing line. The options for drawing lines include the following methods.

Drawing a single line. The simplest method to draw a line is by entering coordinates or picking location for the two endpoints. As you move the screen cursor away from the first point when picking the location, a blinking or dotted "rubberband" line may stretch out from the first endpoint. This allows you to visualize the line better. Once the second point is selected, the line will "become" solid.

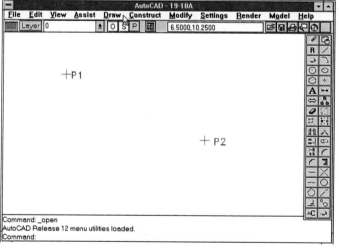
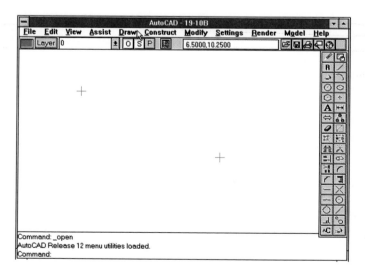

Fig. 19-10. Placing points on the display screen.

Drawing lines at an angle. Lines at an angle can best be drawn by typing in the second point as a polar coordinate. Pick the first endpoint using the screen cursor, then enter the polar coordinates.

Drawing a series of connected lines. With connected lines, the second endpoint of a previous line becomes the first endpoint of another line, Fig. 19-11A. Connected lines are usually the default option of the Line command. After picking the two endpoints of the first line, continue to pick the second endpoints of subsequent lines.

Drawing a series of single lines. To draw a series of single lines, you usually must select one of the Line options. This allows you to pick the first and second endpoints without leaving the Line command. See Fig. 19-11B.

Drawing horizontal and vertical lines. The Horizontal and Vertical options of the Line command allow you to draw lines parallel to the "X" and "Y" axes, Fig. 19-11C. After picking the first endpoint, the line will extend either horizontally or vertically, depending on the option you selected. Some CADD programs call this the Orthogonal option, which limits any drawn line to either horizontal or vertical. You do not need to select individual Horizontal or Vertical options.

Drawing lines at an angle to a given line. To draw a line at an angle to a given line, you must first pick the existing line. Then, enter an angle, pick the first endpoint, and enter the line length, Fig. 19-11D. Some CADD programs ask you to pick both endpoints on the screen. As you locate the second endpoint, though, the line extends only at the entered angle.

Drawing a line parallel to a given line. Drawing parallel lines is easy with CADD. Simply pick the existing line, then pick the endpoints of the new line. The first endpoint of the new line determines the distance away from the existing line. See Fig. 19-11E. Some systems ask you to type in the distance between the two lines.

Drawing a line perpendicular to a given line. To draw a line perpendicular to a given, first pick the existing line, then pick the first endpoint of the new line. As you locate the second endpoint, the line extends at 90° to the existing line, Fig. 19-11F. Some systems have you first pick one endpoint, then the line to which the new line should be perpendicular.

Drawing tangent lines. New lines can be drawn tangent to circles, arcs, and curves. After selecting the Line command's Tangent option, pick the circle or arc to which the new line should be tangent. Then pick a second endpoint at the appropriate location. The second endpoint might also be made tangent by choosing the Tangent option again, Fig. 19-11G.

Some systems offer the Parallel and Perpendicular options with the Tangent command. This allows you to place a line both tangent to a circle and parallel or perpendicular to an existing line.

Drawing double lines. Double lines are parallel lines drawn at the same time. When entering or picking the endpoints, the double lines can be placed left-justified, right-justified, or center-justified in relation to points you pick. This command is especially useful in architectural drafting for drawing walls on a floor plan. See Fig. 19-1H.

Drawing polygons

Any type of polygon can be drawn by connecting several line segments. However, CADD systems offer additional specialized polygon drawing options for rectangles and other regular polygons.

Drawing a rectangle. The Rectangle command allows you to pick the opposite corners of a rectangle. As you locate the second corner, a

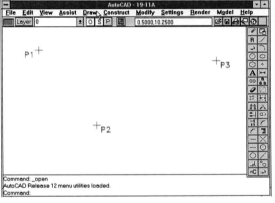
BEFORE

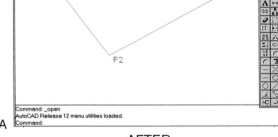
AFTER

Fig. 19-11. Drawing lines. (Continued)

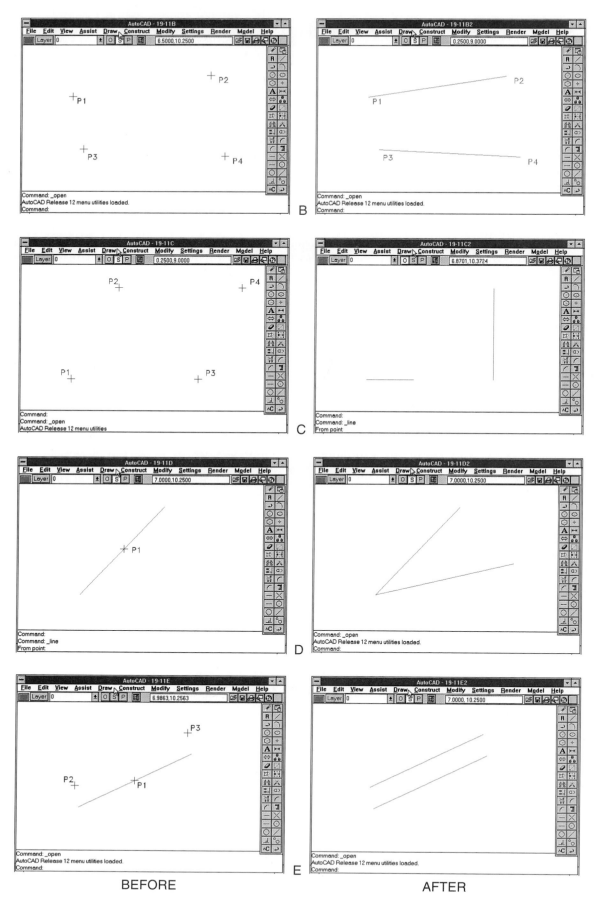

BEFORE   AFTER

Fig. 19-11.   Drawing lines. (Continued)

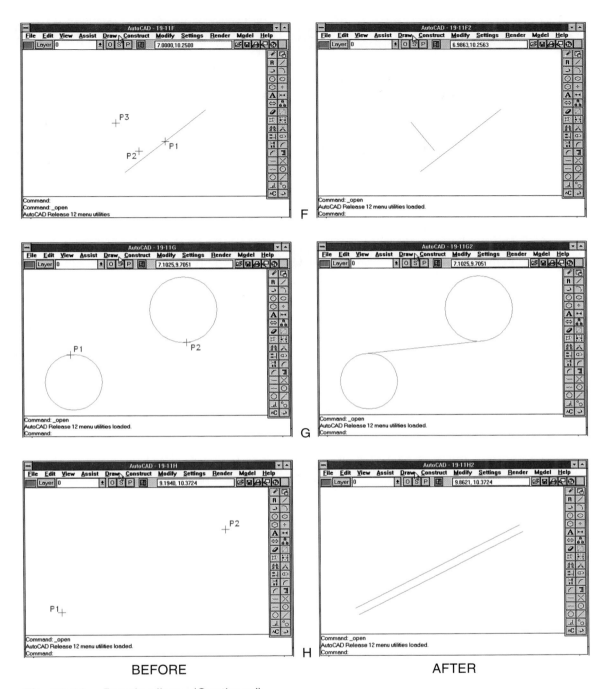

BEFORE

AFTER

Fig. 19-11. Drawing lines. (Continued)

temporary rectangle appears to stretch out from the first corner to help you visualize the size of the rectangle better, Fig. 19-12. This method of drawing a rectangle is much easier than drawing four connected line segments.

Drawing a regular polygon. Regular polygons would be difficult to draw if it were not for the Polygon command. This command generally offers three options:

- Center and vertex.
- Center and edge.
- Vertex and opposite vertex.

Before picking the two points, you must enter the number of sides for the polygon. The results of using the three options to draw a hexagon are shown in Fig. 19-13. As you locate the second point, the polygon may stretch out and rotate with the screen cursor to help you locate its position.

Some CADD software allows you to choose whether you would like the polygon inscribed in or circumscribed about a circle. A polygon is "inscribed" when it is drawn inside a circle and its corners touch the circle. A polygon is "circumscribed" when they are drawn outside of the circle, and the sides are tangent with the circle.

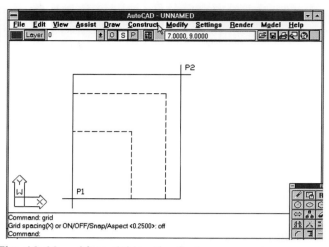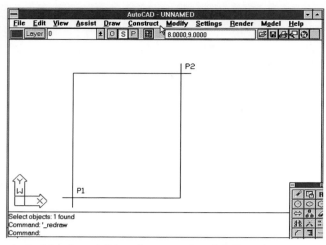

Fig. 19-12.   After picking the first corner, a temporary rectangle "rubberbands" to allow you to pick the size.

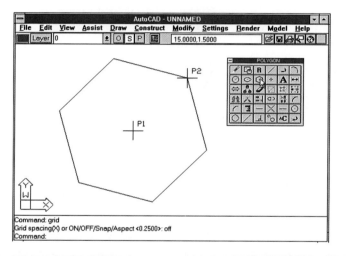

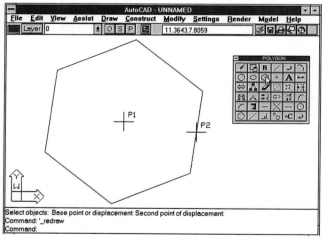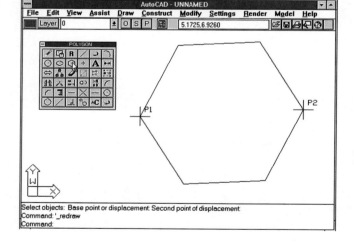

Fig. 19-13.   Drawing a polygon.

Drawing circles

A circle can be drawn by several methods. Choose the appropriate method for the situation. When using your CADD software, selecting the Circle command probably accesses a list of options. Each of these options allow you to draw a circle using a different method of selecting circle attributes, such as center, radius, diameter, and circumference. Fig. 19-14 illustrates the options that follow.

Drawing a circle by center point and radius. After selecting the Radius or other appropriate option, type in the radius, and then pick or enter coordinates for the center location, Fig. 19-14A. Some systems allow you to pick the center point and then "drag" the circle to its size. In this manner, you can visualize the circle as you place it in its final location.

Drawing a circle by center point and diameter. After selecting the Diameter or other appropriate option, type in the diameter, and then pick or enter coordinates for the center location. See Fig. 19-14B.

Drawing a circle using two points. When drawing a circle by two points (2P option), you are actually picking two points along the circumference of

the circle. After you have selected the second point, a complete circle is drawn on screen. See Fig. 19-14C.

Drawing a circle through three points. In certain situations, you must place a circle which touches three other entities. The Three-Point (3P) option allows you to pick three points. The circle is calculated and drawn to pass through each point, Fig. 19-14D.

Drawing circles tangent to other entities. Just as you can draw a line tangent to a circle, you can also draw a circle tangent to a line, arc, another

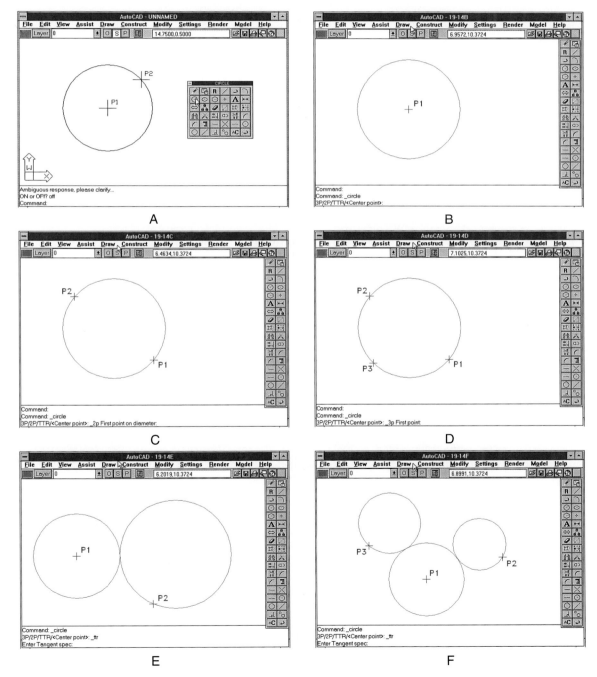

Fig. 19-14. A circle can be drawn using a variety of techniques.

circle, or any other entity. You can construct circles tangent to one item or to several items, depending on the CADD software you are using. Usually, you must pick the circle center and then the entity to which the circle should be tangent. Some systems allow you to pick two entities to which the circle should be tangent. The computer calculates the size of the circle. See Fig. 19-14E and F.

Drawing arcs

Arcs are partial circles which can be defined by their center point, start point, endpoint, radius, and included angle. Selecting the proper Arc command brings up a list of options. These options allow you to draw an arc using different aspects of an arc. Arcs are typically drawn counterclockwise; thus, pick your start and endpoints carefully. See Fig. 19-15 for an explanation of the options that follow.

Three points on an arc. Pick the start point, endpoint, and point along the arc, Fig. 19-15A.

Center, start point, and endpoint. Pick the center, start point, and endpoint of the desired arc, Fig. 19-15B.

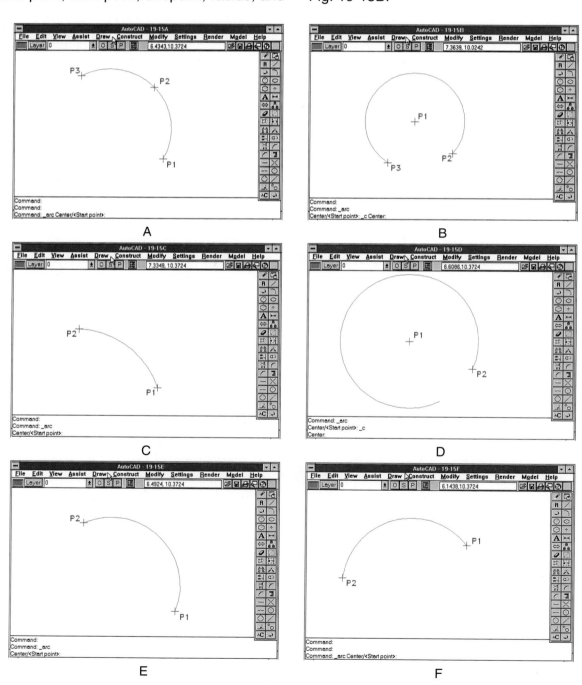

Fig. 19-15. Arcs are generally drawn in a counterclockwise direction. Pick points carefully to ensure that your arc is drawn in the proper direction.

Start point, endpoint, and radius. Type in the radius, then pick the start point and endpoint. Select your start point and endpoint carefully to obtain the desired results. See Fig. 19-15C.

Center, start point, and included angle. Type in the included angle, then pick the center point and start point to draw an arc, Fig. 19-15D.

Start point, endpoint, and included angle. Type in the included angle. Then, pick the start point and endpoint carefully, Fig. 19-15E.

Two points. Pick the two points indicating the diameter of the arc. This results in a 180° arc, or semicircle. See Fig. 19-15F.

Drawing splines

Splines are smooth curves that pass through a series of points. The Spline command will either make a smooth curve out of existing connected lines or request that you pick points to create the curve, Fig. 19-16.

Drawing ellipses

Ellipses can be drawn by several methods. One method is to locate the two axes by selecting two endpoints of one axis and one endpoint of the other axis, Fig. 19-17A. Another method is to place the ellipse's center, then give one endpoint of each axis, Fig. 19-17B. A third method is to pick the ellipse's major axis endpoints and then enter a rotation angle, Fig. 19-17C.

Selecting Linetype, Linewidth, and Color

To comply with standard line conventions, CADD programs allow you to select the linetype and linewidth. For example, this might be done by selecting the Linetype command and then entering HIDDEN. In some CADD programs, the available linetypes and linewidths are shown on the screen. Simply move the cursor over the desired linetype or linewidth and pick it.

CADD systems with color display screens allow you to assign colors to entities. For example, all hidden lines could be drawn red, visible lines in white, and dimensions in blue. Separating items by color helps you to identify different parts of the drawing.

Drawing Aids

Drawing aids are helpful functions which allow you to locate position on the screen. The three common types of drawing aids are discussed here: grids, object snap, and construction lines.

Grid

A **grid** is a pattern of dots on the screen used much like graph paper in traditional drafting, Fig. 19-18. When adding entities, use the grid as a reference to help you locate position. The grid is not a permanent part of the drawing. Even if the grid is displayed during plotting, the grid dots are not plotted.

It is often difficult to position the cursor precisely over a grid point. The screen resolution may be too low to clearly see the dots or the cursor may appear to float or jump. To help you, CADD systems offer "snap" options. A grid snap option locks the drawing cursor to grid dots as it moves across the screen. You cannot pick a point on the screen that does not coincide with a grid dot if the grid snap feature is turned on.

You can set the "X"-axis and "Y"-axis spacing of the grid dots at any value. Set the spacing to the smallest value you will need. For example, if most of the dimensions of your drawing are multiples of .25″, then set your grid at .25″. They can be reset any number of times to meet your needs.

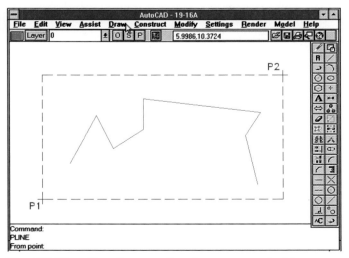
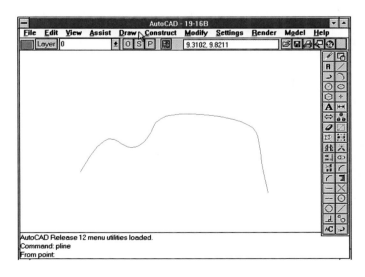

Fig. 19-16. Creating a spline from connected lines.

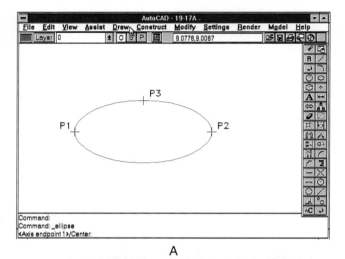

A

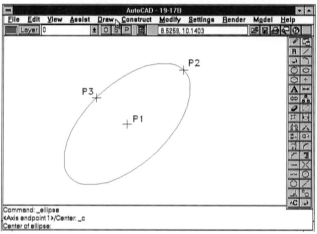

B

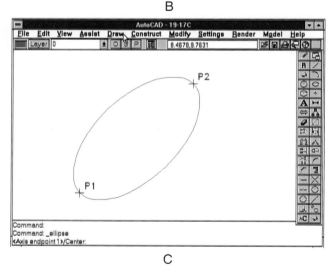

C

Fig. 19-17.   A–An ellipse can be drawn by identifying the major and minor axes, B–the center point and an endpoint on each of the axes, or C–endpoints of the major axis and a rotation angle.

Grids do not have to be displayed. The grid can be turned on when needed and off when it hinders viewing the drawing. The grid snap can be in effect even if the grid is not displayed.

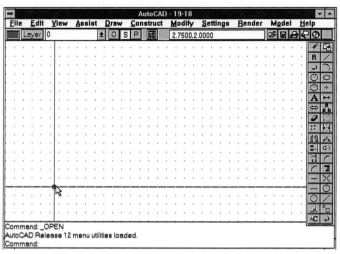

Fig. 19-18.   Grids are similar to graph paper used in traditional drafting. Set the grid to a value appropriate for your drawing.

Several CADD programs allow you to rotate the grid. This is helpful when drawing an auxiliary or other rotated view. Enter the angle of the grid as needed. The grid will then appear at the desired angle.

Object snap

Your ability to precisely pick point locations depends on the resolution of the display and the accuracy of the input device. However, suppose you want to locate the endpoint of a line at the corner of a rectangle. This might be difficult to do accurately. What happens when you need to locate an endpoint at the center point of a circle? How do you find the center point? This is done with object snap.

Object snap allows you to precisely locate a position on an existing entity. After entering a drawing command, select one of the object snap options. They might be found in a screen menu along the side of the screen or along the top of a screen. Some CADD software provides an object snap menu at your cursor location. Available object snap options, shown in Fig. 19-19, allow you to place points at:

- The endpoint of a line, curve, or arc.
- The midpoint of a line.
- The intersection of two entities.
- The center of a circle or arc.
- Predefined positions on an entity.
- Any point along an entity.
- The nearest point on an entity.

Construction lines

Construction lines are temporary lines placed for reference. These lines are not a permanent part of the drawing. Construction lines can be drawn horizontally, vertically, or at an angle. Construction lines can be placed on a separate layer. The layer can then be turned off to view the drawing without

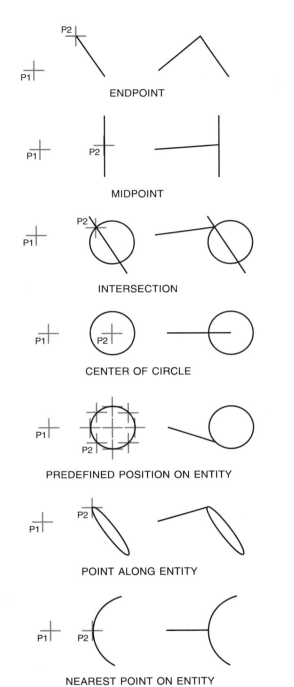

Fig. 19-19. Object snap modes provide an accurate means for attaching to existing objects.

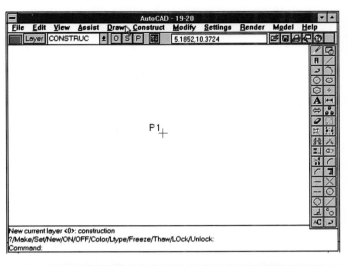

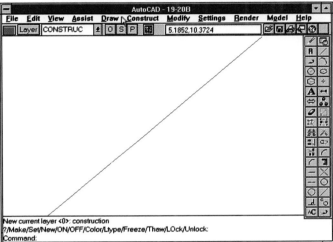

Fig. 19-20. Construction lines are temporary reference lines.

the construction lines. Simply enter the angle and pick the base point, Fig. 19-20. This is the pivot around which the construction line is rotated. After placing a construction line, you can "snap" to it using any of the object snap options.

Editing a Drawing

Editing commands allow you to modify, erase, and manipulate previously drawn entities.

Editing is the primary function of the drafter. In fact, many complex drawings are made by manipulating (working with) just a few drawn objects. CADD software provides a variety of editing functions to suit every situation. You should be able to make the desired change using one of the commands found in this section.

When editing, you will be required to select objects to edit. You might pick an individual item, pick several items, or group objects to edit by picking corners of a window around them or passing a line through them. The way you select items depends on the command. Usually, a message will appear at the bottom of the screen telling you how to select items to perform the editing function.

Erasing and unerasing entities

The Erase or Delete command removes an entity or group of objects from the drawing. You might select entities individually, by picking corners of a window around them, or by entering the Last option to erase the last drawn item. See Fig. 19-21.

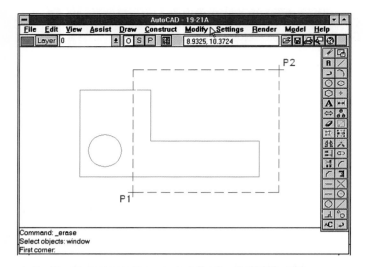

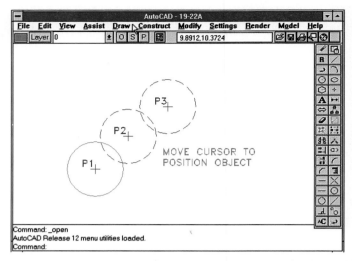

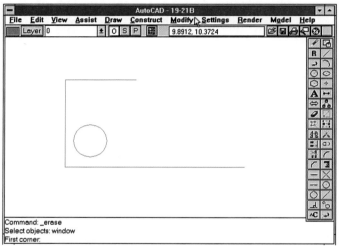

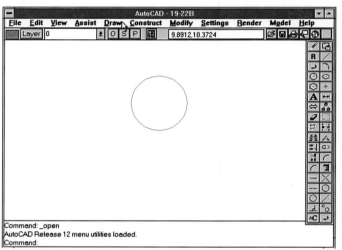

Fig. 19-21. A window may be used to select a group of entities on a drawing. In this case, any line entirely enclosed within the window is selected.

Fig. 19-22. Moving an object by "dragging" it to the new position.

In many cases, CADD programs are forgiving. If you accidentally erase the wrong object(s), select the Unerase command, often called the Oops command, to restore the last erased object or a number of erased objects.

Moving entities

Moving objects requires that you first select the object to be moved, pick a reference point, and then pick the new location. When you pick a single item, it may "drag" along with the cursor to the new position, Fig. 19-22.

In some CADD programs, one of the Move options allows you to move a vertex of a polygon. Lines that start or end at that point are shortened or lengthened, Fig. 19-23.

Copying entities

Copying entities is much like moving them. However, the original entities remain unchanged and the copy is positioned in place, Fig. 19-24.

In many CADD programs, the Copy command also includes array functions. An *array* is a pattern of copies, placed in a rectangular or circular design. When making a rectangular array, you must select the original object(s), enter the number of copies in the "X" and "Y" directions, and enter the distance between copies, Fig. 19-25.

When making a circular array, you must select the object(s), pick a pivot point, enter the number of copies, and enter the angle between copies, Fig. 19-26. Some CADD programs allow you to rotate the objects as they are arrayed.

Mirroring entities

Mirroring creates a reflected image of one or several entities on the other side of a mirror line, Fig. 19-27. One option of the Mirror command allows you to delete the original entities after mirroring them.

The mirroring function is especially helpful for symmetrical objects. Draw one half and then mirror it to create the entire object. This cuts down the

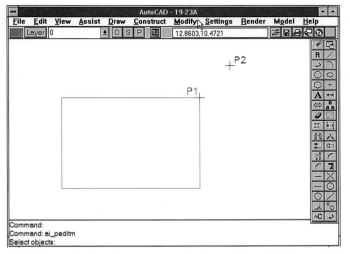

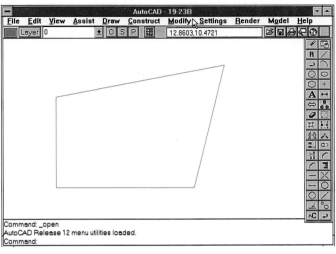

Fig. 19-23. Moving the vertex of a polygon affects all lines which start or end at that point.

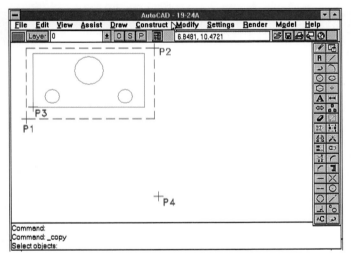

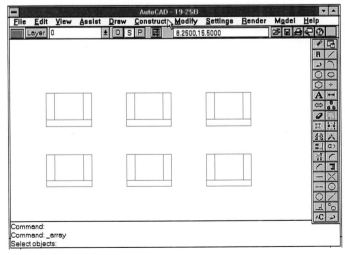

Fig. 19-24. The Copy command.

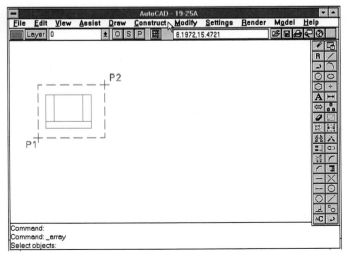

Fig. 19-25. A rectangular array created from a single object. Note that the object is drawn only once, then arrayed.

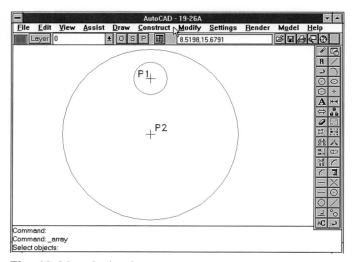

Fig. 19-26.   A circular array.

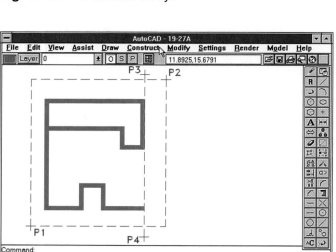

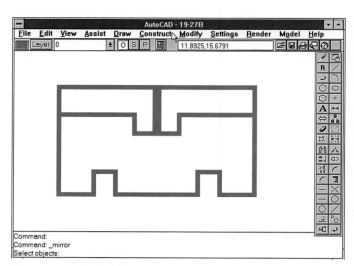

Fig. 19-27.   The Mirror command is commonly used to create symmetrical objects.

drawing time, and greatly enhances productivity. Remember, with a CADD system you should never draw the same entity twice. Simply use the appropriate editing function to modify an existing entity.

Rotating entities

Rotating allows you to revolve an entity or group of entities around a desired pivot point. Select the object(s) to be rotated, enter a rotation angle, and pick a pivot point. The objects are rotated about the pivot point at the specified angle, Fig. 19-28.

Trimming an entity

The Trim command is used to shorten a line, curve, or other entity to its intersection with an existing entity. You must pick the reference (boundary) object and the entity to be trimmed, Fig. 19-29. Check the user's manual for your CADD software for the order in which to pick the reference object and entity to be trimmed.

Extending entities

The Extend command is the opposite of the Trim command. It allows you to lengthen an entity to meet with another reference entity, Fig. 19-30. This command is particularly useful to extend entities that fall just short of an intended intersection.

Stretching an object

The Stretch command allows you to move a selected portion of the drawing while retaining all connections between entities. The portion of the object to stretch is chosen by picking corners of a window around it. Then, a reference point is picked, followed by picking a new position for the reference point, Fig. 19-31.

Scaling objects

The Scale command allows you to enlarge or reduce the size of an entity or group of entities. After selecting items to scale, specify a magnification factor, Fig. 19-32. Some systems allow you to

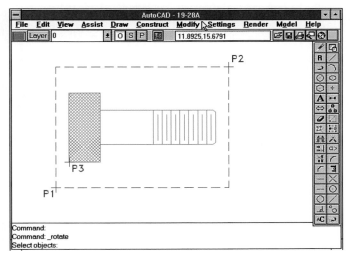

Fig. 19-28. The Rotate command.

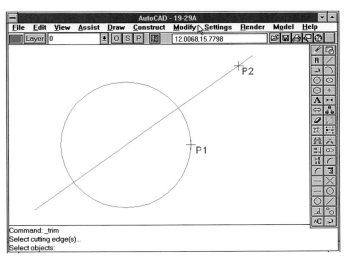

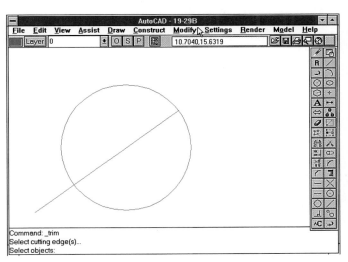

Fig. 19-29. The Trim command shortens an entity to an intersection with another entity.

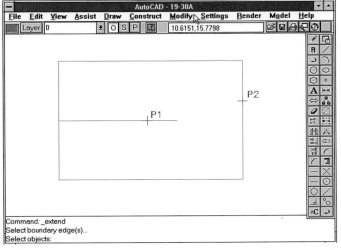

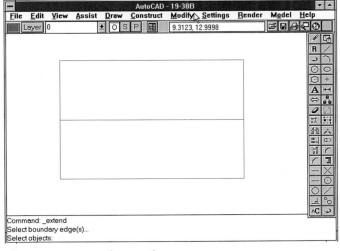

Fig. 19-30. The Extend command allows you to lengthen a line to meet another entity.

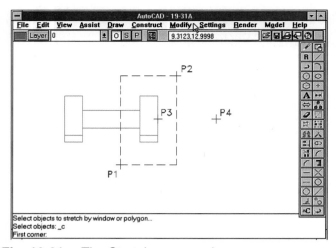

Fig. 19-31. The Stretch command.

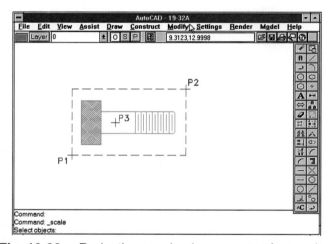

Fig. 19-32. Reductions and enlargements of an entity or group of entities is performed with the Scale command.

scale objects unproportionately (different "X" and "Y" direction scaling factors).

Joining entities

The Join command allows you to combine two entities. It may be easier to edit or change entities joined together rather than edit them separately. To do so, the two items must end at a common meeting point. (This can also be done by trimming each.)

Breaking an entity

The Break command allows you to remove a section of a line, circle, or complex object. This can be done by selecting two points on an entity, breaking a fenced portion, or selecting two objects, Fig. 19-33. When breaking with a fence, all portions of entities which are within or intersecting the fence are removed. "Breaking by object" is a method of breaking a portion of two objects where they intersect.

Exploding an entity

The Explode command breaks an entity into its component parts. A polygon, dimension, or symbol is converted into its component lines. Each item

can then be edited individually. The exploded entity will not change in appearance.

Creating fillets

The Fillet command creates fillets and rounds of a specified radius. Although there are several different options offered by the Fillet command, the simplest technique is to pick the two lines to be filleted. The fillet that is created is an arc tangent to both entities, Fig. 19-34. Most CADD programs then trim the entities to the fillet arc. For some programs, the two entities do not have to intersect. The Fillet command will extend lines to meet the fillet arc.

Creating chamfers

The Chamfer command connects two lines with a chamfer. The chamfer is placed equidistant from the intersection of the lines unless you enter different chamfer distances, Fig. 19-35.

Changing entity properties

Properties refer to linetype, linewidth, layer, and color of lines, circles, arcs, and other entities. You are able to change these properties using editing

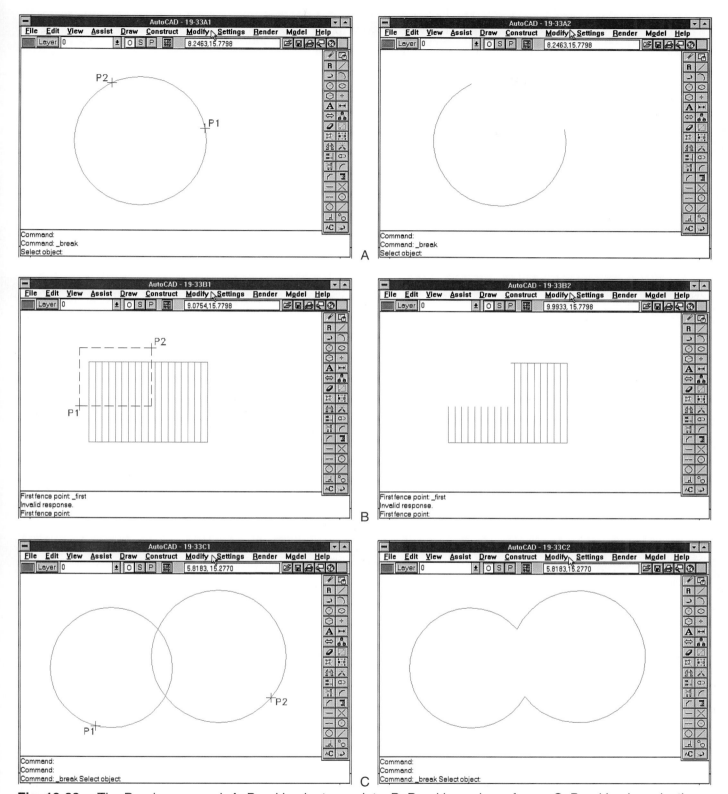

Fig. 19-33. The Break command. A–Breaking by two points. B–Breaking using a fence. C–Breaking by selecting two objects.

functions, usually found under Edit Properties. Simply enter the property to change, select the entities to change, and enter a new value. If you draw a line using the wrong linetype, you will not need to re-draw it. Simply use the Edit Properties command to modify the linetype, Fig. 19-36. The edit properties command can also be used to change the color of selected entities.

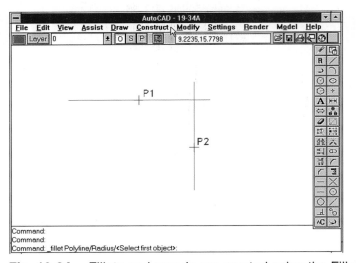
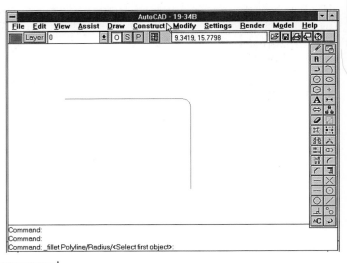

Fig. 19-34. Fillets and rounds are created using the Fillet command.

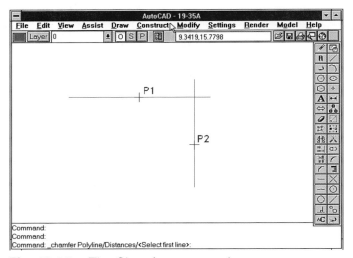
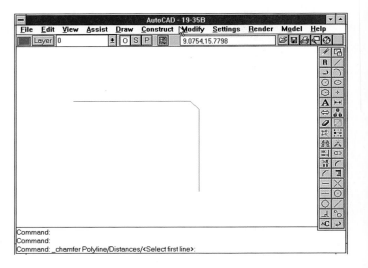

Fig. 19-35. The Chamfer command.

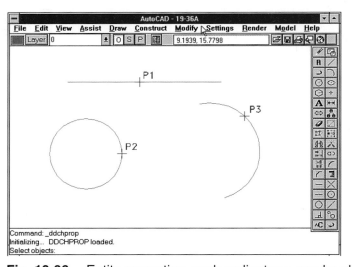
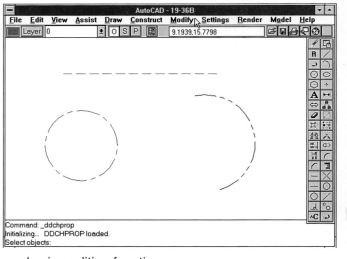

Fig. 19-36. Entity properties, such as linetype, can be changed using editing functions.

Display Controls

Display controls are commands that determine how much and what parts of your drawing are shown on the display screen. Remember that your drawing is held as data in computer memory. Therefore, the CADD program can manipulate the data to show you the entire drawing or only a part of it.

Zoom commands

Zoom options determine how much of the drawing you see on the display screen. They do not change the actual size of the drawing. All object sizes and shapes remain proportional; they are simply magnified or reduced in size temporarily for better viewing of the drawing. Fig. 19-37 illustrates two of the Zoom options: Zoom Window and Zoom Factor.

Zoom window. This option magnifies a portion of the drawing within a specified window. The windowed section enlarges to fill the screen.

Zoom out. This option reduces the current view to fit within a specified window. The rest of the drawing is redrawn on the screen accordingly.

Zoom factor. The Zoom Factor option allows you to enlarge or reduce the drawing by a magnifying factor. You must also select a center to become the center of the magnified view. Entering a magnifying factor greater than one makes the drawing appear larger. A value less than one reduces the size so that more of the drawing fits on the display.

Zoom extents. This option returns the displayed view to the full drawing area. No matter how large your drawing area is (even 1,000,000' wide), it is displayed in its entirety.

Zoom full. The Zoom Full option calculates how big the drawing should be to just fit on the display. It is different than the Zoom Extents option in that it does not show the drawing area defined by the boundaries. Instead, it calculates the smallest screen window which contains your entire drawing.

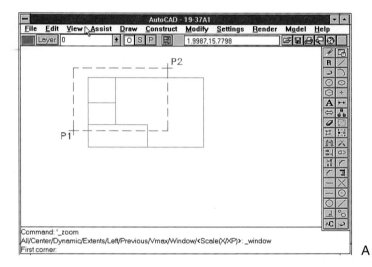

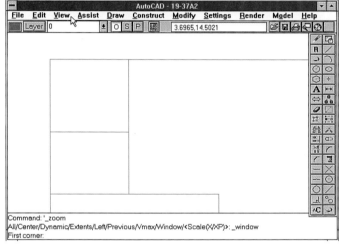

A

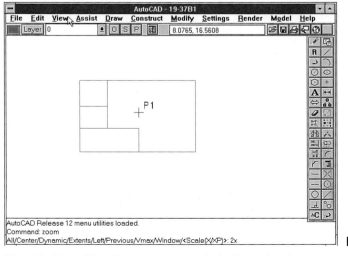

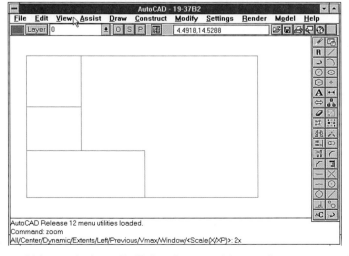

B

Fig. 19-37. The Zoom command. A–Zooming in on an area within a window. B–Enlarging an object using a specified magnification factor.

Panning

When a portion of the drawing is magnified, it might be necessary to see an object which is "just off" the screen. This is done by "panning" across the enlarged view. Pick two points. The view then moves the distance between the two points in the direction of the second point, Fig. 19-38.

Panning is accomplished on some systems by specifying which direction to pan. For example, you might choose Pan Left, Pan Right, Pan Up, or Pan Down. The windowed section of the drawing then moves that direction.

Display last

The Display Last command sequence returns the screen to the previous view of the drawing. You might use this to return to a reduced view after having "zoomed in" to do detail work.

Save and recall view

Using Display Save and Display Recall functions, you can save and recall current views of the drawing for future reference. Save commonly used views so that it is easier to return to these portions of the drawing for more work.

Redraw

When editing, "holes" (unlit pixels) from deleted entities may be left. The Redraw command renews the view of the drawing, clearing off the current display and then redrawing the same view. This "cleans up" the drawing.

PROBLEMS AND ACTIVITIES

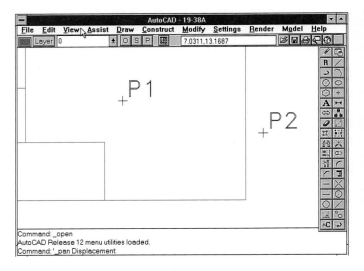

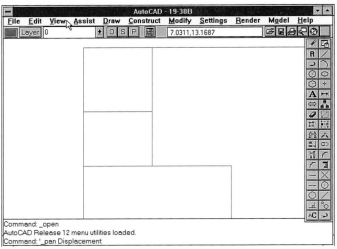

Fig. 19-38.   Panning across a drawing to see a part of an object which is off the screen.

Using standard drafting practices and a CADD system, create a drawing for each of the problems shown in Fig. 19-39. The drawing name should be P19-(problem number). For example, name Problem 1 as P19-1. Set up the drawing using decimal inches as the unit of measurement. Enter a drawing area or select a sheet size large enough to contain the drawing. Set grids and use other drawing aids to your advantage. Do not add dimensions or text to the drawing at this time.

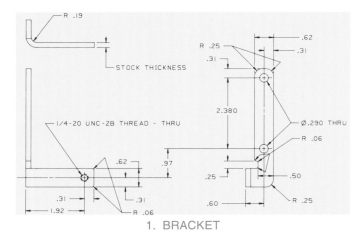

1. BRACKET

Fig. 19-39.   Computer-assisted design and drafting problems. (Continued)

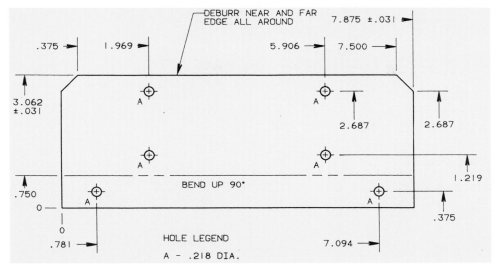

DEBURR NEAR AND FAR EDGE ALL AROUND

7.875 ±.031

.375

1.969

5.906

7.500

3.062 ±.031

2.687

2.687

1.219

BEND UP 90°

.750

0

0

.781

.375

HOLE LEGEND

7.094

A - .218 DIA.

2. BATTERY BRACKET

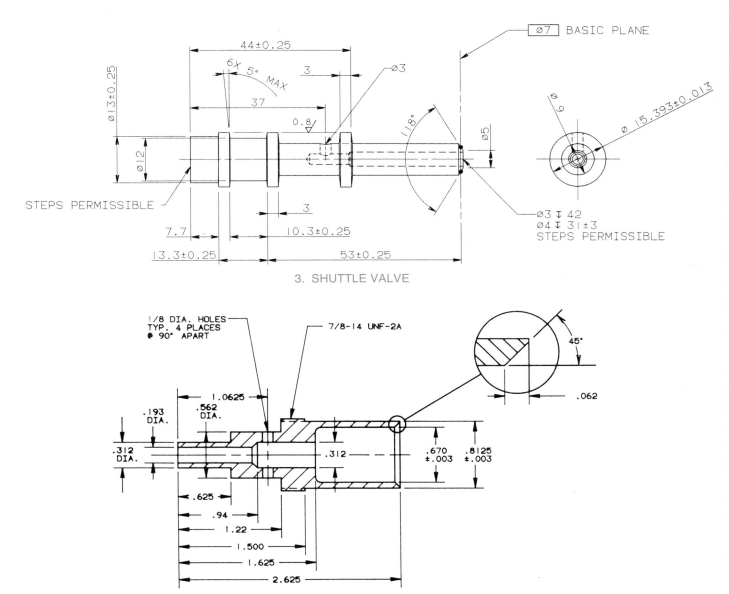

Ø7 BASIC PLANE

44±0.25

6X 5° MAX

3

Ø3

Ø13±0.25

37

0.8

118°

Ø5

Ø12

Ø 15.393±0.013

STEPS PERMISSIBLE

3

Ø3 ⫌ 42
Ø4 ⫌ 31±3
STEPS PERMISSIBLE

7.7

10.3±0.25

13.3±0.25

53±0.25

3. SHUTTLE VALVE

1/8 DIA. HOLES
TYP. 4 PLACES
@ 90° APART

7/8-14 UNF-2A

45°

1.0625

.562 DIA.

.062

.193 DIA.

.312 DIA.

.312

.670 ±.003

.8125 ±.003

.625

.94

1.22

1.500

1.625

2.625

4. TORCH FITTING ADAPTER

Fig. 19-39. Computer-assisted design and drafting problems. (Continued)

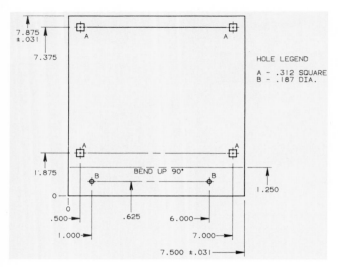

5. PC CARD BRACKET

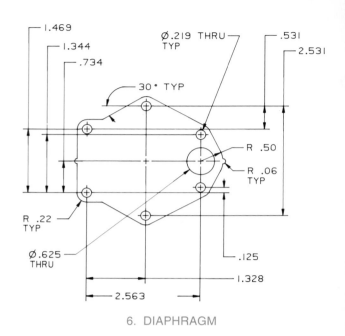

6. DIAPHRAGM

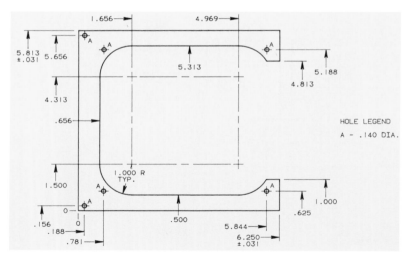

7. COVERBRACKET

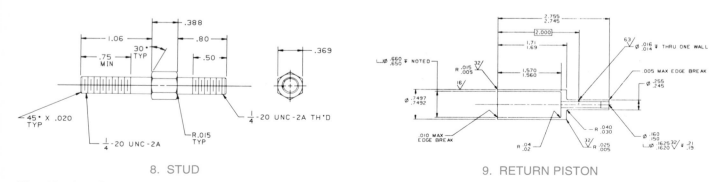

8. STUD

9. RETURN PISTON

Fig. 19-39.   Computer-assisted design and drafting problems. (Continued)

20 Detailing and Advanced CADD Functions

KEY CONCEPTS

- ☐ Layering techniques are used to separate graphic entities onto "overlays."
- ☐ Uniform, consistent text can be easily added to a drawing.
- ☐ Dimensioning commands allow a design-drafter to dimension drawing according to the most recent standard.
- ☐ Section lines can be added to a view using hatching functions of the CADD software.
- ☐ The Inquiry function allows a design-drafter to obtain information about an entire drawing or any components of it.
- ☐ Symbols can be easily created, selected, and stored using a CADD program.
- ☐ Plotting is one method of obtaining a hardcopy of the drawing.

This chapter covers CADD functions that reduce many of the tedious tasks found in traditional drafting. Since a great amount of time typically is spent detailing engineering drawings, CADD systems automate much of this process. Automatic functions for dimensioning, adding section lines, and placing text decrease the amount of drawing time required. In addition, this chapter introduces some advanced functions, such as layering, using symbol libraries, and inquiry functions.

LAYERS

Layering commands allow you to separate different parts of a drawing into levels, much like using layers of film in overlay drafting. This is important to understand before you begin adding text and dimensions to a drawing.

In manual drafting, information on complex drawings is separated onto different sheets of vellum or film, each containing specific, yet related details. For example, in architectural drafting, the floor plan is drawn on one sheet, the foundation plan on another, and the plumbing plan on still another sheet. This is done with CADD by specifying different layers, or levels. Each layer can be thought of as a plastic overlay. Related details of a drawing are drawn on the same layer. This allows the drafter to efficiently organize information such as object views, dimensions, text, and specifications, Fig. 20-1.

Active Layer

When drawing, you are adding entities to only one layer. This is called the **active** or **current layer.** For example, suppose you are currently adding object lines, but now plan to add dimensions. You may wish to choose a new active layer on which to draw the dimensions.

Displaying Layers

All or only selected layers can be displayed. For example, when drawing a plumbing plan, you might show the floor plan, but "hide" layers containing electrical, foundation, and other plans. Entities on layers turned off (not displayed) are not deleted; they are just invisible. They appear again when you turn the layer back on.

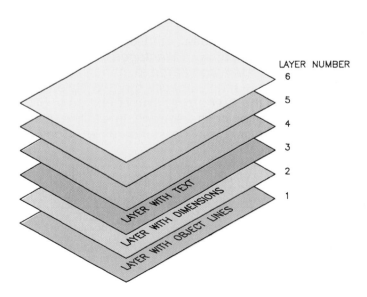

Fig. 20-1. Layers allow you to efficiently organize information about a drawing.

Layer Numbers and Names

When specifying an active layer(s), you most often enter numbers. However, several systems allow you to assign names to layers. For example, the layer containing dimensions might be layer "10", or be named "DIMEN". Names make it easier for you to identify the information contained on each layer.

Layer Color and Linetype

Colors and linetypes may be assigned to layers. Thus, any entity drawn on that layer automatically assumes the color and linetype of that layer. Most systems allow you to individually change the entity color and linetype if necessary.

Erasing Layers

Erasing or deleting layers should be done with caution. This function deletes all entities on the specified layer. Once erased, they generally cannot be brought back.

Changing the Layer of Entities

You may want to change the layer associated with an entity. For example, suppose you added a dimension to the layer containing only object views. The Layer Change function allows you to change the layer associated with the dimension to the proper value.

ADDING TEXT

Text commands allow the drafter to place labels, specifications, notes, and other text on the drawing. There are many advantages to adding text with a CADD system compared to manual lettering. With CADD, you simply specify the text style (features), type in the **text string** (line of text), and pick the desired position on the drawing. In addition, computer-generated text can be standardized. Compa-

nies often determine what style of text is to be used. All drawings then follow those guidelines. This reduces the chance of drafters choosing their own style of lettering, increasing drawing readability.

The steps taken to place text on a drawing are much the same among CADD systems. See Fig. 20-2. Although the exact command names may differ slightly, the procedure usually is:

1. Select the Draw Text command.
2. Specify the various text features, including text font, height, slant, width, and rotation.
3. Select the text justification. This determines the position of the text relative to the location point you pick.
4. Pick the text position on the drawing.
5. Type in the text to be placed on the drawing.

Selecting Text Features

Text features affect the appearance of the lettering on the drawing. As shown in Fig. 20-3, they include the following alternatives.

Font

The **font** refers to the appearance of the text. A font may look like the characters printed in this text, consist of simple line strokes, or be very fancy. Bold

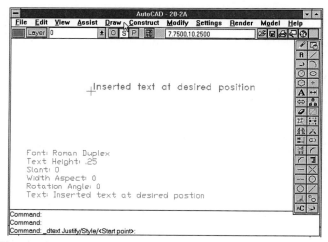

Fig. 20-2.   Placing text on a drawing.

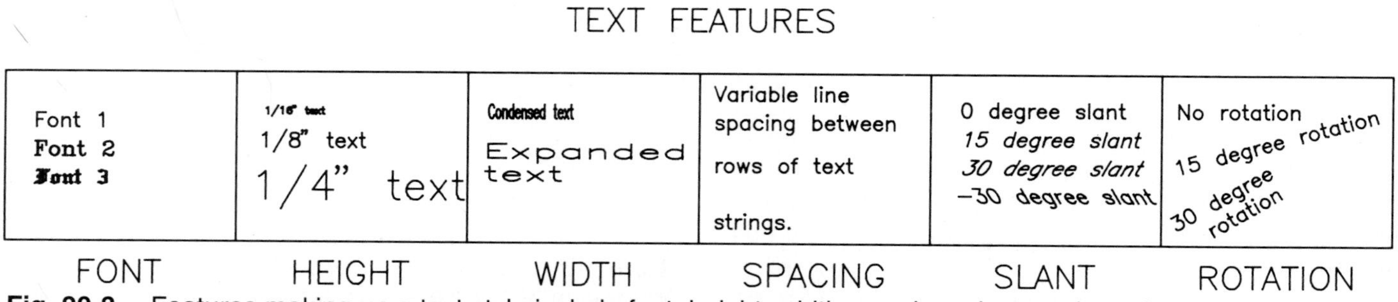

TEXT FEATURES

FONT	HEIGHT	WIDTH	SPACING	SLANT	ROTATION
Font 1 Font 2 *Font 3*	1/16" text 1/8" text 1/4" text	Condensed text Expanded text	Variable line spacing between rows of text strings.	0 degree slant 15 degree slant 30 degree slant -30 degree slant	No rotation 15 degree rotation 30 degree rotation

Fig. 20-3.   Features making up a text style include font, height, width, spacing, slant, and rotation.

fonts are used for titles and section labels. Fancy fonts are used by graphic artists or for business graphics and presentations. For drafting, choose a font that is easy to read, usually composed of simple line strokes.

Height

The **height** is the distance from the bottom to top of a text character. It is usually based on the height of uppercase (capital) letters. Lowercase letters will be smaller than the set height. Choose a height compatible with school or company standards.

Width

Whenever you change the text height, the width automatically adjusts proportionately. However, depending on your CADD system, you might also be able to set the width. A larger width value makes the text look expanded; a smaller width value makes the text look condensed. Some systems use an **aspect value,** which is the ratio of text width to height. For example, entering an aspect value of "2" means the width of a text character will be twice its height.

Spacing

The **spacing** is the distance from the bottom of one line of text to the bottom of the next. The spacing is generally just more than the text height. A spacing option is common on systems that allow you to type more than one text string at a time.

Slant

The **slant,** or **obliquing angle,** is the angle of each character in the text string. You can create italic text using a 15° slant angle. A 0° slant angle makes text characters vertical.

Rotation

Rotation is the angle for entire text strings. Rotation is measured from horizontal in a counterclockwise manner. If you enter a negative rotation angle, the text string is rotated clockwise.

Text features are usually set to default values when you start the CADD system. **Default values** are parameters built into the software, which the computer assigns to the text string until you change them. For example, the default text height may be .125″. You may not wish to use the default value, and can set them to school or company standards during drawing setup. If you change the text style, text previously entered is not affected by the newly set features.

Selecting justification

Justification determines how the text string will be placed in relation to the location point you pick. You may be required to select a justification before or after typing in the text. Justification options, illustrated in Fig. 20-4, include:

Left justified. The left side of the text string is placed at the location point.

Right justified. The right side of the text string is placed at the location point.

Center justified. The text string is evenly spaced on both the left and right sides of the location point.

Middle justified. The text is spaced vertically and horizontally around the location point.

Vertical text. The text string extends vertically downward from the location point. This is not rotation because all of the text characters remain horizontal.

Aligned text. The aligned justification option allows you to position the text string by picking two base points. The text is placed at the angle between the points, and is left justified from the first point selected.

Selecting Text Position

Once you have selected the desired text features and justification, pick the position on the drawing. The on-screen cursor will indicate the desired position.

Typing in Text

Most systems permit you to enter single text strings one at a time. The text string is considered a

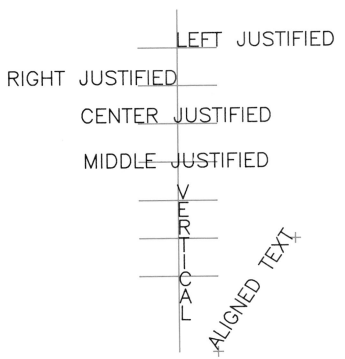

Fig. 20-4. Text justification determines how a text string is placed in relation to the insertion point.

single unit. Thus, if you edit the text string by moving, copying, or deleting, you affect the entire string. Systems that allow you to enter multiple lines consider each line, not the entire paragraph, as a text string.

Some CADD programs offer a Quick Text option. The text is represented on screen with a box. This allows the computer to quickly display the position of the text. When the drawing is printed or plotted, the actual text appears.

Editing Text

Standard editing commands–Move, Copy, Rotate, etc.–affect a text string just as they do graphic entities. However, to revise the text itself, you must select the Edit Text command. Pick the text string to edit, and make the necessary changes to the text string or its features.

DIMENSIONING

Dimensioning commands let you automatically place dimensions on the drawing. In traditional drafting, dimension lines are laid out by hand and measurements are made using a scale. With CADD, the drafter needs only to pick several points. The computer then adds the proper dimension lines, extension lines, leaders, and also calculates and places the dimension. The drafter may have to accept or reject the dimension, a function which allows him/her the chance to make changes before the dimension is placed on the drawing.

Dimensions typically are placed on a separate layer using a different color. This distinguishes them from object lines. Set the active layer and color before dimensioning.

There are four basic methods used to dimension geometric shapes. These include linear, angular, radial (diameter and radius), and leader dimensions. Remember that drafting standards for dimensioning apply equally to computer-generated dimensions as much as they do to manual drafting.

Dimensioning Parameters

When placing dimensions, there are a number of dimensioning parameters available. Some systems include as many as 30 different variables you might set. These parameters might include parameters for dual dimensioning, text size, scale factor, and arrow size. Check your user's manual to research these options. These are set before adding dimensions to the drawing. Common parameters, as illustrated in Fig. 20-5, include the following:

- *Break.* The Break option determines whether the measurement is placed above or within a break in the dimension line.

- *Leading Zero.* The Leading Zero option determines whether a zero precedes measurements less than one. Remember, when using metric dimensions, a leading zero is always used for values less than one.

- *Unidirectional/Aligned Systems.* The Align option determines whether the dimension remains horizontal or aligns with the direction of the dimension line.

- *Terminator.* The Arrow or Terminating Symbol option determines whether an arrow, tick mark, or dot is used at the intersection of dimension and extension lines.

- *Extension Line Offset.* The Extension Line Offset option determines the distance between the extension line and the object.

- *Tolerances.* The Tolerance option allows you to automatically add plus/minus or limit tolerances.

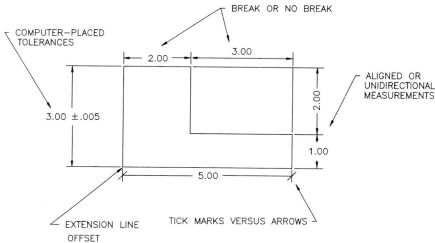

Fig. 20-5.   Dimensioning parameters can be set in a CADD program.

Linear Dimensioning

Linear dimensioning commands are used to dimension straight distances. They include measuring horizontal and vertical distance, and distances along an entity (called *aligned dimensions*). To place a linear dimension, pick two points or select a line. The system will prompt you for the placement of the dimension line. Select the desired placement. The dimension line, extension lines, and dimension value are then placed automatically. This process is shown in Fig. 20-6. The system may allow you to edit the dimension text before entering it on the drawing.

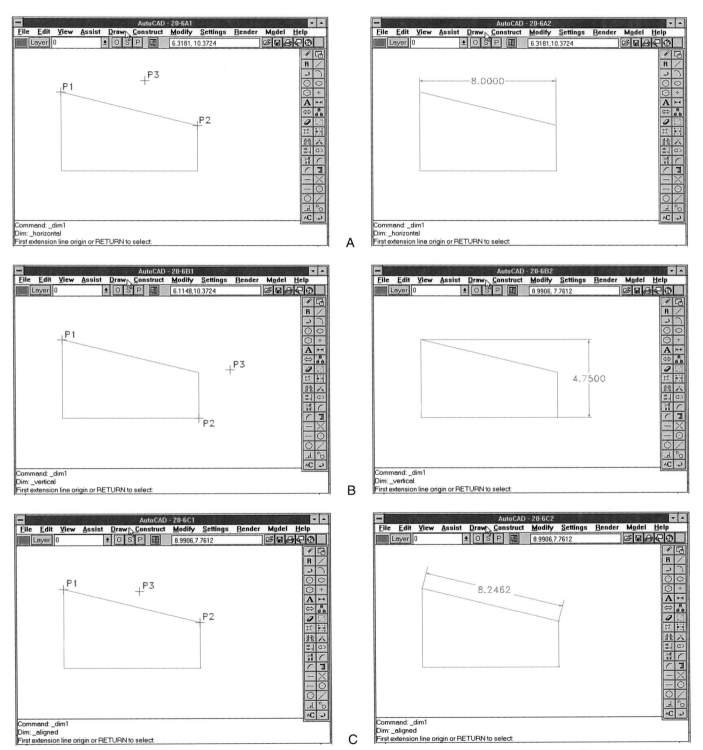

Fig. 20-6. Linear dimensioning. A–Horizontal linear dimensions. B–Vertical linear dimensions. C–Aligned linear dimensions.

Most systems automatically place the dimension value outside the extension lines if the dimension will not fit. However, some systems have you pick a third point (to indicate the dimension line) outside of the dimension area. The dimension will be placed outside the actual measured dimension and the arrows will point in, Fig. 20-7.

Linear dimensioning also includes chained (continued) and baseline (datum) dimensioning functions. These require that you first pick several locations to measure. Next you must pick the position of the first dimension line. The appropriate dimensions are then added automatically, as shown in Fig. 20-8.

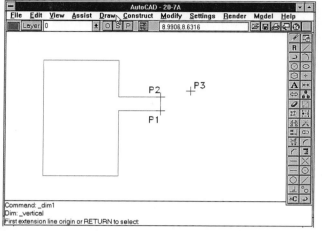

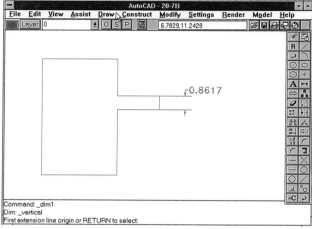

Fig. 20-7. Placing dimensions that are too large to fit between the extension lines.

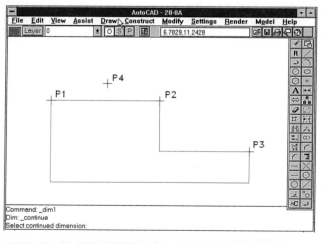

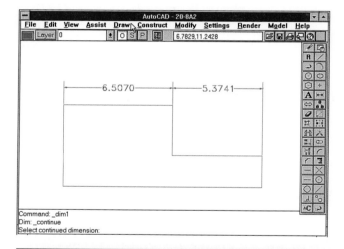

A

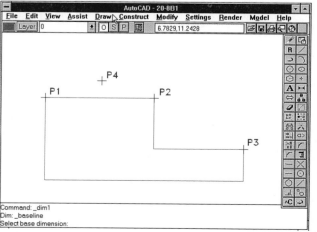

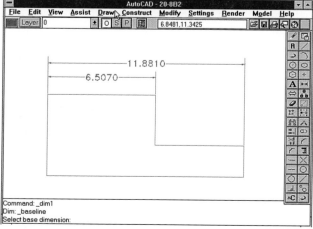

B

Fig. 20-8. Linear dimensioning. A—Chained dimensions. B—Baseline (datum) dimensions.

Angular Dimensioning

Angular dimensions define the angle between two nonparallel lines. Two methods are used by CADD systems for dimensioning angular dimensions: between entities and three points.

Angular dimensioning between entities

With this method, select two lines that form the angle. Then pick the location of the dimension line. See Fig. 20-9.

Angular dimensioning using three points

This method requires that you pick four points on the drawing, Fig. 20-10. The points are the vertex, the two points that form the angle, and the position of the dimension line. The vertex might be a corner, intersection of two lines, or some other point. (It depends on what you are going to dimension.) The two endpoints might be the endpoints of two intersecting lines.

Radial Dimensioning

Radial dimensioning includes diameter dimensioning and radius dimensioning. The steps are simple. Select the Dimension Radius or Dimension Diameter command, and pick the circle or arc to dimension. The dimension line and dimension value are placed automatically, Fig. 20-11.

Diameter and radius dimensioning can be done with or without a leader. Without a leader, the dimension is placed within the circle or arc. With a leader, the dimension is placed outside the feature at the end of a leader pointing toward the feature. Placement of the leader is done one of two ways.

With semiautomatic dimensioning, you pick the dimension either inside or outside the circle or arc. If you pick inside the curve, the dimension is placed inside. If you pick outside the feature, a leader is used. Then you must pick the location of the dimension. See Fig. 20-12.

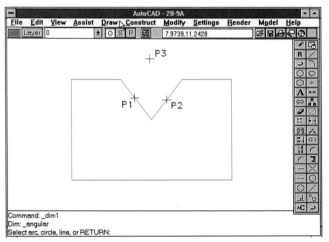

Fig. 20-9. Angular dimensioning between entities.

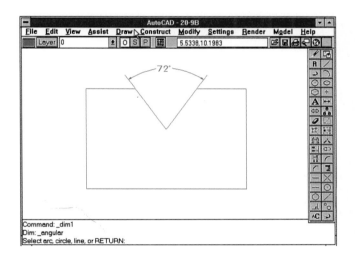

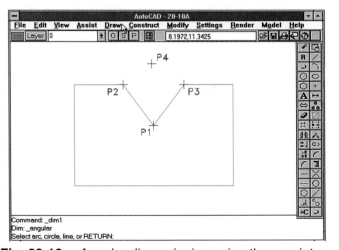

Fig. 20-10. Angular dimensioning using three points.

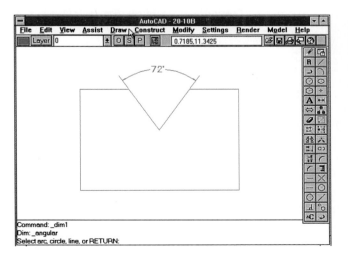

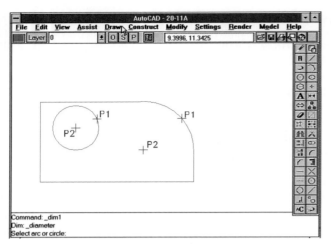
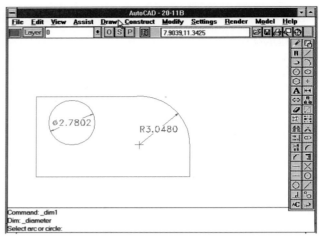

Fig. 20-11. Radial dimensioning includes diameter and radius dimensioning.

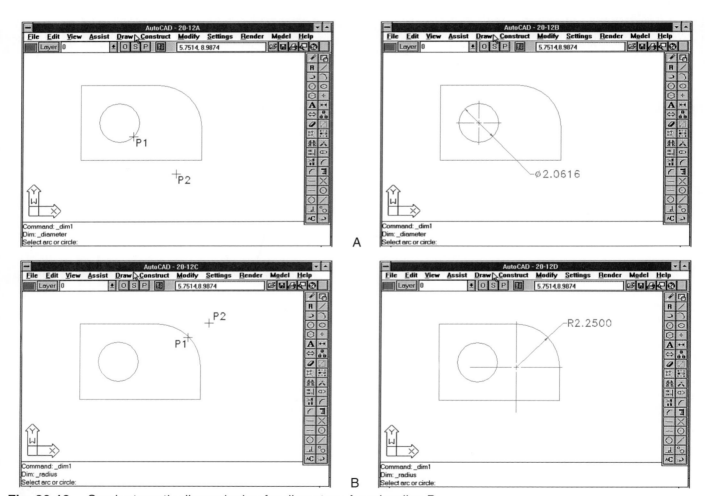

Fig. 20-12. Semiautomatic dimensioning for diameters A and radius B.

CADD systems that use automatic dimensioning determine whether or not to use a leader. These systems check whether the diameter dimension will fit in the circle. If so, the measurement is centered within a dimension line inside the circle and the dimension is complete. If the measurement will not fit, the system asks for the placement length of the leader outside the circle or arc. You must manually enter this information.

Many systems add the proper symbol—diameter (ø) or radius (R)—preceding the dimension. If not, you must add the feature manually. This can be done when the system allows you to edit the dimension before it is placed with the dimension.

Leaders

As previously mentioned, leaders are automatically added when dimensioning arcs and circles. Most CADD systems also offer a separate command to add leaders. After entering the Dimension Leader command, pick two points. The first point marks the feature to which the leader applies. The second point marks the location of the note. Then type in the text. The leader, arrow, and note are then added to the drawing. See Fig. 20-13.

Computer-placed Tolerances

Tolerance values typically can be placed automatically with the dimension value. You must specify whether plus/minus or limit tolerances are used, and the tolerance values. A plus/minus tolerance is appended to the dimension. With a limit tolerance, the computer calculates and shows the upper and lower limits.

HATCHING SECTION VIEWS

One of the most tedious tasks in drafting is adding section lines to a section view. CADD makes drawing section lines quicker by providing automatic hatching (section lining). After selecting the Hatch command, the steps taken to hatch an area include selecting the pattern and picking the entities that make up the boundary. Some CADD programs allow you to pick closed areas, and then automatically fill the area with the desired hatching.

Select the Hatch Pattern

Several standard hatch patterns representing various materials are included with most CADD

programs, Fig. 20-14. You might enter the pattern name, such as ANSI31 (American National Standards Institute 31), or select the pattern from a screen menu. Some systems request that you enter a user-defined pattern. Enter the linetype, angle, and spacing between the section lines.

Define a Hatch Boundary

The *hatch boundary* is the lines, circles, and other entities that border the area to be hatched. Once you select these items, the hatch pattern fills the area within the boundary, Fig. 20-15. Hatching may take several seconds, especially if the pattern is complex or if the area is large.

Entities that form the boundary are chosen individually or by picking corners of a window around them. On most systems, the area must be totally enclosed. In addition, several CADD programs re-

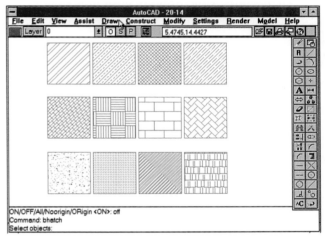

Fig. 20-14. A variety of hatch patterns are available in most CADD programs. Placing hatch patterns with CADD does not require the tedious, time-consuming procedure used with traditional equipment.

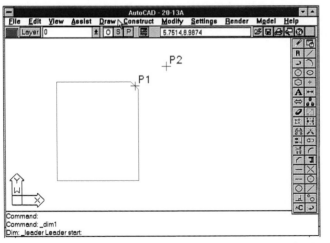

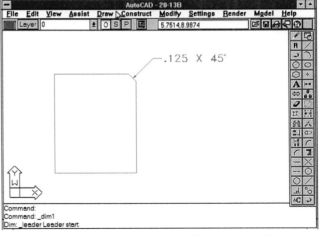

Fig. 20-13. Placing leaders on a drawing. Point P1 indicates the feature to where the leader is to be applied. Point P2 indicates the location of the note.

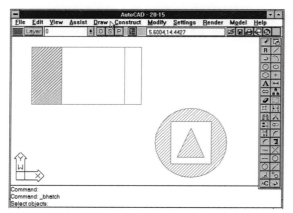

Fig. 20-15. The hatch pattern fills the area within the hatch boundary.

quire that the entities forming the hatch boundary meet, but not extend beyond the boundary area. On these systems, hatching entities which do not form a perfectly enclosed boundary may cause unpredictable results. Edit entities that fall beyond the boundary area using the Break or Trim commands.

INQUIRY COMMANDS

Inquiry commands let you ask questions about your drawing, such as distance, perimeter, area, and drawing status. Unlike manual drafting, with CADD you cannot place a scale on the display screen to measure distance. This is done with a select group of commands.

Entity

The Inquire Entity function lists information about a single object. Access the Inquire Entity function and pick the entity in question. The returned list may include length, diameter, radius, color, linetype, and location.

Distance

To measure distance, select the Inquire Distance function and pick two points on the drawing. These points may be the endpoints of a line or any distance across the drawing area. If you are selecting specific points on an entity, use drawing aids to your advantage.

Angle

To measure an angle formed by two non-intersecting lines, select the Inquire Angle function and pick the lines in question. The angle given indicates the angle between the first and second lines you picked.

Area

Area is the amount of surface enclosed by a circle, ellipse, polygon, or irregular shape composed of several entities. You must pick the entities that totally enclose the shape to be calculated. The Inquire Area command will result in an error message if an attempt is made to measure the area of a region not totally enclosed.

Drawing Status

Selecting the Inquire Status function displays an assortment of information about the current drawing. This might include:

* Drawing boundaries.
* Extent of boundaries used by drawing.
* Portion of drawing shown on display.
* Grid setting.
* Current layer.
* Layers displayed.
* Current color.
* Current linetype.
* Amount of memory left (for microcomputers).
* Amount of data storage space left (for microcomputers).
* Elapsed drawing time.

CREATING AND USING SYMBOL LIBRARIES

Symbols are widely used in industry to represent standard parts or assemblies in diagrams, schematics, and drawings. With manual techniques, symbols are drawn by hand or by using a plastic template. This is not only time-consuming, but the quality may vary and there may not be a template made for a company's unique symbols. With CADD, symbols are drawn once and saved. They can then be inserted as many times as needed in an infinite number of drawings. Some CADD programs come equipped with standard symbols for mechanical, architectural, and electrical drawings.

Inserting a Symbol

As with most CADD functions, a standard procedure must be used to insert symbols. In Fig. 20-16, a bolt symbol is being placed into position. The following steps were used to insert the bolt symbol.

1. Select the active symbol library that contains the symbol you need. A *symbol library* is a special directory on the data storage device

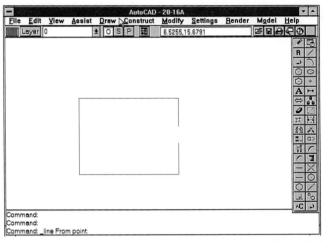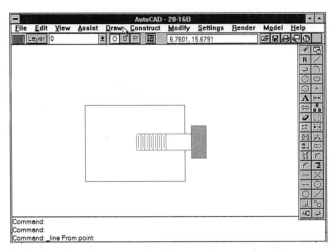

Fig. 20-16. Computer-generated symbols reduce the amount of drawing time required to produce a drawing.

where similar symbol types—mechanical, electrical, architectural, welding, and others—are stored together. See Fig. 20-17. You can choose symbols from only the current library, called the ***active library.***

2. Select the symbol you need. A prompt may ask for the symbol name or number, or a screen menu showing the available symbols could appear. With some CADD systems, you can pick symbols from the tablet menu.

3. Pick the symbol's location on the screen.

4. Enter values for the symbol scale (size factor) and/or rotation angle. These options are not editing commands. They alter only the copy of the symbol being inserted. The functions allow you to resize, rotate, or mirror the symbol before it is added. The default insertion point and size for the symbol remain the same.

Drawing and Storing Symbols

You are not limited to using symbols that come with your system. Creating company-specific symbols for commonly used parts and assemblies may be one of your main tasks. The steps taken to create and store symbols vary slightly among systems. Some CADD programs do not distinguish a drawing from a symbol. (Thus you can insert any existing drawing as a symbol into the current drawing.) However, most systems require that you use a special series of commands that save a group of entities in a symbol library. Fig. 20-18 and the following steps describe the procedure needed to store a symbol.

1. Begin a new drawing and draw the symbol to size.

2. Select the Library Symbol Add command.

3. Enter the name of the symbol library where the symbol is to be stored.

4. Enter the name or number for the symbol. Be as descriptive as possible with your naming convention.

5. Select the entities forming the symbol by picking them individually or by windowing them.

6. Pick an insertion point that determines where the symbol is placed relative to the location point you pick when later inserting the symbol. The insertion point also becomes the reference point for rotating and scaling. Be as accurate as possible when picking the insertion point by using drawing aids to your advantage.

This method of storing symbols has an advantage over systems that use entire drawings as symbols. Suppose that while making a drawing, you find that you need a certain shape many times. Rather than using the Copy command, follow the previous procedure to save the shape as a symbol. Then continue on the drawing, inserting the symbol wherever needed.

Editing Symbols

Symbols are considered to be a single item. Thus, if you select the Erase command and pick the symbol, all entities that make up the symbol are erased. To edit the symbol, you must either:

1. Redraw and resave the symbol. The Replace function (if available) updates all instances of the old symbol with the revised symbol. In addition, all future instances of the symbol reflect the changes made.

2. Select the Explode command to break the symbol into its component parts. This is done when you need to alter a symbol for that particular situation only. Some CADD programs allow you to explode the symbol as it is inserted into the drawing.

Fig. 20-17.   Symbol libraries. (SoftSource, Inc.)

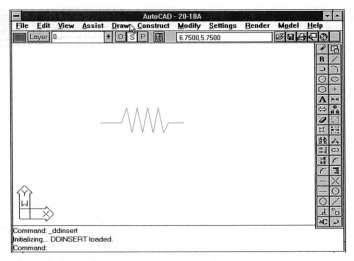

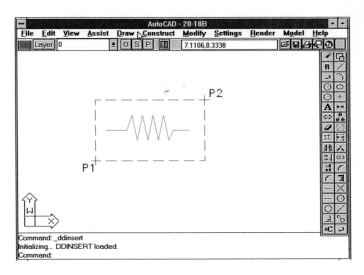

Fig. 20-18. Adding a symbol to the library.

Symbol Attributes

Symbols can be assigned attribute information that can be compiled later when the drawing is complete. Attributes might include part number, name, material, size, weight, cost, and manufacturer. The printed report lists totals for each symbol type and can, therefore, determine the total number of parts or amount of materials required in the design.

PLOTTING DRAWINGS ┼┼┼┼┼┼┼

Plotting is the process of making a hardcopy of a drawing. At any time, a drafter may make a check plot for an update of the drawing status. Check plots are produced by devices that are fast, but often lack quality–such as dot matrix printers, Fig. 20-19A. Final plots are high-quality outputs for reproducing and later distributing the design. Most industries continue to use pen plotters to produce final plots. Pen plotters have high-resolution, quality linework, and the ability to plot on almost any medium, Fig. 20-19B.

The plotting procedure includes entering plot specifications, inserting the paper and proper pens, and selecting the command to begin the plot.

Entering Plot Specifications

Plot specifications are a series of values that specify three things: what part of the drawing is plotted, how it appears on the paper, and to what scale it is plotted. These values include paper size, plot area, plot rotation, plot scale, and pen values. Each of these values is necessary to produce a high-quality plot. Most CADD software is programmed with default specifications that it uses unless you enter new values.

The Size option allows you to select paper sizes supported by your plotter. The Area option determines what part of the drawing to plot. You can plot the entire drawing, pick the corners of a window around the portion to plot, or specify a previously saved view to plot. The Rotation option determines whether the plot is output in a *landscape* (longest paper dimension horizontal) or *portrait* format (longest paper dimension vertical), Fig. 20-20. The Scale option determines the plotted drawing size as a ratio to the actual size of the drawing. This is a very powerful function. It allows you to use full-size dimensions when drawing, yet plot to an exact scale so that the drawing fits on the selected paper size.

The final plot specification–pen values–is needed for pen plotters. Pen specifications determine which pens plot which colored, numbered, or layered entities, and may include pen pressure, pen speed, and pen width. Pen specifications are often shown as a pen table for you to edit, while others simply inform you of the choices you have made, Fig. 20-21.

Plotting the Drawing

Once the specifications are entered, the data can be sent to the plotter. First, prepare the plotter by loading the paper, inserting pens, and readying the plotter. Check the plotter's operating manual for the proper way to align the paper and insert pens into the gripper, rack, or carousel. To ready the plotter, make sure that the command panel on the front of the plotter indicates the remote function. This means that data is received from the computer rather than locally. Local commands allow you to move the paper and pen by pressing buttons on the control panel of the plotter.

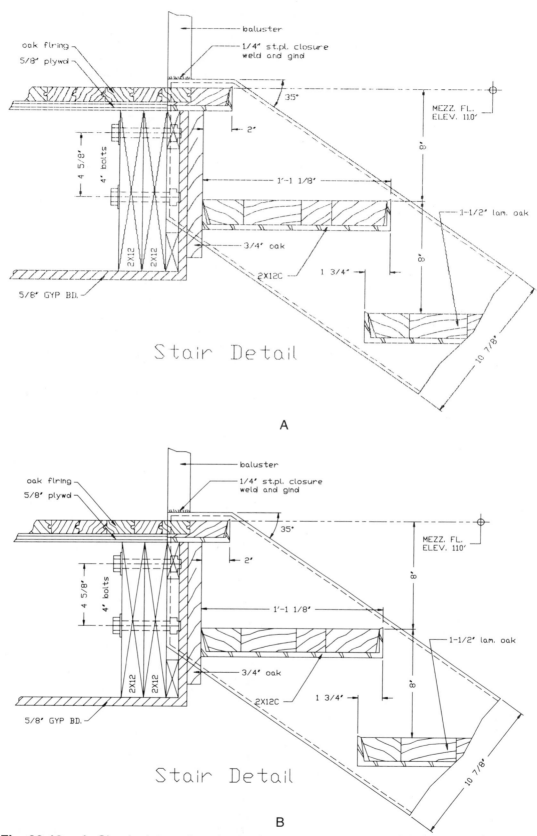

Fig. 20-19.   A—Check plot produced on a dot matrix printer. B—Final plot produced on a pen plotter.

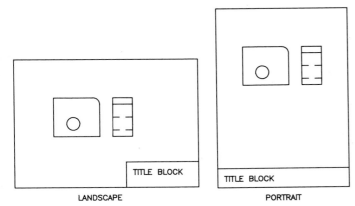

Fig. 20-20. Landscape and portrait formats.

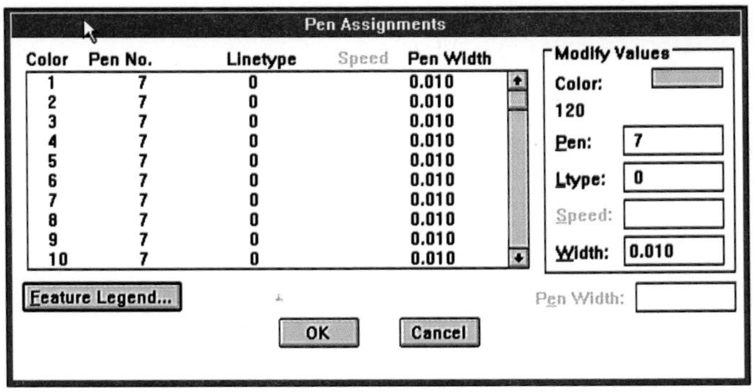

Fig. 20-21. A–Plotter pen specifications. In this example, the pen speed is "grayed out" since the output device is a laser plotter, which does not use pens. B–The feature legend indicates the linetypes used in the drawing.

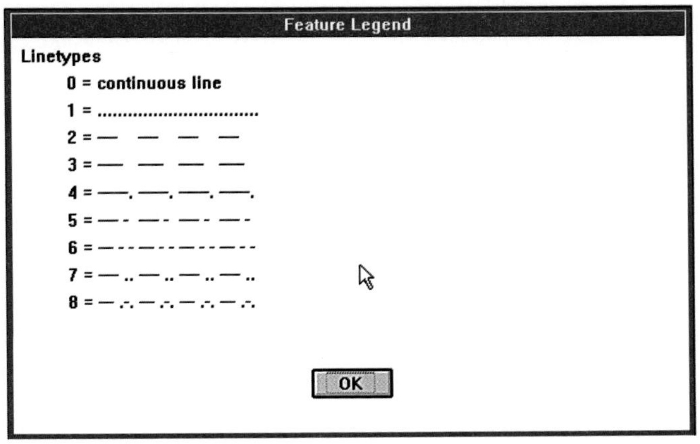

Once the plotter is prepared, select the Plot function to begin sending the drawing data from the computer to the plotter. With single-pen plotters, the plotter to allow you to insert another pen. Multi-pen plotters automatically select pens in numerical order according to the pen specifications previously mentioned. When the plot is finished, remove the paper and tightly recap all pens.

Check Plots

When outputting with a pen plotter, the specifications must be properly set for a high-quality plot. Check plots, on the other hand, generally ignore most of the specifications. Most often, the entire drawing or view shown on the screen is transferred directly to the device.

Plotting Supplies

Plotting supplies include media, pens, adapters, and cleaning items. Carefully consider the compatibility of the media and pen type before you plot. Inks are specially made for film, vellum, gloss bond, and transparency products. Choose pen types based on the quality of plot required and the medium used. Tungsten-tip, steel-tip, and ceramic-tip liquid ink pens are best. Disposable liquid ink, plastic-tip, and fiber-tip pens can also be used, but they do not produce high-quality plots. Note whether an adapter must be screwed to the pen before it is inserted in the plotter. Make test plots of portions of the drawing if you are not sure how the pens will draw on the medium.

Troubleshooting

Plotting problems will arise no matter how carefully you select pens for the medium. Troubles may be caused by worn pens, unacceptable pen speeds, improper pen pressure, or environmental problems. Check your plotter manual for causes and solutions for common pen plotting problems.

PROBLEMS AND ACTIVITIES

Using standard drafting practices and a CADD system, create a drawing for each of the problems in Fig. 20-22. The drawing name should be "P20-(problem number)". For example, name Problem 1 as "P20-1". Set up the drawing using the proper units of measurement. Enter a drawing area or select a sheet size large enough for the drawing. Set grids and use other drawing aids, as necessary, to your advantage. Separate object views, dimensions, and text using different layers. Follow the text and dimensioning standards used to create the problem drawing. Create symbols for geometric tolerancing frames and symbols used on the drawing. Add text with the tolerancing symbols as necessary. After completing the design, plot the entire drawing on the largest paper size supported by your plotter.

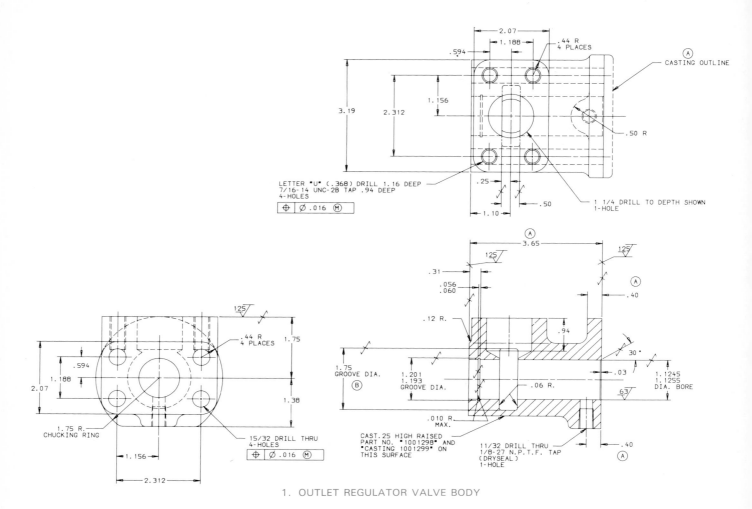

1. OUTLET REGULATOR VALVE BODY

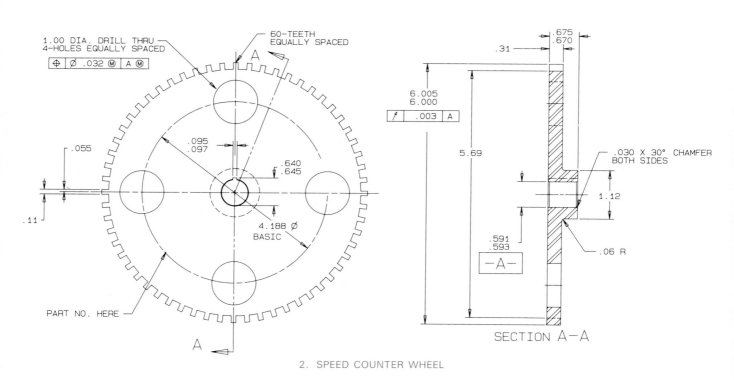

2. SPEED COUNTER WHEEL

Fig. 20-22. Computer-assisted drafting problems. (Continued)

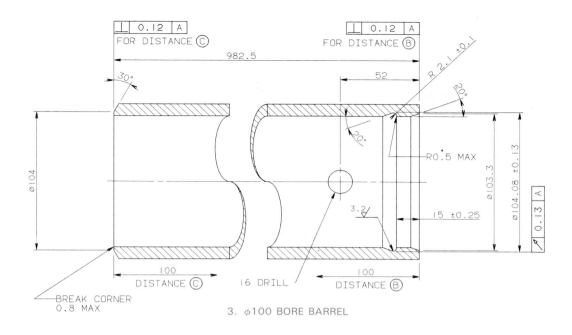

3. φ100 BORE BARREL

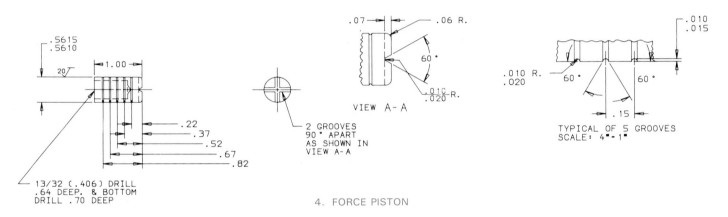

4. FORCE PISTON

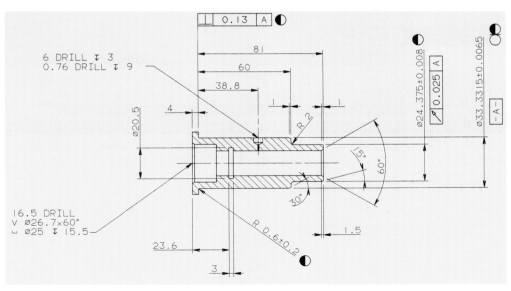

5. IDLER GEAR SHAFT

Fig. 20-22. (Continued) Computer-assisted drafting problems.

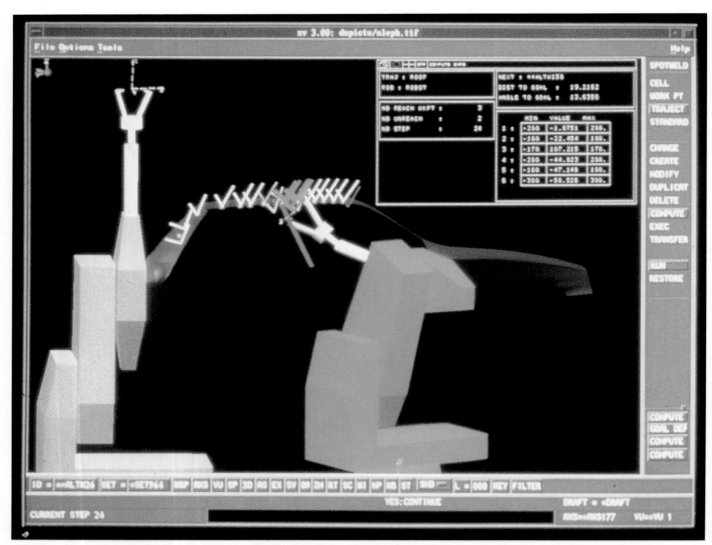

CADD systems not only allow complex objects to be created and manipulated easily, but machine and factory set-ups can be tested on the computer and any modifications made. (IBM)

Part V
Drafting and Design Applications

Drafting and Design Applications builds on the basics developed in preceding chapters to tie the representation of objects (through technical size and shape description) to the manufacturing process. This part of the text covers *Working Drawings; Manufacturing Processes; Linking Design and Manufacturing; Threads and Fastening Devices; Cams, Gears, and Splines,* and *Welding Drawings.*

The Space Shuttle Endeavor climbs into space atop twin pillars of fire from the launch pad at Cape Canaveral, Florida. This accomplishment required the cooperation of thousands of individuals in the design and manufacture of the complicated vehicle and its support systems. (NASA)

Nearly every manufactured part requires a working drawing. CADD greatly reduces the time required to produce these drawings. By linking CADD and CAM, the manufacturing process as a whole is simplified. (Emerson Electronics)

Working Drawings

KEY CONCEPTS

- [ ] A working drawing must contain the information necessary to manufacture, construct, assemble, or install a product or structure.

- [ ] There are a number of types of working drawings. Each type serves a distinct purpose.

- [ ] On all formal industrial drawings, essential information is recorded in the title, materials, and change blocks.

- [ ] National standards govern the amount of information on a drawing and how it is presented. Company or industry standards may modify or expand upon the national standards.

- [ ] The working drawing is the vital link of communication in industry. It makes possible production of individual parts of a machine in widely separated plants.

- [ ] Functional drawing involves producing a drawing that conveys the necessary information with minimum time and effort without sacrificing quality and accuracy.

Working drawings are those that provide all the necessary information to manufacture, con-struct, assemble, or install a machine or structure. Usually, a working drawing is the product of a team effort: engineers or architects, designers, technicians, and drafters all add their special talents to the solution of production problems, Fig. 21-1. Literally thousands of hours go into the preparation of industrial drawings used in modern industrial production.

TYPES OF WORKING DRAWINGS

Working drawings may be divided into a number of subtypes, depending on their use. The first type of working drawing is a freehand sketch; the remainder are instrument drawings to serve various purposes.

Freehand Sketches of Design Prototypes

The engineer-designer will often make a *freehand sketch* on sectional paper that will serve as the working drawing for tooling, for jigs and fixtures, or for test equipment setups, Fig. 21-2. Such drawings are also sometimes used for prototypes, or for experimental or research parts and assemblies.

Fig. 21-1. A typical industrial drafting room where the efforts of many specialists are combined to produce the working drawings essential to manufacturing and construction. (Standard Oil of California)

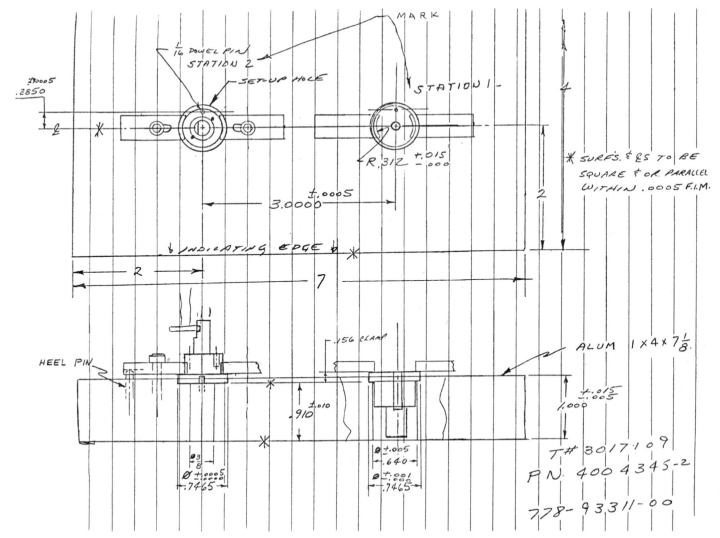

Fig. 21-2. A freehand sketch used as a working drawing in the production of a tooling fixture. (Sperry Flight Systems Div.)

Modifications to the drawing may be made during the construction by the designer or technician. Because of the nature and use of these sketches, only the basic information for part or assembly fabrication is included. It should be emphasized that such working drawings are for limited use; they are not released for general production.

Detail Drawings

A *detail drawing* describes a single part that is to be made from one piece of material. Information is provided through views and by notes, dimensions, tolerances, material, and finish specifications. Also included is any other information needed to fabricate, finish, and inspect the part.

Views of the part usually are in orthographic projection as normal views or sections. Pictorial views may be included for clarification when necessary. An example of a detail working drawing of a machine part is shown in Fig. 21-3. When parts are closely related and space permits, some industries permit detailing of several parts on one detail drawing.

Before beginning on a detail drawing of a part, the drafter should consider the methods by which the part will be processed before it is finally assembled on a machine or product. The drawing should include information sufficient to purchase or make the part and to design the tools used for its manufacture. The drafter must decide how many views will be necessary, and locate those views to allow plenty of space for dimensions and necessary notes.

Tabulated Drawing

A *tabulated drawing* is a type of detailed working drawing that provides information needed to fabricate two or more items which are basically identical but vary in a few characteristics, Fig. 21-4A. These variable characteristics typically involve

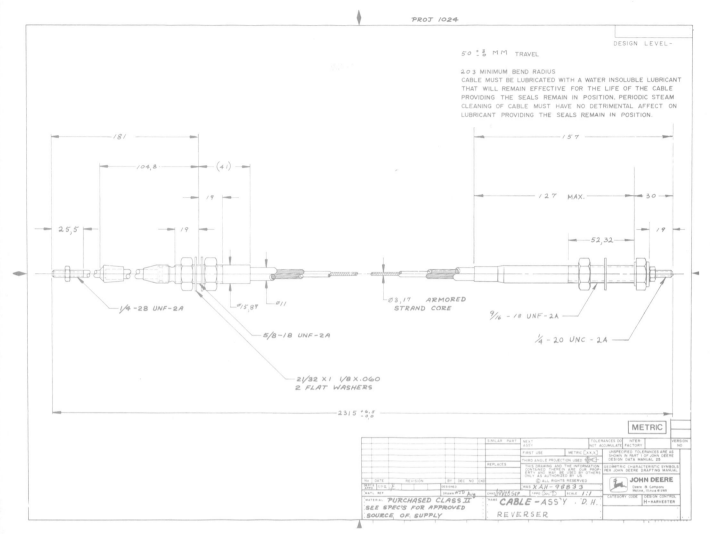

PROJ 1024

DESIGN LEVEL-

50 ±³₋₀ MM TRAVEL

203 MINIMUM BEND RADIUS
CABLE MUST BE LUBRICATED WITH A WATER INSOLUBLE LUBRICANT
THAT WILL REMAIN EFFECTIVE FOR THE LIFE OF THE CABLE
PROVIDING THE SEALS REMAIN IN POSITION. PERIODIC STEAM
CLEANING OF CABLE MUST HAVE NO DETRIMENTAL AFFECT ON
LUBRICANT PROVIDING THE SEALS REMAIN IN POSITION.

181

104,8   (41)

19

25,5   19   157

127 MAX.   30

19

52,32   19

1/4 -28 UNF-2A

Ø15,87   Ø11

Ø3,17 ARMORED
STRAND CORE

9/16 - 18 UNF-2A

5/8 -18 UNF-2A

1/4 - 20 UNC - 2A

21/32 X 1 1/8 X .060
2 FLAT WASHERS

2315 ±⁶,⁵₋₀,₀

METRIC

JOHN DEERE

MATERIAL PURCHASED CLASS II
SEE SPEC'S FOR APPROVED
SOURCE OF SUPPLY

NAME CABLE - ASS'Y . D.H.
REVERSER

Fig. 21-3.   A detail working drawing provides complete information necessary to fabricate, finish, and inspect the part. (John Deere & Company)

dimensions, material, or finish. The ***fixed characteristics*** (those that remain the same for all parts involved) should be detailed only once, either on the body of the drawing, in the material block, or in the tabulation block. ***Variable characteristics***, such as dimensions that change from part to part, are expressed on the drawing with letter symbols. The different values for each symbol are given in the tabulation block. Sizes of stock for the parts are given in a materials list, Fig. 21-4B.

A tabulated drawing eliminates the need to prepare separate drawings of parts that are basically alike.

Assembly Drawings

An ***assembly drawing*** depicts the assembled relationship or positions of two or more detail parts, or of the parts and subassemblies that comprise a unit. As shown in Fig. 21-5, the views of the object are usually orthographic. Isometric or other pictorial views are permissible if needed for clarity. Use only those views, sections, and details necessary to adequately describe the assembly. A list of parts is detailed in tabulated form in the materials block on the drawing, or on a separate sheet, and referenced to the parts in the assembly by numbers.

Assembly drawings should be developed by referring to the detail drawings of the respective parts. This provides an excellent check for fits, clearances, and interferences. Only those operations performed on the assembly in the condition shown should be specified. No detail dimensions should be shown on assembly drawings except to cover operations performed during assembly or to locate detail parts in an adjustable assembly.

A special type of assembly drawing is the ***exploded assembly*** drawing, Fig. 21-6. This type of drawing is most useful when assembling a number

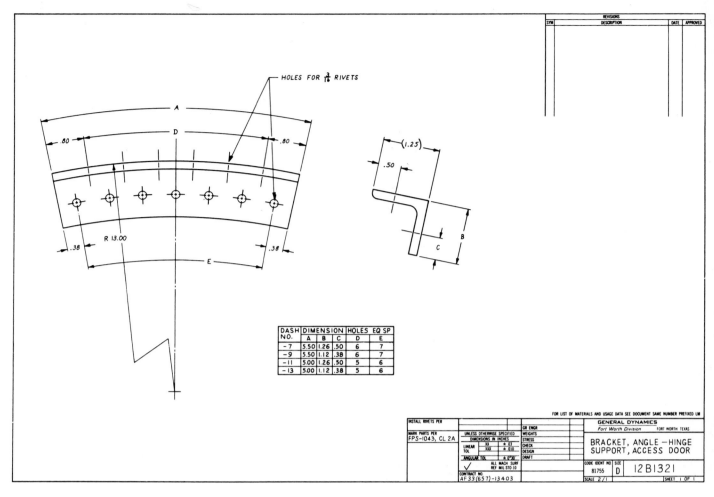

- HOLES FOR 3/16 RIVETS

A
D
.80   .80
R 13.00
.38   .38
E

(1.25)
.50
B
C

DASH NO.	DIMENSION			HOLES EQ SP	
	A	B	C	D	E
-7	5.50	1.26	.50	6	7
-9	5.50	1.12	.38	6	7
-11	5.00	1.26	.50	5	6
-13	5.00	1.12	.38	5	6

FOR LIST OF MATERIALS AND USAGE DATA SEE DOCUMENT SAME NUMBER PREFIXED LM

INSTALL RIVETS PER
MARK PARTS PER FPS-1043, CL 2A

GENERAL DYNAMICS
Fort Worth Division   FORT WORTH, TEXAS
BRACKET, ANGLE—HINGE SUPPORT, ACCESS DOOR
CODE IDENT NO 81755   SIZE D   12B1321
SCALE 2/1   SHEET 1 OF 1
CONTRACT NO. AF 33(657)-13403

A

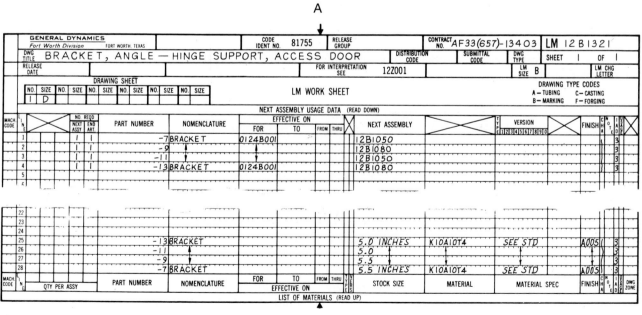

GENERAL DYNAMICS — LM WORK SHEET
BRACKET, ANGLE — HINGE SUPPORT, ACCESS DOOR

MACH CODE			NO. REQD	PART NUMBER	NOMENCLATURE	EFFECTIVE ON				NEXT ASSEMBLY		VERSION		FINISH		
			NEXT ASSY / END ART			FOR	TO	FROM	THRU							
1			1 1	-7	BRACKET	0124B001				12B1050					3	
2			1 1	-9						12B1080					3	
3			1 1	-11						12B1050					3	
4			1 1	-13	BRACKET	0124B001				12B1080					3	

	PART NUMBER	NOMENCLATURE			STOCK SIZE	MATERIAL	MATERIAL SPEC	FINISH	
25	-13	BRACKET			5.0 INCHES	K10A10T4	SEE STD	A005	3
26	-11				5.0				3
27	-9				5.5				3
28	-7	BRACKET			5.5 INCHES	K10A10T4	SEE STD	A005	3

B

Fig. 21-4. A tabulated working drawing is used for the fabrication of parts which are nearly identical. (Convair Aerospace Div., General Dynamics)

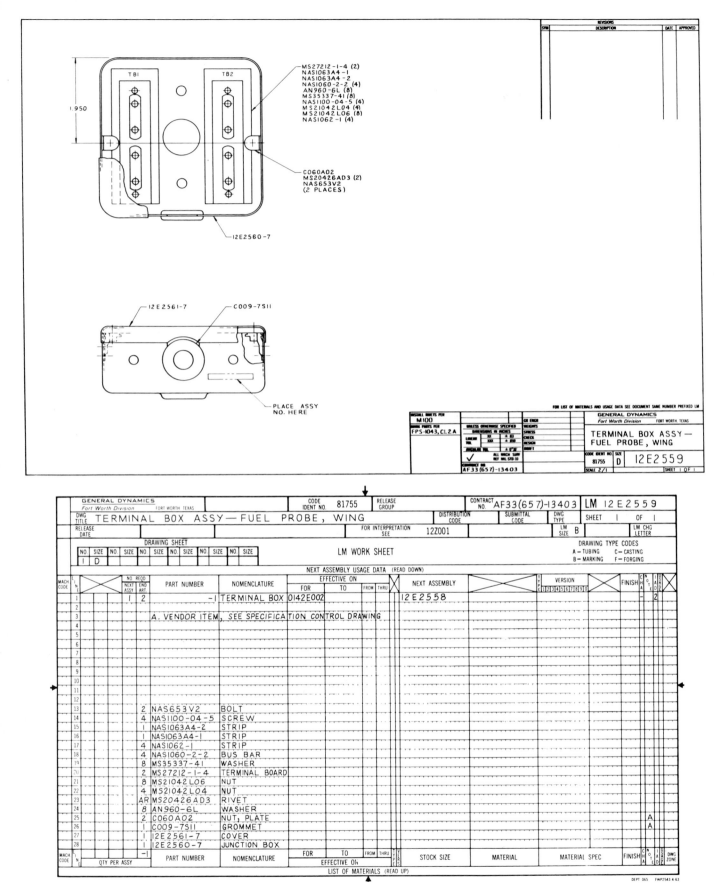

Fig. 21-5. An assembly working drawing and materials list. (Convair Aerospace Div., General Dynamics)

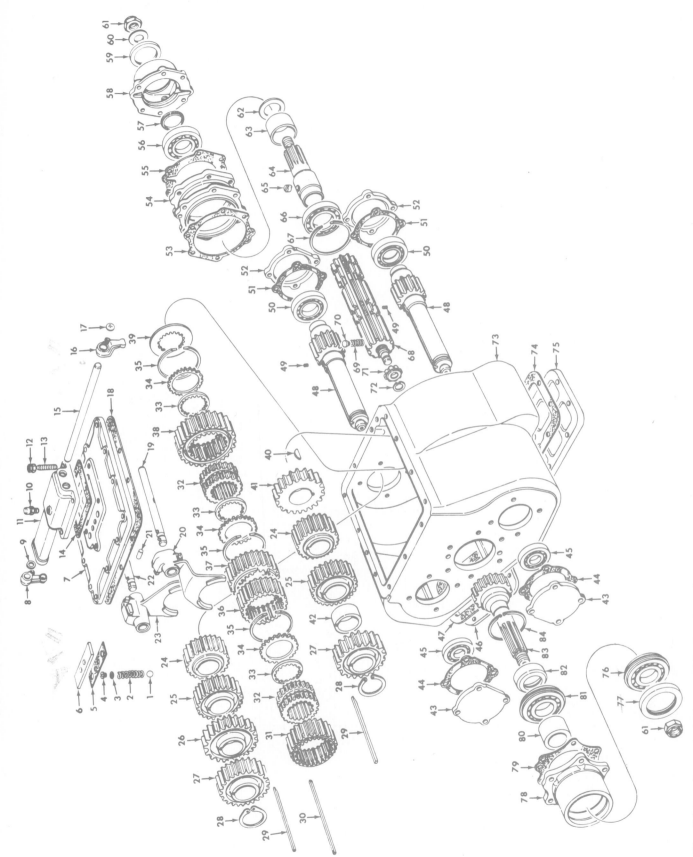

Fig. 21-6.   An exploded assembly drawing of a truck transmission. (International Harvester Co.)

of components. Components are usually drawn in pictorial form, with an axis line showing the sequence of assembly.

Another type of special assembly drawing is the **outline assembly** drawing, Fig. 21-7. These drawings are used for the installation of units and provide overall dimensions to show size and location of points necessary in locating and fastening each unit in place. The outline assembly drawing provides the dimensions essential to making electrical, air, and other connections, as well as the amount of clearance required to operate and service the unit.

Notes may be included to indicate the weight of the unit, the electrical and cooling requirements, and any special notes of caution. These drawings are sometimes termed *installation* drawings.

Process or Operation Drawings

In addition to the types of working drawings already discussed there is another type that concerns production methods. These **process drawings** or **operation drawings** usually provide information for only one step or operation in the making of a part, as shown in Fig. 21-8.

This type of drawing is used by a machine operator when making a particular machine setup and performing single operations, such as drilling a hole or milling a slot. Process or operation drawings usually are accompanied by the machine setup

specifications and specific steps for performing the operation in the machine shop. These drawings are usually prepared by a person knowledgeable in drafting and machining operations.

Layout Drawings

A *layout drawing* is often the original concept for a machine design or for placement of units. It is not a production drawing, but rather serves to record developing design concepts. It is used to obtain approval of a particular design or to check clearances and relationships of component parts, Fig. 21-9. Layout drawings are used by experimental shops when constructing models or prototypes, and by design drafters as a reference when preparing detail drawings of various parts.

Although layout drawings may look like assembly drawings, the purposes of the two are entirely different. Layout drawings are used in the early concept and product design stage; assembly drawings are used near the end for fabrication of the product.

FORMATS FOR WORKING DRAWINGS

All formal industrial drawings include areas in which essential information is recorded in an

DIMENSIONS

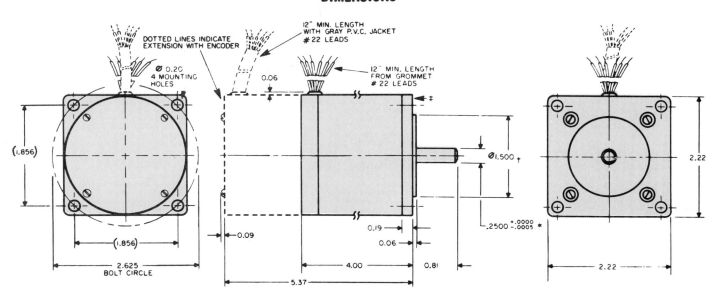

M063 MOTORS

Fig. 21-7. An outline assembly drawing which provides the necessary dimensions for the installation of a motor on a numerical control machine. (Superior Electric Co.)

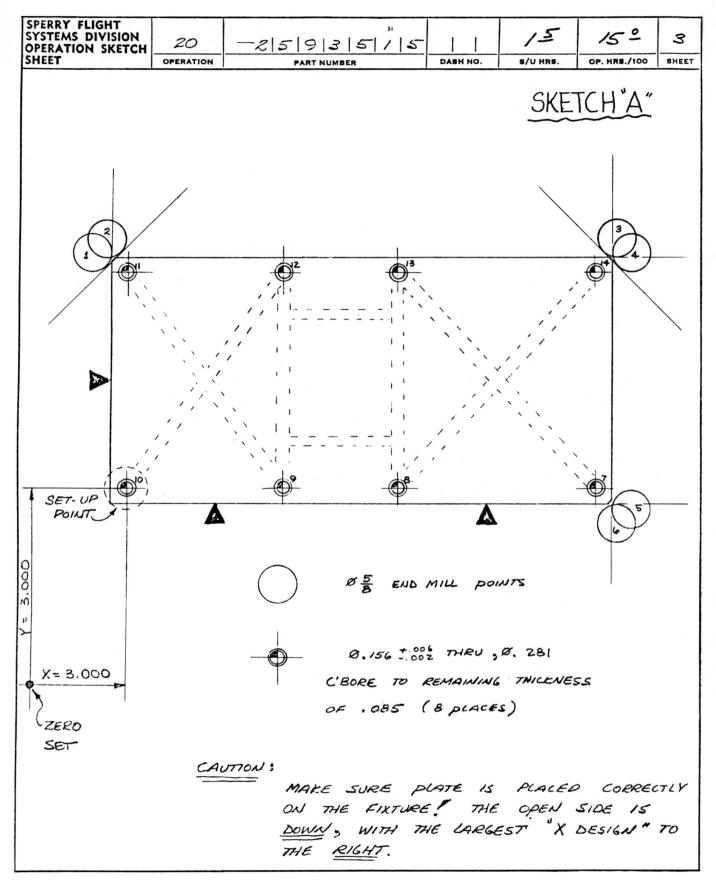

Fig. 21-8. An operation drawing for use with an CNC milling fixture. (Sperry Flight Systems Div.)

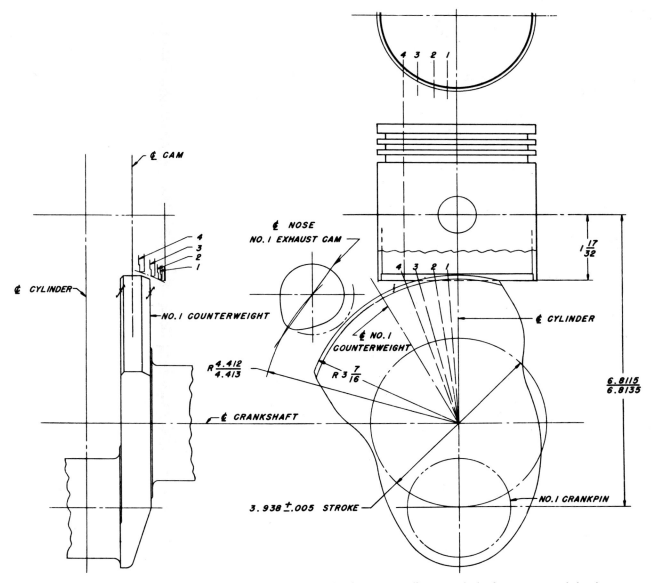

Fig. 21-9. An industry layout drawing to determine the largest radius crankshaft counterweight that can be used and still maintain clearance between the counterweight and the cam and piston. (General Motors Engineering Standards)

organized manner: a title block, a materials block, and a change block. Certain basic information is common to these blocks in nearly all industries, although the style and location of the block on the drawing may vary. The following sections discuss the information usually recorded in these blocks.

Title Block

The *title block,* Fig. 21-10, usually is placed in the lower right-hand corner of the drawing. Included in this block are the title or name of the part or assembly, the drawing number (which is the same as the part number), names of the drafter and checker and signatures of the individuals who are responsible for approvals (engineering, materials,

and production). Other items usually shown in the title block are general tolerances, specifications (material, heat treatment, finish), and an application block that provides additional information.

Materials Block

The *materials block* is a tabular form that usually appears immediately above the title block on assembly and installation drawings, Fig. 21-11. This block is called various names, including *Parts List, List of Materials, Bill of Materials,* or *Schedule of Parts.* It lists the different parts that go into the assembly shown on the drawing, the quantity of each part needed, name or description of the part, and any material specifications. If the part is to be

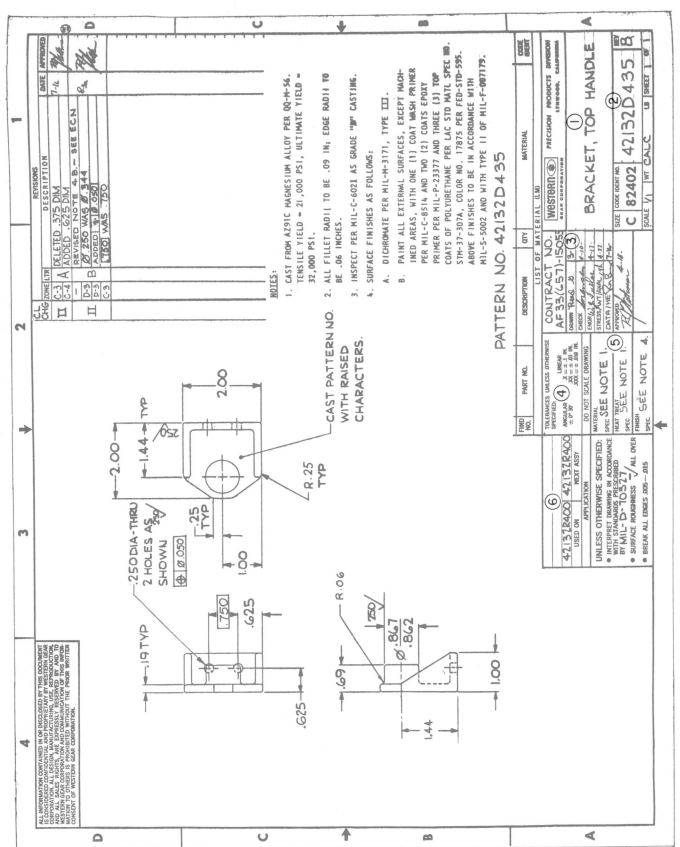

Fig. 21-10.   Elements of a drawing title block are: (1) the title or name of the part or assembly, (2) the drawing number, (3) names of the drafter and checker and approval signatures, (4) general tolerances, (5) specifications information, and (6) application block. (Western Gear Corp.)

REQ'D	REQ'D	ITEM	PART NO.	DESCRIPTION	SPECIFICATION	NOTES
REF		11	600040-1	WIRE LIST (DSTS)		
122		10	220193-1	CLIP, TERMINAL		
AR		9	220226-9	WIRE, ELECTRICAL		
X	1	8	750098-2	ELECTRONIC CMPNT ASSY, CONTROL		
	1	7	803239-1	CKT CARD ASSY, BCD TO DEC CONV		
	1	6	803235-1	CKT CARD ASSY, BCD TO DEC CONV		
	1	5	803231-1	CKT CARD ASSY, BCD TO DEC CONV		
	1	4	803230-1	CKT CARD ASSY, FIXED DATA CONV		
	4	3	MS51957-21	SCREW, MACHINE-PAN HEAD	FF-S-92	
1		2	750044-1	CONNECTOR, PLATE ASSY		
	1	1	850064-1	BRACKET ASSY, CONTROL		
-2	-1	LIST OF PARTS				

Fig. 21-11. A materials block lists all of the parts required to complete the assembly shown on the drawing.

purchased, the supplier should be identified by name or identification code in the materials block.

Change Block

After prints of a drawing have been released to production, it is sometimes necessary to make changes for various reasons: design improvement, production problems, or errors found in the drawing. All changes to the original drawing must be approved by the proper authority.

When a change has been approved and made on the drawing, it is recorded in the ***change block,*** Fig. 21-12. The entry typically consists of a brief description of the change, an identifying letter referencing it to the specific location on the drawing, and the date. The initials or signature of drafter making the change, and those of the person approving the change, are required in most organizations.

LAYOUT OR REF. NO. AM-89303#44, AM-94524#44				
CHK.	DATE	CHANGE	REV.	RELEASE NO.
CP	14OCT	X-6643-4062 DIM 'A' WAS	A	27990 T
46	8, DI	M 'D' WAS 3.4, PROD		
		N° ADDED		
CP	10AUG	WERE C1; B DIMS WERE	B	65030-A
8.0 W/	4" &	8.9 W/5" PIPE & D DIMS WERE		"
		3.0 W/4" & 3.2 W/5" PIPE		"

Fig. 21-12. The change block is a record of changes made to the original drawing. (International Harvester Co.)

The ***change block*** is usually located in the upper right-hand corner of the drawing, and is sometimes titled *Alterations, Notice of Change,* or *Revisions.*

INDUSTRIAL STANDARDS FOR DRAFTING

All industrial drawings should conform to the ***National Standards*** published for the given industry. These standards are available from the American National Standards Institute and are prepared by various industrial and professional organizations. The American Society of Mechanical Engineers has prepared the Y14. series of standards for drafting. Most companies have their own standards, contained in a Drafting Room Manual. These standards conform generally with the National Standards, but are modified or added to so that they meet the needs of that company or a particular industry.

Today, there is a strong trend toward the adoption of certain international standards and symbols (see Reference Section), particularly by those companies engaged in multinational manufacturing and trade activities. "Worldwide sourcing," a method involving assembly of a product from components made in several different countries, has become common. Often, this means that drawings made in one country must be easily read and used to make parts in another.

THE INDUSTRIAL DESIGN AND DRAFTING PROCESS

The production of an industrial drawing begins with an expressed need for something to be produced, Fig. 21-13. This need is given to the design

THE DESIGN DRAFTING PROCESS IN INDUSTRY

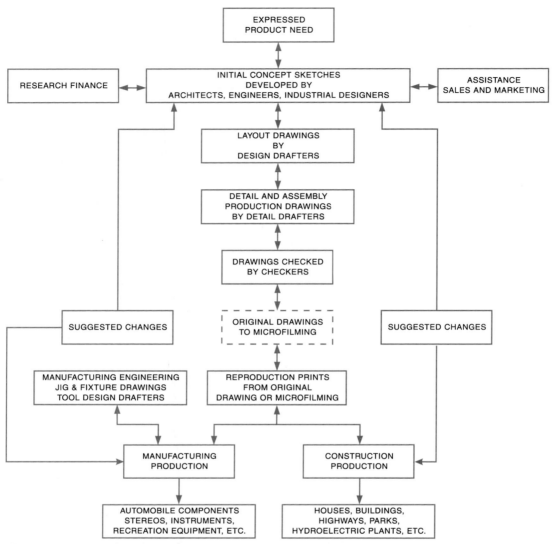

Fig. 21-13.   The industrial drafting process from product conception to production.

department, where the concept is developed and researched. Original designs of the product are sketched or drawn. The designers give the sketches to drafters for further development of the design ideas in layout drawings, or for preparation of detail and assembly drawings. Prints of these original drawings are carefully checked by experienced design drafters for correctness of all details.

After the drawing is approved by design engineers, cost analysts, production engineers, and management, it is released to the reproduction department, where prints are made.

These prints are sent to production, where tool designers prepare drawings for the jigs and fixtures used with the machines that will produce the product. Once prints of the production drawings are in use, drawing changes may be made only by the design department under authority of the project engineer for the product.

APPLICATIONS OF WORKING DRAWINGS

The working drawing is the vital link of communication in industry, making it possible to produce individual parts of a machine in widely separated plants. Some applications of working drawings in major industries are discussed in the following sections.

Aerospace Drafting

Because of the product manufactured, aerospace drawings are perhaps the most elaborate of those of any industry. Precision dimensioning, tolerancing, and rigid specifications characterize these drawings, since many of the parts and components are made by subcontractors and brought together for assembly by the prime contractor.

Literally thousands of detail and assembly working drawings are used in the production of one model of an aerospace vehicle or airplane. Every type of drawing is employed in the aerospace industry, Fig. 21-14.

Automotive Drafting

Probably the largest industrial group in America consists of the automotive manufacturing companies and their subcontractors. It can be argued that more has been done in this industry than any other to perfect the drafting process. As in the aerospace industry, every type of drawing is employed in the production of automobiles and trucks. Both of these industries have produced extensive drafting room procedures to supplement to the National Standards. These drafting procedures are designed to perfect the communication process.

The automotive industry is also one of the leaders in the use of **computer-aided manufacturing (CAM),** Fig. 21-15, which requires special drafting procedures.

Architectural Drafting

Architects, engineers, technicians, and construction workers rely heavily on working drawings in planning and constructing a residential or commercial building. The nature and types of architectural drawings are discussed in Chapter 28. This is a field in which creative design and individuality are expressed in nearly every well-planned structure. To achieve the design solution planned by the architect to meet the desires of the owner, careful working drawings and specifications are necessary.

Electrical-Electronics Drafting

Working drawings in the electrical industry differ from those in other industries, as discussed and illustrated in Chapter 28. Instead of detail and assembly drawings of machine parts, various diagrams and line drawings with numbers of graphic symbols are used. The need for drawings in the electrical/electronics industries has increased due to the more complex circuitry of many household appliances and such electronics applications as computers, process controllers, and aerospace vehicles.

Forging and Casting Drawings

Forging is the forming of heated metal by a hammering or squeezing action, as discussed in Chapter 22. A forged part should be shown on one drawing with the outline drawn in phantom lines,

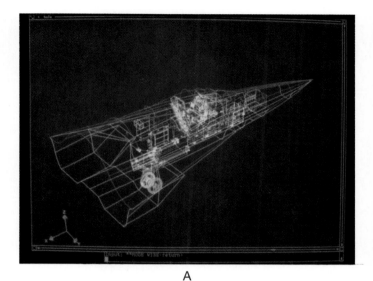

A

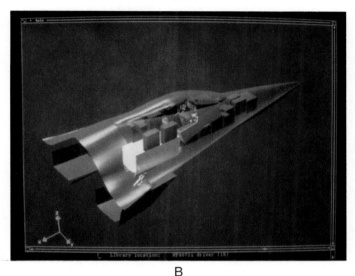

B

C

Fig. 21-14. Space vehicles can be developed entirely on a computer system. A–Initial sketch. C–complete three-dimensional mock-up. The same computer data and commands can then control manufacturing tools. (Lockheed Aircraft Corp.)

Fig. 21-15. A computer-integrated manufacturing system controls movement of these car bodies along the assembly line. To program these robotic welders and other production machines requires a skilled programmer and the use of many detail working drawings. (Cincinnati Milacron)

Fig. 21-16. Unless the amount of finish cannot be controlled by the machining symbol, forging outlines for machining should not be dimensioned, other than a note indicating the allowance. Where the forging is complex, and where the outline of the rough forging must be maintained for tooling purposes, separate dimensioned drawings for the forged and machined piece should be made. Material specification and heat treatment must be called out on the drawing, and the part number indicated and located on the piece.

In the casting drawing, the rough and machined versions of the casting should be combined in the same views on one drawing. The material to be removed by machining is shown in phantom, Fig. 21-17. For complex castings, where the combined rough and machined castings in a single view would cause confusion, make separate dimensioned drawings showing the rough casting and machined piece. The two drawings should be placed on one sheet, when possible. A detail drawing of a casting should give complete information on the following.

- Material specification.
- Hardness specification, if required.
- Machining allowances.

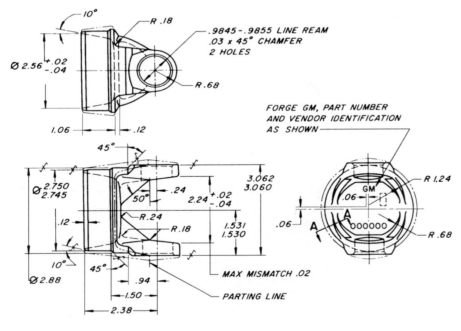

Fig. 21-16. A composite forging drawing with machining dimensions given and the rough forging shown in phantom lines. (General Motors Corp.)

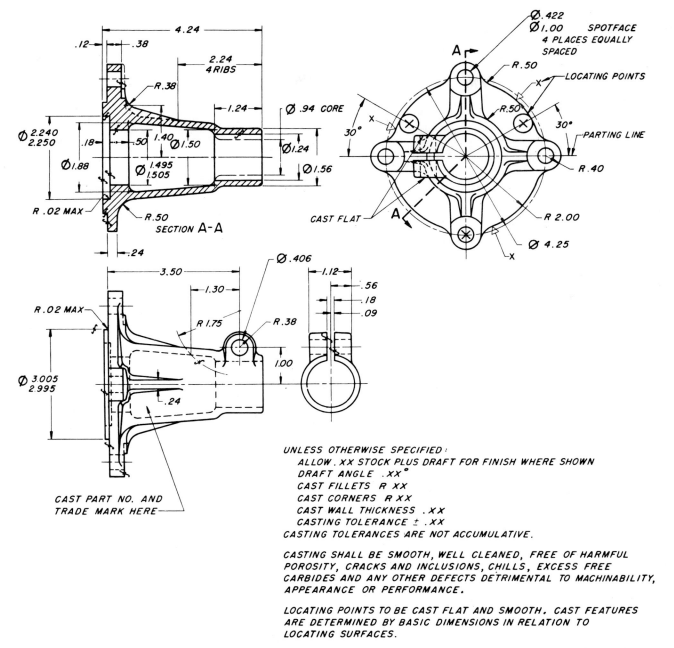

Ø.422
Ø1.00   SPOTFACE
4 PLACES EQUALLY
SPACED

R.50

LOCATING POINTS

R.50

30°

PARTING LINE

R.40

R 2.00

Ø 4.25

CAST FLAT

Ø .94 CORE

Ø 2.240
2.250

Ø1.88

.18

.50 1.40 Ø1.50

Ø 1.495
1.505

R .02 MAX

R.50

SECTION A-A

.24

4.24

.12   .38

2.24
4 RIBS

1.24

R.38

Ø1.24

Ø1.56

3.50

1.30

Ø .406

R 1.75

R .38

R .02 MAX

Ø 3.005
2.995

.24

1.00

1.12

.56

.18

.09

CAST PART NO. AND
TRADE MARK HERE

UNLESS OTHERWISE SPECIFIED:
  ALLOW .XX STOCK PLUS DRAFT FOR FINISH WHERE SHOWN
  DRAFT ANGLE .XX°
  CAST FILLETS R XX
  CAST CORNERS R XX
  CAST WALL THICKNESS .XX
  CASTING TOLERANCE ± .XX
CASTING TOLERANCES ARE NOT ACCUMULATIVE.

CASTING SHALL BE SMOOTH, WELL CLEANED, FREE OF HARMFUL
POROSITY, CRACKS AND INCLUSIONS, CHILLS, EXCESS FREE
CARBIDES AND ANY OTHER DEFECTS DETRIMENTAL TO MACHINABILITY,
APPEARANCE OR PERFORMANCE.

LOCATING POINTS TO BE CAST FLAT AND SMOOTH. CAST FEATURES
ARE DETERMINED BY BASIC DIMENSIONS IN RELATION TO
LOCATING SURFACES.

Fig. 21-17. A composite casting drawing with finished machining dimensions given and locating points identified. (General Motors Corp.)

- Kind of finish.
- Draft angles.
- Limits on draft surfaces that must be controlled.
- Locating points for checking the casting.
- Parting line.
- Part number and trademark.

Piping Drawings

Piping drawings are a type of assembly drawing that represent the piping layout using symbols and either double-line or single-line drawings. To represent the various fittings, either pictorial or graphic symbols may be used. The usual practice is to show piping layouts as single-line drawings in isometric or oblique, because of the difficulty of reading orthographic projection views. A typical piping layout is shown in Fig. 21-18.

Structural Steel Drawings

Structural steel drawings are of two types: design drawings, and shop ("working") drawings. Design drawings show the overall design and dimensions of the structure and specify sizes and

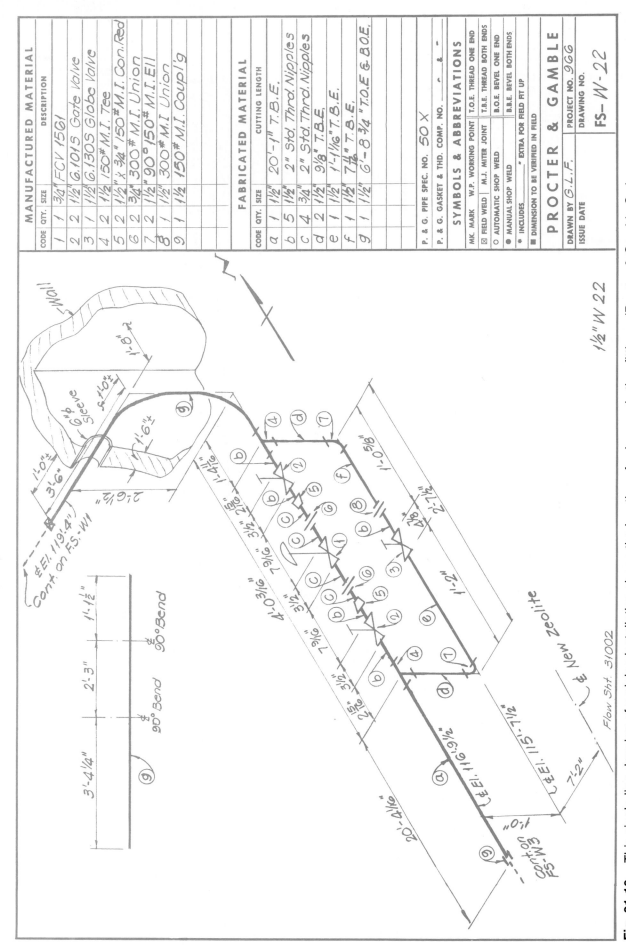

Fig. 21-18. This single-line drawing of a piping installation shows the location of valves and other fittings. (Proctor & Gamble Co.)

types of material used. They are prepared by structural engineers. Working drawings for the actual fabrication of steel members are prepared by the fabricator under the direction and approval of the design engineer. Working drawings that are sent to the job site for erection purposes must detail connections to be made in the field.

Units shipped to the job site from the fabricator's facility are called **subassemblies,** Fig. 21-19. These are fastened together on the job by rivets or bolts or are welded in place, Fig. 21-20.

Reinforced Concrete Drawings

Reinforced concrete construction is achieved by placing steel rods or beams strategically in the forms and pouring the concrete around these reinforcements, Fig. 21-21. This type of construction is very strong structurally and also is fire-resistant.

Drawings for reinforced concrete construction must show the size and location of the reinforcing steel, as well as the dimensions and shapes of the concrete members, Fig. 21-22. The rods are repre-

Fig. 21-21. Steel reinforcing rods, like those exposed here during a bridge deck repair project, are embedded in concrete to provide strength. In addition to paving use, reinforced concrete is widely used for structures. (Jack Klasey)

sented by long dashes in the elevation view and as darkened circles in the sectioned view. The American Concrete Institute publishes an approved *Manual of Standard Practice for Detailing Reinforced Concrete Structures* which is helpful in preparing drawings for reinforced concrete structures, such as buildings, bridges, and walls.

Welding Drawings

The **welding drawing** is a type of assembly drawing that shows the components of an assembly in position to be welded, rather than as separate parts. Specification of the type of welds to be used on various joints has become standard procedure on welding drawings. A series of welding symbols has been prepared by the American Welding Society. These symbols and the procedures used to prepare welding drawings are discussed in Chapter 26.

Fig. 21-19. These large steel intake tubes for a hydroelectric generating plant were fabricated as a subassembly in the shop and shipped to the job site.

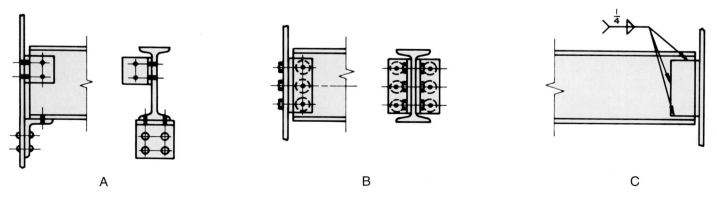

A B C

Fig. 21-20. Methods of illustrating steel beam connections. A–Rivets. B–Bolts. C–Welds.

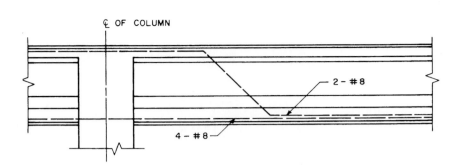
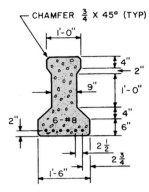

Fig. 21-22.   A detail of a reinforced concrete member showing size and location of steel reinforcing rods.

Patent Drawings

When a patent is sought for a machine or other device which lends itself to illustration by a drawing, the Patent Office requires the applicant to furnish the drawing. Very specific instructions on properly preparing patent drawings are available from the Superintendent of Documents, Washington, DC. These instructions include the type of paper and ink, drawing size, line density, and other requirements, such as the manner in which the drawings are to be sent to the Patent Office.

Patent drawings follow the American Drafting Standards, but the views do not have to appear in a strict orthographic projection relation to one another or even be placed on the same sheet. Isometric or other pictorial views are acceptable if they clearly represent the object for which a patent is desired.

FUNCTIONAL DRAFTING TECHNIQUES

Functional drafting may be defined as making a drawing that includes just those lines, views, symbols, notes, and dimensions needed to completely clarify the construction of an object or part. Application of this technique should not reduce the accuracy of the drawing or the quality of lettering and line delineation.

Every industry continually seeks ways to improve drafting communications. This effort arises from the amount of drafting time involved in planning and development of most industrial projects, and the time spent in reading and interpreting drawings. Functional drafting techniques have successfully reduced drafting time, and thus have been welcomed in most industries.

The following functional drafting techniques have been discussed in other sections of this text as *standard drafting practices.* They are listed here as a summary of standard practices.

- Use the minimum number of views.

- Use a partial view, where adequate.
- Eliminate a view whenever a thickness note will suffice.
- Use symmetry to reduce drawing time.
- Eliminate superfluous detail; use schematic or simplified forms whenever possible.
- Omit drawing of standard parts such as bolts, nuts, and rivets. Locate them conventionally or by a note, and list them in the materials block.
- Avoid unnecessary repetition of detail.
- Omit cross hatching, except where clarity demands its use; use outline sectioning.
- Use standard graphic symbols (such as piping and welding symbols) whenever possible.
- Use templates to draw ellipses, circles, and symbols.
- Use mechanical lead holders, pencil pointers, and thin-lead holders as timesavers.
- Photograph usable sections of drawings to be reworked and start with this.
- Use photo-drafting of models, electronic circuits, and other assemblies to which notes can be added to speed the drafting process.
- Use adhesive-backed appliques for information used repeatedly on drawings such as change blocks and gear data blocks.
- Use tabulated semi-complete drawings of common-shaped items which require only the addition of dimensions and/or specifications.
- Use a general tolerance note in title block for indicating tolerances on drawing when possible.
- Avoid freehand lettering of parts lists and other material that can be typed on adhesive-backed appliques.

PROBLEMS AND ACTIVITIES

The problems in this chapter are to be drawn and should include all dimensions, specifications, and notes required to release the drawing to production.

Detail Working Drawings

1. Make detail working drawings of those parts in shown in Fig. 21-23 as assigned by your instructor. Change all dimensions to decimal limit dimensions with the following tolerances: .XXX = ± .003; .XX = ± .010. Delete all unnecessary dimensions; indicate all flat surfaces as 125 microinches (3.2 micrometers) and all bored and counterbored holes as 63 microinches (1.6 micrometers) in texture.

2. Prepare a three-view detail working drawing of the "BODY PITOT OVERRIDE" shown in Fig. 21-24. Draw one view as a section to clarify interior detail. Dimension the drawing using geometric dimensioning and tolerancing. Delete the notes replaced by geometric dimensioning symbols and add the remaining necessary notes.

3. Draw two views (one of them a section) of the "HYDRAULIC DECHUCK PISTON" shown in Fig. 21-24. Change the number of equally spaced holes from 24 to 18 and dimension for CNC machining. See the Reference Section for the chart used to calculate equally spaced locations. Dimension the drawing, using geometric feature control symbols.

Precision Sheet Metal Drawings

4. Refer to Fig. 21-25 and Fig. 21-26. Calculate the flat pattern size. Make flat pattern dimensioned drawings of those parts shown.

Assembly Working Drawings

5. Make a detail working drawing of the "ROLLER FOR BRICK ELEVATOR" shown in Fig. 21-27. Add bilateral tolerances to those parts requiring fits, as indicated in the notes.

6. Make an exploded assembly drawing of the "ROLLER FOR BRICK ELEVATOR" shown in Fig. 21-27.

7. Make detail drawings of the component parts and an assembly drawing of the "WHEEL, IMPELLER" shown in Fig. 21-28. Include all notes and material list.

8. Prepare detail working drawings of the various parts of the "STEERING QUADRANT ASSEMBLY" shown in Fig. 21-29. Add material and revision items to the individual detail drawings, either as callouts or notes. The drawings may be placed on one sheet or on separate sheets.

9. Make an assembly drawing of the "STEERING QUADRANT ASSEMBLY" shown in Fig. 21-29. Add material and revision blocks and callouts to your drawing.

10. Make detail drawings of component parts (except standard bolts, nuts, and washers) and an assembly drawing of the "WIRE STRAIGHTENER" shown in Fig. 21-30 and Fig. 21-31. Include all notes and a material list.

Design Problems

The following are suggested problems that can be solved by the design method that was discussed in Chapter 1. With the approval of your instructor, you may select any of these problems, or one of your own for which you desire a solution. Prepare detail and assembly working drawings of the solutions.

- Stereo component cabinets.
- A quick-action, easy-to-operate automobile tire jack.
- A water safety device useful when hunting or fishing.
- A caddy to help organize paper clips, rubber bands, pens, pencils, notes, and other items used at your desk.
- A storage device for your sporting equipment.
- A jig or fixture for holding a workpiece in a machine tool.
- A special tool or clamp which combines the function of two or more separate tools.

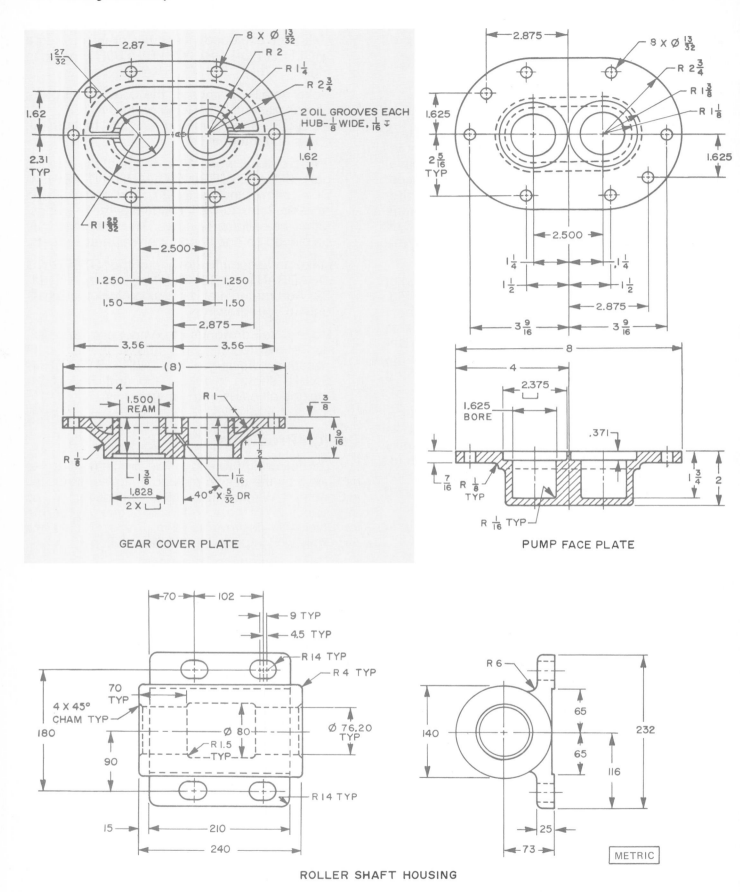

GEAR COVER PLATE

PUMP FACE PLATE

ROLLER SHAFT HOUSING

METRIC

Fig. 21-23.   Parts for which detail working drawings are to be prepared.

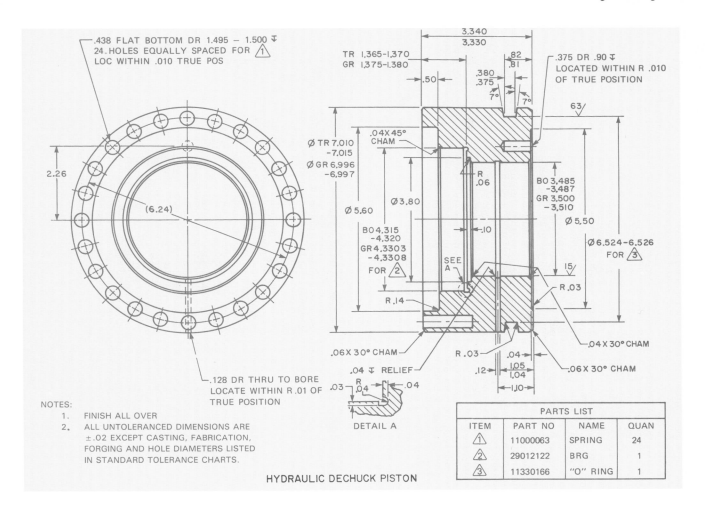

NOTES:
1. FINISH ALL OVER
2. ALL UNTOLERANCED DIMENSIONS ARE
±.02 EXCEPT CASTING, FABRICATION,
FORGING AND HOLE DIAMETERS LISTED
IN STANDARD TOLERANCE CHARTS.

DETAIL A

HYDRAULIC DECHUCK PISTON

ITEM	PART NO	NAME	QUAN
1	11000063	SPRING	24
2	29012122	BRG	1
3	11330166	"O" RING	1

PARTS LIST

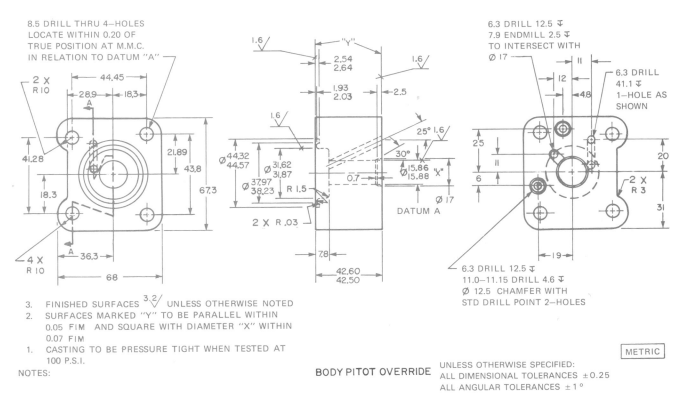

3. FINISHED SURFACES ³·²∕ UNLESS OTHERWISE NOTED
2. SURFACES MARKED "Y" TO BE PARALLEL WITHIN
0.05 FIM AND SQUARE WITH DIAMETER "X" WITHIN
0.07 FIM
1. CASTING TO BE PRESSURE TIGHT WHEN TESTED AT
100 P.S.I.

NOTES:

BODY PITOT OVERRIDE

UNLESS OTHERWISE SPECIFIED:
ALL DIMENSIONAL TOLERANCES ±0.25
ALL ANGULAR TOLERANCES ±1°

METRIC

Fig. 21-24. Parts for which detail working drawings are to be prepared.

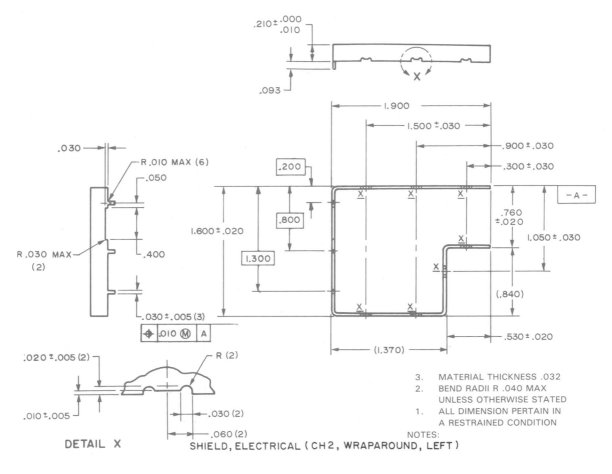

DETAIL X

SHIELD, ELECTRICAL (CH2, WRAPAROUND, LEFT)

3. MATERIAL THICKNESS .032
2. BEND RADII R .040 MAX UNLESS OTHERWISE STATED
1. ALL DIMENSION PERTAIN IN A RESTRAINED CONDITION
NOTES:

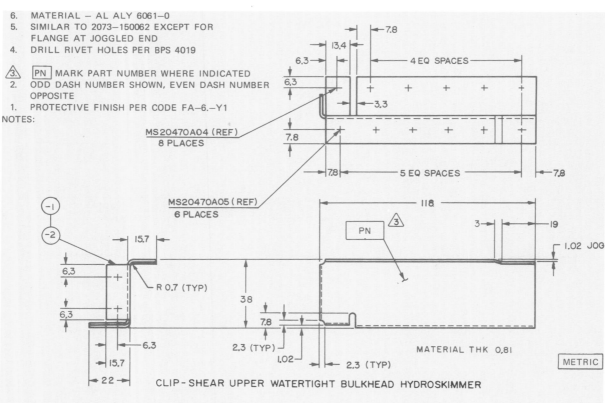

6. MATERIAL – AL ALY 6061–0
5. SIMILAR TO 2073–150062 EXCEPT FOR FLANGE AT JOGGLED END
4. DRILL RIVET HOLES PER BPS 4019
3. PN MARK PART NUMBER WHERE INDICATED
2. ODD DASH NUMBER SHOWN, EVEN DASH NUMBER OPPOSITE
1. PROTECTIVE FINISH PER CODE FA–6.–Y1
NOTES:

CLIP – SHEAR UPPER WATERTIGHT BULKHEAD HYDROSKIMMER

Fig. 21-25. Precision sheet metal parts for which flat pattern dimensioned drawings are to be prepared.

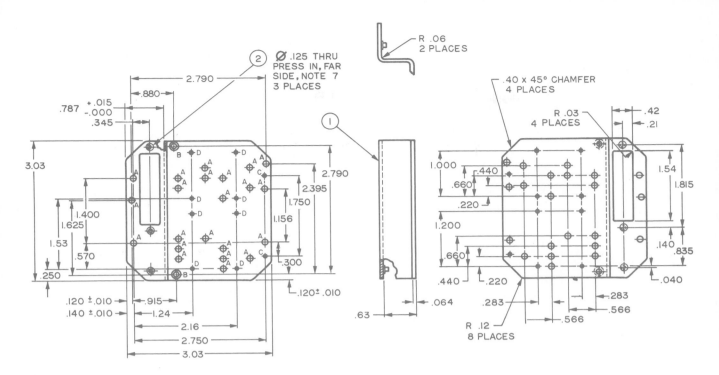

3	X	MIL-C-5541, GR C, CL 1	M690278-1	CHEMICAL FILM TRTMT
2			1752101-11	NUT, SELF LOCKING
1	1		4005071-2	CHASSIS
ITEM NO.	QTY REQD		STANDARD	NAME OR DESCRIPTION
LIST OF MATERIALS				

HOLE TABLE		
HOLE CODE	HOLE SIZE	QTY REQD
A	Ø .125 THRU	22
B	Ø .125 THRU 100° CSK TO Ø .204	2
C	Ø .096 $^{+.004}_{-.001}$ THRU	2
D	Ø .116 $^{+.004}_{-.001}$ THRU	8

NOTES:
1. INTERPRET DRAWING PER ANSI Y 14.5M—1982
2. REMOVE BURRS AND SHARP EDGES, UNLESS OTHERWISE SPECIFIED
3. FINISH ALL OVER 125
4. DIMENSIONS, TOLERANCES AND SURFACE FINISH VALUES APPLY BEFORE THE APPLICATION OF THE FINISH
5. DO NOT APPLY PIECE MARK
6. FINISH AS FOLLOWS:
    6.1   CHEMICAL FILM TREATMENT, ITEM 1 ONLY, PER ITEM 3
7. DO NOT CHAMFER OR BREAK ENTRANCE EDGE OF MOUNTING HOLE RECEIVING INSERT. PRESS IN UNDER STEADY LOAD UNTIL UNDERSIDE OF SHOULDER IS FLUSH WITHIN .005 OF ADJACENT SURFACE
8. TOLERANCES, UNLESS OTHERWISE SPECIFIED:
    2 PLACE DECIMALS = ±.02
    3 PLACE DECIMALS = ±.005
    ANGLES = ±2°
9. MATERIAL THICKNESS .064

Fig. 21-26. Precision sheet metal part for which a flat pattern dimensioned drawing is to be prepared.

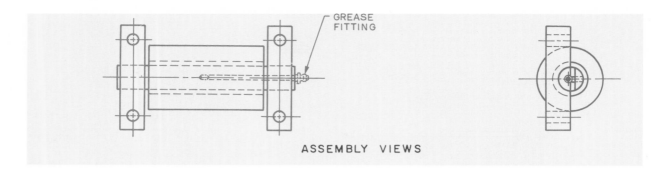

ASSEMBLY VIEWS

GREASE FITTING

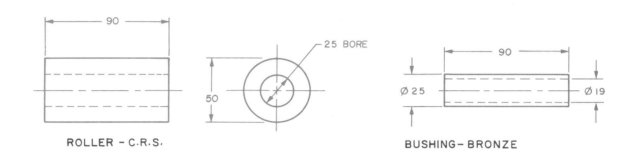

90

25 BORE

50

ROLLER – C.R.S.

90

Ø 25   Ø 19

BUSHING– BRONZE

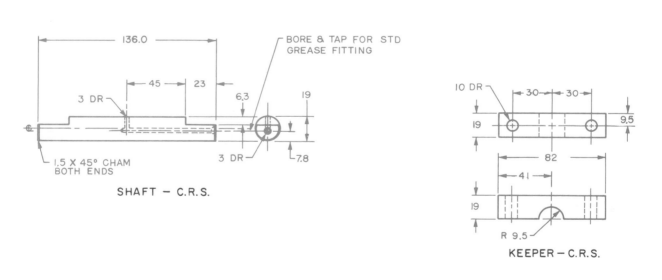

SHAFT – C.R.S.

136.0

45   23

3 DR

1.5 X 45° CHAM BOTH ENDS

6.3   19

3 DR   7.8

BORE & TAP FOR STD GREASE FITTING

KEEPER – C.R.S.

10 DR   30   30

19   9.5

82

41

19

R 9.5

NOTES:
2.   FINISH 125/ ALL OVER.
1.   BUSHING TO BE A LIGHT DRIVE FIT
     IN ROLLER & RUNNING FIT ON SHAFT

METRIC

ROLLER FOR BRICK ELEVATOR

Fig. 21-27.   A subassembly for which an assembly drawing and an exploded assembly drawing are to be prepared.

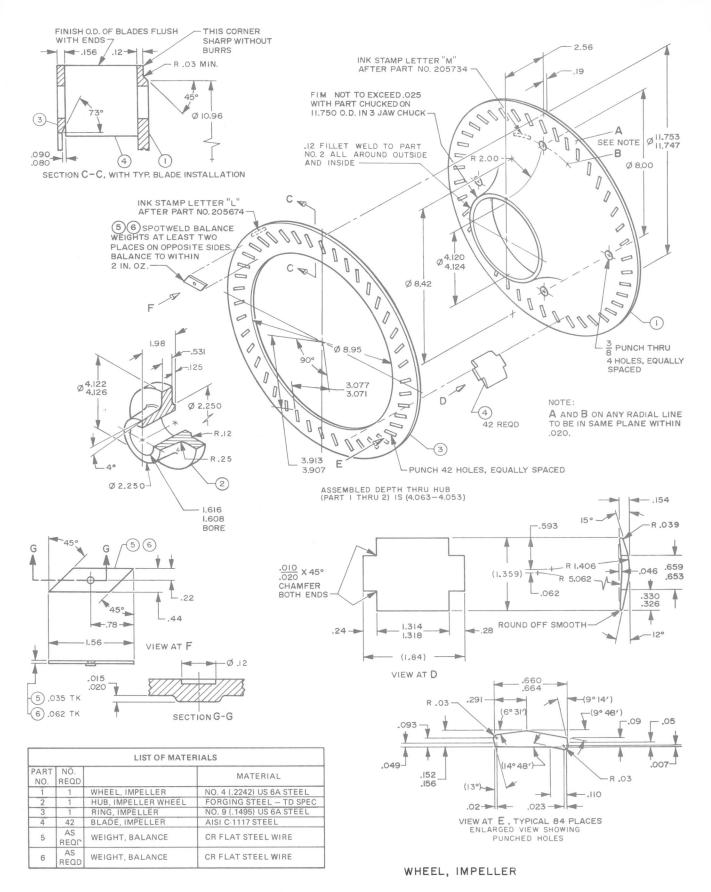

FINISH O.D. OF BLADES FLUSH WITH ENDS
THIS CORNER SHARP WITHOUT BURRS
R .03 MIN.
.156 .12
45°
73°
Ø 10.96
.090 .080
③ ④ ①
SECTION C–C, WITH TYP. BLADE INSTALLATION

INK STAMP LETTER "M" AFTER PART NO. 205734
FIM NOT TO EXCEED .025 WITH PART CHUCKED ON 11.750 O.D. IN 3 JAW CHUCK
.12 FILLET WELD TO PART NO. 2 ALL AROUND OUTSIDE AND INSIDE
2.56
.19
A SEE NOTE
B
Ø 11.753 / 11.747
Ø 8.00
R 2.00
Ø 4.120 / 4.124
Ø 8.42
①
3/8 PUNCH THRU 4 HOLES, EQUALLY SPACED

INK STAMP LETTER "L" AFTER PART NO. 205674
⑤⑥ SPOTWELD BALANCE WEIGHTS AT LEAST TWO PLACES ON OPPOSITE SIDES. BALANCE TO WITHIN 2 IN. OZ.
C
C
F
Ø 8.95
90°
3.077 / 3.071
D
④ 42 REQD

NOTE: A AND B ON ANY RADIAL LINE TO BE IN SAME PLANE WITHIN .020.

1.98
.531
.125
Ø 4.122 / 4.126
Ø 2.250
R .12
R .25
4°
Ø 2.250
1.616 / 1.608 BORE

3.913 / 3.907
E
③
PUNCH 42 HOLES, EQUALLY SPACED

ASSEMBLED DEPTH THRU HUB (PART 1 THRU 2) IS (4.063–4.053)

G
G
45°
45°
⑤ ⑥
.22
.78
.44
1.56
VIEW AT F
⑤ .035 TK
⑥ .062 TK
.015 .020
Ø .12
SECTION G–G

.010 / .020 X 45° CHAMFER BOTH ENDS
.593
15°
R .039
R 1.406
R 5.062
.046
.659 / .653
.330 / .326
12°
(1.359)
.062
.24
1.314 / 1.318
.28
ROUND OFF SMOOTH
(1.84)
VIEW AT D

.660 / .664
.291
(9° 14')
(6° 31')
(9° 48')
R .03
.09
.05
.093
(14° 48')
.007
.049
.152 / .156
R .03
.110
(13°)
.02
.023
VIEW AT E , TYPICAL 84 PLACES ENLARGED VIEW SHOWING PUNCHED HOLES

WHEEL, IMPELLER

LIST OF MATERIALS			
PART NO.	NO. REQD		MATERIAL
1	1	WHEEL, IMPELLER	NO. 4 (.2242) US 6A STEEL
2	1	HUB, IMPELLER WHEEL	FORGING STEEL – TD SPEC
3	1	RING, IMPELLER	NO. 9 (.1495) US 6A STEEL
4	42	BLADE, IMPELLER	AISI C-1117 STEEL
5	AS REQD	WEIGHT, BALANCE	CR FLAT STEEL WIRE
6	AS REQD	WEIGHT, BALANCE	CR FLAT STEEL WIRE

Fig. 21-28. Exploded assembly view of "IMPELLER WHEEL."

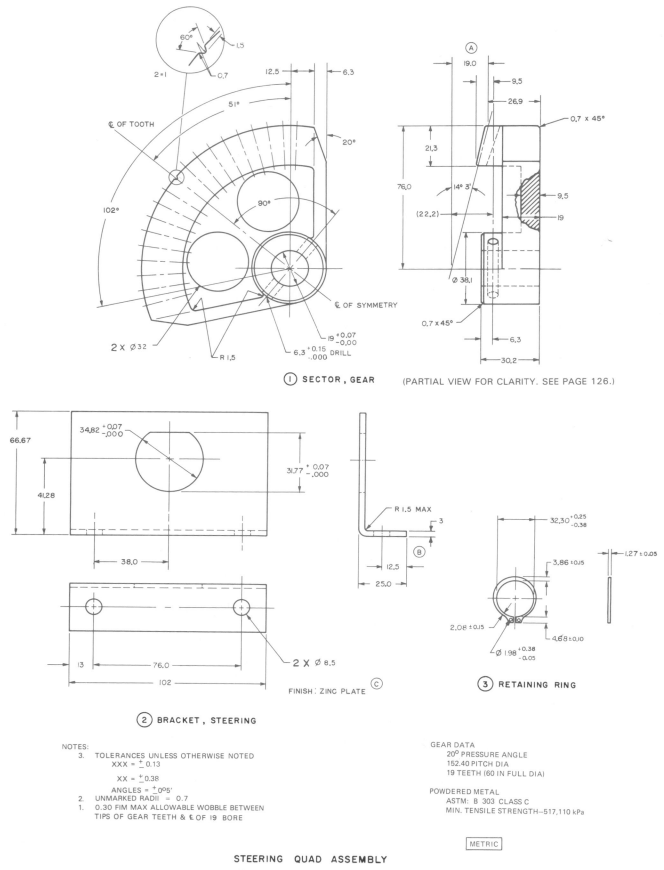

① SECTOR, GEAR   (PARTIAL VIEW FOR CLARITY. SEE PAGE 126.)

② BRACKET, STEERING

③ RETAINING RING

FINISH : ZINC PLATE ©

NOTES:
3. TOLERANCES UNLESS OTHERWISE NOTED
    XXX = ± 0.13
    XX = ± 0.38
    ANGLES = ± 0°5'
2. UNMARKED RADII = 0.7
1. 0.30 FIM MAX ALLOWABLE WOBBLE BETWEEN
    TIPS OF GEAR TEETH & ₵ OF 19 BORE

GEAR DATA
    20° PRESSURE ANGLE
    152.40 PITCH DIA
    19 TEETH (60 IN FULL DIA)

POWDERED METAL
    ASTM : B 303 CLASS C
    MIN. TENSILE STRENGTH—517,110 kPa

METRIC

STEERING QUAD ASSEMBLY

Fig. 21-29. An assembly for which detail working drawings and an assembly drawing are to be prepared.

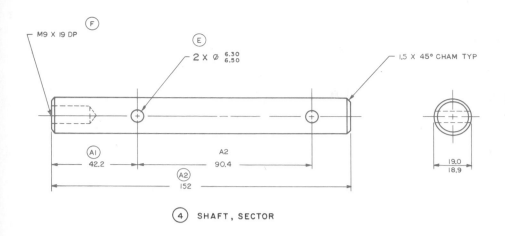

M9 X 19 DP

2 X Ø 6.30 / 6.50

1.5 X 45° CHAM TYP

A1 42.2
A2 90.4
A2 152

19.0 / 18.9

④ SHAFT, SECTOR

REVISIONS			
LTR	DESCRIPTION	DATE	APPD
A	DIM CORRECTED		
B	WAS 11		
C	WAS PAINT		
D	MATL SPEC ADDED		
E	WAS 6.30 DIA		
F	2B ADDED		

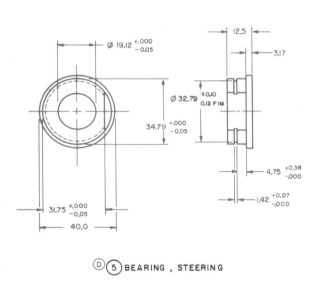

Ø 19.12 +.000 / -.05

12.5

3.17

Ø 32.79

±0.10 0.12 FIM

34.79 +.000 / -.05

4.75 +0.38 / -.000

1.42 +0.07 / -.000

31.75 +.000 / -.05

40.0

Ⓓ⑤ BEARING , STEERING

60.5

37°

Ø 25

R 25

Ø 95 +.015 / -.000

R 17.5

19.0 +0.07 / -.000 REAM

6.30 / 6.50 DRILL AFTER WELDING

30.0

24

12.5

31.7

FINISH: ZINC PLATE ©

⑥ STEERING ARM ASSY, SECTOR SHAFT

METRIC

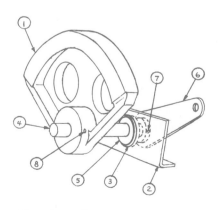

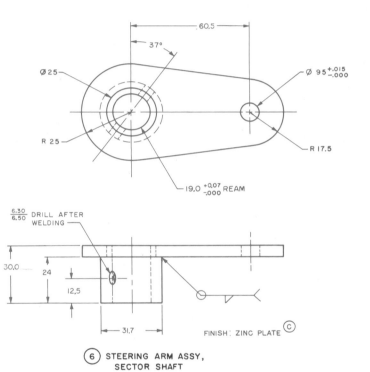

8	1	456722	ROLL PIN (PURCHASE)
7	1	454565	ROLL PIN (PURCHASE)
6	1	39563	ARM, 19 HRS; HUB C1018 STEEL
5	1	39580	ASTM B–202–60T, TYPE 2 CLASS B SINTERED IRON, OIL IMPREGNATED
4	1	39564	19 O.D. C1018 CRS
3	1	39492	TRU–ARC NO. 5100-137
2	1	40200	11GA H.R.P & O
1	1	41088	POWERED METAL ASTM: B 303 CLASS C
ITEM	REQD	PART NO.	MATERIAL
LIST OF PARTS			

Fig. 21-29. (Continued)

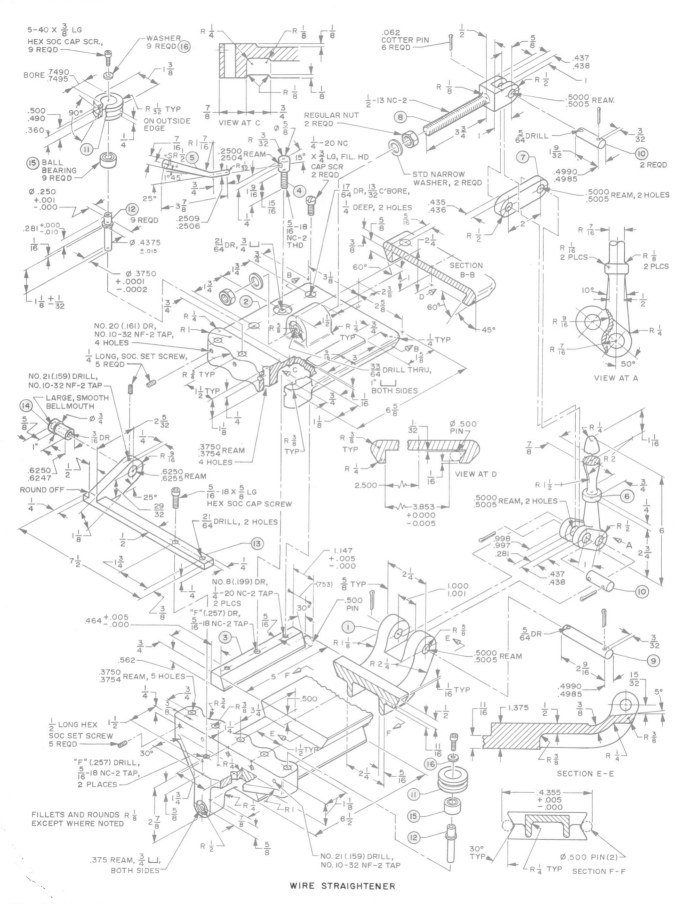

WIRE STRAIGHTENER

Fig. 21-30. Exploded assembly view of "WIRE STRAIGHTENER."

MATERIAL LIST			
PART NO.	NO. REQD.	PART NAME	MATERIAL
1	1	LOWER ADJUSTER BRACKET	CAST IRON
2	1	LEFT ADJUSTER SLIDE BRACKET	CAST IRON
3	1	GIB FOR SLIDE	AISI C-1018 CRS
4	1	CLAMP SCREW FOR SLIDE GIB	AISI C-1018 CRS
5	1	HANDLE FOR SLIDE GIB CLAMP SCREW	AISI C-1018 CRS
6	1	WIRE SET CLAMP HANDLE	AISI C-1018 CRS
7	1	LINK FOR HANDLE	AISI C-1018 CRS
8	1	ADJUSTING FORK SCREW	AISI C-1018 CRS
9	1	PIVOT PIN	AISI C-1018 CRS
10	2	PIVOT PIN	AISI C-1018 CRS
11	9	WIRE STRAIGHTENER ROLLER	CARBON HDN
12	9	WIRE STRAIGHTENER ROLLER SHAFT	CARBON HDN
13	1	STOCK GUIDE BUSHING HOLDER	CARBON HDN
14	1	STOCK GUIDE BUSHING	SAE W1 TOOL STL HDN 60-62R_c
15	9	BALL BEARING	NEW DEPARTURE
16	9	WASHER	MILD STEEL

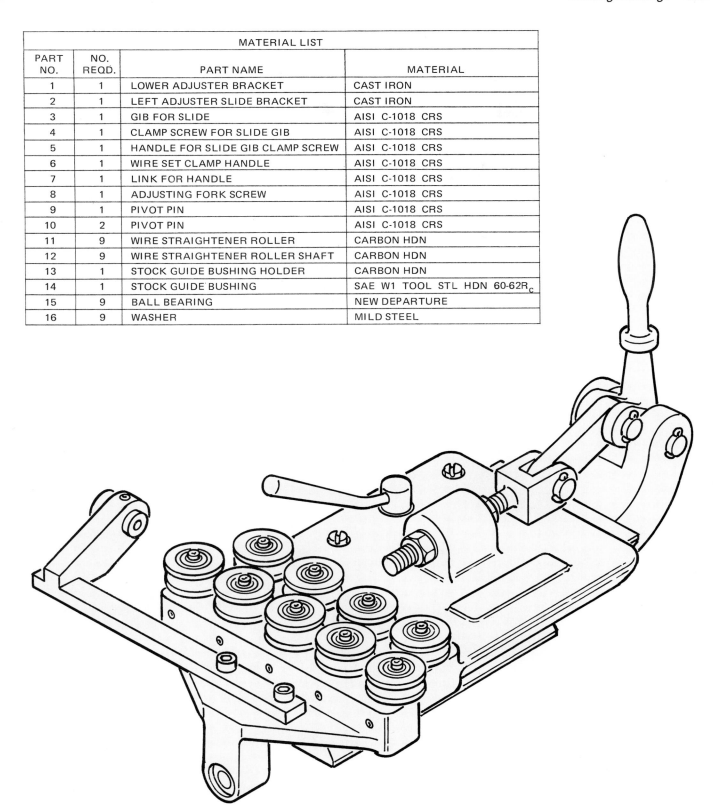

Fig. 21-31. Pictorial view and material list for the "WIRE STRAIGHTENER."

Robots are an integral part of many modern manufacturing facillities. (Garrett)

Manufacturing Processes

KEY CONCEPTS

- ☐ Features to be machined a certain way must have the machine process specified on the drawing.
- ☐ Computer numerical control machining involves programming instructions for tool movement and machine operations.
- ☐ Absolute positioning involves specification of all distances and directions from a zero point.
- ☐ Incremental positioning is a system in which distances and directions are specified from a previous position, rather than a fixed zero point.
- ☐ The American Surface Texture symbol conveys information about the roughness, waviness, and lay of a finished surface.
- ☐ In precision sheet metal work, the drafter must determine the bend allowance to properly dimension a flat pattern drawing.
- ☐ The drafter should be aware of heat treating processes and of forming processes that shape metal by means other than cutting action.
- ☐ Nontraditional metalworking processes, such as electrical discharge machining, chemical milling, and high energy rate forming, are necessary to machine the tougher alloys developed in recent years as a result of the space program.
- ☐ Plastics have become a significant industrial material, and often are worked with processes that are closely similar to those used for metals.

An important aspect of the work done by the drafter and designer in preparing a drawing is specifying the features on a machine part. This is done by notes and/or symbols known as *callouts.* This chapter presents most of the frequently used manufacturing processes.

MACHINE PROCESSES

A feature on a drawing, such as a hole, may simply be dimensioned by giving its diameter, Fig. 22-1A. However, for those features that are to be machined in a certain way (to produce a desired surface texture or to hold a certain tolerance), the machine process must be specified. Following are common machine processes and their callouts.

Drilling

Drilled holes are usually produced by a drill bit chucked in a drill press or portable power drill (depending on the nature of piece and accuracy required). Note that the specification of the drilled hole may be entirely by a callout or by a callout and a dimension for depth on the feature, Fig. 22-1B.

Spotfacing

Spotfacing is a cutting process used to clean up or level the surface around a hole to provide a bearing for a bolt head or nut, Fig. 22-2A. A spotface may be specified by note only and need not be shown on the drawing.

Counterboring

Counterboring involves cutting deeper than spotfacing to allow fillister and socket head screws to be seated below the surface. See Fig. 22-2B.

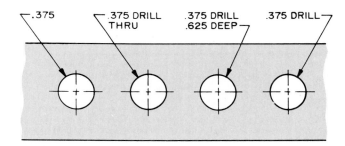

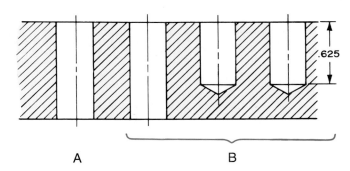

Fig. 22-1. Methods of representing and dimensioning drilled holes.

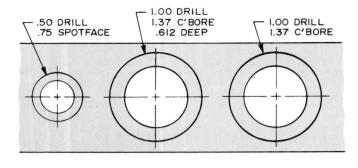

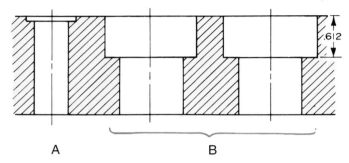

Fig. 22-2. Spotfaced and counterbored holes as represented and dimensioned on drawings.

Counterdrilling

Smaller holes may be partially drilled with a larger drill *(counterdrilling)* to allow room for a fastener or feature of a mating part, as shown in Fig. 22-3B.

Boring

When an extremely accurate hole with a smooth surface texture is required, *boring* is usually specified as the machine process, Fig. 22-4A. This may be done on a lathe or on a boring mill.

Reaming

After a hole is drilled, it may be reamed for greater accuracy and a smoother surface texture, Fig. 22-4B. The hole is drilled slightly undersize, and then reamed to the desired diameter. A reaming tool can only be used on an existing hole.

Broaching

Broaching is the process of pulling or pushing a tool over or through the workpiece to form irregular or unusual shapes. The broach is a long tapered tool with cutting teeth that get progressively larger so that at the completion of a single stroke, the work is finished.

A very simple type of broach is used to cut a keyway on the inside of a pulley or gear hub. A more complex broach is used to cut an internal spline. See Fig. 22-5.

Countersinking

Countersinking is done by cutting a beveled edge (chamfer) in a hole so that a flat head screw will seat flush with the surface, Fig. 22-3A. The outside diameter of the countersunk feature on the surface of the part and the angle of the countersink are dimensioned.

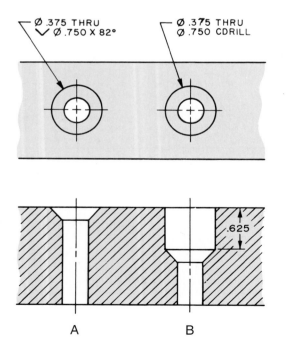

Fig. 22-3. Dimensioning of countersunk and counterdrilled holes.

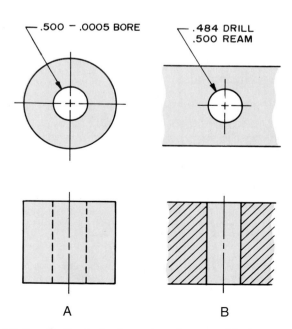

Fig. 22-4. Callouts for bored and reamed holes.

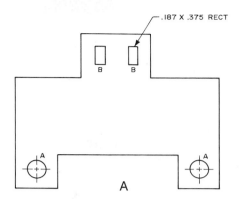

.187 X .375 RECT

A

DESCRIPTION OF HOLES		
SIZE	DESCRIPTION	QTY
A	φ .375 THRU	2
B	.187 x .375 RECT	2

B

Fig. 22-7. Blanking features (whether regular or irregular in shape) is common practice on light gauge metals. Features may be: dimensioned directly, A–placed in a callout, or B–specified in a dimension table.

Fig. 22-5. The internal splines of a pinion are being broached on this 40-ton horizontal broaching machine. (Milwaukee Gear)

Stamping

Circular or irregular holes and other features are often prepared by using dies in a punch press, Fig. 22-6. Stamping operations include perforation, blanking, shearing, bending, and forming. These are normally used on sheet metals. The features may be dimensioned directly, placed in a callout, or specified in a tabular form on the drawing. See Fig. 22-7.

Knurling

Knurling is the process of forming straight-line or diagonal-line (diamond) serrations on a part to provide a better hand grip or interference fit. See Fig. 22-8A. The diametral pitch (DP) type, grade (coarse, medium, and fine), and length of knurl should be specified, Fig. 22-8B. The knurled surface may be fully or partially drawn, as shown in Fig. 22-8C. It may

be omitted from the drawing entirely, since the callout provides a clear description.

Necking and Undercutting

It is sometimes necessary on machine parts to provide a groove on a shaft to terminate a thread (a *neck*) or to cut a recess at a point where the shaft changes size and mating parts such as a pulley must fit flush against a shoulder (an *undercut*). These necks and undercuts are specified as noted in Fig. 22-9A. When too small to detail on the part itself, they should be drawn as an enlarged detail. See Fig. 22-9B and Fig. 22-9C.

Chamfering

Small bevels are usually cut on the ends of holes, shafts, and threaded fasteners to facilitate assembly. These bevels, or *chamfers* are dimensioned as shown in Fig. 22-10. When the chamfer angle is 45°, the dimension should be noted as in Fig. 22-10A. The use of the word "chamfer" is optional. Angles other than 45° must be included in the dimension note, Fig. 22-10B. Internal chamfers are dimensioned as shown in Fig. 22-10C.

Tapers, Conical and Flat

A cone-shaped section of a shaft or a hole is called a *conical taper,* Fig. 22-11. Standard machine

Fig. 22-6. Large mechanical presses, like this 75-ton model, are used to punch and form sheet metal. (Rockwell International Corp.)

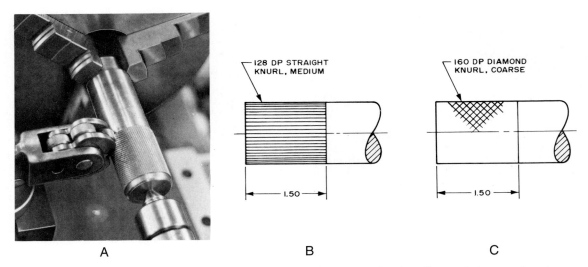

Fig. 22-8.   A—Knurling operation. B & C—Methods of representing knurling patterns on drawings.

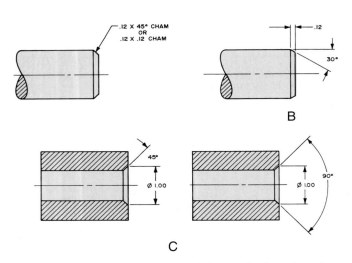

Fig. 22-9.   Specifying necks and undercuts on drawings. Sometimes, an enlarged detail drawing is necessary for clarity.

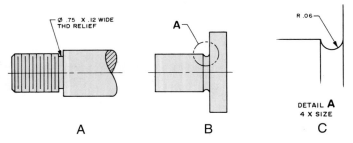

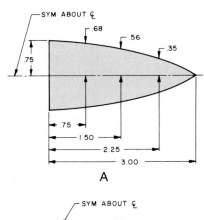

A

B

Fig. 22-11.   Method used to dimension tapers. (American National Standards Institute)

Fig. 22-10.   Representing and dimensioning chamfers on a drawing. Note that angles must be given.

Grinding, Honing, and Lapping

The process of removing metal by means of abrasives is known as ***grinding.*** This is usually a finishing operation, as shown in Fig. 22-12A. Some actual shaping of parts is also done by grinding, which is then referred to as ***abrasive machining,*** Fig. 22-12B. Wheels used for most grinding operations come in a variety of sizes, shapes, and abrasive coarseness grades.

To produce a finished surface, grinding may be done on a surface grinder for flat work, or on a horizontal spindle machine for precision tool and die work. For internal or external grinding of cylindrical parts, a lathe or cylindrical grinder is used,

tapers are used on various machine tool spindles, with mating tapers on the drill bits and tool shanks that fit into them.

A ***flat taper*** increases or decreases in size at a uniform rate to assume a wedge shape, similar to a door stop.

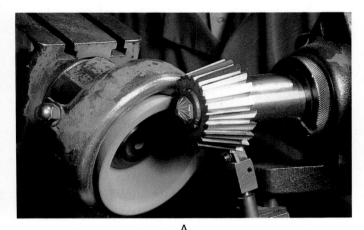

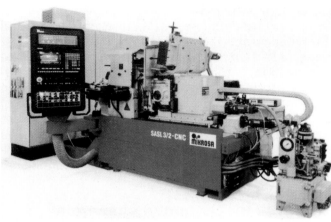

Fig. 22-13. This automatic centerless external cylindrical grinder is used for high-speed production grinding of cylindrical workpieces up to nearly 4 inches in diameter. It operates under computer numerical control. (Mikrosa)

Fig. 22-12. A–Finish grinding is used to produce a smooth surface on these gear teeth. B–Abrasive machining, involves actual shaping of parts with the use of grinders. (Norton Co.)

Fig. 22-13. The surface texture of a machine part is usually produced with a grinding operation, particularly finer finishes.

Honing is done with blocks of very fine abrasive materials under light pressure against the work surface (such as inside of a cylinder). They are rotated rather slowly and are moved backwards and laterally. *Lapping* is quite similar to honing, except a lapping plate or block is used with a very fine paste or liquid abrasive between the metal lap and work surface.

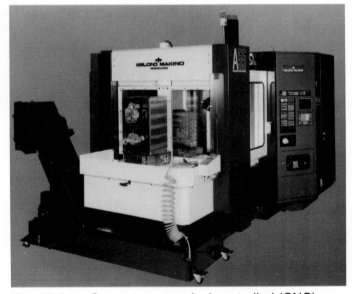

Fig. 22-14. Computer numerical controlled (CNC) machines are a part of modern manufacturing today. This horizontal machining center automatically performs a variety of operations needed to completely process aluminum, steel, or cast iron components from rough castings to finished parts. Workpieces are mounted on square pallets (foreground) and rotate through the machine. Up to 128 cutting tools are stored in the tool changer and automaticly mounted as needed. (LeBlond Makino)

COMPUTER NUMERICAL CONTROL MACHINING

Computer numerical control (CNC) machining is a means of controlling machine tools, Fig. 22-14. It has been applied extensively to milling, drilling, lathe, punch press work, and wire wrapping.

Computer numerical control machining is very flexible and can be used for machining long- or short-run production items. There is a great reduction in conventional tooling and fixturing made possible by the programmed instructions.

The term *computer numerical control (CNC)* is used when a machine is operated by its own computer. These systems are operated from computer software, rather than from the perforated paper tape traditionally used with NC machines. When several machines are controlled by a central computer directly wired to the machines, the system is called **direct numerical control (DNC).** The abbreviation *DNC* is also used to stand for **distributed numerical control**. This system, used in large manufacturing situations, places a number of smaller intermediate computers between the central computer and the CNC machine tools. The distributed control method provides greater flexibility and more rapid response to changing conditions.

Drawings for Computer Numerical Control Machining

There are two reference point systems, *incremental* and *absolute,* used to position the cutting tool of a computer numerical control machine for work on a part. Drawings used in programming NC machines are much the same as those used for more traditional machining. However, the dimensioning system used on the drawing must be compatible with the reference point system of the CNC machine.

Absolute Positioning

Many CNC machines use the **absolute positioning**, or reference point, system to position the cutting tool. In this system, all locations are given as distances and directions from a **zero point.** That

is, each move the tool makes is given as a distance and direction from the zero point, Fig. 22-15. The first "X" dimension is 1.0 + .625 or +1.625. The "Y" dimension is 1.0 + .625 or +1.625. The location for the second hole is "X" = +1.625; "Y" = +5.875 (1.625 + 4.250). The remaining holes are located in a similar manner.

Absolute dimensions on drawings for CNC machining should be of the rectangular datum dimensioning type. Two types, coordinate and ordinate, are shown in Fig. 22-16. The coordinate type is typical of rectangular dimensioning, but each dimension is measured from a datum plane. Ordinate dimensions are also measured from datums and are shown on extension lines without the use of dimension lines or arrowheads. A drawing dimensioned in a manner suitable for CNC machining with the absolute reference point system of positioning is shown in Fig. 22-17.

Incremental Positioning

The **incremental positioning** (continuous path) system of the cutting tool in relation to the workpiece is based upon programming the machine a specific distance and direction from its *current position* rather than from a fixed *zero reference point,* as in absolute positioning. That is, each move the tool must make is given as a distance and direction from the previous location or point.

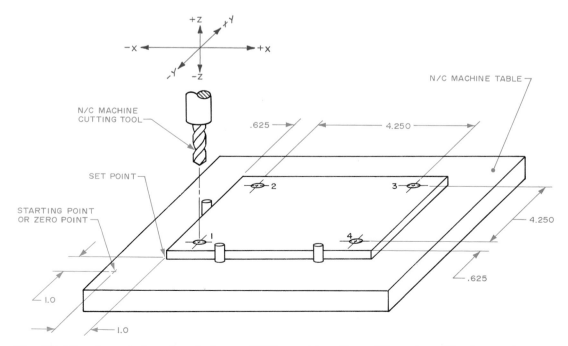

Fig. 22-15. A workpiece located on a CNC machine. Two different positioning systems are used to control the direction and distance the cutting tool moves. In the *absolute* system, each move is referenced to the zero point at lower left. In the *incremental* system, distance and direction are referenced to the end point of the last move.

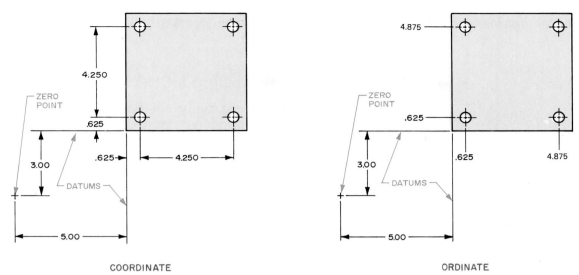

Fig. 22-16. Coordinate or ordinate dimensions measured from datum are used for dimensioning drawings for CNC machines that use the absolute positioning system.

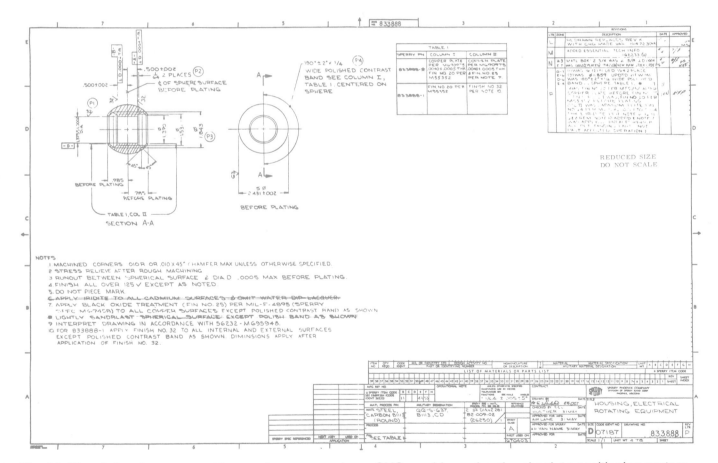

Fig. 22-17. A drawing prepared for use in programming a CNC machine using the absolute positioning system. Note the datum dimensioning and zero point. (Sperry Flight Systems Div.)

The first dimension is given as a distance and direction from the starting (zero) point to the first location where the tool will perform its work. Refer again to Fig. 22-15. The distance and direction from the starting point to the first hole is "X" = 1.625 and "Y" = 1.625. The second hole is "X" = 0.0 and "Y" = 4.250 from the previous location; the third, "X" = 4.250 and "Y" = 0.0; the fourth, "X" = 0.0 and "Y" = 4.250. The programming to return the CNC tool to its zero point would be "X" = 5.875 and "Y" = 1.625.

Incremental dimensions should be applied as successive (chain) dimensions, Fig. 22-18. The programmer can read these directly without having to calculate individual settings for preparing the documents needed to punch the tape or write the program that feeds information into the CNC machine control unit. These dimensions are the same as *basic* dimensions in that they are untoleranced. The tolerances that can be held between features in CNC machining are built into the machine. Toleranced dimensions on the drawing would not change the machined part.

Program Sheet and Machine Setup Sheet

The program sheet for CNC machining is prepared from the drawing by a technician called a *programmer,* Fig. 22-19. The programmer must be able to read and interpret prints made from drawings and be thoroughly familiar with machine processes and capabilities. The result may be a computer program file, or a paper tape prepared by a tape punch machine operator working from the program sheet. Tapes may also be prepared on a tape punch machine connected to a digitizer, a measuring device that converts a drawing to a series of digits or coordinate points.

After the program sheet has been carefully written, a machine setup sheet is prepared. This sheet indicates how the part is to be located and secured on the machine worktable and lists the tools which are to be loaded onto the machine. Data from the program sheet is then input to the CNC tape or computer. When properly programmed, CNC machining centers like those shown in Fig. 22-20, can complete a number of operations in sequence to perform all the necessary machining on a complex part. This is considerably more efficient than moving the part to several different machines, each performing a single operation.

Interpreting a CNC Program

It will help you in drawing for CNC machining if you understand how a CNC machine responds to commands and moves the tool or the workpiece to the desired location.

A CNC machine is wired either for a *fixed zero setpoint* or a *floating zero setpoint.* In the fixed zero system, the machine refers to the established point as zero; parts to be machined are located with reference to this point. In the floating zero system, the CNC programmer may establish zero at any convenient point by coding it into the program.

Dimensional instructions for a CNC program are given on a two- or three-dimensional coordinate plane called the *Cartesian coordinate system,* Fig. 22-21. When the operator is facing a vertical spindle machine, table movement to the left or right is the "X" direction or axis. Table movement away from the operator or toward the operator is the "Y" direction or axis. Vertical movement of the working tool is called the "Z" axis. Machining of some parts requires the use of only the "X" and "Y" axes; others require "X," "Y," and "Z."

It is easier to understand the direction of movement if you assume the table remains stationary and the tool moves over the work. When the work is located on the table with the datum zero point at the zero point of the machine, movement of the tool along the "X" axis to the right is in +"X" direction and to the left is in the -"X" direction. Movement of the tool into the work away from the operator is the +"Y" direction and toward the operator is the -"Y" direction. Tool movement down into the work is -"Z" and up from the work is +"Z."

Tool movements are assumed to be in the *positive* (+) direction unless marked *negative* (-). It is therefore desirable to establish the datum zero point on a drawing in the lower left-hand corner or at a point just off the part to be machined. Refer to Fig. 22-18.

When the zero point is located in this manner, all datum dimensions are + dimensions and do not need to be indicated as such. This also eliminates the possibility of errors in working with + and - dimensions.

SURFACE TEXTURE

Measuring the smoothness of a surface, or finish, is done by using an instrument called a *profilometer.* This instrument measures the roughness of a surface in microinches or microme-

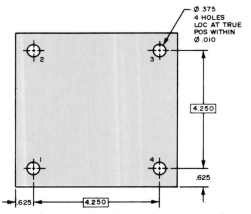

Fig. 22-18. Successive dimensions, sometimes called chain dimensions, are used for dimensioning drawings for CNC machines that use the incremental positioning system.

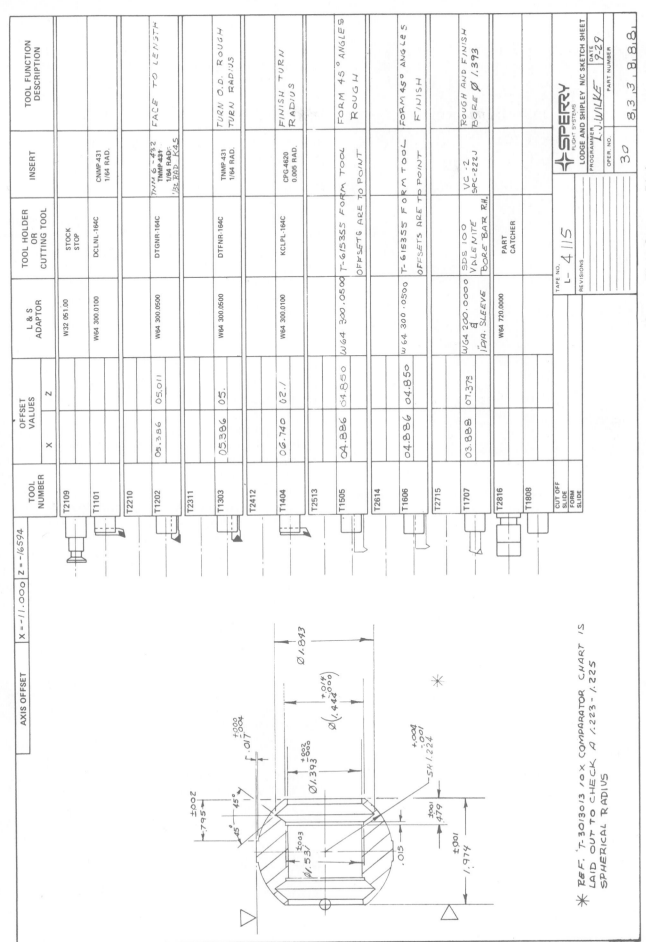

Fig. 22-19. A CNC program sheet and machine setup sheet prepared from a drawing by a programmer. (Sperry Flight Systems Div.)

Fig. 22-20. This computer-controlled *machining cell* includes two high-speed CNC horizontal machining centers and pallet delivery system to move workpieces from one machine to another as needed for additional operations. By switching to another computer program, such machining centers can quickly be set up to process a different component. (LeBlond Makino)

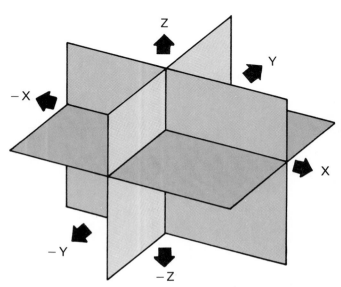

Fig. 22-21. Cartesian coordinates are used to describe movements of a tool or the workpiece on a CNC machine. Movements left or right are on the "X" axis; movements toward or away from the operator are on the "Y" axis. Up and down movements are on the "Z" axis.

ters. Refer to Fig. 22-22. The symbol used on drawings for micro (millionths) is the Greek letter μ (microinch appears as "μ in" and micrometer appears as "μ m"). The surface texture of a machine part should be specified on the drawing as part of the design specifications and not left to the discretion of the machine operator. The surface texture value is used in conjunction with the American Standard Surface Texture symbol.

Machine processes such as milling, shaping, and turning can produce surface textures in the order of 125 to 8 μ inches (3.2 to 0.2 μ meters). Grinding operations can produce surface textures in the range of 64 to 4 μ inches. This depends on the coarseness of the wheel and rate of feed.

Honing and lapping remove only very small amounts of metal and surface textures as fine as 2 μ inches are possible.

Surface Texture Symbol

The **American Standard Surface Texture symbol** is used to designate the classifications of roughness, waviness, and lay.

Roughness refers to the finer irregularities in a surface. Included are those that result from the machine production process, such as traverse feed marks. Surface roughness is measured for height and width.

Surface roughness height can deviate from 1 to 1000 microinches, as measured by a profilometer along a nominal centerline, Fig. 22-23. The preferred series of roughness height values, ranging from "very rough" to "extremely smooth machine finish," are shown in Fig. 22-24. Surface roughness height is designated above the "vee" in the surface texture symbol, Fig. 22-25. A horizontal extension bar is added to the symbol where values other than roughness are specified.

Fig. 22-22. This sophisticated analyzer can measure up to 50 different surface finish parameters and display the results on its screen. (Federal Products Co.)

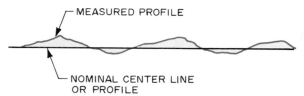

Fig. 22-23. The nominal profile (assumed true surface) and the measured profile, as measured by a profilometer (greatly exaggerated).

Roughness width is the distance between successive peaks or ridges of the predominant pattern of roughness. This characteristic is measured in inches or millimeters. Refer to Fig. 22-25. The roughness-width cutoff is the greatest spacing of irregularities in the measurement of roughness height.

Waviness is the widest-spaced component of the surface; it covers a greater horizontal distance than the roughness-width cutoff. Roughness may be thought of as occurring on a "wavy" surface. **Waviness width** is the spacing from one wave peak to the next, and waviness height is the distance from peak to valley, measured in inches or millimeters. Refer to Fig. 22-25.

Lay is the direction of the predominant surface pattern (such as parallel, perpendicular, or angular to line representing the surface to which the symbol is applied). Refer to Fig. 22-26.

SHEET METAL FABRICATION ┼┼┼

The fabrication of sheet metal varies considerably from the construction industry to the electronics and instrumentation industries. The difference is primarily in the degree of tolerances held when fabricating objects.

Sheet Metal Work in the Construction Industry

The development of pattern layouts for sheet metal products in the construction industry was discussed in Chapter 15. The flat-pattern layouts for these products vary from quite simple to very complex. Since these objects usually have considerable margin in fit, tolerances are not held closely, nor is bend allowance normally figured in laying out patterns. Some common types of sheet metal hems and joints used in air conditioning duct work are shown in Fig. 22-27.

Sheet Metal Work in Electronic and Instrumentation Industries

Sheet metal work in the electronic and instrumentation industries is frequently referred to as pre-

ROUGHNESS HEIGHT RATING		SURFACE DESCRIPTION	PROCESS
MICROMETERS	MICROINCHES		
25.2	1000	Very rough	Saw and torch cutting, forging or sand casting.
12.5	500	Rough machining	Heavy cuts and coarse feeds in turning, milling, and boring.
6.3	250	Coarse	Very coarse surface grind, rapid feeds in turning, planning, milling, boring and filing.
3.2	125	Medium	Machining operations with sharp tools, high speeds, fine feeds, and light cuts.
1.6	63	Good machine finish	Sharp tools, high speeds, extra fine feeds, and cuts.
0.8	32	High grade machine finish	Extremely fine feeds and cuts on lathe, mill and shapers required. Easily produced by centerless, cylindrical, and surface grinding.
0.4	16	High quality machine finish	Very smooth reaming or fine cylindrical or surface grinding, or course hone or lapping of surface.
0.2	8	Very fine machine finish	Fine honing and lapping of surface.
0.05 0.1	2-4	Extremely smooth machine finish	Extra fine honing and lapping of surface.

Fig. 22-24. Description of roughness height values used in conjunction with the American Standard Surface Texture symbol.

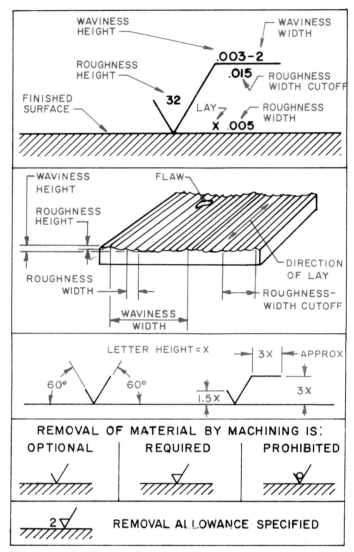

Fig. 22-25. The surface texture symbol is used to show classifications of roughness, waviness, and lay of a surface.

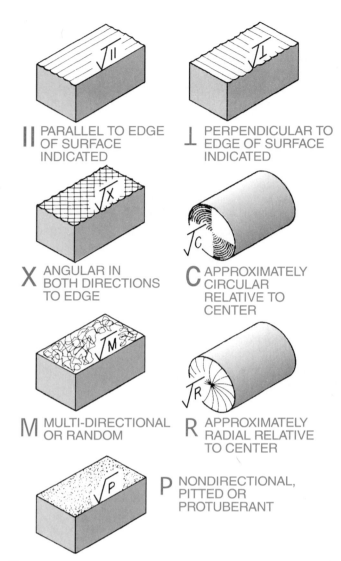

II PARALLEL TO EDGE OF SURFACE INDICATED

⊥ PERPENDICULAR TO EDGE OF SURFACE INDICATED

X ANGULAR IN BOTH DIRECTIONS TO EDGE

C APPROXIMATELY CIRCULAR RELATIVE TO CENTER

M MULTI-DIRECTIONAL OR RANDOM

R APPROXIMATELY RADIAL RELATIVE TO CENTER

P NONDIRECTIONAL, PITTED OR PROTUBERANT

Fig. 22-26. The lay symbols and their meaning. Lay symbols are located beneath the horizontal bar on the surface texture symbol.

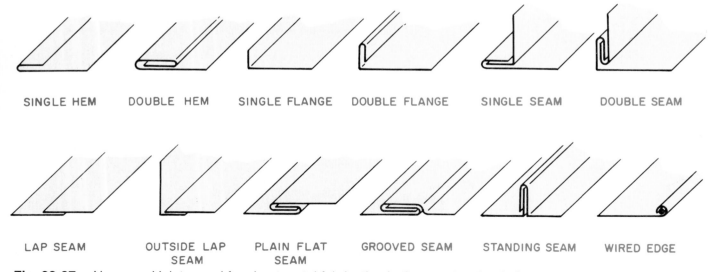

SINGLE HEM   DOUBLE HEM   SINGLE FLANGE   DOUBLE FLANGE   SINGLE SEAM   DOUBLE SEAM

LAP SEAM   OUTSIDE LAP SEAM   PLAIN FLAT SEAM   GROOVED SEAM   STANDING SEAM   WIRED EDGE

Fig. 22-27. Hems and joints used for sheet metal fabrication in the construction industry.

precision sheet metal work because of the close tolerances to which parts are held. **Precision sheet metal** may be defined as "working light-gauge metal to machine shop tolerances."

Precision sheet metal parts are machined in the flat, then folded to shape, and must hold the tolerances between related features, Fig. 22-28. The design drafter must calculate the bend allowance or obtain the information from charts to properly dimension the layout for a flat pattern.

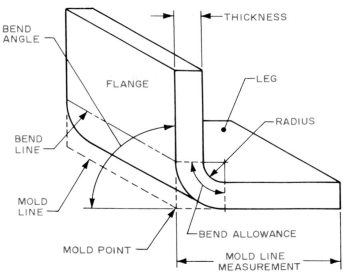

Fig. 22-29. Terms used in describing precision sheet metal.

Precision Sheet Metal Terms

- **Bend allowance:** Length of material required for a bend, measured from bend line to bend line, Fig. 22-29.

- **Bend angle:** Full angle through which sheet metal is bent; not to be confused with angle between flange and adjacent leg.

- **Bend line:** Tangent line where bend changes to a flat surface. Each bend has two bend lines.

- **Blank:** A flat sheet metal piece of approximately the correct size, on which a pattern has been laid out ready for machining and forming.

- **Center line of bend:** A radial line, passing through bend radius, which bisects the included angle between bend lines.

- **Developed length:** Length of the flat pattern layout. This length is always shorter than sum of mold line dimensions on part.

- **Flat pattern:** Pattern used to lay out the sheet metal part on a blank.

- **Mold line:** Line of intersection formed by projection of two flat surfaces.

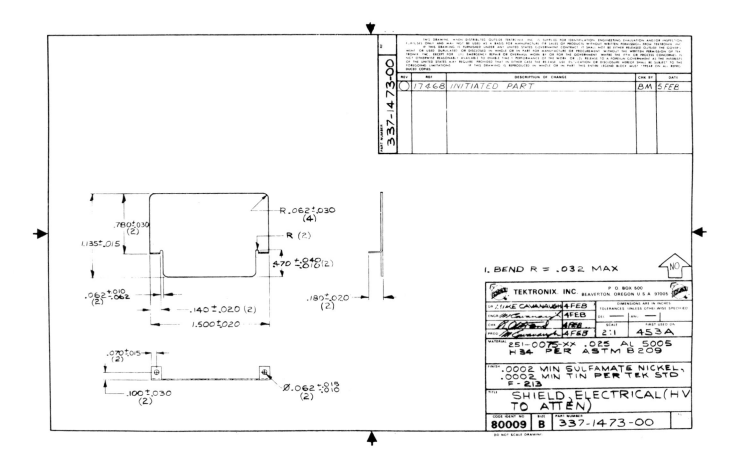

Fig. 22-28. A precision sheet metal drawing for an instrument housing.

- ***Set-back:*** Amount of deduction in length resulting from a bend developed in a flat pattern.

Bend Length Formulas

To calculate the lineal length of the bend on precision sheet metal parts one of two formulas is used. The formula chosen will depend on the size of the bend radius and the thickness of the metal. See Fig. 22-30. After finding the lineal length, the developed length (length in the flat) can be calculated.

To find A (lineal length of a 90° bend) when R (inside radius) is *less than twice the stock thickness,* use the following formula.

$$A = 1/2\pi(R + .4T)$$

Example: To find A, where R = 1/16 or .0625 and T = .064 (inside radius is *less* than twice stock thickness):

$$
\begin{aligned}
A &= 1/2\pi(R + .4T) \\
&= 1/2 \times 3.1416\,(.0625 + .4 \times .064) \\
&= 1.5708\,(.0625 + .0256) \\
&= 1.5708\,(.0881) \\
A &= .1384 = \text{Lineal length of bend}
\end{aligned}
$$

When the bend radius is *more than twice the stock thickness,* the following formula should be used to find A:

$$A = 1/2\pi(R + .5T)$$

To find the developed length of the part after the lineal bend length (A) has been found, refer to Fig. 22-30 and use this formula:

Developed length = X + Y + A - (2R + 2T)

Where: X = Outside distance of one side
Y = Outside distance of other side
A = Lineal length of bend
R = Bend radius
T = Material thickness

Developed length = 1.00 + .75 + .1384 - (.125 + .128)
= 1.8884 - .253
= 1.6354 or 1.64

Set-back Charts

Set-back charts are available in most industries where precision sheet metal work is done. A portion of a chart for 90° bends is shown in Fig. 22-31. A more complete chart is shown in the Reference Section. These charts will save time and errors when making calculations in the shop.

The set-back figure is found by following across the row representing the bend radius until it meets the vertical column representing the thickness of the sheet metal to be bent. For example, a bend radius of 1/8″ on metal .040″ thick would require a set-back figure of .106″. The diagram in Fig. 22-32, and the formula that follows, show how the set-back figure is used to calculate the *developed length* (length in the flat to produce desired folded size) of a precision sheet metal part.

Using the set-back chart:

Developed length = X + Y - Z
Where: X = Outside distance of one side
Y = Outside distance of other side
Z = Set-back allowance for the 90° bend (from chart, Fig. 22-31)

Example, Fig. 22-32:

Developed length = X + Y - Z
= 1.00 + .75 - .110
= 1.75 - .110
Developed length = 1.64

PRECISION SHEET METAL SET-BACK CHART							
MATERIAL THICKNESS							
	.016	.020	.025	.032	.040	.051	.064
1/32	.034	.039	.046	.055	.065	.081	.097
3/64	.041	.046	.053	.062	.072	.086	.104
1/16	.048	.053	.059	.068	.079	.093	.110
5/64	.054	.060	.066	.075	.086	.100	.117
3/32	.061	.066	.073	.082	.092	.107	.124
7/64	.068	.073	.080	.089	.099	.113	.130
1/8	.075	.080	.086	.095	.106	.120	.137
9/64	.081	.087	.093	.102	.113	.127	.144
5/32	.088	.093	.100	.109	.119	.134	.150
11/64	.095	.100	.107	.116	.126	.140	.157
3/16	.102	.107	.113	.122	.133	.147	.164

(Left column label: 90 DEG BEND RADIUS)

Fig. 22-31. Precision sheet metal set-back chart for 90° bends.

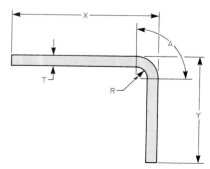

Fig. 22-30. Diagram for use with formula when calculating developed length.

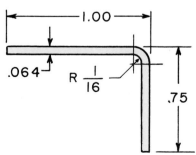

Fig. 22-32. Diagram for calculating developed length from set-back chart.

Note that the same 90° bend precision sheet metal part was used in both the formula calculation and set-back chart examples. The same answer was obtained for the developed length in each example. However, the process is much shorter using the set-back chart.

Allowances for bends of angles other than 90° may be easily calculated. This is done by multiplying the developed length for 90° bends by a factor representing the number of degrees in the desired bend over 90. Suppose, for instance, that the bend in the above example had been 60°. The bend length for the 60° bend would be calculated by taking the reading for the 90° bend from the set-

back chart and multiplying it by $\frac{60}{90}$.

60° bend length = reading for 90° + $\frac{60}{90}$

60° bend length = .110 + $\frac{60}{90}$ = .073

Variations in Bend Lengths of Different Metals

When sheet metal is bent, the **median line** along the interior of the metal remains true length throughout the bending process, Fig. 22-33. The metal on the outside of this line is stretched and the metal on the inside of the bend is compressed. For most metals, the median line is approximately 44% of the distance from the interior face to the outer face.

The location of the median line forms the basis for bend allowance calculations. The harder the

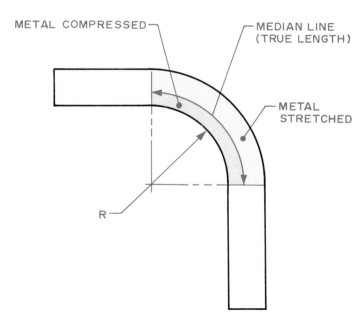

Fig. 22-33. When bent, metal will stretch on the outside portion of the bend and compress on the inside. The median line will remain true length.

metal, the greater the bend length. In actual practice, some experimentation may need to be done with the different metals and different shipments of the same metal alloy to arrive at the correct allowance. Most industries will lay out the flat pattern on two identical blanks. One blank will be formed and measurements checked. Any adjustments needed can then be laid out on a third blank, using the unformed blank as a reference.

Bend Relief Cutouts

Wherever sheet metal bends intersect, a **relief** (notch) must be cut out, as shown in Fig. 22-34, to prevent the part from buckling or wrinkling. The size and shape of the relief cutout may vary as long as it extends at least .03″ beyond the intersection of the bend lines. The cutout usually has a radius, although one is not required for proper bend results.

Precision Sheet Metal Drawings

Most precision sheet metal work will be machined on computer numerical control equipment. Therefore, drawings for this type of work should be datum dimensioned. Usually, a flat pattern and a pictorial of the folded pattern are drawn, showing toleranced dimensions that must be held. The direction of bend on a flange or leg is indicated by the words "bend up" or "bend down" shown on the pattern for the part, Fig. 22-35.

Undimensioned Drawings

An undimensioned drawing is a precise, full-scale drawing of a flat pattern on an environmentally stable material. It is used as a direct pattern for layout or machining.

METAL FORMING AND HEAT TREATMENT PROCESSES

Manufacturing processes in which metals are formed or shaped by actions other than cutting, as well as the various methods of heat treatment, should also be understood by the design drafter. Most drawings refer to materials that have been formed or must be heat treated by one of these processes. A comprehensive treatment of these processes is beyond the scope of this text, but the brief descriptions that follow will help you become familiar with them. A more intensive study can be conducted by consulting manufacturing processes texts.

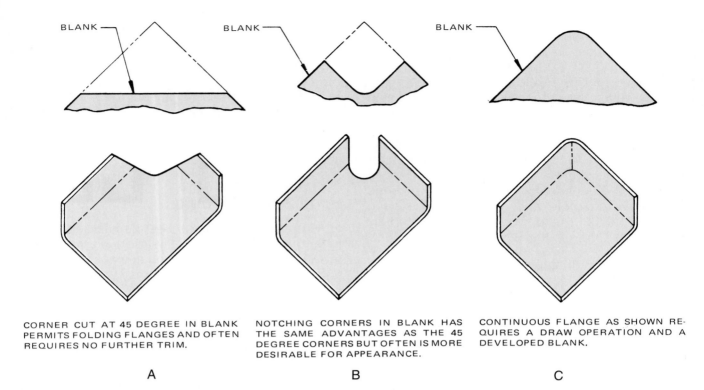

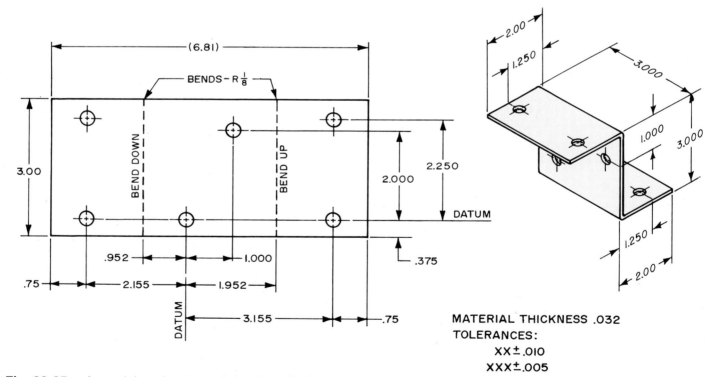

Fig. 22-34. Cutouts relieve the stress in sheet metal at points where bends intersect. Without cutouts, the bends would cause buckling and wrinkling. (General Motors Drafting Standards)

Fig. 22-35. A precision sheet metal drawing of a flat pattern layout.

Casting

Casting, as the term is used in the metal industries, is the process of pouring molten metal into a mold, where it hardens into the desired form as it cools. There are several different casting processes. Casting processes differ primarily in the materials used for molds and in the type of molds. The principal casting processes and their uses are discussed in the following sections.

Sand casting

Metal objects of any size, from small parts weighing a few ounces to large pieces weighing many tons, may be cast by the *sand casting* process, Fig. 22-36. The ability to cast intricate details and low cost are the advantages of this process. Disadvantages are the rough surface and low accuracy of castings.

Shell-mold casting

The *shell-mold casting* process is used mainly for smaller castings (up to 50 pounds in weight), although some larger castings are being poured. The molds are made in the form of thin shells, Fig. 22-37. Sharp reproduction of details on the original pattern, smooth finish, and fairly close accuracy are possible. Nearly all castable alloys can be shell-molded. Molds are more expensive than those for sand casting, but this is offset by time and labor savings where the casting has to be machined.

Plaster-mold casting

Nonferrous metals (aluminum, brass, and bronze) can be cast with a very smooth finish and good accuracy by using *plaster-mold casting.* See Fig. 22-38. Small castings from a fraction of an ounce to 10 pounds are normal, with pieces up to 200 pounds possible. The Antioch process uses a mixture of plaster and sand, providing excellent results with large and complex pieces such as tire molds, wave guides for the electronics industry, and torque converters. Ceramicast® is a special ceramic material that can be used to make molds for casting certain ferrous alloys, such as carbon and stainless steels.

Permanent mold casting

Permanent molds are made from cast iron or steel and used for casting lower-melting-point alloys, such as aluminum, magnesium, zinc, tin, lead, and some of the copper base alloys, Fig. 22-39. Accuracy is exceptionally good in such castings. *Permanent mold casting* is best-suited to producing medium or large quantities of small or medium-sized parts.

Slush-mold casting (a variation of permanent mold casting) uses zinc, lead, or tin alloys. The alloy is poured into the permanent mold and left just long enough to form a thin shell on the inside of the mold. The remaining molten metal is poured out, resulting in a thin-walled hollow casting.

Centrifugal casting

Tubular or cylindrical parts, such as aluminum or cast iron pipe, are cast in whirling metal molds in a process called *centrifugal casting.* The whirling forces the metal against the wall of the mold, leaving a hole in the center, Fig. 22-40.

Fig. 22-36. A mechanized molding line used to cast automotive components. Molds are precisely set by the mechanized handling equipment in the background. (Central Foundry Division, GMC)

Fig. 22-37. A ceramic shell mold used to cast turbine blades. (Garrett-AiResearch Casting Div.)

Fig. 22-38. These aluminum alloy turbocharger wheels were cast in plaster molds made from rubber patterns. (Garrett-AiResearch Casting Div.)

Fig. 22-39. Molten aluminum is poured into a permanent mold made from iron. Such molds are used over and over. (Garrett-AiResearch Casting Div.)

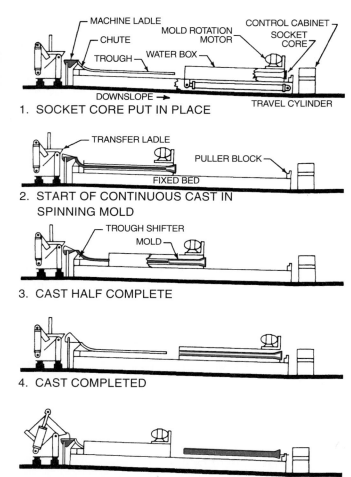

1. SOCKET CORE PUT IN PLACE

2. START OF CONTINUOUS CAST IN SPINNING MOLD

3. CAST HALF COMPLETE

4. CAST COMPLETED

5. PIPE REMOVED-LADLE REFILLED

Fig. 22-40. Method used to cast pipe in centrifugal molds. (United States Pipe and Foundry Co.)

Investment casting

The **investment casting** process, sometimes called "lost wax" or precision casting, is used primarily for small, intricate parts that require high accuracy and an excellent finish, Fig. 22-41. The process is relatively expensive, but the time and labor saved in machining help offset the cost.

Fig. 22-41. Small, intricate parts can be cast effectively using the investment casting process. (Garrett-AiResearch Casting Div.)

Die casting

Sometimes called "pressure casting," the *die casting* process is particularly well-suited to mass production, Fig. 22-42. Molten aluminum, zinc, or magnesium is forced under pressure into a metal die or mold and allowed to harden. The mold then opens, the part is ejected, and the cycle repeats.

Very good accuracy and an excellent finish are obtainable with this process. Since it involves expensive dies, however, it is suited only to large-quantity production.

Forging

The working of heated (but not molten) metal into shape by means of pressure (usually a hammering or squeezing action) is known as *forging,* Fig. 22-43. This process develops the greatest strength and toughness possible in steel, bronze, brass, copper, aluminum, and magnesium parts. Huge mechanically operated presses are used to forge machine parts that, for some applications, are not otherwise attainable.

Extrusion of Metals

Extrusion is the process of producing long lengths of rod, tubing, molding, and other shapes from aluminum, brass, copper, and magnesium. Extrusions are formed by placing a billet of hot metal in the cylinder of an extruding press and forcing it through a die with an opening of the desired shape, Fig. 22-44.

Tremendous pressure is created by a hydraulic ram that squeezes the heat-softened metal out through the die and onto a conveyer to cool. The extrusions are cut to length and straightened by a

A

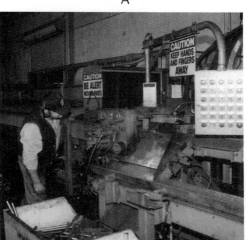

B

C

Fig. 22-43. Automatic drop forging eliminates operator handling of steel bars or billets. A–Following a temperature check, the hot stock is moved to the end of the furnace conveyer, and located for pickup by the impact hammer's stock transfer device. The stock is then positioned vertically, ready for transfer to the pallet feed device. B–The bar is fed through the shear until it contacts the length gage. After shearing, the billet is discharged to the heater conveyer. Two 65-ton trimming presses are located at the exit end of the forging cell. C–The production rate of the impact process is such that two persons are required to hot trim the continuous flow of forging platters. (Klein Tools, Inc./Chambersburg Engineering Co.)

Fig. 22-42. Die casting produces items of high accuracy and excellent finish. (Fisher Body Division, GMC)

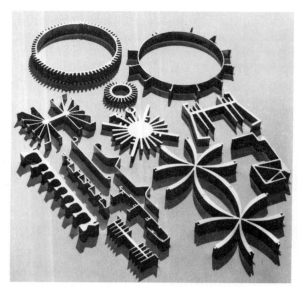

Fig. 22-44. The extrusion process is used to form aluminum and other metals into various shapes. (Aluminum Association)

Fig. 22-45. Threaded fasteners are made in these cold heading machines. (ELCO Industries, Inc.)

stretching operation. Extrusions are used extensively in the aerospace, automotive, and structural industries.

Cold Heading

If sufficient pressure is applied and the correct type of die is used, metals may be formed while cold. *Cold heading* (also called chipless machining, cold extrusion, cold forging, or impact forging) is the process of applying tremendous pressure to metal, forcing it to flow upward into a narrow space around a punch and die. Cold heading is often used to form an enlarged head or flange on a rod or bar.

A widespread use of this process is making bolts, screws, and other fasteners from large rolls of wire stock that are fed to automatic cold heading machines, Fig. 22-45.

Heat Treating and Case Hardening

Heat treating involves heating metal to a high temperature, then cooling it at various rates to produce qualities of hardness, ductility, and strength. *Annealing* is a form of heat treatment that reduces the hardness of a metal to make it machine or form more easily. *Normalizing* is a form of heat treatment aimed at relieving stresses caused by previous hot or cold working of a part.

Case hardening forms a hard outer layer on a piece, leaving the inner core more ductile. This produces a part that is very hard, long-wearing, and resistant to breaking under impact. Casehardening is generally applied to lower-carbon steels and used for such parts as automobile wrist pins and

races for ball and roller bearings. It is sometimes called *carburizing,* since the part is usually heated to a high temperature for an extended period in contact with materials from which the steel absorbs more carbon or nitrogen.

Flame hardening is similar to casehardening. It is a method of producing surface or localized hardening by directly heating a surface, then immediately quenching the piece before the heat has had a chance to penetrate far below the surface. This process is widely used to harden gears, splines, and ratchets, Fig. 22-46.

Hardness Testing

The hardness of a piece of metal is tested by measuring the indentation impression or the height of rebound on machines designed for the purpose,

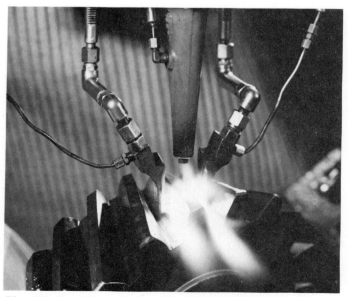

Fig. 22-46. Flame hardening of gear teeth.

Fig. 22-47. Metals must be carefully selected to give the desired qualities of strength and hardness.

OTHER METALWORKING PROCESSES

The space program and technologies resulting from it have created the need for new metal alloys that are lighter, stronger, and tougher than any used before. These new alloys are also far more difficult to work than any used before, which has lead industry to develop a number of sophisticated metalworking techniques and processes. A number of these "nontraditional" metalworking processes are described on the following pages.

Electrical Discharge Machining (EDM)

The working of metals by eroding the material away with an electric spark is known as *electrical discharge machining (EDM),* Fig. 22-48. The process is a magnified and controlled version of the pitting or burning that occurs when a charged electrical wire momentarily contacts a grounded piece of metal.

The electrode in an EDM machine, usually graphite or a brass material, is the cutting tool. It is surrounded with a coolant that serves as a dielectric barrier between the electrode and the workpiece at the arc gap. A servo-mechanism accurately controls movement of the electrode and maintains the proper gap. The electrical discharge occurs at a rate of 20,000 to 30,000 times per second.

Fig. 22-47. Measuring the hardness of a piece of steel in a hardness tester. (Wilson Instrument Div. ACCO)

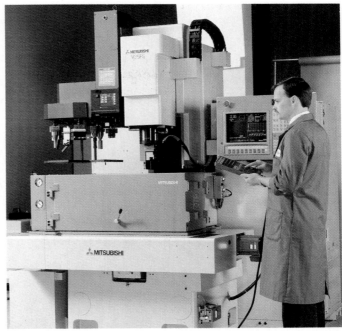

Fig. 22-48. This electrical discharge machine operates under computer numerical control. The operator is using a multi-function remote control for the unit. (Mitsubishi)

Each spark erodes a small amount of metal, while the dielectric flushes the eroded particles away from the cutting action and keeps the electrode and workpiece cool. This machining process is very effective on metals that would be difficult to machine in any other way. Accuracies within .0005 inch are possible, with a very fine surface texture.

Electrochemical Machining (ECM)

The process of *electrochemical machining (ECM)* is the reverse of electroplating. In the electroplating process, a thin layer of metal is added to another metal by means of a direct current through a liquid solution. In electrochemical machining, a very high current density is used. The workpiece is the anode or positive part of the circuit; the electrode is the cathode or negative part.

The electrolyte, or solution, flows at high pressure between the shaped electrode and workpiece. This results in the metal being electrolytically eroded from the workpiece and washed away, rather than being deposited on the electrode, Fig. 22-49.

The electrochemical method of machining metal can be used on materials that are very difficult to machine by other methods. No strains are set up in the workpiece, since no contact occurs between it and the electrode. Considerable work has been done in perfecting the shape of electrodes to produce desired shapes on workpieces. ECM is a

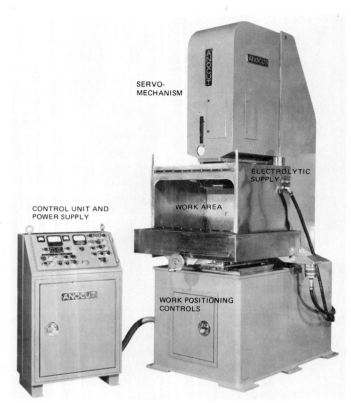

Fig. 22-49. Equipment used for electrochemical machining of metals.

much faster method of machining than EDM and is used more in production operations.

Chemical Milling

A method of removing material by etching with a chemical is called **chemical milling.** The process, which can be used on plastics and glass as well as metals, is also referred to as "chem milling" or "contour etching." Chemical milling works on that portion of the material which is not protected by a mask.

On large pieces, such as aircraft structural members that are chem milled to reduce their weight, the masking is done by dipping in a plastic material. The portion to be etched is then exposed by cutting away the plastic covering. This process is also used extensively with parts made from thin metals, printed circuits, and integrated circuits. Refer to Fig. 22-50. The masking for these is usually done by using a photosensitive resist material.

Chemical milling of metal has several advantages. It is relatively inexpensive and produces a surface texture in the range of 30 to 125 μ inches. Tolerances can be held within a few thousandths, the metal is not stressed as it would be in blanking, and heat-treated or hardened parts can be worked without affecting the characteristics of the metal.

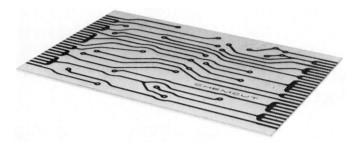

Fig. 22-50. A printed circuit board which was produced by chem milling. (Chemcut Corp.)

Ultrasonic Machining

Certain materials, such as alloys of nickel, change dimensions slightly when subjected to a strong magnetic field. **Ultrasonic machining** is based on this property. When placed in a strong magnetic field that fluctuates rapidly, a nickel-alloy rod changes its length about .004″. An alternating current is used to reverse the magnetic field about 25,000 to 30,000 times per second. The "cutting tool," a rod of brass or soft steel of the desired shape, is attached to the nickel rod.

In operation, the cutting tool rests on the workpiece and a waterborne abrasive flows between the points of contact. The abrasive is driven into the workpiece by ultrasonic vibration of the cutting tool. The shaped-rod tool cuts its way through the workpiece leaving a clean hole of the desired shape, Fig. 22-51.

Tungsten carbide, quartz, glass, and ceramic materials can be machined with amazing speed. Tolerances as close as .0005″ and surface textures

Fig. 22-51. Ultrasonic machining is done at frequencies of 25,000 to 30,000 cycles per second. These frequencies are far above those that can be heard as sounds by the human ear. (Raytheon Co.)

of 10 to 15 μ inches may be obtained. The process is used for specialized cutting operations and for making carbide dies for extrusion, drawing, and stamping.

Laser Applications

A device that has gained wide acceptance in science, research, machining, welding, and measurement is the laser, Fig. 22-52. **Laser** is an acronym formed from the initial letters of the words, *l*ight *a*mplification by *s*timulated *e*mission of *r*adiation. The laser generates a narrow beam of monochromatic light of extremely high intensity in very short pulses. Its principle industrial applications are in metal removal, welding, and measurement. Because of its capability of producing an extremely narrow beam of light and temperatures up to 75,000°F, the laser can be used for perforating holes in stainless steel, carbide, ceramics, and even diamonds. Actually, the holes are not drilled: the material is melted and vaporized with each burst of energy from the laser beam. The laser machine can work to very close tolerances.

Fig. 22-52. This CNC laser cutting center can precisely cut steel as thick as 3/4 inch. Unlike many laser cutting systems in which the workpiece moves beneath a stationary laser, this unit features a stationary workpiece and moving cutting head, allowing simpler programming. (Trumpf, Inc.)

High Energy Rate Forming (HERF)

The forming of large-diameter sheet metal parts requires tremendous power as well as a press of sufficient size. Often neither are available. However, an ingenious and inexpensive method has been developed to process these kinds of jobs. It is known as **High Energy Rate Forming (HERF)**, and is sometimes called "explosive forming." A die of the desired shape is prepared and the sheet metal piece is cut or fabricated, then clamped in place with a retaining ring. The area around and above the die is filled with water and a vacuum pulled between the piece and the die. Finally, a carefully calculated amount of an explosive material is set off above the piece of sheet metal to be formed, Fig. 22-53. The piece is formed in a fraction of a second and retains its new shape within acceptable tolerances. A piece formed by HERF is shown in Fig. 22-54.

Electrohydraulic Forming

Two methods used to form sheet metal parts and to bulge metal tubular parts by **electrohydraulic forming** are the **hydrospark process** and the **exploding bridge wire.** The hydrospark process discharges an electric spark under water to produce a high velocity shock wave that forms the part, Fig. 22-55.

The shock wave is created by the release of stored electrical energy generated by a high-voltage power supply and a coaxial electrode. The force of the shock wave can be changed by varying the voltage.

The exploding bridge wire method is similar to the electrospark process, except that a wire is used between the electrodes. The released electrical energy vaporizes the wire, and the shock wave and expanding vapor exert force on the surrounding

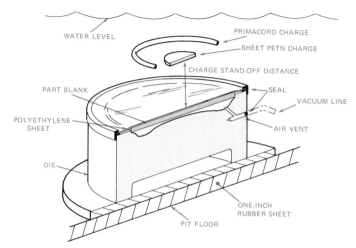

Fig. 22-53. The operating principle of High Energy Rate Forming. (Grumman Aerospace Corp.)

Fig. 22-54. The part on the left was formed in a fraction of a second by the explosive force used in High Energy Rate Forming. The part on the right was hot-spun and required considerably more time. (Lockheed Aircraft Corp.)

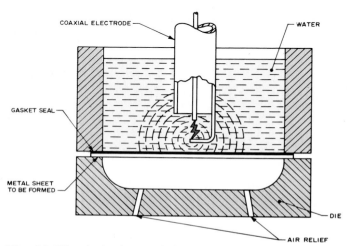

COAXIAL ELECTRODE
WATER
GASKET SEAL
METAL SHEET TO BE FORMED
DIE
AIR RELIEF

Fig. 22-55. In hydrospark forming, shock waves generated by an electrical spark are used to form a sheet metal part.

water. The water transmits the pressure to the sheet metal, forcing it into the die.

The electrospark method is faster, since the wire need not be replaced as it must be in the exploding bridge wire method. The wire method does lend itself to better control, however, since the wire can be shaped to fit the cavity. The electrohydraulic forming method is safer, more precise, and lower in cost than conventional hydraulic forming.

PLASTIC PRODUCTION PROCESSES

Plastics are one of the significant materials of modern industry, used for everything from electronic components to machine tool parts. The broad field of plastics can be divided into two general groups, thermoplastics and thermosetting plastics.

Thermoplastic materials become soft when heated and harden as they cool. They can be softened repeatedly by heating. Included in this group are the styrenes, vinyls, acrylics, polyethylenes, and nylons.

Thermosetting plastics cannot be softened by reheating, since they change chemically during the curing process. Included in the thermosets are the phenolics, epoxies, ureas, melamines, and polyesters.

Many of the processes for working plastics are the same as those used for metalworking. However, there are some processes that are specific to the processing of plastics. These processes are discussed in the following sections.

Injection Molding

Many different processes and techniques are used to form plastics into useful products. One of the most widely used is ***injection molding.*** Plastic granules are loaded into the hopper of the injection-molding machine, then softened by heating. The softened plastic is forced through a nozzle into the mold to form and cool. After cooling, the part is ejected and the feeder gates are trimmed to finish the part. Refer to Fig. 22-56.

Injection molding lends itself to a wide variety of products, some of them quite complex in design, Fig. 22-57. Thermoplastics are primarily used with the injection molding process.

Extrusion of Plastics

The extrusion of plastics is similar to that of metals. Continuous lengths of pipe, rod, special shapes, and sheets can be formed through dies of the machine. Thermoplastic resin is heated and forced through the die. The finished shape is picked up on a conveyer belt to cool as it leaves the machine, Fig. 22-58.

Blow Molding

Thin-walled hollow plastic parts are formed by ***blow molding,*** Fig. 22-59. A tube or cylinder of

Fig. 22-56. An injection molding machine used to form plastic parts. (Union Carbide Corp.)

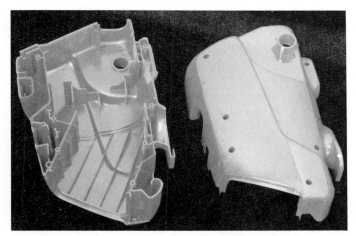

Fig. 22-57. The two halves of this plastic part for a home appliance, with an internal structure featuring stiffening ribs and sockets for fasteners, were injection molded. (Mitsubishi)

Fig. 22-58. Modern plastic extrusion machines are computer controlled for efficient operation. Operation is monitored on a screen that graphically represents each step in the extrusion process. (Rohm and Haas)

Fig. 22-59. An experimental blow molding machine designed to produce large components, such as automobile instrument panels and assemblies, or components used in building construction. (GE Plastics, Inc.)

heated thermoplastic resin, called a **parison,** is extruded and placed between halves of the split mold. The mold closes, pinching the ends of the plastic tube, then a blast of air forces the thermoplastic against the mold. When the plastic cools, it becomes stable and retains the shape of the mold.

The technology of plastic blow molding makes it possible to produce many different types of products rapidly and at relatively low cost. Typical blow-molded objects include fuel tanks for automobiles and boats, and containers for soap, medicine, food products, instruments, and tools.

Compression Molding

Compression molding is one of the most common processes used in forming thermosetting plastics. A measured amount of plastic resin is placed in the open heated mold, then the mold is closed. Pressure is applied to force the plastic into the shape of the mold cavity. The plastic first becomes a liquid, then solidifies and cures into a permanently hard material.

Compression molding presses come in varying sizes. Those commonly used generate temperatures of 270°F to 360°F, and pressures from 300 psi to 8000 psi. Refer to Fig. 22-60.

Transfer Molding

The plastic resin used in *transfer molding* is not fed directly into the mold as it is in compression

Fig. 22-60. Removing a newly formed plastic part from a compression molding machine. (Union Carbide Corp.)

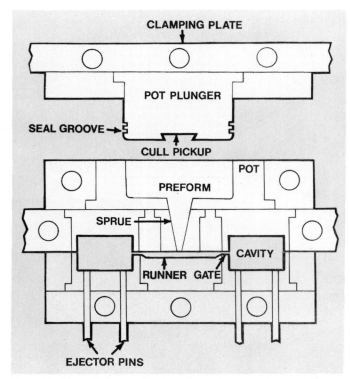

Fig. 22-61. Schematic drawing of a transfer mold showing the separate chamber or *pot*, where the resin is melted. The runner and gates channel plastic into the mold cavity when pressure is applied by the pot plunger. (Durez Div.)

molding. Instead, the resin is placed in a separate chamber and heated under the pressure of a plunger until molten. Higher pressures are then exerted, forcing the softened resin through runners and gates into the mold cavities, Fig. 22-61.

Calendering

Plastic sheet and film stock are produced by a process known as *calendering,* Fig. 22-62. The plastic resin is fed from a hopper, heated, and passed through a series of rollers. The rollers reduce the resin sheet to the desired thickness. After cooling, the finished sheet is trimmed to width. Various materials can be mixed with the plastic resin to give it the color and other qualities desired.

Rotational Molding

In *rotational molding,* plastic resin in the form of powder or liquid is placed in a mold of the desired form. The mold rotates, spreading the powder or liquid evenly over its interior surface. As the resin melts, a solid coating is formed on the mold's surface to produce the required shape. Some products of rotational molding are plastic refuse cans, ice chests, footballs, and sports helmets.

Post-forming of Thermoplastic Sheets

In the *post-forming* process, thermoplastic sheets from .010″ to 5/8 ″ or more can be formed

Fig. 22-62. This calendering machine is forming plastic sheet material. (Union Carbide Corp.)

into desired shapes by one of several methods. Sheets may be heated and forced into a pair of matched molds; heated and forced with air pressure over or into a half mold, or heated and vacuum-formed over male or female molds, Fig. 22-63.

The vacuum process is most useful with thinner sheets. Air pressure and matched molds are used for heavier forming such as aircraft noses, domes, and similar products.

Casting

Plastics may be cast from thermoplastic or thermosetting resins by heating to convert them to a liquid state. The materials may be hardened into final form by heating or by chemical action (curing) in the case of thermosets, or by cooling (for thermoplastics). Sheets, rods, and special shapes may be formed by this method. Casting is suitable for small production runs or for making parts for a prototype. Molds for casting plastic may be made from such inexpensive materials as lead, plaster, or plastisols. Pressure is not required, but a vacuum is sometimes used to eliminate bubbles and voids in castings.

Plastisols

Plastisols are a mixture of plastic resins with **plasticizers** (flexibility agents) that improve the workability of the mixture and reduce brittleness after curing. Plastisols are cured by heating to about 350°F, to become tough, flexible, solid materials. They are generally used in slush molding and dip molding processes. Common uses are coatings for such articles as wire dish racks and electroplating racks, and for the inside surfaces of drums and tanks.

Fig. 22-63. This plastic window shutter was post-formed from plastic sheet by the vacuum-forming process. (B. F. Goodrich Chemical Co.)

Slush Molding and Dip Molding

Slush molding of plastics is similar to the slush molding of metals except the mold is heated and filled with plastisol. After a short period, a layer of the plastisol forms on the inside surfaces of the mold, and then the remainder is poured out. The mold is then placed in an oven to cure the plastic. Products such as toy doll parts, syringe bulbs, and spark plug covers are molded by this process.

Dip molding is similar to slush molding, except that a heated male mold is used. The plastisol fuses to the outside surface of the mold, Fig. 22-64. This is further cured in an oven, and then stripped from the mold. The detail of the male mold is reproduced on the interior of the plastisol piece. Typical products are toy doll parts, boots, and spark plug covers. Tool handles may also be coated by this process. Sometimes the dip mold product is stripped and used as a mold for casting other plastic parts.

Pressure Laminates

The product resulting from bonding two or more layers of material together is known as a **laminate.** The individual layers making up the laminate may be sheet plastic, cloth, paper, or wood. The layers are impregnated with a plastic resin, then heat and pressure are applied by a laminating press. When the pressures exceed 1000 psi, the process is known as high-pressure laminating. Typical high pressure laminates are those used for kitchen countertops, cabinets, and furniture. Refer to Fig. 22-65. Low-pressure laminating is used to form boat hulls, automobile bodies, luggage, credit card plates, and component housings of various types.

Foamed Plastic

Plastic resins to which air or gas are added to form a sponge-like substance are known as **foamed plastics.** The foam may be either rigid or flexible, depending upon the resin used. Plastic resins commonly used for foamed plastics are the styrenes, phenolics, polyurethane, epoxies, and silicones.

The rigid type is used as core material for sandwich panels in aircraft and construction work. The panels are strong and have good heat- and sound-insulating qualities. This type of foamed plastic can be formed, removed from the mold, and used as packing for machine parts or in slab form. It can also be made to adhere to the inside of refrigerator doors, aircraft assemblies, or other areas where its insulating qualities are desirable.

Fig. 22-64.   A plastisol dip molding system for making insulating "boots" for high-voltage switchgear. Thickness of the cured plastic is .125″. (W. S. Rockwell Co.)

Fig. 22-65.   High-pressure laminates are widely used for countertops and other surfaces in kitchens and bathrooms. (Kohler Co.)

Flexible foamed plastics resemble foam rubber and are widely used for weatherstripping and sound-proofing products. They are also used as cushioning for the shipping of fragile items or expensive precision equipment. These flexible foams are being used in many applications to replace rubber.

QUESTIONS FOR DISCUSSION

1. Which machine process, drilling or reaming, is likely to produce a more accurate hole? Why?

2. How does boring differ from drilling or reaming?

3. Explain the difference between counterboring and spotfacing. Make a sketch to illustrate the difference between the two processes.

4. What is broaching? Give examples of machine parts or objects that have been produced by broaching.

5. How does chamfering differ from countersinking? What is the function of each process?

6. Write a definition of computer numerical control machining in terms that your classmates will understand.

7. Does a drawing to be used in preparing computer program for CNC machining differ from a standard drawing? Why or why not?

8. What other documents must be prepared from the drawing for CNC machining prior to programming? Who prepares these?

9. Using simple sketches, explain the difference between the *absolute* and *incremental* positioning systems used in CNC work.

10. How do honing and lapping differ from grinding?

11. Explain the meaning of surface texture and state the purpose of designating this feature on a drawing. How does surface texture differ from surface finish?

12. There are two general types of sheet metal drafting. Explain how these two differ.

13. Give some examples of metal parts that have been cast. Which process of casting was used (or likely used) for each example?

14. What is forging? Give some examples of forged parts.

15. How is the extrusion of metal parts done? List some examples of parts that you believe were produced by the extrusion process.

16. Explain the difference between annealing and case hardening.

17. Explain the process of electrical discharge machining. How does electrochemical machining differ from EDM?

18. What is chemical milling? Give one or more examples of where is it used.

19. What is ultrasonic machining? How does it work? Where is it used?

20. List some uses for the laser in modern industrial work.

21. Explain how high energy rate forming is used to produce parts. How does this differ from electrohydraulic forming?

22. What are the two general types of plastics? What is the major difference between them?

23. Describe the injection molding process of forming plastics.

24. What is blow molding? List some examples of plastic products produced by this process.

25. How does transfer molding of plastics differ from compression molding?

26. What plastic product is produced by the process called "calendering"?

27. What are plastisols? What uses do they have?

28. Explain the difference between slush molding and dip molding of plastics.

29. What are plastic laminates? Give some examples of uses for both high-pressure and low-pressure laminated plastics.

30. Name the two types of foamed plastics. List several uses for each type.

Many designers worked as a team to produce this advanced-concept prototype for an all terrain vehicle. Once the manufacture begins, the designers and engineers must work together to create the final produced vehicle. (Isuzu)

23 Linking Design and Manufacturing

Increased national and international competition in manufacturing is causing industrial leaders to rethink their production plans. Recent developments in computers and manufacturing have prompted these leaders to look at new strategies for remaining competitive and improving the quality of their products. Developments in the computer industry are having a profound impact on manufacturing at all levels from design to machine processing. The management and marketing components of manufacturing companies are also affected.

The purpose of this chapter is to help you develop a knowledge and understanding of these new strategies in manufacturing. Also covered in this chapter is the effect these strategies are likely to have on the design-drafting component of industry, both now and in the future. The impact of computer technology on design and manufacturing is presented first as it relates to design-documentation (computer-assisted design and drafting, or CADD). Then, the area of computer-assisted manufacturing (CAM) and how it is linked to CADD will be covered. Other innovative manufacturing plans, such as flexible manufacturing systems (FMS) and computer-integrated manufacturing (CIM) will be examined, as well.

COMPUTER-ASSISTED DESIGN AND DRAFTING (CADD)

CADD refers to the computer-assisted design and/or drafting process. The term *CAD* (computer-aided design) is also used at times. It is assumed that the documentation of the design process will result in some computer-generated drawings (drafting) and other documents essential to the entire manufacturing cycle, Fig. 23-1.

Fig. 23-1. Computer-generated drawings are essential to the design process. (Hewlett-Packard)

Development of CADD

In the early stages of CADD, the principal uses of the technology were producing and maintaining drawings. There was considerable value even in the generation of drawings, since considerable time-savings could be realized through the ready application of symbols, dimensioning elements, and projection of views. But it was realized that CADD had much more to offer in the design of a product. Related data such as alternate designs for the product, costs, and materials analysis (including stress analysis) could be stored in a database, ready for instant recall. In addition, information generated in the design process could be extended for use in manufacturing as the concept of the automated factory developed.

Need for a database in today's manufacturing market

In today's highly competitive manufacturing atmosphere, the designer does not have the luxury of redesigning or reworking of a product once the manufacturing process has begun. The design selected for manufacture must be the best among several that have been proven by thoroughly analyzing alternate designs, material options, machine processes, and labor costs. Anything less than top performance in product design contributes to problems in a number of areas. These include manufacturability, cost overruns, product failure in service, and lack of customer confidence in the company. Poorly designed products eventually contribute to a company's failure.

Perfecting the manufacturability of a product requires the capability of thoroughly examining design alternatives. Knowledge gathered from previous experience can be stored in the computer database for ready use in design work. This information will be available to help the designer make wise decisions in the design process.

How CADD works as a subsystem of CIM

CADD is one subsystem of **computer-integrated manufacturing (CIM)**, Fig. 23-2. With the assistance of CADD equipment, designers are able to analyze, test, and discuss each design decision, Fig. 23-3. Specialists in materials, tooling, manufacturing processing, sales, and marketing also provide input to the design process. Once the de-

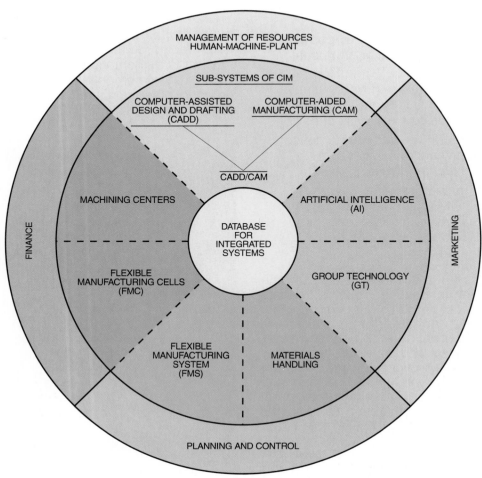

Fig. 23-2. Components of computer-integrated manufacturing. These components of computer integrated manufacturing (CIM) are described in this chapter.

Fig. 23-3. Today, traditional "board" drafting and CADD often are used side-by-side. CADD allows designer-drafters to try multiple design solutions, confer with other designers or specialists, and quickly make modifications to the drawing. (IBM)

sign decision has been made, information is entered into the central database for use and adaptation to other subsystems in the manufacturing process, Fig. 23-4. When the accepted design for a product leaves the CADD department, it is assumed to meet all requirements for manufacturability and meeting customer needs.

Advantages of CADD

CADD provides many advantages for the design-drafting department and those who work there. Following are a few of these advantages.

- It removes the need for tedious calculations by designers and drafters.
- It saves valuable time by generating notations, bills of materials, and symbols to be placed on the drawing.
- It eliminates many of the time-consuming tasks of manual drafting, such as drawing lines, geometric shapes, and measuring distances.
- It provides time and essential data to review alternate design solutions.
- It requires input from other departments, such as manufacturing and sales, thus eliminating later problems.
- It provides more reliability in design work by making relevant information available to design personnel.
- It generates a CADD database on the design and documentation of the product, which can be used in other subsystems of CIM.

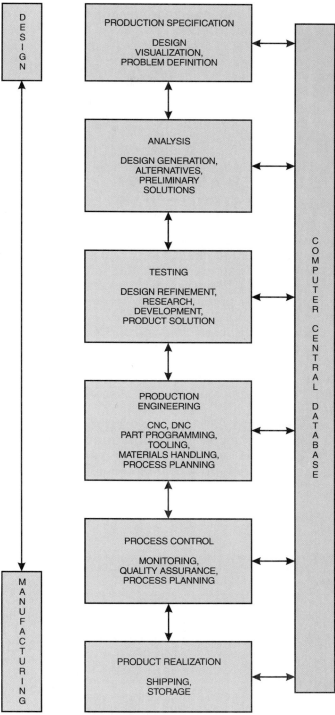

Fig. 23-4. In a CADD/CAM system, the central computer database links design and manufacturing.

- It reduces the number of drawings required by providing the ability to retrieve design models whenever needed.

CADD Designer-Drafter Qualifications

Today, the design-drafter has available more assistance in the way of design information, analysis, and testing than at any previous time. To take full advantage of these capabilities, the designer-drafter must acquire the knowledge and skills needed to make effective use of the CADD systems that will be in the design departments of tomorrow, Fig. 23-5. In Chapters 17 through 20, the fundamentals of computer-assisted design and drafting are presented to assist you in getting started in computer graphics.

COMPUTER-ASSISTED MANUFACTURING (CAM)

Computer-assisted manufacturing (CAM) facility is a natural extension of CADD technology. CAM can be defined as a manufacturing method that uses mills, lathes, drills, punches, and other programmable production equipment under computer control, Fig. 23-6. These machines are known as ***computer numerical control (CNC)*** machines. They can be programmed to perform a wide variety of machine processes at great speed, while holding close tolerances. ***Robots*** are another type of programmable equipment that is essential to a CAM environment.

CADD and CAM are linked together by the database developed in the design of the product, Fig. 23-7. This same database is used by production engineering to program computer-assisted manufacturing equipment.

Computers are also used in a CAM facility to control production scheduling and quality control,

Fig. 23-6.   Computer numerical control (CNC) allows precise automated machining of complex components. (Ford Motor Company)

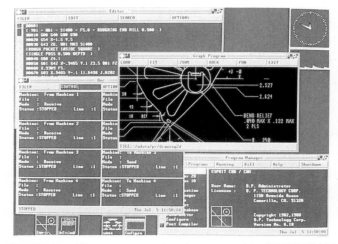

Fig. 23-7.   The common database used in CADD/CAM allows users of this manufacturing software to use multiple window panels on the computer screen to view various types of information. The upper window on the left displays the NC code used for a machining operation on the part shown in the drawing detail displays at right center. Also displayed is a panel showing the status of various machines in the manufacturing cell. (DP Technology, Inc.)

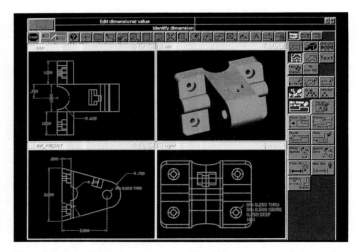

Fig. 23-5.   CADD programs like this one can integrate traditional drafting views with three-dimensional surface modeling, as shown here, as well as three-dimensional wireframe modeling, three-dimensional solids modeling, and geometric properties analysis. It can interface with a number of other design, analysis, tooling, and manufacturing programs. (Intergraph Corporation)

and such manufacturing business functions such as purchasing, financial planning, and marketing.

Numerically Controlled (NC) Machines

When numerical control (NC) was first introduced as a method of programming and controlling production machine tools, instructions (or "programs") were on perforated **punch cards.** Later, punched **paper tape** was used to store and "play back" a program. Since these methods of numerical control were subject to damage in a hostile machine tool environment, the paper tape was replaced by more durable plastic (Mylar®) tape.

Now, computers control NC machines. These machines are called **computer numerical controlled machines (CNC).** In CNC, a computer is used to write, store, edit, and execute a program. One method of computer control used in manufacturing is **direct numerical control (DNC).** In this method, the computer serves as the control unit for one or more NC machines. Refer to Fig. 23-8. A more accelerated stage is called **distributed numerical control (DNC).** In this method, a main computer controls several intermediate computers that are coupled to certain machine tools, robots, and inspection stations.

Fig. 23-8. A machine operator uses a DNC computer keyboard to load the appropriate program for a part to be processed in a machining center. The computer on the machine is linked with the factory's main computer to make use of the information in a central database. (DLoG-Remex)

Robots

For a number of years, robots have been performing tasks of varying difficulty in industry, Fig. 23-9. Single-purpose devices that cannot be reprogrammed to perform other tasks (such as those that merely transfer a part from one machine to another) do not fit the accepted definition of "robot." The Robotic Industries Association defines a robot as "a *programmable multifunctional manipulator* designed to move material, parts, tools, or specific devices through variable motions for the performance of a variety of tasks." This defines a tool that is flexible and capable of functioning in a number of industrial settings just like any other programmable machine.

Advantages of CAM

Many of the advantages claimed for CAM come from its linkage with CADD as **CADD/CAM.** As manufacturing becomes more computer-based and moves toward full integration in all facets of design, production engineering, process control, and marketing, the advantages of CADD/CAM will become even more pronounced. Advantages claimed for CAM include:

• Communications are improved by the direct transfer of documentation from design to manufacturing.

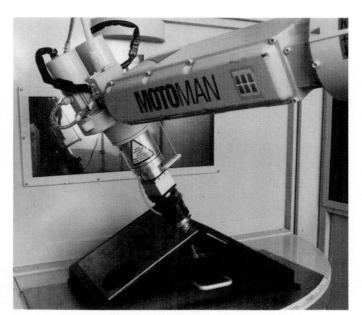

Fig. 23-9. Robotics technology is widely used in industry today, especially for painting, cutting, welding, and assembly tasks. This robot arm is moving a laser cutting head, under CNC control, as it removes a precisely dimensioned circle of material from a metal plate. Fiber optic "light piping" channels the light beam from the laser generator to the cutting head. (Motoman)

- Production is more efficient and output is increased.
- Errors are reduced with design and manufacturing sharing the same database.
- Materials handling and machine processing are more efficient.
- Quality control is improved.
- Lead times are reduced, improving market response.
- Work environment is safer.

CAM works as a subsystem of CIM

The scope of CAM may be limited to a single machining cell or may be expanded to include an entire department or facility, achieving what is often referred to as **CIM,** or **computer-integrated manufacturing.** The following sections describe components usually found in a more comprehensive CIM installation.

Machining Centers

A **machining center** is a CNC machine tool that is capable of performing a variety of material removal operations on a part such as drilling, milling, or boring. Usually, these machines are equipped with automatic tool changing and storage capabilities and part delivery or shuttle mechanisms. CNC turning centers, CNC grinding centers, and other CNC machines are also available for stand-alone machining or for systems integration. See Fig. 23-10.

Operations scheduled for machining centers are numerically controlled by computers and sensors are built into the system to protect the equipment from overload and maintain product quality. These sensors enable the controller to monitor the plant, process, and product, Fig. 23-11. The flow of lubricants and coolants, tool life, and tool breakage are also monitored.

Machining centers require a minimum of operator supervision; work in process is limited only by pallet storage and the number of tools stored in the tool magazine. Machining centers are usually installed as integral parts of flexible manufacturing systems (FMS). The machining center is considered the smallest building block of FMS.

Flexible Manufacturing Cells (FMC)

A **flexible manufacturing cell (FMC)** (sometimes referred to as a "flexible manufacturing *center*") consists of a grouping of machine tools organized into a working unit. Cells are usually configured to perform virtually all of the machining processes needed to produce a part or family of parts. Another form of the FMC is a grouping of like machines dedicated to a particular type of machin-

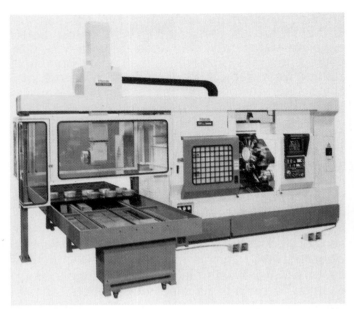

Fig. 23-10. This dual-spindle turning center can work on both ends of a part simultaneously, as well as perform secondary operations such as drilling or tapping. Note the tool-changing turret next to the orange-colored safety cover (cover has been opened to show turret). Loading of parts to be processed is done automatically by the gantry-type robotic loader at the left end of the machine. (Mazak Corporation)

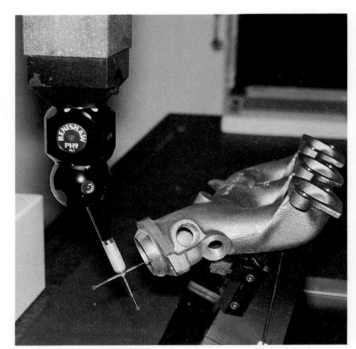

Fig. 23-11. Automatic gauging of a finished part allows immediate adjustment of the manufacturing process if variations from specifications are found. (Ford Motor Company)

ing process, such as a small group of horizontal machining centers. See Fig. 23-12. The equipment

Fig. 23-12. This FMC consists of two horizontal machining centers served by a rail-guided pallet transporter. A load-unload station and a remote staging terminal are used with the 13 parts-holding pallets that help ensure continuous flow of material through the machining centers. (Cincinnati Milacron)

in an FMC can be linked by automated material handling equipment, such as a robot, a conveyor and pallet system, or ***automated guided vehicles (AGVs).*** Operation of the FMC will often be directed by a device called a "cell controller," which may be a computer or a ***programmable logic controller (PLC).*** The PLC is a simpler device than a computer, and is programmed using a step-by-step method called ***ladder logic***.

Flexible Manufacturing Systems (FMS)

A ***flexible manufacturing system (FMS)*** is a production approach that consists of highly automated and computer-controlled machines (machining cells, robots, and inspection equipment) connected through the use of integrated materials handling and storage systems. The automated manufacturing operation is monitored and controlled from a central location, Fig. 23-13.

The term "flexible" refers to the system's ability to process a variety of similar products and to reroute or reschedule production in the event of equipment failure. These systems are designed to deliver quality output in a cost-effective manner and to respond quickly to changing production demands.

Advantages of FMS

The FMS has many advantages, and is the form of manufacturing that most closely approaches

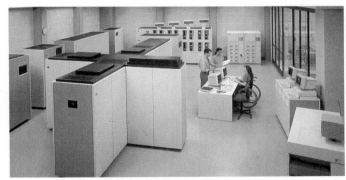

Fig. 23-13. A large automated manufacturing operation is typically controlled from a central location, such as this mainframe computer. (IBM)

computer-integrated manufacturing (CIM). The FMS offers the following advantages:

- Increases volume of production while decreasing costs.
- Manufactures completed single parts or batches of parts in random order, as needed.
- Produces parts "on order," rather than in large lots that must be warehoused, thereby reducing inventory.
- Improves quality control by using 100 percent inspection.
- Decreases hazardous and repetitive work, making the work environment safer and more pleasant.

Group Technology (GT)

Group technology (GT) is a manufacturing philosophy that consists of organizing components into families of parts to be produced in machining cells. These parts are similar in design or in manufacturing requirements. The design characteristics are similar in terms of materials, dimensions and tolerances, shape, and finish. Similarity in manufacturing characteristics might include such factors as tool and machine processes, fixtures required, and sequence of operations. GT is considered an essential element in the implementation of computer-integrated manufacturing (CIM).

Advantages of GT

Although there is expense involved in coding and organizing products for the use of group technology, there are also some distinct advantages.

These include improving product design; providing standardization and families of parts, thereby reducing costs; reducing tooling and set-up costs, and reducing ***work-in-process (WIP)***.

Just-In-Time (JIT) Manufacturing

One of the most widely accepted new concepts introduced to manufacturing in recent years has been the ***Just-In-Time (JIT)*** philosophy. In this method of operation, the goal is to reduce work-in-process to an absolute minimum. This involves the reduction of lead times, reduction of actual WIP inventories, and reduction of set-up times. See Fig. 23-14.

The conventional approach in industry has been to automate existing processes of machining and assembly, which has resulted in isolated cells of production, or "islands of automation." This leads

Fig. 23-14. Application of the JIT philosophy to manufacturing operations, such as this tractor assembly line, emphasizes delivery of parts in the correct quantity, exactly when needed. By avoiding the stockpiling of materials and parts, costs are reduced and efficiency is increased. (John Deere & Company)

to costly periods of waiting time between cells, rather than a smooth efficient flow of material between production processes. The most efficient system moves the product to be manufactured from the firm's suppliers to its customers in a continuous manner with few or no rejects.

The JIT system regards production processes as the only means of adding value to a manufactured product. All other tasks such as transportation, inspection, and storage are defined as "wastes" to be eliminated wherever possible. An efficient production system requires a highly consistent, shortcycled process with minimal inventory in process.

Suppliers to a manufacturer using JIT are expected to function as a coordinated part of the system, delivering materials or parts "just-in-time" for use on the line. Supply and manufacturing both work in terms of small lot sizes, as opposed to purchasing or manufacturing large quantities and storing a portion until needed. Traditional large inventory practices require capital to be invested in nonproductive (at the time) supplies and storage facilities.

Artificial Intelligence (AI) and Expert Systems

Artificial intelligence (AI) is an attempt to program a computer with the information (data) and range of possible responses needed to allow it to identify a problem and make decisions on the best solution to that problem. This is a process that would normally be associated only with human intelligence. Speech recognition, language interpretation, and computer vision (scene interpretation) are among the tools used in this branch of computer science.

Expert systems technology is an application of AI. An expert system's software uses knowledge and inference procedures to solve problems. A fundamental concept involved in operation of an expert system is its ability to "learn" and adapt its responses from information it gathers, rather than function strictly with firm, programmed decisions.

This brief description is an oversimplification of the topics of AI and expert systems. Until further progress is made in these areas, human interaction with automated manufacturing is likely to remain.

COMPUTER-INTEGRATED MANUFACTURING (CIM)

In the strictest definition of the term, *computer-integrated manufacturing (CIM)* is the full automation and joining of all facets of an industrial enterprise: design, documentation, materials selection and handling, machine processing, quality assurance, storing and/or shipping, management, and marketing. There are few, if any, CIM installations at this time that fully meet the definition. There are, however, many partial CIM systems (such as CADD/CAM, FMS, and FMC installation) in operation, especially in the automotive, aircraft production, electronics, and electrical equipment industries, Fig. 23-15. These partial CIM facilities vary widely in size and complexity.

Advantages of CIM

The advantages of CIM in today's manufacturing enterprises are applicable to some extent to any of the subsystems of CIM. CIM and other steps toward automation improve productivity and efficiency in manufacturing, while reducing costs of production and making industry more competitive. Computer control of production operations also increases the effectiveness of quality control activities, improving product reliability. It also makes the workplace safer and working conditions more pleasant for workers.

Future Developments in CIM

Over a relatively short period of time, computer systems have developed in capacity and speed

Fig. 23-15. This CIM facility is used to manufacture electrical components. (Allen-Bradley Company)

while significantly decreasing in cost. This reinforces the strong trend toward computer-integrated manufacturing. To become fully effective, however, CIM must be improved in its adaptive characteristics, with the capability to self-determine which decision to make. This will be accomplished when computer designers develop systems that can learn and respond with the kind of decision-making skills a worker gains from experience, Fig. 23-16. Progress in the area of artificial intelligence and application of this new computer technology will further enhance CIM. As CIM becomes more and more the standard in manufacturing, industry will increasingly require individuals who understand the role of the computer and the design-manufacturing problems that must be solved.

QUESTIONS
FOR DISCUSSION

1. In the early stages of computer-assisted design and drafting (CADD), the computer was used primarily for generation of drawings. What brought about the expanded use of CADD? What is included in that "expanded use?"

2. How does CADD contribute to an industry's ability to remain competitive? To improve quality in products?

3. Explain the term "central database."

4. List some of the advantages of a CADD system.

5. Define computer-assisted manufacturing (CAM). How is it related to CADD?

6. How does a machining center differ from a flexible manufacturing cell (FMC)?

7. Describe what is meant by the term "flexible manufacturing cell." How does the FMC differ from the "flexible manufacturing system (FMS)?"

8. What is "group technology (GT)?"

9. Explain the meaning of the term "Just-in-time manufacturing."

10. What is meant by "artificial intelligence?"

11. What is computer-integrated manufacturing (CIM)? What are its capabilities? How does CIM relate to the fully automated factory?

Fig. 23-16.   Products are tested on an in-line, in-circuit test machine that provides information used to monitor product quality and make necessary changes in manufacturing processes. Computer system designers are attempting to develop software that will be able to make the types of experience-based decisions employed by human operators. (Allen-Bradley Company)

 **24** # Threads and Fastening Devices

KEY CONCEPTS

☐ There are many different types of fasteners used in industry.

☐ Threads can be represented on drawings as detailed, schematic, or simplified.

☐ There are many different types of bolts and screws.

☐ Washers and retaining rings can serve many different purposes.

☐ There are many different types of nuts used in industry.

☐ Keys and pins are types of fastening devices used in industry.

☐ Springs are represented on drawings in a manner very similar to threads.

The hardware and techniques required by modern industry to join components makes fastening one of the most dynamic and fastest growing technologies, Fig. 24-1. To a nontechnical person, fasteners may appear quite simple. Many fasteners are simple. However, in high volume assembly work, the speed of assembly, holding capabilities, and reliability of fasteners call for many special types of fasteners, Fig. 24-2. Many different industries use fasteners. Some of these industries include the aerospace, appliance, automotive, and electrical industries.

Threads and threaded fasteners are used on most machine assemblies produced in industry. Standard methods of specifying and representing screw threads are presented in the following sections.

Considerable progress has been made jointly by the United States, Canada, and England in standardizing screw threads. The result of this cooperative effort is the **Unified Thread Series.** This is now the American standard for fastening types of screw threads.

Unified threads and the former standard, American National threads, have essentially the same thread form. These threads are mechanically interchangeable. The chief difference in the two types of threads is in the application of allowances, tolerances, pitch diameter, and specification.

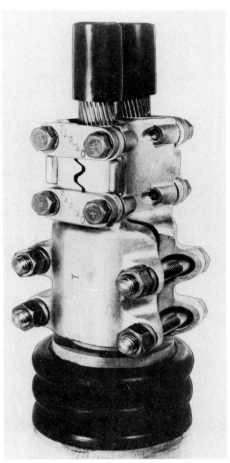

Fig. 24-1. Modern industry depends on fasteners. (ITT Harper, Inc.)

THREAD TERMINOLOGY

The following list of thread terminology includes the more important terms, Fig. 24-3.

• **Major diameter:** The largest diameter on an external or internal screw thread.

• **Minor diameter:** The smallest diameter on an external or internal screw thread.

• **Pitch diameter:** The diameter of an imaginary cylinder passing through the thread profiles at the point where the widths of the thread and groove are equal.

• **Pitch:** The distance from a point on one screw thread to a corresponding point on the next

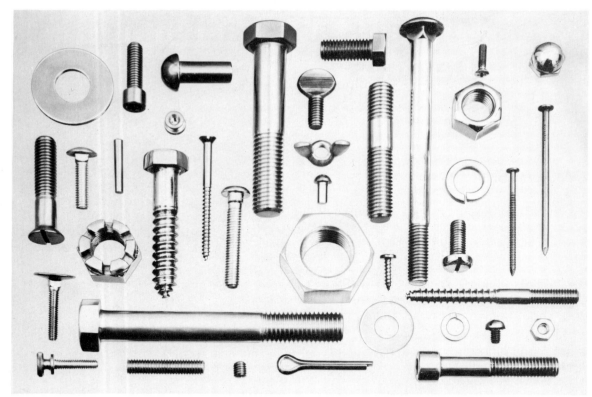

Fig. 24-2. There are many different types of fasteners used in industry. (ITT Harper, Inc.)

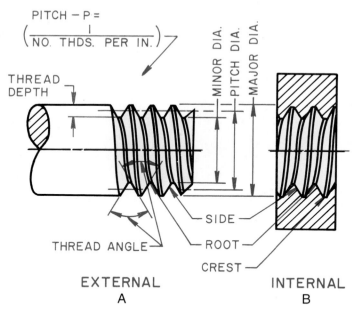

Fig. 24-3. There are specific terms associated with threads.

thread, measured parallel to the axis. The pitch for a particular thread may be calculated mathematically by dividing one inch by the number of threads per inch. Example:

$$\frac{1 \text{ inch}}{10 \text{ thds/inch}} = \text{pitch} = .10 \text{ inch}$$

- **Crest:** The top surface of thread joining two sides (or flanks).
- **Root:** The bottom surface of thread joining two sides (or flanks).
- **Angle of thread:** The included angle between the sides, or flanks, of the thread measured in an axial plane.
- **External thread:** The thread on the outside of a cylinder such as a machine bolt.
- **Internal thread:** The thread on the inside of a cylinder such as a nut.
- **Thread form:** The profile of the thread as viewed on the axial plane. (Refer to Fig. 24-4 for standard thread forms.)
- **Thread series:** The groups of diameter-pitch combinations distinguished from each other by the number of threads per inch applied to a specific diameter.
- **Thread class:** The fit between two mating thread parts with respect to the amount of clearance or interference which is present when they are assembled. "Class 1" represents a loose fit and "class 3" a tight fit.
- **Right-hand thread:** A thread, when viewed in the end view, winds clockwise to assemble. A thread is considered to be right-handed (RH) unless otherwise stated.
- **Left-hand thread:** A thread, when viewed in the end view, winds counterclockwise to

assemble. Left-handed (LH) threads are indicated.

Thread Form

The thread form is the profile of the thread as viewed on the axial plane. There are a number of standard thread forms, Fig. 24-4. However, the Unified form has been agreed upon by the United States, Canada, and Great Britain as the standard for such fasteners as bolts, machine screws, and nuts. The Unified is a combination of the American National and the British Whitworth. The Unified has almost completely replaced the American National form due to fewer difficulties encountered in producing the flat crest and root of the thread.

While the Unified form is used for fasteners, the *Square, Acme, Buttress,* and *Worm* threads are used to transmit motion and power. This is due to the thread profiles. The profiles of these threads are more vertical with their axes. Examples of motion and power transmission are steering gears (worm screw) and lead screws (square) on machine lathes. The *Sharp V* is used where friction is desired, such as setscrews. The *Knuckle* form is for fast assembling of parts such as light bulbs and bottle caps.

Thread Series

The thread series designates the form of thread for a particular application. There are four series of Unified screw threads: coarse, fine, extra-fine, and constant-pitch series.

The *Unified Coarse* series is designated *UNC.* This series is used for bolts, screws, nuts, and threads in cast iron, soft metals, or plastic where fast assembly or disassembly is required.

The *Unified Fine* is labeled *UNF.* This series is used for bolts, screws, and nuts where a higher

tightening force between parts is required. The Fine series is also used where the length of the thread engagement is short and where a small lead angle is desired.

The *Unified Extra-Fine (UNEF)* series is used for even shorter lengths of thread engagements. It is also used for thin-wall tubes, nuts, ferrules, and couplings. It is also used for applications requiring high stress resistance.

The *Unified Constant-pitch* series is designated *UN.* The number of the threads per inch precedes the designation. For example, "8UN" specifies a Unified Constant-pitch thread with eight threads per inch. This series of threads is for special purposes such as high pressure applications. This thread is also used for large diameters where other thread series do not meet the requirements.

The *8-thread series (8UN)* is also used as a substitute for the Coarse-Thread series for diameters larger than 1 inch. The *12-thread series (12UN)* is used as a continuation of the Fine-Thread series for diameters larger than 1 1/2 inches. The *16-thread series (16UN)* is used as a continuation of the Extra-Fine series for diameters larger than 1 11/16 inches. Dimensions for the Unified series of thread are given in the Reference Section.

Thread Classes

The classes of fit for external and internal threads of mating parts are distinguished from each other by the amount of tolerance and allowance permitted for each class. Classes "1A," "2A," and "3A" indicate external threads. Classes "1B," "2B," and "3B" indicate internal threads.

Classes of fit are identified by numeral "1" for applications requiring minimum binding to permit frequent and quick assembly or disassembly of parts. A class "2" fit is for bolts, screws, nuts, and similar fasteners for normal applications in mass

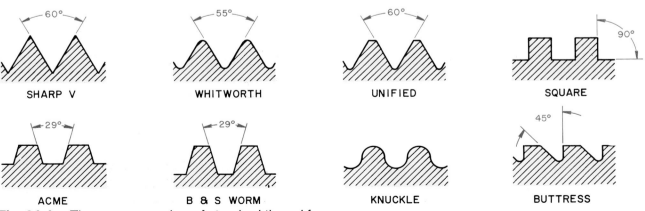

Fig. 24-4. There are a number of standard thread forms.

production. A class "3" fit is for applications requiring closer tolerances than the other classes for a fit to withstand greater stress and vibration.

Single and Multiple Threads

A screw or other threaded machine part may contain single or multiple threads, Fig. 24-5. A *single-threaded screw* will move forward into its mating part a distance equal to its pitch in one complete revolution (360°). In the case of the single-threaded screw, the pitch (P) is equal to the lead. (Refer to Fig. 24-5A.) Notice that the crest line is offset a distance of 1/2P since a single view shows only a 1/2 revolution of a thread.

A *double-threaded screw* has two threads side by side and moves forward into its mating part a distance equal to its lead, or 2P. (Refer to Fig. 24-5B.) The crest line of a double-threaded screw is offset a distance of P in a single view.

A *triple-threaded screw* has three individual threads. A triple-threaded screw moves forward a distance equal to its lead, or 3P. (Refer to Fig. 24-5C.)

Single threads are used where considerable pressure or power is to be exerted in the movement of mating parts. A nut and bolt, or a machinist vise screw and its jaws, are two examples where single threads are used. Multiple threads are used where rapid movement between mating parts is desired. The mating parts of a ballpoint pen, or water faucet valves, are two examples where multiple threads are used.

Thread Specification Notes

Screw threads are specified by a note, Fig. 24-6. The note will provide the following information

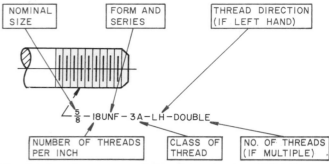

Fig. 24-6. A note specifying threads should list the nominal size, the number of threads per inch, the form and series, and the thread class, in that order. The thread is assumed to be right-handed unless noted. If the thread is left-handed, "LH" should follow the thread class. If the thread is anything other than a single thread, that indication should follow the thread class and the thread direction (if required).

in a standard sequential order: nominal size (major diameter or screw number), number of threads per inch, thread form (UN), and series F grouped together (UNF) and the class of fit (3A). A thread is assumed to be a right-hand, single thread unless noted otherwise. A left-hand thread is indicated by "LH" and multiple threads are indicted by "DOUBLE" or "TRIPLE."

Thread specifications must be included on all threaded parts by a note and a leader to the external thread as shown in Fig. 24-6, or to an internal thread in the circular view, Fig. 24-7A. The length or depth of the threaded part is given as the last item in the specification, or it may be dimensioned directly on the part, Fig. 24-7C. Standard size bolts and nuts may be called out by a letter on the drawing and specified in the materials list. A thread note providing information on the tap drill size, depth, countersinking, and number of holes to be threaded is shown in Fig. 24-8.

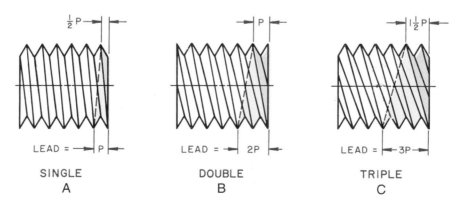

Fig. 24-5. A–A single thread moves into the part a distance equal to the pitch in one revolution. B–A double thread moves into the part a distance equal to twice its pitch in one revolution. C–A triple thread moves into the part a distance equal to three times its pitch in one revolution.

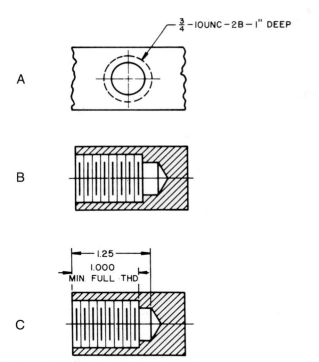

Fig. 24-7. A–An internal thread can be specified with a callout to the circular view. C–An internal thread can also be specified by dimensions in a section view.

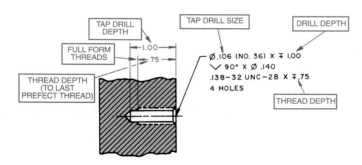

Fig. 24-8. Thread notes may provide additional information such as the tap drill size, the drill depth, and the thread depth.

Metric Translation of Unified Screw Threads

In drawings which are dual-dimensioned, the thread designation shall specify in sequence the nominal size (expressed in decimal inches), the number of threads per inch, thread series symbol, and class symbol. When the pitch diameter (PD) is given, it is presented as follows.

Where the inch system is the primary dimension on dual-dimensioned drawing:

0.375–24 UNF–2A

PD 0.349–0.343

   (8.808–8.713 mm)

Where the metric system is the primary dimension on dual-dimensioned drawing:

0.375–24 UNF–2A

PD 8.808–8.713

   (0.349–0.343 IN)

Metric Threads

Metric threads are designated similar to Unified and American National Standard, but with some slight variations. A diameter and pitch are used to designate the metric series, as in the inch system, with the following modifications.

1. Metric coarse threads are designated by simply giving the prefix "M" and the diameter. In Fig. 24-9, for example, M8 is a coarse thread designation representing a nominal thread diameter of 8 mm with a pitch of 1.25 mm understood. (A thread designation is for a coarse thread unless otherwise noted.)

2. Metric fine threads are designated by listing the pitch as a suffix. A fine thread for a part would be "M8 x 1.0," or 8 mm diameter with a pitch of 1.0 mm. The most common metric thread is coarse. A coarse metric thread generally falls between the coarse and fine series of inch system measurements for a comparable diameter.

In the inch series, the designation of the pitch of a thread is given as the number of threads per inch: "3/4—10UNC" has a pitch of 1/10 inch. In the metric series, the number given for the pitch is the actual pitch: "M x 1.0" has a pitch of 1 mm. Tables showing the International Standards Organization (ISO) metric thread series are in the Reference Section. The basic form of the ISO thread is also shown along with information for thread calculations.

The tolerance and class of fit in metric threads are designated by adding numbers and letters in a certain sequence to the callout. The thread designation in Fig. 24-10 calls for a fine thread of 6 mm diameter, 0.75 mm pitch (no pitch is given in the designation for a coarse thread) with a pitch diameter tolerance grade 6 and an allowance "h", crest diameter tolerance grade 6 and allowance "g".

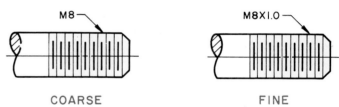

Fig. 24-9. The metric coarse thread designation gives only the diameter, the pitch is understood. The fine thread designation gives the pitch, following the diameter.

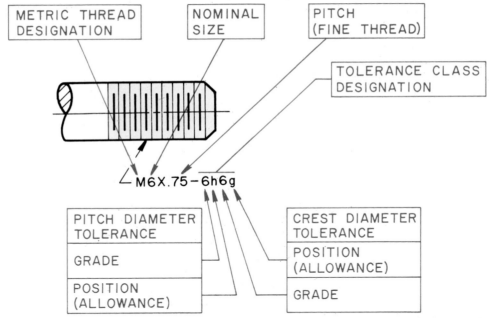

Fig. 24-10. A complete designation for an ISO metric thread will provide all of the information to fully describe the thread.

Thread Representation

There are three conventional methods of representing threads on drawings. The three methods are ***detailed, schematic,*** and ***simplified,*** Fig. 24-11. The detailed convention is a closer representation of the actual thread. It is sometimes used to show the geometry of a thread form as an enlarged detail. However, the schematic and simplified conventions are most commonly used. They save drafting time and, in many instances, produce a clearer drawing.

Detailed representation of V-type threads

The construction of the Sharp V Unified National forms of detail thread representation is shown in Fig. 24-12. The pitch of the thread can be used to lay out thread spacing, Fig. 24-12A. However, the conventional practice, especially on small machine parts, is to approximate the pitch spacing for the thread crests so that they appear natural and in keeping with the size of the part. (Refer to Fig. 24-12A.)

Start the spacing with a half space, since the first thread crest represents only a 1/2 revolution, or

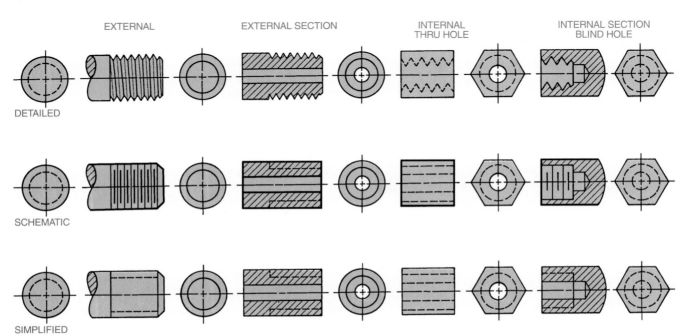

Fig. 24-11. There are conventional representations of screw threads that should be used.

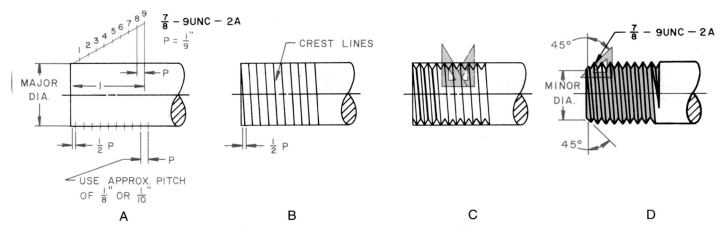

Fig. 24-12. Constructing a detailed representation of a Unified National Thread.

180°. Continue the spacing along the bottom edge of the threaded part. Draw the crest lines by adjusting a triangle along a straightedge to the correct slope, Fig. 24-12B. Actually, the crest and root lines are helix curves (refer to Fig. 6-21), but they are drawn as straight lines.

Notice that the slope for a right-hand thread is shown in Fig. 24-12. The thread advances into its mating part when turned clockwise on its axis. The slope for a left-hand thread would be the opposite.

Draw 60° sides for the thread form, Fig. 24-12C. Join the bottom of these threads to form the root lines. Notice that the root lines are not parallel to the crest lines in this detailed representation. This is due to the difference in diameters. Complete the detail representation by drawing a chamfer on the end of the thread at the minor diameter, Fig. 24-12D. Add the callout (note) to provide the thread specification.

Detailed representation of square threads

The detailed representation of square threads is shown in Fig. 24-13. This is an approximation of the thread since the thread would appear as a helix

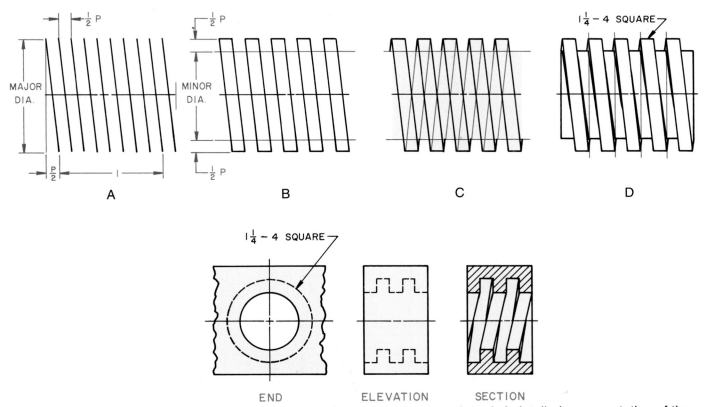

Fig. 24-13. Constructing a detailed representation of an external square thread. A detailed representation of the internal thread is also shown.

curve rather than a straight line. Proceed as follows to draw a detailed thread representation.

1. Construct the major diameter. Draw crest lines 1/2P apart for single threads, Fig. 24-13A. The pitch of square threads is equal to one over number of threads per inch (1/4 in Fig. 24-13).

2. Next draw the lines for the top of the threads, Fig. 24-13B.

3. Draw light lines representing the minor diameter a distance of 1/2P from the major diameter.

4. Draw diagonal construction lines connecting the tops of the thread to represent that portion of the thread on the backside that is visible. Darken the visible thread line outside the minor diameter, Fig. 24-13C.

5. Draw a light construction line from the inside crest lines to locate points on the minor diameter where the root lines meet the minor diameter, Fig. 24-13D.

6. Connect these points with points where the adjacent crest line crosses the centerline to form root lines of the thread. (Refer back to Fig. 24-13D.)

7. Add a note to provide the thread specification.

Detailed representation of the Acme thread

The Acme thread is similar to the square, except the sides of the Acme are drawn to provide an included angle of 30° (actually 29) for the groove and 30° for the thread, Fig. 24-14. The steps in drawing a detailed Acme thread are as follows.

1. Draw a centerline and lay off the length and the major diameter, Fig. 24-14A.

2. Determine the pitch by dividing 1 by the number of threads per inch (1/4 in the example). Draw the minor thread diameter 1/2P from the major diameter.

3. Draw the pitch diameter halfway between the major and minor diameters, Fig. 24-14B. Lay off the thread pitch along one pitch diameter

line by measuring a series of divisions 1/2P (for thread and groove) and projecting these divisions to the opposite pitch diameter.

4. Construct sides of threads by drawing lines 15° with the vertical and through points marked on the pitch diameter, Fig. 24-14C. Draw the crest and root diameters of the thread.

5. Connect the thread crests with lines sloping downward 1/2P to the right for right-hand threads (to the left for left-handed threads).

6. Draw root diameter lines to complete the thread, Fig. 24-14D.

7. Add a note to provide thread specifications.

The construction of detailed internal Acme threads is shown in Fig. 24-15.

Two classes of Acme threads are provided: General Purpose and Centralizing. **General Purpose** classes (2G, 3G, and 4G) provide clearances on all diameters for free movement. The variation in classes has to do with the amount of backlash or end play in the threads. Class 2G is the preferred choice and, if less backlash is desired, classes 3G and 4G are provided.

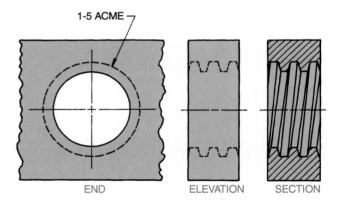

Fig. 24-15. Detailed representation of internal Acme threads.

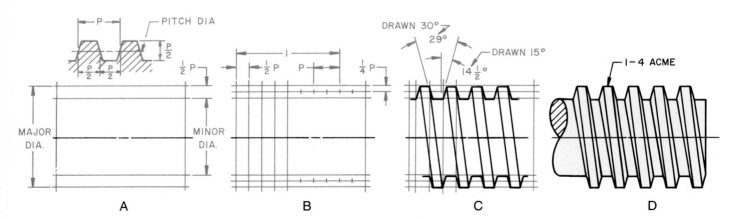

Fig. 24-14. Constructing a detailed representation of an external Acme thread.

Some examples of Acme thread notes specifying General Purpose classes are:

1/2 – 10 ACME– 2G

2 – 4 ACME – 4G

Centralizing classes of Acme threads (2C to 6C) have limited clearances at the major diameters of internal and external threads. This permits a bearing to maintain approximate alignment of the threads and prevents wedging. A class 6C Acme thread is a closer fit than a class 2C. Some examples of Acme thread notes specifying Centralizing classes are:

3/8 – 12 ACME– 5C

4 – 2 ACME – 2C

Schematic representation of threads

Schematic thread symbols are recommended and approved for the representation of all screw threads. These symbols are used by industry, along with the thread specifications, on most drawings to represent threaded parts, Fig. 24-16. Notice that the external thread in the section view is not a schematic symbol but a detailed representation. Also notice that the internal thread in the elevation is the same symbol used for the simplified internal thread in the elevation.

The construction of the schematic thread symbol is as follows.

1. Lay out a centerline and the major diameter of the thread, Fig. 24-17A.
2. Lay off the pitch of the thread by the graphical method or by measurement. The pitch does not need to be true, and can be estimated if laid off uniformly.
3. Draw thin lines across the diameter to represent the crest lines of the thread.
4. To find the minor diameter, lay off a 60° "V" between the crest lines. Draw a light construction line along the threaded length of part, Fig. 24-17B. Repeat this on the opposite side of the piece.
5. Use a heavy line for the root line. This line is drawn between the lines marking the minor

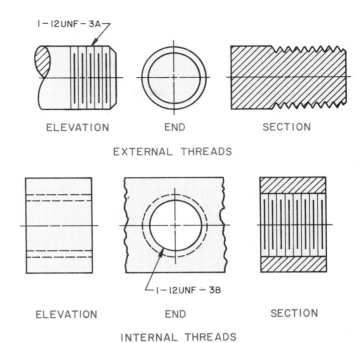

Fig. 24-16. Schematic representation of threads.

diameter and is spaced uniformly between the crest lines, Fig. 24-17C. Do not draw a root line in the first space next to the end.

6. Draw a 45° chamfer from the last full thread. Add a note to provide thread specifications, Fig. 24-17D.

Simplified representation of threads

The use of simplified thread symbols is the fastest way to represent screw threads on a drawing, Fig. 24-18. The major diameter is found by direct measurement. The minor diameter is found by the 60° "V" method, or it is estimated.

Representation of Small Threads

Threaded parts of a small diameter are difficult to draw to true or reduced scale dimensions. The small screw pitch will crowd the crest and root lines

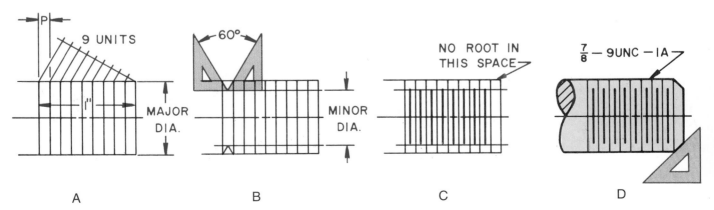

Fig. 24-17. Constructing a schematic thread symbol for an external thread.

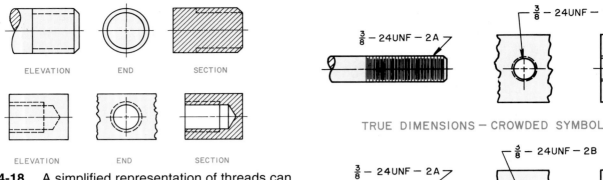

Fig. 24-18.   A simplified representation of threads can provide a clear view of the drawing. These should be used when specific details of the thread do not need to be seen.

TRUE DIMENSIONS — CROWDED SYMBOL

EXAGGERATED SPACING CLARIFIES SYMBOL

Fig. 24-19.   Exaggerated spacing clarifies thread symbols on small size thread drawings.

in the schematic method, Fig. 24-19. In the simplified method, the clarity of the symbol would be impaired by a crowding of the major and minor diameters. The conventional practice is to exaggerate the space between crests and roots, and major and minor diameters, since accuracy is not as important as clarity of the symbol. The note specifying the thread controls the actual thread characteristics.

Pipe Threads

Three forms of American Standard pipe threads are used in industry. These three forms are the **Regular, Dryseal,** and **Aeronautical.** The regular

pipe thread is the standard for the plumbing trade. It is available in *tapered* and *straight* threads.

Tapered pipe threads are cut on a taper of 1 in 16 measured on the diameter. They may be drawn straight or at an angle, since the thread note indicates whether the thread is straight or tapered. When the tapered pipe threads are drawn at an angle, they should be exaggerated by measuring 1 unit in 16. These are measured on the radius rather than the diameter. This is an angle of approximately 3°, Fig. 24-20.

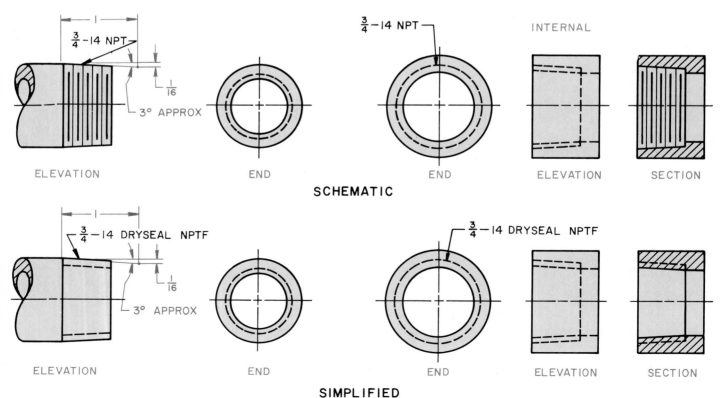

Fig. 24-20.   Tapered pipe threads can be represented in schematic and simplified forms.

The Dryseal pipe thread is standard for automotive, refrigeration, and hydraulic tube and pipe fittings. The general forms and dimensions of these threads are the same as regular pipe threads except for the truncation of the crests and roots. The Dryseal pipe thread form has no clearance since the flats of the crests on the external and internal thread meet, producing a metal-to-metal contact and eliminating the need for a sealer.

Aeronautical pipe thread is the standard in the aerospace industry. In this industry, the internally threaded part is typically made of soft light materials (such as aluminum or magnesium alloys) and the screw is made from high-strength steel. An insert, usually of phosphor bronze, is used as the bearing part of the internal thread, preventing wear on the light alloy thread, Fig. 24-21.

The Regular and Aeronautical pipe thread forms require a sealer to prevent leakage in the joint. The Dryseal form requires no seal.

The specifications for American Standard pipe threads are listed in sequence: nominal size, number of threads per inch, symbols for form, and series. For example, a typical specification is 1/2—14 NPT.

The following symbols are used to designate the more common American Standard pipe threads.

- NPT—American Standard Taper Pipe Thread
- NPTR—American Standard Taper Pipe for Railing Joints
- NPTF—Dryseal American Standard Pipe Thread

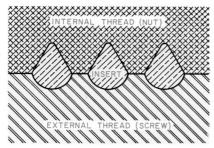

Fig. 24-21. Aeronautical pipe threads have an insert that will bear the load. These threads are used when one of the thread materials is soft and the other is hard.

- NPSF—Dryseal American Standard Fuel Internal Straight Pipe Thread
- NPSI—Dryseal American Standard Intermediate Internal Straight Pipe Thread

TYPES OF BOLTS AND SCREWS

There are many varieties and sizes of bolts, nuts, and screws for all kinds of industrial applications. There are five general types of threaded fasteners that the drafter-designer should be familiar with.

A *bolt* has a head on one end and is threaded on the other end to receive a nut. It is inserted through clearance holes to hold two or more parts together, Fig. 24-22A.

A *cap screw* is similar to a bolt with a head on one end, but usually a greater length of thread on the other. It is screwed into a part with mating internal threads for greater strength and rigidity, Fig. 24-22B.

A *stud* is a rod threaded on both ends to be screwed into a part with mating internal threads. A nut is used on the other end to secure two or more parts together, Fig. 24-22C.

A *machine screw* is similar to a cap screw except it is smaller and has a slotted head, Fig. 24-22D.

A *setscrew* is used to prevent motion between two parts, such as rotation of a collar on a shaft, Fig. 24-22E.

The ranges of sizes and exact dimensions for all of these threaded fasteners are given in the Appendix, starting on page 643.

Drawing Square Bolt Heads and Nuts

The drawing of square bolt heads is identical to the drawing of square nuts, except the nut is usually thicker than the bolt head. The method illustrated in Fig. 24-23 is based on the bolt diameter and is an approximation of the actual projection.

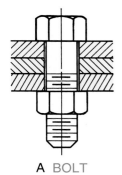 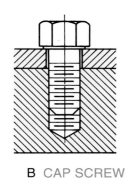 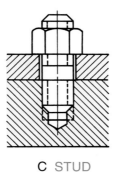 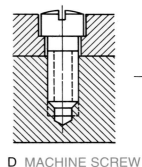 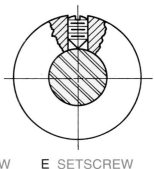

A BOLT   B CAP SCREW   C STUD   D MACHINE SCREW   E SETSCREW

Fig. 24-22. There are five general types of bolts and screws.

A drawing of bolt heads and nuts across their corners is the best representation. This method should be used when a choice is available. This method is shown first.

1. Draw the bolt diameter (nominal size), Fig. 24-23A.
2. Draw the bolt head thickness and diameter.
3. Draw the square head around the diameter at 45° and project it to the front view, Fig. 24-23B.
4. Locate the centers for chamfer arcs in the front view by projecting lines 60° with the horizontal down from the center and outside of the corners of the top surface.
5. Complete the drawing of the square bolt head across the corners by drawing a 30° chamfer line at the outside corners in the front view, Fig. 24-23C.

The regular square head nut shown in Fig. 24-23D is 7/8 D in thickness. The thickness of the heavy duty nut equals the diameter of the bolt it matches. The hidden lines in the front view that represent threads are not normally shown in applications, especially in bolt and nut assemblies. (See Fig. 24-22.)

Drawing Hexagonal Bolt Heads and Nuts

Hexagonal bolt heads and nuts are also best represented when drawn across their corners. For this method, shown in Fig. 24-24, proceed as follows.

1. Draw the bolt diameter (nominal size), Fig. 24-24A.
2. Draw the bolt head thickness and diameter.

3. Lay out the hexagonal head around the diameter at 60° with the horizontal. Project this to the front view, Fig. 24-24B.
4. Locate centers for the center chamfer arc in the front view by projecting lines 60° with the horizontal and down from the outside corners of the top surface.
5. Locate centers for the side chamfer arcs by projecting 60° lines down from the two inside corners to meet the other 60° lines.
6. Complete the drawing of the hexagonal bolt head across the corners by drawing a 30° chamfer line at the outside corners in the front view, Fig. 24-24C.

The construction of a hexagonal nut is shown in Fig. 24-24D. A regular nut is drawn 7/8 D in thickness and a heavy duty hex nut is drawn 1D in thickness. Hidden lines may be omitted unless needed for clarity.

The procedure for drawing hexagonal head bolts and nuts across their flats are illustrated in Fig. 24-25.

Drawing Cap Screws

A drawing of cap screws (like a drawing of bolts) is a proportioned, approximate drawing based on the diameter of the screw. Five types of standard cap screws are shown in Fig. 24-26, together with the dimensions for their construction.

Specific dimensions for assembly and thread lengths should be checked in a standards table. A cap screw may be specified as 7/16—14 UNC—2A x 2 BRASS HEX CAP SCR. If the cap screw is made of steel, the material term is omitted.

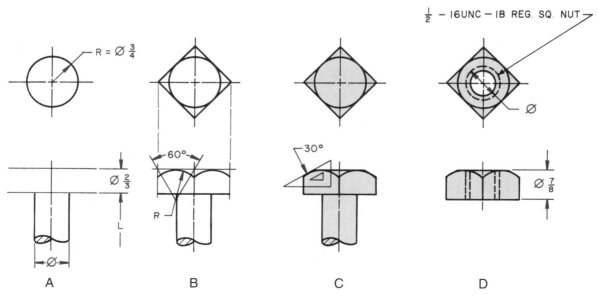

A   B   C   D

Fig. 24-23. Drawing square bolt heads and nuts across their corners gives a close approximation of true projection.

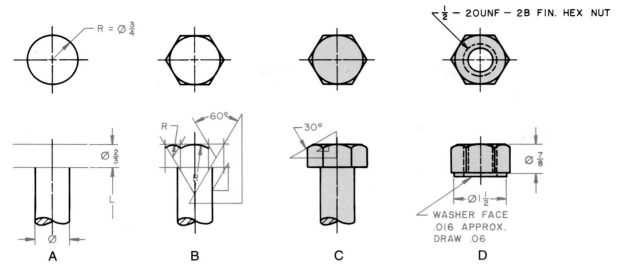

Fig. 24-24. The approximation method can be used to draw hexagonal bolt heads and nuts across corners.

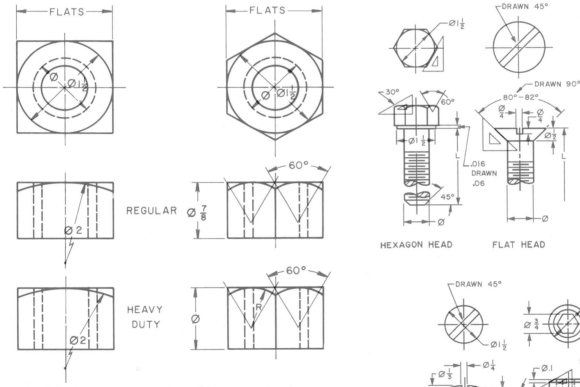

Fig. 24-25. Drawing square and hexagonal nuts across the flats does not give a good representation. This position should be avoided whenever possible.

HEXAGON HEAD   FLAT HEAD   ROUND HEAD

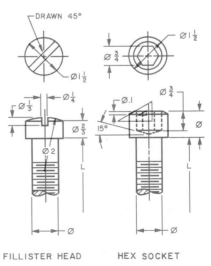

FILLISTER HEAD   HEX SOCKET

Fig. 24-26. There are several different types of cap screws and each is represented in a different manner on drawings.

Drawing Machine Screws

Four common types of machine screws and their construction are shown in Fig. 24-27. Machine screws are similar to cap screws, but usually smaller in diameter.

The threads on machine screws are either Unified Coarse (UNC) or Unified Fine (UNF), Class 2A.

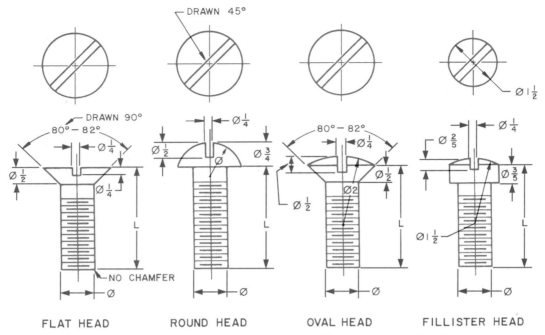

Fig. 24-27. There are several different types of machine screws and each is represented in a different manner on drawings.

Screws that are 2 inches in length or less are threaded to within two threads of the bearing surface. The thread length on screws longer than 2 inches is a minimum of 1 3/4 inches. A machine screw may be specified as 8—24 UN F—2A x 1 OVAL HD MACH SCR.

Drawing Setscrews

Setscrews are made in the standard square head type as well as several headless types, Fig. 24-28. Several styles of points are also available with each type. When a setscrew is used against a

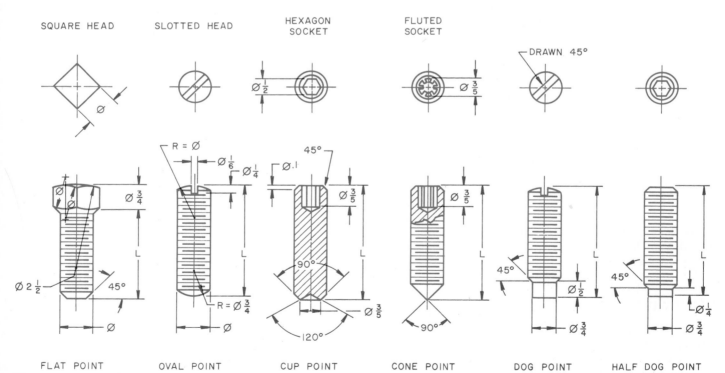

Fig. 24-28. There are several different types of setscrews and each is represented differently on drawings.

round shaft, a cup point is likely to hold best. A flat or dog point is used where a flat spot on a shaft has been machined. The threads for setscrews can be coarse, fine, or 8-thread series, class 2A. An exception is the square-head setscrews. These are normally stocked in the coarse series and size 1/4 inch or larger. A setscrew may be specified as 1/4—28 UNF—2A x 5/8 HEX SOCK CUP PT SET SCR.

Drawing Self-tapping Screws

Time in assembly work is of great importance in many industrial applications. To meet this condition where threaded fasteners are used, self-tapping screws have been developed, Fig. 24-29. Self-tapping screws can be divided into two major kinds.

- **Thread-cutting screws** act like a tap. They cut away material as they enter the hole. These screws have flutes or slots in the point to form a cutting edge. Their thread form is similar to standard Unified Threads, Fig. 24-30. This type screw is suitable for applications in metal as well as plastics. Thread-cutting screws can be removed and reassembled without noticeable loss of holding power.

- **Thread-forming screws** form threads by displacing the material rather than cutting it. These screws are sometimes called **sheet-metal screws.** They are especially suited for thin-gage sheet metal up to .375 inch in thickness, as well as in any soft material such as wood and plastic, Fig. 24-31. The thread form on this screw is a narrow, sharp crest. No chips or waste material is formed in their application.

Fig. 24-29. Self-tapping screws speed the work of industry. (ITT Harper, Inc.)

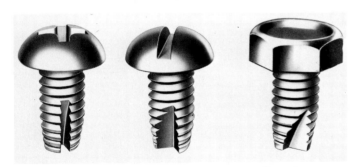

Fig. 24-30. Thread-cutting screws actually cut threads as they are inserted. (ELCO Industries, Inc.)

Fig. 24-31. Thread-forming screws displace the metal to form threads.

Two patented thread-forming screws have a special shape to displace the metal and form a tight fitting thread. These screws are known as Swageform® and Taptite®, Fig. 24-32.

Metallic Drive Screws

Some industrial applications call for permanent fasteners. These fasteners are not expected to be disassembled. Drive screws are designed for this use, Fig. 24-33. They have multiple threads with a large lead angle. They are driven by a force in line with their axis, rather than torque. (Torque is a circular force, or force through a radius.)

Once seated, metallic drive screws cannot be removed and reinserted easily. Economy is one of the main reasons for using this kind of screw where circumstances permit. Safety is another reason. Sometimes it is necessary to assemble components so that they cannot be disassembled without special tools. Drive screws are available for use with a variety of materials. They are typically used with wood, plastic, or metal.

Fig. 24-32. Swageform® and Taptite® threads have special forms to provide a more secure thread engagement. (ELCO Industries, Inc.)

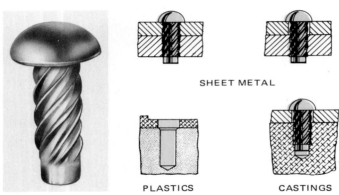

Fig. 24-33. Drive screws are an economical and permanent type of fastener. (ELCO Industries, Inc.)

Wood Screws

Wood screws are standardized with three head types: flat, round, and oval, Fig. 24-34. These are available in slotted or Phillips head drives. The Phillips head is used in most commercially manufactured products where time in assembly is important.

Wood screws range in size from 0 to 32. The diameter varies from .060″ to .372″. They may be specified in a note or in the Materials List as NO. 9 x 1 1/2 OVAL HD WOOD SCR.

Templates for Drawing Threaded Fasteners

A variety of templates are available for drawing threads, bolts, screws, nuts, and head types, Fig. 24-35. Templates are used in industry to speed drafting time and to produce more uniform representations of threaded fasteners.

WASHERS AND RETAINING RINGS

Washers are added to screw assemblies for several different reasons. The following is a list of the more common reasons for using washers.

- Load distribution.
- Surface protection.
- Insulation.
- Spanning an over-size clearance hole.
- Sealing.
- Electrical connections.
- Taking up spring tension.
- Locking.

Lock washers are either split-spring or toothed washers.

A *finishing washer* distributes the load and eliminates the need for a countersunk hole. It is used extensively for attaching fabric coverings. *Flat washers* are used primarily for load distribution, Fig. 24-36.

Retaining rings are inexpensive devices used to provide a shoulder for holding, locking, or positioning components on shafts, pins, studs, or in bores, Fig. 24-37. These are available in a wide variety of designs. They almost always slip or snap into grooves and are sometimes called "snap" rings.

NUTS—COMMON AND SPECIAL

There is a wide variety of nuts, Fig. 24-38. Nuts are discussed in this section under two broad

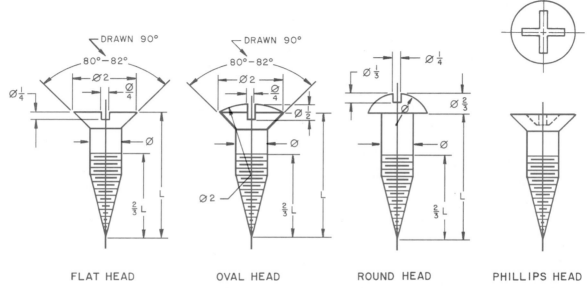

FLAT HEAD   OVAL HEAD   ROUND HEAD   PHILLIPS HEAD

Fig. 24-34. There are several different types of wood screws. Each type is represented differently on drawings.

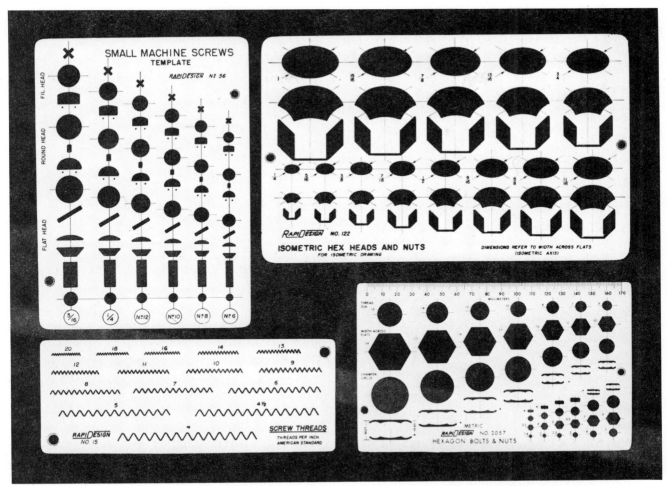

Fig. 24-35. Templates should be used when available to speed the drawing of threaded fasteners. (RapiDesign)

Fig. 24-36. There are several different types of washers used in mechanical assemblies. Some examples include finishing, flat, and lock washers. (ELCO Industries, Inc.)

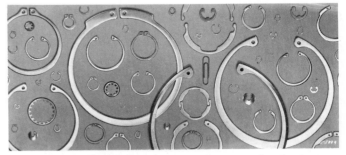

Fig. 24-37. Retaining rings are inexpensive fasteners and can be quickly assembled or removed. (Koh-I-Noor Rapidograph, Inc.)

classes: common and special. Only a few of the special nuts that are available are discussed here. However, those presented indicate the wide variety available. A supplier's catalog would contain many hundreds of special use items.

Common Nuts

Nuts used on bolts for assemblies are known as ***common nuts.*** Common nuts are generally divided between finished and heavy, Fig. 24-39. ***Fin-***

ished nuts are used for close tolerances. ***Heavy nuts*** are used for a looser fit, for large-clearance holes, and for high loads.

Special Nuts

There are several types of special nuts available. ***Special nuts*** are used where an application requires features not found on common nuts. Cap

Fig. 24-38. The variety of nuts used in industry is seemingly limitless. (Standard Pressed Steel Co.)

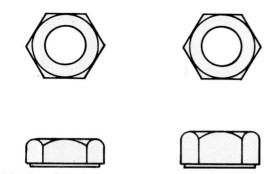

Fig. 24-39. Common nuts are one type of nut.

nuts, single-thread engaging nuts, captive nuts, and locknuts are all types of special nuts.

Cap, wing, and knurled nuts

A *cap nut* is used for appearance, Fig. 24-40A. Cap nuts are sometimes called *acorn nuts. Wing nuts* and *knurled nuts* allow for hand tightening, Fig. 24-40B and Fig. 24-40C.

Single-thread engaging nuts

Nuts formed by stamping a thread engaging impression in a flat piece of metal are called *single-thread engaging nuts.* An example of a single-thread engaging nut is shown in Fig. 24-41. The nut shown has helical prongs that engage and lock on the screw thread root diameter. A protruding truncated cone nut is stamped into the metal. This provides a ramp for the screw to climb as it turns. Single-thread nuts can be formed from nearly any ferrous or nonferrous alloy. Usually, however, they

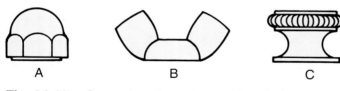

Fig. 24-40. Cap nuts, wing nuts, and knurled nuts are special nuts.

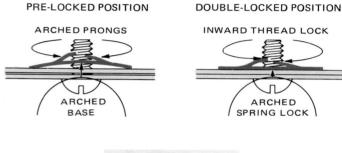

Fig. 24-41. Single-thread engaging nuts are used in applications where very little clearance is present. (Eaton Corp.)

are made of high-carbon steel, hardened and drawn to a "spring" temper. These nuts are often used to reduce assembly costs where lighter-duty applications are involved.

Captive or self-retaining nuts

Captive or *self-retaining nuts* are multiple-threaded nuts that are held in place by a clamp or binding device of light gage metal. They are used for applications where thin materials are used. These nuts are also used where threaded fasteners are needed at inaccessible or blind locations. Assemblies that require repeated assembly and disassembly often use these nuts as well.

Self-retaining nuts may be grouped according to four means of attachment.

- *Plate* or *anchor nuts* have mounting lugs that can be screwed, riveted, or welded to the assembly, Fig. 24-42A.
- *Caged nuts* are held in place by a spring-steel cage that snaps into a hole or clamps over an edge. Fig. 24-42B.
- *Clinch nuts* are designed with a pilot collar clinched or staked into a parent part through a precut hole, Fig. 24-42C.
- *Self-piercing nuts* are held in a strip or a tool and used with a punch press, Fig. 24-42D. The punch press "punches" a hole in the work piece with the hardened nut. The press also clinches the nut in place in the work piece.

Locknuts

Locknuts are a type of special nut that prevent the nut from loosening up once it is properly tightened. There are three groups of locknuts: free-spinning, prevailing-torque, spring-action nuts.

The *free-spinning* type grips tightly only when the nut is seated on a surface or when two mating parts are tightened together, Fig. 24-43. There are several types of free-spinning locknuts. Those with two mating parts clamp the threads of bolt when

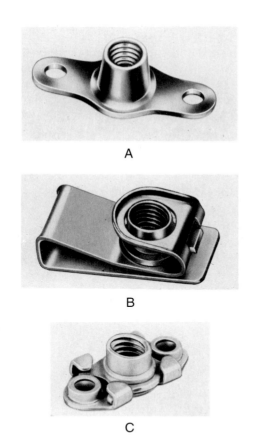

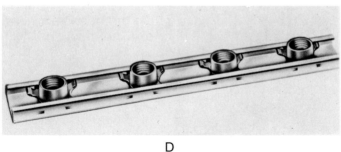

Fig. 24-42. There are several different types of self-retaining nuts. (Esna Corp.)

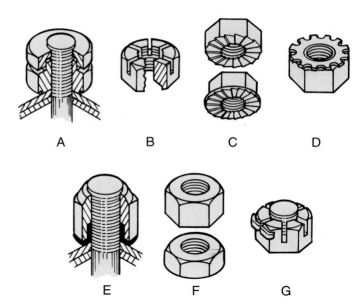

Fig. 24-43. There are many types of free-spinning locknuts.

seated and resist back-off, Fig. 24-43A. Locknuts with a recessed bottom and slotted upper portion cause a spring action when seated, and bind upper threads of the nut, Fig. 24-43B. Those nuts with a deformed bearing surface tend to dig in and remain tight when seated, Fig. 24-43C.

Some locknuts have a lock washer secured to the main nut, Fig. 24-43D. Others have inserts, Fig. 24-43E. The insert tends to flow around the threads when seated, forming a tight lock and seal. Jam nuts are thin nuts used under common nuts, Fig. 24-43F. When seated under pressure, the threads of the jam nut and bolt are elastically deformed. This causes considerable resistance against loosening. Slotted nuts have slots to receive a cotter pin or wire, Fig. 24-43G. The pin or wire passes through a drilled hole in the bolt, locking the nut in place. These nuts look

very similar to "spring action" nuts described above. (Refer back to Fig. 24-43B.)

Prevailing-torque locknuts start freely, then a wrench must be used to get to the final position. This is due to a deformation of the threads or insert in the center or upper portion of the nut. These nuts maintain a constant load against loosening whether seated or not.

Spring-action locknuts are the single-thread type, usually stamped from spring-steel. They lock in place when driven up against a surface. (Refer back to Fig. 24-41.) These nuts are sometimes classed as free-spinning locknuts, but they can be jammed onto a thread without spinning. Frequently in mass production situations, these nuts are jammed onto a thread. (Note that these locknuts are also single-thread engaging nuts.)

RIVETS

The manufacture of many assembled products requires a permanent type of fastener. In these cases, rivets often are the answer. Rivets are typically used for aircraft structures, small appliances, and jewelry. Rivet sizes are indicated by the diameter of the shank and by the length of the shank, if it is an unusual length. Rivets are available in a variety of head styles and are grouped into two general types: standard and blind.

Standard Rivets

Standard rivets come in several styles, depending on strength, methods of application, and other design requirements, Fig. 24-44.

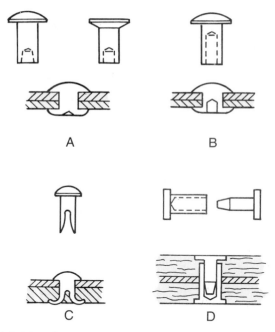

Fig. 24-44. Standard rivets come in a variety of different types.

Semitubular rivets are the most widely standard rivet, Fig. 24-44A. This type of rivet becomes essentially a solid rivet when properly specified and set. Semitubular rivets can be used to pierce very thin light metals, although they are not classified as self-piercing rivets.

Full tubular rivets have deeper shank holes. These rivets can also punch their own holes in fabric, some plastics, and other soft materials, Fig. 24-44B. The shear strength of full tubular rivets is less than that of semitubular rivets.

Bifurcated or ***split rivets*** are punched or sawed to form prongs that enable them to punch their own holes in fiber, wood, plastic, or metal, Fig. 24-44C. These rivets are also called ***self-piercing rivets.***

Compression rivets consist of two parts: a deep-drilled tubular part and a solid part designed for an interference fit when set, Fig. 24-44D. This type of rivet is used when both sides of a work piece must have a finished appearance, such as the handle of a kitchen knife.

Blind Rivets

Blind rivets can be installed in a joint that is accessible from only one side. However, they are increasingly being used in applications where standard rivets were used in the past to simplify assembly, reduce cost, and to improve appearance. Blind rivets are classified by the methods used in setting. They are also available in a variety of head styles.

Pull-mandrel blind rivets are set by inserting the rivet in the joint and pulling a mandrel to upset the blind end of the rivet, Fig. 24-45. These are sometimes called ***pop rivets.*** Some rivets have mandrels that pull through leaving a hole in the rivet, Fig. 24-45A. Others have "break-type" mandrels that break during the pull-through process and plug the hole, Fig. 24-45B. A third "nonbreak-type" mandrel must be trimmed off after the rivet is set, Fig. 24-45C.

The ***threaded-type blind rivet*** consists of an internally threaded rivet that is torqued or pulled to expand and set the rivet, Fig. 24-45D.

The ***drive-pin blind rivet*** is like the mandrel type in reverse. The pin is driven into the body to set the blind side of the rivet, Fig. 24-45E.

Chemically expanded blind rivets have a hollow end filled with an explosive that detonates when heat or an electric current is applied to set the rivet, Fig. 24-45F.

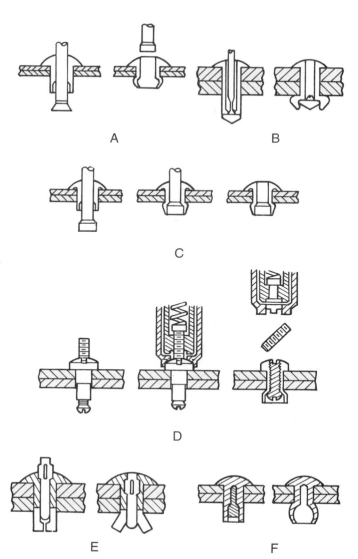

Fig. 24-45. There are many different types of blind rivets.

SPRINGS, PIN FASTENERS, AND KEYS

Keys and pin fasteners are two other ways of joining parts. Each type has an application that it is best suited for. Springs are presented in this section because their graphic representation is very similar to threads.

Pin Fasteners

Where the load is "primarily" shear, **pins** can be an inexpensive and effective means of fastening, Fig. 24-46. The method of representing pins on a drawing is shown in Fig. 24-47. Some of the different types of pins include the following.

- **Hardened dowel.**
- **Ground dowel.**
- **Hardened taper.**
- **Ground taper.**
- **Grooved surface.**
- **Spring (or tubular).**
- **Clevis.**
- **Cotter pins.**

Springs

The steps in laying out a representation of a coil spring on a drawing are similar to those in representing screw threads. A detail drawing of a coil spring from a check valve is shown in Fig. 24-48. A schematic representation of various types of coil springs is shown in Fig. 24-49.

To lay out a coil spring, mark off the pitch distance along the diameter of the coil. (Refer back to Fig. 24-49.) Give the coils a slope of one-half of the pitch for closely wound springs. Note the difference in the representation in a tension and a compression spring and how the different types of ends are drawn in each case. To avoid a repetitious series of coils, phantom lines may be used to represent repeated detail between spring ends.

Keys

Keys are used to prevent rotation between a shaft and a machine part. Some parts that typically use keys are gears and pulleys. The four most common types of keys are **square, gib head, Pratt and Whitney,** and **Woodruff.** These are shown in Fig. 24-50 along with the method of dimensioning them.

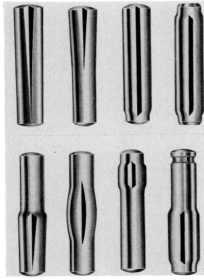

Fig. 24-46. Pin fasteners are often used in assembly work where the load is primary. (Groov-Pin Corp.)

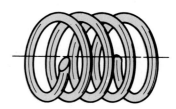

CHECK VALVE SPRING

Fig. 24-48. Detailed representation of coil springs.

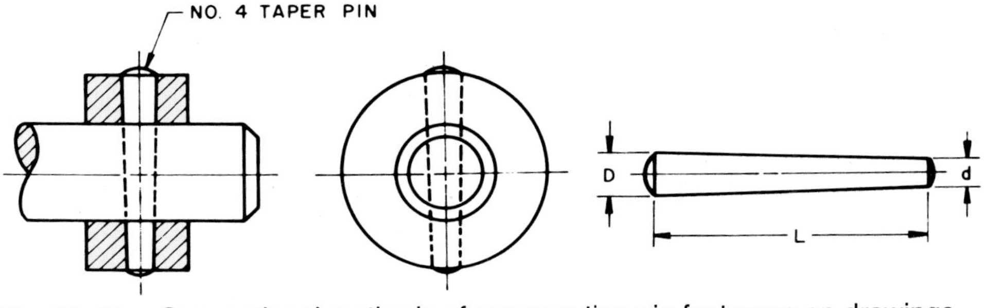

NO. 4 TAPER PIN

D   d

L

Fig. 24-47. Conventional methods of representing pin fasteners on drawings should be followed.

TENSION SPRING   COMPRESSION SPRING   TORSION SPRING

Fig. 24-49. Schematic representation of coil springs.

PROBLEMS AND ACTIVITIES

The following problems will provide you with the opportunity to apply the knowledge of threads and fasteners presented in this chapter. These problems will also allow you to become familiar with the drafting procedures necessary to represent fasteners on drawings. Unless indicated otherwise, use the suggested layout shown in Fig. 24-51.

1. Construct a detail representation of a Unified National Coarse thread showing an external threaded shaft and a sectional view of an internal thread of the same specification. The thread specification is DIA = 1 1/2; 6 THDS/INCH, CLASS 3. Place the thread specification note on the drawing.

2. Make a detailed drawing of an external and internal square thread specified as 1 1/2—3 SQUARE. Dimension the thread with a note.

3. Using the same specifications as for Problem 2, draw an Acme thread, class 2G.

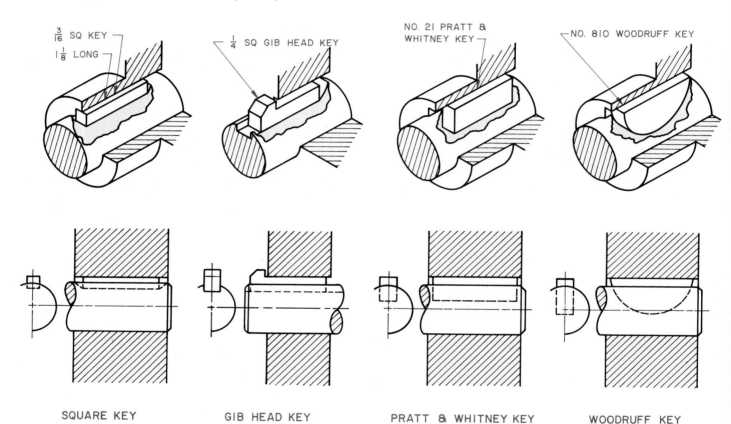

$\frac{3}{16}$ SQ KEY

$1\frac{1}{8}$ LONG

$\frac{1}{4}$ SQ GIB HEAD KEY

NO. 21 PRATT & WHITNEY KEY

NO. 810 WOODRUFF KEY

SQUARE KEY   GIB HEAD KEY   PRATT & WHITNEY KEY   WOODRUFF KEY

Fig. 24-50. There are several different types of standard keys. Keys are used to prevent rotation between a machine part and a shaft.

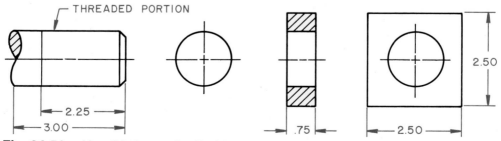

THREADED PORTION

2.25

3.00

.75

2.50

2.50

Fig. 24-51. Use this layout for Problems 1-9 in the "Problems and Activities" section.

4. Draw a schematic representation of threads for a 7/8 class 2 Unified National thread. Check the table in the Reference Section for thread specifications. Dimension the thread with a note.

5. Use the same information from Problem 4, except represent a Square thread with an outside diameter of 3/4 inch.

6. Draw a simplified thread representation of an Acme thread specified as 1—5 ACME—2G—LH—DOUBLE. Dimension the thread with a note.

7. Draw two views, one a sectional view, of a schematic representation of a threaded hole 2″ deep in a 1″ square bar of steel 3″ long. The hole is to be pilot drilled to a depth of 2 1/4″ and threaded with a 9/16 Unified National Extra Fine class 3 thread. Dimension the feature with a note.

8. Make a simplified thread drawing of the threaded pieces in Problem 1.

9. Draw a schematic representation of Unified National Thread Series 8 with an outside diameter of 2″.

10. Draw a schematic representation of an American Standard Taper Pipe thread with a nominal pipe size of 1 1/4″. Check the Reference Section for the thread specifications. Show two views of an external thread and two views, one to be a section, of an internal thread. Dimension the threads with a note.

11. Make a drawing similar to the one in Problem 10 for an American Standard Taper Pipe thread with a nominal pipe size 3/8″, Dryseal.

12. Draw a regular hexagonal and square head bolt and nut of the following specifications: Nominal size 1/2″, length 3″, length of thread 2″. Show the threads in simplified representation. Place a dimensional note on the drawing.

13. Make a drawing of a 3/8″ semifinished hexagonal head bolt 2″ long. The bolt clamps two pieces of 5/8″ steel plate together. The bolt is also held by a jam nut and a semifinished regular nut. Specify the bolt and nuts in a note.

14. Make a two-view drawing of the "SPINDLE BEARING ADJUSTING NUT," Fig. 24-52A. Make the circular view a full section. Use schematic symbol representation for the threads.

15. Draw two views of the "SPECIAL ADJUSTING SCREW," Fig. 24-52B. Show the counterbore and full thread as a broken-out section. Use the schematic symbol to represent the thread and dimension the part. (Check the Reference Section for dimensions not furnished on the drawing.)

16. Draw the necessary views to adequately describe the "CONTROL SHAFT," Fig. 24-52C. Use the simplified symbol representation for the threads. Dimension the drawing.

17. Draw the views required to adequately describe the "SPINDLE RAM SCREW," Fig. 24-52D. Show a detail representation of the thread by drawing two full threads on each end of threaded portion. Indicate the remainder of the thread by using phantom lines at the major diameter. Dimension the part.

18. Draw the necessary views to adequately describe the "PEDESTAL BRACKET SHAFT," Fig. 24-53A. Use schematic thread symbol for threads. Dimension the drawing.

19. Draw a half section of the "SHANK," Fig. 24-53B. Show the threads in schematic form. Dimension the part.

20. Make a front view as shown of the "GEAR SHAFT," Fig. 24-53C. Also make a partial top view sufficient to show both keyseats for the 3/16″ square key at "1" and a No. 404 Woodruff key at "2." (Check the Reference Section for key specifications.) Use schematic symbols for thread representation. Dimension the drawing.

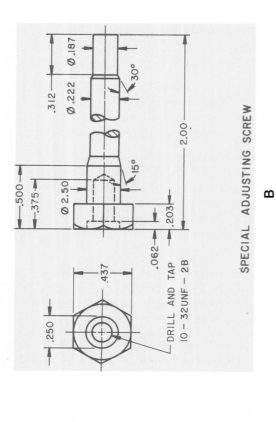

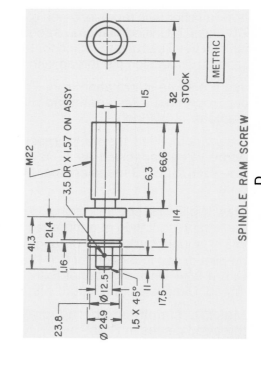

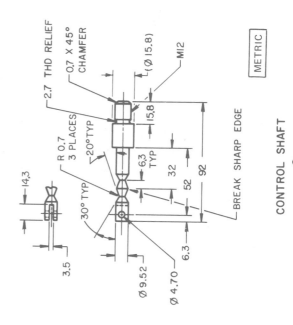

Fig. 24-52.   Use these objects for the problems indicated in the "Problems and Activities" section that involve thread representation.

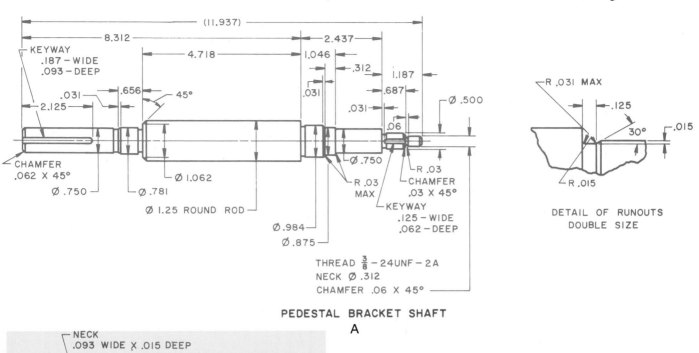

PEDESTAL BRACKET SHAFT

A

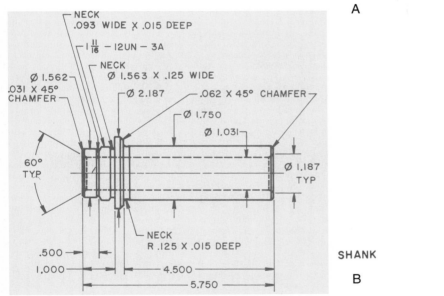

SHANK

B

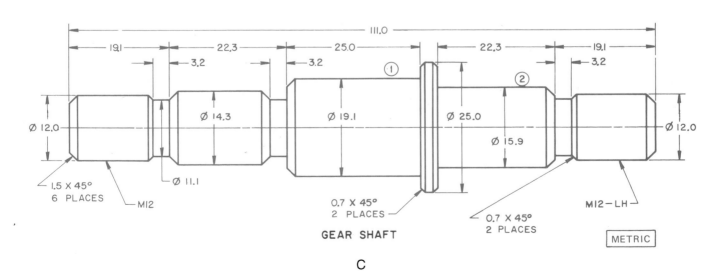

GEAR SHAFT

C

Fig. 24-53. Use these objects for the problems indicated in the "Problems and Activities" section that involve thread representation.

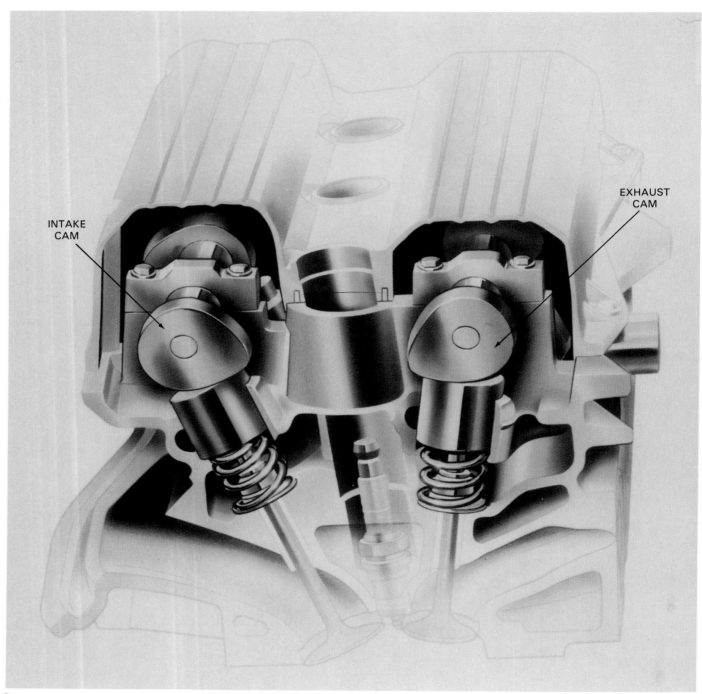

Cams are an essential part of modern engin design.

Cams, Gears, and Splines

KEY CONCEPTS

- [ ] Cams transmit rotary motion into reciprocating motion.
- [ ] Three basic types of cams are plate cams, groove cams, and cylindrical cams.
- [ ] Uniform motion, harmonic motion, uniformly accelerated motion, and combination motion are all types of cam motion.
- [ ] A displacement diagram is typically the first step in designing a cam.
- [ ] Spur gears, bevel, gears, worm gears and worms, are all types of gears.
- [ ] Splines are similar to multiple, parallel keys and prevent rotation between a shaft and its related member.

Modern machines usually require mechanisms to transfer motion and power from one source to another. Most often this transfer has to occur without slippage that might be present with belts. It is also necessary in some instances to convert rotary motion to reciprocal motion (straight-line motion) at a certain rate of speed for related parts. For example, the firing action of a four-stroke internal combustion engine converts reciprocal motion to circular motion, Fig. 25-1.

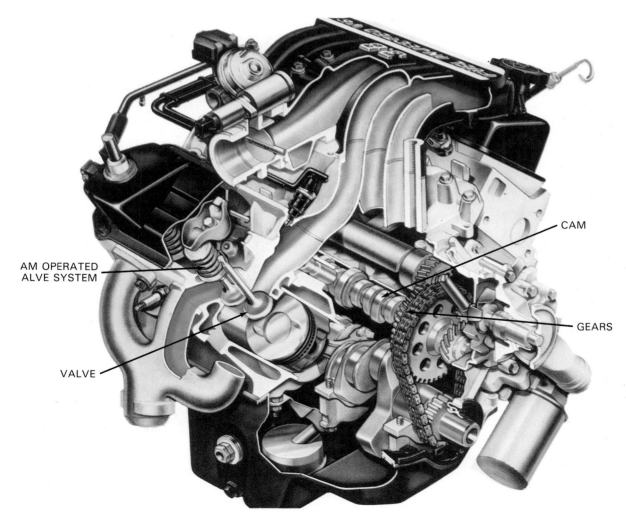

CAM

GEARS

AM OPERATED
ALVE SYSTEM

VALVE

Fig. 25-1. Gears and cams are used to maintain the precise relationship between valves and pistons to produce the intake, compression, power, and exhaust strokes of an internal combustion engine. (Ford Motor Company)

In four-stroke engine operation, there must be a specific timing for opening and closing of the valves in relation to the cycling of the piston. This is achieved with gears and cams. This chapter presents information on some basic types of cams, gears, and splines, and how these features are represented on drawings.

CAMS

A ***cam*** is a mechanical device that changes uniform rotating motion into reciprocating motion of varying speed, Fig. 25-2. Three types of cams are commonly used. These types are ***plate cam,*** ***groove cam,*** and ***cylindrical cam,*** Fig. 25-2.

A ***follower*** makes contact with the surface or groove of the cam. The follower is held against the cam by gravity, spring action, or by a groove in the groove cam, Fig. 25-3. Basic types of cam followers are ***knife edge, flat face,*** and ***roller,*** Fig. 25-3.

Cams may be designed to provide a number of different types of motion and displacement patterns. ***Motion*** refers to the cam follower's rate of speed or movement in relation to the uniform rotation speed of the cam, Fig. 25-4. ***Displacement*** refers to the distance the cam follower moves in relation to the rotation of the cam. The three basic

types of cam follower motion will be discussed later in this chapter.

Cam Displacement Diagrams

A ***displacement diagram*** is a graph or drawing of the displacement (travel) pattern of the cam follower caused by one rotation of the cam, Fig. 25-5. Construction of a displacement diagram is usually the first step in the design of a cam.

In Fig. 25-5, divisions on the abscissa scale, or angular sectors, around the base circle represent time intervals of the revolving cam. (The ***abscissa*** is the horizontal coordinate of a point in a plane Cartesian coordinate system obtained by measuring parallel to the "X" axis.) It has been established that when the speed of rotation of a cam is constant, the time intervals are uniform. Note in Fig. 25-5 that the distance from the base circle on the ordinate scale represents the distance of travel or displacement of the cam follower. When the time intervals are kept constant, the cam follower's rate of speed varies as the angle or incline of the cam changes.

The divisions along the abscissa scale can approximate the actual spaces on the cam base circle since, the displacement diagram is only representative of the motion of the cam. Displacement

Fig. 25-2.   There are three types of cams commonly used in mechanisms. These three types are: A–The plate cam, B–The groove cam, and C–The cylindrical cam. (Ferguson Machine Co.)

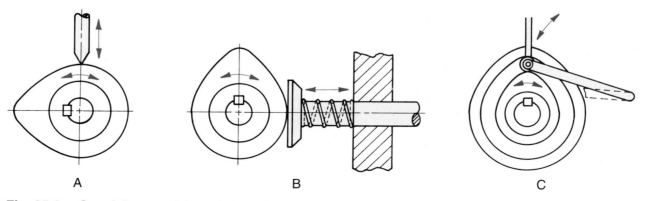

Fig. 25-3.   Cam followers pick up the rotating motion of the cam and change it to reciprocating motion. The three basic types of followers are: A–The knife edge, B–The flat face, and C–The roller.

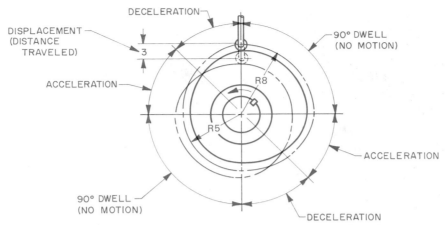

Fig. 25-4. Cams can be designed to provide a variety of motion and displacement patterns.

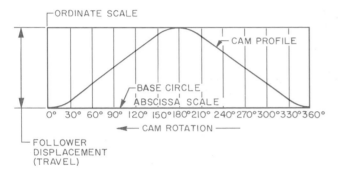

Fig. 25-5. A cam displacement diagram is a graph of the travel pattern of the cam follower caused by one rotation of the cam. Creating this is usually the first step in designing a cam.

on the ordinates must be accurate since these are used in laying off measurements on the cam layout itself.

Types of Cam Follower Motion

The three principal types of motion for cam followers are uniform motion, simple harmonic motion, and uniformly accelerated motion.

Uniform motion is produced when the cam moves the follower at the same rate of speed from the beginning to the end of the displacement cycle. The shape of a uniform motion cam is shown as a straight line in the displacement diagram, Fig. 25-6A.

However, with a straight line cam design, the starting and stopping of the follower would be very abrupt due to instantaneous changes in velocity. So cam shape is usually modified with arcs (R) having a radius of one-fourth to one-half the follower displacement. This smooths out the beginning and ending of the follower stroke. The uniform motion cam is used for machinery operating at a slow rate of speed.

Proceed as follows to lay out a uniform motion cam.

1. Lay out a base circle with a radius equal to the distance from the cam axis to the lowest follower position as shown at the 0° position, Fig. 25-6B.
2. Draw a convenient number of equally spaced radial lines dividing the base circle into intervals representing the angular motion of cam.

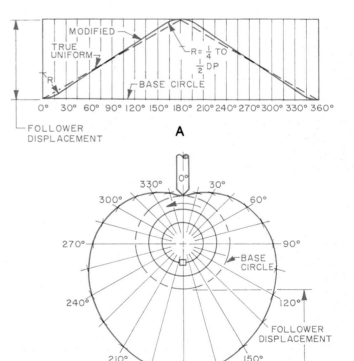

Fig. 25-6. A uniform motion cam produces a constant motion throughout the travel of the follower. The displacement diagram will show the travel as a straight line (no curves).

(The number of divisions must equal the divisions along the base circle of the displacement diagram for the cam.) In Fig. 25-6, 24 increments of 15° have been used.

3. Starting with the 0° position of the displacement diagram, transfer with dividers the distances that the modified cam profile line lies above the base circle on each 15° line to the corresponding line on the cam layout beyond the base circle. Note that the cam rotates counterclockwise and the plotting progresses in opposite direction, Fig. 25-6B.

4. When all points have been located, sketch a smooth curve through the points. Finish with an irregular curve.

Harmonic motion moves the follower in a smooth continuous motion. This movement is based on the successive positions of a point moving at a constant velocity around the circumference of a circle, Fig. 25-7A. The harmonic cam is used for machinery operating at moderate speeds.

Proceed as follows to lay out a harmonic motion cam.

1. Lay out a displacement diagram by constructing a semicircle with a diameter equal to the desired follower displacement. Divide the semicircle into the same number of equal parts as there are angular divisions for one-half of the cam layout. (Refer back to Fig. 25-7A.) Project these divisions to their corresponding angular ordinate. Draw a curve representing the displacement diagram for the harmonic motion cam.

2. Lay out the base circle of the cam with a radius equal to the distance from the cam axis to the lowest follower position, as shown at 0° position in Fig. 25-7B.

3. Draw a convenient number of equally spaced radial lines dividing the base circle into sectors representing the angular motion of the cam. (The number of divisions must equal the divisions along the base circle of the displacement diagram.)

4. Starting with the 0° position of the displacement diagram, transfer the distances that the harmonic curve lies off of the base circle at each ordinate to its corresponding radial line in the cam layout, Fig. 25- 7B. Note that the cam rotates clockwise and the plotting progresses in the opposite direction.

5. When all of the points have been located, sketch a smooth curve through the points. Finish with an irregular curve.

Uniformly accelerated motion is designed into a cam to provide constant acceleration or deceleration of the follower displacement, Fig. 25-8.

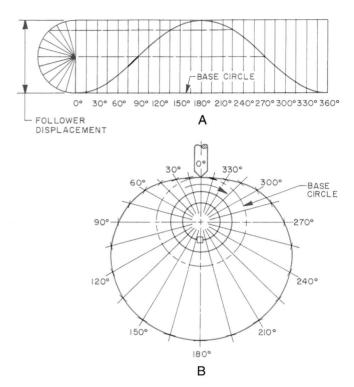

Fig. 25-7. A harmonic motion cam moves the follower at a constant speed, based on a point moving around the circumference of a circle. The displacement diagram will show the travel as a smooth curve.

The displacement on the ordinates varies at the end of successive uniform intervals on the abscissa scale or angular sectors around the base circle such as 0, $1^2 = 1$, $2^2 = 4$, $3^2 = 9$, and so on.

Divisions along the abscissa scale represent equal intervals of time, Fig. 25-8A. Divisions on the ordinate scale represent distances that are the squares of each successive time interval.

During the first one-half revolution of the cam, the follower rises with constant acceleration from 0° to 90°. From 90° to 180°, the cam still rises but with constant deceleration. Note on the displacement diagram that there is a reversal of the ordinate scale at 90° (or midway). (Refer back to Fig. 25-8A.)

During the second half revolution of the cam, the follower falls with constant acceleration from 180° to 270°. It continues to fall from 270° to 360° with constant deceleration, returning the follower to its lowest or "zero" point. (Refer back to Fig. 25-8A.) This type of cam motion is suited for high speed cam operation.

Proceed as follows to lay out a uniformly accelerated motion cam.

1. Lay out a displacement diagram by drawing an inclined line and laying off squares of successive intervals of time. (Refer back to Fig. 25-8A.) Note that the squares of the intervals increase through interval 3 (90°) and decrease

in same manner from interval 3 to the height of the full displacement (180°). Project these divisions to ordinate lines at 0° and from there to their corresponding angular ordinate. Draw a curve representing the displacement diagram for the uniformly accelerated motion cam.

2. Lay out a base circle of a cam with a radius equal to the distance from the cam axis to the lowest follower position as shown at the 0° position, Fig. 25-8B.

3. Draw a convenient number of equally spaced radial lines dividing the base circle into sectors representing the angular motion of cam. (The number of divisions must equal the divisions along the base circle of the displacement diagram.)

4. Starting with the "zero" position of the displacement diagram, transfer the distances the uniformly accelerated curve lies off of the base circle at each ordinate to its corresponding radial line in the cam layout. (Refer back to Fig. 25-8B.) Note that this cam rotates clockwise and the plotting progresses in the opposite direction from the base circle outwards.

5. When all of the points have been located, sketch a smooth curve through the points. Finish with an irregular curve.

Combination motion may be designed for a single cam in order to achieve the follower displacement desired, Fig. 25-9. Note that the cam follower is the roller type, and the center of the roller is assumed to start on the base circle for

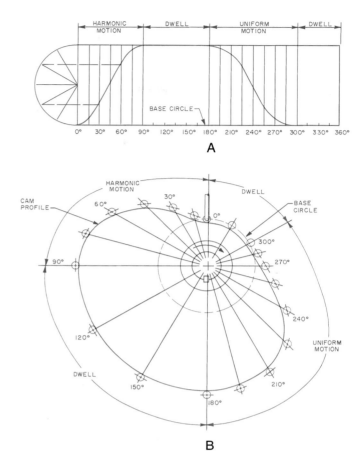

Fig. 25-9. A combination motion cam may provide a combination of the characteristics of the uniform motion cam, the harmonic motion cam, and the uniformly accelerated motion cam.

layout purposes. Transfer displacement distances from the diagram to their respective radial lines in the layout in the usual manner. Lay off an arc with a radius equal to that of the roller from these points. The cam profile is drawn tangent to the roller positions on the radial lines.

Cam with Offset Roller Follower

A uniformly accelerated cam with an offset roller follower is shown in Fig. 25-10. Since the motion is uniformly accelerated throughout, it can be plotted directly from the follower without drawing a displacement diagram.

Note in Fig. 25-10 that the center of the roller follower is located on the base circle. Draw a circle with its center at the center of the base circle and tangent to the extended centerline of the roller follower. Divide this circle into twelve 30° sections and tangents drawn at the section points. Next, transfer distances from the uniform acceleration diagram to these tangent lines that extend from the base circle outward as shown at the 90° radial line. Then draw

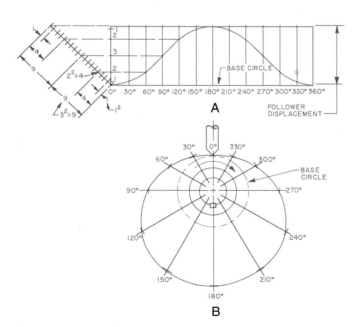

Fig. 25-8. A uniformly accelerated motion cam provides uniform acceleration or deceleration of the follower, resulting in a smooth curve displacement diagram.

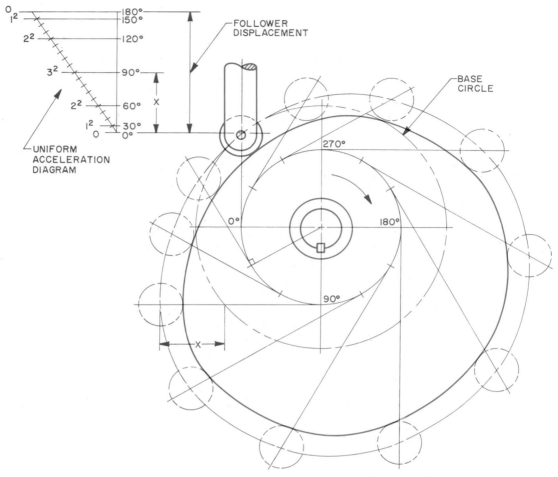

Fig. 25-10.   An offset follower has a center point that is located on the base circle. The surface of the follower that is in contact with the cam is "offset" from the base circle.

circles representing the roller at each of these locations. Draw a smooth curve tangent to the 12 positions of the roller to form the profile of the offset roller uniformly accelerated motion cam.

GEARS

Gears are machine parts used to transmit motion and power by means of successively engaging teeth. **Gear teeth** are shaped so contact between the teeth of mating gears is continually maintained while rotation is occurring, Fig. 25-11. Teeth with the involute curve are the type most commonly used for gears. The purpose of this section is to provide an introduction to the terminology, representation, and specification of basic gear types on drawings.

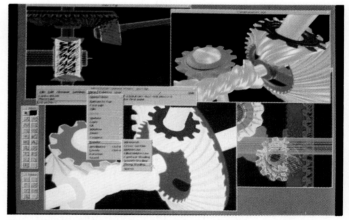

Fig. 25-11.   Gear designs can be easily created, and communicated, using CADD software. (Intergraph Corporation)

Spur Gears

Spur gears are used to transmit rotary motion between two or more parallel shafts, Fig. 25-12. The teeth of a spur gear may be cut parallel to the

gear axis. These gears are called **straight spur gears,** Fig. 25-12. These gears are satisfactory for low or moderate speeds but tend to be noisy at high speeds. The "reverse" gear in a manual transmission car is a straight spur gear. This is why manual transmissions tend to "grind" in reverse. Modifications of the spur gear for heavier loading and higher speeds are achieved through helical and herringbone toothed gears, Fig. 25-13.

When mating spur gears of different size are in mesh, the larger one is called the gear, the smaller one is the pinion. Only straight spur gears are discussed here.

Spur gear terminology

The design drafter must know and understand gear terminology in order to properly specify and represent gears on drawings. Some essential terms are defined here. Formulas are given, when appropriate, for finding various gear measurements.

Number of teeth or threads (N, n). The *number of teeth* in the gear or pinion, or number of

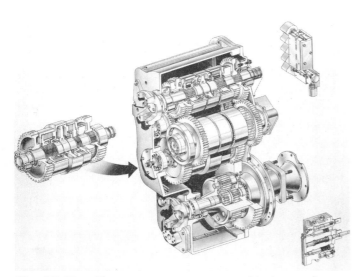

Fig. 25-12. Many spur gears are used to transmit motion and power in manual transmissions. (International Harvester Co.)

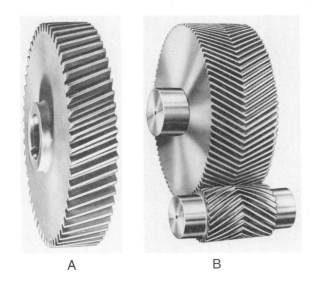

A   B

Fig. 25-13. The teeth of spur gears can be A—helical in shape or B—herringbone.

threads in the worm. This is abbreviated with either "N" or "n."

Diametral pitch (P). The *diametral pitch* is the number of teeth (N) in a gear per inch of pitch diameter. A gear having 48 teeth and a pitch diameter of 3 inches has a diametral pitch of 16.

$$P = \frac{N}{D}$$

Pitch circle. The *pitch circle* is an imaginary circle located approximately half the distance from the roots and tops of the gear teeth. It is tangent to the pitch circle of the mating gear, Fig. 25-14.

Pitch diameter D. The *pitch diameter* is the diameter of the pitch circle.

$$D = \frac{N}{P}$$

Addendum (a). The *addendum* is the radial distance between the pitch circle and the top of the tooth.

$$a = \frac{1}{P} = 0.5 \, (D_0 - D)$$

Dedendum (b). The *dedendum* is the radial distance between the pitch circle and the bottom of the tooth.

$$b = \frac{1.157^*}{P} + 0.5 \, (D - D_R)$$

*A constant for involute gears.

Outside circle or addendum circle. The *outside,* or *addendum, circle* is the diameter of the pitch circle plus twice the addendum (same as outside diameter).

Outside diameter (D_o). The *outside diameter* is the diameter of a circle coinciding with the tops of the teeth of an external gear (same as addendum circle).

$$D_0 = D + 2a = \frac{N}{P} + 2\left(\frac{1}{P}\right) = \frac{N + 2}{P}$$

Root circle or dedendum circle. The *root,* or *dedendum circle* is the circle that coincides with the bottom of the gear teeth.

Root diameter (D_r). The *root diameter* is the diameter of the root circle. It is equal to the pitch diameter minus twice the dedendum.

$$D_r + D - 2b = \frac{N}{P} - \frac{2(1.157)}{P} = \frac{N - 2.314}{P}$$

Center distance C. The *center distance* is the center-to-center distance between the axes of two meshing gears.

$$C = PR_1 + PR_2 = \frac{N_1 + N_2}{2P}$$

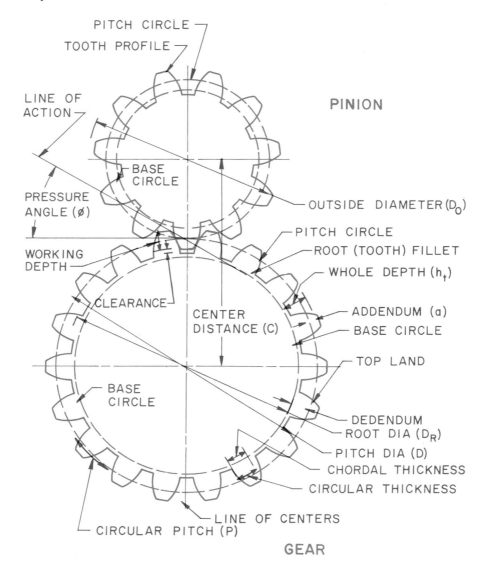

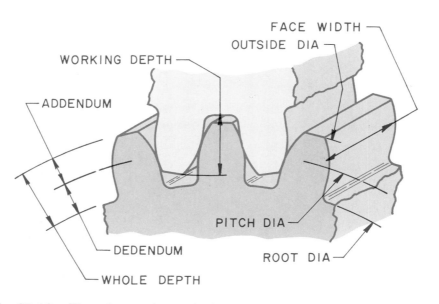

Fig. 25-14. There is certain terminology that applies to spur gears. The drafter/designer must know and understand these terms.

Where PR_1 and PR_2 are the respective pitch radii, and $N1$ and $N2$ are the respective number of teeth of the two meshing gears.

Clearance C. The *clearance* is the radial distance between the top of a tooth and the bottom of the tooth space of a mating gear.

$$c = b - a = \frac{1.157}{P} - \frac{1}{P} = \frac{0.157}{P}$$

Circular pitch (p). The *circular pitch* is the length of the arc along the pitch circle between similar points on adjacent teeth.

$$p = \frac{\pi D}{N} = \frac{\pi}{P}$$

Circular thickness (t). The *circular thickness* is the length of the arc along the pitch circle between the two sides of the tooth.

$$t = \frac{p}{2} = \frac{\pi D}{2N}$$

Face width (F). The *face width* is the width of the tooth measured parallel to the gear axis.

Chordal addendum (a_c). The *chordal addendum* is the radial distance from the top of the tooth to the chord of the pitch circle.

$$a_c = a + \frac{D}{2}\left[1 - \cos\left(\frac{90°}{N}\right)\right]$$

Chordal thickness (t_c). The *chordal thickness* is the length of the chord along the pitch circle between the two sides of the tooth.

$$t_c = D \sin\left(\frac{90°}{N}\right)$$

Whole depth (h_t). The *whole depth* is the total depth of a tooth (addendum plus dedendum).

$$h_t = a + b = \frac{1}{P} + \frac{1.157}{P} = \frac{2.157}{P}$$

Working depth. The *working depth* is the sum of the addendums of two mating gears.

Pressure angle (ø). The *pressure angle* is the angle of pressure between contacting teeth of meshing gears. Two involute systems, the 14 1/2° and 20°, are standard with the 20° gradually replacing the older 14 1/2°. The pressure angle determines the size of the base circle that it is tangent to, Fig. 25-14.

Base circle. The *base circle* is the circle from where the involute profile is generated. The diameter of the base circle is determined by the pressure angle of the gear system.

Spur gear representation

The normal practice in representing gears on industrial drawings is to show the gear teeth in simplified conventional form rather than draw them in detail form, Fig. 25-15.

The circular view may be omitted unless needed for clarity. A table of gear data is included on the drawing to supply the specifications needed to manufacture the gear, Fig. 25-16. Note that a phantom line is used to represent the outside and root diameters and a centerline is used to represent the pitch circle. Where it is necessary to show tooth profiles for clarity, a gear template should be used, Fig. 25-17.

Rack and pinion

A *rack* is a spur gear with its teeth spaced along a straight pitch line, Fig. 25-18. The rack and pinion have a number of uses in machinery and equipment, such as lowering and raising the spindle of a drill press.

Bevel Gears

Bevel gears are used to transmit motion and power between two or more shafts whose axes are at an angle (usually 90°) and would intersect if extended, Fig. 25-19. Bevel gears of the same size and at right angles are called *miter gears*. Straight toothed bevel gears are discussed here, but helical toothed bevel gears are often used for quieter and smoother operation.

Bevel gear terminology

Some of the terms used for bevel gears are the same as for spur gears. These are noted in the list that follows. Also given, where appropriate, are formulas for straight bevel gear measurements.

Diametral pitch (P_d). The *diametral pitch* is the same as for spur gears.

Pitch diameter D. The *pitch diameter* is the diameter of the pitch circle at the base of the pitch cone, Fig. 25-20.

$$D = \frac{N}{P_d}$$

Circular pitch (p). The *circular pitch* is the same as for spur gears.

Circular thickness (t). The *circular thickness* is the same as for spur gears, but measured at large end of tooth.

Outside diameter (D_o). The *outside diameter* is the diameter of the crown circle of the gear teeth.

$$D_o = D + 2_a \cos\Gamma$$

Crown height (χ). The *crown height* is the distance from the cone apex to the crown of the gear tooth measured parallel to the gear axis.

$$\chi = \tfrac{1}{2} D_o / \tan\Gamma_o$$

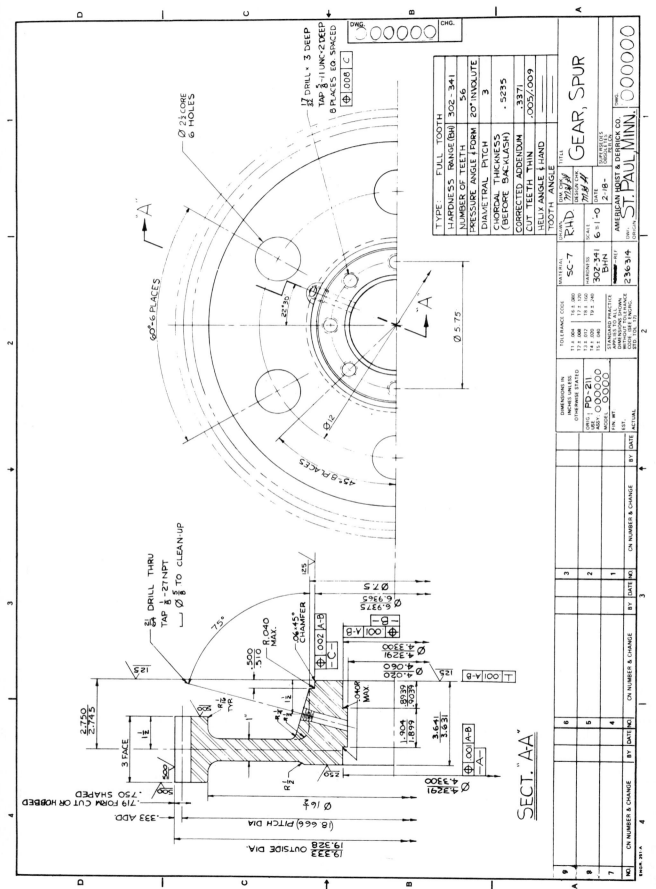

Fig. 25-15. Notice on this industrial drawing of a spur gear that the outside diameter of the gear teeth are represented by phantom lines. (American Hoist & Derrick Co.)

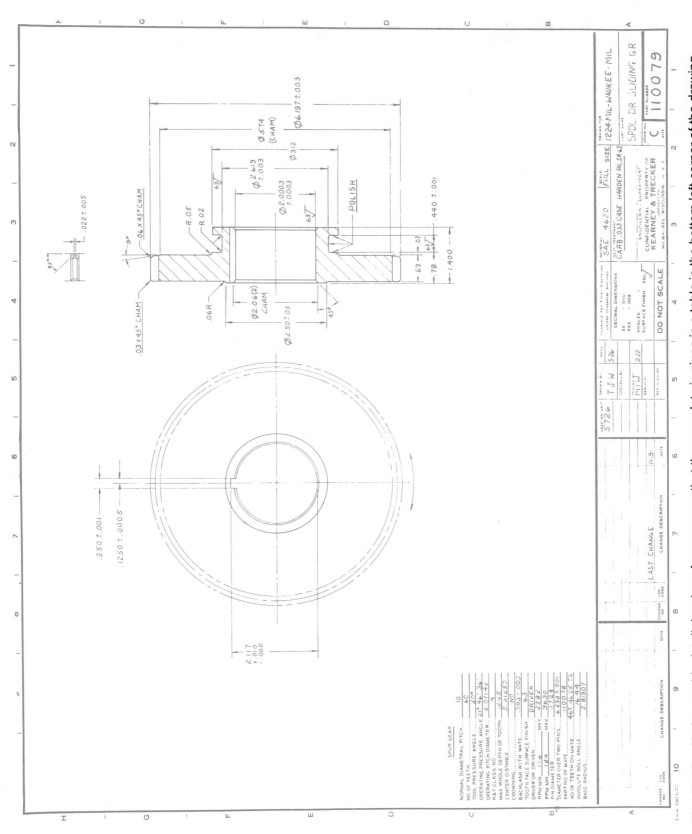

Fig. 25-16. Notice on this detail drawing of a spur gear that the gear data is given in a table in the bottom left corner of the drawing. (Kearney & Trecker Corp.)

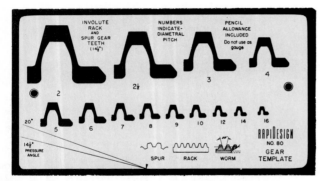

Fig. 25-17. Templates are available for drawing spur rack and spur gear teeth. (RapiDesign)

Fig. 25-18. A rack is a spur gear with teeth spaced along a straight pitch line. The rack is used with a smaller gear, or pinion. (Boston Gear Div.)

Fig. 25-19. The smaller of two bevel gears is called a pinion.

Backing (Y). The *backing* is the distance from the back of the gear hub to the base of the pitch cone measured parallel to the gear axis.

Crown backing (Z). The *crown backing* is the distance from the back of gear hub to the crown of the gear, measured parallel to the gear axis.

$$Z = Y + a \sin\Gamma$$

Mounting distance (MD). The *mounting distance* is the distance from a locating surface of a

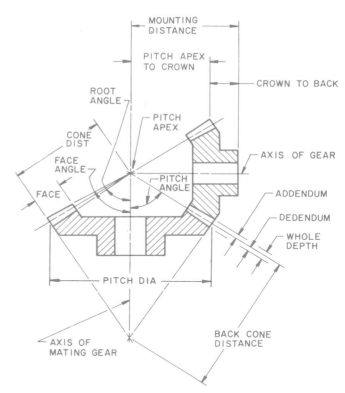

Fig. 25-20. There are several terms associated with bevel gears that the drafter/designer must know and understand.

gear (such as end of hub) to the centerline of its mating gear. It is used for proper assembling of bevel gears.

$$MD = Y + \tfrac{1}{2}D/\tan\Gamma$$

Addendum (a). The *addendum* is the same as for spur gears, but measured at large end of tooth.

Addendum angle (a). The *addendum angle* is the angle between the elements of the face cone and the pitch cone. It is the same for the gear and pinion.

$$a = \tan^{-1}\frac{A}{CD}$$

Dedendum (b). The *dedendum* is the same as for spur gears, but measured at large end of tooth.

Dedendum angle (δ). The *dedendum angle* is the angle between elements of root cone and pitch cone and is the same for gear and pinion.

$$\delta = \tan^{-1}\frac{D}{CD}$$

Face angle (Γ_o or γ_o). The *face angle* is the angle between an element of the face cone and the axis of the gear or pinion.

$$\Gamma_o = \Gamma + \delta_P$$

$$\gamma_o = \gamma + \delta_G$$

Pitch angle (Γ or γ). The *pitch angle* is the angle between an element of the pitch cone and its axis.

Gear: $\Gamma = \tan^{-1}\dfrac{N}{n} = \tan^{-1}\dfrac{D}{d}$

Pinion: $\gamma = \tan^{-1}\dfrac{n}{N} = \tan^{-1}\dfrac{d}{D}$

Root angle (Γ_R or γ_R). The *root angle* is the angle between an element of the root cone and its axis.

Gear: $\Gamma_R = \Gamma - \delta_G$

Pinion: $\gamma_R = \gamma - \delta_P$

Shaft angle (Σ). The *shaft angle* is the angle between the shaft of the two gears, usually 90°.

Pressure angle (ø). The *pressure angle* is the same as for spur gears.

Cone distance (A_o). The *cone distance* is the distance along an element of the pitch cone and is the same for the gear and pinion.

$A_O = \dfrac{D}{2\sin\Gamma}$

Whole depth (h_t). The *whole depth* is the same as for spur gears, but measured at large end of tooth.

Chordal thickness (t_c). The *chordal thickness* is the length of the chord subtending a circular thickness arc.

For Bevel Gear: $t_c = D\sin\left(\dfrac{90° \cos\Gamma}{N}\right)$

For Pinion: $t_c = D\sin\left(\dfrac{90°\cos\gamma}{N}\right)$

Chordal addendum (a_c). The *chordal addendum* is the distance from the top of the tooth to the chord subtending the circular thickness arc.

$$a_c = a + \frac{D}{2\cos\Gamma}\left[1 - \cos\left(\frac{90°\cos\Gamma}{N}\right)\right]$$

Bevel gear representation

The construction of a bevel gear is shown in Fig. 25-21. The teeth are normally drawn in simplified conventional form. Proceed as follows to draw the gear.

1. Lay out the pitch diameters and axes of the gear and the pinion, Fig. 25-21A.
2. Show the whole tooth depth by drawing light construction lines for the addendum and dedendum, Fig. 25-21B.
3. Lay off the face width and other features using the dimensions specified (or dimensions from gear data tables), Fig. 25-21C.
4. Erase construction lines and complete the drawing of the bevel gear and pinion, Fig. 25-21D.

Worm Gear and Worm

The *worm mesh* is a gear type used for transmitting motion and power between nonintersecting shafts usually at 90° to each other, Fig. 25-22. The worm mesh consists of the *worm* and the *worm gear.* The worm mesh is characterized by a high velocity ratio of worm to gear. These gears are capable of carrying greater loads than the cross helical gears. The driving member of the worm mesh is the worm.

The worm is actually an Acme-type thread that in section appears much like a gear rack, Fig. 25-23. To increase the contact of the worm mesh, the worm gear is made in a *throated* (concave) shape to wrap around the worm. (See Fig. 25-22). One revolution of a single-threaded worm advances the worm gear one tooth space. This is called the *lead.* Worms can be either right-hand or left-hand thread, depending on the rotation desired. Worms may have single, double, or triple threads.

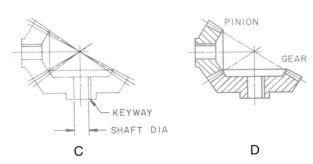

Fig. 25-21. When constructing a bevel gear, the teeth are usually shown in simplified conventional form.

The speed ratio of a worm mesh depends on the number of threads on the worm and the number of teeth on the gear. A worm with a single thread meshed with a gear having 48 teeth must revolve 48 times to rotate the gear 1 time. This is a ratio of 48:1. The same speed reduction with a pair of spur gears would require a gear with 480 teeth and a pinion with 10 teeth. A double-threaded worm would require 24 revolutions to rotate the 48 tooth gear once. This is a ratio of 24:1.

Fig. 25-22.   A worm mesh is a worm gear and a worm.

Worm and worm gear terminology

The following terms are used in reference to worm gears and worms.

Axial pitch (p_x). The *axial pitch* is the distance between corresponding sides of adjacent threads in a worm, Fig. 25-23.

Lead (ℓ). The *lead* is the axial advance of the worm in one complete revolution. The lead is equal to the pitch for single-thread worms, twice the pitch for double-thread, and three times the pitch for triple-thread worms.

Lead angle (λ). The *lead angle* is the angle between a tangent to the helix of the thread at the pitch diameter and a plane perpendicular to the axis of the worm.

$$\lambda = \tan^{-1} \frac{\ell}{\pi D_\omega}$$

Pitch diameter of the worm (D_ω). The *pitch diameter of the worm* is the diameter of the pitch circle of a worm thread. This can be calculated using the formula below. This is a recommended value, but it may be varied.

$$D_\omega = 2.4p_x + 1.1$$

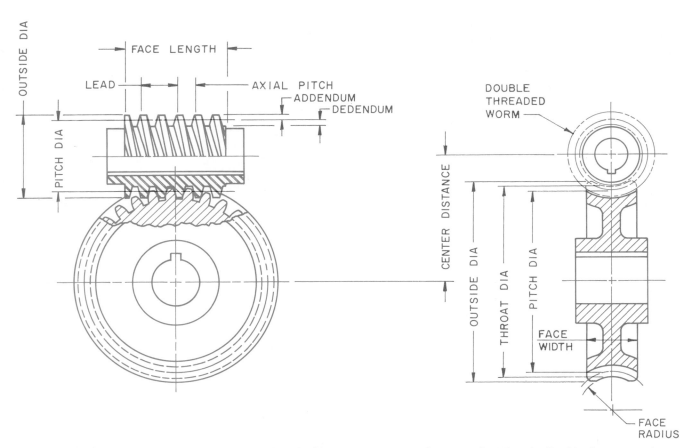

Fig. 25-23.   There are certain terms associated with worm gears and worms that the drafter/designer must know and understand.

Addendum of thread (a_ω). The *addendum of thread* is the same as for spur gears.

$a_\omega = 0.138p_x$

Dedendum of thread (b_ω). The *dedendum of thread* is the same as for spur gears.

$b_\omega = 0.368p_x$

Whole depth of thread ($h_{t\omega}$). The *whole depth of thread* is the same as for spur gears.

$h_{t\omega} = 0.686p_x$

Outside diameter of the worm ($D_{o\omega}$). The *outside diameter of the worm* is the pitch diameter of the worm plus twice the addendum.

$D_{o\omega} = D_\omega + 0.636p_x$

Face length of the worm (F_ω). The *face length of the worm* is the overall length of the worm thread section.

$F_\omega = p_x \left(4.5 + \dfrac{N_{\omega G}}{50} \right)$

Number of teeth on the worm gear ($N_{\omega G}$). The *number of teeth on the worm gear* is determined by the desired speed ratio between the worm and worm gear.

$N_{\omega G} = SR^* \times$ No. of Threads

Circular pitch of the worm gear. The *circular pitch of the worm gear* is the same as for spur gears. It must be the same as the axial pitch of the worm.

Pitch diameter of the worm gear ($D_{\omega G}$). The *pitch diameter of the worm gear* is the same as for spur gears. The following formula is recommended.

$D_G = \dfrac{p_x(N_{\omega G})}{\pi}$

Addendum ($a_{\omega G}$). The *addendum* must equal the addendum of the worm thread.

$a_{\omega G} = 0.318p\chi$

Whole depth ($h_{t\omega G}$). The *whole depth* must equal the whole depth of the worm thread.

$h_{t\omega G} = 0.696\chi$

Throat diameter of the worm gear (D_t). The *throat diameter of the worm gear* is the outside diameter of the worm gear measured at the bottom of the tooth arc. It is equal to the pitch diameter of the gear plus twice the addendum.

$D_t = \dfrac{p_x(N_{\omega G})}{\pi} + 0.636p\chi = p\chi \, \dfrac{N_{\omega G} + 1.113\pi}{\pi}$

Face radius of the worm gear (F_r). The *face radius of the worm gear* is the outside arc radius of the worm gear teeth that curves around the worm.

$F_r = \dfrac{D_\omega}{2} - 0.318p\chi$

Outside diameter of the worm gear ($D_{o\omega G}$). The *outside diameter of the worm gear* is measured at the top of the tooth arc.

$D_{o\omega G} + D_t + 0.477p\chi$

Worm gear and worm representation

The way worm gears and worms are represented on drawings is shown in Fig. 25-24. The gear teeth and worm thread are usually drawn in simplified, conventional form. Specifications for machining the gear and worm are given in table form on the drawing.

SPLINES

Splines are similar to multiple keys on a shaft. Splines prevent rotation between the shaft and its related member. The teeth on a spline may have parallel sides, but splines with involute teeth are increasing in use, Fig. 25-25.

A drawing of an external and internal spline is shown in Fig. 25-26. Note the specifications given for each spline on the drawing. Terminology for involute splines is the same as for spur gears.

PROBLEMS AND ACTIVITIES

In laying out the problems that follow, use B-size sheets. Arrange the required features to make good use of the space available.

Cam Layouts

The following dimensions are standard for all of the cam problems: base circle, 3.50"; shaft diameter, 1"; hub, 1.50" diameter; keyway, 1/8" x 1/16"; knife edge follower, 0.625" round stock; roller follower, 0.875" diameter. The follower is aligned vertically over the center of the base circle and the cam rises in 180° and falls in 180° unless otherwise noted.

1. Make a displacement diagram and cam layout for a modified uniform motion cam with a rise of 1.375". (Use an arc of one-quarter of the rise to modify the uniform motion in the dis-

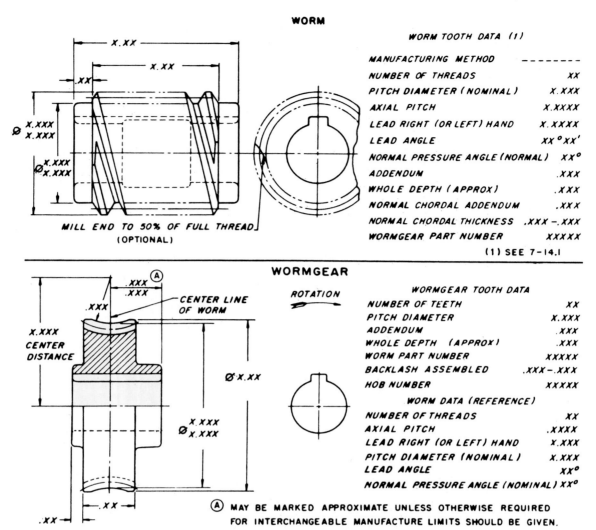

WORM

WORM TOOTH DATA (1)

MANUFACTURING METHOD	--------
NUMBER OF THREADS	XX
PITCH DIAMETER (NOMINAL)	X.XXX
AXIAL PITCH	X.XXXX
LEAD RIGHT (OR LEFT) HAND	X.XXXX
LEAD ANGLE	XX°XX'
NORMAL PRESSURE ANGLE (NORMAL)	XX°
ADDENDUM	.XXX
WHOLE DEPTH (APPROX)	.XXX
NORMAL CHORDAL ADDENDUM	.XXX
NORMAL CHORDAL THICKNESS	.XXX −.XXX
WORMGEAR PART NUMBER	XXXXX

(1) SEE 7-14.1

MILL END TO 50% OF FULL THREAD
(OPTIONAL)

WORMGEAR

ROTATION

CENTER LINE
OF WORM

WORMGEAR TOOTH DATA

NUMBER OF TEETH	XX
PITCH DIAMETER	X.XXX
ADDENDUM	.XXX
WHOLE DEPTH (APPROX)	.XXX
WORM PART NUMBER	XXXXX
BACKLASH ASSEMBLED	.XXX−.XXX
HOB NUMBER	XXXXX

WORM DATA (REFERENCE)

NUMBER OF THREADS	XX
AXIAL PITCH	.XXXX
LEAD RIGHT (OR LEFT) HAND	X.XXX
PITCH DIAMETER (NOMINAL)	X.XXX
LEAD ANGLE	XX°
NORMAL PRESSURE ANGLE (NOMINAL)	XX°

(A) MAY BE MARKED APPROXIMATE UNLESS OTHERWISE REQUIRED
FOR INTERCHANGEABLE MANUFACTURE LIMITS SHOULD BE GIVEN.

Fig. 25-24. Conventions should be followed when representing and specifying worm gears and worms on drawings. (American National Standards Institute)

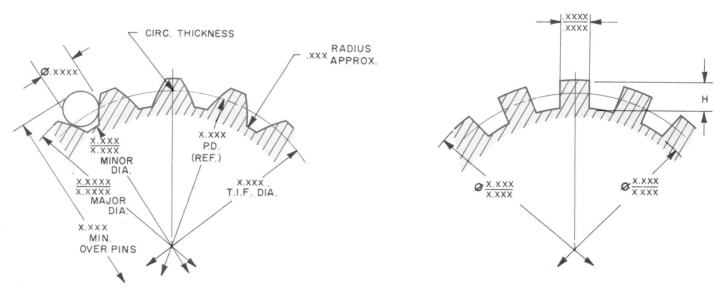

Fig. 25-25. Involute A and parallel B splines are used to prevent rotary motion between a shaft and coupling or gear mounted on the shaft.

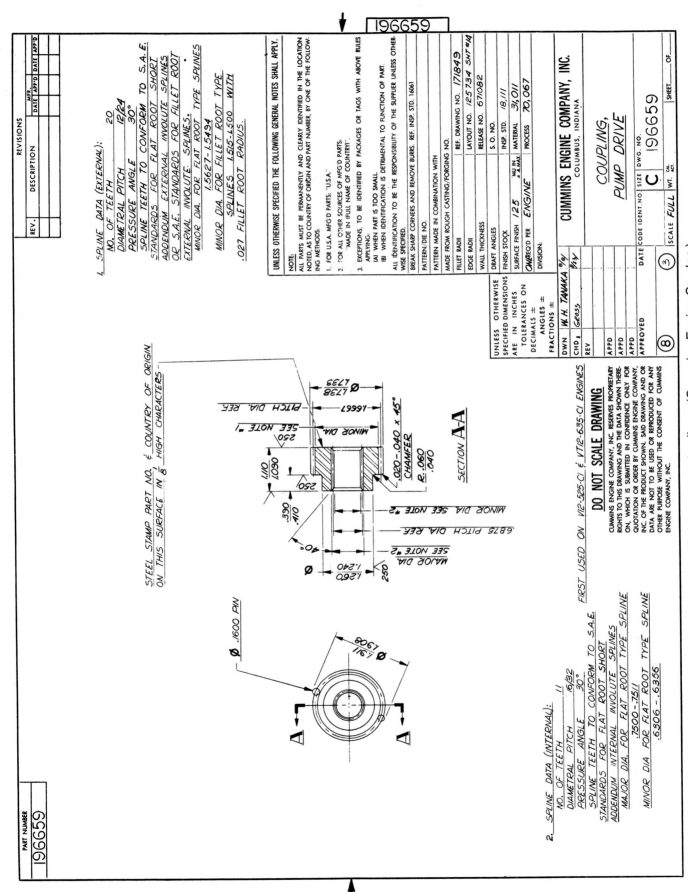

Fig. 25-26 Many industrial drawings use simplified convention to represent splines. (Cummins Engine Co., Inc.)

placement diagram.) The cam rotates clockwise and a knife edge follower is used.

2. Make a displacement diagram and cam layout for a modified uniform motion cam with a rise of 1.250″. (Use an arc of one-third of the rise to modify the uniform motion in the displacement diagram.) The cam rotates counterclockwise and a knife edge follower is used.

3. Make a displacement diagram and cam layout for a harmonic motion cam with a rise of 1.50″. The cam rotates counterclockwise and a knife edge follower is used.

4. Make a displacement diagram and cam layout for a harmonic motion cam with a rise of 1.125″ in 120°, dwell for 90°, fall 1.125″ with harmonic motion in 120° and dwell for 30°. The cam rotates clockwise and a knife edge follower is used.

5. Make a displacement diagram and cam layout for a uniformly accelerated motion cam with a rise of 1.250″. The cam rotates clockwise and a knife edge follower is used.

6. Make a displacement diagram and cam layout for a uniformly accelerated motion cam with a rise of 1.375″ in 90°, dwell for 90°, fall 1.375″ with uniformly decelerated motion in 90° and dwell for 90°. The cam rotates counterclockwise and a roller follower is used.

7. Make a displacement diagram and cam layout for a uniformly accelerated motion cam with a rise of 1.125″ in 120° dwell for 60°, fall 1.125″ in 120° with uniformly decelerated motion and dwell for 60°. The cam rotates counterclockwise. The roller follower is offset .50″ to left of vertical centerline.

8. Make a displacement diagram and cam layout for a uniformly accelerated motion cam with a rise of 1.50″ in 180°, dwell for 60° and fall 1.50″ with harmonic motion in 120°. The cam rotates clockwise and has a roller follower.

Cam Design Problems

9. Design a cam that will open and close a valve on an automatic hot-wax spray at a car wash in one revolution. To open the valve, the cam follower must move 1.125″. The valve is to open in 20° of cam rotation, remain open for 320°, close in 10°, and remain closed for 10°. The cam operates at moderate speed. You are to select the appropriate cam motion, size of base circle, and type of cam follower. Make a

full-size working drawing of the displacement diagram and the cam.

10. Design a cam that will raise a control lever, permitting a work piece to be fed to a machine. The lever must be raised a distance of 1″, remain open, and close in equal segments of cam revolution. The cam operates at a relatively high speed with moderate pressure on the cam follower. The cam follower must be offset to the right of center .75″. Select the appropriate cam motion, size of base circle, and type of cam follower. Make a full-size working drawing of the displacement diagram and the cam.

Gear Problems

11. Make a working drawing of a spur gear having 40 teeth, a diametral pitch of 8, a pressure angle of 20°, a shaft diameter of .75″, a hub diameter of 1.5″, a hub width of 1.00″, a face width of .50″, and a keyway that is 1/8″ x 1/16″. Compute values for the pitch diameter, circular thickness, and whole depth. Include these in a table on drawing. One view should be a sectional view.

12. Make an assembly drawing of a spur gear having 48 teeth, a pitch diameter of 3.00″, a shaft diameter of .625″, a face width of .75″, and a keyway that is 1/8″ x 1/16″. Also draw a pinion having 24 teeth and a pitch diameter of 1.250″. Other dimensions of the pinion are the same as for the spur gear. The pressure angle of the gear and pinion is 20°. Include a table of specifications for the gear and pinion on the drawing.

13. Make a detail drawing of a bevel gear having 36 teeth, a diametral pitch of 12, a pressure angle of 20°, a face width of .53″, a shaft diameter of 1.00″, a whole length of 1.25″, a mounting distance of 1.875″, a hub diameter of 2.125″, and a 1/8″ x 1/16″ keyway. Make a sectional view for clarity. Compute values for the pitch diameter, circular pitch, whole depth, addendum, and dedendum. Include these in a table on the drawing.

14. Make an assembly drawing of a 64 tooth bevel gear and a 16 tooth pinion assembled at a 90° shaft angle. The diametral pitch is 16, the pressure angle is 20°, and the face width is .48″. For the gear, the shaft size is .625″; the hub diameter is 2.250″; the keyway is 1/8″ x 1/16″; and the mounting distance (MD) is 1.375″. For the pinion, the shaft size is .375″; the hub diameter is .8125″; the keyway is 1/8″ x 3/64″;

and the mounting distance is 1.50″. Draw the assembly in a section. Compute values for the pitch diameter, circular pitch, and whole depth. Include these in a table on the drawing.

15. Make an assembly drawing of a worm mesh that has following specifications. For the gear, the pitch diameter is 5.80″; the pressure angle is 20°; the number of teeth is 29; the face width is 1.375″; the shaft diameter is 1.250″; the hub diameter is 2.750″; the keyway is 1/4″ x 1/8″; and the outside diameter is 6.40″. For the worm, the pitch diameter is 2.30″; the face length is 3.0″; the shaft diameter is 1.125″; the hub diameter is 1.837″; and the keyway is 1/4″ x 1/8″.

Note that the axial pitch of the worm can be found **by computing** the circular pitch of the gear.

The other values can be found by using the formulas given in the section on worm gears and worms. Show the noncircular view as a sectional view. Include the following specifications either in table form or as direct dimensions on the drawing.

For the gear, include the number of teeth, the pressure angle, the pitch diameter, the outside diameter, and the face width.

For the worm, include the lead, the pitch diameter, the outside diameter, the length of face, and the whole depth of thread.

16. Design a gear assembly involving two gears, or a worm gear and worm, to achieve a definite ratio. Obtain basic specifications for gears from a machinist's handbook or from a gear catalog. Make an assembly drawing of the gears. Add the necessary dimensions and specifications.

Many manufactured items require welding during their fabrication. Welding drawings specify exactly how and where the weld should be placed. (Metal-Fab, Inc.)

Welding Drawings

Welding has become one of industry's principal means of fastening parts together, Fig. 26-1. Welding can also be used to build up the surface of a part. Modern technology has developed welding

Fig. 26-1. Modern industry depends upon welding processes for many jobs. A drawing must clearly specify the engineering designer's intent for each weld if the part is to be properly fabricated. (Lincoln Electric)

processes and materials to meet nearly any metal fabricating need. This capability has placed a major responsibility on design and drafting departments to adequately specify welds required for a particular structure or machine part.

WELDING PROCESSES

Numerous welding processes have been developed to meet the need for joining different types of metals. The processes that are commonly used and those developed for welding the "exotic" metals of the aerospace industry are discussed in the following section.

Brazing

Brazing is the process of joining metals by adhesion with a low melting point alloy. This process does not melt the parent metal. A copper base with tin, zinc, and/or lead is commonly used.

Fusion Welding

The basic types of *fusion welding* processes include oxyacetylene, arc, TIG, and MIG welding. The heat generated by a flame or arc causes the parent metal and a feeder metal rod to melt and "fuse" into one piece. This type of welding joins the metals by cohesion, Fig. 26-2.

TIG stands for tungsten inert gas welding. It is a gas-shielded arc welding process. The tungsten electrode maintains an intense heat and a metal filler rod may or may not be added, depending on the requirements of the joint. An inert gas (one that does not chemically combine with the weld) surrounds the weld and produces a clean weld. The gas typically used is a combination of argon and helium.

The primary use of the TIG welding process is in joining lightweight (less than 1/4″ thick) nonferrous metal including aluminum, magnesium, silicon-bronze, copper and nickel alloys, stainless steel, and precious metals. The gas-shielded arc gives an unobstructed view of the slag-free weld.

MIG is the abbreviation for metal inert gas welding. It is a gas-shielded arc welding process similar to TIG welding. In MIG welding, the electrode is a filler wire that is fed into the weld automatically, Fig. 26-3. MIG is used for welding metals 1/4″ thick or thicker.

Resistance Welding

Resistance welding is an effective and economical means of fastening metal parts, Fig. 26-4. An electric current is the source of heat. Pressure is applied to bring the parts together at the point of weld.

Resistance welding is based on the principle that resistance to current flow causes metal to become hot. Resistance is greatest at the joint between the pieces. Therefore, when the current is properly adjusted, the metal pieces melt and fuse at the joint.

Fig. 26-2.   Arc welding is classified as fusion welding. In fusion welding, metal is added to the parent materials being joined. (Lincoln Electric)

Fig. 26-3.   MIG welding uses a "shield" of inert gas to protect the area being welded. A metal filler wire is automatically fed into the weld area. (Lincoln Electric)

Fig. 26-4. A spot welder is a type of resistance welder. (Arco Automation Systems, Inc.)

Types of resistance welding are:

- **Spot welding,** where the metal is fluxed only in the contact spots.
- **Butt** or **seam welding** of an entire joint or seam.

- **Flash welding,** where the ends of two metal parts are brought together under pressure and resistance welded.

Induction Welding

Induction welding is similar to resistance welding. However, in induction welding the heat generated for the weld is produced by the resistance of the metal parts to the flow of an induced electric current. The welding action may occur with or without pressure.

Electron Beam Welding (EBW)

The source of heat in **electron beam welding** is a high-intensity beam of electrons focused in a small area at the surface to be welded, Fig. 26-5A. Although this electron beam welding is a fusion welding process, it is unlike the common fusion welding processes that are often used.

Electron beam welding is done in a vacuum. This practically eliminates contamination of the weld from the atmosphere, Fig. 26-5B. There is minimum distortion of the work piece because the heat is concentrated in a small area. No rod, gas, or flux is needed in electron beam welding. EBW is used in welding metals such as titanium, beryllium, and zirconium. These metals are common to the aerospace industry and are difficult to weld by other welding processes.

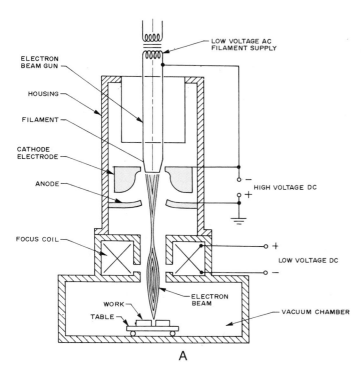

A

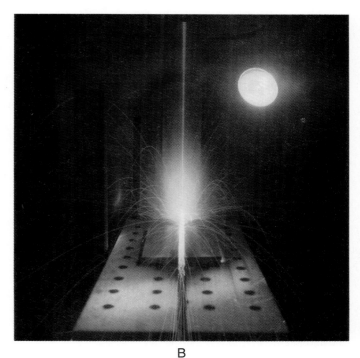

B

Fig. 26-5. An electron beam welder uses a concentrated beam of electrons. The welding is done inside of a vacuum chamber. (United Aircraft Corp.)

TYPES OF WELDED JOINTS

The welding process lends itself to a variety of joints in fastening metal parts. There are five types of joints commonly used. These five are the **butt joint, corner joint, tee joint, lap joint,** and **edge joint.** These joints, and the welds applicable to each type, are shown in Fig. 26-6.

TYPES OF WELDS

The term "weld" refers to the basic design of the weld itself, Fig. 26-7. Design selection is basically determined by the thickness of the metals to be joined. The design selection is also determined by the penetration of the weld into the joint for the strength required. The type of metal also has a bearing on the weld design selected.

BASIC WELD SYMBOLS

The American Welding Society (AWS) has developed a set of standard symbols for use in specifying types of fusion and resistance welds on drawings, Fig. 26-8. These weld symbols should be understood by designers, drafters, welders, and all persons in industries using welding processes. Weld symbols should be used only as a part of the welding symbol discussed in the next section.

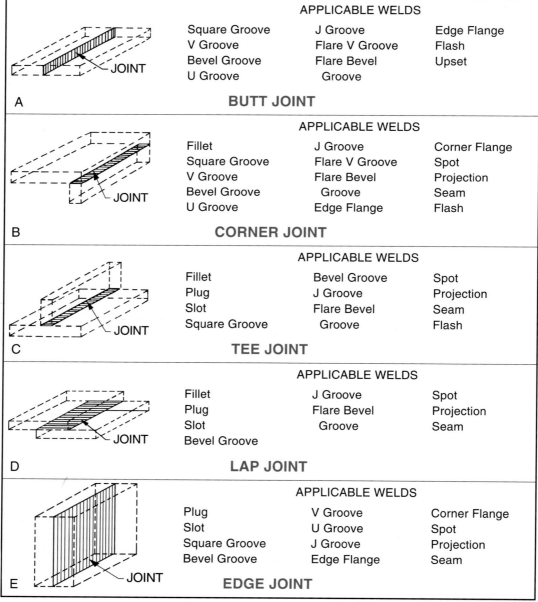

Fig. 26-6.   There are five basic types of joints used in welding. (American Welding Society)

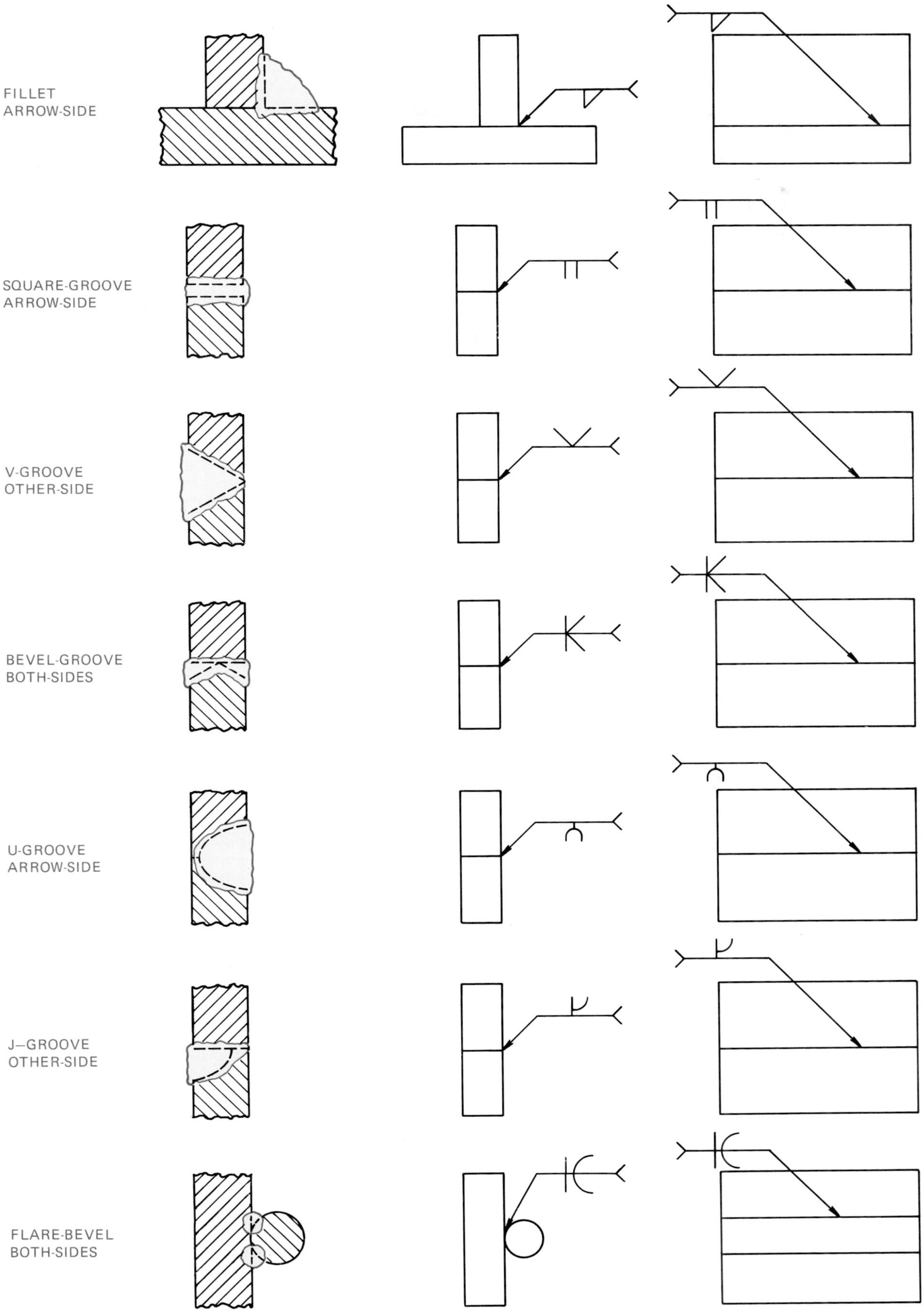

FILLET
ARROW-SIDE

SQUARE-GROOVE
ARROW-SIDE

V-GROOVE
OTHER-SIDE

BEVEL-GROOVE
BOTH-SIDES

U-GROOVE
ARROW-SIDE

J—GROOVE
OTHER-SIDE

FLARE-BEVEL
BOTH-SIDES

Fig. 26-7. There are several types of welds. Each one is represented differently on a drawing.

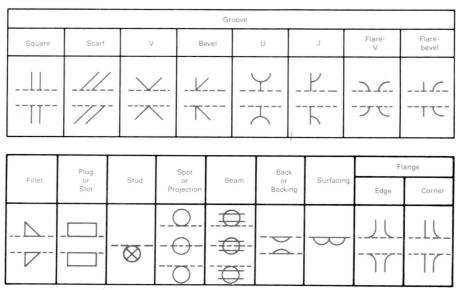

Fig. 26-8. Basic weld symbols specify the type of weld to be performed. (American Welding Society)

Standard Symbol

The standard welding symbol is a composite symbol that carries all pertinent information for a particular weld. It indicates the type of weld, the size, the location, and the welding process (if specified), Fig. 26-9. Note that the elements along the reference line of the symbol remain the same when the tail and arrow are reversed, Fig. 26-10. A template for use in preparing standard welding symbols is shown in Fig. 26-11.

Weld symbols attached to the reference line are shown in an "upright" position when on the far side (top side) of the line. Weld symbols are in an "up-side-down" position when on the near side (lower side) of the line. The weld symbols are never

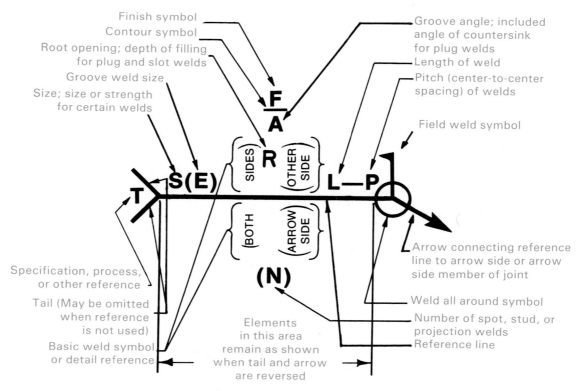

Fig. 26-9. The elements of the welding symbol have standard locations that should be followed. This makes sure that anyone who picks up the drawing and who is familiar with the standard can accurately read the drawing.

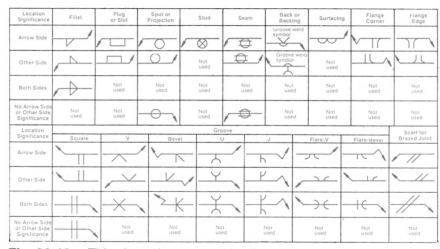

Location Significance	Fillet	Plug or Slot	Spot or Projection	Stud	Seam	Back or Backing	Surfacing	Flange Corner	Flange Edge
Arrow Side	[symbol]	[symbol]	[symbol]	[symbol]	[symbol]	Groove weld symbol [symbol]	[symbol]	[symbol]	[symbol]
Other Side	[symbol]	[symbol]	[symbol]	Not used	[symbol]	Groove weld symbol [symbol]	Not used	[symbol]	[symbol]
Both Sides	[symbol]	Not used	Not used	Not used	Not used	Not used	Not used	Not used	Not used
No Arrow Side or Other Side Significance	Not used	Not used	[symbol]	Not used	[symbol]	Not used	Not used	Not used	Not used

Location Significance	Groove							Scarf for Brazed Joint
	Square	V	Bevel	U	J	Flare-V	Flare-Bevel	
Arrow Side	[symbol]	[symbol]	[symbol]	[symbol]	[symbol]	[symbol]	[symbol]	[symbol]
Other Side	[symbol]	[symbol]	[symbol]	[symbol]	[symbol]	[symbol]	[symbol]	[symbol]
Both Sides	[symbol]	[symbol]	[symbol]	[symbol]	[symbol]	[symbol]	[symbol]	[symbol]
No Arrow Side or Other Side Significance	[symbol]	Not used	Not used	Not used	Not used	Not used	Not used	Not used

Fig. 26-10. This chart shows some basic weld symbols and their location significance.

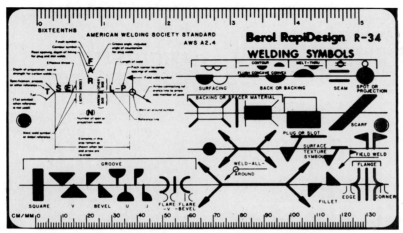

Fig. 26-11. Templates speed the application of welding symbols to drawings. (RapiDesign)

reversed. For example, the perpendicular leg of the fillet and groove weld symbols always are shown on the left. When no specification, welding process, or other reference is given, the tail section of the symbol may be omitted.

The location of welds with respect to a joint is controlled by the placement of the weld symbol on the reference line of the welding symbol. Welds that are to be located on the arrow side of the joint are shown by placing the weld symbol on the side of the reference line toward the reader, Fig. 26-12A. Welds that are to be on the side opposite the arrow are considered to be on the other side of the joint, so the weld symbol is shown on the side of the reference line away from the reader, Fig. 26-12B.

When the joint is to be welded on both sides, the weld symbol is shown on both sides of the reference line, Fig. 26-12C. Note in the second example of Fig. 26-12C that a different weld may be called out for each side of the joint and that a combination of welds may also be specified.

Supplementary Symbols

The dimensions of welds, the contour of the weld surface, welds that are to melt through, and other criteria can be specified by adding the appropriate information or symbol to the welding symbol. Refer to the welding symbols chart in the Reference Section, and to the bulletin *Standard Symbols for Welding, Brazing, and Nondestructive Examination (ANSI/AWS A2.4-86)* published by the American Welding Society.

SYMBOL   INTERPRETATION

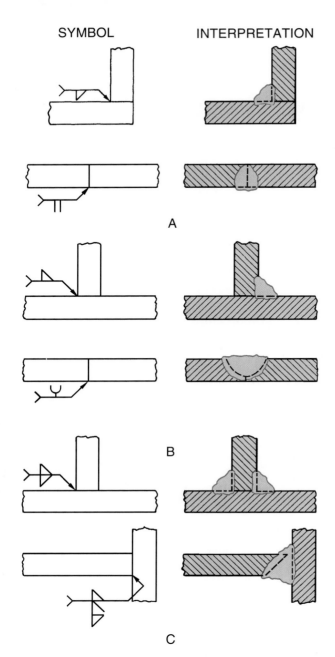

A

B

C

Fig. 26-12. Arrows in the welding symbol have a specific meaning in locating welds.

QUESTIONS FOR DISCUSSION

1. How does brazing of a metal joint differ from a fusion welded joint?

2. Check a welding text or reference book for the meaning and interpretation of the terms "adhesion" and "cohesion." Report the findings to your class.

3. What are "TIG" and "MIG" welding? How do they differ from regular electric arc welding?

4. What is the principle that resistance welding is based on? How does it work? Give some examples of products that have been resistance welded.

5. How does induction welding operate?

6. What is electron beam welding? What metals is it used on? Where is it done?

7. Differentiate between a "weld symbol" and "the welding symbol."

8. Of what significance is the arrow in the welding symbol in determining the location of the weld?

PROBLEMS AND ACTIVITIES

1. Make working drawings, including the specification of welds, for the objects shown in Fig. 26-13, Fig. 26-14, and Fig. 26-15, as assigned by your instructor.

2. Design a piece of furniture requiring welded parts. Use the design method (Chapter 1) to arrive at the final design. Make a working drawing of the piece.

3. Using the design method, design a tool, jig, or fixture requiring welded parts for some problem that needs solving around home, school, or your place of work. Make a working drawing and construct a scaled model or prototype of the item.

SELECTED ADDITIONAL READING

Symbols for Welding, Brazing, and Nondestructive Examination, ANSI/AWS A2.4-91, American Welding Society, 2501 N.W. Seventh St., Miami, FL 33125.

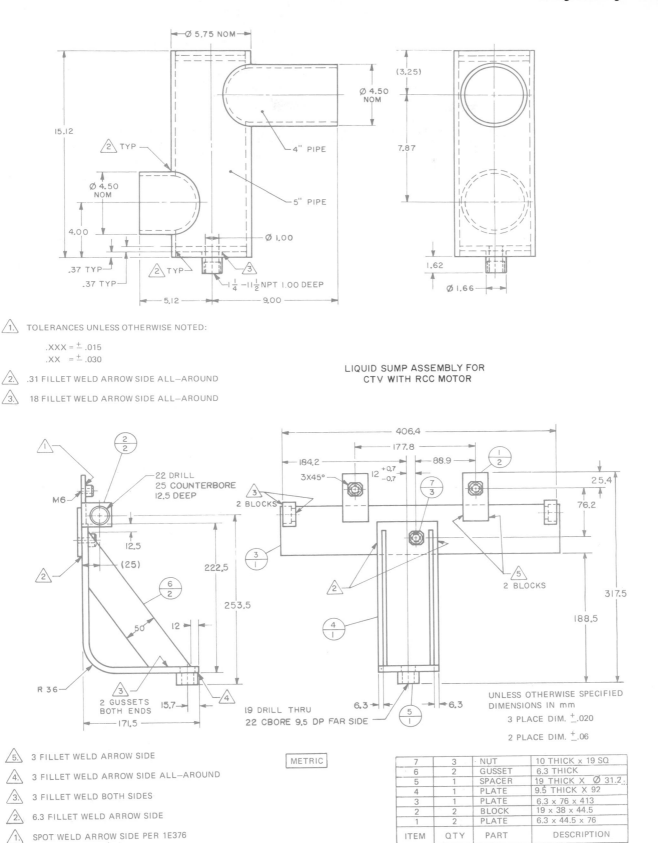

Fig. 26-13. Use these objects to prepare working drawings. Include the specification of welded joints.

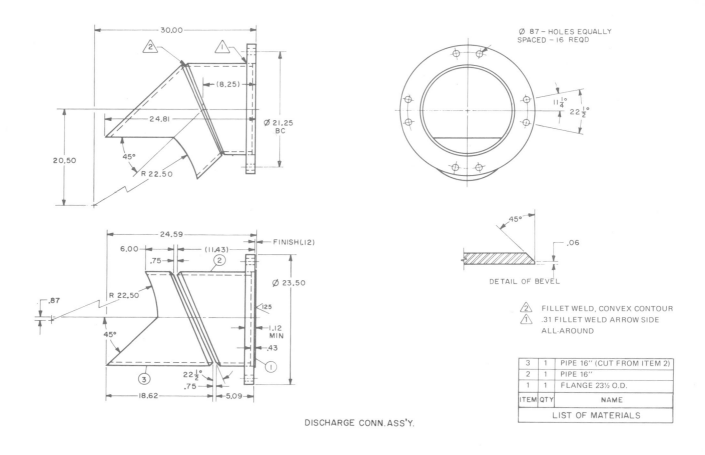

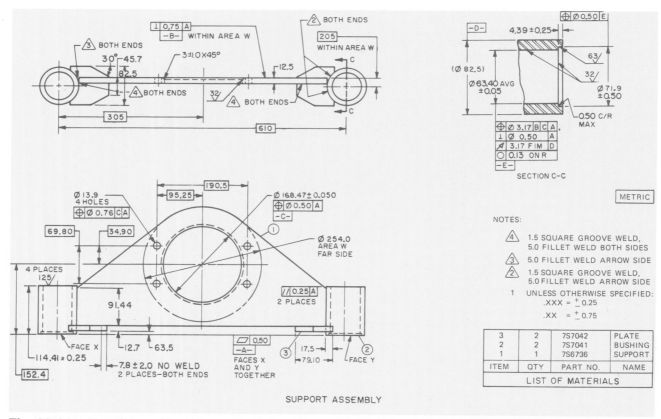

Fig. 26-14. Use these objects to prepare working drawings. Include the specification of welded joints.

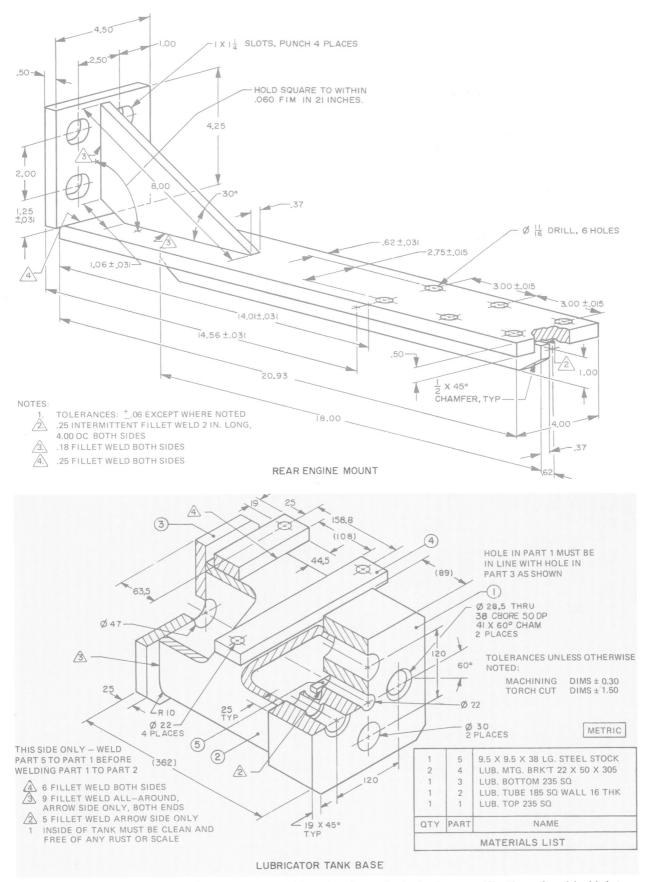

REAR ENGINE MOUNT

NOTES:
1. TOLERANCES: ±.06 EXCEPT WHERE NOTED
2. .25 INTERMITTENT FILLET WELD 2 IN. LONG, 4.00 OC BOTH SIDES
3. .18 FILLET WELD BOTH SIDES
4. .25 FILLET WELD BOTH SIDES

I X I¼ SLOTS, PUNCH 4 PLACES

HOLD SQUARE TO WITHIN .060 FIM IN 21 INCHES.

Ø 11/16 DRILL, 6 HOLES

½ X 45° CHAMFER, TYP

HOLE IN PART 1 MUST BE IN LINE WITH HOLE IN PART 3 AS SHOWN

Ø 28.5 THRU
38 CBORE 50 DP
41 X 60° CHAM
2 PLACES

TOLERANCES UNLESS OTHERWISE NOTED:
MACHINING DIMS ± 0.30
TORCH CUT DIMS ± 1.50

Ø 22

Ø 30
2 PLACES

METRIC

THIS SIDE ONLY – WELD PART 5 TO PART 1 BEFORE WELDING PART 1 TO PART 2

4 6 FILLET WELD BOTH SIDES
3 9 FILLET WELD ALL–AROUND, ARROW SIDE ONLY, BOTH ENDS
2 5 FILLET WELD ARROW SIDE ONLY
1 INSIDE OF TANK MUST BE CLEAN AND FREE OF ANY RUST OR SCALE

1	5	9.5 X 9.5 X 38 LG. STEEL STOCK
2	4	LUB. MTG. BRK'T 22 X 50 X 305
1	3	LUB. BOTTOM 235 SQ
1	2	LUB. TUBE 185 SQ WALL 16 THK
1	1	LUB. TOP 235 SQ
QTY	PART	NAME
	MATERIALS LIST	

LUBRICATOR TANK BASE

Fig. 26-15. Use these objects to prepare working drawings. Include the specification of welded joints.

Drafting is important in many aspects of road construction. Maps are needed, civil engineers create drawings, even the machinery used to build the roads requires drawings such as welding, assembly, and engineering drawings.

Part VI
Careers and Opportunities

The preceding chapters of this text have presented in detail the many aspects of the "graphic language," describing the practices associated with the production of technical drawings, and touching upon the many jobs, roles, and positions linked to the broad field of drafting and design.

This part of the text covers in more detail the careers and opportunities associated with drafting and design. The *Drafting Careers* chapter describes the various types of drafting-related careers, the typical duties performed by persons in those fields, and the training required to qualify for them. The *Drafting Specialties* chapter provides glimpses of several specific areas that draw upon the skills and knowledge developed in technical drawing. Topics covered include: architectural drawings, electrical and electronics drawings, maps and survey drawings, technical illustration, and graphs and charts.

The modern drafter must be able to use common CADD software to perform drafting and design functions, as well as being able to make freehand sketches and produce drawings using traditional equipment. (Altium)

Cartographers rely on accurate information obtained from surveyors. (The Lietz Company)

Drafting Careers

KEY CONCEPTS

☐ Drafting is a method used to effectively communicate ideas and plans.

☐ There are many possible careers in drafting and related technical fields.

☐ The educational and other requirements for careers in drafting and related fields vary widely, as do the duties of the different career areas.

The fundamentals of drafting are the same, regardless of the type of drawing. The symbols used, the lettering styles, or the general arrangement of the drawing may vary from mechanical to electronic or architectural to map drafting. However, the standards used, and the fundamental processes involved in producing, checking, and reproducing the finished drawings are much the same. Once the basics of drafting are understood, a person with a continuing interest in drafting may select an area of specialization to gain additional experience.

CAREERS IN DRAFTING

Job titles and duties in the field of drafting vary from industry to industry, and the nature of activities may vary among industries under the same general classification. However, the following career activities in drafting are typical for the job levels and industries discussed. These should be supplemented by researching career information concerning specific industries of your choice.

The following career fields do not require a college degree. However preparation in a technical school or community college will give you skills that may be advantageous.

Drafting Trainee

At the time of employment, a **drafting trainee** is expected to have a basic understanding of drafting instruments or software and skill in their use, a knowledge of procedures for representing views of objects, and the ability to produce neat freehand sketches and lettering. Generally, a trainee will work under the close supervision of senior drafting personnel.

Typical duties of a drafting trainee include revising of drawings, redrawing or repairing damaged drawings, and gathering information from reference sources needed to detail components (for themselves or other drafters), Fig. 27-1. The work-training program will involve drawing detail and sectional views, dimensioning and preparing tables, and developing working drawings. The trainee must become familiar with the company's drafting standards, as expressed in its drafting room manual.

Taking courses in drafting, mathematics, science, electronics, metals, manufacturing, and architecture is recommended. Additionally, courses in the use of computer-aided design and drafting (CADD) would be essential for work in a growing number of companies.

Detail Drafter

Detail drafters are well-informed in the fundamentals of drafting, and have gained proficiency and speed in handling instruments or software. They usually work as **detailers** in the preparation of working drawings for manufacturing or construction. Primarily, they will revise drawings and bills of material, prepare detail drawings, and work on simple assembly drawings, wiring or circuit diagrams, charts, and graphs.

Detail drafters should be thoroughly familiar with drafting standards and symbols, and must be able to make basic calculations in their area of drafting. They also should have a practical knowledge of either engineering or architectural materials and procedures.

Fig. 27-1. A drafting trainee will gain experience in a number of drawing areas and activities. Later on in their career, they will draw on this experience to solve many different problems presented to them.

Layout Drafter

It is the job of the *layout drafter* to prove out the product design, using sketches and models and a scaled layout drawing. Design layout helps to determine the manufacturing feasibility of a product.

This is an exacting type of drafting, requiring a knowledge of the field and the products being drawn. It may include preparation of some original layouts and studies to determine proper fits or clearances. It also could involve making some changes in the design after consulting with the engineer in charge.

Since layout drafters may be required to make dimensional computations and allowances, a knowledge of machine shop practices and materials is essential. The ability to research reference manuals is also necessary.

Design Drafter

The *design drafter* is a senior level drafter, representing the highest level of drafting skill. After acquiring considerable experience in the drafting field, a design drafter will do layout work and prepare complex detail and assembly drawings of machines, equipment, structures, wiring diagrams, piping diagrams, and construction drawings. A design drafter works from basic data supplied by architects, engineers, or industrial designers.

The design drafter must possess a sound knowledge of good engineering and drafting procedures, shop practices, mathematics, and science. He or she may make design changes when required, in consultation with the design engineer or architect, Fig. 27-2. The drafter must then follow through on changes made to see that they are reflected on other drawings involved. The design drafter may prepare cost estimates based on the materials and parts list, according to the design problem. Instructing and supervising other drafters in assigned tasks or other related drawing problems is usually another responsibility of the design drafter.

Checker

After a drawing is finished by a drafter, it must be reviewed for accuracy, completeness, clarity, and manufacturing feasibility. This examination is made by an experienced drafter called a *checker,* Fig. 27-3.

Checkers are persons who understand manufacturing processes and are thoroughly familiar with both the drafting practices in their particular industry and the national drafting standards set by ANSI (American National Standards Institute). Usually, checkers have reached the level of design drafters. As such,

Fig. 27-2.   A team effort involving engineer-designer, drafter, and manufacturing personnel is vital to the manufacturing of goods. (Motorola, Inc.)

Fig. 27-3.   The skills and knowledge of an experienced drafter are needed to check a drawing for accuracy and completeness and make any necessary corrections.

they may suggest modifications in design or specifications, or other changes to facilitate production. The approval signature of a checker normally will appear in the title block of the drawing.

Technical Illustrator

Technical illustration is the drawing of objects (usually machine parts, assemblies, or mechanisms) in pictorial form. These illustrations may be enhanced by line weight variations, shading, or colors to give them a more realistic appearance, Fig. 27-4.

A *technical illustrator* should thoroughly understand drafting fundamentals, including the con-

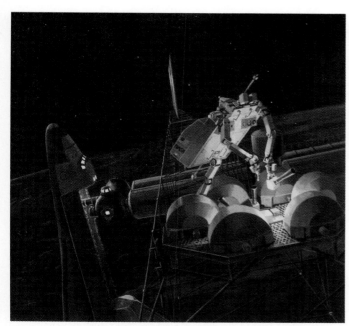

Fig. 27-4. Technical illustrators present images of things that cannot be easily photographed, such as a robotic servicer at work repairing a component of a space station. (Martin-Marietta Corporation)

struction of pictorial views. The illustrator should be able to read and interpret blueprints and must have some background in art, industrial design, and manufacturing processes.

Technical illustrations are used in manufacturing and production to assist workers in interpreting blueprints. Such illustrations also aid workers who are unable to read blueprints, by helping them visualize the object and its construction. A more "artistic" type of technical illustration is often used in marketing and advertising literature.

CAREERS RELATED TO DRAFTING

There are a number of related career opportunities for persons who have talent in the drafting field. All require a background in drafting, since it is used on the job. Many of these related career fields require preparation at the college level.

Architect

Architects plan, design, and oversee the construction of residential, commercial, and industrial building projects. They often are involved in such fields as city planning and *landscape architecture* (the design of parks, golf courses, and other outdoor facilities). Because the field is so broad, architects tend to specialize in one type of structure, such as residences, churches, schools, factories, or office buildings.

The architect's duties begin with a study of the client's needs and desires. They then progress through preliminary plans and sketches, finished drawings, and often, presentation renderings and scale models, Fig. 27-5. The architect also must develop cost estimates, prepare specifications, and supervise construction of the project.

Training for a career in architecture usually consists of four or five years of college, including drafting experience. An architect should have a sound understanding of mathematics and science, as well as the arts and humanities.

Industrial Designer

The profession of *industrial design* is concerned with the development of solutions to three-dimensional problems that involve esthetics, materials, manufacturing processes, human factors, and creativity. The industrial designer's task is working with scientific ideas and discoveries, endeavoring to develop these into products and services useful for humans.

Products such as office machines, furniture, systems for controlling forest fires, and special equipment to enable handicapped persons to lead fuller and more productive lives are goals of the industrial designer's efforts, Fig. 27-6.

There are two areas of emphasis in the four-year college preparatory program for this career. One is product design; the other is mechanical design.

The *product designer* works primarily with those problems where the user interacts with the product (design of a telephone, automobile steering wheel, or furniture for a library). The *mechanical designer* addresses problems where there is a machine-to-machine relationship, and no direct human

Fig. 27-5. Scale models are a tool that an architect might use to help a client better understand a design. (Urban Investment and Development Co.)

Fig. 27-6.   Industrial designers often develop products to make the lives of people easier or more satisfying. This folding travel wheelchair, designed to maneuver easily in airplane, train, or bus aisles, won a design excellence award from the Industrial Designers Society of America. It weighs only 16 pounds and folds down to briefcase size, as shown at left, for storage. (SEATCASE, Inc.)

Fig. 27-7.   The mechanical designer deals with systems and products that might be very complex. A material handling system used in a large distribution warehouse might involve a very complex conveyor arrangement. A mechanical designer is needed to design a solution to this type of problem. (SI Handling Systems, Inc.)

Fig. 27-8.   Professional engineers must have a good understanding of drafting procedures to be able to effectively communicate with other members of the technical team. Many engineers today are fully competent using CADD systems, as well as manual drafting practices. (IBM)

interaction is involved (design of an automobile transmission, a machine tool, or an improved conveyer system for materials handling), Fig. 27-7.

Although an industrial designer may not be working as a drafter, she or he should have a good background in drafting, mathematics, and science. The designer should be creative and have a thorough understanding of design problem-solving techniques.

The design drafting process is not a one-person job. It involves a team of engineers, designers, and drafters.

Engineer

Like industrial designers, professional engineers are also concerned with creative design solutions, Fig. 27-8. They usually have a strong background in science and mathematics, since they must be able to apply these principles in searching out practical problem solutions.

The engineer's field of interest and knowledge is broad, even though she or he may specialize in one particular area. A good understanding of drafting procedures is needed so that the engineer can communicate with other members of the technical team. The chief means of expressing ideas to others is by original freehand sketches. Engineers also review drawings prepared by drafters and make suggestions for their alteration.

There are a number of areas of specialization within the broad field of engineering, but all require four to five years of college education. Some of the better-known areas are presented here.

Aerospace engineering

This specialization deals with the design and development of all types of conventional and experimental aircraft and aerospace vehicles, Fig. 27-9. An *aerospace engineer* usually concentrates on one area, such as aerodynamics, propulsion systems, structures, instrumentation, or manufacturing.

Agricultural engineering

The problems of production, handling, and processing of food and fiber for the benefit of soci-

Fig. 27-9. Aerospace engineers are involved in all phases of the U.S. space program. The development and testing of the "piggyback" method of transporting a space shuttle from California to Florida atop a Boeing 747 aircraft required many aerospace engineers to solve many different problems. (NASA)

ety is the province of the *agricultural engineer.* More specifically, the agricultural engineer's specialty is the design and development of farm machinery, farm structures, processing equipment, and the control and conservation of water resources.

Ceramic engineering

Ceramic engineers are concerned with the research, design, and development of nonmetallic materials into useful products. Examples are as widely varied as glassware, joint replacement in humans, electrical insulators, and the fusing of refractory materials as a protective coating for metals. Ceramic engineers developed the protective shielding tiles applied to the space shuttles to control heat during reentry to the earth's atmosphere. Similar materials have been used to coat metal signs, cooking utensils, and sinks with a protective layer.

Chemical engineering

The branch of engineering that processes materials to undergo chemical change is called chemical engineering. *Chemical engineers* design and develop the processes and equipment that convert raw materials into useful products such as petrochemicals, plastics, synthetic fibers, and medicines.

Civil engineering

The design and development of transportation systems, including highways, railroads, and airports, is done by the *civil engineer.* This field of engineering also includes the design and construction of water systems, waste disposal systems, marine harbors, pipelines, buildings, dams, and bridges, Fig. 27-10.

Fig. 27-10. One of the greatest civil engineering projects in history was the building of the Panama Canal to connect the Atlantic and Pacific oceans. There are six locks on the 51-mile long waterway. To accomplish this construction, many civil engineers had to solve a variety of problems that this task presented. (Panama Canal Commission)

Electrical engineering

The field of electrical engineering has three major branches: electrical power, electronics, and computer engineering. The electrical power area involves the generation, transmission, and utilization of electrical energy. ***Electrical engineers*** specializing in this field design major projects, such as power transmission lines or the electrical distribution systems for large buildings or industries. They are also involved in the design of such equipment as electrical generators and motors, Fig. 27-11.

Engineers working in electronics are concerned with communication systems, especially radio and television broadcasting. They also work with industrial electronics, designing automated control systems for processing equipment.

Engineers working in the area of computer engineering design computers to control equipment and devices, such as automobiles, aerospace vehicles, and manufacturing machinery.

Industrial engineering

The form of engineering concerned with the design, operation, and management of systems is called industrial engineering. ***Industrial engineers*** design plant layouts and devise improved methods of manufacturing and processing. They also work with quality control, production control, and cost analysis, Fig. 27-12. In these activities, industrial engineers work with engineers in other areas of specialization and with personnel managers in developing and coordinating these systems.

Mechanical engineering

The design and development of mechanical devices ranging from extremely small machine components to large earthmoving equipment is carried out by ***mechanical engineers,*** Fig. 27-13.

A mechanical engineer will usually specialize in a given area, such as machinery, automobiles, ships, turbines, jet engines, or manufacturing facilities.

Fig. 27-11. Electrical engineers design and develop a wide range of devices and systems, from huge power plants to small DC wound-field motor. Electrical engineers also work in the electronics and computer fields. (Baldor)

Fig. 27-12. Industrial engineers discuss process improvements with floor supervisors. (IBM)

Fig. 27-13. Earthmoving equipment must be designed and engineered to withstand rough and heavy work. (Jack Klasey)

Metallurgical engineering

The ***metallurgical engineer*** is involved in the location, extraction, and refining of metals. A metallurgical engineer may specialize in one general area of the total field, such as mining and extraction, refining, or the welding of metals. An engineer's responsibilities, for example, could include altering the structure of a metal through alloying or other processes to produce a material with certain characteristics to perform a special purpose.

Nuclear engineering

Research, design, and development in the field of nuclear energy is the responsibility of the

nuclear engineer. This type of engineering includes the design and operation of nuclear-fueled electrical generating plants used in ships, submarines, or locomotives. This field of engineering offers many challenges in the area of power systems, as well as in chemistry, biology, and medicine.

Petroleum engineering

The **petroleum engineer** is involved in the location and recovery of petroleum resources and the development and transportation of petroleum products, Fig. 27-14. Considerable research has been done on the use and conservation of petroleum resources, since petroleum-based products are in greater demand each year. New sources of petroleum and gases need to be located by petroleum engineers while the search for substitute materials continues.

PROBLEMS AND ACTIVITIES

The following problems and activities are designed to help you understand the nature of drafting and its opportunities.

1. Select an object around your home or school, such as window trim or the molding around a door, and write a description of it, using words alone.

2. Make a freehand sketch of the object you described in No. 1. Which description was easier to prepare? Which description conveys information with more clarity?

3. Review current issues of magazines, such as Aztlan, Ebony, Entrepreneur, Popular Mechanics, Scientific American, or Working Woman and prepare a report on an individual who is distinguished in a drafting or a related career. Present your report to the class.

4. Use magazines, newspapers, and brochures to find as many different types and uses made of drafting as you can. These may include drawings, sketches, graphs, charts, or diagrams where drafting procedures were used. Mount your collection on notebook paper and be prepared to show them in class. Preserve the collection for later reference.

5. Interview an architect, industrial designer, drafter, or engineer on the nature of their work and its rewards. Find out as much as you can about educational requirements and any specialized training needed, the nature of the drafting work, pay compared to other types of work, and opportunities for employment and advancement.

6. Interview several persons not directly engaged in technical work. This might include your par-

Fig. 27-14. Petroleum engineers design and supervise the construction of refineries and gas processing plants.

ents, neighbors, or the parents of your friends. Obtain their opinions on the value of drafting to the average citizen. Find out what activities they have done where a knowledge of drafting was (or would have been) helpful.

7. Try to obtain an actual print of a house plan or some manufactured machine part and bring it to class. Discuss its features and details with your classmates.

8. Select a person who has become well-known in a technical field related to drafting. Staff members at your school or community library can help you find material. Prepare a report on how that person became involved in this work and the contributions that they have made. Present your report to the class.

SELECTED ADDITIONAL READING

U.S. Bureau of Labor Statistics, Occupational Outlook Handbook, Government Printing Office, Washington, DC, latest edition.

28 Drafting Specialties

KEY CONCEPTS

☐ Advanced drawing skills and knowledge of the subject field are necessary to producing acceptable drawings used in the architecture, electronics, mapping and surveying fields.

☐ Technical illustrations, usually drawn with three-dimensional effect, are used to supplement working drawings and to clarify complex assembly and operational procedures.

☐ Drafting departments are sometimes called upon to produce charts and graphs for informational presentations.

Drafting applications relate to the many areas in which advanced skill in drawing and knowledge of the field are essential to preparing acceptable technical drawings for industry. This chapter covers architectural drawings, electrical and electronics drawings, map and survey drawings, technical illustration, and graphs and charts.

ARCHITECTURAL DRAWINGS

Architectural drafting is the area of design and drafting that specializes in the preparation of drawings for the building of structures, such as houses, churches, schools, shopping centers, bridges, roads, airports, commercial buildings and factories, Fig. 28-1.

Most architectural drafting applications require a "set of drawings." Basic drawings included in

Fig. 28-1. Architects prepare the drawings for the construction of houses, restaurants, public buildings, and commercial structures.

such a set are typically the ***site/plot plans, foundation plan, floor plan, elevations, plumbing plan, electrical plan, HVAC plan,*** and ***construction details. Presentation plans*** or ***models*** are usually provided, as well. Some projects require additional drawings and analyses such as traffic flow, land use, landscaping, parking, cut and fill, program analysis, and environmental impact statements. The specific structure to be built usually dictates the type and number of drawings required.

Types of Drawings

Architectural drawings generally build upon the theory and procedures learned in technical drawing, but there are basic differences. The most noticeable of these differences is the more "artistic" style of lettering, Fig. 28-2. Architectural drafting also makes greater use of presentation drawings. Architectural drawings also make use of a significant number of detailed symbols that are not found on technical drawings of machine parts.

Site/plot plans

Site plans describe the basic features of the site: its shape and size, natural features, existing structures, and easements. ***Plot plans*** show the location of the major structure and any other buildings on the site, site dimensions and topographical features, meridian arrow, drives, walks, and utilities, Fig. 28-3. Sometimes, the roof plan is shown on this drawing if it is complicated or otherwise not clear. Scale of the plot plan is generally 1″ = 10′ 0″, 1″ = 20′ 0″, or 1″ = 30′ 0″, depending on the size of the site.

Foundation/basement plans

The ***foundation plan*** shows the size and materials of the foundation. The foundation plan includes information on excavation, waterproofing, and supporting structures, Fig. 28-4. This plan is prepared primarily for the excavation contractor and the masons, carpenters, and cement workers who build the foundation.

A specialized foundation plan, called a ***basement plan,*** is required when the structure has a basement. It differs from a typical foundation plan in that interior walls, doors, windows, and stairs are included in the excavated area. This drawing looks much more like a floor plan than a foundation plan,

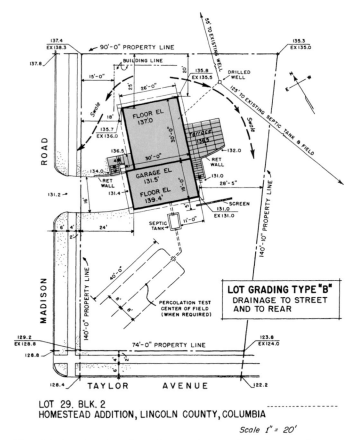

LOT 29. BLK. 2
HOMESTEAD ADDITION, LINCOLN COUNTY, COLUMBIA

Scale 1″ = 20′

Fig. 28-3. A plot plan shows the roof outline and location of the principal features on the lot. (FHA)

Fig. 28-5. The scale of the foundation and basement plan is generally 1/4″ = 1′ 0″ for residential building and 1/8″ = 1′ 0″ for commercial structures.

Floor plans

A ***floor plan*** shows all exterior and interior walls, doors, windows, patios, walks, decks, fireplaces, mechanical equipment, built-in cabinets, and appliances, Fig. 28-6. A separate plan view is drawn for each floor of the building. The floor plan is the heart of a set of construction drawings. It is the one plan to which all tradeworkers refer. The floor plan is actually a section drawing "cut" about half-way up the wall of each floor. Sometimes, when the structure is not complex, the floor plan may include information that would ordinarily be found on other drawings. For example, the electrical switches and outlets, furnace, and plumbing features may be shown on this plan for a simple structure. The scale of floor plans is usually 1/4″ = 1′ 0″ for residential buildings and 1/8″ = 1′ 0″ for commercial buildings.

Elevations

Elevations are drawn for each side of the structure, Fig. 28-7. These drawings are typically orthographic projections that show the exterior features of the building. Elevations show placement of

Fig. 28-2. The style of architectural lettering is not as formal as the type used on other drawings.

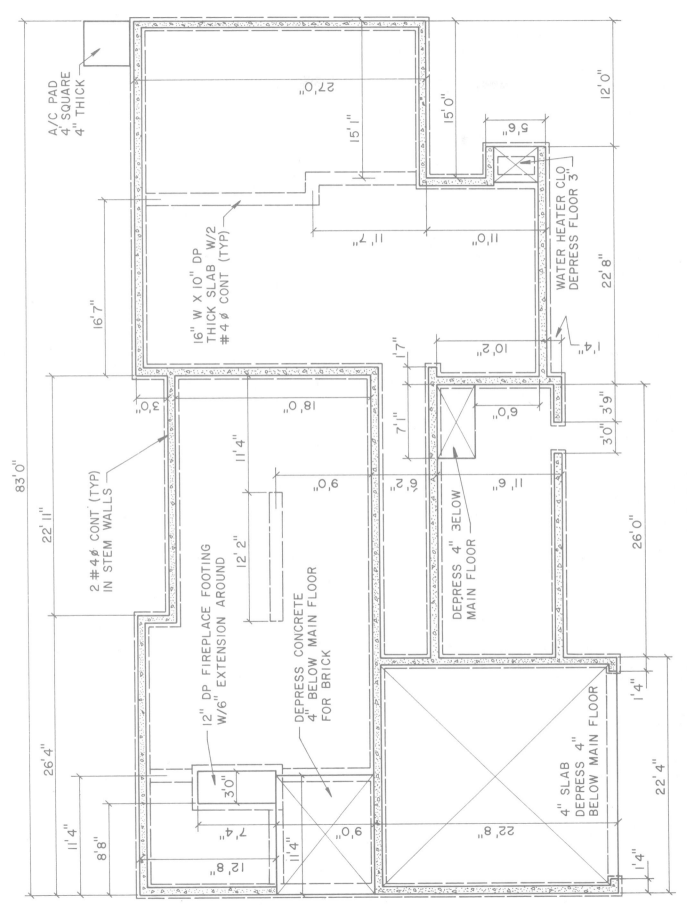

Fig. 28-4. A foundation plan shows the foundation walls and the footings.

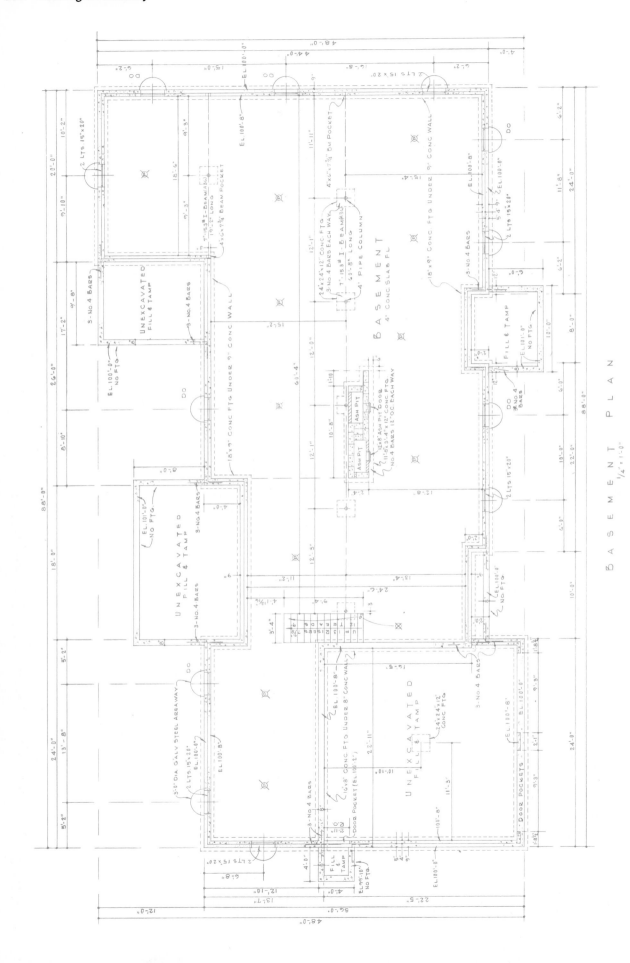

Fig. 28-5.   A basement floor plan includes the interior walls, doors, windows, and stairs. (Garlinghouse, Inc.)

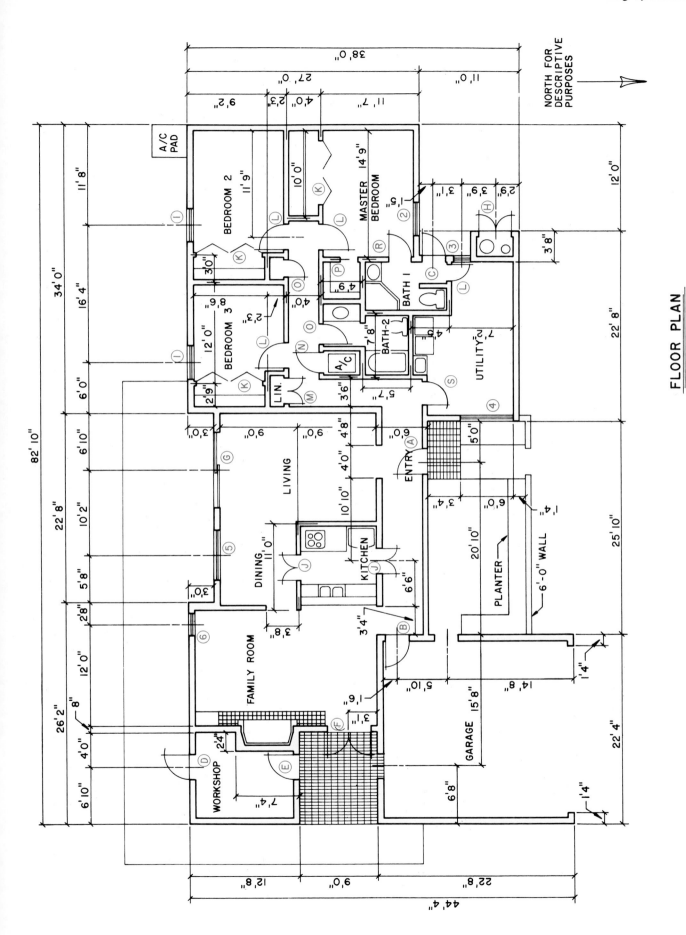

FLOOR PLAN

LIVING AREA   2320 SQ. FT.

Fig. 28-6. An architectural floor plan shows the basic arrangement of rooms in a building.

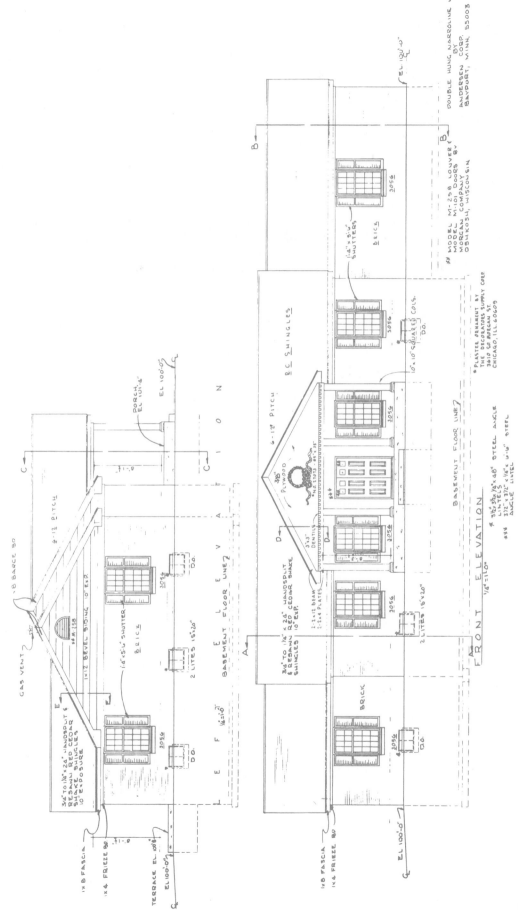

Fig. 28-7. Front and side elevations show a building as if a person was standing and looking at it. (Garlinghouse, Inc.)

windows and doors, types of exterior materials, steps, chimneys, rooflines, grade lines, finished floor and ceiling levels, and other exterior details. An elevation drawing also includes the vertical height dimensions of basic features of the structure that cannot be shown very well on other drawings. The scale of a building elevation is the same used for the floor plan and the foundation plan.

Plumbing plan

The plumbing plan is a plan view drawing, usually traced from the floor plan, that shows the plumbing system. This plan shows water supply lines, waste disposal lines, and fixtures, Fig. 28-8. It describes the sizes and types of all piping and fittings used in the system. Gas lines and built-in vacuum systems, if required, are also included on the plumbing plan. A plumbing fixture schedule is frequently included on the plumbing plan.

Electrical plans

The **electrical plan** is a plan view drawing in section, similar to the floor and foundation plans,

Fig. 28-9. It is usually traced from the floor plan and is used to show the electrical meter, distribution panel box, electrical outlets, switches, and special electrical features. It identifies the number and types of circuits in the building and may include a schedule of lighting fixtures as well.

HVAC plan

The **HVAC plan** is sometimes called the **climate control plan,** Fig. 28-10. It is a plan view drawing that shows the location, size, and type of all HVAC (heating, ventilating, and air conditioning) equipment. It also shows any humidification or air cleaning devices. Other information usually included on this plan is the location of distribution pipes or ducts, thermostats, registers, or baseboard convectors. An equipment schedule and heat-loss calculations may be included, as well.

Construction details

Construction detail drawings are usually produced when more information is needed to fully describe how the construction is to be done.

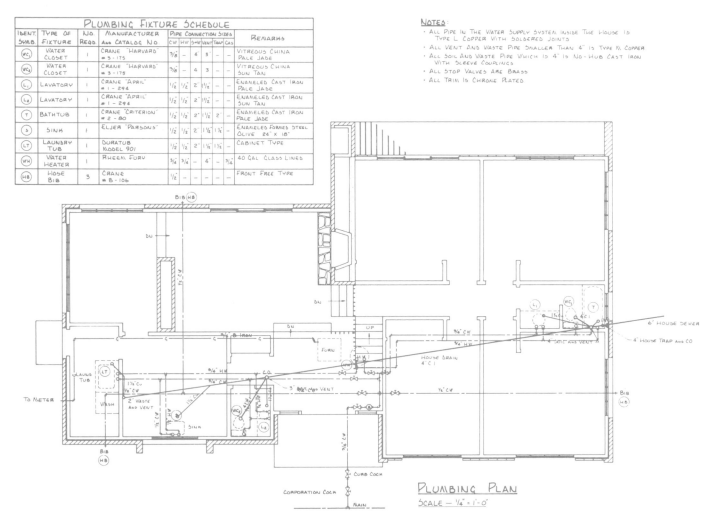

Fig. 28-8. A typical residential plumbing plan shows the water supply system and waste water removal system for a house.

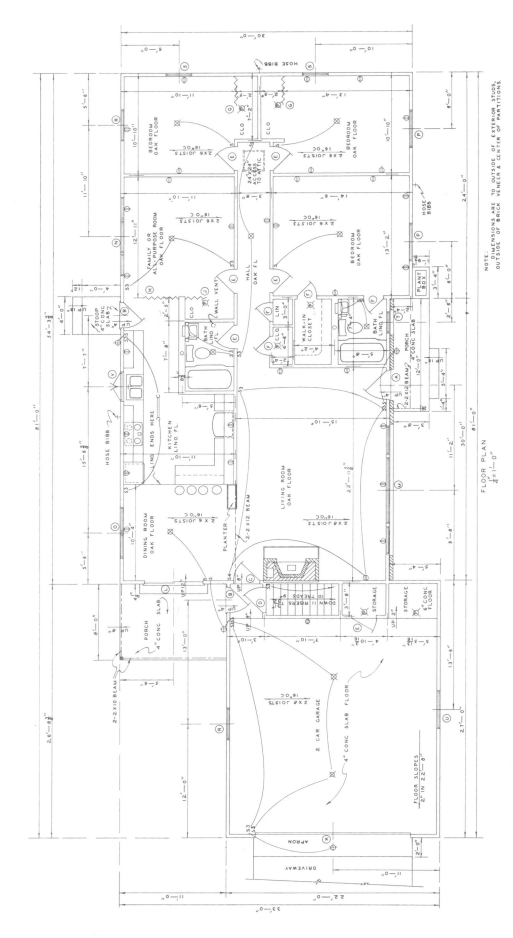

Fig. 28-9. An electrical plan shows the location of switches, outlets, and other important electrical devices that must be installed.

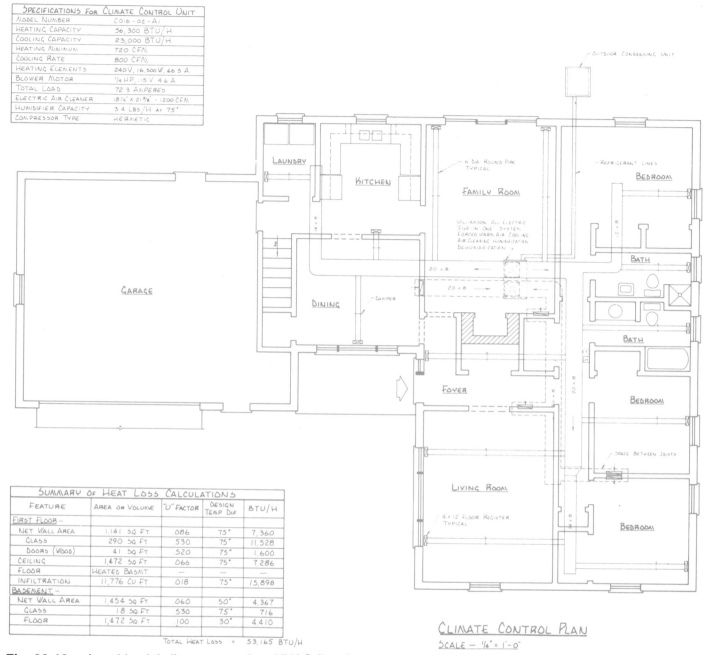

Specifications For Climate Control Unit	
Model Number	C016-02-A1
Heating Capacity	56,300 BTU/H
Cooling Capacity	23,000 BTU/H
Heating Minimum	720 CFM
Cooling Rate	800 CFM
Heating Elements	240 V, 16,500 W, 66.5 A
Blower Motor	¼ HP, 115 V, 4.6 A
Total Load	72.3 Amperes
Electric Air Cleaner	18½" x 21⅝" - 1200 CFM
Humidifier Capacity	3.4 Lbs/H at 75°
Compressor Type	Hermetic

Summary of Heat Loss Calculations				
Feature	Area or Volume	"U" Factor	Design Temp Dif	BTU/H
First Floor –				
Net Wall Area	1,141 Sq Ft	.086	75°	7,360
Glass	290 Sq Ft	.530	75°	11,528
Doors (Wood)	41 Sq Ft	.520	75°	1,600
Ceiling	1,472 Sq Ft	.066	75°	7,286
Floor	Heated Basmt	–	–	–
Infiltration	11,776 Cu Ft	.018	75°	15,898
Basement –				
Net Wall Area	1,454 Sq Ft	.060	50°	4,367
Glass	18 Sq Ft	.530	75°	716
Floor	1,472 Sq Ft	.100	30°	4,410
		Total Heat Loss =		53,165 BTU/H

CLIMATE CONTROL PLAN
SCALE – ¼" = 1'-0"

Fig. 28-10. A residential climate control, or HVAC (heating, ventilating, and air conditioning), plan show the location of the vents, ducts, and furnace and/or air conditioner.

Typical detail drawings include those for kitchens, Fig. 28-11. Detail drawings for chimneys and fireplaces are included, Fig. 28-12. Foundation walls also need detail drawings, Fig. 28-13. Stairs, windows and doors, complex framing, and other items of special construction all need detail drawings to fully explain information.

Presentation plans and models

A **presentation plan** (drawing) or **model** is frequently developed for a proposed construction project to help communicate the scope and appearance of the structure to prospective clients, funding agencies, boards, and other interested parties. The type and number of presentation plans for a given structure will depend on the complexity of the project and purpose of the presentations.

Several types of presentation drawings are commonly used. Exterior and interior perspectives are presentation drawings, Fig. 28-14. Rendered elevations are also presentation drawings, Fig. 28-15. Plot plans and floor plans are often used as presentation drawings, Fig. 28-16. Landscape plans, structural sections, traffic flow diagrams, and land use presentations all use presentation drawings to communicate information to a specific audience, Fig. 28-17.

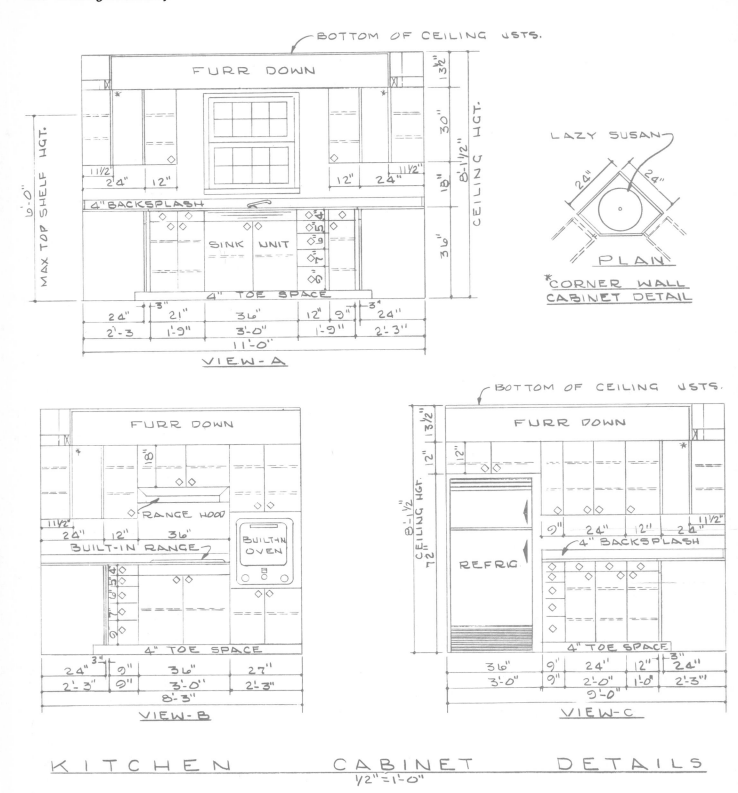

Fig. 28-11. Typical detail drawings may show elevations of kitchen cabinets and their arrangement.

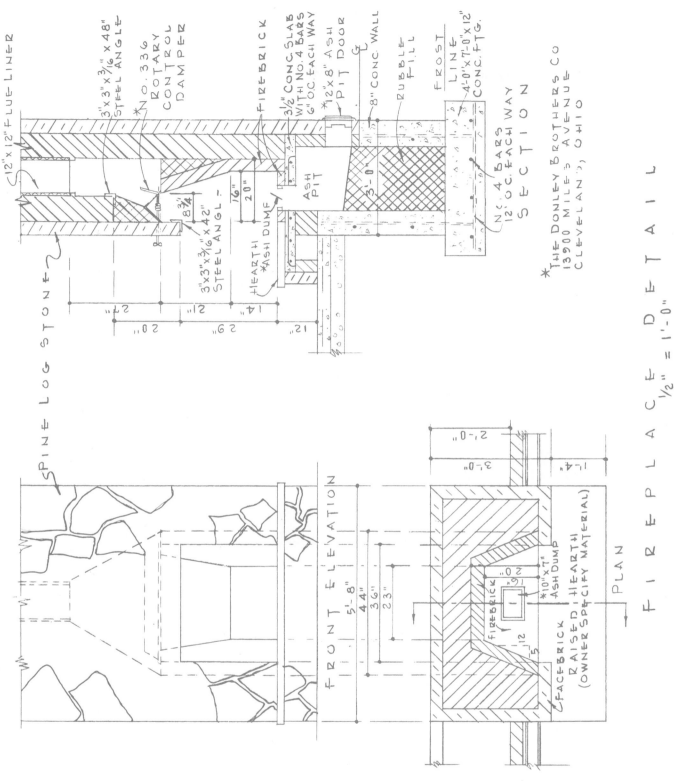

Fig. 28-12. Plan views provide location and specifications for items in a building such as a fireplace. Sometimes detail sections are needed to fully explain the feature.

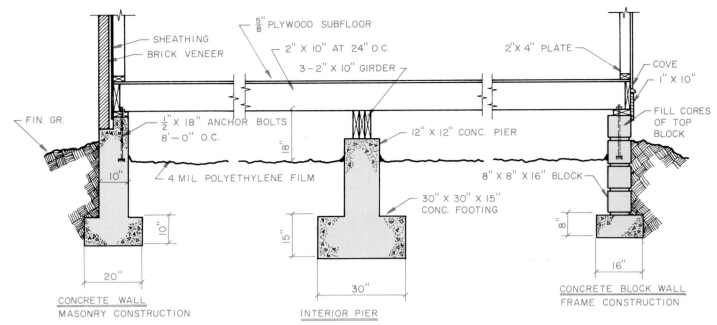

SHEATHING
BRICK VENEER

$\frac{5}{8}''$ PLYWOOD SUBFLOOR

2" X 10" AT 24" O.C.

3 – 2" X 10" GIRDER

2"X 4" PLATE

COVE

1" X 10"

FILL CORES OF TOP BLOCK

FIN. GR.

$\frac{1}{2}''$ X 18" ANCHOR BOLTS 8'—0" O.C.

10"

18"

12" X 12" CONC. PIER

4 MIL POLYETHYLENE FILM

8" X 8" X 16" BLOCK

30" X 30" X 15" CONC. FOOTING

15"

8"

10"

20"

30"

16"

CONCRETE WALL
MASONRY CONSTRUCTION

INTERIOR PIER

CONCRETE BLOCK WALL
FRAME CONSTRUCTION

Fig. 28-13. Detail drawing can show sections such as poured concrete foundations.

Fig. 28-14. An architectural rendering of a house in perspective helps the client to better visualize the finished building. (Ken Hawk)

NORTH

Fig. 28-15. Presentation elevations, rather than a perspective, are sometimes used to represent a structure. (Larry Campbell)

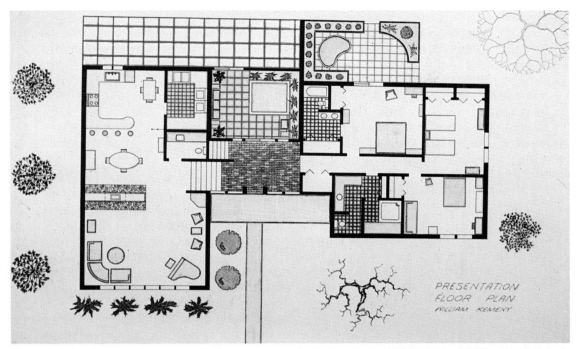

Fig. 28-16. A rendered presentation floor plan increases communication. (William Kemeny)

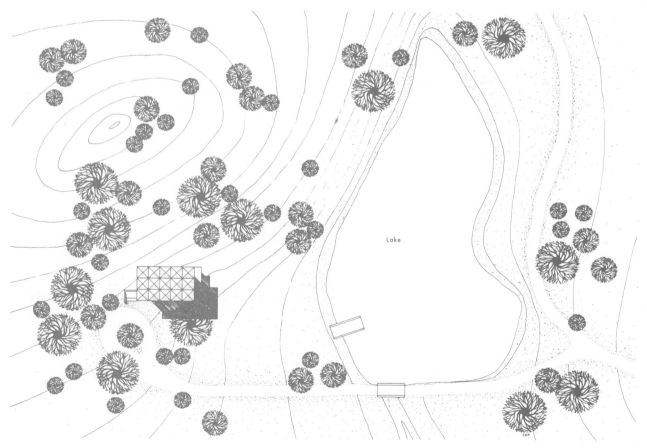

Fig. 28-17. Land use presentations can be rendered in ink so that they will reproduce cleanly. Other drawings are inked for this reason as well.

In addition to communicating the scope and appearance of a project to a prospective client, models are used to study a basic design, check for relationship between parts of the structure, and for advertising purposes. Three basic types of models are generally used: small-scale solid models, structural models, and presentation models. (Refer to Fig. 28-18 for an example of a structural model.)

The purpose of the **small-scale solid model** is to show how a building will relate to surrounding structures. Scales used range from 1/32″ = 1′ 0″ to 1/8″ = 1′ 0″. Very little detail is shown on those models.

Structural models are used to show the structural elements of a building–how it is constructed. These models range in size from 1/2″ = 1′ 0″ to 1″ = 1′ 0″. The purpose of a **presentation model** is to show the finished appearance of the building as realistically as possible, Fig. 28-19.

ELECTRICAL AND ELECTRONICS DRAWINGS

Growth of the electronics industry in recent years has brought increased demand for drafters who are capable of preparing electrical and electronic circuit drawings. **Electrical and electronics drafting** involves the same basic principles used in other types of drawings. The difference is in the special symbols that have been developed to represent electrical circuits and wiring devices.

Major requirements for electrical and electronics drawings have been standardized by industry through the American National Standards Institute (ANSI) and by the military through the Military Standard Publications. These two sets of standards are almost identical.

The components used in electricity and electronics vary in size from large transformers at an electrical generating plant to microscopic integrated semiconductor circuits, Fig. 28-20. Because of the complex devices involved and their relationship to each other in electrical circuits, you must acquire a basic knowledge of electricity/electronics and understand how an electrical or electronic circuit operates if you wish to specialize in this type of drafting.

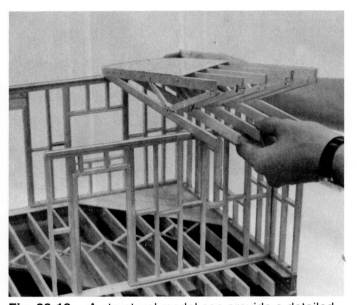

Fig. 28-18. A structural model can provide a detailed view of such features as the framing of a building.

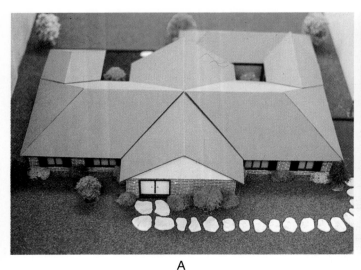

A

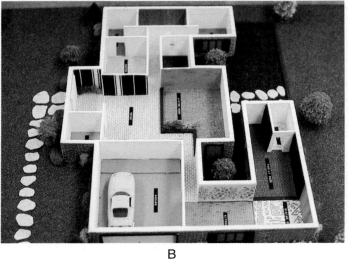

B

Fig. 28-19. Small scale architectural presentation models allow a customer to visualize what the finished house will look like.

Fig. 28-20. The first full 32-bit microprocessor was a single microchip, smaller than a dime, and containing 150,000 transistors. It was made possible by the use of computer graphics and microelectronics. (Bell Laboratories)

Pictorial and Photo Drawings

Pictorial drawings using symbols are sometimes drawn to illustrate component parts in electrical and electronics drawings. Pictorials are particularly useful for assembly line workers, do-it-yourself hobbyists, and others who are not trained in reading graphic symbols on electrical and electronics drawings. A pictorial drawing of several electrical components and their relative positions on a circuit board is illustrated in Fig. 28-21.

Another drafting technique used in electronics work is the *photodrawing.* This type of drawing is produced by photographing electronic components or assemblies and adding line work to complete the drawing, Fig. 28-22.

Single-Line Diagrams

Single-line diagrams are simplified representations of complex circuits or entire systems. The ANSI Standard for Electrical and Electronics Diagrams states that a single line or a one-line diagram is one that: "... shows, by means of single lines and graphic symbols, the course of an electric circuit or system of circuits and the component devices or parts used therein."

Single-line diagrams are used primarily in the electrical power and industrial control areas, with some limited applications in electronics and communications. A single-line diagram is usually one of the first drawings made in the design of a large electrical power system, because it contains the basic information that will serve as a guide in the preparation of more detailed plans.

A typical single-line diagram used in the electrical power field is shown in Fig. 28-23. The thick connecting lines on the drawing indicate primary circuits, and the medium lines indicate connections to current or potential sources. In either case, a single line is used to represent a multiconductor circuit.

In single-line diagrams, it is standard practice to use either horizontal or vertical connecting lines with the highest voltages at the top or left of the drawing and successively lower voltages toward the bottom or right of the drawing. When constructing such a diagram, try to maintain a logical se-

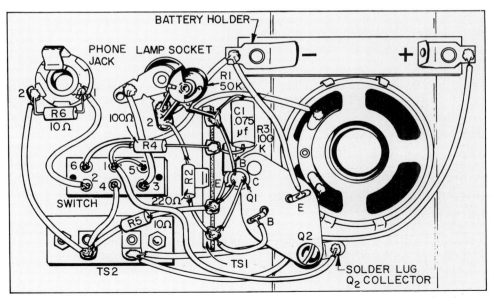

Fig. 28-21. Pictorial drawings can show the parts and their relationships in a realistic manner.

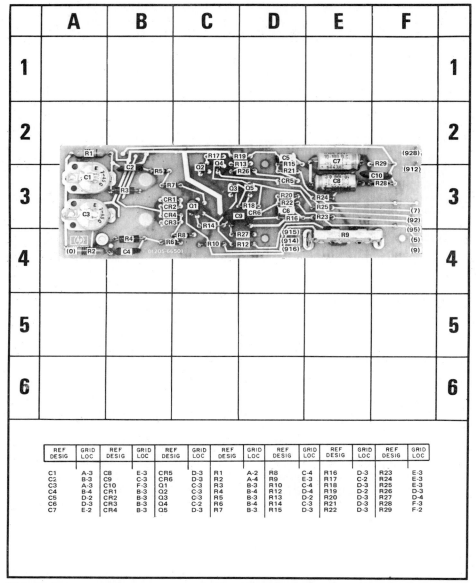

Fig. 28-22.   A photo that has been superimposed on a line-art grid is called a photodrawing. A photodrawing is designed to aid in identifying components by code number and grid location. (Hewlett-Packard Co.)

quence while avoiding an excessive number of line crossings.

Block Diagrams

Block diagrams are closely related to single-line diagrams: each contains basic information that presents an overview of a system in its simplest form. Squares and rectangles are primarily used on block diagrams, but an occasional triangle or circle may be used for emphasis. Graphic symbols are rarely used, except to represent input and output devices.

The blocks should be arranged in a definite pattern of rows and columns, with the main signal path progressing from left to right whenever possi-

ble, Fig. 28-24. Auxiliary units, such as power supply or oscillator circuits, should be placed below the main diagram. Each block should contain a brief description or function of the stage it represents. Additional information may be placed elsewhere on the drawing. The block that requires the greatest amount of lettering usually determines the size of all the blocks. However, the use of two block sizes on one drawing is not objectionable.

A heavy line should be used to represent the signal path. In a complex circuit or system, more than one line may lead into or away from a block, Fig. 28-25. Arrows should be used to show the direction of the signal flow. The overall appearance of the diagram should be a consistent and organized pattern that is well-balanced and easy to read.

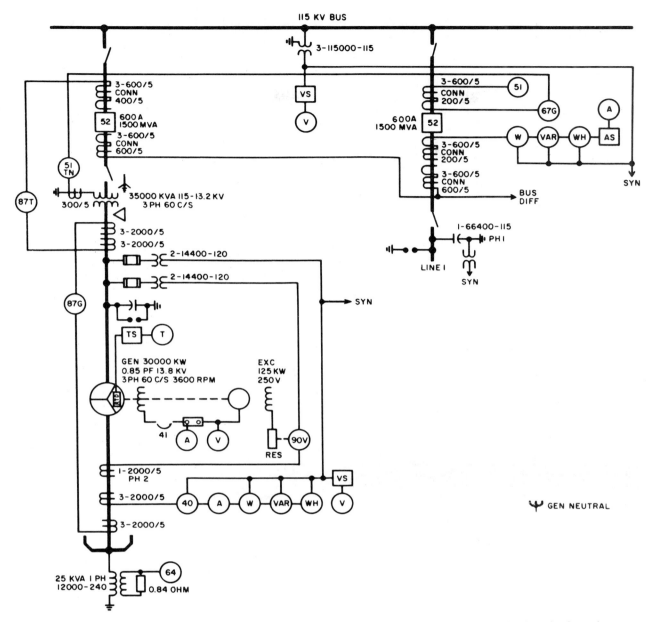

Fig. 28-23. A typical single-line diagram shows power switchgear and complete device designations. (American National Standards Institute)

Schematic or Elementary Diagrams

As defined in the ANSI Standard for Electrical and Electronics Diagrams, a schematic or elementary diagram is one that: "… shows, by means of graphic symbols, the electrical connections and functions of a specific circuit arrangement. The schematic diagram facilitates tracing the circuit and its functions without regard to the actual physical size, shape, or location of the component device or parts."

The most frequently used drawing in the electronics field is the **schematic diagram,** Fig. 28-26. It serves as the master drawing for production drawings, parts lists, and component specifica-tions. It is used by engineering groups for circuit design and analysis, and by technical personnel for installation and maintenance of the finished product. Elements of a schematic diagram may be combined with other types of diagrams to provide a more comprehensive and useful drawing, Fig. 28-27.

Connection and Interconnection Wiring Diagrams

Connection and interconnection wiring diagrams are drawings that supplement schematic diagrams. These drawings contain information used in the manufacture, installation, and mainte-

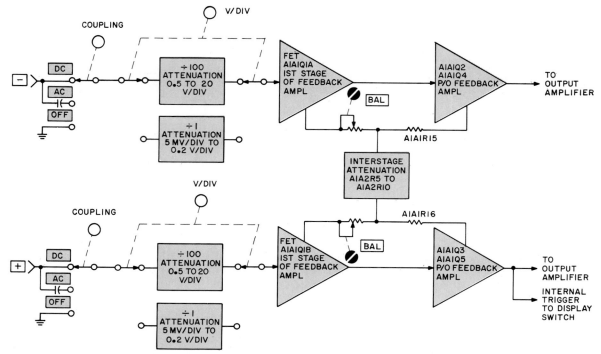

Fig. 28-24. A typical block diagram has blocks arranged in rows and columns. (Hewlett-Packard Co.)

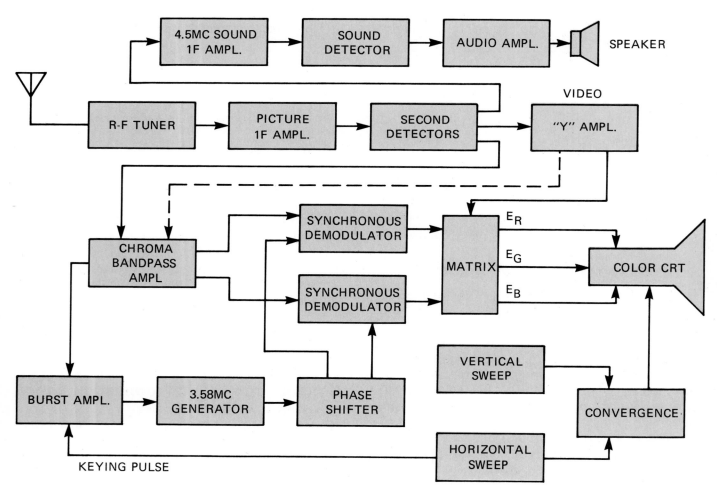

Fig. 28-25. Arrows are used in typical block diagrams to show direction of flow.

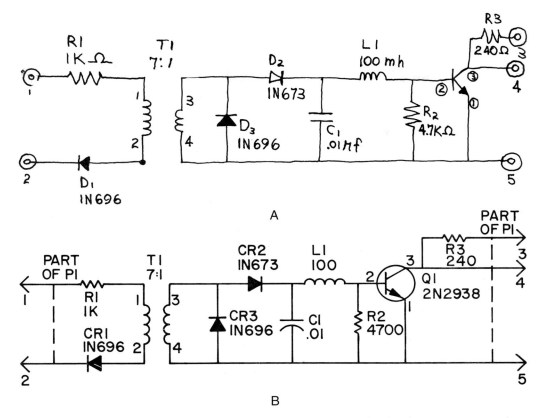

Fig. 28-26. A–An engineer's sketch of a circuit will have the basic appearance of a schematic, only less refined. B–A drafter will take the sketch and produce a finished schematic diagram.

nance of electrical and electronic equipment. They graphically represent the conducting paths (wires or cables) between component devices.

Connection diagrams and interconnection diagrams are very similar. The ANSI definitions may help to differentiate between the two types:

The **connection or wiring diagram** is one that "... shows the connections of an installation or its component devices or parts. It may cover internal or external connections, or both, and contains such detail as is needed to make trace connections that are involved. The connection diagram usually shows general physical arrangement of the component devices or parts."

An **interconnection diagram** is "... a form of connection or wiring diagram which shows only external connections between unit assemblies or equipment. The internal connections of the unit assemblies or equipment are usually omitted."

Wiring diagrams are divided into three major classifications: **continuous-line, interrupted-line,** and **tabular.** (Refer to Fig 28-28 for an example of continuous-line. Refer to Fig. 28-29 for an example of interrupted-line. Refer to Fig. 28-30 for an example of tabular.) Each type indicates the method used to show the connections between component parts or devices. The component parts in a connec-

tion diagram are represented by pictorial, block, or graphic symbols with only outline and terminal circles shown.

A type of continuous-line diagram that is often used to show the point-to-point connections of a device is illustrated in Fig. 28-28. The diagram in Fig. 28-27 is more complicated, but it is a point-to-point connection diagram, as well. An interrupted-line connection diagram may contain symbols to indicate the connecting paths between component parts or devices. The tabular type of connection diagram is sufficient for many wiring operations. Basically, it is a simple "from-to" list, but it may be expanded to show additional information such as wire lengths, sizes, or types.

Solid-State Electronics

The greatest development in the field of electronics in recent years has been in solid-state electronics. The accompanying problems of miniaturization have placed unusual demands upon the drafter. These have led to the increased use of sophisticated drafting equipment such as artwork generators, automatic drafting machines, and computer-assisted design equipment.

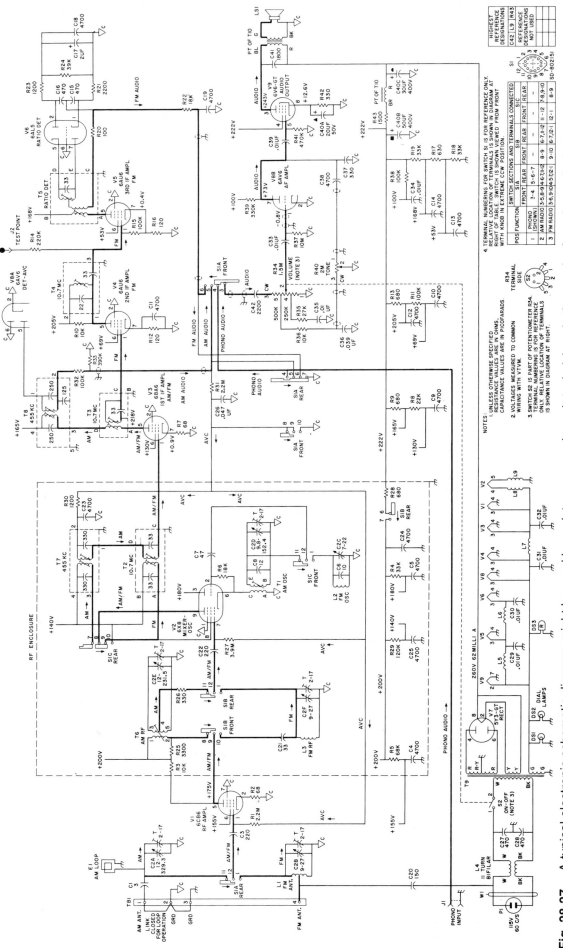

Fig. 28-27. A typical electronic schematic diagram might be used by maintenance and repair technicians in the repair of an electronic device. (American National Standards Institute)

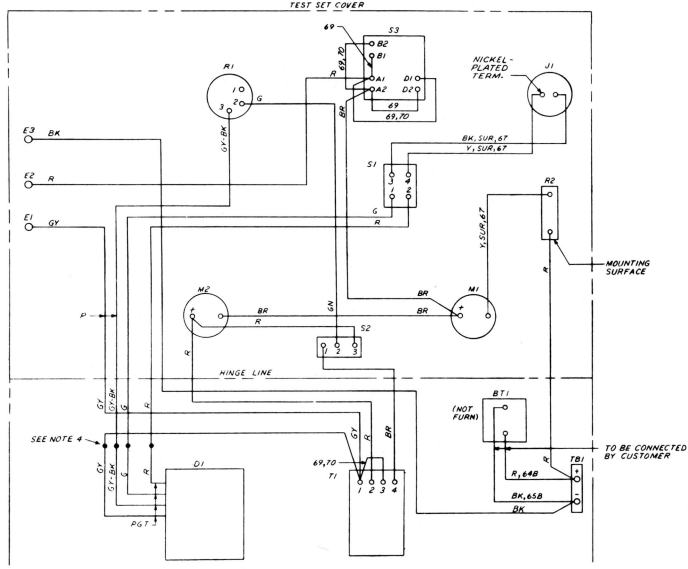

Fig. 28-28. A typical continuous-line connection diagram, sometimes referred to as a point-to-point wiring diagram, will show all of the components and how they fit together, but not the actual physical locations of the parts. (American National Standards Institute)

NOTES:

1. UNLESS OTHERWISE SPECIFIED, ALL WIRES ARE INCLUDED IN THE CABLE ASSEMBLY XXXXX.

2. ITEM NUMBERS REFERRED TO ARE SHOWN IN PARTS LIST OF ASSEMBLY DRAWING XXXXX.

3. ALL SOLDERING SHALL BE IN ACCORDANCE WITH QQ-S-524 METHOD C.

4. SPLICE AND SOLDER AND WRAP WITH ONE LAYER OF TAPE ITEM 58 AND TWO LAYERS OF TAPE ITEM 60.

5. SUR-WIRING-WIRE TO BE DRESSED BACK AND RUN ALONG THE MOUNTING SURFACES IN THE MOST CONVENIENT MANNER.

6. PGT - LEADS FURNISHED WITH PART.

Solid-state electronics began with the invention of the transistor and the subsequent development of the printed circuit board. Both involve an etching process where very precise masks, or windows, are used to control the areas that will be eaten or etched away by a chemical solution.

The next step in the development of solid-state electronics was to construct thin film circuits in which passive components and connections were deposited as extremely thin conductive films on a supporting **substrate** (base). At the present time,

there are numerous active devices that can be interconnected on a supporting substrate, Fig. 28-31.

The current phase of development in solid-state technology is known as ***very large scale integration (VLSI)*** of complex circuits and functions. These have made possible "computer-on-a-chip" devices within a single package, Fig. 28-32.

The drawings for solid-state devices are usually photographically reduced to as much as 1/40 original size, and successively aligned with other reduced drawings to an accuracy of one ten-thousandth of an

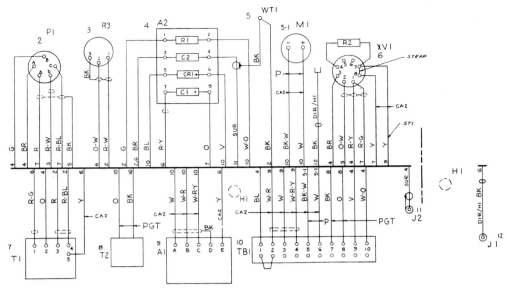

Fig. 28-29.   A typical interrupted line connection diagram, also called a base-line diagram, may contain symbols to indicate connections between components or devices.

REV		WIRE				FROM				TO			
SYM	TRAN	COLOR	AWG	SYMBOL	METHOD OR PATH	NOTE	AREA LOC	TERMINAL	LEVEL	NOTES	AREA LOC	TERMINAL	LEVEL	NOTES
–		W–R		ST1	CA2			TB1	2			A1	B		
–		MS1			CA1			TB1	2			TB1	1		
–		W		ST1	CA2			TB1	3			A1	A		
–		W–R–Y		ST1	CA2			TB1	4			A1	C		
–		BK–W		P1	CA2			TB1	5			M1	NEG		
–		W		P1	CA2			TB1	6			M1	PØS		
–		BK			PGT			TB1	7			T2			
–		Ø			PGT			TB1	8			T2			
–		V			CA1			TB1	9			A2	6		
–		W–Ø			CA1			TB1	10			A2	2		
–		R–G		SS4	CA1			T1	1			XV1	1		
–		Ø			CA1			T1	2			A2	8		
–		R		SP1	CA1			T1	3			P1	A•		
–		R–BL		SP1	CA1			T1	4			P1	C		
–		Y			CA2			T1	5			XV1	8		
–		BK			PGT			T2				TB1	7		
–		Ø			PGT			T2				TB1	8		
–		BK		H1	DIR			V1	CAP			J1			
–		BK			SUR			WT1				A2	4$		
–		BK			CA1			WT1				P1	C$		
–		R–Y		SS3	CA1			XV1	1			A2	7		
–		R–G		SS4	CA1			XV1	1			T1	1		
–		CØM			CA1			XV1	1$			XV1	1$		
–		CØM			CA1			XV1	1$			XV1	1$		
–		CØM			CA1			XV1	1$			XV1	2$		
–		Ø–W		SS2	CA1			XV1	2			R3	2		
–		CØM			CA1			XV1	2$			XV1	1$		
–		BR			CA1			XV1	3			A2	3		
–		R2			PGT			XV1	4			XV1	7		
–								XV1	5						
–								XV1	6						
–		R2			PGT			XV1	7			XV1	4		
–		STRAP						XV1	7			XV1	8		
–		Y		ST1	CA2			XV1	7(ST1			A1	E		
–		W			CA2			A1	A			TB1	3		
–		W–R		ST1	CA2			A1	B			TB1	2		

Fig. 28-30.   A typical tabular-type connection diagram presents wiring connections in the form of a table. (American National Standards Institute)

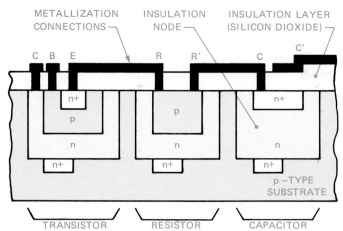

METALLIZATION CONNECTIONS — INSULATION NODE — INSULATION LAYER (SILICON DIOXIDE) —

Fig. 28-31. A cross-sectional view of a typical integrated circuit shows that the transistors, capacitors, and resistors contained on a single chip have different constructions.

Fig. 28-32. With very large scale integration (VLSI), the components of an entire microcomputer can be contained on a single, tiny chip. (Motorola, Inc.)

inch (.0001). They are typically plotted on a special film base, Fig. 28-33. This method has virtually replaced the older manual procedure that involved cutting a peelable film to form an opaque overlay of the circuit pattern.

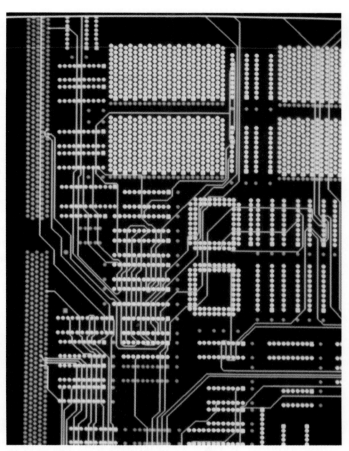

Fig. 28-33. A solid state schematic can be plotted on a film base. This film is then used in the miniaturization of the circuit. (Lockheed Aircraft Corp.)

MAP AND SURVEY DRAWINGS

Cartography is the science of map making, and special kinds of drafting are required in the preparation of copy for maps. Drafting techniques used for some of the more common types of maps are presented in this section.

Mapping and Surveying Terms

The following terms are basic to an understanding of surveying and map drafting.

Azimuth: The angle that a line makes with a north-south line, measured clockwise from the north, Fig. 28-34.

Bearing: An angle having a value of 0° to 90°, measured from either the north or south. A line with a bearing of 20° to the west of south would be stated as South 20° West.

Contour: The irregularly-shaped lines found on topographic and other plan-view maps to indicate changes in terrain elevation. On any single contour line, every point is at the same elevation.

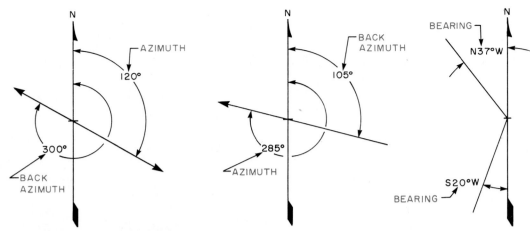

Fig. 28-34.   The azimuth of a line describes a line in degrees from North. The bearing of a line describes a line in terms of degrees from one or more of the four compass points (North, South, East, and West).

Horizontal curve: A change of direction in the horizontal or plan view that is achieved by means of a curve.

Interpolation: A technique used to locate, by proportion, intermediate points between grid data given in contour plotting problems.

Mosaic: A series of aerial photographs or radar images of adjacent land areas, taken with intentional overlaps and fitted together to produce a larger picture.

North: The direction normally indicated on the top of a map. **Magnetic north,** as indicated by a magnetic compass, is satisfactory for most maps, but is subject to local deflection errors affecting the magnetic compass. **True north,** as determined by sighting on Polaris (the "North Star") is considered to be the most accurate.

Plat: A plan that shows land ownership, boundaries, and subdivisions.

Survey: The use of linear and angular measurements and calculations to determine boundaries, position, elevation, or profile, of a part of the earth's or other planets' surfaces.

Stations: The turning points in a map traverse, Fig. 28-35. In highway construction surveys, points at 100 foot intervals on the centerline in the plan view are also called stations. They are located by stakes with station numbers on them.

Traverse: Measuring or laying out a line, such as a property line, by means of angular and linear measurements, Fig. 28-35. A **closed traverse** is one that returns to its point of origin in a previously identified point. An **open traverse** neither returns nor ends at a previously identified point.

Vertical curve: A change of direction in the grade, shown in a profile view, that is achieved by means of a curve (usually a parabolic curve).

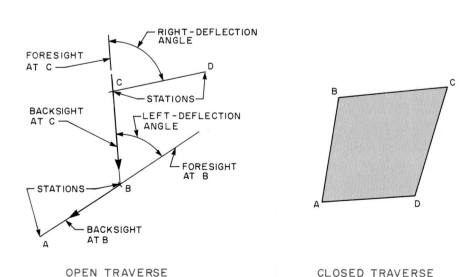

OPEN TRAVERSE   CLOSED TRAVERSE

Fig. 28-35.   There are different terms used in laying out open and closed traverses when mapping.

Types of Maps

All maps are representations, on flat surfaces, of a part of the surface of the earth (or any other planet). Although all types of maps have common features, they may be classified into types based upon their intended use.

Geographic maps

The **geographic map** is familiar to most students as the type contained in social studies textbooks. It illustrates, by color variation or other technique, such elements as climate, soil, vegetation, land use, rivers, population, cities, and topographic features. The geographic map normally represents a large area and must be drawn to a very small scale.

Geologic maps

Geology is the study of the earth's surface, its outer crust and interior structure, and the changes that have taken and are taking place. **Geologic maps** report this information pictorially. Maps showing the geologic surface are used, Fig. 28-36. Maps that show geologic cross-sections of the subsurface are also used, Fig. 28-37.

Topographic maps

In contrast to a geographic map, a **topographic map** gives a detailed description of a relatively small area. Depending on their intended use, topographic maps may include natural features, boundaries, cities, roads, pipelines, electric lines, houses, and vegetation, Fig. 28-38. Contour lines

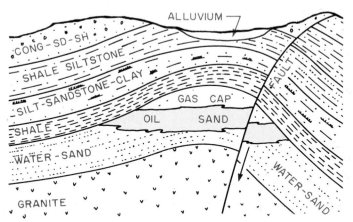

Fig. 28-37. A geologic cross-section map shows a section of the earth's interior structure.

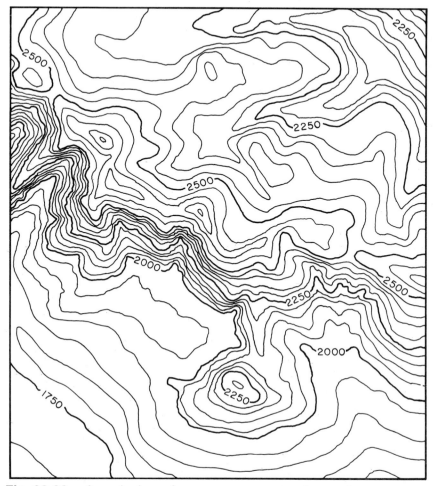

Fig. 28-36. A geologic surface map can cover a large area.

are normally used to show elevation. Standard map symbols may be employed to show natural or man-made features.

Cadastral maps

Cadastral maps are drawn to a scale large enough to accurately show locations of streets, property lines, buildings, etc. They are also used in the control and transfer of property, Fig. 28-39. Cadastral maps are used to show plats of additions to a city and to identify property owners along a road right-of-way.

Engineering construction maps

Maps under this category range from a simple plot plan for a residence to such major engineering projects as commercial buildings, industrial plants, electrical transmission lines, bridges, and hydroelectric dams. An **engineering map** for a highway construction project is shown in Fig. 28-40. The horizontal curve is plotted in the aerial or plan view. The vertical curve is shown on the same sheet in the profile view. The two are referenced by common check points.

An engineering map that requires careful study and detailing is one that locates a dam in relation to the elevation and configuration of the surrounding land. The location and cross section of the Morrow Point Dam in Colorado is shown in Fig. 28-41.

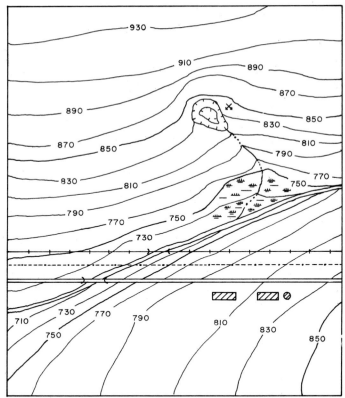

Fig. 28-38.   A topographic map shows contour elevations, natural features, and manmade features.

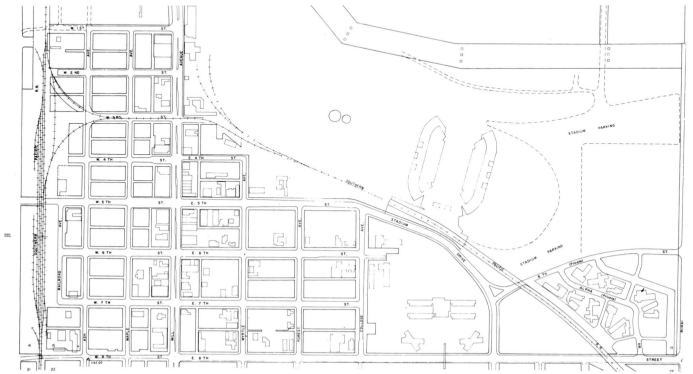

Fig. 28-39.   A cadastral map aids in locating property lines. It is drawn to a scale large enough to show individual buildings.

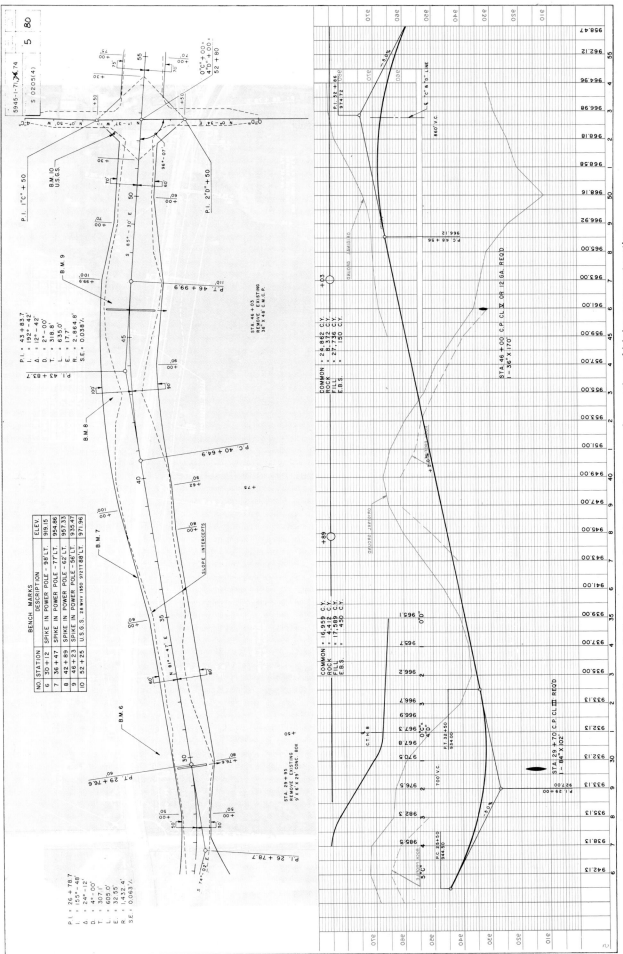

Fig. 28-40. A civil engineering map might show both the horizontal and vertical curves of a section of highway construction. (Wisconsin Department of Transportation)

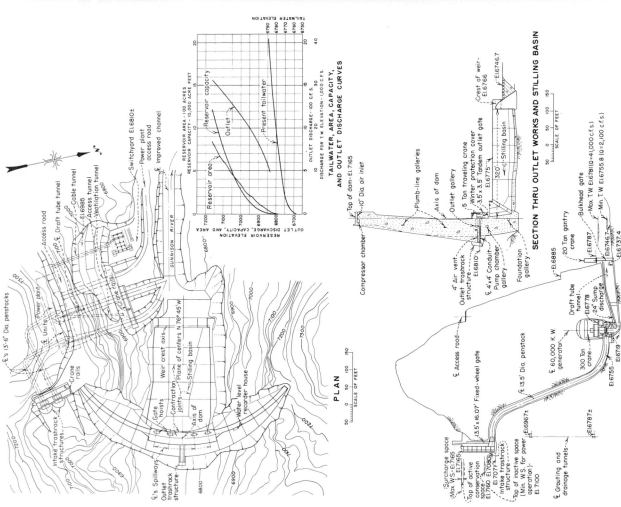

Fig. 28-41. The site for construction of the Morrow Point Dam in Colorado was located by means of an engineering map. (U.S. Department of the Interior)

Map Format

There are perhaps as many different layouts for maps as there are types of maps. However, each map has a title, scale, lettering and notes, symbols, or other standard data for which certain guidelines are recognized.

Title

The **title** of a map states what the map is, its location, when it was prepared, and the individual, company, or government agency for whom it was prepared. The title is placed in the lower right-hand corner of the sheet when possible. If this is not possible, it is placed in any area that gives clarity to the drawing.

Scale

The **scale** of a map should be indicated just below the title. Most maps are laid out on a **base ten** ratio with a civil engineer's scale. The map scales used range from very large (1″ = 1′ 0″), such as on a plot plan for a residence or commercial building, to greatly reduced scales (1″ = 400 miles) on geographic maps.

Scales also may be indicated as a ratio, such as 1:250,000. In this case, 1 inch equals 250,000 inches, or nearly 4 miles. A decimal scale, using the side marked "50," could be used for this scale by letting each major unit equal 50,000 similar units.

For a scale of 1″ = 400′, the "40" side of a decimal scale could be used with the smallest subdivision equal to 10 feet. Some maps use the graphic scale which is quickly and easily interpreted.

Lettering and notes

On engineering maps, **lettering** is done with single-stroke capital letters, either vertical or inclined. The two lettering types are never used on the same map. Titles of maps are sometimes done in Roman-style letters, providing a little more flair than the single-stroke Gothic. All lettering and notes are placed to read from the bottom or right-hand side of the sheet.

Symbols

Because of the very small scales used on many maps, not all features can be shown. Many that can be shown must be represented by **symbols,** Fig. 28-42. Regardless of the scale of the map, symbols are drawn essentially the same size.

North indication

Maps must be properly oriented to be useful. This is done with the **direction arrow,** which indicates *true north* unless otherwise stated. The main feature of the arrow should be its body line, with the arrowhead clearly indicating north.

Gathering map data

Surveying is the means of collecting data for use in making maps. This is accomplished in a variety of ways.

Field survey crews

For years, survey crews equipped with the surveyor's chain, transit, and level have gathered data in the field for use in making maps. This is a

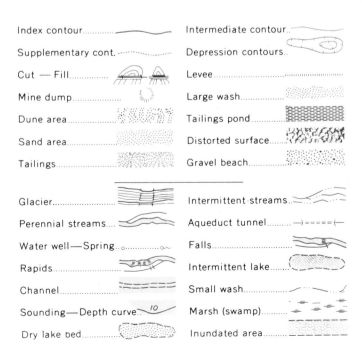

Fig. 28-42. Standard symbols are used on maps to conserve space and make them easier to understand. (U.S. Department of the Interior)

time-consuming and laborious task, particularly where the equipment has to be carried over rough terrain.

Today, many of these surveying devices have been replaced, particularly on large-scale projects, with faster and more accurate instruments. However, transit-equipped survey crews are still used to gather map-making information on small tracts of land, in highway construction, and on geological exploration projects, Fig. 28-43.

Photogrammetry

The tremendous expansion in state and interstate highway programs has caused a need for new and improved techniques of gathering survey information. One of these techniques, photogrammetry, which had been in limited use for a number of years, is now used extensively.

Photogrammetry is the use of photography, either aerial or land-based, to produce useful data for the preparation of contour topographic, planimetric, and orthophoto cross-section profile maps.

Once the area to be mapped has been identified, control points are placed to be used in controlling photographic stereo models (three-dimension viewing). Next, ground control surveys are made as checks. Aerial photographs are then taken for translation into photomaps, orthophoto cross-section maps, and topographic maps. This is accomplished by means of a stereoplotter, a device for reading elevations from a flat surface, Fig. 28-44.

Automated digitizers for recording horizontal coordinates and control points are used, along with computers. The data are returned to drafters and a map is drawn from the survey data. The data also may be processed and fed to a plotter in the form of a computer program. The plotter then produces the maps. Photogrammetry represents a considerable savings of time over field survey methods in the collection of survey data.

Radar imagery

A recent development in gathering information for map drafting is ***radar imagery***, Fig. 28-45. This is accomplished by a high-resolution, side-looking, airborne radar that records a "photo-like" image of the terrain. The radar is called ***side-looking*** because the electronic signals are sent out at right angles to the aircraft's path.

Fig. 28-44. A stereoplotter is a device that produces vertical readings from a flat surface. (Virginia Dept. of Highways)

Fig. 28-43. Despite the introduction of newer technology, traditional surveying tools are still in wide use. The transit theodolite is an instrument used by surveyors to take accurate measurements of land features. (Colorado Department of Highways)

Fig. 28-45. Highly detailed radar images can show very detailed images such as the terrain of a volcanic island in southeast Asia.
(Goodyear Aerospace Corp. and International Aero Service Corp.)

Radar imagery mapping can be done day or night, in any kind of weather, even when heavy clouds block out the use of aerial photography. Unlike aerial photography, the scale of the radar image is constant and without distortion, regardless of the range or altitude of the aircraft. This makes it ideal for mosaics and other mapping purposes.

The radar imagery method of surveying has applications in the fields of geology, geophysics, hydrology, topographic mapping, highway construction, and agriculture. Output is planimetric, but three-dimensional qualities are provided when prepared in duplicate, overlapping sets.

TECHNICAL ILLUSTRATION

Industrial processes and products have become so technical that many items sold are accompanied with one or more technical illustrations explaining their operation and use. *Technical illus-*

tration is the preparation of drawings, usually pictorial, with three-dimensional effect. Usually, they are shaded or finished in multiple colors to give more realism to the object or process and to improve understanding.

Technical illustrations are used to supplement working drawings and to clarify complex assembly and operational procedures. They are indispensable to nontechnical personnel and to those who use technical equipment but have difficulty reading working drawings.

Technical illustrations are widely used in industry, in service manuals, and in do-it-yourself kits.

Types of Technical Illustrations

Technical illustration work may be classified into two general fields: engineering-production illustrations and publication illustrations.

Engineering-production illustrations are used for engineering design, contract proposals, and production work. More emphasis is given to technical accuracy than to their styling.

Publication illustrations are used in service manuals, parts catalogs, operational handbooks, and in sales and advertising brochures and catalogs, Fig. 28-46. Publication illustrations may involve the use of cartoon sketches, shading effects, life drawings, and other commercial art techniques. In both the engineering-production and the publication fields, the following types of illustrations are used.

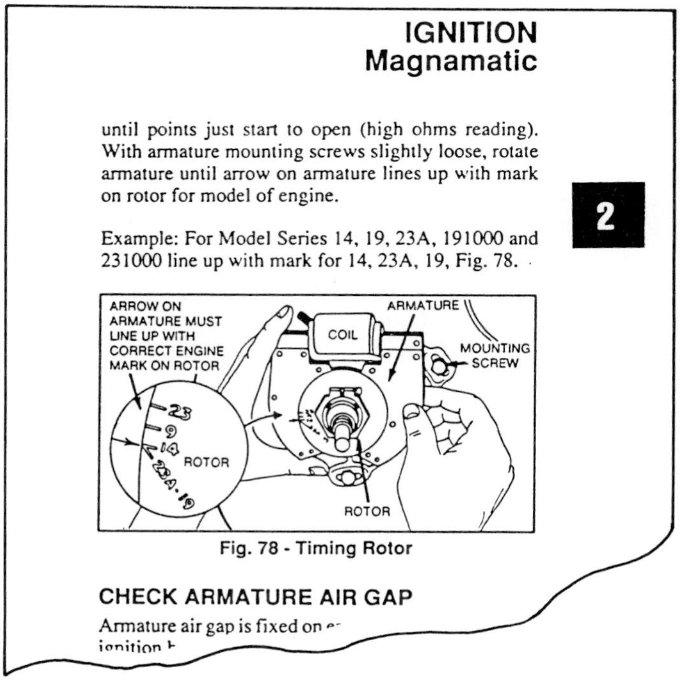

IGNITION
Magnamatic

until points just start to open (high ohms reading). With armature mounting screws slightly loose, rotate armature until arrow on armature lines up with mark on rotor for model of engine.

Example: For Model Series 14, 19, 23A, 191000 and 231000 line up with mark for 14, 23A, 19, Fig. 78.

2

ARROW ON ARMATURE MUST LINE UP WITH CORRECT ENGINE MARK ON ROTOR

ARMATURE

COIL

MOUNTING SCREW

ROTOR

ROTOR

Fig. 78 - Timing Rotor

CHECK ARMATURE AIR GAP

Armature air gap is fixed on ignition

Fig. 28-46. Technical illustrations aid understanding. They are a vital part of equipment service manuals. (Briggs & Stratton Corporation)

Pictorial

The ***pictorial drawing*** is perhaps the most elementary type of technical illustration. A pictorial drawing may be a line drawing in any one of the standard pictorial projections: isometric, dimetric, trimetric, oblique, or perspective. In other cases, the drawing may be a sophisticated rendering of a drawing that closely resembles a photograph of an object, Fig. 28-47.

Cutaway assembly

The ***cutaway assembly drawing*** helps to clarify multiview drawings of complex assemblies, Fig. 28-48. Cutaway assembly drawings are used in production operations and frequently are found in service manuals.

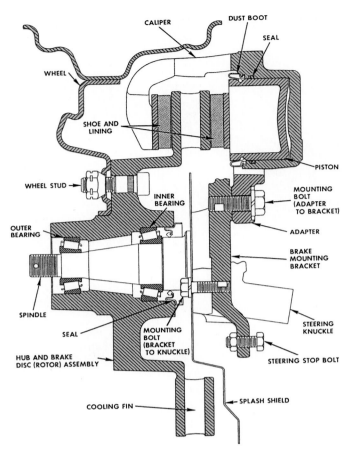

Fig. 28-48. A cutaway assembly drawing shows the relationship of various parts after they are assembled. (International Harvester Co.)

Exploded assembly

A type of technical illustration that is used frequently in manufacturing assembly, service manuals, and customer product instructions is the ***exploded assembly,*** Fig. 28-49. Few illustrations are as effective in clarifying a procedure for assembly. To leave no doubt about the order of assembly, a centerline is often shown joining the parts.

Hidden and telltale sections

Hidden and telltale sections permit the illustrator to display sheetmetal relationships when it is desirable to show the outside of a piece in its entirety, Fig. 28-50.

Peeled section

Several layers of material in a built-up part may be shown by use of a ***peeled section,*** Fig. 28-51. The use of this type of section calls for artistic judgment on the part of the illustrator and checker.

Basic Illustration Techniques

The following techniques are widely used in industry to prepare all types of technical illustrations.

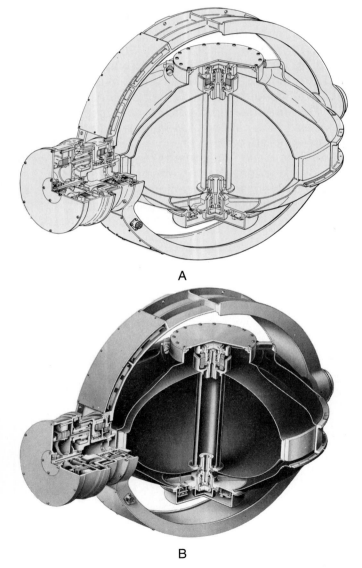

Fig. 28-47. A–The pictorial drawing is a basic type of technical illustration. B–Pictorial drawings can be rendered with relative ease. Renderings are often used as presentation drawings. (Sperry Flight Systems Div.)

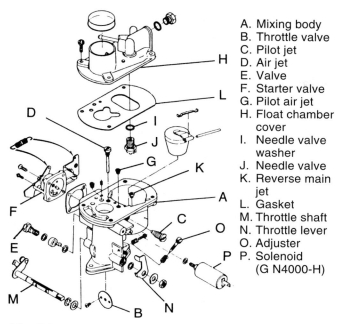

A. Mixing body
B. Throttle valve
C. Pilot jet
D. Air jet
E. Valve
F. Starter valve
G. Pilot air jet
H. Float chamber cover
I. Needle valve washer
J. Needle valve
K. Reverse main jet
L. Gasket
M. Throttle shaft
N. Throttle lever
O. Adjuster
P. Solenoid (G N4000-H)

Fig. 28-49. An exploded assembly illustration shows how different parts are assembled together. These drawings make assembly operations easy to follow, especially when performing repair or maintenance tasks. Such illustrations are widely used in service manuals. (Kubota, Ltd.)

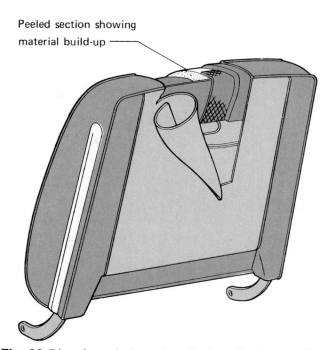

Peeled section showing material build-up

Fig. 28-51. A peeled section displays the layers of a built-up product like an automobile seat. (General Motors Corp.)

Locating shadows of geometric forms

Locating shadows of geometric forms helps in understanding the shading of these forms in technical illustrations. Four geometric figures are shown in Fig. 28-52, along with the procedures used in locating their shadows. Note the direction of the light source and how it affects the shading on the

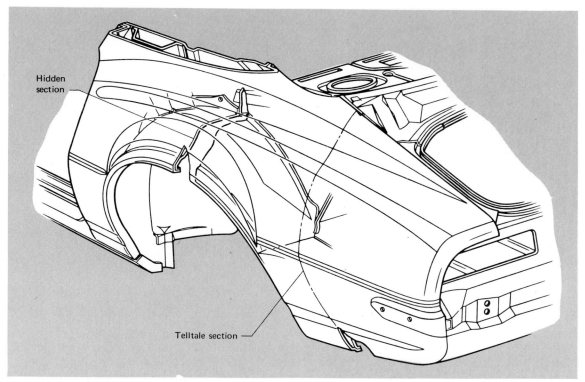

Hidden section

Telltale section

Fig. 28-50. Hidden and telltale sections are used to show internal details of sheetmetal parts. (General Motors Corp.)

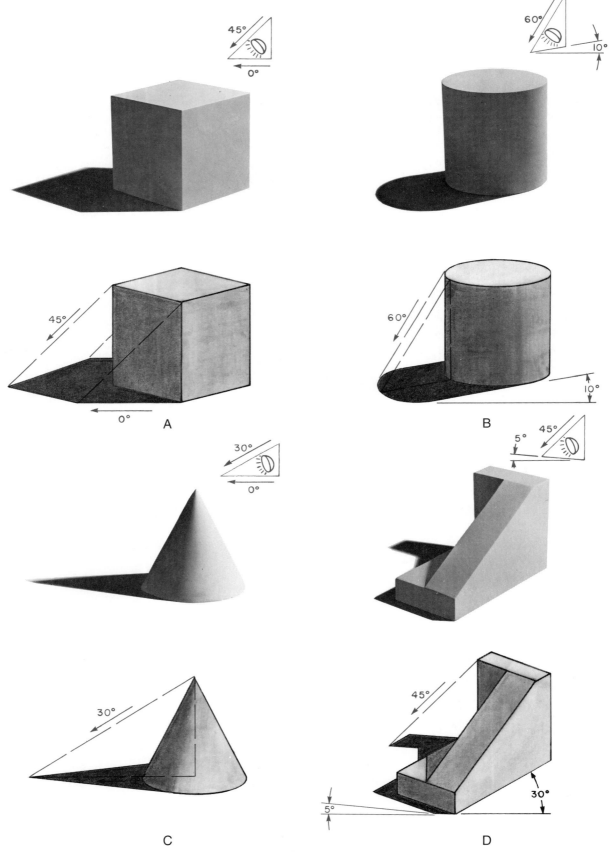

Fig. 28-52. Shadows of geometric figures will appear differently for light sources coming from different directions.

objects. Carefully study the projection of the shadows and how the light source affects the direction that the shadows are projected.

Outline shading

A multiview or a pictorial drawing may be given more realism by a technique called **outline shading,** Fig. 28-53. The light direction is assumed, and lines that are on the side away from the light source are considered to be in the shade. For holes, the side nearest the light is assumed to be in the shade.

The remaining lines outlining the part (those not in the shade) are done in normal visible line weight. The lines inside may be drawn either in normal weight or in lighter weight for greater emphasis.

In multiview drawings, the line is widened on the side away from the light to approximately three times its normal width. Notice that cylindrical features are shifted on an axis parallel to the light source.

In pictorial drawings, the line is widened on *both* sides to approximately one and one-half

times its normal width, for a total of three line widths. Notice that cylindrical features, which appear as ellipses in pictorial views, are shifted along the axis which most nearly aligns with the light source.

Line shading

Shading to produce a three-dimensional effect on an object may be accomplished by using lines of varying spacing and length, Fig. 28-54. A light source should be assumed. The shading lines are then drawn more closely together on the portions of a surface furthest from the light or those shaded from the light. Note how the lines may be shortened or omitted on surfaces where the light seems to fall most strongly.

On external cylindrical surfaces, note that the lines appear closer together as they move to the shaded side of the piece. Where the light falls on the far side on *internal* cylindrical surface, fewer lines are drawn. It is best to arrange the light so that it does not fall in the center of a cylindrical surface.

Shaded flat surfaces are darkest in the foreground, where they contrast sharply with lighted surfaces, and tend to lighten in the background. **Lighted** flat surfaces are lightest, or void of line shading, in the foreground and tend to darken somewhat in the background.

Fillets and rounds

Fillets and **rounds** usually catch the light source, and thus, tend to appear light. When line shading is used, fillets and rounds may be treated either with curved lines as shown in Fig. 28-54 or with straight lines as shown in Fig. 28-55. Straight lines are preferred because they may be applied more readily and present a neater appearance.

Threads

Threads are represented by a series of ellipses, unless they appear in the frontal plane of an

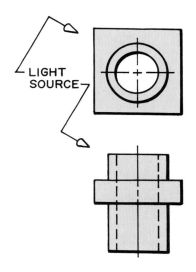

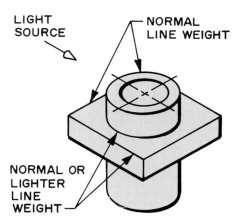

Fig. 28-53. Outline shading technique is a technique for making drawings more realistic. Outline shading can be applied to multiview, pictorial, and many other types of drawings.

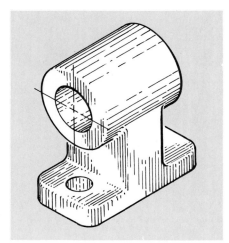

Fig. 28-54. Curved-line shading is used to give a three-dimensional effect to drawings.

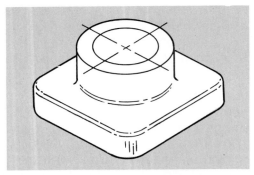

Fig. 28-55.   The straight-line shading technique can be used to shade fillets and rounds.

oblique drawing. In that case, they are represented as circles. Threads may be shaded solid except for the highlights, Fig. 28-56.

Smudge shading

When a tone shading that flows smoothly from dark to light and back to dark is desired, ***smudge shading*** is commonly used, Fig. 28-57. This is done by going over the area to be shaded with a soft lead (3B to HB) pencil. Then, to produce the desired tone, the graphite is rubbed into the texture

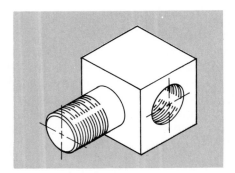

Fig. 28-56.   Different methods can be used for shading external and internal threads.

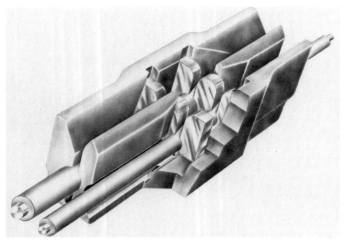

Fig. 28-57.   Smudge shading is used to give a realistic appearance to drawings. (Bell Laboratories)

of the paper with a stub of paper, a piece of soft cloth, or a finger. Where a darker portion is needed, more graphite may be added with the pencil and rubbed. A protective spray coating should be applied to the rendering when it is completed.

Pencil shading

A soft lead pencil, 3B to HB, used on paper that has a slight texture, is an effective means of shading a drawing. Variations in rendering are achieved by using the side of the lead as well as the point, and by using leads of different degrees of hardness. This technique produces soft tones that are especially suited to architectural renderings, Fig. 28-58. A spray coating should be applied to the rendering to protect it from smears during handling and storage.

Inking

Technical illustrations planned for publication are usually inked to achieve desired reproduction qualities. Not only is ink well suited for line work, but it is adaptable to stippling (for shading) with a fine pen point or sponge.

Sponge shading is done by using a small piece of sponge with a stamp pad or film of printer's ink, Fig. 28-59. Light touches should be used at first, with darker areas reworked to the desired tone. Protect surrounding areas of the illustration with a paper overlay.

Fig. 28-58.   Architectural renderings can be effectively done in pencil.

Fig. 28-59. Sponge shading with ink can produce very realistic renderings. (Bell Laboratories)

Shading with an airbrush

One of the most effective tools for illustration work is the **airbrush,** Fig. 28-60. The airbrush operates by using compressed air or carbon dioxide to spray a mist of ink or watercolor onto the drawing.

Parts of the drawing that are not to be sprayed should be protected by a paper template or a **frisket,** which is a special paper with an adhesive back. The frisket is laid over the entire drawing, then "windows" are cut out of the material over any areas to be airbrushed.

To develop skill in the use of the airbrush, some practice is needed. Once the technique is mastered, however, illustrations more effective than a photograph are possible, Fig. 28-61.

Photo retouching

A photograph is a fast and inexpensive means of producing a technical illustration. However, a photograph often fails to bring out the detail that is desired. Sometimes, only a part of the object photographed is to be emphasized, while the rest is subdued. The process of reworking photographs to emphasize or sharpen certain details and hold others back is known as **photo retouching.** As on any other illustration, retouching can be done by hand or with an airbrush. Lines may be added by **scribing** (scratching) the negative or removed by using a special lacquer. A photograph before and after retouching to bring out the detail is shown in Fig. 28-62.

Overlay film

A wide assortment of **overlay films** for shading or color work in technical illustration is available commercially. These films come in glossy finish for illustrations that are to be reproduced by the **diazo** (dark line) print process or by photography. When the original artwork is to serve as the finished illustration, overlay film with a matte finish should be used.

Overlay film has an adhesive back and can be applied to almost any working surface. The film is laid in place and lightly pressed, then carefully cut to the desired shape. The part that is to serve as the overlay is left in place, and the remainder is removed. Some films require burnishing to set them in place, others do not.

Fig. 28-61. An illustration prepared with an airbrush can be more effective than an actual photograph. (Cincinnati Milacron)

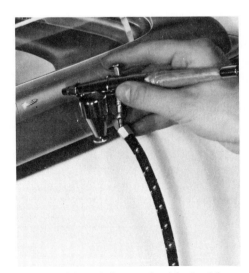

Fig. 28-60. An airbrush is a valuable tool for many types of technical illustration projects. (Paasche Airbrush Co.)

A   B

Fig. 28-62. A–Retouching can eliminate defects and improve the visibility of parts of a photograph. The photograph may have areas that appear damaged or dirty. B–Retouching the photo can eliminate these problems, making the object appear better. (Mack Trucks, Inc.)

Transfers and tapes

Where a number of standard parts or a number of like components is to be shown, adhesive-backed or pressure sensitive **transfers** and **tapes** should be used. These are available commercially in a wide variety of symbols and other useful devices, and can achieve a considerable savings in time for the technical illustrator, Fig. 28-63.

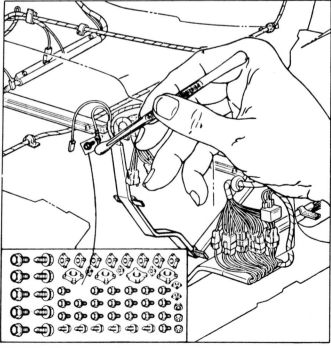

Fig. 28-63. Transfer sheets and tapes containing symbols and standard parts are a valuable aid to the technical illustrator. (General Motors Corp.)

Computer-assisted illustration

The computer has become a valuable tool for the technical illustrator. The speed and ease of using the computer, combined with the great diversity of available programs, open to the illustrator many design possibilities not attainable by traditional methods, Fig. 28-64.

As in computer graphics applications for standard drafting procedures, the computer provides immediate access to a stored library of symbols and illustrations of parts that can be used to create new illustrations. Changes and revisions can be completed quickly and easily, leaving the illustrator more time for creativity.

GRAPHS AND CHARTS

Drafting departments are called on from time to time to prepare graphs and charts. These visual devices are excellent means of presenting data in graphic form for contract proposals, analysis, and marketing. As one familiar with drafting procedures, you possess many of the skills for this work.

Graphs are diagrams that show the relationships between two or more factors. *Charts* may be defined as a means of presenting information in a tabulated form, or sometimes in graphic form as a line diagram.

Types of Graphs and Charts

Many forms of graphs and charts are used to analyze and clarify data. The major types are presented in this section.

Fig. 28-64. Computer graphics can be used by the technical illustrator to produce pictorial drawings of exceptional quality. (Intergraph Corporation and Lockheed)

Line graph

One of the simplest graphs to construct is the *line graph.* Line graphs are used to show relationships of quantities to a time span. The horizontal axis, called the *abscissa* or "X-axis," usually contains the time element. The vertical axis, called the *ordinate* or "Y-axis," expresses the other factor in terms of numbers or percentages.

After the data are plotted in the line graph, the line can be a *broken line curve* (drawn point-to-point), Fig. 28-65. The line can also be a *smooth curve,* Fig. 28-66. When it is desired to show an actual condition or status for each of the time periods, use the broken line curve. However, when the change is continuous, the smooth curve approximating the actual data for each time interval is more meaningful.

Symbols at each of the data points or variations in the form of the line itself are sometimes used to represent differences in factors or methods of treatment, Fig. 28-67.

The line graph may also be used to compare design factors of a material, Fig. 28-68. A set of characteristic curves for an electronic transistor can also be illustrated with a line graph, Fig. 28-69.

Bar graphs

Bar graphs are a popular form of presenting statistical data, because they are easily understood by laypersons. They are used to show relationships between two or more variables, Fig. 28-70. However, bar graphs have fewer plotted values for each variable than a line graph.

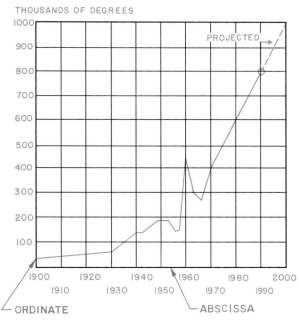

Fig. 28-65. A line graph can show the relationships of quantities over a period of time.

The data presented in bar graphs are usually for a total period of time rather than for various periods such as those in a line graph. For example, the bar graph shows production per hour, day, or year, but usually not successive periods of production.

There are two basic types of bar graphs: index bar and range bar. *Index bar graphs* have a com-

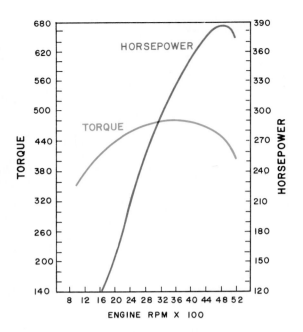

The data represented by these curves were obtained through Standard Test No. 20 of the General Motors Automotive Engine Test Code. This test establishes a uniform method of determining the gross power output of the bare engine. It is run with distributor and carburetor adjusted for maximum power at each speed. Data are corrected to 60°F. using an SAE correction factor.

Fig. 28-66. Smooth curves in a line graph reflect a continuous change, rather than a sharp change for each interval. Data are gathered for each interval and plotted. The curve approximates the plotted points.

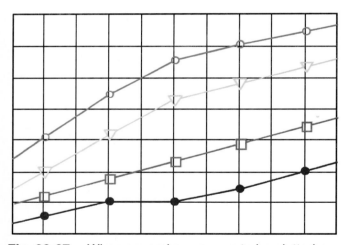

Fig. 28-67. When several curves are to be plotted on the same graph, symbols can be used to differentiate each item. Color can also be used, alone or in combination with symbols, to set curves apart.

mon base where the bars originate. A number of variations are possible with index bar graphs, Fig. 28-70. **Range bar graphs** are individual bars, representing segments of the whole, that are plotted within the range of the total project time schedule.

Vertical and horizontal bar graphs are the most common types of index bar graphs. However,

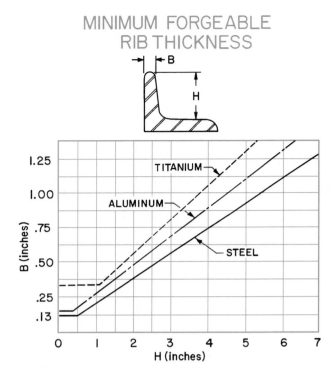

Fig. 28-68. Line graphs can be used to compare dimensional features of different metal stocks.

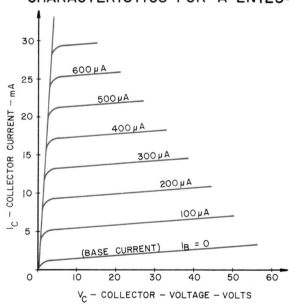

Fig. 28-69. Graphs can display hypothetical output curves for different parameters of an electronic component.

the **grouped bar graph** permits the inclusion of other variables in an effective manner. When the grouped bar method is used, an identical sequence of elements should be maintained, and each element should be distinctively shaded or colored.

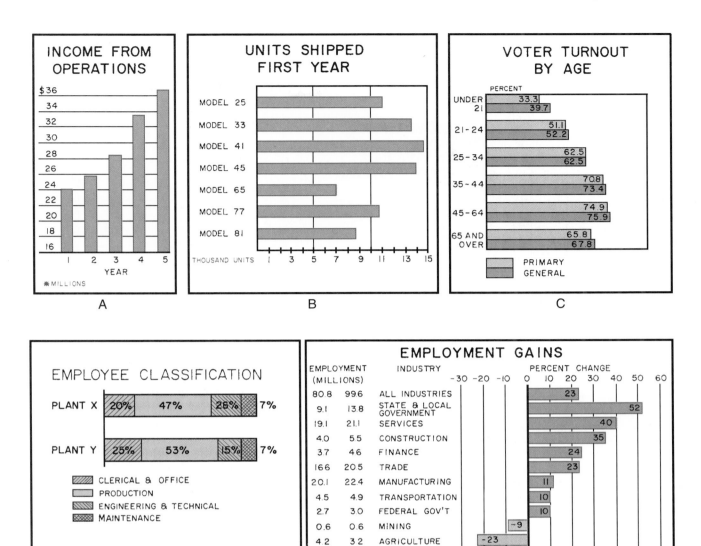

Fig. 28-70. There are several variations in preparing index bar graphs. Common index bar graphs include: A–vertical bar, B–horizontal bar, C–grouped bar, D–subdivided bar, and E–deviation bar.

The **subdivided bar graph** is effective when there are fewer than five divisions. The graph loses its value when too many divisions make it difficult to appraise the relative value of each.

When the subdivided bar is used, the most important or sizable element should be plotted first (next to index line). Follow this with the item next in importance, and so on until the graph is completed. The same order of elements should be retained when two or more subdivided bars are used, regardless of the variation in importance or size in successive bars.

The **percentage bar graph** is a type of subdivided bar which is particularly easy to read when the percentages are included, Fig. 28-70.

The **paired bar graph** is useful when comparing two sets of factors on different scales. A typical use is to show total value of raw products pur-

chased, as contrasted with the percentage of this amount purchased from a single supplier.

The **deviation bar graph** provides a comparison between a number of factors and their deviation from a "break-even" point. This type of graph lends itself well to such comparisons as profit and loss or increase and decrease. On a horizontal deviation bar graph, the bars are drawn from a zero index line with positive values running to the right and negative values to the left. On a vertical deviation bar graph, positive values should appear above the index line and negative values below.

The **range bar graph** variation normally plots the items against a time line. A typical example would be to plot a production schedule in which each phase would be plotted as a time range within the time schedule shown for the entire project, Fig. 28-71. The range bar graph also may be used to show progress.

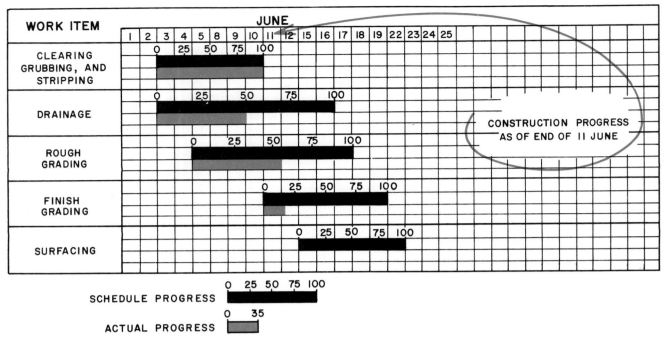

WORK ITEM	JUNE																						
	1	2	3	4	5	8	9	10	11	12	15	16	17	18	19	22	23	24	25				

CONSTRUCTION PROGRESS
AS OF END OF 11 JUNE

SCHEDULE PROGRESS   0 25 50 75 100

ACTUAL PROGRESS   0   35

Fig. 28-71. A range bar graph can be used to compare the scheduled progress and actual progress on a construction project.

Surface or area graphs

A **surface graph** or **area graph** is an adaptation of the line or bar graph. The area between the curve and the abscissa axis is shaded for emphasis, Fig. 28-72. Variations in the surface graph are **shaded-zone graph,** and **pictorial-surface graph.**

Pie graphs

Pie graphs, sometimes called circle or sector graphs, are frequently used to contrast individual segments (parts) with the whole. A typical example of this is the graph shown in Fig. 28-73 A. This graph depicts the distribution of the labor force in one area.

The pie graph can be varied by drawing it as a pictorial, Fig.28-73 B. This type of graph is sometimes used to represent cost expenditures. The pictorial graph shown resembles a silver dollar divided into the various categories of expense.

Flow charts

Flow charts are a graphic means of depicting the sequence of technical processes that would be difficult to describe in narrative form, Fig. 28-74. The

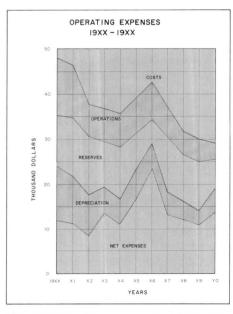

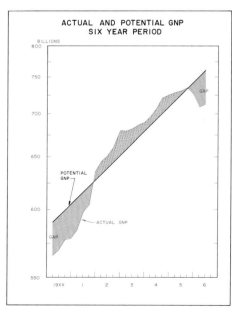

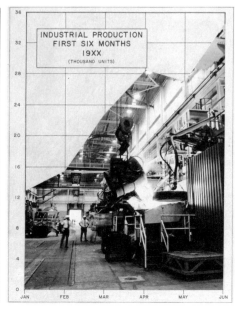

Fig. 28-72. There are several types of surface or area graphs.

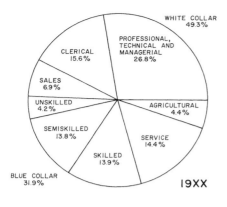

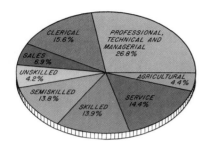

EMPLOYMENT
BY OCCUPATIONAL GROUP

Fig. 28-73. A–Information can be presented in the form of a pie graph. B–A pictorial pie graph is a variation of a pie graph.

flow of various processes and materials in the line of manufacturing production can be clearly detailed in a flow chart. Pictures, symbols, and diagrams should be used when they aid in understanding.

Organizational charts

Organizational charts do for personnel what flow charts do for processes and materials. They show relationships between individuals and departments within an organization and the operations or services each performs, Fig. 28-75.

The organizational chart also shows the relationship between line personnel and staff personnel within an organization. **Line personnel,** such as supervisors or department heads, have authority to direct an operation or a group. Lines in the chart clearly show this authority.

Staff personnel, such as consultants for numerical control machines, may suggest and recommend procedures and types of equipment to the manufacturing manager. However, they cannot direct that the suggestions or recommendations be carried out. The lines on the organizational chart indicate these responsibilities.

Nomographs

A type of graph useful in solving a succession of nearly identical problems is the **nomograph,** Fig. 28-76. This graph usually contains three parallel

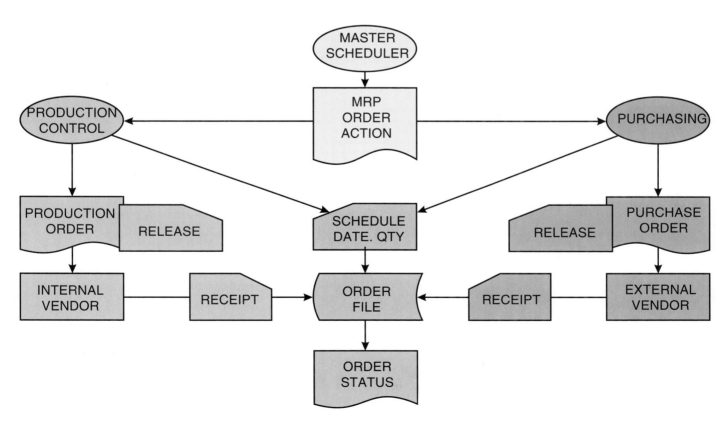

Fig. 28-74. Flow charts can be used to show how material flows into the manufacturing process through the purchasing and production control functions.

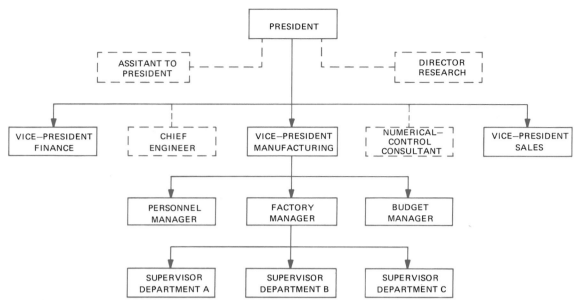

Fig. 28-75. An organizational chart shows the lines of authority and responsibility. It can also distinguish between those who serve in staff and support positions.

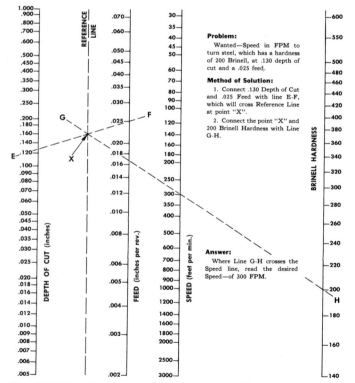

Fig. 28-76. Nomographs can be used to make quick determinations within given parameters. A nomograph might be created for a specific carbide cutting tool. This would allow different cutting speeds to be determined quickly and easily depending on factors of feed, depth of cut, and material hardness. (Carboloy Div., GE Co.)

scales that are graduated for different variables. When a straight line connects values of any two scales, the related value may be read directly from the third at the point intersected by the line.

PROBLEMS AND ACTIVITIES

The following problems are divided into different sections based on the principles applied. Each section is designed to further your understanding in an area presented in this chapter.

Problems in Architectural Drafting

The following problems are planned to give you experience with the basics of architectural drafting by making various types of drawings used in architectural work and to try your problem-solving skills in the field of architectural design.

1. Prepare a scaled working drawing of the floor plan of the house shown on the following page. Include all necessary dimensions and notes.

2. Draw a foundation plan for the house used in Problem 1.

3. Trace the floor plan from Problem 1 and prepare an electrical plan. Check your local electrical code for the requirements on spacing wall outlets. Show lines to switches on all outlets controlled by switches.

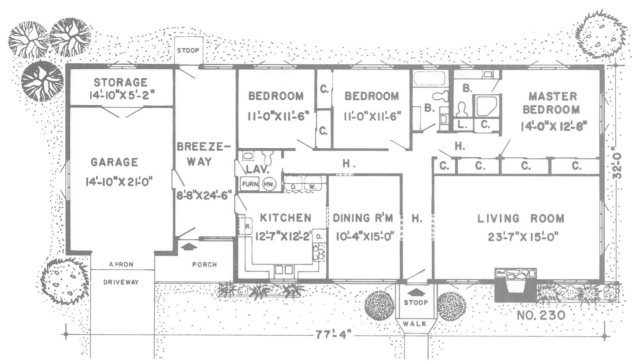

Problem 1

4. Draw a front elevation of the house in Problem 1 above. Also make a wall section to show details of construction. Add necessary dimensions and notes to the elevation and wall section.

5. Prepare a plan view and elevations of the kitchen cabinets and develop a window and door schedule.

Problems in Electrical and Electronics Drafting

6. Make a pictorial drawing of a small transistor radio or similar electronic device.

7. Draw and label the following component symbols: (Use a template or draw each component to same relative size.)
   A. Battery, 9 volts, BT_1.
   B. Switch, single-pole, single-throw, S_1.
   C. Ammeter, M_1.
   D. Resistor, 4700 ohms, R_1.
   E. Lamp, incandescent, dial lamp, DS_1.

8. Replace the blocks with the correct single-line symbols in the following diagram of an electrical power system.

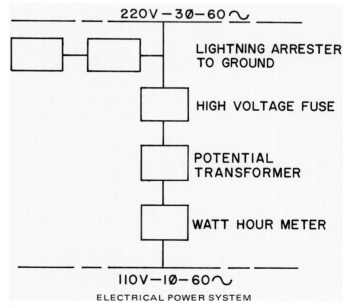

ELECTRICAL POWER SYSTEM

Problem 8

9. Draw a block diagram of a noise level meter with the following stages:
   A. Input microphone.
   B. Audio amplifier, Q_1.
   C. Audio amplifier, Q_2.

D. Audio amplifier, Q_3.

E. Decibel meter.

10. Complete the following elementary diagram so that the operation will cease after a complete sequence of operations.

11. Redraw the sketch below and add the information listed. Avoid crowding and wasted space.

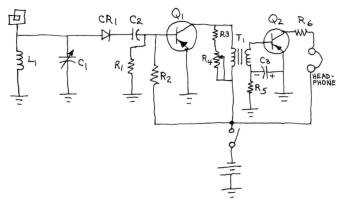

Problem 11

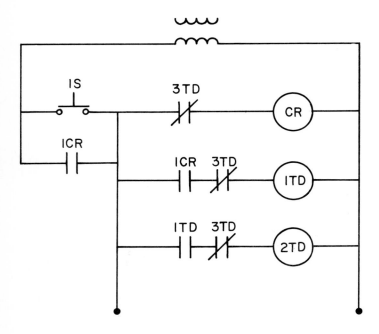

Problem 10

12. Redraw the continuous line connection diagram shown below, as an interrupted line connection diagram. For each lead, show the subassembly number, the terminal to which it is going, and the color code. For example, "Lead 3" on "Unit A1" would be labeled "A4/4-GN-BK," meaning it is a green lead with a black tracer stripe that goes to "Unit A4, Terminal 4."

Problems in Map Drafting

The following problems are designed to provide you with the opportunity to apply knowledge

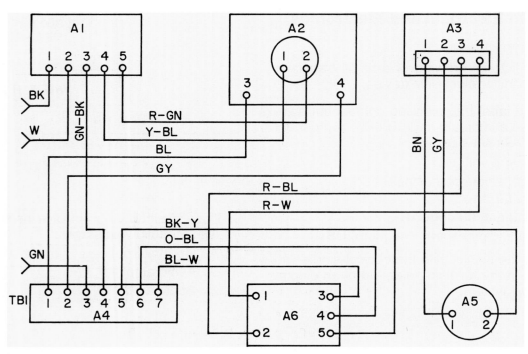

Problem 12

gained in your study of map drafting and to help you become familiar with the procedures used.

13. Select an appropriate scale and contour interval and plot the contours for the map shown below.

14. Select an appropriate scale and contour interval and plot the contours, as well as the natural and constructed features, for the map shown on the following page.

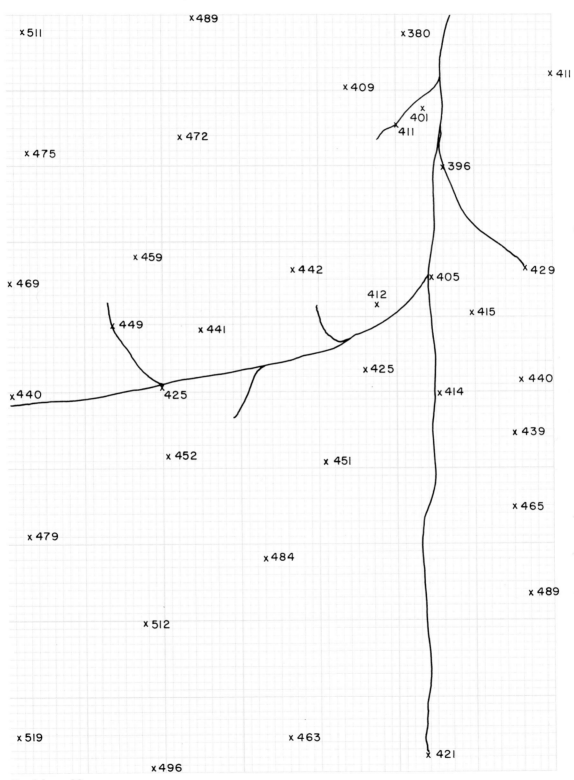

x 511   x 489   x 380   x 411

x 409

x 472   x 401   x 411

x 475   x 396

x 459   x 442   x 429

x 469   x 405   x 415

x 449   x 441   412 x

x 425   x 440

x 440   425   x 414   x 439

x 452   x 451   x 465

x 479

x 484   x 489

x 512

x 519   x 463   x 465

x 496   x 421

Problem 13

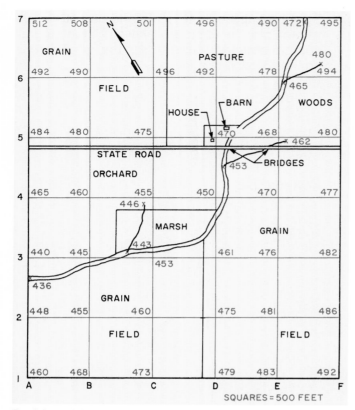

Problem 14

15. Draw a profile map showing the shape of the terrain at lines "1" and "4" through the map section shown in Problem 14. Use the same scale as used for the contour map.

16. Lay out the map traverse shown below. Indicate a North point on the map, then orient the first station and backsight line with it.

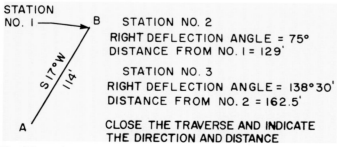

STATION NO. 1 ──────► B   STATION NO. 2
                            RIGHT DEFLECTION ANGLE = 75°
                            DISTANCE FROM NO. 1 = 129'

                            STATION NO. 3
                            RIGHT DEFLECTION ANGLE = 138°30'
                            DISTANCE FROM NO. 2 = 162.5'

S 17° W   114'

A   CLOSE THE TRAVERSE AND INDICATE
           THE DIRECTION AND DISTANCE

Problem 16

Problems in Technical Illustration

The following problems will give you opportunities to try several of the techniques used in technical illustration. With the guidance and approval of

your instructor, select as many types of illustrations and techniques as time and equipment permit. Then prepare the technical illustrations.

17. Select an object from Fig. 7-40 and prepare a multiview illustration using outline shading in pencil.

18. Select an object from Fig. 11-63 and prepare an isometric illustration using outline shading in ink.

19. Prepare a smudge shading illustration of an object from Fig. 11-64. The illustration is to be a dimetric drawing, using one of the axes suggested in Fig. 11-32. Select the axes that will present the best view of the object's features.

20. Prepare a pencil shading rendering an object of your choice.

Problems in Graphs and Charts

The following problems are planned to improve your understanding and skills in the use of basic drafting techniques related to constructing graphs and charts.

21. Collect data on a subject of interest to you and use the information to construct a line graph. Magazines in your school, public, or drafting library are good sources of data.

22. Collect data on a subject of interest to you and prepare a bar graph of a type not yet used.

23. Construct a pictorial surface graph illustrating traffic safety and accident statistics. Gather data from driver training classes, safety classes, or an insurance company. Select an appropriate photo from a magazine or take one yourself.

24. Gather budget information from an annual report (local or state government, or a private organization or business). Use it to prepare a pie graph to graphically contrast the budgeted dollars.

25. Construct a flow chart illustrating the sequence followed in an industrial process, such as anodizing aluminum, or making an electronic-circuit board, or preparing and spray painting a metal surface.

26. Gather information on your city government and prepare an organizational chart showing line and staff organization.

Reference Section

Sheet Metal and Wire Gage Designation

Precision Sheet Metal Set-back Chart

Coordinates for Locating Equally Spaced Holes in Jig Boring

Graphic Symbols for Drawings

Drawing Sheet Layouts

Standard Abbreviations for Use on Drawings

A

Abrasive	ABRSV
Accessory	ACCESS
Accumulator	ACCUMR
Acetylene	ACET
Actual	ACT
Actuator	ACTR
Addendum	ADD
Adhesive	ADH
Adjust	ADJ
Advance	ADV
Aeronautic	AERO
Alclad	CLAD
Alignment	ALIGN
Allowance	ALLOW
Alloy	ALY
Alteration	ALT
Alternate	ALT
Alternating Current	AC
Aluminum	AL
American National Standards Institute	ANSI
American Wire Gage	AWG
Ammeter	AMM
Amplifier	AMPL
Anneal	ANL
Anodize	ANOD
Antenna	ANT
Approved	APPD
Approximate	APPROX
Arrangement	ARR
Asbestos	ASB
As Required	AR
Assemble	ASSEM
Assembly	ASSY
Attenuation, Attenuator	ATTEN
Audio Frequency	AF
Automatic	AUTO
Automatic Frequency Control	AFC
Automatic Gain Control	AGC
Auxiliary	AUX
Average	AVG

B

Babbit	BAB
Base Line	BL
Battery	BAT
Bearing	BRG

Beat-Frequency Oscillator	BFO
Bend Radius	BR
Bevel	BEV
Bill of Material	B/M
Blueprint	BP or B/P
Bolt Circle	BC
Bracket	BRKT
Brass	BRS
Brazing	BRZG
Brinnell Hardness Number	BHN
Bronze	BRZ
Brown & Sharpe (Gage)	B&S
Burnish	BNH
Bushing	BUSH

C

Cabinet	CAB
Calculated	CACL
Cancelled	CANC
Capacitor	CAP
Capacity	CAP
Carburize	CARB
Case Harden	CH
Casting	CSTG
Cast Iron	CI
Cathode-Ray Tube	CRT
Center	CTR
Center to Center	C to C
Centigrade	C
Centimeter	CM
Centrifugal	CENT
Chamfer	CHAM
Check Valve	CV
Chrome Vanadium	CR VAN
Circuit	CKT
Circular	CIR
Circumference	CIRC
Clearance	CL
Clockwise	CW
Closure	CLOS
Coated	CTD
Cold-Drawn Steel	CDS
Cold-Rolled Steel	CRS
Color Code	CC
Commercial	COMM
Concentric	CONC
Condition	COND
Conductor	CNDCT
Contour	CTR

Control	CONT
Copper	COP
Counterbore	CBORE
Counterclockwise	CCW
Counter-Drill	CDRILL
Countersink	CSK
Coupling	CPLG
Cubic	CU
Cylinder	CYL

D

Datum	DAT
Decimal	DEC
Decrease	DECR
Degree	DEG
Detail	DET
Detector	DET
Developed Length	DL
Developed Width	DW
Deviation	DEV
Diagonal	DIAG
Diagram	DIAG
Diameter	DIA
Diameter Bolt Circle	DBC
Diametral Pitch	DP
Dimension	DIM
Direct Current	DC
Disconnect	DISC
Double-Pole Double-Throw	DPDT
Double-Pole Single-Throw	DPST
Dowel	DWL
Draft	DFT
Drafting Room Manual	DRM
Drawing	DWG
Drawing Change Notice	DCN
Drill	DR
Drop Forge	DF
Duplicate	DUP

E

Each	EA
Eccentric	ECC
Effective	EFF
Electric	ELEC
Electrolytic	ELCTLT
Enclosure	ENCL
Engine	ENG
Engineer	ENGR

Engineering	ENGRG	Inch	IN	Measure	MEAS
Engineering Change Order	ECO	Inclined	INCL	Mechanical	MECH
Engineering Order	EO	Include, Including, Inclusive	INCL	Medium	MED
Equal	EQ	Increase	INCR	Meter	MTR
Equivalent	EQUIV	Independent	INDEP	Middle	MID
Estimate	EST	Indicator	IND	Military	MIL
		Information	INFO	Millimeter	MM
F		Inside Diameter	ID	Minimum	MIN
Fabricate	FAB	Installation	INSTL	Miscellaneous	MISC
Fillet	FIL	Intermediate Frequency	IF	Modification	MOD
Finish	FIN	International Standards Organization	ISO	Mold Line	ML
Finish All Over	FAO	Interrupt	INTER	Motor	MOT
Fitting	FTG			Mounting	MTG
Fixed	FXD			Multiple	MULT
Fixture	FIX	**J**			
Flange	FLG	Joggle	JOG	**N**	
Flat Head	FHD	Junction	JCT	Nickel Steel	NS
Flat Pattern	F/P			Nomenclature	NOM
Flexible	FLEX	**K**		Nominal	NOM
Fluid	FL	Keyway	KWY	Normalize	NORM
Forged Steel	FST			Not to Scale	NTS
Forging	FORG	**L**		Number	NO.
Furnish	FURN	Laboratory	LAB		
		Lacquer	LAQ	**O**	
G		Laminate	LAM	Obsolete	OBS
Gage	GA	Left-Hand	LH	Opposite	OPP
Gallon	GAL	Length	LG	Oscillator	OSC
Galvanized	GALV	Letter	LTR	Oscilloscope	SCOPE
Gasket	GSKT	Limited	LTD	Ounce	OZ
Generator	GEN	Limit Switch	LS	Outside Diameter	OD
Grind	GRD	Linear	LIN	Over-All	OA
Ground	GRD	Liquid	LIQ		
		List of Material	L/M	**P**	
H		Long	LG	Package	PKG
Half-Hard	1/2H	Low Carbon	LC	Parting Line (Castings)	PL
Handle	HDL	Low Frequency	LF	Parts List	P/L
Harden	HDN	Low Voltage	LV	Pattern	PATT
Head	HD	Lubricate	LUB	Piece	PC
Heat Treat	HTTR			Pilot	PLT
Hexagon	HEX	**M**		Pitch	P
High Carbon Steel	HCS	Machine(ing)	MACH	Pitch Circle	PC
High Frequency	HF	Magnaflux	M	Pitch Diameter	PD
High Speed	HS	Magnesium	MAG	Plan View	PV
Horizontal	HOR	Maintenance	MAINT	Plastic	PLSTC
Hot-Rolled Steel	HRS	Major	MAJ	Plate	PL
Hour	HR	Malleable	MALL	Pneumatic	PNEU
Housing	HSG	Malleable Iron	MI	Port	P
Hydraulic	HYD	Manual	MAN	Positive	POS
Hydrostatic	HYDRO	Manufacture(ing)(ed)(er)	MFG	Potentiometer	POT
		Mark	MK	Pounds Per Square Inch	PSI
I		Master Switch	MS	Pounds Per Square Inch Gage	PSIG
Identification	IDENT	Material	MATL	Power Amplifier	PA
Impregnate	IMPG	Maximum	MAX		

Power Supply	PWR SPLY	Unified Screw Thread Coarse	UNC	Thread	THD
Pressure	PRESS	Unified Screw Thread Fine	UNF	Through	THRU
Primary	PRI			Tolerance	TOL
Process, Procedure	PROC	Unified Screw Thread Extra Fine	UNEF	Tool Steel	TS
Product, Production	PROD			Torque	TOR

Q

Quality	QUAL	Unified Screw Thread 8 Thread	8UN	Total Indicator Reading	TIR
Quantity	QTY	Section	SECT	Transceiver	XCVR
Quarter-Hard	1/4H	Sequence	SEQ	Transformer	XFMR

R

		Serial	SER	Transistor	XSTR
Radar	RDR	Serrate	SERR	Transmitter	XMTR
Radio	RAD	Sheathing	SHTHG	True Involute Form	TIF
Radio Frequency	RF	Sheet	SH	Tungsten	TU
Radius	RAD or R	Silver Solder	SILS	Typical	TYP
Ream	RM	Single-Pole Double-Throw	SPDT		
Receptacle	RECP				

U

Reference	REF	Single-Pole Double-Throw	SPDT	Ultra-High Frequency	UHF
Regular	REG	Single-Pole Single-Throw	SPST	Unit	U
Regulator	REG	Society of Automotive Engineers	SAE	Universal	UNIV
Release	REL			Unless Otherwise Specified	UOS
Required	REQD	Solder	SLD		
Resistor	RES	Solenoid	SOL		

V

Revision	REV	Speaker	SPKR	Vacuum	VAC
Revolutions Per Minute	RPM	Special	SPL	Vacuum Tube	VT
Right-Hand	RH	Specification	SPEC	Variable	VAR
Rivet	RIV	Spot Face	SF	Vernier	VER
Rockwell Hardness	RH	Spring	SPG	Vertical	VERT
Round	RD	Square	SQ	Very High Frequency	VHF
		Stainless Steel	SST	Vibrate	VIB

S

		Standard	STD	Video	VD
Schedule	SCH	Steel	STL	Void	VD
Schematic	SCHEM	Stock	STK	Volt	V
Screw	SCR	Support	SUP	Volume	VOL
Screw Threads		Switch	SW		
American National Coarse	NC	Symbol	SYM		

W

American National Fine	NF	Symmetrical	SYM	Washer	WASH
American National Extra Fine	NEF	System	SYS	Watt	W

T

American National 8 Pitch	8N			Watt Hour	WH
		Tabulate	TAB	Wattmeter	WM
American Standard Taper Pipe	NTP	Tangent	TAN	Weatherproof	WP
		Tapping	TAP	Weight	WT
American Standard Straight Pipe	NPSC	Technical Manual	TM	Wide, Width	W
		Teeth	T	Wire Wound	WW
American Standard Taper (Dryseal)	NPTF	Television	TV	Wood	WD
		Temper	TEM	Wrought Iron	WI
American Standard Straight (Dryseal)	NPSF	Temperature	TEM		

Y

		Tensile Strength	TS	Yield Point (PSI)	YP
		Thick	THK	Yield Strength (PSI)	YS

Electrical and Electronics Reference Designations

(Partial List)

Amplifier	AR	Crystal Unit, Piezoelectric	Y	Plug	P
Antenna	E	Delay Line	DL	Potentiometer	R
Assembly	A	Electron Tube	V	Power Supply	PS
Attenuator	AT	Filter	FL	Receiver, Radio	RE
Audible Signaling Device	DS	Fuse	F	Recorder, Sound	A
Ballast Tube or Lamp	RT	Generator	G	Regulator, Voltage	VR
Battery	BT	Handset	HS	Relay	K
Bell	DS	Hardware	H	Reproducer, Sound	A
Buzzer	DS	Indicator	DS	Resistor	R
Capacitor	C	Inductor	L	Rheostat	R
Circuit Breaker	CB	Jack	J	Semiconductor Diode	CR
Clock	M	Lamp	DS	Socket	X
Coil	L	Loudspeaker	LS	Switch	S
Computer	A	Meter	M	Terminal Board	TB
Connector, Plug	P	Microphone	MK	Transformer	T
Connector, Receptacle	J	Motor	B	Transistor	Q
Contact, Electrical	E	Oscilloscope	M	Transmitter, Radio	TR
Counter, Electrical	M	Pickup	PU		

Courtesy of American National Standards Institute, Reference Designations for Electrical and Electronics Parts and Equipment, Y32.16-1968.

Running and Sliding Fits

Limits are in thousandths of an inch.

Limits for hole and shaft are applied algebraically to the basic size to obtain the limits of size for the parts.

Data in bold face are in accordance with ABC agreements.

Symbols H5, g5, etc., are Hole and Shaft designations used in ABC System

Nominal Size Range Inches		Class RC 1			Class RC 2			Class RC 3			Class RC 4		
		Limits of Clearance	Standard Limits		Limits of Clearance	Standard Limits		Limits of Clearance	Standard Limits		Limits of Clearance	Standard Limits	
Over	To		Hole H5	Shaft g4		Hole H6	Shaft g5		Hole H7	Shaft f6		Hole H8	Shaft f7
0	− 0.12	0.1 0.45	+ 0.2 0	− 0.1 − 0.25	0.1 0.55	+ 0.25 0	− 0.1 − 0.3	0.3 0.95	+ 0.4 0	− 0.3 − 0.55	0.3 1.3	+ 0.6 0	− 0.3 − 0.7
0.12	− 0.24	0.15 0.5	+ 0.2 0	− 0.15 − 0.3	0.15 0.65	+ 0.3 0	− 0.15 − 0.35	0.4 1.2	+ 0.5 0	− 0.4 − 0.7	0.4 1.6	+ 0.7 0	− 0.4 − 0.9
0.24	− 0.40	0.2 0.6	+ 0.25 0	− 0.2 − 0.35	0.2 0.85	+ 0.4 0	− 0.2 − 0.45	0.5 1.5	+ 0.6 0	− 0.5 − 0.9	0.5 2.0	+ 0.9 0	− 0.5 − 1.1
0.40	− 0.71	0.25 0.75	+ 0.3 0	− 0.25 − 0.45	0.25 0.95	+ 0.4 0	− 0.25 − 0.55	0.6 1.7	+ 0.7 0	− 0.6 − 1.0	0.6 2.3	+ 1.0 0	− 0.6 − 1.3
0.71	− 1.19	0.3 0.95	+ 0.4 0	− 0.3 − 0.55	0.3 1.2	+ 0.5 0	− 0.3 − 0.7	0.8 2.1	+ 0.8 0	− 0.8 − 1.3	0.8 2.8	+ 1.2 0	− 0.8 − 1.6
1.19	− 1.97	0.4 1.1	+ 0.4 0	− 0.4 − 0.7	0.4 1.4	+ 0.6 0	− 0.4 − 0.8	1.0 2.6	+ 1.0 0	− 1.0 − 1.6	1.0 3.6	+ 1.6 0	− 1.0 − 2.0
1.97	− 3.15	0.4 1.2	+ 0.5 0	− 0.4 − 0.7	0.4 1.6	+ 0.7 0	− 0.4 − 0.9	1.2 3.1	+ 1.2 0	− 1.2 − 1.9	1.2 4.2	+ 1.8 0	− 1.2 − 2.4
3.15	− 4.73	0.5 1.5	+ 0.6 0	− 0.5 − 0.9	0.5 2.0	+ 0.9 0	− 0.5 − 1.1	1.4 3.7	+ 1.4 0	− 1.4 − 2.3	1.4 5.0	+ 2.2 0	− 1.4 − 2.8
4.73	− 7.09	0.6 1.8	+ 0.7 0	− 0.6 − 1.1	0.6 2.3	+ 1.0 0	− 0.6 − 1.3	1.6 4.2	+ 1.6 0	− 1.6 − 2.6	1.6 5.7	+ 2.5 0	− 1.6 − 3.2
7.09	− 9.85	0.6 2.0	+ 0.8 0	− 0.6 − 1.2	0.6 2.6	+ 1.2 0	− 0.6 − 1.4	2.0 5.0	+ 1.8 0	− 2.0 − 3.2	2.0 6.6	+ 2.8 0	− 2.0 − 3.8
9.85	−12.41	0.8 2.3	+ 0.9 0	− 0.8 − 1.4	0.7 2.8	+ 1.2 0	− 0.7 − 1.6	2.5 5.7	+ 2.0 0	− 2.5 − 3.7	2.2 7.2	+ 3.0 0	− 2.2 − 4.2
12.41	−15.75	1.0 2.7	+ 1.0 0	− 1.0 − 1.7	0.7 3.1	+ 1.4 0	− 0.7 − 1.7	3.0 6.6	+ 2.2 0	− 3.0 − 4.4	2.5 8.2	+ 3.5 0	− 2.5 − 4.7
15.75	−19.69	1.2 3.0	+ 1.0 0	− 1.2 − 2.0	0.8 3.4	+ 1.6 0	− 0.8 − 1.8	4.0 8.1	+ 2.5 0	− 4.0 − 5.6	2.8 9.3	+ 4.0 0	− 2.8 − 5.3
19.69	−30.09	1.6 3.7	+ 1.2 0	− 1.6 − 2.5	1.6 4.8	+ 2.0 0	− 1.6 − 2.8	5.0 10.0	+ 3.0 0	− 5.0 − 7.0	5.0 13.0	+ 5.0 0	− 5.0 − 8.0
30.09	−41.49	2.0 4.6	+ 1.6 0	− 2.0 − 3.0	2.0 6.1	+ 2.5 0	− 2.0 − 3.6	6.0 12.5	+ 4.0 0	− 6.0 − 8.5	6.0 16.0	+ 6.0 0	− 6.0 −10.0
41.49	−56.19	2.5 5.7	+ 2.0 0	− 2.5 − 3.7	2.5 7.5	+ 3.0 0	− 2.5 − 4.5	8.0 16.0	+ 5.0 0	− 8.0 −11.0	8.0 21.0	+ 8.0 0	− 8.0 −13.0
56.19	−76.39	3.0 7.1	+ 2.5 0	− 3.0 − 4.6	3.0 9.5	+ 4.0 0	− 3.0 − 5.5	10.0 20.0	+ 6.0 0	−10.0 −14.0	10.0 26.0	+10.0 0	−10.0 −16.0
76.39	−100.9	4.0 9.0	+ 3.0 0	− 4.0 − 6.0	4.0 12.0	+ 5.0 0	− 4.0 − 7.0	12.0 25.0	+ 8.0 0	−12.0 −17.0	12.0 32.0	+12.0 0	−12.0 −20.0
100.9	−131.9	5.0 11.5	+ 4.0 0	− 5.0 − 7.5	5.0 15.0	+ 6.0 0	− 5.0 − 9.0	16.0 32.0	+10.0 0	−16.0 −22.0	16.0 42.0	+16.0 0	−16.0 −26.0
131.9	−171.9	6.0 14.0	+ 5.0 0	− 6.0 − 9.0	6.0 19.0	+ 8.0 0	− 6.0 −11.0	18.0 38.0	+12.0 0	−18.0 −26.0	18.0 50.0	+20.0 0	−18.0 −30.0
171.9	−200	8.0 18.0	+ 6.0 0	− 8.0 −12.0	8.0 22.0	+10.0 0	− 8.0 −12.0	22.0 48.0	+16.0 0	−22.0 −32.0	22.0 63.0	+25.0 0	−22.0 −38.0

(ANSI)

Continued

Running and Sliding Fits (Continued)

Limits are in thousandths of an inch.

Limits for hole and shaft are applied algebraically to the basic size to obtain the limits of size for the parts

Data in bold face are in accordance with ABC agreements

Symbols H8, e7, etc., are Hole and Shaft designations used in ABC System

Class RC 5			Class RC 6			Class RC 7			Class RC 8			Class RC 9			Nominal Size Range Inches	
Limits of Clearance	Hole H8	Shaft e7	Limits of Clearance	Hole H9	Shaft e8	Limits of Clearance	Hole H9	Shaft d8	Limits of Clearance	Hole H10	Shaft c9	Limits of Clearance	Hole H11	Shaft	Over	To
0.6	+0.6	-0.6	0.6	+1.0	-0.6	1.0	+1.0	-1.0	2.5	+1.6	-2.5	4.0	+2.5	-4.0	0	0.12
1.6	-0	-1.0	2.2	-0	-1.2	2.6	0	-1.6	5.1	0	-3.5	8.1	0	-5.6		
0.8	+0.7	-0.8	0.8	+1.2	-0.8	1.2	+1.2	-1.2	2.8	+1.8	-2.8	4.5	+3.0	-4.5	0.12	0.24
2.0	-0	-1.3	2.7	-0	-1.5	3.1	0	-1.9	5.8	0	-4.0	9.0	0	-6.0		
1.0	+0.9	-1.0	1.0	+1.4	-1.0	1.6	+1.4	-1.6	3.0	+2.2	-3.0	5.0	+3.5	-5.0	0.24	0.40
2.5	-0	-1.6	3.3	-0	-1.9	3.9	0	-2.5	6.6	0	-4.4	10.7	0	-7.2		
1.2	+1.0	-1.2	1.2	+1.6	-1.2	2.0	+1.6	-2.0	3.5	+2.8	-3.5	6.0	+4.0	-6.0	0.40	0.71
2.9	-0	-1.9	3.8	-0	-2.2	4.6	0	-3.0	7.9	0	-5.1	12.8	-0	-8.8		
1.6	+1.2	-1.6	1.6	+2.0	-1.6	2.5	+2.0	-2.5	4.5	+3.5	-4.5	7.0	+5.0	-7.0	0.71	1.19
3.6	-0	-2.4	4.8	-0	-2.8	5.7	0	-3.7	10.0	0	-6.5	15.5	0	-10.5		
2.0	+1.6	-2.0	2.0	+2.5	-2.0	3.0	+2.5	-3.0	5.0	+4.0	-5.0	8.0	+6.0	-8.0	1.19	1.97
4.6	-0	-3.0	6.1	-0	-3.6	7.1	0	-4.6	11.5	0	-7.5	18.0	0	-12.0		
2.5	+1.8	-2.5	2.5	+3.0	-2.5	4.0	+3.0	-4.0	6.0	+4.5	-6.0	9.0	+7.0	-9.0	1.97	3.15
5.5	-0	-3.7	7.3	-0	-4.3	8.8	0	-5.8	13.5	0	-9.0	20.5	0	-13.5		
3.0	+2.2	-3.0	3.0	+3.5	-3.0	5.0	+3.5	-5.0	7.0	+5.0	-7.0	10.0	+9.0	-10.0	3.15	4.73
6.6	-0	-4.4	8.7	-0	-5.2	10.7	0	-7.2	15.5	0	-10.5	24.0	0	-15.0		
3.5	+2.5	-3.5	3.5	+4.0	-3.5	6.0	+4.0	-6.0	8.0	+6.0	-8.0	12.0	+10.0	-12.0	4.73	7.09
7.6	-0	-5.1	10.0	-0	-6.0	12.5	0	-8.5	18.0	0	-12.0	28.0	0	-18.0		
4.0	+2.8	-4.0	4.0	+4.5	-4.0	7.0	+4.5	-7.0	10.0	+7.0	-10.0	15.0	+12.0	-15.0	7.09	9.85
8.6	-0	-5.8	11.3	0	-6.8	14.3	0	-9.8	21.5	0	-14.5	34.0	0	-22.0		
5.0	+3.0	-5.0	5.0	+5.0	-5.0	8.0	+5.0	-8.0	12.0	+8.0	-12.0	18.0	+12.0	-18.0	9.85	12.41
10.0	0	-7.0	13.0	0	-8.0	16.0	0	-11.0	25.0	0	-17.0	38.0	0	-26.0		
6.0	+3.5	-6.0	6.0	+6.0	-6.0	10.0	+6.0	-10.0	14.0	+9.0	-14.0	22.0	+14.0	-22.0	12.41	15.75
11.7	0	-8.2	15.5	0	-9.5	19.5	0	-13.5	29.0	0	-20.0	45.0	0	-31.0		
8.0	+4.0	-8.0	8.0	+6.0	-8.0	12.0	+6.0	-12.0	16.0	+10.0	-16.0	25.0	+16.0	-25.0	15.75	19.69
14.5	0	-10.5	18.0	0	-12.0	22.0	0	-16.0	32.0	0	-22.0	51.0	0	-35.0		
10.0	+5.0	-10.0	10.0	+8.0	-10.0	16.0	+8.0	-16.0	20.0	+12.0	-20.0	30.0	+20.0	-30.0	19.69	30.09
18.0	0	-13.0	23.0	0	-15.0	29.0	0	-21.0	40.0	0	-28.0	62.0	0	-42.0		
12.0	+6.0	-12.0	12.0	+10.0	-12.0	20.0	+10.0	-20.0	25.0	+16.0	-25.0	40.0	+25.0	-40.0	30.09	41.49
22.0	0	-16.0	28.0	0	-18.0	36.0	0	-26.0	51.0	0	-35.0	81.0	0	-56.0		
16.0	+8.0	-16.0	16.0	+12.0	-16.0	25.0	+12.0	-25.0	30.0	+20.0	-30.0	50.0	+30.0	-50.0	41.49	56.19
29.0	0	-21.0	36.0	0	-24.0	45.0	0	-33.0	62.0	0	-42.0	100	0	-70.0		
20.0	+10.0	-20.0	20.0	+16.0	-20.0	30.0	+16.0	-30.0	40.0	+25.0	-40.0	60.0	+40.0	-60.0	56.19	76.39
36.0	0	-26.0	46.0	0	-30.0	56.0	0	-40.0	81.0	0	-56.0	125	0	-85.0		
25.0	+12.0	-25.0	25.0	+20.0	-25.0	40.0	+20.0	-40.0	50.0	+30.0	-50.0	80.0	+50.0	-80.0	76.39	100.9
45.0	0	-33.0	57.0	0	-37.0	72.0	0	-52.0	100	0	-70.0	160	0	-110		
30.0	+16.0	-30.0	30.0	+25.0	-30.0	50.0	+25.0	-50.0	60.0	+40.0	-60.0	100	+60.0	-100	100.9	131.9
56.0	0	-40.0	71.0	0	-46.0	91.0	0	-66.0	125	0	-85.0	200	0	-140		
35.0	+20.0	-35.0	35.0	+30.0	-35.0	60.0	+30.0	-60.0	80.0	+50.0	-80.0	130	+80.0	-130	131.9	171.9
67.0	0	-47.0	85.0	0	-55.0	110.0	0	-80.0	160	0	-110	260	0	-180		
45.0	+25.0	-45.0	45.0	+40.0	-45.0	80.0	+40.0	-80.0	100	+60.0	-100	150	+100	-150	171.9	200
86.0	0	-61.0	110.0	0	-70.0	145.0	0	-105.0	200	0	-140	310	0	-210		

End of Table

Locational Clearance Fits

Limits are in thousandths of an inch.
Limits for hole and shaft are applied algebraically to the basic size to obtain the limits of size for the parts.
Data in bold face are in accordance with ABC agreements.
Symbols H6, h5, etc., are Hole and Shaft designations used in ABC System

Nominal Size Range Inches Over / To	Class LC 1 Limits of Clearance	Class LC 1 Hole H6	Class LC 1 Shaft h5	Class LC 2 Limits of Clearance	Class LC 2 Hole H7	Class LC 2 Shaft h6	Class LC 3 Limits of Clearance	Class LC 3 Hole H8	Class LC 3 Shaft h7	Class LC 4 Limits of Clearance	Class LC 4 Hole H10	Class LC 4 Shaft h9	Class LC 5 Limits of Clearance	Class LC 5 Hole H7	Class LC 5 Shaft g6
0 — 0.12	0 / 0.45	+0.25 / −0	+0 / −0.2	0 / 0.65	+0.4 / −0	+0 / −0.25	0 / 1	+0.6 / −0	+0 / −0.4	0 / 2.6	+1.6 / −0	+0 / −1.0	0.1 / 0.75	+0.4 / −0	−0.1 / −0.35
0.12 — 0.24	0 / 0.5	+0.3 / −0	+0 / −0.2	0 / 0.8	+0.5 / −0	+0 / −0.3	0 / 1.2	+0.7 / −0	+0 / −0.5	0 / 3.0	+1.8 / −0	+0 / −1.2	0.15 / 0.95	+0.5 / −0	−0.15 / −0.45
0.24 — 0.40	0 / 0.65	+0.4 / −0	+0 / −0.25	0 / 1.0	+0.6 / −0	+0 / −0.4	0 / 1.5	+0.9 / −0	+0 / −0.6	0 / 3.6	+2.2 / −0	+0 / −1.4	0.2 / 1.2	+0.6 / −0	−0.2 / −0.6
0.40 — 0.71	0 / 0.7	+0.4 / −0	+0 / −0.3	0 / 1.1	+0.7 / −0	+0 / −0.4	0 / 1.7	+1.0 / −0	+0 / −0.7	0 / 4.4	+2.8 / −0	+0 / −1.6	0.25 / 1.35	+0.7 / −0	−0.25 / −0.65
0.71 — 1.19	0 / 0.9	+0.5 / −0	+0 / −0.4	0 / 1.3	+0.8 / −0	+0 / −0.5	0 / 2	+1.2 / −0	+0 / −0.8	0 / 5.5	+3.5 / −0	+0 / −2.0	0.3 / 1.6	+0.8 / −0	−0.3 / −0.8
1.19 — 1.97	0 / 1.0	+0.6 / −0	+0 / −0.4	0 / 1.6	+1.0 / −0	+0 / −0.6	0 / 2.6	+1.6 / −0	+0 / −1	0 / 6.5	+4.0 / −0	+0 / −2.5	0.4 / 2.0	+1.0 / −0	−0.4 / −1.0
1.97 — 3.15	0 / 1.2	+0.7 / −0	+0 / −0.5	0 / 1.9	+1.2 / −0	+0 / −0.7	0 / 3	+1.8 / −0	+0 / −1.2	0 / 7.5	+4.5 / −0	+0 / −3	0.4 / 2.3	+1.2 / −0	−0.4 / −1.1
3.15 — 4.73	0 / 1.5	+0.9 / −0	+0 / −0.6	0 / 2.3	+1.4 / −0	+0 / −0.9	0 / 3.6	+2.2 / −0	+0 / −1.4	0 / 8.5	+5.0 / −0	+0 / −3.5	0.5 / 2.8	+1.4 / −0	−0.5 / −1.4
4.73 — 7.09	0 / 1.7	+1.0 / −0	+0 / −0.7	0 / 2.6	+1.6 / −0	+0 / −1.0	0 / 4.1	+2.5 / −0	+0 / −1.6	0 / 10	+6.0 / −0	+0 / −4	0.6 / 3.2	+1.6 / −0	−0.6 / −1.6
7.09 — 9.85	0 / 2.0	+1.2 / −0	+0 / −0.8	0 / 3.0	+1.8 / −0	+0 / −1.2	0 / 4.6	+2.8 / −0	+0 / −1.8	0 / 11.5	+7.0 / −0	+0 / −4.5	0.6 / 3.6	+1.8 / −0	−0.6 / −1.8
9.85 — 12.41	0 / 2.1	+1.2 / −0	+0 / −0.9	0 / 3.2	+2.0 / −0	+0 / −1.2	0 / 5	+3.0 / −0	+0 / −2.0	0 / 13	+8.0 / −0	+0 / −5	0.7 / 3.9	+2.0 / −0	−0.7 / −1.9
12.41 — 15.75	0 / 2.4	+1.4 / −0	+0 / −1.0	0 / 3.6	+2.2 / −0	+0 / −1.4	0 / 5.7	+3.5 / −0	+0 / −2.2	0 / 15	+9.0 / −0	+0 / −6	0.7 / 4.3	+2.2 / −0	−0.7 / −2.1
15.75 — 19.69	0 / 2.6	+1.6 / −0	+0 / −1.0	0 / 4.1	+2.5 / −0	+0 / −1.6	0 / 6.5	+4 / −0	+0 / −2.5	0 / 16	+10.0 / −0	+0 / −6	0.8 / 4.9	+2.5 / −0	−0.8 / −2.4
19.69 — 30.09	0 / 3.2	+2.0 / −0	+0 / −1.2	0 / 5.0	+3 / −0	+0 / −2	0 / 8	+5 / −0	+0 / −3	0 / 20	+12.0 / −0	+0 / −8	0.9 / 5.9	+3.0 / −0	−0.9 / −2.9
30.09 — 41.49	0 / 4.1	+2.5 / −0	+0 / −1.6	0 / 6.5	+4 / −0	+0 / −2.5	0 / 10	+6 / −0	+0 / −4	0 / 26	+16.0 / −0	+0 / −10	1.0 / 7.5	+5.0 / −0	−1.0 / −3.5
41.49 — 56.19	0 / 5.0	+3.0 / −0	+0 / −2.0	0 / 8.0	+5 / −0	+0 / −3	0 / 13	+8 / −0	+0 / −5	0 / 32	+20.0 / −0	+0 / −12	1.2 / 9.2	+5.0 / −0	−1.2 / −4.2
56.19 — 76.39	0 / 6.5	+4.0 / −0	+0 / −2.5	0 / 10	+6 / −0	+0 / −4	0 / 16	+10 / −0	+0 / −6	0 / 41	+25.0 / −0	+0 / −16	1.2 / 11.2	+6.0 / −0	−1.2 / −5.2
76.39 — 100.9	0 / 8.0	+5.0 / −0	+0 / −3.0	0 / 13	+8 / −0	+0 / −5	0 / 20	+12 / −0	+0 / −8	0 / 50	+30.0 / −0	+0 / −20	1.4 / 14.4	+8.0 / −0	−1.4 / −6.4
100.9 — 131.9	0 / 10.0	+6.0 / −0	+0 / −4.0	0 / 16	+10 / −0	+0 / −6	0 / 26	+16 / −0	+0 / −10	0 / 65	+40.0 / −0	+0 / −25	1.6 / 17.6	+10.0 / −0	−1.6 / −7.6
131.9 — 171.9	0 / 13.0	+8.0 / −0	+0 / −5.0	0 / 20	+12 / −0	+0 / −8	0 / 32	+20 / −0	+0 / −12	0 / 80	+50.0 / −0	+0 / −30	1.8 / 21.8	+12.0 / −0	−1.8 / −9.8
171.9 — 200	0 / 16.0	+10.0 / −0	+0 / −6.0	0 / 26	+16 / −0	+0 / −10	0 / 41	+25 / −0	+0 / −16	0 / 100	+60.0 / −0	+0 / −40	1.8 / 27.8	+16.0 / −0	−1.8 / −11.8

(ANSI)

Continued

Locational Clearance Fits (Continued)

Limits are in thousandths of an inch.

Limits for hole and shaft are applied algebraically to the basic size to obtain the limits of size for the parts.

Data in bold face are in accordance with ABC agreements.

Symbols H9, f8, etc., are Hole and Shaft designations used in ABC System

Class LC 6			Class LC 7			Class LC 8			Class LC 9			Class LC 10			Class LC 11			Nominal Size Range Inches	
Limits of Clearance	Hole H9	Shaft f8	Limits of Clearance	Hole H10	Shaft e9	Limits of Clearance	Hole H10	Shaft d9	Limits of Clearance	Hole H11	Shaft c10	Limits of Clearance	Hole H12	Shaft	Limits of Clearance	Hole H13	Shaft	Over	To
0.3 / 1.9	+1.0 / 0	−0.3 / −0.9	0.6 / 3.2	+1.6 / 0	−0.6 / −1.6	1.0 / 3.6	+1.6 / −0	−1.0 / −2.0	2.5 / 6.6	+2.5 / −0	−2.5 / −4.1	4 / 12	+4 / −0	−4 / −8	5 / 17	+6 / −0	−5 / −11	0	0.12
0.4 / 2.3	+1.2 / 0	−0.4 / −1.1	0.8 / 3.8	+1.8 / 0	−0.8 / −2.0	1.2 / 4.2	+1.8 / −0	−1.2 / −2.4	2.8 / 7.6	+3.0 / −0	−2.8 / −4.6	4.5 / 14.5	+5 / −0	−4.5 / −9.5	6 / 20	+7 / −0	−6 / −13	0.12	0.24
0.5 / 2.8	+1.4 / 0	−0.5 / −1.4	1.0 / 4.6	+2.2 / 0	−1.0 / −2.4	1.6 / 5.2	+2.2 / −0	−1.6 / −3.0	3.0 / 8.7	+3.5 / −0	−3.0 / −5.2	5 / 17	+6 / −0	−5 / −11	7 / 25	+9 / −0	−7 / −16	0.24	0.40
0.6 / 3.2	+1.6 / 0	−0.6 / −1.6	1.2 / 5.6	+2.8 / 0	−1.2 / −2.8	2.0 / 6.4	+2.8 / −0	−2.0 / −3.6	3.5 / 10.3	+4.0 / −0	−3.5 / −6.3	6 / 20	+7 / −0	−6 / −13	8 / 28	+10 / −0	−8 / −18	0.40	0.71
0.8 / 4.0	+2.0 / 0	−0.8 / −2.0	1.6 / 7.1	+3.5 / 0	−1.6 / −3.6	2.5 / 8.0	+3.5 / −0	−2.5 / −4.5	4.5 / 13.0	+5.0 / −0	−4.5 / −8.0	7 / 23	+8 / −0	−7 / −15	10 / 34	+12 / −0	−10 / −22	0.71	1.19
1.0 / 5.1	+2.5 / 0	−1.0 / −2.6	2.0 / 8.5	+4.0 / 0	−2.0 / −4.5	3.0 / 9.5	+4.0 / −0	−3.0 / −5.5	5 / 15	+6 / −0	−5 / −9	8 / 28	+10 / −0	−8 / −18	12 / 44	+16 / −0	−12 / −28	1.19	1.97
1.2 / 6.0	+3.0 / 0	−1.2 / −3.0	2.5 / 10.0	+4.5 / 0	−2.5 / −5.5	4.0 / 11.5	+4.5 / −0	−4.0 / −7.0	6 / 17.5	+7 / −0	−6 / −10.5	10 / 34	+12 / −0	−10 / −22	14 / 50	+18 / −0	−14 / −32	1.97	3.15
1.4 / 7.1	+3.5 / 0	−1.4 / −3.6	3.0 / 11.5	+5.0 / 0	−3.0 / −6.5	5.0 / 13.5	+5.0 / −0	−5.0 / −8.5	7 / 21	+9 / −0	−7 / −12	11 / 39	+14 / −0	−11 / −25	16 / 60	+22 / −0	−16 / −38	3.15	4.73
1.6 / 8.1	+4.0 / 0	−1.6 / −4.1	3.5 / 13.5	+6.0 / 0	−3.5 / −7.5	6 / 16	+6 / −0	−6 / −10	8 / 24	+10 / −0	−8 / −14	12 / 44	+16 / −0	−12 / −28	18 / 68	+25 / −0	−18 / −43	4.73	7.09
2.0 / 9.3	+4.5 / 0	−2.0 / −4.8	4.0 / 15.5	+7.0 / 0	−4.0 / −8.5	7 / 18.5	+7 / −0	−7 / −11.5	10 / 29	+12 / −0	−10 / −17	16 / 52	+18 / −0	−16 / −34	22 / 78	+28 / −0	−22 / −50	7.09	9.85
2.2 / 10.2	+5.0 / 0	−2.2 / −5.2	4.5 / 17.5	+8.0 / 0	−4.5 / −9.5	7 / 20	+8 / −0	−7 / −12	12 / 32	+12 / −0	−12 / −20	20 / 60	+20 / −0	−20 / −40	28 / 88	+30 / −0	−28 / −58	9.85	12.41
2.5 / 12.0	+6.0 / 0	−2.5 / −6.0	5.0 / 20.0	+9.0 / 0	−5 / −11	8 / 23	+9 / −0	−8 / −14	14 / 37	+14 / −0	−14 / −23	22 / 66	+22 / −0	−22 / −44	30 / 100	+35 / −0	−30 / −65	12.41	15.75
2.8 / 12.8	+6.0 / 0	−2.8 / −6.8	5.0 / 21.0	+10.0 / 0	−5 / −11	9 / 25	+10 / −0	−9 / −15	16 / 42	+16 / −0	−16 / −26	25 / 75	+25 / −0	−25 / −50	35 / 115	+40 / −0	−35 / −75	15.75	19.69
3.0 / 16.0	+8.0 / 0	−3.0 / −8.0	6.0 / 26.0	+12.0 / −0	−6 / −14	10 / 30	+12 / −0	−10 / −18	18 / 50	+20 / −0	−18 / −30	28 / 88	+30 / −0	−28 / −58	40 / 140	+50 / −0	−40 / −90	19.69	30.09
3.5 / 19.5	+10.0 / 0	−3.5 / −9.5	7.0 / 33.0	+16.0 / −0	−7 / −17	12 / 38	+16 / −0	−12 / −22	20 / 61	+25 / −0	−20 / −36	30 / 110	+40 / −0	−30 / −70	45 / 165	+60 / −0	−45 / −105	30.09	41.49
4.0 / 24.0	+12.0 / 0	−4.0 / −12.0	8.0 / 40.0	+20.0 / −0	−8 / −20	14 / 46	+20 / −0	−14 / −26	25 / 75	+30 / −0	−25 / −45	40 / 140	+50 / −0	−40 / −90	60 / 220	+80 / −0	−60 / −140	41.49	56.19
4.5 / 30.5	+16.0 / 0	−4.5 / −14.5	9.0 / 50.0	+25.0 / −0	−9 / −25	16 / 57	+25 / −0	−16 / −32	30 / 95	+40 / −0	−30 / −55	50 / 170	+60 / −0	−50 / −110	70 / 270	+100 / −0	−70 / −170	56.19	76.39
5.0 / 37.0	+20.0 / 0	−5 / −17	10.0 / 60.0	+30.0 / −0	−10 / −30	18 / 68	+30 / −0	−18 / −38	35 / 115	+50 / −0	−35 / −65	50 / 210	+80 / −0	−50 / −130	80 / 330	+125 / −0	−80 / −205	76.39	100.9
6.0 / 47.0	+25.0 / 0	−6 / −22	12.0 / 67.0	+40.0 / −0	−12 / −27	20 / 85	+40 / −0	−20 / −45	40 / 140	+60 / −0	−40 / −80	60 / 260	+100 / −0	−60 / −160	90 / 410	+160 / −0	−90 / −250	100.9	131.9
7.0 / 57.0	+30.0 / 0	−7 / −27	14.0 / 94.0	+50.0 / −0	−14 / −44	25 / 105	+50 / −0	−25 / −55	50 / 180	+80 / −0	−50 / −100	80 / 330	+125 / −0	−80 / −205	100 / 500	+200 / −0	−100 / −300	131.9	171.9
7.0 / 72.0	+40.0 / 0	−7 / −32	14.0 / 114.0	+60.0 / −0	−14 / −54	25 / 125	+60 / −0	−25 / −65	50 / 210	+100 / −0	−50 / −110	90 / 410	+160 / −0	−90 / −250	125 / 625	+250 / −0	−125 / −375	171.9	200

End of table

Locational Transition Fits

Limits are in thousandths of an inch.

Limits for hole and shaft are applied algebraically to the basic size to obtain the limits of size for the mating parts.

Data in bold face are in accordance with ABC agreements.

"Fit" represents the maximum interference (minus values) and the maximum clearance (plus values).

Symbols H7, js6, etc., are Hole and Shaft designations used in ABC System

Nominal Size Range Inches Over — To	Class LT 1 Fit	Hole H7	Shaft js6	Class LT 2 Fit	Hole H8	Shaft js7	Class LT 3 Fit	Hole H7	Shaft k6	Class LT 4 Fit	Hole H8	Shaft k7	Class LT 5 Fit	Hole H7	Shaft n6	Class LT 6 Fit	Hole H7	Shaft n7
0 – 0.12	−0.10 +0.50	+0.4 −0	+0.10 −0.10	−0.15 +0.65	+0.6 −0	+0.2 −0.2							−0.5 +0.15	+0.4 −0	+0.5 +0.25	−0.65 +0.15	+0.4 −0	+0.65 +0.25
0.12 – 0.24	−0.15 +0.65	+0.5 −0	+0.15 −0.15	−0.25 +0.95	+0.7 −0	+0.25 −0.25							−0.6 +0.2	+0.5 −0	+0.6 +0.3	−0.8 +0.2	+0.5 −0	+0.8 +0.3
0.24 – 0.40	−0.2 +0.8	+0.6 −0	+0.2 −0.2	−0.3 +1.2	+0.9 −0	+0.3 −0.3	−0.5 +0.5	+0.6 −0	+0.5 +0.1	−0.7 +0.8	+0.9 −0	+0.7 +0.1	−0.8 +0.2	+0.6 −0	+0.8 +0.4	−1.0 +0.2	+0.6 −0	+1.0 +0.4
0.40 – 0.71	−0.2 +0.9	+0.7 −0	+0.2 −0.2	−0.35 +1.35	+1.0 −0	+0.35 −0.35	−0.5 +0.6	+0.7 −0	+0.5 +0.1	−0.8 +0.9	+1.0 −0	+0.8 +0.1	−0.9 +0.2	+0.7 −0	+0.9 +0.5	−1.2 +0.2	+0.7 −0	+1.2 +0.5
0.71 – 1.19	−0.25 +1.05	+0.8 −0	+0.25 −0.25	−0.4 +1.6	+1.2 −0	+0.4 −0.4	−0.6 +0.7	+0.8 −0	+0.6 +0.1	−0.9 +1.1	+1.2 −0	+0.9 +0.1	−1.1 +0.2	+0.8 −0	+1.1 +0.6	−1.4 +0.2	+0.8 −0	+1.4 +0.6
1.19 – 1.97	−0.3 +1.3	+1.0 −0	+0.3 −0.3	−0.5 +2.1	+1.6 −0	+0.5 −0.5	−0.7 +0.9	+1.0 −0	+0.7 +0.1	−1.1 +1.5	+1.6 −0	+1.1 +0.1	−1.3 +0.3	+1.0 −0	+1.3 +0.7	−1.7 +0.3	+1.0 −0	+1.7 +0.7
1.97 – 3.15	−0.3 +1.5	+1.2 −0	+0.3 −0.3	−0.6 +2.4	+1.8 −0	+0.6 −0.6	−0.8 +1.1	+1.2 −0	+0.8 +0.1	−1.3 +1.7	+1.8 −0	+1.3 +0.1	−1.5 +0.4	+1.2 −0	+1.5 +0.8	−2.0 +0.4	+1.2 −0	+2.0 +0.8
3.15 – 4.73	−0.4 +1.8	+1.4 −0	+0.4 −0.4	−0.7 +2.9	+2.2 −0	+0.7 −0.7	−1.0 +1.3	+1.4 −0	+1.0 +0.1	−1.5 +2.1	+2.2 −0	+1.5 +0.1	−1.9 +0.4	+1.4 −0	+1.9 +1.0	−2.4 +0.4	+1.4 −0	+2.4 +1.0
4.73 – 7.09	−0.5 +2.1	+1.6 −0	+0.5 −0.5	−0.8 +3.3	+2.5 −0	+0.8 −0.8	−1.1 +1.5	+1.6 −0	+1.1 +0.1	−1.7 +2.4	+2.5 −0	+1.7 +0.1	−2.2 +0.4	+1.6 −0	+2.2 +1.2	−2.8 +0.4	+1.6 −0	+2.8 +1.2
7.09 – 9.85	−0.6 +2.4	+1.8 −0	+0.6 −0.6	−0.9 +3.7	+2.8 −0	+0.9 −0.9	−1.4 +1.6	+1.8 −0	+1.4 +0.2	−2.0 +2.6	+2.8 −0	+2.0 +0.2	−2.6 +0.4	+1.8 −0	+2.6 +1.4	−3.2 +0.4	+1.8 −0	+3.2 +1.4
9.85 – 12.41	−0.6 +2.6	+2.0 −0	+0.6 −0.6	−1.0 +4.0	+3.0 −0	+1.0 −1.0	−1.4 +1.8	+2.0 −0	+1.4 +0.2	−2.2 +2.8	+3.0 −0	+2.2 +0.2	−2.6 +0.6	+2.0 −0	+2.6 +1.4	−3.4 +0.6	+2.0 −0	+3.4 +1.4
12.41 – 15.75	−0.7 +2.9	+2.2 −0	+0.7 −0.7	−1.0 +4.5	+3.5 −0	+1.0 −1.0	−1.6 +2.0	+2.2 −0	+1.6 +0.2	−2.4 +3.3	+3.5 −0	+2.4 +0.2	−3.0 +0.6	+2.2 −0	+3.0 +1.6	−3.8 +0.6	+2.2 −0	+3.8 +1.6
15.75 – 19.69	−0.8 +3.3	+2.5 −0	+0.8 −0.8	−1.2 +5.2	+4.0 −0	+1.2 −1.2	−1.8 +2.3	+2.5 −0	+1.8 +0.2	−2.7 +3.8	+4.0 −0	+2.7 +0.2	−3.4 +0.7	+2.5 −0	+3.4 +1.8	−4.3 +0.7	+2.5 −0	+4.3 +1.8

(ANSI)

Force and Shrink Fits

Limits are in thousandths of an inch.

Limits for hole and shaft are applied algebraically to the basic size to obtain the limits of size for the parts.

Data in bold face are in accordance with ABC agreements.

Symbols H7, s6, etc., are Hole and Shaft designations used in ABC System

Nominal Size Range Inches (Over – To)	Class FN 1 Limits of Interference	Class FN 1 Hole H6	Class FN 1 Shaft	Class FN 2 Limits of Interference	Class FN 2 Hole H7	Class FN 2 Shaft s6	Class FN 3 Limits of Interference	Class FN 3 Hole H7	Class FN 3 Shaft t6	Class FN 4 Limits of Interference	Class FN 4 Hole H7	Class FN 4 Shaft u6	Class FN 5 Limits of Interference	Class FN 5 Hole H8	Class FN 5 Shaft x7
0 – 0.12	0.05 / 0.5	+0.25 / −0	+0.5 / +0.3	0.2 / 0.85	+0.4 / −0	+0.85 / +0.6				0.3 / 0.95	+0.5 / −0	+0.95 / +0.7	0.3 / 1.3	+0.6 / −0	+1.3 / +0.9
0.12 – 0.24	0.1 / 0.6	+0.3 / −0	+0.6 / +0.4	0.2 / 1.0	+0.5 / −0	+1.0 / +0.7				0.4 / 1.2	+0.5 / −0	+1.2 / +0.9	0.5 / 1.7	+0.7 / −0	+1.7 / +1.2
0.24 – 0.40	0.1 / 0.75	+0.4 / −0	+0.75 / +0.5	0.4 / 1.4	+0.6 / −0	+1.4 / +1.0				0.6 / 1.6	+0.6 / −0	+1.6 / +1.2	0.5 / 2.0	+0.9 / −0	+2.0 / +1.4
0.40 – 0.56	0.1 / 0.8	+0.4 / −0	+0.8 / +0.5	0.5 / 1.6	+0.7 / −0	+1.6 / +1.2				0.7 / 1.8	+0.7 / −0	+1.8 / +1.4	0.6 / 2.3	+1.0 / −0	+2.3 / +1.6
0.56 – 0.71	0.2 / 0.9	+0.4 / −0	+0.9 / +0.6	0.5 / 1.6	+0.7 / −0	+1.6 / +1.2				0.7 / 1.8	+0.7 / −0	+1.8 / +1.4	0.8 / 2.5	+1.0 / −0	+2.5 / +1.8
0.71 – 0.95	0.2 / 1.1	+0.5 / −0	+1.1 / +0.7	0.6 / 1.9	+0.8 / −0	+1.9 / +1.4				0.8 / 2.1	+0.8 / −0	+2.1 / +1.6	1.0 / 3.0	+1.2 / −0	+3.0 / +2.2
0.95 – 1.19	0.3 / 1.2	+0.5 / −0	+1.2 / +0.8	0.6 / 1.9	+0.8 / −0	+1.9 / +1.4	0.8 / 2.1	+0.8 / −0	+2.1 / +1.6	1.0 / 2.3	+0.8 / −0	+2.3 / +1.8	1.3 / 3.3	+1.2 / −0	+3.3 / +2.5
1.19 – 1.58	0.3 / 1.3	+0.6 / −0	+1.3 / +0.9	0.8 / 2.4	+1.0 / −0	+2.4 / +1.8	1.0 / 2.6	+1.0 / −0	+2.6 / +2.0	1.5 / 3.1	+1.0 / −0	+3.1 / +2.5	1.4 / 4.0	+1.6 / −0	+4.0 / +3.0
1.58 – 1.97	0.4 / 1.4	+0.6 / −0	+1.4 / +1.0	0.8 / 2.4	+1.0 / −0	+2.4 / +1.8	1.2 / 2.8	+1.0 / −0	+2.8 / +2.2	1.8 / 3.4	+1.0 / −0	+3.4 / +2.8	2.4 / 5.0	+1.6 / −0	+5.0 / +4.0
1.97 – 2.56	0.6 / 1.8	+0.7 / −0	+1.8 / +1.3	0.8 / 2.7	+1.2 / −0	+2.7 / +2.0	1.3 / 3.2	+1.2 / −0	+3.2 / +2.5	2.3 / 4.2	+1.2 / −0	+4.2 / +3.5	3.2 / 6.2	+1.8 / −0	+6.2 / +5.0
2.56 – 3.15	0.7 / 1.9	+0.7 / −0	+1.9 / +1.4	1.0 / 2.9	+1.2 / −0	+2.9 / +2.2	1.8 / 3.7	+1.2 / −0	+3.7 / +3.0	2.8 / 4.7	+1.2 / −0	+4.7 / +4.0	4.2 / 7.2	+1.8 / −0	+7.2 / +6.0
3.15 – 3.94	0.9 / 2.4	+0.9 / −0	+2.4 / +1.8	1.4 / 3.7	+1.4 / −0	+3.7 / +2.8	2.1 / 4.4	+1.4 / −0	+4.4 / +3.5	3.6 / 5.9	+1.4 / −0	+5.9 / +5.0	4.8 / 8.4	+2.2 / −0	+8.4 / +7.0
3.94 – 4.73	1.1 / 2.6	+0.9 / −0	+2.6 / +2.0	1.6 / 3.9	+1.4 / −0	+3.9 / +3.0	2.6 / 4.9	+1.4 / −0	+4.9 / +4.0	4.6 / 6.9	+1.4 / −0	+6.9 / +6.0	5.8 / 9.4	+2.2 / −0	+9.4 / +8.0
4.73 – 5.52	1.2 / 2.9	+1.0 / −0	+2.9 / +2.2	1.9 / 4.5	+1.6 / −0	+4.5 / +3.5	3.4 / 6.0	+1.6 / −0	+6.0 / +5.0	5.4 / 8.0	+1.6 / −0	+8.0 / +7.0	7.5 / 11.6	+2.5 / −0	+11.6 / +10.0
5.52 – 6.30	1.5 / 3.2	+1.0 / −0	+3.2 / +2.5	2.4 / 5.0	+1.6 / −0	+5.0 / +4.0	3.4 / 6.0	+1.6 / −0	+6.0 / +5.0	5.4 / 8.0	+1.6 / −0	+8.0 / +7.0	9.5 / 13.6	+2.5 / −0	+13.6 / +12.0
6.30 – 7.09	1.8 / 3.5	+1.0 / −0	+3.5 / +2.8	2.9 / 5.5	+1.6 / −0	+5.5 / +4.5	4.4 / 7.0	+1.6 / −0	+7.0 / +6.0	6.4 / 9.0	+1.6 / −0	+9.0 / +8.0	9.5 / 13.6	+2.5 / −0	+13.6 / +12.0
7.09 – 7.88	1.8 / 3.8	+1.2 / −0	+3.8 / +3.0	3.2 / 6.2	+1.8 / −0	+6.2 / +5.0	5.2 / 8.2	+1.8 / −0	+8.2 / +7.0	7.2 / 10.2	+1.8 / −0	+10.2 / +9.0	11.2 / 15.8	+2.8 / −0	+15.8 / +14.0
7.88 – 8.86	2.3 / 4.3	+1.2 / −0	+4.3 / +3.5	3.2 / 6.2	+1.8 / −0	+6.2 / +5.0	5.2 / 8.2	+1.8 / −0	+8.2 / +7.0	8.2 / 11.2	+1.8 / −0	+11.2 / +10.0	13.2 / 17.8	+2.8 / −0	+17.8 / +16.0
8.86 – 9.85	2.3 / 4.3	+1.2 / −0	+4.3 / +3.5	4.2 / 7.2	+1.8 / −0	+7.2 / +6.0	6.2 / 9.2	+1.8 / −0	+9.2 / +8.0	10.2 / 13.2	+1.8 / −0	+13.2 / +12.0	13.2 / 17.8	+2.8 / −0	+17.8 / +16.0
9.85 – 11.03	2.8 / 4.9	+1.2 / −0	+4.9 / +4.0	4.0 / 7.2	+2.0 / −0	+7.2 / +6.0	7.0 / 10.2	+2.0 / −0	+10.2 / +9.0	10.0 / 13.2	+2.0 / −0	+13.2 / +12.0	15.0 / 20.0	+3.0 / −0	+20.0 / +18.0
11.03 – 12.41	2.8 / 4.9	+1.2 / −0	+4.9 / +4.0	5.0 / 8.2	+2.0 / −0	+8.2 / +7.0	7.0 / 10.2	+2.0 / −0	+10.2 / +9.0	12.0 / 15.2	+2.0 / −0	+15.2 / +14.0	17.0 / 22.0	+3.0 / −0	+22.0 / +20.0
12.41 – 13.98	3.1 / 5.5	+1.4 / −0	+5.5 / +4.5	5.8 / 9.4	+2.2 / −0	+9.4 / +8.0	7.8 / 11.4	+2.2 / −0	+11.4 / +10.0	13.8 / 17.4	+2.2 / −0	+17.4 / +16.0	18.5 / 24.2	+3.5 / −0	+24.2 / +22.0
13.98 – 15.75	3.6 / 6.1	+1.4 / −0	+6.1 / +5.0	5.8 / 9.4	+2.2 / −0	+9.4 / +8.0	9.8 / 13.4	+2.2 / −0	+13.4 / +12.0	15.8 / 19.4	+2.2 / −0	+19.4 / +18.0	21.5 / 27.2	+3.5 / −0	+27.2 / +25.0
15.75 – 17.72	4.4 / 7.0	+1.6 / −0	+7.0 / +6.0	6.5 / 10.6	+2.5 / −0	+10.6 / +9.0	9.5 / 13.6	+2.5 / −0	+13.6 / +12.0	17.5 / 21.6	+2.5 / −0	+21.6 / +20.0	24.0 / 30.5	+4.0 / −0	+30.5 / +28.0
17.72 – 19.69	4.4 / 7.0	+1.6 / −0	+7.0 / +6.0	7.5 / 11.6	+2.5 / −0	+11.6 / +10.0	11.5 / 15.6	+2.5 / −0	+15.6 / +14.0	19.5 / 23.6	+2.5 / −0	+23.6 / +22.0	26.0 / 32.5	+4.0 / −0	+32.5 / +30.0

Locational Interference Fits

Limits are in thousandths of an inch.

Limits for hole and shaft are applied algebraically to the basic size to obtain the limits of size for the parts.

Data in bold face are in accordance with ABC agreements,

Symbols H7, p6, etc., are Hole and Shaft designations used in ABC System

Nominal Size Range Inches (Over – To)	Class LN 1 Limits of Interference	Class LN 1 Hole H6	Class LN 1 Shaft n5	Class LN 2 Limits of Interference	Class LN 2 Hole H7	Class LN 2 Shaft p6	Class LN 3 Limits of Interference	Class LN 3 Hole H7	Class LN 3 Shaft r6
0 – 0.12	0 / 0.45	+0.25 / −0	+0.45 / +0.25	0 / 0.65	+0.4 / −0	+0.65 / +0.4	0.1 / 0.75	+0.4 / −0	+0.75 / +0.5
0.12 – 0.24	0 / 0.5	+0.3 / −0	+0.5 / +0.3	0 / 0.8	+0.5 / −0	+0.8 / +0.5	0.1 / 0.9	+0.5 / −0	+0.9 / +0.6
0.24 – 0.40	0 / 0.65	+0.4 / −0	+0.65 / +0.4	0 / 1.0	+0.6 / −0	+1.0 / +0.6	0.2 / 1.2	+0.6 / −0	+1.2 / +0.8
0.40 – 0.71	0 / 0.8	+0.4 / −0	+0.8 / +0.4	0 / 1.1	+0.7 / −0	+1.1 / +0.7	0.3 / 1.4	+0.7 / −0	+1.4 / +1.0
0.71 – 1.19	0 / 1.0	+0.5 / −0	+1.0 / +0.5	0 / 1.3	+0.8 / −0	+1.3 / +0.8	0.4 / 1.7	+0.8 / −0	+1.7 / +1.2
1.19 – 1.97	0 / 1.1	+0.6 / −0	+1.1 / +0.6	0 / 1.6	+1.0 / −0	+1.6 / +1.0	0.4 / 2.0	+1.0 / −0	+2.0 / +1.4
1.97 – 3.15	0.1 / 1.3	+0.7 / −0	+1.3 / +0.8	0.2 / 2.1	+1.2 / −0	+2.1 / +1.4	0.4 / 2.3	+1.2 / −0	+2.3 / +1.6
3.15 – 4.73	0.1 / 1.6	+0.9 / −0	+1.6 / +1.0	0.2 / 2.5	+1.4 / −0	+2.5 / +1.6	0.6 / 2.9	+1.4 / −0	+2.9 / +2.0
4.73 – 7.09	0.2 / 1.9	+1.0 / −0	+1.9 / +1.2	0.2 / 2.8	+1.6 / −0	+2.8 / +1.8	0.9 / 3.5	+1.6 / −0	+3.5 / +2.5
7.09 – 9.85	0.2 / 2.2	+1.2 / −0	+2.2 / +1.4	0.2 / 3.2	+1.8 / −0	+3.2 / +2.0	1.2 / 4.2	+1.8 / −0	+4.2 / +3.0
9.85 – 12.41	0.2 / 2.3	+1.2 / −0	+2.3 / +1.4	0.2 / 3.4	+2.0 / −0	+3.4 / +2.2	1.5 / 4.7	+2.0 / −0	+4.7 / +3.5
12.41 – 15.75	0.2 / 2.6	+1.4 / −0	+2.6 / +1.6	0.3 / 3.9	+2.2 / −0	+3.9 / +2.5	2.3 / 5.9	+2.2 / −0	+5.9 / +4.5
15.75 – 19.69	0.2 / 2.8	+1.6 / −0	+2.8 / +1.8	0.3 / 4.4	+2.5 / −0	+4.4 / +2.8	2.5 / 6.6	+2.5 / −0	+6.6 / +5.0
19.69 – 30.09		+2.0 / −0		0.5 / 5.5	+3 / −0	+5.5 / +3.5	4 / 9	+3 / −0	+9 / +7
30.09 – 41.49		+2.5 / −0		0.5 / 7.0	+4 / −0	+7.0 / +4.5	5 / 11.5	+4 / −0	+11.5 / +9
41.49 – 56.19		+3.0 / −0		1 / 9	+5 / −0	+9 / +6	7 / 15	+5 / −0	+15 / +12
56.19 – 76.39		+4.0 / −0		1 / 11	+6 / −0	+11 / +7	10 / 20	+6 / −0	+20 / +16
76.39 – 100.9		+5.0 / −0		1 / 14	+8 / −0	+14 / +9	12 / 25	+8 / −0	+25 / +20
100.9 – 131.9		+6.0 / −0		2 / 18	+10 / −0	+18 / +12	15 / 31	+10 / −0	+31 / +25
131.9 – 171.9		+8.0 / −0		4 / 24	+12 / −0	+24 / +16	18 / 38	+12 / −0	+38 / +30
171.9 – 200		+10.0 / −0		4 / 30	+16 / −0	+30 / +20	24 / 50	+16 / −0	+50 / +40

(ANSI)

Drill Size Decimal Equivalents-Metric and Inch

NUMBER AND LETTER DRILLS.

Drill No.	Frac	Deci
80		.0135
79		.0145
	1/64	.0156
78		.0160
77		.0180
76		.0200
75		.0210
74		.0225
73		.0240
72		.0250
71		.0260
70		.0280
69		.0292
68		.0310
	1/32	.0313
67		.0320
66		.0330
65		.0350
64		.0360
63		.0370
62		.0380
61		.0390
60		.0400
59		.0410
58		.0420
57		.0430
56		.0465
	3/64	.0469
55		.0520
54		.0550
53		.0595
	1/16	.0625
52		.0635
51		.0670
50		.0700
49		.0730
48		.0760
	5/64	.0781
47		.0785
46		.0810
45		.0820
44		.0860
43		.0890
42		.0935
	3/32	.0938
41		.0960
40		.0980
39		.0995
38		.1015
37		.1040
36		.1065
	7/64	.1094
35		.1100
34		.1110
33		.1130
32		.116
31		.120
	1/8	.125
30		.129
29		.136

Drill No.	Frac	Deci
28		.140
	9/64	.141
27		.144
26		.147
25		.150
24		.152
23		.154
	5/32	.156
22		.157
21		.159
20		.161
19		.166
18		.170
	11/64	.172
17		.173
16		.177
15		.180
14		.182
13		.185
	3/16	.188
12		.189
11		.191
10		.194
9		.196
8		.199
7		.201
	13/64	.203
6		.204
5		.206
4		.209
3		.213
	7/32	.219
2		.221
1		.228
	15/64	.234
A		.234
B		.238
C		.242
D		.246
	1/4	.250
E		.250
F		.257
G		.261
H		.266
	17/64	.266
I		.272
J		.277
K		.281
	9/32	.281
L		.290
M		.295
	19/64	.297
N		.302
	5/16	.313
O		.316
P		.323
	21/64	.328
Q		.332
R		.339
	11/32	.344

Drill No.	Frac	Deci
S		.348
T		.358
	23/64	.359
U		.368
	3/8	.375
V		.377
W		.386
	25/64	.391
X		.397
Y		.404
	13/32	.406
Z		.413
	27/64	.422
	7/16	.438
	29/64	.453
	15/32	.469
	31/64	.484
	1/2	.500
	33/64	.516
	17/32	.531
	35/64	.547
	9/16	.562
	37/64	.578
	19/32	.594
	39/64	.609
	5/8	.625
	41/64	.641
	21/32	.656
	43/64	.672
	11/16	.688
	45/64	.703
	23/32	.719
	47/64	.734
	3/4	.750
	49/64	.766
	25/32	.781
	51/64	.797
	13/16	.813
	53/64	.828
	27/32	.844
	55/64	.859
	7/8	.875
	57/64	.891
	29/32	.906
	59/64	.922
	15/16	.938
	61/64	.953
	31/32	.969
	63/64	.984
	1	1.000

METRIC DRILLS.

MM	DEC.	MM	DEC.	MM	DEC.	MM	DEC.
1.	.0394	3.2	.1260	6.3	.2480	9.5	.3740
1.05	.0413	3.25	.1280	6.4	.2520	9.6	.3780
1.1	.0433	3.3	.1299	6.5	.2559	9.7	.3819
1.15	.0453	3.4	.1339	6.6	.2598	9.75	.3839
1.2	.0472	3.5	.1378	6.7	.2638	9.8	.3858
1.25	.0492	3.6	.1417	6.75	.2657	9.9	.3898
1.3	.0512	3.7	.1457	6.8	.2677	10.	.3937
1.35	.0531	3.75	.1476	6.9	.2717	10.5	.4134
1.4	.0551	3.8	.1496	7.	.2756	11.	.4331
1.45	.0571	3.9	.1535	7.1	.2795	11.5	.4528
1.5	.0591	4.	.1575	7.2	.2835	12.	.4724
1.55	.0610	4.1	.1614	7.25	.2854	12.5	.4921
1.6	.0630	4.2	.1654	7.3	.2874	13.	.5118
1.65	.0650	4.25	.1673	7.4	.2913	13.5	.5315
1.7	.0669	4.3	.1693	7.5	.2953	14.	.5512
1.75	.0689	4.4	.1732	7.6	.2992	14.5	.5709
1.8	.0709	4.5	.1772	7.7	.3031	15.	.5906
1.85	.0728	4.6	.1811	7.75	.3051	15.5	.6102
1.9	.0748	4.7	.1850	7.8	.3071	16.	.6299
1.95	.0768	4.75	.1870	7.9	.3110	16.5	.6496
2.	.0787	4.8	.1890	8.	.3150	17.	.6693
2.05	.0807	4.9	.1929	8.1	.3189	17.5	.6890
2.1	.0827	5.	.1968	8.2	.3228	18.	.7087
2.15	.0846	5.1	.2008	8.25	.3248	18.5	.7283
2.2	.0866	5.2	.2047	8.3	.3268	19.	.7480
2.25	.0886	5.25	.2067	8.4	.3307	19.5	.7677
2.3	.0906	5.3	.2087	8.5	.3346	20.	.7874
2.35	.0925	5.4	.2126	8.6	.3386	20.5	.8071
2.4	.0945	5.5	.2165	8.7	.3425	21.	.8268
2.45	.0965	5.6	.2205	8.75	.3445	21.5	.8465
2.5	.0984	5.7	.2244	8.8	.3465	22.	.8661
2.6	.1024	5.75	.2264	8.9	.3504	22.5	.8858
2.7	.1063	5.8	.2283	9.	.3543	23.	.9055
2.75	.1083	5.9	.2323	9.1	.3583	23.5	.9252
2.8	.1102	6.	.2362	9.2	.3622	24.	.9449
2.9	.1142	6.1	.2402	9.25	.3642	24.5	.9646
3.	.1181	6.2	.2441	9.3	.3661	25.	.9843
3.1	.1220	6.25	.2461	9.4	.3701		

TAP DRILL SIZES FOR UNIFIED STANDARD SCREW THREADS

Screw Thread Major Diameter	Threads Per Inch	Tap Drill Size Or Number	Screw Thread Major Diameter	Threads Per Inch	Tap Drill Size Or Number
0	80	3/64	3/8	16	5/16
1	64	53	3/8	24	Q
1	72	53	7/16	14	U
2	56	50	7/16	20	25/64
2	64	50	1/2	13	27/64
3	48	47	1/2	20	29/64
3	56	45	9/16	12	31/64
4	40	43	9/16	18	33/64
4	48	42	5/8	11	17/32
5	40	38	5/8	18	37/64
5	44	37	3/4	10	21/32
6	32	36	3/4	16	11/16
6	40	33	7/8	9	49/64
8	32	29	7/8	14	13/16
8	36	29	1	8	7/8
10	24	25	1	12	59/64
10	32	21	1 1/8	7	63/64
12	24	16	1 1/8	12	1 3/64
12	28	14	1 1/4	7	1 7/64
1/4	20	7	1 1/4	12	1 11/64
1/4	28	3	1 3/8	6	1 7/32
5/16	18	F	1 3/8	12	1 19/64
5/16	24	I	1 1/2	6	1 11/32
			1 1/2	12	1 27/64

TAP DRILL SIZES FOR ISO METRIC THREADS

Nominal Size mm	Series — Coarse		Series — Fine	
	Pitch mm	Tap Drill	Pitch mm	Tap Drill
10	1.5	8.5	1.25	8.75
12	1.75	10.25	1.25	10.50
14	2	12.00	1.5	12.50
16	2	14.00	1.5	14.50
18	2.5	15.50	1.5	16.50
20	2.5	17.50	1.5	18.50
22	2.5	19.50	1.5	20.50
24	3	21.00	2	22.00
27	3	24.00	2	25.00

TAP DRILL SIZES FOR ISO METRIC THREADS

Nominal Size mm	Series — Coarse		Series — Fine	
	Pitch mm	Tap Drill	Pitch mm	Tap Drill
1.4	0.3	1.1	—	—
1.6	0.35	1.25	—	—
2	0.4	1.6	—	—
2.5	0.45	2.05	—	—
3	0.5	2.5	—	—
4	0.7	3.3	—	—
5	0.8	4.2	—	—
6	1.0	5.0	—	—
8	1.25	6.75	1	7.0

Unified Standard Screw Thread Series

Sizes Primary	Sizes Secondary	Basic Major Diameter	Coarse UNC	Fine UNF	Extra fine UNEF	4UN	6UN	8UN	12UN	16UN	20UN	28UN	32UN	Sizes
0		0.0600	—	80	—	—	—	—	—	—	—	—	—	0
	1	0.0730	64	72	—	—	—	—	—	—	—	—	—	1
2		0.0860	56	64	—	—	—	—	—	—	—	—	—	2
	3	0.0990	48	56	—	—	—	—	—	—	—	—	—	3
4		0.1120	40	48	—	—	—	—	—	—	—	—	—	4
5		0.1250	40	44	—	—	—	—	—	—	—	—	—	5
6		0.1380	32	40	—	—	—	—	—	—	—	—	UNC	6
8		0.1640	32	36	—	—	—	—	—	—	—	—	UNC	8
10		0.1900	24	32	—	—	—	—	—	—	—	—	UNF	10
	12	0.2160	24	28	32	—	—	—	—	—	—	UNF	UNEF	12
1/4		0.2500	20	28	32	—	—	—	—	—	UNC	UNF	UNEF	1/4
5/16		0.3125	18	24	32	—	—	—	—	—	20	28	UNEF	5/16
3/8		0.3750	16	24	32	—	—	—	—	UNC	20	28	UNEF	3/8
7/16		0.4375	14	20	28	—	—	—	—	16	UNF	UNEF	32	7/16
1/2		0.5000	13	20	28	—	—	—	—	16	UNF	UNEF	32	1/2
9/16		0.5625	12	18	24	—	—	—	UNC	16	20	28	32	9/16
5/8		0.6250	11	18	24	—	—	—	12	16	20	28	32	5/8
	11/16	0.6875	—	—	24	—	—	—	12	16	20	28	32	11/16
3/4		0.7500	10	16	20	—	—	—	12	UNF	UNEF	28	32	3/4
	13/16	0.8125	—	—	20	—	—	—	12	16	UNEF	28	32	13/16
7/8		0.8750	9	14	20	—	—	—	12	16	UNEF	28	32	7/8
	15/16	0.9375	—	—	20	—	—	—	12	16	UNEF	28	32	15/16
1		1.0000	8	12	20	—	—	UNC	UNF	16	UNEF	28	32	1
	1 1/16	1.0625	—	—	18	—	—	8	12	16	20	28	—	1 1/16
1 1/8		1.1250	7	12	18	—	—	8	UNF	16	20	28	—	1 1/8
	1 3/16	1.1875	—	—	18	—	—	8	12	16	20	28	—	1 3/16
1 1/4		1.2500	7	12	18	—	—	8	UNF	16	20	28	—	1 1/4
	1 5/16	1.3125	—	—	18	—	—	8	12	16	20	28	—	1 5/16
1 3/8		1.3750	6	12	18	—	UNC	8	UNF	16	20	28	—	1 3/8
	1 7/16	1.4375	—	—	18	—	6	8	12	16	20	28	—	1 7/16
1 1/2		1.5000	6	12	18	—	UNC	8	UNF	16	20	28	—	1 1/2
	1 9/16	1.5625	—	—	18	—	6	8	12	16	20	—	—	1 9/16
1 5/8		1.6250	—	—	18	—	6	8	12	16	20	—	—	1 5/8
	1 11/16	1.6875	—	—	18	—	6	8	12	16	20	—	—	1 11/16
1 3/4		1.7500	5	—	—	—	6	8	12	16	20	—	—	1 3/4
	1 13/16	1.8125	—	—	—	—	6	8	12	16	20	—	—	1 13/16
1 7/8		1.8750	—	—	—	—	6	8	12	16	20	—	—	1 7/8
	1 15/16	1.9375	—	—	—	—	6	8	12	16	20	—	—	1 15/16
2		2.0000	4 1/2	—	—	—	6	8	12	16	20	—	—	2
	2 1/8	2.1250	—	—	—	—	6	8	12	16	20	—	—	2 1/8
2 1/4		2.2500	4 1/2	—	—	—	6	8	12	16	20	—	—	2 1/4
	2 3/8	2.3750	—	—	—	—	6	8	12	16	20	—	—	2 3/8
2 1/2		2.5000	4	—	—	UNC	6	8	12	16	20	—	—	2 1/2
	2 5/8	2.6250	—	—	—	4	6	8	12	16	20	—	—	2 5/8
2 3/4		2.7500	4	—	—	UNC	6	8	12	16	20	—	—	2 3/4
	2 7/8	2.8750	—	—	—	4	6	8	12	16	20	—	—	2 7/8
3		3.0000	4	—	—	UNC	6	8	12	16	20	—	—	3
	3 1/8	3.1250	—	—	—	4	6	8	12	16	—	—	—	3 1/8
3 1/4		3.2500	4	—	—	UNC	6	8	12	16	—	—	—	3 1/4
	3 3/8	3.3750	—	—	—	4	6	8	12	16	—	—	—	3 3/8
3 1/2		3.5000	4	—	—	UNC	6	8	12	16	—	—	—	3 1/2
	3 5/8	3.6250	—	—	—	4	6	8	12	16	—	—	—	3 5/8
3 3/4		3.7500	4	—	—	UNC	6	8	12	16	—	—	—	3 3/4
	3 7/8	3.8750	—	—	—	4	6	8	12	16	—	—	—	3 7/8
4		4.0000	4	—	—	UNC	6	8	12	16	—	—	—	4
	4 1/8	4.1250	—	—	—	4	6	8	12	16	—	—	—	4 1/8
4 1/4		4.2500	—	—	—	4	6	8	12	16	—	—	—	4 1/4
	4 3/8	4.3750	—	—	—	4	6	8	12	16	—	—	—	4 3/8
4 1/2		4.5000	—	—	—	4	6	8	12	16	—	—	—	4 1/2
	4 5/8	4.6250	—	—	—	4	6	8	12	16	—	—	—	4 5/8
4 3/4		4.7500	—	—	—	4	6	8	12	16	—	—	—	4 3/4
	4 7/8	4.8750	—	—	—	4	6	8	12	16	—	—	—	4 7/8
5		5.0000	—	—	—	4	6	8	12	16	—	—	—	5
	5 1/8	5.1250	—	—	—	4	6	8	12	16	—	—	—	5 1/8
5 1/4		5.2500	—	—	—	4	6	8	12	16	—	—	—	5 1/4
	5 3/8	5.3750	—	—	—	4	6	8	12	16	—	—	—	5 3/8
5 1/2		5.5000	—	—	—	4	6	8	12	16	—	—	—	5 1/2
	5 5/8	5.6250	—	—	—	4	6	8	12	16	—	—	—	5 5/8
5 3/4		5.7500	—	—	—	4	6	8	12	16	—	—	—	5 3/4
	5 7/8	5.8750	—	—	—	4	6	8	12	16	—	—	—	5 7/8
6		6.0000	—	—	—	4	6	8	12	16	—	—	—	6

ISO Metric Screw Thread Standard Series

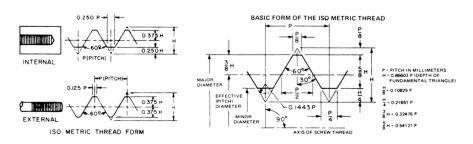

INTERNAL
EXTERNAL
ISO METRIC THREAD FORM
BASIC FORM OF THE ISO METRIC THREAD

Nominal Size Diam. (mm) Column a			Pitches (mm)														Nominal Size Diam. (mm)
			Series With Graded Pitches		Series With Constant Pitches												
1	2	3	Coarse	Fine	6	4	3	2	1.5	1.25	1	0.75	0.5	0.35	0.25	0.2	(mm)
0.25			0.075	—	—	—	—	—	—	—	—	—	—	—	—	—	0.25
0.3			0.08	—	—	—	—	—	—	—	—	—	—	—	—	—	0.3
	0.35		0.09	—	—	—	—	—	—	—	—	—	—	—	—	—	0.35
0.4			0.1	—	—	—	—	—	—	—	—	—	—	—	—	—	0.4
	0.45		0.1	—	—	—	—	—	—	—	—	—	—	—	—	—	0.45
0.5			0.125	—	—	—	—	—	—	—	—	—	—	—	—	—	0.5
	0.55		0.125	—	—	—	—	—	—	—	—	—	—	—	—	—	0.55
0.6			0.15	—	—	—	—	—	—	—	—	—	—	—	—	—	0.6
	0.7		0.175	—	—	—	—	—	—	—	—	—	—	—	—	—	0.7
0.8			0.2	—	—	—	—	—	—	—	—	—	—	—	—	—	0.8
	0.9		0.225	—	—	—	—	—	—	—	—	—	—	—	—	—	0.9
1			0.25	—	—	—	—	—	—	—	—	—	—	—	—	0.2	1
	1.1		0.25	—	—	—	—	—	—	—	—	—	—	—	—	0.2	1.1
1.2			0.25	—	—	—	—	—	—	—	—	—	—	—	—	0.2	1.2
	1.4		0.3	—	—	—	—	—	—	—	—	—	—	—	—	0.2	1.4
1.6			0.35	—	—	—	—	—	—	—	—	—	—	—	—	0.2	1.6
	1.8		0.35	—	—	—	—	—	—	—	—	—	—	—	—	0.2	1.8
2			0.4	—	—	—	—	—	—	—	—	—	—	—	0.25	—	2
	2.2		0.45	—	—	—	—	—	—	—	—	—	—	—	0.25	—	2.2
2.5			0.45	—	—	—	—	—	—	—	—	—	—	0.35	—	—	2.5
3			0.5	—	—	—	—	—	—	—	—	—	—	0.35	—	—	3
	3.5		0.6	—	—	—	—	—	—	—	—	—	—	0.35	—	—	3.5
4			0.7	—	—	—	—	—	—	—	—	—	0.5	—	—	—	4
	4.5		0.75	—	—	—	—	—	—	—	—	—	0.5	—	—	—	4.5
5			0.8	—	—	—	—	—	—	—	—	—	0.5	—	—	—	5
		5.5	—	—	—	—	—	—	—	—	—	—	0.5	—	—	—	5.5
6			1	—	—	—	—	—	—	—	—	0.75	—	—	—	—	6
		7	1	—	—	—	—	—	—	—	—	0.75	—	—	—	—	7
8			1.25	1	—	—	—	—	—	—	1	0.75	—	—	—	—	8
		9	1.25	—	—	—	—	—	—	—	1	0.75	—	—	—	—	9
10			1.5	1.25	—	—	—	—	—	1.25	1	0.75	—	—	—	—	10
		11	1.5	—	—	—	—	—	—	—	1	0.75	—	—	—	—	11
12			1.75	1.25	—	—	—	—	1.5	1.25	1	—	—	—	—	—	12
	14		2	1.5	—	—	—	—	1.5	1.25b	1	—	—	—	—	—	14
		15	—	—	—	—	—	—	1.5	—	1	—	—	—	—	—	15
16			2	1.5	—	—	—	—	1.5	—	1	—	—	—	—	—	16
		17	—	—	—	—	—	—	1.5	—	1	—	—	—	—	—	17
	18		2.5	1.5	—	—	—	2	1.5	—	1	—	—	—	—	—	18
20			2.5	1.5	—	—	—	2	1.5	—	1	—	—	—	—	—	20
	22		2.5	1.5	—	—	—	2	1.5	—	1	—	—	—	—	—	22
24			3	2	—	—	—	2	1.5	—	1	—	—	—	—	—	24
		25	—	—	—	—	—	2	1.5	—	1	—	—	—	—	—	25
		26	—	—	—	—	—	—	1.5	—	1	—	—	—	—	—	26
	27		3	2	—	—	—	2	1.5	—	1	—	—	—	—	—	27
		28	—	—	—	—	—	2	1.5	—	1	—	—	—	—	—	28
30			3.5	2	—	—	(3)	2	1.5	—	1	—	—	—	—	—	30
		32	—	—	—	—	—	2	1.5	—	—	—	—	—	—	—	32
	33		3.5	2	—	—	(3)	2	1.5	—	—	—	—	—	—	—	33
		35c	—	—	—	—	—	—	1.5	—	—	—	—	—	—	—	35c
36			4	3	—	—	—	2	1.5	—	—	—	—	—	—	—	36
		38	—	—	—	—	—	—	1.5	—	—	—	—	—	—	—	38
	39		4	3	—	—	—	2	1.5	—	—	—	—	—	—	—	39
		40	—	—	—	—	3	2	1.5	—	—	—	—	—	—	—	40
42			4.5	3	—	4	3	2	1.5	—	—	—	—	—	—	—	42
	45		4.5	3	—	4	3	2	1.5	—	—	—	—	—	—	—	45

a Thread diameter should be selected from columns 1, 2 or 3; with preference being given in that order.
b Pitch 1.25 mm in combination with diameter 14 mm has been included for spark plug applications.
c Diameter 35 mm has been included for bearing locknut applications.
The use of pitches shown in parentheses should be avoided wherever possible.
The pitches enclosed in the bold frame, together with the corresponding nominal diameters in Columns 1 and 2, are those combinations which have been established by ISO Recommendations as a selected "coarse" and "fine" series for commercial fasteners. Sizes 0.25 mm through 1.4 mm are covered in ISO Recommendation R 68 and, except for the 0.25 mm size, in AN Standard ANSI B1.10.

(ANSI)

Inch-Metric Thread Comparison

INCH SERIES			METRIC			
Size	Dia. (In.)	TPI	Size	Dia. (In.)	Pitch (MM)	TPI (Approx)
			M1.4	.055	.3 / .2	85 / 127
#0	.060	80				
			M1.6	.063	.35 / .2	74 / 127
#1	.073	64 / 72				
			M2	.079	.4 / .25	64 / 101
#2	.086	56 / 64				
			M2.5	.098	.45 / .35	56 / 74
#3	.099	48 / 56				
#4	.112	40 / 48				
			M3	.118	.5 / .35	51 / 74
#5	.125	40 / 44				
#6	.138	32 / 40				
			M4	.157	.7 / .5	36 / 51
#8	.164	32 / 36				
#10	.190	24 / 32				
			M5	.196	.8 / .5	32 / 51
			M6	.236	1.0 / .75	25 / 34
1/4	.250	20 / 28				
5/16	.312	18 / 24				
			M8	.315	1.25 / 1.0	20 / 25
3/8	.375	16 / 24				
			M10	.393	1.5 / 1.25	17 / 20
7/16	.437	14 / 20				
			M12	.472	1.75 / 1.25	14.5 / 20
1/2	.500	13 / 20				
			M14	.551	2 / 1.5	12.5 / 17
5/8	.625	11 / 18				
			M16	.630	2 / 1.5	12.5 / 17
			M18	.709	2.5 / 1.5	10 / 17
3/4	.750	10 / 16				
			M20	.787	2.5 / 1.5	10 / 17
			M22	.866	2.5 / 1.5	10 / 17
7/8	.875	9 / 14				
			M24	.945	3 / 2	.8.5 / 12.5
1''	1.000	8 / 12				
			M27	1.063	3 / 2	8.5 / 12.5

(Standard Pressed Steel Co.)

Acme Screw Threads

1	2	3	4	5	6	7	8	9	10	11	12
Identification		Basic Diameters			Thread Data						
Nominal Sizes (All Classes)	Threads per Inch,* n	Classes 2G, 3G, and 4G			Pitch, p	Thickness at Pitch Line, $t = p/2$	Basic Height of Thread, $h = p/2$	Basic Width of Flat, $F = 0.3707p$	Lead Angle at Basic Pitch Diameter* Classes 2G, 3G, and 4G, λ	Shear Area† Class 3G	Stress Area‡ Class 3G
		Major Diameter, D	Pitch Diameter,§ $E = D - h$	Minor Diameter, $K = D - 2h$					Deg Min		
¼	16	0.2500	0.2188	0.1875	0.06250	0.03125	0.03125	0.0232	5 12	0.350	0.0285
5⁄16	14	0.3125	0.2768	0.2411	0.07143	0.03571	0.03571	0.0265	4 42	0.451	0.0474
3⁄8	12	0.3750	0.3333	0.2917	0.08333	0.04167	0.04167	0.0309	4 33	0.545	0.0699
7⁄16	12	0.4375	0.3958	0.3542	0.08333	0.04167	0.04167	0.0309	3 50	0.660	0.1022
½	10	0.5000	0.4500	0.4000	0.10000	0.05000	0.05000	0.0371	4 3	0.749	0.1287
5⁄8	8	0.6250	0.5625	0.5000	0.12500	0.06250	0.06250	0.0463	4 3	0.941	0.2043
¾	6	0.7500	0.6667	0.5833	0.16667	0.08333	0.08333	0.0618	4 33	1.108	0.2848
7⁄8	6	0.8750	0.7917	0.7083	0.16667	0.08333	0.08333	0.0618	3 50	1.339	0.4150
1	5	1.0000	0.9000	0.8000	0.20000	0.10000	0.10000	0.0741	4 3	1.519	0.5354
1 ⅛	5	1.1250	1.0250	0.9250	0.20000	0.10000	0.10000	0.0741	3 33	1.751	0.709
1 ¼	5	1.2500	1.1500	1.0500	0.20000	0.10000	0.10000	0.0741	3 10	1.983	0.907
1 ⅜	4	1.3750	1.2500	1.1250	0.25000	0.12500	0.12500	0.0927	3 39	2.139	1.059
1 ½	4	1.5000	1.3750	1.2500	0.25000	0.12500	0.12500	0.0927	3 19	2.372	1.298
1 ¾	4	1.7500	1.6250	1.5000	0.25000	0.12500	0.12500	0.0927	2 48	2.837	1.851
2	4	2.0000	1.8750	1.7500	0.25000	0.12500	0.12500	0.0927	2 26	3.301	2.501
2 ¼	3	2.2500	2.0833	1.9167	0.33333	0.16667	0.16667	0.1236	2 55	3.643	3.049
2 ½	3	2.5000	2.3333	2.1667	0.33333	0.16667	0.16667	0.1236	2 36	4.110	3.870
2 ¾	3	2.7500	2.5833	2.4167	0.33333	0.16667	0.16667	0.1236	2 21	4.577	4.788
3	2	3.0000	2.7500	2.5000	0.50000	0.25000	0.25000	0.1853	3 19	4.786	5.27
3 ½	2	3.5000	3.2500	3.0000	0.50000	0.25000	0.25000	0.1853	2 48	5.73	7.50
4	2	4.0000	3.7500	3.5000	0.50000	0.25000	0.25000	0.1853	2 26	6.67	10.12
4 ½	2	4.5000	4.2500	4.0000	0.50000	0.25000	0.25000	0.1853	2 9	7.60	13.13
5	2	5.0000	4.7500	4.5000	0.50000	0.25000	0.25000	0.1853	1 55	8.54	16.53

* All other dimensions are given in inches.

§ British: Effective Diameter.

† Per inch length of engagement of the external thread in line with the minor diameter crests of the internal thread. Computed from this formula: Shear Area $= \pi K_n [0.5 + h \tan 14\frac{1}{2}° (E_s - K_n)]$. Figures given are the minimum shear area based on max K_n and min E_s.

‡ Figures given are the minimum stress area based on the mean of the minimum minor and pitch diameters of the external thread.

(ANSI)

Square Screw Threads

SIZE	THREADS PER INCH	SIZE	THREADS PER INCH	SIZE	THREADS PER INCH
3/8	12	7/8	5	2	2 1/2
7/16	10	1	5	2 1/4	2
1/2	10	1 1/8	4	2 1/2	2
9/16	8	1 1/4	4	2 3/4	2
5/8	8	1 1/2	3	3	1 1/2
3/4	6	1 3/4	2 1/2	3 1/4	1 1/2

(ANSI)

Standard Taper Pipe Threads

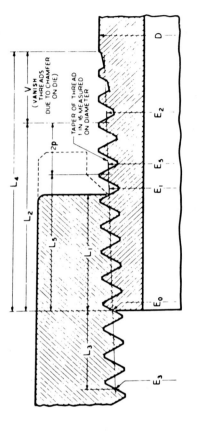

Basic Dimensions of USA (American) Standard Taper Pipe Thread, NPT[1]

Nominal[8] Pipe Size	Outside Diameter of Pipe, D	Threads per inch, n	Pitch of Thread, p	Pitch Diameter at beginning of External Thread, E_0	Handtight Engagement			Effective Thread, External		
					Length[2], L_1		Dia[3], E_1	Length[4], L_2		Dia[3], E_2
					In.	Thds.		In.	Thds.	
1	2	3	4	5	6	7	8	9	10	11
1/16	0.3125	27	0.03704	0.27118	0.160	4.32	0.28118	0.2611	7.05	0.28750
1/8	0.405	27	0.03704	0.36351	0.1615	4.36	0.37360	0.2639	7.12	0.38000
1/4	0.540	18	0.05556	0.47739	0.2278	4.10	0.49163	0.4018	7.23	0.50250
3/8	0.675	18	0.05556	0.61201	0.240	4.32	0.62701	0.4078	7.34	0.63750
1/2	0.840	14	0.07143	0.75843	0.320	4.48	0.77843	0.5337	7.47	0.79179
3/4	1.050	14	0.07143	0.96768	0.339	4.75	0.98887	0.5457	7.64	1.00179
1	1.315	11.5	0.08696	1.21363	0.400	4.60	1.23863	0.6828	7.85	1.25630
1 1/4	1.660	11.5	0.08696	1.55713	0.420	4.83	1.58338	0.7068	8.13	1.60130
1 1/2	1.900	11.5	0.08696	1.79609	0.420	4.83	1.82234	0.7235	8.32	1.84130
2	2.375	11.5	0.08696	2.26902	0.436	5.01	2.29627	0.7565	8.70	2.31630
2 1/2	2.875	8	0.12500	2.71953	0.682	5.46	2.76216	1.1375	9.10	2.79062
3	3.500	8	0.12500	3.34062	0.766	6.13	3.38850	1.2000	9.60	3.41562
3 1/2	4.000	8	0.12500	3.83750	0.821	6.57	3.88881	1.2500	10.00	3.91562
4	4.500	8	0.12500	4.33438	0.844	6.75	4.38712	1.3000	10.40	4.41562
5	5.563	8	0.12500	5.39073	0.937	7.50	5.44929	1.4063	11.25	5.47862
6	6.625	8	0.12500	6.44609	0.958	7.66	6.50597	1.5125	12.10	6.54062
8	8.625	8	0.12500	8.43359	1.063	8.50	8.50003	1.7125	13.70	8.54062
10	10.750	8	0.12500	10.54531	1.210	9.68	10.62094	1.9250	15.40	10.66562
12	12.750	8	0.12500	12.53281	1.360	10.88	12.61781	2.1250	17.00	12.66562
14 OD	14.000	8	0.12500	13.77500	1.562	12.50	13.87262	2.2500	18.00	13.91562
16 OD	16.000	8	0.12500	15.76250	1.812	14.50	15.87575	2.4500	19.60	15.91562
18 OD	18.000	8	0.12500	17.75000	2.000	16.00	17.87500	2.6500	21.20	17.91562
20 OD	20.000	8	0.12500	19.73750	2.125	17.00	19.87031	2.8500	22.80	19.91562
24 OD	24.000	8	0.12500	23.71250	2.375	19.00	23.86094	3.2500	26.00	23.91562

[1] The basic dimensions of the USA (American) Standard Taper Pipe Thread are given in inches to four or five decimal places. While this implies a greater degree of precision than is ordinarily attained, these dimensions are the basis of gage dimensions and are so expressed for the purpose of eliminating errors in computations.
[2] Also length of thin ring gage and length from gaging notch to small end of plug gage.
[3] Also pitch diameter at gaging notch (handtight plane.)
[4] Also length of plug gage.
[5] The length L_5 from the end of the pipe determines the plane beyond which the thread form is incomplete at the crest. The next two threads are complete at the root. At this plane the cone formed by the crests of the thread intersects the cylinder forming the external surface of the pipe. $l_5 = l_2 - 2p$. (See Appendix E.)
[6] Given as information for use in selecting tap drills.
[7] Military Specification MIL-P-7105 gives the wrench makeup as three threads for 3 in. and smaller. The L_3 dimensions are as follows: Size 2.5 in. 2.69609 and size 3 in. 3.31719.
[8] Designated, for example, as 3/8 NPT or 0.675 NPT.

(ANSI)

Metric-Inch Equivalents

INCHES (FRACTIONS)	DECIMALS	MILLIMETERS	INCHES (FRACTIONS)	DECIMALS	MILLIMETERS
	.00394	.1	15/32	.46875	11.9063
	.00787	.2		.47244	12.00
	.01181	.3	31/64	.484375	12.3031
1/64	.015625	.3969	1/2	.5000	12.70
	.01575	.4		.51181	13.00
	.01969	.5	33/64	.515625	13.0969
	.02362	.6	17/32	.53125	13.4938
	.02756	.7	35/64	.546875	13.8907
1/32	.03125	.7938		.55118	14.00
	.0315	.8	9/16	.5625	14.2875
	.03543	.9	37/64	.578125	14.6844
	.03937	1.00		.59055	15.00
3/64	.046875	1.1906	19/32	.59375	15.0813
1/16	.0625	1.5875	39/64	.609375	15.4782
5/64	.078125	1.9844	5/8	.625	15.875
	.07874	2.00		.62992	16.00
3/32	.09375	2.3813	41/64	.640625	16.2719
7/64	.109375	2.7781	21/32	.65625	16.6688
	.11811	3.00		.66929	17.00
1/8	.125	3.175	43/64	.671875	17.0657
9/64	.140625	3.5719	11/16	.6875	17.4625
5/32	.15625	3.9688	45/64	.703125	17.8594
	.15748	4.00		.70866	18.00
11/64	.171875	4.3656	23/32	.71875	18.2563
3/16	.1875	4.7625	47/64	.734375	18.6532
	.19685	5.00		.74803	19.00
13/64	.203125	5.1594	3/4	.7500	19.05
7/32	.21875	5.5563	49/64	.765625	19.4469
15/64	.234375	5.9531	25/32	.78125	19.8438
	.23622	6.00		.7874	20.00
1/4	.2500	6.35	51/64	.796875	20.2407
17/64	.265625	6.7469	13/16	.8125	20.6375
	.27559	7.00		.82677	21.00
9/32	.28125	7.1438	53/64	.828125	21.0344
19/64	.296875	7.5406	27/32	.84375	21.4313
5/16	.3125	7.9375	55/64	.859375	21.8282
	.31496	8.00		.86614	22.00
21/64	.328125	8.3344	7/8	.875	22.225
11/32	.34375	8.7313	57/64	.890625	22.6219
	.35433	9.00		.90551	23.00
23/64	.359375	9.1281	29/32	.90625	23.0188
3/8	.375	9.525	59/64	.921875	23.4157
25/64	.390625	9.9219	15/16	.9375	23.8125
	.3937	10.00		.94488	24.00
13/32	.40625	10.3188	61/64	.953125	24.2094
27/64	.421875	10.7156	31/32	.96875	24.6063
	.43307	11.00		.98425	25.00
7/16	.4375	11.1125	63/64	.984375	25.0032
29/64	.453125	11.5094	1	1.0000	25.4001

Square Bolts

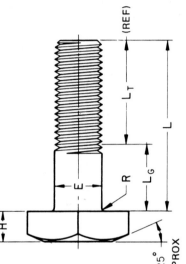

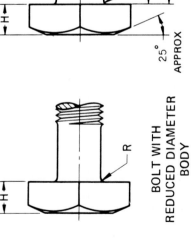

BOLT WITH
REDUCED DIAMETER
BODY

25° APPROX

Dimensions of Square Bolts

Nominal Size or Basic Product Dia		E Body Dia. Max	F Width Across Flats			G Width Across Corners		H Height			R Radius of Fillet		L_T Thread Length For Bolt Lengths	
			Basic	Max	Min	Max	Min	Basic	Max	Min	Max	Min	6 in. and shorter Basic	Over 6 in. Basic
1/4	0.2500	0.260	3/8	0.375	0.362	0.530	0.498	11/64	0.188	0.156	0.03	0.01	0.750	1.000
5/16	0.3125	0.324	1/2	0.500	0.484	0.707	0.665	13/64	0.220	0.186	0.03	0.01	0.875	1.125
3/8	0.3750	0.388	9/16	0.562	0.544	0.795	0.747	1/4	0.268	0.232	0.03	0.01	1.000	1.250
7/16	0.4375	0.452	5/8	0.625	0.603	0.884	0.828	19/64	0.316	0.278	0.03	0.01	1.125	1.375
1/2	0.5000	0.515	3/4	0.750	0.725	1.061	0.995	21/64	0.348	0.308	0.03	0.01	1.250	1.500
5/8	0.6250	0.642	15/16	0.938	0.906	1.326	1.244	27/64	0.444	0.400	0.06	0.02	1.500	1.750
3/4	0.7500	0.768	1 1/8	1.125	1.088	1.591	1.494	1/2	0.524	0.476	0.06	0.02	1.750	2.000
7/8	0.8750	0.895	1 5/16	1.312	1.269	1.856	1.742	19/32	0.620	0.568	0.06	0.02	2.000	2.250
1	1.0000	1.022	1 1/2	1.500	1.450	2.121	1.991	21/32	0.684	0.628	0.09	0.03	2.250	2.500
1 1/8	1.1250	1.149	1 11/16	1.688	1.631	2.386	2.239	3/4	0.780	0.720	0.09	0.03	2.500	2.750
1 1/4	1.2500	1.277	1 7/8	1.875	1.812	2.652	2.489	27/32	0.876	0.812	0.09	0.03	2.750	3.000
1 3/8	1.3750	1.404	2 1/16	2.062	1.994	2.917	2.738	29/32	0.940	0.872	0.09	0.03	3.000	3.250
1 1/2	1.5000	1.531	2 1/4	2.250	2.175	3.182	2.986	1	1.036	0.964	0.09	0.03	3.250	3.500

Note: L_G is the grip gaging length (nominal bolt length minus the basic thread length L_T).
L_T is the basic thread length and is the distance from the extreme end of the bolt to the last complete (full form) thread.
Bold type indicates products unified dimensionally with British and Canadian standards.
For additional requirements see ANSI B18.2.1

(ANSI)

Hex Bolts

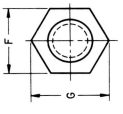

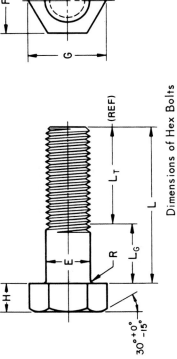

Dimensions of Hex Bolts

Nominal Size or Basic Product Dia		E Body Dia Max	E Body Dia Basic	F Width Across Flats Max	F Width Across Flats Min	G Width Across Corners Max	G Width Across Corners Min	H Height Basic	H Height Max	H Height Min	R Radius of Fillet Max	R Radius of Fillet Min	L_T Thread Length For Bolt Lengths 6 in. and Shorter Basic	L_T Thread Length For Bolt Lengths Over 6 in. Basic
1/4	0.2500	0.260	7/16	0.438	0.425	0.505	0.484	11/64	0.188	0.150	0.03	0.01	0.750	1.000
5/16	0.3125	0.324	1/2	0.500	0.484	0.577	0.552	7/32	0.235	0.195	0.03	0.01	0.875	1.125
3/8	0.3750	0.388	9/16	0.562	0.544	0.650	0.620	1/4	0.268	0.226	0.03	0.01	1.000	1.250
7/16	0.4375	0.452	5/8	0.625	0.603	0.722	0.687	19/64	0.316	0.272	0.03	0.01	1.125	1.375
1/2	0.5000	0.515	3/4	0.750	0.725	0.866	0.826	11/32	0.364	0.302	0.03	0.01	1.250	1.500
5/8	0.6250	0.642	15/16	0.938	0.906	1.083	1.033	27/64	0.444	0.378	0.06	0.02	1.500	1.750
3/4	0.7500	0.768	1 1/8	1.125	1.088	1.299	1.240	1/2	0.524	0.455	0.06	0.02	1.750	2.000
7/8	0.8750	0.895	1 5/16	1.312	1.269	1.516	1.447	37/64	0.604	0.531	0.06	0.02	2.000	2.250
1	1.0000	1.022	1 1/2	1.500	1.450	1.732	1.653	43/64	0.700	0.591	0.09	0.03	2.250	2.500
1 1/8	1.1250	1.149	1 11/16	1.688	1.631	1.949	1.859	3/4	0.780	0.658	0.09	0.03	2.500	2.750
1 1/4	1.2500	1.277	1 7/8	1.875	1.812	2.165	2.066	27/32	0.876	0.749	0.09	0.03	2.750	3.000
1 3/8	1.3750	1.404	2 1/16	2.062	1.994	2.382	2.273	29/32	0.940	0.810	0.09	0.03	3.000	3.250
1 1/2	1.5000	1.531	2 1/4	2.250	2.175	2.598	2.480	1	1.036	0.902	0.09	0.03	3.250	3.500
1 3/4	1.7500	1.785	2 5/8	2.625	2.538	3.031	2.893	1 5/32	1.196	1.054	0.12	0.04	3.750	4.000
2	2.0000	2.039	3	3.000	2.900	3.464	3.306	1 11/32	1.388	1.175	0.12	0.04	4.250	4.500
2 1/4	2.2500	2.305	3 3/8	3.375	3.262	3.897	3.719	1 1/2	1.548	1.327	0.19	0.06	4.750	5.000
2 1/2	2.5000	2.559	3 3/4	3.750	3.625	4.330	4.133	1 21/32	1.708	1.479	0.19	0.06	5.250	5.500
2 3/4	2.7500	2.827	4 1/8	4.125	3.988	4.763	4.546	1 13/16	1.869	1.632	0.19	0.06	5.750	6.000
3	3.0000	3.081	4 1/2	4.500	4.350	5.196	4.959	2	2.060	1.815	0.19	0.06	6.250	6.500
3 1/4	3.2500	3.335	4 7/8	4.875	4.712	5.629	5.372	2 3/16	2.251	1.936	0.19	0.06	6.750	7.000
3 1/2	3.5000	3.589	5 1/4	5.250	5.075	6.062	5.786	2 5/16	2.380	2.057	0.19	0.06	7.250	7.500
3 3/4	3.7500	3.858	5 5/8	5.625	5.437	6.495	6.198	2 1/2	2.572	2.241	0.19	0.06	7.750	8.000
4	4.0000	4.111	6	6.000	5.800	6.928	6.612	2 11/16	2.764	2.424	0.19	0.06	8.250	8.500

Note: L_G is the grip gaging length (nominal bolt length minus the basic thread length L_T).
L_T is the basic thread length and is the distance from the extreme end of the bolt to the last complete (full form) thread.
Bold type indicates products unified dimensionally with British and Canadian standards.
For additional requirements see ANSI B18.2.1.

(ANSI)

Finished Hex Bolts

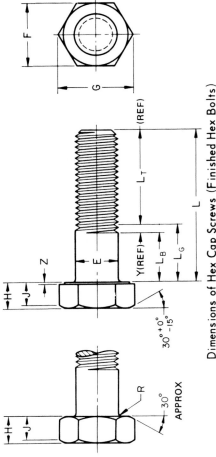

APPROX

30° + 0° − 15°

Dimensions of Hex Cap Screws (Finished Hex Bolts)

Nominal Size or Basic Product Dia		E Body Dia		F Width Across Flats			G Width Across Corners		H Height			J Wrenching Height	L_T Thread Length For Screw Lengths		Y Transition Thread Length For Screw Lengths		Z Runout of Bearing Surface FIR
		Max	Min	Basic	Max	Min	Max	Min	Basic	Max	Min	Min	6 in. and Shorter Basic	Over 6 in. Basic	6 in. and Shorter Max	Over 6 in. Max	Max
1/4	0.2500	0.2500	0.2450	7/16	0.438	0.428	0.505	0.488	5/32	0.163	0.150	0.106	0.750	1.000	0.400	0.650	0.010
5/16	0.3125	0.3125	0.3065	1/2	0.500	0.489	0.577	0.557	13/64	0.211	0.195	0.140	0.875	1.125	0.417	0.667	0.011
3/8	0.3750	0.3750	0.3690	9/16	0.562	0.551	0.650	0.628	15/64	0.243	0.226	0.160	1.000	1.250	0.438	0.688	0.012
7/16	0.4375	0.4375	0.4305	5/8	0.625	0.612	0.722	0.698	9/32	0.291	0.272	0.195	1.125	1.375	0.464	0.714	0.013
1/2	0.5000	0.5000	0.4930	3/4	0.750	0.736	0.866	0.840	5/16	0.323	0.302	0.215	1.250	1.500	0.481	0.731	0.014
9/16	0.5625	0.5625	0.5545	13/16	0.812	0.798	0.938	0.910	23/64	0.371	0.348	0.250	1.375	1.625	0.750	0.750	0.015
5/8	0.6250	0.6250	0.6170	15/16	0.938	0.922	1.083	1.051	25/64	0.403	0.378	0.269	1.500	1.750	0.773	0.773	0.017
3/4	0.7500	0.7500	0.7410	1 1/8	1.125	1.100	1.299	1.254	15/32	0.483	0.455	0.324	1.750	2.000	0.800	0.800	0.020
7/8	0.8750	0.8750	0.8660	1 5/16	1.312	1.285	1.516	1.465	35/64	0.563	0.531	0.378	2.000	2.250	0.833	0.833	0.023
1	1.0000	1.0000	0.9900	1 1/2	1.500	1.469	1.732	1.675	39/64	0.627	0.591	0.416	2.250	2.500	0.875	0.875	0.026
1 1/8	1.1250	1.1250	1.1140	1 11/16	1.688	1.631	1.949	1.859	11/16	0.718	0.658	0.461	2.500	2.750	0.929	0.929	0.029
1 1/4	1.2500	1.2500	1.2390	1 7/8	1.875	1.812	2.165	2.066	25/32	0.813	0.749	0.530	2.750	3.000	0.929	0.929	0.033
1 3/8	1.3750	1.3750	1.3630	2 1/16	2.062	1.994	2.382	2.273	27/32	0.878	0.810	0.569	3.000	3.250	1.000	1.000	0.036
1 1/2	1.5000	1.5000	1.4880	2 1/4	2.250	2.175	2.598	2.480	15/16	0.974	0.902	0.640	3.250	3.500	1.000	1.000	0.039
1 3/4	1.7500	1.7500	1.7380	2 5/8	2.625	2.538	3.031	2.893	1 3/32	1.134	1.054	0.748	3.750	4.000	1.100	1.100	0.046
2	2.0000	2.0000	1.9880	3	3.000	2.900	3.464	3.306	1 7/32	1.263	1.175	0.825	4.250	4.500	1.167	1.167	0.052
2 1/4	2.2500	2.2500	2.2380	3 3/8	3.375	3.262	3.897	3.719	1 3/8	1.423	1.327	0.933	4.750	5.000	1.167	1.167	0.059
2 1/2	2.5000	2.5000	2.4880	3 3/4	3.750	3.625	4.330	4.133	1 17/32	1.583	1.479	1.042	5.250	5.500	1.250	1.250	0.065
2 3/4	2.7500	2.7500	2.7380	4 1/8	4.125	3.988	4.763	4.546	1 11/16	1.744	1.632	1.151	5.750	6.000	1.250	1.250	0.072
3	3.0000	3.0000	2.9880	4 1/2	4.500	4.350	5.196	4.959	1 7/8	1.935	1.815	1.290	6.250	6.500	1.250	1.250	0.079

Note: L_B is the body length from the underside of the head to the last scratch of thread.
L_G is the grip gaging length (nominal bolt length minus the basic thread length L_T).
L_T is the basic thread length and is the distance from the extreme end of the bolt to the last complete (full form) thread.
Bold type indicates products unified dimensionally with British and Canadian standards.
For additional requirements see ANSI B18.2.1.

(ANSI)

Square Nuts

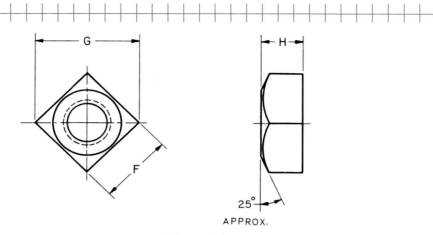

Dimensions of Square Nuts

Nominal Size or Basic Major Dia of Thread		F Width Across Flats			G Width Across Corners		H Thickness		
		Basic	Max	Min	Max	Min	Basic	Max	Min
1/4	0.2500	7/16	0.438	0.425	0.619	0.584	7/32	0.235	0.203
5/16	0.3125	9/16	0.562	0.547	0.795	0.751	17/64	0.283	0.249
3/8	0.3750	5/8	0.625	0.606	0.884	0.832	21/64	0.346	0.310
7/16	0.4375	3/4	0.750	0.728	1.061	1.000	3/8	0.394	0.356
1/2	0.5000	13/16	0.812	0.788	1.149	1.082	7/16	0.458	0.418
5/8	0.6250	1	1.000	0.969	1.414	1.330	35/64	0.569	0.525
3/4	0.7500	1 1/8	1.125	1.088	1.591	1.494	21/32	0.680	0.632
7/8	0.8750	1 5/16	1.312	1.269	1.856	1.742	49/64	0.792	0.740
1	1.0000	1 1/2	1.500	1.450	2.121	1.991	7/8	0.903	0.847
1 1/8	1.1250	1 11/16	1.688	1.631	2.386	2.239	1	1.030	0.970
1 1/4	1.2500	1 7/8	1.875	1.812	2.652	2.489	1 3/32	1.126	1.062
1 3/8	1.3750	2 1/16	2.062	1.994	2.917	2.738	1 13/64	1.237	1.169
1 1/2	1.5000	2 1/4	2.250	2.175	3.182	2.986	1 5/16	1.348	1.276
See Notes	8	3							

(ANSI)   (ANSI)

Hex Flat Nuts and Hex Flat Jam Nuts

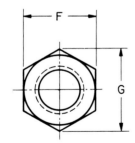  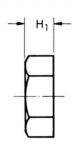

Dimensions of Hex Flat Nuts and Hex Flat Jam Nuts

Nominal Size or Basic Major Dia of Thread		F Width Across Flats			G Width Across Corners		H Thickness Hex Flat Nuts			H_1 Thickness Hex Flat Jam Nuts		
		Basic	Max	Min	Max	Min	Basic	Max	Min	Basic	Max	Min
1 1/8	1.1250	1 11/16	1.688	1.631	1.949	1.859	1	1.030	0.970	5/8	0.655	0.595
1 1/4	1.2500	1 7/8	1.875	1.812	2.165	2.066	1 3/32	1.126	1.062	3/4	0.782	0.718
1 3/8	1.3750	2 1/16	2.062	1.994	2.382	2.273	1 13/64	1.237	1.169	13/16	0.846	0.778
1 1/2	1.5000	2 1/4	2.250	2.175	2.598	2.480	1 5/16	1.348	1.276	7/8	0.911	0.839
See Notes	10	4										

(ANSI)

Hex Nuts and Hex Jam Nuts

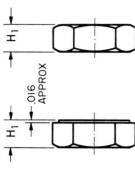

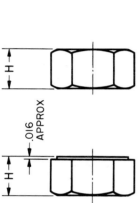

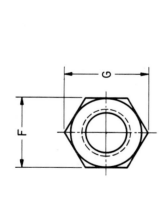

Dimensions of Hex Nuts and Hex Jam Nuts

Nominal Size or Basic Major Dia of Thread		F Width Across Flats			G Width Across Corners		H Thickness Hex Nuts			H₁ Thickness Hex Jam Nuts			Hex Nuts Specified Proof Load — Runout of Bearing Face, FIR Max		Jam Nuts All Strength Levels
		Basic	Max	Min	Max	Min	Basic	Max	Min	Basic	Max	Min	Up to 150,000 psi	150,000 psi and Greater	
1/4	0.2500	7/16	0.438	0.428	0.505	0.488	7/32	0.226	0.212	5/32	0.163	0.150	0.015	0.010	0.015
5/16	0.3125	1/2	0.500	0.489	0.577	0.557	17/64	0.273	0.258	3/16	0.195	0.180	0.016	0.011	0.016
3/8	0.3750	9/16	0.562	0.551	0.650	0.628	21/64	0.337	0.320	7/32	0.227	0.210	0.017	0.012	0.017
7/16	0.4375	11/16	0.688	0.675	0.794	0.768	3/8	0.385	0.365	1/4	0.260	0.240	0.018	0.013	0.018
1/2	0.5000	3/4	0.750	0.736	0.866	0.840	7/16	0.448	0.427	5/16	0.323	0.302	0.019	0.014	0.019
9/16	0.5625	7/8	0.875	0.861	1.010	0.982	31/64	0.496	0.473	5/16	0.324	0.301	0.020	0.015	0.020
5/8	0.6250	15/16	0.938	0.922	1.083	1.051	35/64	0.559	0.535	3/8	0.387	0.363	0.021	0.016	0.021
3/4	0.7500	1 1/8	1.125	1.088	1.299	1.240	41/64	0.665	0.617	27/64	0.446	0.398	0.023	0.018	0.023
7/8	0.8750	1 5/16	1.312	1.269	1.516	1.447	3/4	0.776	0.724	31/64	0.510	0.458	0.025	0.020	0.025
1	1.0000	1 1/2	1.500	1.450	1.732	1.653	55/64	0.887	0.831	35/64	0.575	0.519	0.027	0.022	0.027
1 1/8	1.1250	1 11/16	1.688	1.631	1.949	1.859	31/32	0.999	0.939	39/64	0.639	0.579	0.030	0.025	0.030
1 1/4	1.2500	1 7/8	1.875	1.812	2.165	2.066	1 1/16	1.094	1.030	23/32	0.751	0.687	0.033	0.028	0.033
1 3/8	1.3750	2 1/16	2.062	1.994	2.382	2.273	1 11/64	1.206	1.138	25/32	0.815	0.747	0.036	0.031	0.036
1 1/2	1.5000	2 1/4	2.250	2.175	2.598	2.480	1 9/32	1.317	1.245	27/32	0.880	0.808	0.039	0.034	0.039

Note: Bold type indicates products unified dimensionally with British and Canadian standards.
For additional requirements see ANSI B18.2.1.
(ANSI)

Slotted Flat Countersunk Head Cap Screws

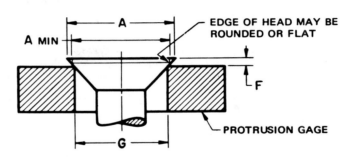

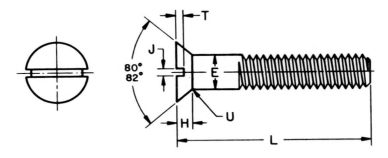

Dimensions of Slotted Flat Countersunk Head Cap Screws

Nominal Size[1] or Basic Screw Diameter		E Body Diameter		A Head Diameter		H[2] Head Height	J Slot Width		T Slot Depth		U Fillet Radius	F[3] Protrusion Above Gaging Diameter		G[3] Gaging Diameter
		Max	Min	Max, Edge Sharp	Min, Edge Rounded or Flat	Ref	Max	Min	Max	Min	Max	.Max	Min	
1/4	0.2500	0.2500	0.2450	0.500	0.452	0.140	0.075	0.064	0.068	0.045	0.100	0.046	0.030	0.424
5/16	0.3125	0.3125	0.3070	0.625	0.567	0.177	0.084	0.072	0.086	0.057	0.125	0.053	0.035	0.538
3/8	0.3750	0.3750	0.3690	0.750	0.682	0.210	0.094	0.081	0.103	0.068	0.150	0.060	0.040	0.651
7/16	0.4375	0.4375	0.4310	0.812	0.736	0.210	0.094	0.081	0.103	0.068	0.175	0.065	0.044	0.703
1/2	0.5000	0.5000	0.4930	0.875	0.791	0.210	0.106	0.091	0.103	0.068	0.200	0.071	0.049	0.756
9/16	0.5625	0.5625	0.5550	1.000	0.906	0.244	0.118	0.102	0.120	0.080	0.225	0.078	0.054	0.869
5/8	0.6250	0.6250	0.6170	1.125	1.020	0.281	0.133	0.116	0.137	0.091	0.250	0.085	0.058	0.982
3/4	0.7500	0.7500	0.7420	1.375	1.251	0.352	0.149	0.131	0.171	0.115	0.300	0.099	0.068	1.208
7/8	0.8750	0.8750	0.8660	1.625	1.480	0.423	0.167	0.147	0.206	0.138	0.350	0.113	0.077	1.435
1	1.0000	1.0000	0.9900	1.875	1.711	0.494	0.188	0.166	0.240	0.162	0.400	0.127	0.087	1.661
1 1/8	1.1250	1.1250	1.1140	2.062	1.880	0.529	0.196	0.178	0.257	0.173	0.450	0.141	0.096	1.826
1 1/4	1.2500	1.2500	1.2390	2.312	2.110	0.600	0.211	0.193	0.291	0.197	0.500	0.155	0.105	2.052
1 3/8	1.3750	1.3750	1.3630	2.562	2.340	0.665	0.226	0.208	0.326	0.220	0.550	0.169	0.115	2.279
1 1/2	1.5000	1.5000	1.4880	2.812	2.570	0.742	0.258	0.240	0.360	0.244	0.600	0.183	0.124	2.505

[1] Where specifying nominal size in decimals, zeros preceding decimal and in the fourth decimal place shall be omitted.

[2] Tabulated values determined from formula for maximum H, Appendix III.

[3] No tolerance for gaging diameter is given. If the gaging diameter of the gage used differs from tabulated value, the protrusion will be affected accordingly and the proper protrusion values must be recalculated using the formulas shown in Appendix II.

(ANSI)

Slotted Round Head Cap Screws

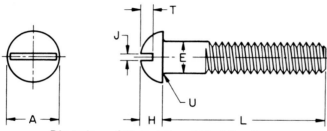

Dimensions of Slotted Round Head Cap Screws

Nominal Size[1] or Basic Screw Diameter		E Body Diameter		A Head Diameter		H Head Height		J Slot Width		T Slot Depth		U Fillet Radius	
		Max	Min	Max	Min	Max	Min	Max	Min	Max	Min	Max	Min
1/4	0.2500	0.2500	0.2450	0.437	0.418	0.191	0.175	0.075	0.064	0.117	0.097	0.031	0.016
5/16	0.3125	0.3125	0.3070	0.562	0.540	0.245	0.226	0.084	0.072	0.151	0.126	0.031	0.016
3/8	0.3750	0.3750	0.3690	0.625	0.603	0.273	0.252	0.094	0.081	0.168	0.138	0.031	0.016
7/16	0.4375	0.4375	0.4310	0.750	0.725	0.328	0.302	0.094	0.081	0.202	0.167	0.047	0.016
1/2	0.5000	0.5000	0.4930	0.812	0.786	0.354	0.327	0.106	0.091	0.218	0.178	0.047	0.016
9/16	0.5625	0.5625	0.5550	0.937	0.909	0.409	0.378	0.118	0.102	0.252	0.207	0.047	0.016
5/8	0.6250	0.6250	0.6170	1.000	0.970	0.437	0.405	0.133	0.116	0.270	0.220	0.062	0.031
3/4	0.7500	0.7500	0.7420	1.250	1.215	0.546	0.507	0.149	0.131	0.338	0.278	0.062	0.031

[1] Where specifying nominal size in decimals, zeros preceding decimal and in the fourth decimal place shall be omitted.
(ANSI)

Slotted Fillister Head Cap Screws

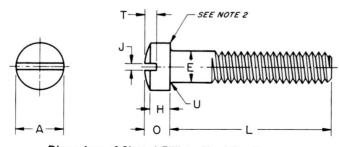

Dimensions of Slotted Fillister Head Cap Screws

Nominal Size[1] or Basic Screw Diameter		E Body Diameter		A Head Diameter		H Head Side Height		O Total Head Height		J Slot Width		T Slot Depth		U Fillet Radius	
		Max	Min	Max	Min	Max	Min	Max	Min	Max	Min	Max	Min	Max	Min
1/4	0.2500	0.2500	0.2450	0.375	0.363	0.172	0.157	0.216	0.194	0.075	0.064	0.097	0.077	0.031	0.016
5/16	0.3125	0.3125	0.3070	0.437	0.424	0.203	0.186	0.253	0.230	0.084	0.072	0.115	0.090	0.031	0.016
3/8	0.3750	0.3750	0.3690	0.562	0.547	0.250	0.229	0.314	0.284	0.094	0.081	0.142	0.112	0.031	0.016
7/16	0.4375	0.4375	0.4310	0.625	0.608	0.297	0.274	0.368	0.336	0.094	0.081	0.168	0.133	0.047	0.016
1/2	0.5000	0.5000	0.4930	0.750	0.731	0.328	0.301	0.413	0.376	0.106	0.091	0.193	0.153	0.047	0.016
9/16	0.5625	0.5625	0.5550	0.812	0.792	0.375	0.346	0.467	0.427	0.118	0.102	0.213	0.168	0.047	0.016
5/8	0.6250	0.6250	0.6170	0.875	0.853	0.422	0.391	0.521	0.478	0.133	0.116	0.239	0.189	0.062	0.031
3/4	0.7500	0.7500	0.7420	1.000	0.976	0.500	0.466	0.612	0.566	0.149	0.131	0.283	0.223	0.062	0.031
7/8	0.8750	0.8750	0.8660	1.125	1.098	0.594	0.556	0.720	0.668	0.167	0.147	0.334	0.264	0.062	0.031
1	1.0000	1.0000	0.9900	1.312	1.282	0.656	0.612	0.803	0.743	0.188	0.166	0.371	0.291	0.062	0.031

[1] Where specifying nominal size in decimals, zeros preceding decimal and in the fourth decimal place shall be omitted.
[2] A slight rounding of the edges at periphery of head shall be permissible provided the diameter of the bearing circle is equal to no less than 90 per cent of the specified minimum head diameter.
(ANSI)

Slotted Headless Set Screws

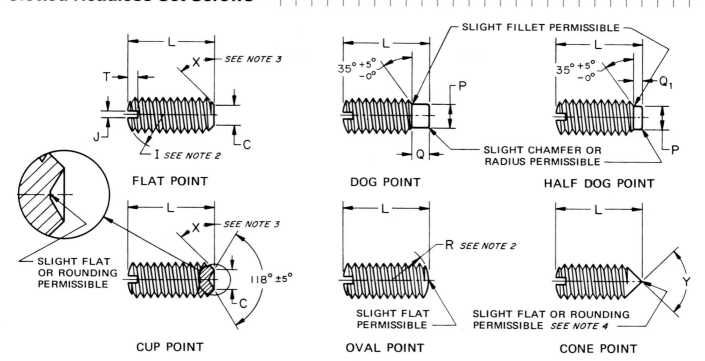

FLAT POINT   DOG POINT   HALF DOG POINT

CUP POINT   OVAL POINT   CONE POINT

Dimensions of Slotted Headless Set Screws

Nominal Size[1] or Basic Screw Diameter		I^2 Crown Radius	J Slot Width		T Slot Depth		C Cup and Flat Point Diameters		P Dog Point Diameters		Q Point Length Dog		Q_1 Point Length Half Dog		R^2 Oval Point Radius	Y Cone Point Angle $90° \pm 2°$ For These Nominal Lengths or Longer; $118° \pm 2°$ For Shorter Screws
		Basic	Max	Min	Max	Min	Max	Min	Max	Min	Max	Min	Max	Min	Basic	
0	0.0600	0.060	0.014	0.010	0.020	0.016	0.033	0.027	0.040	0.037	0.032	0.028	0.017	0.013	0.045	5/64
1	0.0730	0.073	0.016	0.012	0.020	0.016	0.040	0.033	0.049	0.045	0.040	0.036	0.021	0.017	0.055	3/32
2	0.0860	0.086	0.018	0.014	0.025	0.019	0.047	0.039	0.057	0.053	0.046	0.042	0.024	0.020	0.064	7/64
3	0.0990	0.099	0.020	0.016	0.028	0.022	0.054	0.045	0.066	0.062	0.052	0.048	0.027	0.023	0.074	1/8
4	0.1120	0.112	0.024	0.018	0.031	0.025	0.061	0.051	0.075	0.070	0.058	0.054	0.030	0.026	0.084	5/32
5	0.1250	0.125	0.026	0.020	0.036	0.026	0.067	0.057	0.083	0.078	0.063	0.057	0.033	0.027	0.094	3/16
6	0.1380	0.138	0.028	0.022	0.040	0.030	0.074	0.064	0.092	0.087	0.073	0.067	0.038	0.032	0.104	3/16
8	0.1640	0.164	0.032	0.026	0.046	0.036	0.087	0.076	0.109	0.103	0.083	0.077	0.043	0.037	0.123	1/4
10	0.1900	0.190	0.035	0.029	0.053	0.043	0.102	0.088	0.127	0.120	0.095	0.085	0.050	0.040	0.142	1/4
12	0.2160	0.216	0.042	0.035	0.061	0.051	0.115	0.101	0.144	0.137	0.115	0.105	0.060	0.050	0.162	5/16
1/4	0.2500	0.250	0.049	0.041	0.068	0.058	0.132	0.118	0.156	0.149	0.130	0.120	0.068	0.058	0.188	5/16
5/16	0.3125	0.312	0.055	0.047	0.083	0.073	0.172	0.156	0.203	0.195	0.161	0.151	0.083	0.073	0.234	3/8
3/8	0.3750	0.375	0.068	0.060	0.099	0.089	0.212	0.194	0.250	0.241	0.193	0.183	0.099	0.089	0.281	7/16
7/16	0.4375	0.438	0.076	0.068	0.114	0.104	0.252	0.232	0.297	0.287	0.224	0.214	0.114	0.104	0.328	1/2
1/2	0.5000	0.500	0.086	0.076	0.130	0.120	0.291	0.270	0.344	0.334	0.255	0.245	0.130	0.120	0.375	9/16
9/16	0.5625	0.562	0.096	0.086	0.146	0.136	0.332	0.309	0.391	0.379	0.287	0.275	0.146	0.134	0.422	5/8
5/8	0.6250	0.625	0.107	0.097	0.161	0.151	0.371	0.347	0.469	0.456	0.321	0.305	0.164	0.148	0.469	3/4
3/4	0.7500	0.750	0.134	0.124	0.193	0.183	0.450	0.425	0.562	0.549	0.383	0.367	0.196	0.180	0.562	7/8

[1] Where specifying nominal size in decimals, zeros preceding decimal and in the fourth decimal place shall be omitted.

[2] Tolerance on radius for nominal sizes up to and including 5 (0.125 in.) shall be plus 0.015 in. and minus 0.000, and for larger sizes, plus 0.031 in. and minus 0.000. Slotted ends on screws may be flat at option of manufacturer.

[3] Point angle X shall be 45° plus 5°, minus 0°, for screws of nominal lengths equal to or longer than those listed in Column Y, and 30° minimum for screws of shorter nominal lengths.

[4] The extent of rounding or flat at apex of cone point shall not exceed an amount equivalent to 10 per cent of the basic screw diameter.

(ANSI)

Square Head Set Screws

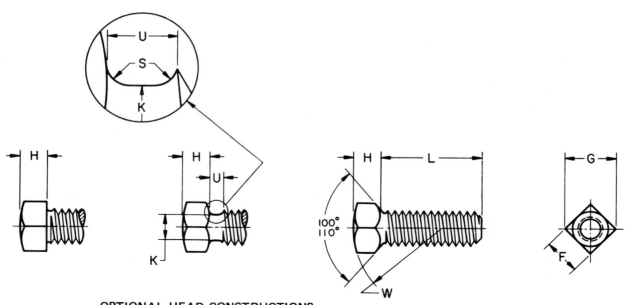

OPTIONAL HEAD CONSTRUCTIONS

Dimensions of Square Head Set Screws

Nominal Size[1] or Basic Screw Diameter		F Width Across Flats		G Width Across Corners		H Head Height		K Neck Relief Diameter		S Neck Relief Fillet Radius	U Neck Relief Width	W Head Radius
		Max	Min	Max	Min	Max	Min	Max	Min	Max	Min	Min
10	0.1900	0.188	0.180	0.265	0.247	0.148	0.134	0.145	0.140	0.027	0.083	0.48
1/4	0.2500	0.250	0.241	0.354	0.331	0.196	0.178	0.185	0.170	0.032	0.100	0.62
5/16	0.3125	0.312	0.302	0.442	0.415	0.245	0.224	0.240	0.225	0.036	0.111	0.78
3/8	0.3750	0.375	0.362	0.530	0.497	0.293	0.270	0.294	0.279	0.041	0.125	0.94
7/16	0.4375	0.438	0.423	0.619	0.581	0.341	0.315	0.345	0.330	0.046	0.143	1.09
1/2	0.5000	0.500	0.484	0.707	0.665	0.389	0.361	0.400	0.385	0.050	0.154	1.25
9/16	0.5625	0.562	0.545	0.795	0.748	0.437	0.407	0.454	0.439	0.054	0.167	1.41
5/8	0.6250	0.625	0.606	0.884	0.833	0.485	0.452	0.507	0.492	0.059	0.182	1.56
3/4	0.7500	0.750	0.729	1.060	1.001	0.582	0.544	0.620	0.605	0.065	0.200	1.88
7/8	0.8750	0.875	0.852	1.237	1.170	0.678	0.635	0.731	0.716	0.072	0.222	2.19
1	1.0000	1.000	0.974	1.414	1.337	0.774	0.726	0.838	0.823	0.081	0.250	2.50
1 1/8	1.1250	1.125	1.096	1.591	1.505	0.870	0.817	0.939	0.914	0.092	0.283	2.81
1 1/4	1.2500	1.250	1.219	1.768	1.674	0.966	0.908	1.064	1.039	0.092	0.283	3.12
1 3/8	1.3750	1.375	1.342	1.945	1.843	1.063	1.000	1.159	1.134	0.109	0.333	3.44
1 1/2	1.5000	1.500	1.464	2.121	2.010	1.159	1.091	1.284	1.259	0.109	0.333	3.75

[1]Where specifying nominal size in decimals, zeros preceding decimal and in the fourth decimal place shall be omitted.
(ANSI)

Continued

Square Head Set Screws (Continued)

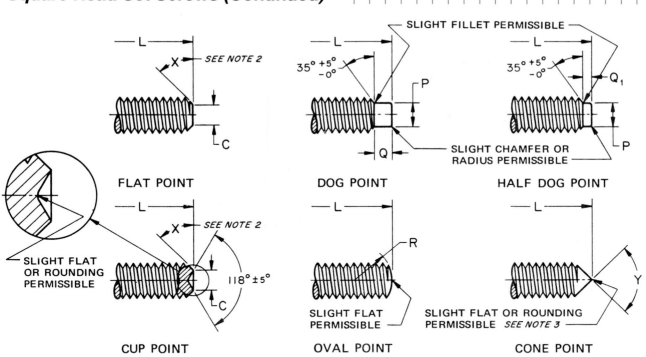

FLAT POINT   DOG POINT   HALF DOG POINT

CUP POINT   OVAL POINT   CONE POINT

Dimensions of Square Head Set Screws (continued)

Nominal Size[1] or Basic Screw Diameter		C Cup and Flat Point Diameters		P Dog and Half Dog Point Diameters		Q Point Length Dog		Q_1 Point Length Half Dog		R Oval Point Radius +0.031 −0.000	Y Cone Point Angle 90° ±2° For These Nominal Lengths or Longer; 118° ±2° For Shorter Screws
		Max	Min	Max	Min	Max	Min	Max	Min		
10	0.1900	0.102	0.088	0.127	0.120	0.095	0.085	0.050	0.040	0.142	1/4
1/4	0.2500	0.132	0.118	0.156	0.149	0.130	0.120	0.068	0.058	0.188	5/16
5/16	0.3125	0.172	0.156	0.203	0.195	0.161	0.151	0.083	0.073	0.234	3/8
3/8	0.3750	0.212	0.194	0.250	0.241	0.193	0.183	0.099	0.089	0.281	7/16
7/16	0.4375	0.252	0.232	0.297	0.287	0.224	0.214	0.114	0.104	0.328	1/2
1/2	0.5000	0.291	0.270	0.344	0.334	0.255	0.245	0.130	0.120	0.375	9/16
9/16	0.5625	0.332	0.309	0.391	0.379	0.287	0.275	0.146	0.134	0.422	5/8
5/8	0.6250	0.371	0.347	0.469	0.456	0.321	0.305	0.164	0.148	0.469	3/4
3/4	0.7500	0.450	0.425	0.562	0.549	0.383	0.367	0.196	0.180	0.562	7/8
7/8	0.8750	0.530	0.502	0.656	0.642	0.446	0.430	0.227	0.211	0.656	1
1	1.0000	0.609	0.579	0.750	0.734	0.510	0.490	0.260	0.240	0.750	1 1/8
1 1/8	1.1250	0.689	0.655	0.844	0.826	0.572	0.552	0.291	0.271	0.844	1 1/4
1 1/4	1.2500	0.767	0.733	0.938	0.920	0.635	0.615	0.323	0.303	0.938	1 1/2
1 3/8	1.3750	0.848	0.808	1.031	1.011	0.698	0.678	0.354	0.334	1.031	1 5/8
1 1/2	1.5000	0.926	0.886	1.125	1.105	0.760	0.740	0.385	0.365	1.125	1 3/4

[1] Where specifying nominal size in decimals, zeros preceding decimal and in the fourth decimal place shall be omitted.

[2] Point angle X shall be 45° plus 5°, minus 0°, for screws of nominal lengths equal to or longer than those listed in Column Y, and 30° minimum for screws of shorter nominal lengths.

[3] The extent of rounding or flat at apex of cone point shall not exceed an amount equivalent to 10 per cent of the basic screw diameter.

End of table

Slotted Flat Countersunk Head Machine Screws

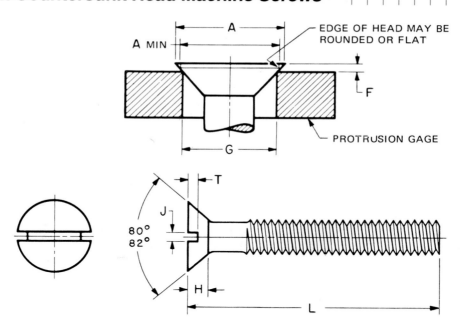

Dimensions of Slotted Flat Countersunk Head Machine Screws

Nominal Size[1] or Basic Screw Diameter		L[2] These Lengths or Shorter are Undercut .	A Head Diameter Max, Edge Sharp	A Head Diameter Min, Edge Rounded or Flat	H[3] Head Height Ref	J Slot Width Max	J Slot Width Min	T Slot Depth Max	T Slot Depth Min	F[4] Protrusion Above Gaging Diameter Max	F[4] Protrusion Above Gaging Diameter Min	G[4] Gaging Diameter
0000	0.0210	—	0.043	0.037	0.011	0.008	0.004	0.007	0.003	*	*	*
000	0.0340	—	0.064	0.058	0.016	0.011	0.007	0.009	0.005	*	*	*
00	0.0470	—	0.093	0.085	0.028	0.017	0.010	0.014	0.009	*	*	*
0	0.0600	1/8	0.119	0.099	0.035	0.023	0.016	0.015	0.010	0.026	0.016	0.078
1	0.0730	1/8	0.146	0.123	0.043	0.026	0.019	0.019	0.012	0.028	0.016	0.101
2	0.0860	1/8	0.172	0.147	0.051	0.031	0.023	0.023	0.015	0.029	0.017	0.124
3	0.0990	1/8	0.199	0.171	0.059	0.035	0.027	0.027	0.017	0.031	0.018	0.148
4	0.1120	3/16	0.225	0.195	0.067	0.039	0.031	0.030	0.020	0.032	0.019	0.172
5	0.1250	3/16	0.252	0.220	0.075	0.043	0.035	0.034	0.022	0.034	0.020	0.196
6	0.1380	3/16	0.279	0.244	0.083	0.048	0.039	0.038	0.024	0.036	0.021	0.220
8	0.1640	1/4	0.332	0.292	0.100	0.054	0.045	0.045	0.029	0.039	0.023	0.267
10	0.1900	5/16	0.385	0.340	0.116	0.060	0.050	0.053	0.034	0.042	0.025	0.313
12	0.2160	3/8	0.438	0.389	0.132	0.067	0.056	0.060	0.039	0.045	0.027	0.362
1/4	0.2500	7/16	0.507	0.452	0.153	0.075	0.064	0.070	0.046	0.050	0.029	0.424
5/16	0.3125	1/2	0.635	0.568	0.191	0.084	0.072	0.088	0.058	0.057	0.034	0.539
3/8	0.3750	9/16	0.762	0.685	0.230	0.094	0.081	0.106	0.070	0.065	0.039	0.653
7/16	0.4375	5/8	0.812	0.723	0.223	0.094	0.081	0.103	0.066	0.073	0.044	0.690
1/2	0.5000	3/4	0.875	0.775	0.223	0.106	0.091	0.103	0.065	0.081	0.049	0.739
9/16	0.5625	—	1.000	0.889	0.260	0.118	0.102	0.120	0.077	0.089	0.053	0.851
5/8	0.6250	—	1.125	1.002	0.298	0.133	0.116	0.137	0.088	0.097	0.058	0.962
3/4	0.7500	—	1.375	1.230	0.372	0.149	0.131	0.171	0.111	0.112	0.067	1.186

[1] Where specifying nominal size in decimals, zeros preceding decimal and in the fourth decimal place shall be omitted.
[2] Screws of these lengths and shorter shall have undercut heads as shown in Table 5.
[3] Tabulated values determined from formula for maximum H, Appendix V.
[4] No tolerance for gaging diameter is given. If the gaging diameter of the gage used differs from tabulated value, the protrusion will be affected accordingly and the proper protrusion values must be recalculated using the formulas shown in Appendix I.
*Not practical to gage.
(ANSI)

Cross Recessed Flat Countersunk Head Machine Screws

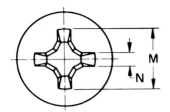

This type of recess has a large center opening, tapered wings, and blunt bottom, with all edges relieved or rounded.

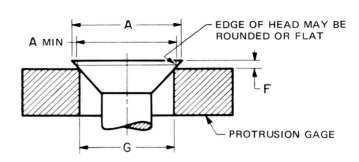

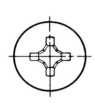

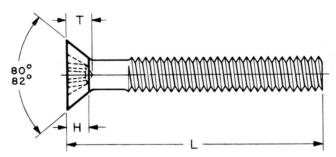

Dimensions of Type I Cross Recessed Flat Countersunk Head Machine Screws

Nominal Size[1] or Basic Screw Diameter		L[2] These Lengths or Shorter are Undercut	A Head Diameter Max, Edge Sharp	A Head Diameter Min, Edge Rounded or Flat	H[3] Head Height Ref	M Recess Diameter Max	M Recess Diameter Min	T Recess Depth Max	T Recess Depth Min	N Recess Width Min	Driver Size	Recess Penetration Gaging Depth Max	Recess Penetration Gaging Depth Min	F[4] Protrusion Above Gaging Diameter Max	F[4] Protrusion Above Gaging Diameter Min	G[4] Gaging Diameter
0	0.0600	1/8	0.119	0.099	0.035	0.069	0.056	0.043	0.027	0.014	0	0.036	0.020	0.026	0.016	0.078
1	0.0730	1/8	0.146	0.123	0.043	0.077	0.064	0.051	0.035	0.015	0	0.044	0.028	0.028	0.016	0.101
2	0.0860	1/8	0.172	0.147	0.051	0.102	0.089	0.063	0.047	0.017	1	0.056	0.040	0.029	0.017	0.124
3	0.0990	1/8	0.199	0.171	0.059	0.107	0.094	0.068	0.052	0.018	1	0.061	0.045	0.031	0.018	0.148
4	0.1120	3/16	0.225	0.195	0.067	0.128	0.115	0.089	0.073	0.018	1	0.082	0.066	0.032	0.019	0.172
5	0.1250	3/16	0.252	0.220	0.075	0.154	0.141	0.086	0.063	0.027	2	0.075	0.052	0.034	0.020	0.196
6	0.1380	3/16	0.279	0.244	0.083	0.174	0.161	0.106	0.083	0.029	2	0.095	0.072	0.036	0.021	0.220
8	0.1640	1/4	0.332	0.292	0.100	0.189	0.176	0.121	0.098	0.030	2	0.110	0.087	0.039	0.023	0.267
10	0.1900	5/16	0.385	0.340	0.116	0.204	0.191	0.136	0.113	0.032	2	0.125	0.102	0.042	0.025	0.313
12	0.2160	3/8	0.438	0.389	0.132	0.268	0.255	0.156	0.133	0.035	3	0.139	0.116	0.045	0.027	0.362
1/4	0.2500	7/16	0.507	0.452	0.153	0.283	0.270	0.171	0.148	0.036	3	0.154	0.131	0.050	0.029	0.424
5/16	0.3125	1/2	0.635	0.568	0.191	0.365	0.352	0.216	0.194	0.061	4	0.196	0.174	0.057	0.034	0.539
3/8	0.3750	9/16	0.762	0.685	0.230	0.393	0.380	0.245	0.223	0.065	4	0.225	0.203	0.065	0.039	0.653
7/16	0.4375	5/8	0.812	0.723	0.223	0.409	0.396	0.261	0.239	0.068	4	0.241	0.219	0.073	0.044	0.690
1/2	0.5000	3/4	0.875	0.775	0.223	0.424	0.411	0.276	0.254	0.069	4	0.256	0.234	0.081	0.049	0.739
9/16	0.5625	—	1.000	0.889	0.260	0.454	0.431	0.300	0.278	0.073	4	0.280	0.258	0.089	0.053	0.851
5/8	0.6250	—	1.125	1.002	0.298	0.576	0.553	0.342	0.316	0.079	5	0.309	0.283	0.097	0.058	0.962
3/4	0.7500	—	1.375	1.230	0.372	0.640	0.617	0.406	0.380	0.087	5	0.373	0.347	0.112	0.067	1.186

[1] Where specifying nominal size in decimals, zeros preceding decimal and in the fourth decimal place shall be omitted.
[2] Screws of these lengths and shorter shall have undercut heads as shown in Table 6.
[3] Tabulated values determined from formula for maximum H, Appendix V.
[4] No tolerance for gaging diameter is given. If the gaging diameter of the gage used differs from tabulated value, the protrusion will be affected accordingly and the proper protrusion values must be recalculated using the formulas shown in Appendix I.

(ANSI)

Slotted and Countersunk Head Machine Screws

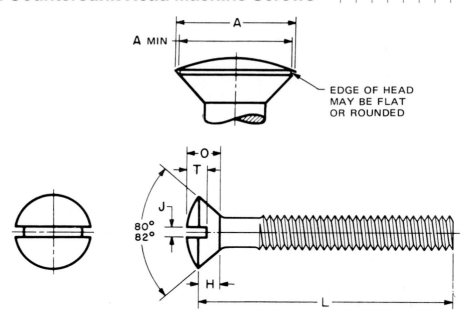

EDGE OF HEAD
MAY BE FLAT
OR ROUNDED

Dimensions of Slotted Oval Countersunk Head Machine Screws

Nominal Size[1] or Basic Screw Diameter		L[2] These Lengths or Shorter are Undercut	A Head Diameter		H[3] Head Side Height	O Total Head Height		J Slot Width		T Slot Depth	
			Max, Edge Sharp	Min, Edge Rounded or Flat	Ref	Max	Min	Max	Min	Max	Min
00	0.0470	—	0.093	0.085	0.028	0.042	0.034	0.017	0.010	0.023	0.016
0	0.0600	1/8	0.119	0.099	0.035	0.056	0.041	0.023	0.016	0.030	0.025
1	0.0730	1/8	0.146	0.123	0.043	0.068	0.052	0.026	0.019	0.038	0.031
2	0.0860	1/8	0.172	0.147	0.051	0.080	0.063	0.031	0.023	0.045	0.037
3	0.0990	1/8	0.199	0.171	0.059	0.092	0.073	0.035	0.027	0.052	0.043
4	0.1120	3/16	0.225	0.195	0.067	0.104	0.084	0.039	0.031	0.059	0.049
5	0.1250	3/16	0.252	0.220	0.075	0.116	0.095	0.043	0.035	0.067	0.055
6	0.1380	3/16	0.279	0.244	0.083	0.128	0.105	0.048	0.039	0.074	0.060
8	0.1640	1/4	0.332	0.292	0.100	0.152	0.126	0.054	0.045	0.088	0.072
10	0.1900	5/16	0.385	0.340	0.116	0.176	0.148	0.060	0.050	0.103	0.084
12	0.2160	3/8	0.438	0.389	0.132	0.200	0.169	0.067	0.056	0.117	0.096
1/4	0.2500	7/16	0.507	0.452	0.153	0.232	0.197	0.075	0.064	0.136	0.112
5/16	0.3125	1/2	0.635	0.568	0.191	0.290	0.249	0.084	0.072	0.171	0.141
3/8	0.3750	9/16	0.762	0.685	0.230	0.347	0.300	0.094	0.081	0.206	0.170
7/16	0.4375	5/8	0.812	0.723	0.223	0.345	0.295	0.094	0.081	0.210	0.174
1/2	0.5000	3/4	0.875	0.775	0.223	0.354	0.299	0.106	0.091	0.216	0.176
9/16	0.5625	—	1.000	0.889	0.260	0.410	0.350	0.118	0.102	0.250	0.207
5/8	0.6250	—	1.125	1.002	0.298	0.467	0.399	0.133	0.116	0.285	0.235
3/4	0.7500	—	1.375	1.230	0.372	0.578	0.497	0.149	0.131	0.353	0.293

[1] Where specifying nominal size in decimals, zeros preceding decimal and in the fourth decimal place shall be omitted.
[2] Screws of these lengths and shorter shall have undercut heads as shown in Table 24.
[3] Tabulated values determined from formula for maximum H, Appendix V.

(ANSI)

Slotted Pan Head Machine Screws

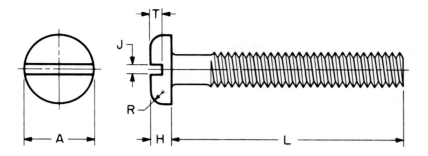

Dimensions of Slotted Pan Head Machine Screws

Nominal Size[1] or Basic Screw Diameter		A Head Diameter		H Head Height		R Head Radius	J Slot Width		T Slot Depth	
		Max	Min	Max	Min	Max	Max	Min	Max	Min
0000	0.0210	0.042	0.036	0.016	0.010	0.007	0.008	0.004	0.008	0.004
000	0.0340	0.066	0.060	0.023	0.017	0.010	0.012	0.008	0.012	0.008
00	0.0470	0.090	0.082	0.032	0.025	0.015	0.017	0.010	0.016	0.010
0	0.0600	0.116	0.104	0.039	0.031	0.020	0.023	0.016	0.022	0.014
1	0.0730	0.142	0.130	0.046	0.038	0.025	0.026	0.019	0.027	0.018
2	0.0860	0.167	0.155	0.053	0.045	0.035	0.031	0.023	0.031	0.022
3	0.0990	0.193	0.180	0.060	0.051	0.037	0.035	0.027	0.036	0.026
4	0.1120	0.219	0.205	0.068	0.058	0.042	0.039	0.031	0.040	0.030
5	0.1250	0.245	0.231	0.075	0.065	0.044	0.043	0.035	0.045	0.034
6	0.1380	0.270	0.256	0.082	0.072	0.046	0.048	0.039	0.050	0.037
8	0.1640	0.322	0.306	0.096	0.085	0.052	0.054	0.045	0.058	0.045
10	0.1900	0.373	0.357	0.110	0.099	0.061	0.060	0.050	0.068	0.053
12	0.2160	0.425	0.407	0.125	0.112	0.078	0.067	0.056	0.077	0.061
1/4	0.2500	0.492	0.473	0.144	0.130	0.087	0.075	0.064	0.087	0.070
5/16	0.3125	0.615	0.594	0.178	0.162	0.099	0.084	0.072	0.106	0.085
3/8	0.3750	0.740	0.716	0.212	0.195	0.143	0.094	0.081	0.124	0.100
7/16	0.4375	0.863	0.837	0.247	0.228	0.153	0.094	0.081	0.142	0.116
1/2	0.5000	0.987	0.958	0.281	0.260	0.175	0.106	0.091	0.161	0.131
9/16	0.5625	1.041	1.000	0.315	0.293	0.197	0.118	0.102	0.179	0.146
5/8	0.6250	1.172	1.125	0.350	0.325	0.219	0.133	0.116	0.197	0.162
3/4	0.7500	1.435	1.375	0.419	0.390	0.263	0.149	0.131	0.234	0.192

[1] Where specifying nominal size in decimals, zeros preceding decimal and in the fourth decimal place shall be omitted.

(ANSI)

Slotted Fillister Head Machine Screws

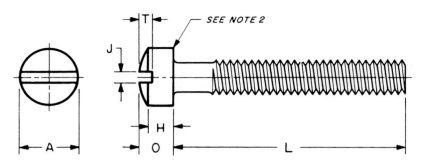

Dimensions of Slotted Fillister Head Machine Screws

Nominal Size[1] or Basic Screw Diameter		A Head Diameter		H Head Side Height		O Total Head Height		J Slot Width		T Slot Depth	
		Max	Min	Max	Min	Max	Min	Max	Min	Max	Min
0000	0.0210	0.038	0.032	0.019	0.011	0.025	0.015	0.008	0.004	0.012	0.006
000	0.0340	0.059	0.053	0.029	0.021	0.035	0.027	0.012	0.006	0.017	0.011
00	0.0470	0.082	0.072	0.037	0.028	0.047	0.039	0.017	0.010	0.022	0.015
0	0.0600	0.096	0.083	0.043	0.038	0.055	0.047	0.023	0.016	0.025	0.015
1	0.0730	0.118	0.104	0.053	0.045	0.066	0.058	0.026	0.019	0.031	0.020
2	0.0860	0.140	0.124	0.062	0.053	0.083	0.066	0.031	0.023	0.037	0.025
3	0.0990	0.161	0.145	0.070	0.061	0.095	0.077	0.035	0.027	0.043	0.030
4	0.1120	0.183	0.166	0.079	0.069	0.107	0.088	0.039	0.031	0.048	0.035
5	0.1250	0.205	0.187	0.088	0.078	0.120	0.100	0.043	0.035	0.054	0.040
6	0.1380	0.226	0.208	0.096	0.086	0.132	0.111	0.048	0.039	0.060	0.045
8	0.1640	0.270	0.250	0.113	0.102	0.156	0.133	0.054	0.045	0.071	0.054
10	0.1900	0.313	0.292	0.130	0.118	0.180	0.156	0.060	0.050	0.083	0.064
12	0.2160	0.357	0.334	0.148	0.134	0.205	0.178	0.067	0.056	0.094	0.074
1/4	0.2500	0.414	0.389	0.170	0.155	0.237	0.207	0.075	0.064	0.109	0.087
5/16	0.3125	0.518	0.490	0.211	0.194	0.295	0.262	0.084	0.072	0.137	0.110
3/8	0.3750	0.622	0.590	0.253	0.233	0.355	0.315	0.094	0.081	0.164	0.133
7/16	0.4375	0.625	0.589	0.265	0.242	0.368	0.321	0.094	0.081	0.170	0.135
1/2	0.5000	0.750	0.710	0.297	0.273	0.412	0.362	0.106	0.091	0.190	0.151
9/16	0.5625	0.812	0.768	0.336	0.308	0.466	0.410	0.118	0.102	0.214	0.172
5/8	0.6250	0.875	0.827	0.375	0.345	0.521	0.461	0.133	0.116	0.240	0.193
3/4	0.7500	1.000	0.945	0.441	0.406	0.612	0.542	0.149	0.131	0.281	0.226

[1] Where specifying nominal size in decimals, zeros preceding decimal and in the fourth decimal place shall be omitted.

[2] A slight rounding of the edges at periphery of head shall be permissible provided the diameter of the bearing circle is equal to no less than 90 per cent of the specified minimum head diameter.

(ANSI)

Plain and Slotted Hex Washer Head Machine Screws

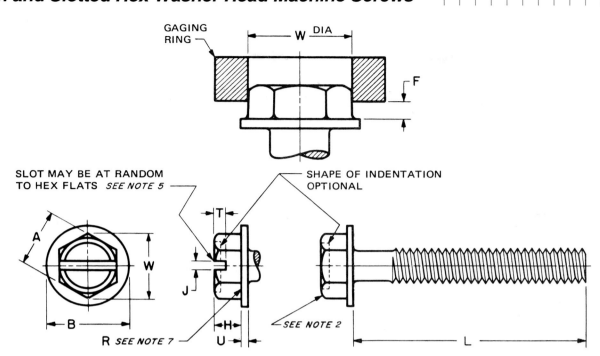

Dimensions of Plain and Slotted Hex Washer Head Machine Screws

Nominal Size[1] or Basic Screw Diameter		A[3] Width Across Flats		W[3,4] Width Across Corners	H Head Height		B Washer Diameter		U Washer Thickness		J[5] Slot Width		T[5,6] Slot Depth		F[4] Protrusion Beyond Gaging Ring
		Max	Min	Min	Max	Min	Max	Min	Max	Min	Max	Min	Max	Min	Min
2	0.0860	0.125	0.120	0.134	0.050	0.040	0.166	0.154	0.016	0.010	—	—	—	—	0.024
3	0.0990	0.125	0.120	0.134	0.055	0.044	0.177	0.163	0.016	0.010	—	—	—	—	0.026
4	0.1120	0.188	0.181	0.202	0.060	0.049	0.243	0.225	0.019	0.011	0.039	0.031	0.042	0.025	0.029
5	0.1250	0.188	0.181	0.202	0.070	0.058	0.260	0.240	0.025	0.015	0.043	0.035	0.049	0.030	0.035
6	0.1380	0.250	0.244	0.272	0.093	0.080	0.328	0.302	0.025	0.015	0.048	0.039	0.053	0.033	0.048
8	0.1640	0.250	0.244	0.272	0.110	0.096	0.348	0.322	0.031	0.019	0.054	0.045	0.074	0.052	0.058
10	0.1900	0.312	0.305	0.340	0.120	0.105	0.414	0.384	0.031	0.019	0.060	0.050	0.080	0.057	0.063
12	0.2160	0.312	0.305	0.340	0.155	0.139	0.432	0.398	0.039	0.022	0.067	0.056	0.103	0.077	0.083
1/4	0.2500	0.375	0.367	0.409	0.190	0.172	0.520	0.480	0.050	0.030	0.075	0.064	0.111	0.083	0.103
5/16	0.3125	0.500	0.489	0.545	0.230	0.208	0.676	0.624	0.055	0.035	0.084	0.072	0.134	0.100	0.125
3/8	0.3750	0.562	0.551	0.614	0.295	0.270	0.780	0.720	0.063	0.037	0.094	0.081	0.168	0.131	0.162

[1] Where specifying nominal size in decimals, zeros preceding decimal and in the fourth decimal place shall be omitted.

[2] A slight rounding of all edges and corners of the hex surfaces shall be permissible.

[3] Dimensions across flats and across corners of the head shall be measured at the point of maximum metal. Taper of sides of hex (angle between one side and the axis) shall not exceed 2 deg or 0.004 in., whichever is greater, the specified width across flats being the large dimension.

[4] The rounding due to lack of fill on all six corners of the head shall be reasonably uniform and the width across corners of the head shall be such that when a sharp ring having an inside diameter equal to the specified minimum width across corners is placed on the top of the head, the hex portion of the head shall protrude by an amount equal to, or greater than, the F value tabulated. See Appendix II for Across Corners Gaging of Hex Heads.

[5] Unless otherwise specified by purchaser, hex washer head machine screws are not slotted.

[6] Slot depth beyond bottom of indentation shall not be less than 1/3 of the minimum slot depth specified.

[7] Fillet radius R at junction of sides of hex and top of washer shall not exceed 0.15 times the basic screw diameter.

(ANSI)

Plain Washers

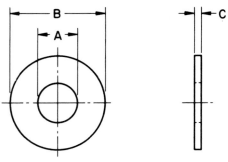

Dimensions of Preferred Sizes of Type A Plain Washers**

Nominal Washer Size***			Inside Diameter A			Outside Diameter B			Thickness C		
			Basic	Tolerance		Basic	Tolerance		Basic	Max	Mi.
				Plus	Minus		Plus	Minus			
—	—		0.078	0.000	0.005	0.188	0.000	0.005	0.020	0.025	0.016
—	—		0.094	0.000	0.005	0.250	0.000	0.005	0.020	0.025	0.016
—	—		0.125	0.008	0.005	0.312	0.008	0.005	0.032	0.040	0.025
No. 6	0.138		0.156	0.008	0.005	0.375	0.015	0.005	0.049	0.065	0.036
No. 8	0.164		0.188	0.008	0.005	0.438	0.015	0.005	0.049	0.065	0.036
No. 10	0.190		0.219	0.008	0.005	0.500	0.015	0.005	0.049	0.065	0.036
$\frac{3}{16}$	0.188		0.250	0.015	0.005	0.562	0.015	0.005	0.049	0.065	0.036
No. 12	0.216		0.250	0.015	0.005	0.562	0.015	0.005	0.065	0.080	0.051
$\frac{1}{4}$	0.250	N	0.281	0.015	0.005	0.625	0.015	0.005	0.065	0.080	0.051
$\frac{1}{4}$	0.250	W	0.312	0.015	0.005	0.734*	0.015	0.007	0.065	0.080	0.051
$\frac{5}{16}$	0.312	N	0.344	0.015	0.005	0.688	0.015	0.007	0.065	0.080	0.051
$\frac{5}{16}$	0.312	W	0.375	0.015	0.005	0.875	0.030	0.007	0.083	0.104	0.064
$\frac{3}{8}$	0.375	N	0.406	0.015	0.005	0.812	0.015	0.007	0.065	0.080	0.051
$\frac{3}{8}$	0.375	W	0.438	0.015	0.005	1.000	0.030	0.007	0.083	0.104	0.064
$\frac{7}{16}$	0.438	N	0.469	0.015	0.005	0.922	0.015	0.007	0.065	0.080	0.051
$\frac{7}{16}$	0.438	W	0.500	0.015	0.005	1.250	0.030	0.007	0.083	0.104	0.064
$\frac{1}{2}$	0.500	N	0.531	0.015	0.005	1.062	0.030	0.007	0.095	0.121	0.074
$\frac{1}{2}$	0.500	W	0.562	0.015	0.005	1.375	0.030	0.007	0.109	0.132	0.086
$\frac{9}{16}$	0.562	N	0.594	0.015	0.005	1.156*	0.030	0.007	0.095	0.121	0.074
$\frac{9}{16}$	0.562	W	0.625	0.015	0.005	1.469*	0.030	0.007	0.109	0.132	0.086
$\frac{5}{8}$	0.625	N	0.656	0.030	0.007	1.312	0.030	0.007	0.095	0.121	0.074
$\frac{5}{8}$	0.625	W	0.688	0.030	0.007	1.750	0.030	0.007	0.134	0.160	0.108
$\frac{3}{4}$	0.750	N	0.812	0.030	0.007	1.469	0.030	0.007	0.134	0.160	0.108
$\frac{3}{4}$	0.750	W	0.812	0.030	0.007	2.000	0.030	0.007	0.148	0.177	0.122
$\frac{7}{8}$	0.875	N	0.938	0.030	0.007	1.750	0.030	0.007	0.134	0.160	0.108
$\frac{7}{8}$	0.875	W	0.938	0.030	0.007	2.250	0.030	0.007	0.165	0.192	0.136
1	1.000	N	1.062	0.030	0.007	2.000	0.030	0.007	0.134	0.160	0.108
1	1.000	W	1.062	0.030	0.007	2.500	0.030	0.007	0.165	0.192	0.136
$1\frac{1}{8}$	1.125	N	1.250	0.030	0.007	2.250	0.030	0.007	0.134	0.160	0.108
$1\frac{1}{8}$	1.125	W	1.250	0.030	0.007	2.750	0.030	0.007	0.165	0.192	0.136
$1\frac{1}{4}$	1.250	N	1.375	0.030	0.007	2.500	0.030	0.007	0.165	0.192	0.136
$1\frac{1}{4}$	1.250	W	1.375	0.030	0.007	3.000	0.030	0.007	0.165	0.192	0.136
$1\frac{3}{8}$	1.375	N	1.500	0.030	0.007	2.750	0.030	0.007	0.165	0.192	0.136
$1\frac{3}{8}$	1.375	W	1.500	0.045	0.010	3.250	0.045	0.010	0.180	0.213	0.153
$1\frac{1}{2}$	1.500	N	1.625	0.030	0.007	3.000	0.030	0.007	0.165	0.192	0.136
$1\frac{1}{2}$	1.500	W	1.625	0.045	0.010	3.500	0.045	0.010	0.180	0.213	0.153
$1\frac{5}{8}$	1.625		1.750	0.045	0.010	3.750	0.045	0.010	0.180	0.213	0.153
$1\frac{3}{4}$	1.750		1.875	0.045	0.010	4.000	0.045	0.010	0.180	0.213	0.153
$1\frac{7}{8}$	1.875		2.000	0.045	0.010	4.250	0.045	0.010	0.180	0.213	0.153
2	2.000		2.125	0.045	0.010	4.500	0.045	0.010	0.180	0.213	0.153
$2\frac{1}{4}$	2.250		2.375	0.045	0.010	4.750	0.045	0.010	0.220	0.248	0.193
$2\frac{1}{2}$	2.500		2.625	0.045	0.010	5.000	0.045	0.010	0.238	0.280	0.210
$2\frac{3}{4}$	2.750		2.875	0.065	0.010	5.250	0.065	0.010	0.259	0.310	0.228
3	3.000		3.125	0.065	0.010	5.500	0.065	0.010	0.284	0.327	0.249

*The 0.734 in., 1.156 in., and 1.469 in. outside diameters avoid washers which could be used in coin operated devices.
**Preferred sizes are for the most part from series previously designated "Standard Plate" and "SAE." Where common sizes existed in the two series, the SAE size is designated "N" (narrow) and the Standard Plate "W" (wide). These sizes as well as all other sizes of Type A Plain Washers are to be ordered by ID, OD, and thickness dimensions.
***Nominal washer sizes are intended for use with comparable nominal screw or bolt sizes.

(ANSI)

Regular Helical Spring Lock Washers

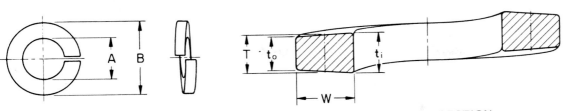

ENLARGED SECTION

Dimensions of Regular Helical Spring Lock Washers[1]

Nominal Washer Size		A Inside Diameter		B Outside Diameter	T Mean Section Thickness $\left(\dfrac{t_i + t_o}{2}\right)$	W Section Width
		Max	Min	Max[2]	Min	Min
No. 2	0.086	0.094	0.088	0.172	0.020	0.035
No. 3	0.099	0.107	0.101	0.195	0.025	0.040
No. 4	0.112	0.120	0.114	0.209	0.025	0.040
No. 5	0.125	0.133	0.127	0.236	0.031	0.047
No. 6	0.138	0.148	0.141	0.250	0.031	0.047
No. 8	0.164	0.174	0.167	0.293	0.040	0.055
No. 10	0.190	0.200	0.193	0.334	0.047	0.062
No. 12	0.216	0.227	0.220	0.377	0.056	0.070
¼	0.250	0.262	0.254	0.489	0.062	0.109
5/16	0.312	0.326	0.317	0.586	0.078	0.125
⅜	0.375	0.390	0.380	0.683	0.094	0.141
7/16	0.438	0.455	0.443	0.779	0.109	0.156
½	0.500	0.518	0.506	0.873	0.125	0.171
9/16	0.562	0.582	0.570	0.971	0.141	0.188
⅝	0.625	0.650	0.635	1.079	0.156	0.203
11/16	0.688	0.713	0.698	1.176	0.172	0.219
¾	0.750	0.775	0.760	1.271	0.188	0.234
13/16	0.812	0.843	0.824	1.367	0.203	0.250
⅞	0.875	0.905	0.887	1.464	0.219	0.266
15/16	0.938	0.970	0.950	1.560	0.234	0.281
1	1.000	1.042	1.017	1.661	0.250	0.297
1 1/16	1.062	1.107	1.080	1.756	0.266	0.312
1⅛	1.125	1.172	1.144	1.853	0.281	0.328
1 3/16	1.188	1.237	1.208	1.950	0.297	0.344
1¼	1.250	1.302	1.271	2.045	0.312	0.359
1 5/16	1.312	1.366	1.334	2.141	0.328	0.375
1⅜	1.375	1.432	1.398	2.239	0.344	0.391
1 7/16	1.438	1.497	1.462	2.334	0.359	0.406
1½	1.500	1.561	1.525	2.430	0.375	0.422

[1] Formerly designated Medium Helical Spring Lock Washers.
[2] The maximum outside diameters specified allow for the commercial tolerances on cold drawn wire.

(ANSI)

Internal Tooth Lock Washers

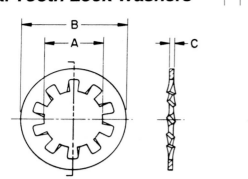

TYPE A

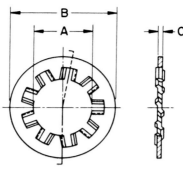

TYPE B

Dimensions of Internal Tooth Lock Washers

Nominal Washer Size		A Inside Diameter		B Outside Diameter		C Thickness	
		Max	Min	Max	Min	Max	Min
No. 2	0.086	0.095	0.089	0.200	0.175	0.015	0.010
No. 3	0.099	0.109	0.102	0.232	0.215	0.019	0.012
No. 4	0.112	0.123	0.115	0.270	0.255	0.019	0.015
No. 5	0.125	0.136	0.129	0.280	0.245	0.021	0.017
No. 6	0.138	0.150	0.141	0.295	0.275	0.021	0.017
No. 8	0.164	0.176	0.168	0.340	0.325	0.023	0.018
No. 10	0.190	0.204	0.195	0.381	0.365	0.025	0.020
No. 12	0.216	0.231	0.221	0.410	0.394	0.025	0.020
¼	0.250	0.267	0.256	0.478	0.460	0.028	0.023
5/16	0.312	0.332	0.320	0.610	0.594	0.034	0.028
3/8	0.375	0.398	0.384	0.692	0.670	0.040	0.032
7/16	0.438	0.464	0.448	0.789	0.740	0.040	0.032
½	0.500	0.530	0.512	0.900	0.867	0.045	0.037
9/16	0.562	0.596	0.576	0.985	0.957	0.045	0.037
5/8	0.625	0.663	0.640	1.071	1.045	0.050	0.042
11/16	0.688	0.728	0.704	1.166	1.130	0.050	0.042
¾	0.750	0.795	0.769	1.245	1.220	0.055	0.047
13/16	0.812	0.861	0.832	1.315	1.290	0.055	0.047
7/8	0.875	0.927	0.894	1.410	1.364	0.060	0.052
1	1.000	1.060	1.019	1.637	1.590	0.067	0.059
1⅛	1.125	1.192	1.144	1.830	1.799	0.067	0.059
1¼	1.250	1.325	1.275	1.975	1.921	0.067	0.059

(ANSI)

External Tooth Lock Washers

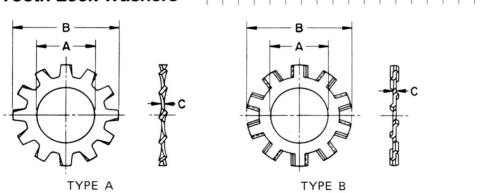

TYPE A   TYPE B

Dimensions of External Tooth Lock Washers

Nominal Washer Size		A Inside Diameter		B Outside Diameter		C Thickness	
		Max	Min	Max	Min	Max	Min
No. 3	0.099	0.109	0.102	0.235	0.220	0.015	0.012
No. 4	0.112	0.123	0.115	0.260	0.245	0.019	0.015
No. 5	0.125	0.136	0.129	0.285	0.270	0.019	0.014
No. 6	0.138	0.150	0.141	0.320	0.305	0.022	0.016
No. 8	0.164	0.176	0.168	0.381	0.365	0.023	0.018
No. 10	0.190	0.204	0.195	0.410	0.395	0.025	0.020
No. 12	0.216	0.231	0.221	0.475	0.460	0.028	0.023
¼	0.250	0.267	0.256	0.510	0.494	0.028	0.023
5/16	0.312	0.332	0.320	0.610	0.588	0.034	0.028
⅜	0.375	0.398	0.384	0.694	0.670	0.040	0.032
7/16	0.438	0.464	0.448	0.760	0.740	0.040	0.032
½	0.500	0.530	0.513	0.900	0.880	0.045	0.037
9/16	0.562	0.596	0.576	0.985	0.960	0.045	0.037
⅝	0.625	0.663	0.641	1.070	1.045	0.050	0.042
11/16	0.688	0.728	0.704	1.155	1.130	0.050	0.042
¾	0.750	0.795	0.768	1.260	1.220	0.055	0.047
13/16	0.812	0.861	0.833	1.315	1.290	0.055	0.047
⅞	0.875	0.927	0.897	1.410	1.380	0.060	0.052
1	1.000	1.060	1.025	1.620	1.590	0.067	0.059

(ANSI)

Countersunk Head Semi-Tubular Rivets

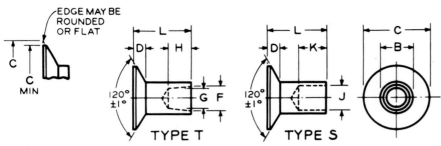

Dimensions of 120 Deg Countersunk Head Semi-Tubular Rivets
(General Purpose)

B		C		D	Type-T			Type-S		
					Taper Hole Rivets			Straight Hole Rivets		
					F	G	H	J		K
Shank Diameter		Head Diameter		Head Thick- ness	Hole Dia at End of Rivet		Hole Dia at Bottom of Hole	Hole Depth to Start of Apex	Hole Dia at End of Rivet		Hole Depth to Start of Apex*
		Max Sharp†	Min, Edge Rounded or Flat								
Min	Max			Ref	Min	Max	Min	Min	Min	Max	Nom
0.085	0.089	0.223	0.203	0.030	0.064	0.068	0.050	0.057	0.062	0.068	0.064
0.118	0.123	0.271	0.245	0.037	0.091	0.095	0.079	0.082	0.084	0.090	0.094
0.141	0.146	0.337	0.307	0.045	0.106	0.112	0.085	0.104	0.100	0.107	0.126
0.182	0.188	0.404	0.369	0.065	0.139	0.145	0.110	0.135	0.134	0.141	0.155
0.210	0.217	0.472	0.430	0.071	0.158	0.166	0.136	0.151	0.155	0.163	0.189
0.244	0.252	0.540	0.493	0.087	0.181	0.191	0.150	0.183	0.176	0.184	0.219

†Head diameter "Max Sharp" is a theoretical dimension determined by projection.
*For rivets having a length tolerance of ±0.015 in. or greater, the straight hole nominal depth shall be increased 0.010 in. over the depth specified.

(ANSI)

Full Tubular Rivets

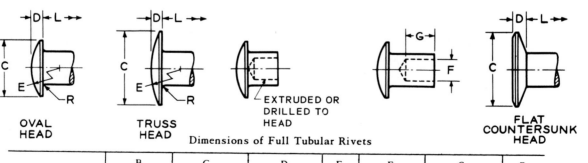

Dimensions of Full Tubular Rivets

Head Style	B		C		D		E	F		G		R
	Shank Diameter		Head Diameter		Head Thickness		Head Radius (Ref)	Diameter of Hole		Depth of Hole*		Fillet Radius
	Min	Max	Basic	Tol	Basic	Tol	Min	Min	Max	Max	Min	Max
Oval	0.141	0.146	0.234	±0.005	0.040	±0.005	0.27	0.100	0.107	0.375	To Head	0.020
Truss	0.141	0.146	0.312	±0.006	0.040	±0.005	0.45	0.100	0.107	0.375	To Head	0.020
	0.182	0.188	0.375	±0.006	0.060	±0.005	0.53	0.134	0.141	0.375	To Head	0.025
Flat Countersunk†	0.141	0.146	0.312	±0.005	0.045	±0.005	——	0.100	0.107	0.375	To Head	——
	0.182	0.188	0.358	±0.006	0.054	±0.006	——	0.134	0.141	0.375	To Head	——

*Full tubular rivets having nominal length of ⅜ in. or shorter shall be drilled or extruded to head.
†Angle of head not specified since it is assumed this type of rivet would generally be used in soft materials and therefore form its own countersink.

(ANSI)

Flat Countersunk Head Rivets

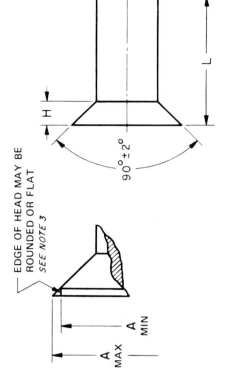

EDGE OF HEAD MAY BE ROUNDED OR FLAT
SEE NOTE 3

90°±2°

Dimensions of Flat Countersunk Head Rivets

Nominal Size[1] or Basic Shank Diameter	E Shank Diameter		A Head Diameter		H Head Height
	Max	Min	Max[2]	Min[3]	Ref[4]
1/16 0.062	0.064	0.059	0.118	0.110	0.027
3/32 0.094	0.096	0.090	0.176	0.163	0.040
1/8 0.125	0.127	0.121	0.235	0.217	0.053
5/32 0.156	0.158	0.152	0.293	0.272	0.066
3/16 0.188	0.191	0.182	0.351	0.326	0.079
7/32 0.219	0.222	0.213	0.413	0.384	0.094
1/4 0.250	0.253	0.244	0.469	0.437	0.106
9/32 0.281	0.285	0.273	0.528	0.491	0.119
5/16 0.312	0.316	0.304	0.588	0.547	0.133
11/32 0.344	0.348	0.336	0.646	0.602	0.146
3/8 0.375	0.380	0.365	0.704	0.656	0.159
13/32 0.406	0.411	0.396	0.763	0.710	0.172
7/16 0.438	0.443	0.428	0.823	0.765	0.186

[1] Where specifying nominal size in decimals, zeros preceding decimal shall be omitted.
[2] Sharp edged head. Tabulated maximum values calculated on basic diameter of rivet and 92° included angle extended to a sharp edge.
[3] Rounded or flat edged irregular shaped head. See Paragraph 2.1 of General Data.
[4] Head height, H, is given for reference purposes only. Variations in this dimension are controlled by the head and shank diameters and the included angle of the head.

(ANSI)

Tinners Rivets

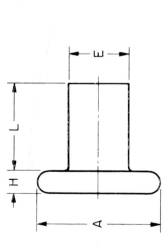

Dimensions of Tinners Rivets

Rivet Size Number[1]	E Shank Diameter		A Head Diameter		H Head Height		L Rivet Length	
	Max	Min	Max	Min	Max	Min	Max	Min
6 oz	0.081	0.075	0.213	0.193	0.028	0.016	0.135	0.115
8 oz	0.091	0.085	0.225	0.205	0.036	0.024	0.166	0.146
10 oz	0.097	0.091	0.250	0.230	0.037	0.025	0.182	0.162
12 oz	0.107	0.101	0.265	0.245	0.037	0.025	0.198	0.178
14 oz	0.111	0.105	0.275	0.255	0.038	0.026	0.198	0.178
1 lb	0.113	0.107	0.285	0.265	0.040	0.028	0.213	0.193
1 1/4 lb	0.122	0.116	0.295	0.275	0.045	0.033	0.229	0.209
1 1/2 lb	0.132	0.126	0.316	0.294	0.046	0.034	0.244	0.224
1 3/4 lb	0.136	0.130	0.331	0.309	0.049	0.035	0.260	0.240
2 lb	0.146	0.140	0.341	0.319	0.050	0.036	0.276	0.256
2 1/2 lb	0.150	0.144	0.311	0.289	0.069	0.055	0.291	0.271
3 lb	0.163	0.154	0.329	0.303	0.073	0.059	0.323	0.303
3 1/2 lb	0.168	0.159	0.348	0.322	0.074	0.060	0.338	0.318
4 lb	0.179	0.170	0.368	0.342	0.076	0.062	0.354	0.334
5 lb	0.190	0.181	0.388	0.362	0.084	0.070	0.385	0.365
6 lb	0.206	0.197	0.419	0.393	0.090	0.076	0.401	0.381
7 lb	0.223	0.214	0.431	0.405	0.094	0.080	0.416	0.396
8 lb	0.227	0.218	0.475	0.445	0.101	0.085	0.448	0.428
9 lb	0.241	0.232	0.490	0.460	0.103	0.087	0.463	0.443
10 lb	0.241	0.232	0.505	0.475	0.104	0.088	0.479	0.459
12 lb	0.263	0.251	0.532	0.498	0.108	0.090	0.510	0.490
14 lb	0.288	0.276	0.577	0.543	0.113	0.095	0.525	0.505
16 lb	0.304	0.292	0.597	0.563	0.128	0.110	0.541	0.521
18 lb	0.347	0.335	0.706	0.668	0.156	0.136	0.603	0.583

[1] Size numbers in ounces and pounds refer to the approximate weight of 1000 rivets.

(ANSI)

Pan Head Rivets

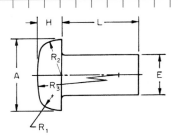

Dimensions of Pan Head Rivets

Nominal Size[1] or Basic Shank Diameter		E Shank Diameter		A Head Diameter		H Head Height		R1 Head Corner Radius	R2 Head Side Radius	R3 Head Crown Radius
		Max	Min	Max	Min	Max	Min	Approx	Approx	Approx
1/16	0.062	0.064	0.059	0.118	0.098	0.040	0.030	0.019	0.052	0.217
3/32	0.094	0.096	0.090	0.173	0.153	0.060	0.048	0.030	0.080	0.326
1/8	0.125	0.127	0.121	0.225	0.205	0.078	0.066	0.039	0.106	0.429
5/32	0.156	0.158	0.152	0.279	0.257	0.096	0.082	0.049	0.133	0.535
3/16	0.188	0.191	0.182	0.334	0.308	0.114	0.100	0.059	0.159	0.641
7/32	0.219	0.222	0.213	0.391	0.365	0.133	0.119	0.069	0.186	0.754
1/4	0.250	0.253	0.244	0.444	0.414	0.151	0.135	0.079	0.213	0.858
9/32	0.281	0.285	0.273	0.499	0.465	0.170	0.152	0.088	0.239	0.963
5/16	0.312	0.316	0.304	0.552	0.518	0.187	0.169	0.098	0.266	1.070
11/32	0.344	0.348	0.336	0.608	0.570	0.206	0.186	0.108	0.292	1.176
3/8	0.375	0.380	0.365	0.663	0.625	0.225	0.205	0.118	0.319	1.286
13/32	0.406	0.411	0.396	0.719	0.675	0.243	0.221	0.127	0.345	1.392
7/16	0.438	0.443	0.428	0.772	0.728	0.261	0.239	0.137	0.372	1.500

1 Where specifying nominal size in decimals, zeros preceding decimal shall be omitted.

(ANSI)

Key Size versus Shaft Diameter

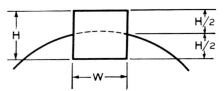

Key Size Versus Shaft Diameter

NOMINAL SHAFT DIAMETER		NOMINAL KEY SIZE			NOMINAL KEYSEAT DEPTH	
Over	To (Incl)	Width, W	Height, H		H/2	
			Square	Rectangular	Square	Rectangular
5/16	7/16	3/32	3/32		3/64	
7/16	9/16	1/8	1/8	3/32	1/16	3/64
9/16	7/8	3/16	3/16	1/8	3/32	1/16
7/8	1-1/4	1/4	1/4	3/16	1/8	3/32
1-1/4	1-3/8	5/16	5/16	1/4	5/32	1/8
1-3/8	1-3/4	3/8	3/8	1/4	3/16	1/8
1-3-4	2-1/4	1/2	1/2	3/8	1/4	3/16
2-1/4	2-3/4	5/8	5/8	7/16	5/16	7/32
2-3/4	3-1/4	3/4	3/4	1/2	3/8	1/4
3-1/4	3-3/4	7/8	7/8	5/8	7/16	5/16
3-3/4	4-1/2	1	1	3/4	1/2	3/8
4-1/2	5-1/2	1-1/4	1-1/4	7/8	5/8	7/16
5-1/2	6-1/2	1-1/2	1-1/2	1	3/4	1/2
6-1/2	7-1/2	1-3/4	1-3/4	1-1/2*	7/8	3/4
7-1/2	9	2	2	1-1/2	1	3/4
9	11	2-1/2	2-1/2	1-3/4	1-1/4	7/8
11	13	3	3	2	1-1/2	1
13	15	3-1/2	3-1/2	2-1/2	1-3/4	1-1/4
15	18	4		3		1-1/2
18	22	5		3-1/2		1-3/4
22	26	6		4		2
26	30	7		5		2-1/2

* Some key standards show 1-1/4". Prefered size is 1-1/2".

(ANSI)

Parallel and Taper Keys

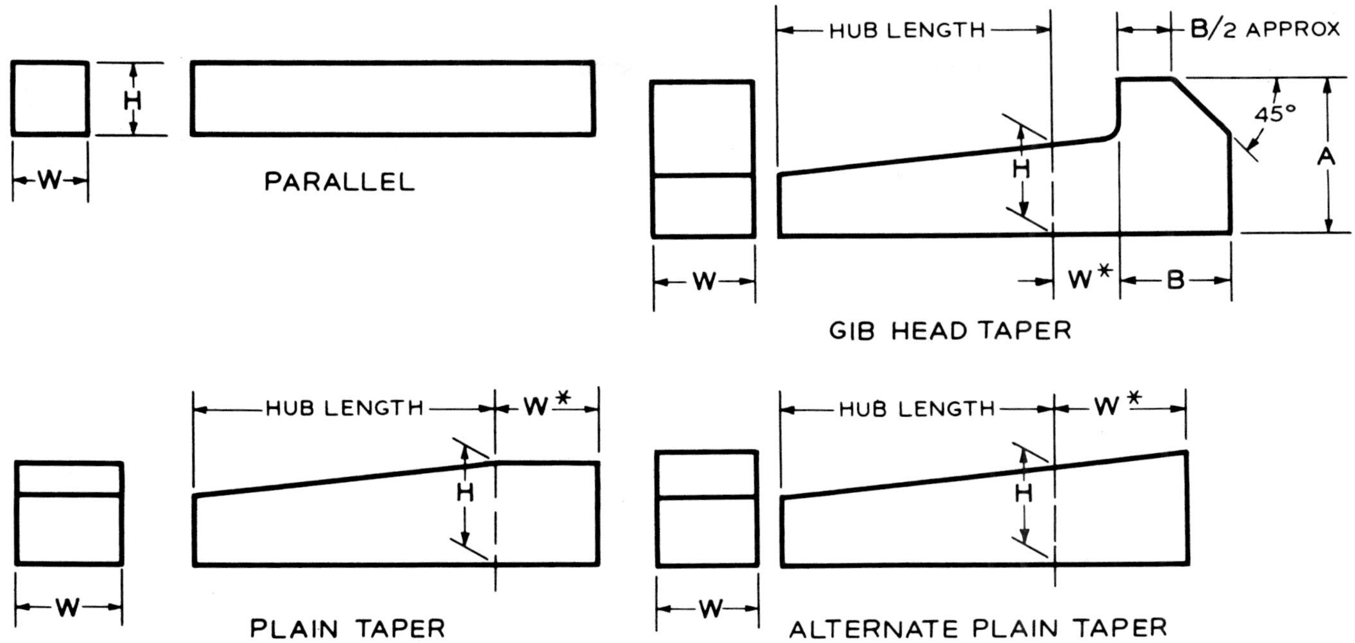

PARALLEL

GIB HEAD TAPER

PLAIN TAPER

ALTERNATE PLAIN TAPER

Plain and Gib Head Taper Keys Have a 1/8″ Taper in 12″

Key Dimensions and Tolerances

KEY			NOMINAL KEY SIZE		TOLERANCE	
			Width, W		Width, W	Height, H
			Over	To (Incl)		
Parallel	Square	Bar Stock	— 3/4 1-1/2 2-1/2	3/4 1-1/2 2-1/2 3-1/2	+0.000 −0.002 +0.000 −0.003 +0.000 −0.004 +0.000 −0.006	+0.000 −0.002 +0.000 −0.003 +0.000 −0.004 +0.000 −0.006
		Keystock	— 1-1/4 3	1-1/4 3 3-1/2	+0.001 −0.000 +0.002 −0.000 +0.003 −0.000	+0.001 −0.000 +0.002 −0.000 +0.003 −0.000
	Rectangular	Bar Stock	— 3/4 1-1/2 3 4 6	3/4 1-1/2 3 4 6 7	+0.000 −0.003 +0.000 −0.004 +0.000 −0.005 +0.000 −0.006 +0.000 −0.008 +0.000 −0.013	+0.000 −0.003 +0.000 −0.004 +0.000 −0.005 +0.000 −0.006 +0.000 −0.008 +0.000 −0.013
		Keystock	— 1-1/4 3	1-1/4 3 7	+0.001 −0.000 +0.002 −0.000 +0.003 −0.000	+0.005 −0.005 +0.005 −0.005 +0.005 −0.005
Taper	Plain or Gib Head Square or Rectangular		— 1-1/4 3	1-1/4 3 7	+0.001 −0.000 +0.002 −0.000 +0.003 −0.000	+0.005 −0.000 +0.005 −0.000 +0.005 −0.000

*For locating position of dimension H. Tolerance does not apply.
 See Table 2A for dimensions on gib heads.
 All dimensions given in inches.

(ANSI)

Woodruff Keys

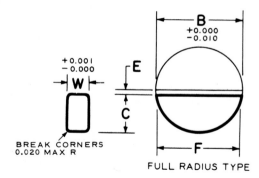

FULL RADIUS TYPE

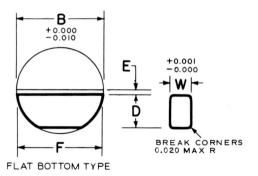

FLAT BOTTOM TYPE

Woodruff Keys

Key No.	Nominal Key Size W × B	Actual Length F +0.000-0.010	Height of Key				Distance Below Center E
			C		D		
			Max	Min	Max	Min	
202	1/16 × 1/4	0.248	0.109	0.104	0.109	0.104	1/64
202.5	1/16 × 5/16	0.311	0.140	0.135	0.140	0.135	1/64
302.5	3/32 × 5/16	0.311	0.140	0.135	0.140	0.135	1/64
203	1/16 × 3/8	0.374	0.172	0.167	0.172	0.167	1/64
303	3/32 × 3/8	0.374	0.172	0.167	0.172	0.167	1/64
403	1/8 × 3/8	0.374	0.172	0.167	0.172	0.167	1/64
204	1/16 × 1/2	0.491	0.203	0.198	0.194	0.188	3/64
304	3/32 × 1/2	0.491	0.203	0.198	0.194	0.188	3/64
404	1/8 × 1/2	0.491	0.203	0.198	0.194	0.188	3/64
305	3/32 × 5/8	0.612	0.250	0.245	0.240	0.234	1/16
405	1/8 × 5/8	0.612	0.250	0.245	0.240	0.234	1/16
505	5/32 × 5/8	0.612	0.250	0.245	0.240	0.234	1/16
605	3/16 × 5/8	0.612	0.250	0.245	0.240	0.234	1/16
406	1/8 × 3/4	0.740	0.313	0.308	0.303	0.297	1/16
506	5/32 × 3/4	0.740	0.313	0.308	0.303	0.297	1/16
606	3/16 × 3/4	0.740	0.313	0.308	0.303	0.297	1/16
806	1/4 × 3/4	0.740	0.313	0.308	0.303	0.297	1/16
507	5/32 × 7/8	0.866	0.375	0.370	0.365	0.359	1/16
607	3/16 × 7/8	0.866	0.375	0.370	0.365	0.359	1/16
707	7/32 × 7/8	0.866	0.375	0.370	0.365	0.359	1/16
807	1/4 × 7/8	0.866	0.375	0.370	0.365	0.359	1/16
608	3/16 × 1	0.992	0.438	0.433	0.428	0.422	1/16
708	7/32 × 1	0.992	0.438	0.433	0.428	0.422	1/16
808	1/4 × 1	0.992	0.438	0.433	0.428	0.422	1/16
1008	5/16 × 1	0.992	0.438	0.433	0.428	0.422	1/16
1208	3/8 × 1	0.992	0.438	0.433	0.428	0.422	1/16
609	3/16 × 1 1/8	1.114	0.484	0.479	0.475	0.469	5/64
709	7/32 × 1 1/8	1.114	0.484	0.479	0.475	0.469	5/64
809	1/4 × 1 1/8	1.114	0.484	0.479	0.475	0.469	5/64
1009	5/16 × 1 1/8	1.114	0.484	0.479	0.475	0.469	5/64

All dimensions given are in inches.

The key numbers indicate nominal key dimensions. The last two digits give the nominal diameter B in eighths of an inch and the digits preceding the last two give the nominal width W in thirty-seconds of an inch.

Example:
No. 204 indicates a key 2/32 × 4/8 or 1/16 × 1/2.
No. 808 indicates a key 8/32 × 8/8 or 1/4 × 1.
No. 1212 indicates a key 12/32 × 12/8 or 3/8 × 1 1/2.

(ANSI)

Woodruff Keyseats

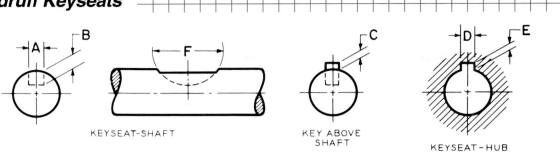

KEYSEAT-SHAFT   KEY ABOVE SHAFT   KEYSEAT-HUB

Keyseat Dimensions

Key Number	Nominal Size Key	Keyseat — Shaft					Key Above Shaft	Keyseat — Hub	
		Width A*		Depth B	Diameter F		Height C	Width D	Depth E
		Min	Max	+0.005 -0.000	Min	Max	+0.005 -0.005	+0.002 -0.000	+0.005 -0.000
202	1/16 × 1/4	0.0615	0.0630	0.0728	0.250	0.268	0.0312	0.0635	0.0372
202.5	1/16 × 5/16	0.0615	0.0630	0.1038	0.312	0.330	0.0312	0.0635	0.0372
302.5	3/32 × 5/16	0.0928	0.0943	0.0882	0.312	0.330	0.0469	0.0948	0.0529
203	1/16 × 3/8	0.0615	0.0630	0.1358	0.375	0.393	0.0312	0.0635	0.0372
303	3/32 × 3/8	0.0928	0.0943	0.1202	0.375	0.393	0.0469	0.0948	0.0529
403	1/8 × 3/8	0.1240	0.1255	0.1045	0.375	0.393	0.0625	0.1260	0.0685
204	1/16 × 1/2	0.0615	0.0630	0.1668	0.500	0.518	0.0312	0.0635	0.0372
304	3/32 × 1/2	0.0928	0.0943	0.1511	0.500	0.518	0.0469	0.0948	0.0529
404	1/8 × 1/2	0.1240	0.1255	0.1355	0.500	0.518	0.0625	0.1260	0.0685
305	3/32 × 5/8	0.0928	0.0943	0.1981	0.625	0.643	0.0469	0.0948	0.0529
405	1/8 × 5/8	0.1240	0.1255	0.1825	0.625	0.643	0.0625	0.1260	0.0685
505	5/32 × 5/8	0.1553	0.1568	0.1669	0.625	0.643	0.0781	0.1573	0.0841
605	3/16 × 5/8	0.1863	0.1880	0.1513	0.625	0.643	0.0937	0.1885	0.0997
406	1/8 × 3/4	0.1240	0.1255	0.2455	0.750	0.768	0.0625	0.1260	0.0685
506	5/32 × 3/4	0.1553	0.1568	0.2299	0.750	0.768	0.0781	0.1573	0.0841
606	3/16 × 3/4	0.1863	0.1880	0.2143	0.750	0.768	0.0937	0.1885	0.0997
806	1/4 × 3/4	0.2487	0.2505	0.1830	0.750	0.768	0.1250	0.2510	0.1310
507	5/32 × 7/8	0.1553	0.1568	0.2919	0.875	0.895	0.0781	0.1573	0.0841
607	3/16 × 7/8	0.1863	0.1880	0.2763	0.875	0.895	0.0937	0.1885	0.0997
707	7/32 × 7/8	0.2175	0.2193	0.2607	0.875	0.895	0.1093	0.2198	0.1153
807	1/4 × 7/8	0.2487	0.2505	0.2450	0.875'	0.895	0.1250	0.2510	0.1310
608	3/16 × 1	0.1863	0.1880	0.3393	1.000	1.020	0.0937	0.1885	0.0997
708	7/32 × 1	0.2175	0.2193	0.3237	1.000	1.020	0.1093	0.2198	0.1153
808	1/4 × 1	0.2487	0.2505	0.3080	1.000	1.020	0.1250	0.2510	0.1310
1008	5/16 × 1	0.3111	0.3130	0.2768	1.000	1.020	0.1562	0.3135	0.1622
1208	3/8 × 1	0.3735	0.3755	0.2455	1.000	1.020	0.1875	0.3760	0.1935
609	3/16 × 1 1/8	0.1863	0.1880	0.3853	1.125	1.145	0.0937	0.1885	0.0997
709	7/32 × 1 1/8	0.2175	0.2193	0.3697	1.125	1.145	0.1093	0.2198	0.1153
809	1/4 × 1 1/8	0.2487	0.2505	0.3540	1.125	1.145	0.1250	0.2510	0.1310
1009	5/16 × 1 1/8	0.3111	0.3130	0.3228	1.125	1.145	0.1562	0.3135	0.1622

* Width A values were set with the maximum keyseat (shaft) width as that figure which will receive a key with the greatest amount of looseness consistent with assuring the key's sticking in the keyseat (shaft). Minimum keyseat width is that figure permitting the largest shaft distortion acceptable when assembling maximum key in minimum keyseat.

Dimensions A, B, C, D are taken at side intersection.

(ANSI)

Straight Pins

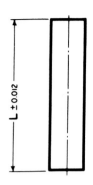

CHAMFERED

L ± 0.012

B ± 0.010

25°

SQUARE END

L ± 0.012

Dimensions of Straight Pins

Nominal Diameter	Diameter A		Chamfer B
	Max	Min	
0.062	0.0625	0.0605	0.015
0.094	0.0937	0.0917	0.015
0.109	0.1094	0.1074	0.015
0.125	0.1250	0.1230	0.015
0.156	0.1562	0.1542	0.015
0.188	0.1875	0.1855	0.015
0.219	0.2187	0.2167	0.015
0.250	0.2500	0.2480	0.015
0.312	0.3125	0.3095	0.030
0.375	0.3750	0.3720	0.030
0.438	0.4375	0.4345	0.030
0.500	0.500	0.4970	0.030

All dimensions are given in inches.
These pins must be straight and free from burrs or any other defects that will affect their serviceability.
(ANSI)

Gib Head Keys

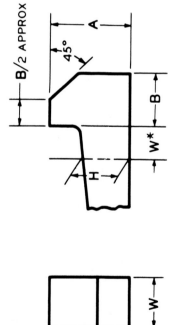

B/2 APPROX

A

45°

B

W*

H

W

Gib Head Nominal Dimensions

Nominal Key Size Width, W	SQUARE			RECTANGULAR		
	H	A	B	H	A	B
1/8	1/8	1/4	1/4	3/32	3/16	1/8
3/16	3/16	5/16	5/16	1/8	1/4	1/4
1/4	1/4	7/16	3/8	3/16	5/16	5/16
5/16	5/16	1/2	7/16	1/4	7/16	3/8
3/8	3/8	5/8	1/2	1/4	7/16	3/8
1/2	1/2	7/8	5/8	3/8	5/8	1/2
5/8	5/8	1	3/4	7/16	3/4	9/16
3/4	3/4	1-1/4	7/8	1/2	7/8	5/8
7/8	7/8	1-3/8	1	5/8	1	3/4
1	1	1-5/8	1-1/8	3/4	1-1/4	7/8
1-1/4	1-1/4	2	1-7/16	7/8	1-3/8	
1-1/2	1-1/2	2-3/8	1-3/4	1	1-5/8	1-1/8
1-3/4	1-3/4	2-3/4	2	1-1/2	2-3/8	1-3/4
2	2	3-1/2	2-1/4	1-1/2	2-3/8	1-3/4
2-1/2	2-1/2	4	3	1-3/4	2-3/4	2
3	3	5	3-1/2	2	3-1/2	2-1/4
3-1/2	3-1/2	6	4	2-1/2	4	3

*For locating position of dimension H.

For larger sizes the following relationships are suggested as guides for establishing A and B.

$$A = 1.8 H \qquad B = 1.2 H$$

All dimensions given in inches.
(ANSI)

Hardened and Ground Dowel Pins

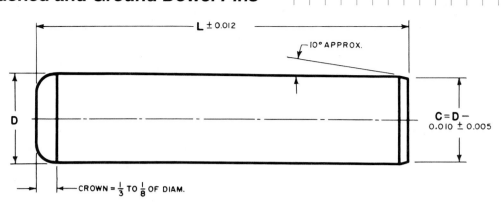

Dimensions of Hardened and Ground Dowel Pins

Length, L	Nominal Diameter D									
	1/8	3/16	1/4	5/16	3/8	7/16	1/2	5/8	3/4	7/8
	Diameter Standard Pins ±0.0001									
	0.1252	0.1877	0.2502	0.3127	0.3752	0.4377	0.5002	0.6252	0.7502	0.8752
	Diameter Oversize Pins ±0.0001									
	0.1260	0.1885	0.2510	0.3135	0.3760	0.4385	0.5010	0.6260	0.7510	0.8760
1/2	X	X	X	X						
5/8	X	X	X	X						
3/4	X	X	X	X	X					
7/8	X	X	X	X	X	X				
1	X	X	X	X	X	X				
1 1/4		X	X	X	X	X	X	X		
1 1/2		X	X	X	X	X	X	X	X	
1 3/4		X	X	X	X	X	X	X	X	
2			X	X	X	X	X	X	X	X
2 1/4				X	X	X	X			
2 1/2				X	X	X	X	X	X	X
3							X	X	X	X
3 1/2							X	X		
4							X	X	X	X
4 1/2								X	X	X
5									X	X
5 1/2									X	X

All dimensions are given in inches.

These pins are extensively used in the tool and machine industry and a machine reamer of nominal size may be used to produce the holes into which these pins tap or press fit. They must be straight and free from any defects that will affect their serviceability.

(ANSI)

Taper Pins

Dimensions of Taper Pins

Length, L	7/0 0.0625	6/0 0.0780	5/0 0.0940	4/0 0.1090	3/0 0.1250	2/0 0.1410	0 0.1560	1 0.1720	2 0.1930	3 0.2190	4 0.2500	5 0.2890	6 0.3410	7 0.4090	8 0.4920	9 0.5910	10 0.7060
0.375	X	X															
0.500	X	X	X	X													
0.625	X	X	X	X	X												
0.750		X	X	X	X	X	X										
0.875			X	X	X	X	X	X	X								
1.000				X	X	X	X	X	X	X							
1.250					X	X	X	X	X	X	X	X					
1.500						X	X	X	X	X	X	X	X				
1.750							X	X	X	X	X	X	X				
2.000								X	X	X	X	X	X				
2.250								X	X	X	X	X	X	X			
2.500									X	X	X	X	X	X	X		
2.750										X	X	X	X	X	X		
3.000										X	X	X	X	X	X		
3.250													X	X	X	X	
3.500													X	X	X	X	X
3.750													X	X	X	X	X
4.000														X	X	X	X
4.250														X	X	X	X
4.500														X	X	X	X
4.750															X	X	X
5.000															X	X	X
5.250															X	X	X
5.500																X	X
5.750																X	X
6.000																	X

All dimensions are given in inches.   Standard reamers are available for pins given above the line.

Pins Nos. 11 (size 0.8600), 12 (size 1.032), 13 (size 1.241), and 14 (1.523) are special sizes—hence their lengths are special.

To find small diameter of pin, multiply the length by 0.02083 and subtract the result from the large diameter.

TYPES	COMMERCIAL TYPE	PRECISION TYPE
Sizes	7/0 to 14	7/0 to 10
Tolerance on Diameter	(+0.0013, −0.0007)	(+0.0013, −0.0007)
Taper	$\frac{1}{4}$ In. per Ft	$\frac{1}{4}$ In. per Ft
Length Tolerance	(±0.030)	(±0.030)
Concavity Tolerance	None	0.0005 up to 1 in. long 0.001 1 $\frac{1}{16}$ to 2 in. long 0.002 2 $\frac{1}{16}$ and longer

(ANSI)

Sheet Metal and Wire Gage Designation

GAGE NO.	AMERICAN OR BROWN & SHARPE'S A.W.G. OR B. & S.	BIRMING-HAM OR STUBS WIRE B.W.G.	WASHBURN & MOEN OR AMERICAN S.W.G.	UNITED STATES STANDARD	MANU-FACTURERS' STANDARD FOR SHEET STEEL	GAGE NO.
0000000	- - - -	- - - -	.4900	.500	- - - -	0000000
000000	.5800	- - - -	.4615	.469	- - - -	000000
00000	.5165	- - - -	.4305	.438	- - - -	00000
0000	.4600	.454	.3938	.406	- - - -	0000
000	.4096	.425	.3625	.375	- - - -	000
00	.3648	.380	.3310	.344	- - - -	00
0	.3249	.340	.3065	.312	- - - -	0
1	.2893	.300	.2830	.281	- - - -	1
2	.2576	.284	.2625	.266	- - - -	2
3	.2294	.259	.2437	.250	.2391	3
4	.2043	.238	.2253	.234	.2242	4
5	.1819	.220	.2070	.219	.2092	5
6	.1620	.203	.1920	.203	.1943	6
7	.1443	.180	.1770	.188	.1793	7
8	.1285	.165	.1620	.172	.1644	8
9	.1144	.148	.1483	.156	.1495	9
10	.1019	.134	.1350	.141	.1345	10
11	.0907	.120	.1205	.125	.1196	11
12	.0808	.109	.1055	.109	.1046	12
13	.0720	.095	.0915	.0938	.0897	13
14	.0642	.083	.0800	.0781	.0747	14
15	.0571	.072	.0720	.0703	.0673	15
16	.0508	.065	.0625	.0625	.0598	16
17	.0453	.058	.0540	.0562	.0538	17
18	.0403	.049	.0475	.0500	.0478	18
19	.0359	.042	.0410	.0438	.0418	19
20	.0320	.035	.0348	.0375	.0359	20
21	.0285	.032	.0317	.0344	.0329	21
22	.0253	.028	.0286	.0312	.0299	22
23	.0226	.025	.0258	.0281	.0269	23
24	.0201	.022	.0230	.0250	.0239	24
25	.0179	.020	.0204	.0219	.0209	25
26	.0159	.018	.0181	.0188	.0179	26
27	.0142	.016	.0173	.0172	.0164	27
28	.0126	.014	.0162	.0156	.0149	28
29	.0113	.013	.0150	.0141	.0135	29
30	.0100	.012	.0140	.0125	.0120	30
31	.0089	.010	.0132	.0109	.0105	31
32	.0080	.009	.0128	.0102	.0097	32
33	.0071	.008	.0118	.00938	.0090	33
34	.0063	.007	.0104	.00859	.0082	34
35	.0056	.005	.0095	.00781	.0075	35
36	.0050	.004	.0090	.00703	.0067	36
37	.0045	- - - -	.0085	.00624	.0064	37
38	.0040	- - - -	.0080	.00625	.0060	38
39	.0035	- - - -	.0075	- - - - -	- - - -	39
40	.0031	- - - -	.0070	- - - - -	- - - -	40
41	.0028	- - - -	.0066	- - - - -	- - - -	41
42	.0025	- - - -	.0062	- - - - -	- - - -	42
43	.0022	- - - -	.0060	- - - - -	- - - -	43
44	.0020	- - - -	.0058	- - - - -	- - - -	44
45	.0018	- - - -	.0055	- - - - -	- - - -	45
46	.0016	- - - -	.0052	- - - - -	- - - -	46
47	.0014	- - - -	.0050	- - - - -	- - - -	47
48	.0012	- - - -	.0048	- - - - -	- - - -	48

Precision Sheet Metal Set-back Chart

MATERIAL THICKNESS (T)

90 DEG BEND RADIUS (R)	.016	.020	.025	.032	.040	.051	.064	.072	.078	.081	.091	.102	.125	.129	.156	.162	.187	.250
1/32	.034	.039	.046	.055	.065	.081	.102	.113	.121	.125	.139							
3/64	.041	.046	.053	.062	.072	.090	.108	.119	.127	.131	.145							
1/16	.048	.053	.059	.068	.079	.093	.110	.122	.134	.138	.152							
5/64	.054	.060	.066	.075	.086	.100	.117	.127	.138	.144	.158							
3/32	.061	.066	.073	.082	.092	.107	.124	.134	.142	.146	.160							
7/64	.068	.073	.080	.089	.099	.113	.130	.141	.148	.153	.167	.181						
1/8	.075	.080	.086	.095	.106	.120	.137	.147	.155	.159	.172	.186	.216	.221				
9/64	.081	.087	.093	.102	.113	.127	.144	.154	.162	.166	.179	.193	.223	.228	.263			
5/32	.088	.093	.100	.109	.119	.134	.150	.161	.169	.173	.186	.200	.230	.235	.270	.278		
11/64	.095	.100	.107	.116	.126	.140	.157	.168	.175	.179	.192	.207	.236	.242	.277	.284	.317	
3/16	.102	.107	.113	.122	.133	.147	.164	.174	.182	.186	.199	.213	.243	.248	.283	.291	.324	.405
13/64	.108	.114	.120	.129	.140	.154	.171	.181	.189	.193	.206	.220	.250	.255	.290	.298	.330	.412
7/32	.115	.120	.127	.136	.146	.161	.177	.188	.196	.199	.212	.227	.257	.262	.297	.305	.337	.419
15/64	.122	.127	.134	.143	.153	.167	.184	.195	.202	.206	.219	.233	.263	.269	.304	.311	.344	.426
1/4	.129	.134	.140	.149	.160	.174	.191	.201	.209	.213	.226	.240	.270	.275	.310	.318	.351	.432
17/64	.135	.141	.147	.156	.166	.181	.198	.208	.216	.220	.233	.247	.277	.282	.317	.325	.357	.439
9/32	.142	.147	.154	.163	.173	.187	.204	.215	.223	.226	.239	.254	.284	.289	.324	.332	.364	.446
5/16	.156	.161	.167	.176	.187	.201	.218	.228	.236	.240	.253	.267	.297	.302	.337	.345	.378	.459
11/32	.169	.174	.181	.190	.200	.214	.231	.242	.250	.253	.266	.281	.311	.316	.351	.359	.391	.473
3/8	.183	.188	.194	.203	.214	.228	.245	.255	.263	.267	.280	.294	.324	.329	.364	.372	.404	.486

(STOCK THICKNESS) T

Z = SET–BACK ALLOWANCE FROM CHART

DEVELOPED LENGTH = X + Y – Z

Coordinates for Locating Equally Spaced Holes in Jig Boring

The constants in the table are multiplied by the diameter of the bolt-hole pitch circle to obtain the longitudinal and lateral adjustments of the right-angle slides of the jig borer, in boring equally spaced holes. While holes may be located by these right-angular measurements, an auxiliary rotary table provides a more direct method. With a rotary table, the holes are spaced by precise angular movements after adjustment to the required radius.

MULTIPLY VALUES SHOWN BY DIAMETER OF PITCH CIRCLE.

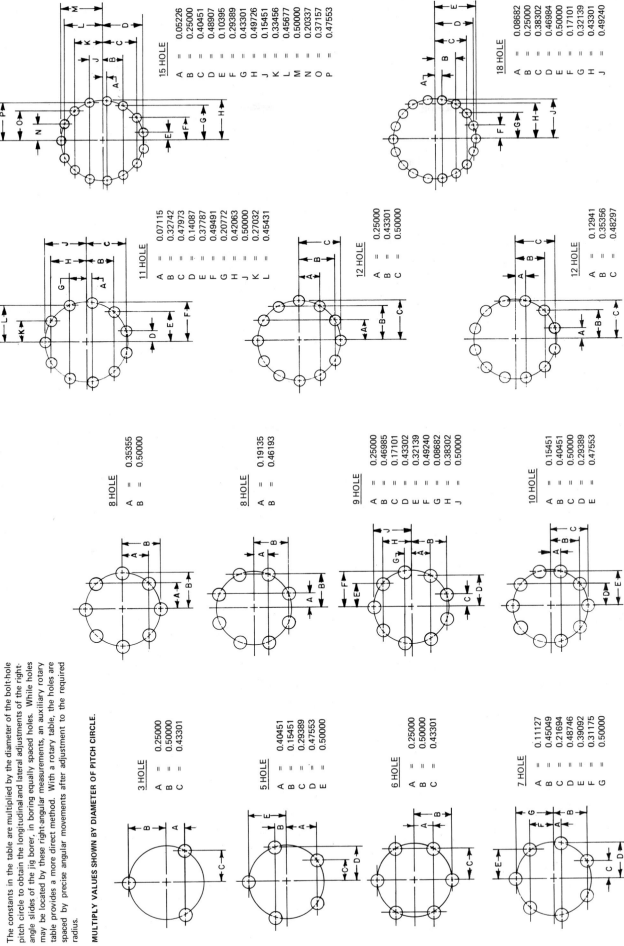

15 HOLE

A	=	0.05226
B	=	0.25000
C	=	0.40451
D	=	0.48907
E	=	0.10395
F	=	0.29389
G	=	0.43301
H	=	0.49726
J	=	0.15451
K	=	0.33456
L	=	0.45677
M	=	0.50000
N	=	0.20337
O	=	0.37157
P	=	0.47553

18 HOLE

A	=	0.08682
B	=	0.25000
C	=	0.38302
D	=	0.46984
E	=	0.50000
F	=	0.17101
G	=	0.32139
H	=	0.43301
J	=	0.49240

11 HOLE

A	=	0.07115
B	=	0.32742
C	=	0.47973
D	=	0.14087
E	=	0.37787
F	=	0.49491
G	=	0.20772
H	=	0.42063
J	=	0.50000
K	=	0.27032
L	=	0.45431

12 HOLE

A	=	0.25000
B	=	0.43301
C	=	0.50000

12 HOLE

A	=	0.12941
B	=	0.35356
C	=	0.48297

8 HOLE

A	=	0.35355
B	=	0.50000

8 HOLE

A	=	0.19135
B	=	0.46193

9 HOLE

A	=	0.25000
B	=	0.46985
C	=	0.17101
D	=	0.43302
E	=	0.32139
F	=	0.49240
G	=	0.08682
H	=	0.38302
J	=	0.50000

10 HOLE

A	=	0.15451
B	=	0.40451
C	=	0.50000
D	=	0.29389
E	=	0.47553

3 HOLE

A	=	0.25000
B	=	0.50000
C	=	0.43301

5 HOLE

A	=	0.40451
B	=	0.15451
C	=	0.29389
D	=	0.47553
E	=	0.50000

6 HOLE

A	=	0.25000
B	=	0.50000
C	=	0.43301

7 HOLE

A	=	0.11127
B	=	0.45049
C	=	0.21694
D	=	0.48746
E	=	0.39092
F	=	0.31175
G	=	0.50000

Geometric Tolerancing Symbols

SYMBOL FOR:	ASME Y14.5	ISO
STRAIGHTNESS	———	———
FLATNESS	▱	▱
CIRCULARITY	○	○
CYLINDRICITY	⌀/	⌀/
PROFILE OF A LINE	⌒	⌒
PROFILE OF A SURFACE	⌓	⌓
ALL AROUND PROFILE	⟲○	NONE
ANGULARITY	∠	∠
PERPENDICULARITY	⊥	⊥
PARALLELISM	//	//
POSITION	⊕	⊕
CONCENTRICITY/COAXIALITY	◎	◎
SYMMETRY	⩵	⩵
CIRCULAR RUNOUT	* ↗	* ↗
TOTAL RUNOUT	* ↗↗	* ↗↗
AT MAXIMUM MATERIAL CONDITION	Ⓜ	Ⓜ
AT LEAST MATERIAL CONDITION	Ⓛ	NONE
REGARDLESS OF FEATURE SIZE	NONE	NONE
PROJECTED TOLERANCE ZONE	Ⓟ	Ⓟ
DIAMETER	⌀	⌀
BASIC DIMENSION	30	30
REFERENCE DIMENSION	(30)	(30)
DATUM FEATURE	A◄	* ⫝̸— OR * ⫝̸—A
DATUM TARGET	06/A1	06/A1
TARGET POINT	✕	✕
DIMENSION ORIGIN	⊕►	NONE
FEATURE CONTROL FRAME	⊕ ⌀0.5ⓂABC	⊕ ⌀0.5ⓂABC
CONICAL TAPER	▷	▷
SLOPE	◁	◁
COUNTERBORE/SPOTFACE	⌴	NONE
COUNTERSINK	⌵	NONE
DEPTH/DEEP	↧	NONE
SQUARE (SHAPE)	□	□
DIMENSION NOT TO SCALE	15̲	15̲
NUMBER OF TIMES/PLACES	8X	8X
ARC LENGTH	⌢105	NONE
RADIUS	R	R
SPHERICAL RADIUS	SR	NONE
SPHERICAL DIAMETER	S⌀	NONE

*   May Be Filled In

Electrical and Electronics Diagrams

4.13 Selector or Multiposition Switch

The position in which the switch is shown may be indicated by a note or designation of switch position.

4.13.1 General (for power and control diagrams)

Any number of transmission paths may be shown.

5.3 Connector
Disconnecting Device
Jack F
Plug F

The contact symbol is not an arrowhead. It is larger and the lines are drawn at a 90-degree angle.

5.3.1 Female contact

IEC ———⟨

5.3.2 Male contact

IEC ———→

7.3 Typical Applications

7.3.1 Triode with directly heated filamentary cathode and envelope connection to base terminal

7.3.2 Equipotential-cathode pentode showing use of elongated envelope

7.3.3 Equipotential-cathode twin triode showing use of elongated envelope and rule of item 7.2.3

7.3.4 Cold-cathode gas-filled tube

7.3.4.1 Rectifier; voltage regulator for direct-current operation
See also symbol 11.1.3.2

8.6 Typical Applications, Three- (or more) Terminal Devices

8.6.1 PNP transistor (also PNIP transistor, if omitting the intrinsic region will not result in ambiguity)

See paragraph A4.11 of the Introduction

8.6.1.1 Application: PNP transistor with one electrode connected to envelope (in this case, the collector electrode)

8.6.2 NPN transistor (also NPIN transistor, if omitting the intrinsic region will not result in ambiguity)

See paragraph A4.11 of the Introduction

8.6.8 Unijunction transistor with N-type base

See paragraph A4.11 of the Introduction

8.6.10 Field-effect transistor with N-channel (junction gate and insulated gate)

8.6.10.1 N-channel junction gate

If desired, the junction-gate symbol element may be drawn opposite the preferred source.

See paragraph A4.11 of the Introduction

12.1 Meter
Instrument

NOTE 12.1A: The asterisk is not part of the symbol. Always replace the asterisk by one of the following letter combinations, depending on the function of the meter or instrument, unless some other identification is provided in the circle and explained on the diagram.

*See Note 12.1A

A	Ammeter F IEC
AH	Ampere-hour meter
C	Coulombmeter
CMA	Contact-making (or breaking) ammeter
CMC	Contact-making (or breaking) clock
CMV	Contact-making (or breaking) voltmeter
CRO	Oscilloscope F
	Cathode-ray oscillograph
DB	DB (decibel) meter
	Audio level/meter F
DBM	DBM (decibels referred to 1 milliwatt) meter
DM	Demand meter
DTR	Demand-totalizing relay
F	Frequency meter F
GD	Ground detector
I	Indicating meter
INT	Integrating meter
μA or UA	Microammeter
MA	Milliammeter
NM	Noise meter
OHM	Ohmmeter F
OP	Oil pressure meter
OSCG	Oscillograph, string
PF	Power factor meter
PH	Phasemeter F
PI	Position indicator
RD	Recording demand meter
REC	Recording meter
RF	Reactive factor meter
SY	Synchroscope
t°	Temperature meter
THC	Thermal converter
TLM	Telemeter
TT	Total time meter
	Elapsed time meter
V	Voltmeter F IEC
VA	Volt-ammeter
VAR	Varmeter F
VARH	Varhour meter
VI	Volume indicator
	Audio-level meter F
VU	Standard volume indicator
	Audio-level meter F
W	Wattmeter F IEC
WH	Watthour meter

Standard Welding Symbols

Typical Welding Symbols

Double-Fillet Welding Symbol
- Weld size

Chain Intermittant Fillet Welding Symbol
- Pitch (distance between centers) of increments
- Size (length of leg)
- Length of increments

Staggered Intermittant Fillet Welding Symbol
- Pitch (distance between centers) of increments
- Size (length of leg)
- Length of increments

Plug Welding Symbol
- Include angle of countersink
- Size (diameter of hole at root)
- Depth of filling in inches (omission indicates filling is complete)
- Omission indicates that weld extends between abrupt changes in direction or as dimensioned
- Length
- Pitch (distance between centers) of welds

Back Welding Symbol
- Back weld
- 2nd operation
- 1st operation

Backing Welding Symbol
- Backing weld
- 1st operation
- 2nd operation

Spot Welding Symbol
- Size or strength
- Number of welds
- Pitch
- Process
- RSW

Stud Welding Symbol
- Size
- Number of studs
- Pitch

Seam Welding Symbol
- Increment length
- Size or strength
- Pitch
- Process
- RSEW

Square-Groove Welding Symbol
- Weld size
- Root opening

Single-V Groove Welding Symbol
- Depth of preparation
- Root opening
- Groove angle

Double-Bevel-Groove Welding Symbol
- Weld size
- Arrow points toward member to be prepared

Symbol with Backgouging
- Depth of preparation
- Back gouge

Flare-V Groove Welding Symbol
- Weld size

Flare-Bevel-Groove Welding Symbol
- Weld size

Multiple Reference Lines
- 1st operation on line nearest arrow
- 2nd operation
- 3rd operation

Complete Penetration
- Indicates complete penetration regardless of type of weld or joint preparation
- CJP

Edge Flange Welding Symbol

Flash or Upset Welding Symbol
- Process reference
- FW

Melt-Thru Symbol

Joint with Backing
- Weld size

Joint with Spacer
- With modified groove weld symbol

Convex Contour Symbol

Flush Contour Symbol

Double bevel groove

- Radius
- Height above point of tangency
- R indicates backing removed after welding

Basic Welding Symbols and Their Location Significance

Location Significance	Fillet	Plug or Slot	Spot or Projection	Stud	Seam	Back or Backing	Surfacing	Flange Corner	Flange Edge
Arrow Side						Groove weld symbol			
Other Side				Not used		Groove weld symbol	Not used	Not used	
Both Sides				Not used		Not used	Not used	Not used	Not used
No Arrow Side or Other Side Significance	Not used	Not used		Not used		Not used	Not used	Not used	Not used

Location Significance	Square	V	Bevel	Groove U	J	Flare-V	Flare-Bevel	Scarf for Brazed Joint
Arrow Side								
Other Side								
Both Sides							Not used	Not used
No Arrow Side or Other Side Significance	Not used	Not used	Not used	Not used	Not used	Not used	Not used	Not used

Supplementary Symbols

Weld-All Around	Field Weld	Melt-Thru	Consumable Insert	Backing Spacer	Flush	Contour Convex	Concave

Location of Elements of a Welding Symbol

- Finish symbol
- Contour symbol
- Root opening, depth of filling for plug and slot welds
- Groove weld size
- Size or strength for certain welds
- Depth of preparation; size or strength for certain welds
- Specification, process, or other reference
- Tail (Tail omitted when reference is not used)
- Basic weld symbol or detail reference
- Groove angle; included angle of countersink for plug welds
- Length of weld
- Pitch (center-to-center spacing) of welds
- Field weld symbol
- Weld-all-around symbol
- Arrow connecting reference line to arrow side member of joint or arrow side of joint
- Reference line
- Number of spot, stud, or projection welds
- Elements in this area remain as shown when tail and arrow are reversed
- F, A, R, S(E) — OTHER SIDE, BOTH SIDES, ARROW SIDE (N), T, L-P

Identification of Arrow Side and Other Side of Joint

Butt Joint
- Arrow side of joint
- Other side of joint
- Arrow of welding symbol

Corner Joint
- Arrow side of joint
- Other side of joint
- Arrow of welding symbol

T-Joint
- Arrow side of joint
- Other side of joint
- Arrow of welding symbol

Lap Joint
- Other side member of joint
- Arrow side member of joint
- Arrow of welding symbol

Edge Joint
- Arrow side of joint
- Joint
- Arrow of welding symbol

Process Abbreviations

Where process abbreviations are to be included in the tail of the welding symbol, reference is made to Table 1, Designation of Welding and Allied Processes by Letters, of AWS A2.4-86.

AMERICAN WELDING SOCIETY, INC.
550 N.W. LeJeune Rd., Miami, Florida 33126

Topographic Map

VARIATIONS WILL BE FOUND ON OLDER MAPS

Hard surface, heavy duty road, four or more lanes . . . . . .

Hard surface, heavy duty road, two or three lanes . . . .

Hard surface, medium duty road, four or more lanes . . . .

Hard surface, medium duty road, two or three lanes . . . . .

Improved light duty road . . . .

Unimproved dirt road and trail . . . .

Dual highway, dividing strip 25 feet or less . . . .

Dual highway, dividing strip exceeding 25 feet . . . .

Road under construction . . . .

Railroad, single track and multiple track . . . .

Railroads in juxtaposition . . . .

Narrow gage, single track and multiple track . . . .

Railroad in street and carline . . . .

Bridge, road and railroad . . . .

Drawbridge, road and railroad . . . .

Footbridge . . . .

Tunnel, road and railroad . . . .

Overpass and underpass . . . .

Important small masonry or earth dam . . . .

Dam with lock . . . .

Dam with road . . . .

Canal with lock . . . .

Buildings (dwelling, place of employment, etc.) . . . .

School, church, and cemetery . . . .   Cem

Buildings (barn, warehouse, etc.) . . . .

Power transmission line . . . .

Telephone line, pipeline, etc. (labeled as to type) . . . .

Wells other than water (labeled as to type) . . . . o Oil . . . . o Gas

Tanks; oil, water, etc. (labeled as to type) . . . . • • ● ⊘ Water

Located or landmark object; windmill . . . . o

Open pit, mine, or quarry; prospect . . . . ⤫ x

Shaft and tunnel entrance . . . . ■ . . . . Y

Horizontal and vertical control station:

   Tablet, spirit level elevation . . . . BM △ 5653

   Other recoverable mark, spirit level elevation . . . . △ 5455

Horizontal control station: tablet, vertical angle elevation   VABM △ 9519

   Any recoverable mark, vertical angle or checked elevation   △3775

Vertical control station: tablet, spirit level elevation . . . . .   BM × 957

   Other recoverable mark, spirit level elevation . . . .   × 954

Checked spot elevation . . . .   ×4675

Unchecked spot elevation and water elevation . . . . × 5657 870

(U.S. Geological Survey)

Boundary, national . . . .

   State . . . .

   County, parish, municipio . . . .

   Civil township, precinct, town, barrio . . . .

   Incorporated city, village, town, hamlet . . . .

   Reservation, national or state . . . .

   Small park, cemetery, airport, etc. . . . .

   Land grant . . . .

Township or range line, United States land survey . . . . . .

Township or range line, approximate location . . . .

Section line, United States land survey . . . .

Section line, approximate location . . . .

Township line, not United States land survey . . . .

Section line, not United States land survey . . . .

Section corner, found and indicated . . . . + ... +

Boundary monument: land grant and other . . . . □ . . . . □

United States mineral or location monument . . . . ▲

Index contour . . . .   Intermediate contour . .

Supplementary contour   Depression contours . .

Fill . . . .   Cut . . . .

Levee . . . .   Levee with road . . . . .

Mine dump . . . .   Wash . . . .

Tailings . . . .   Tailings pond . . . .

Strip mine . . . .   Distorted surface . . . . .

Sand area . . . .   Gravel beach . . . .

Perennial streams . . . .   Intermittent streams . .

Elevated aqueduct . . . .   Aqueduct tunnel . . . . . .

Water well and spring . .   Disappearing stream . .

Small rapids . . . .   Small falls . . . .

Large rapids . . . .   Large falls . . . .

Intermittent lake . . . . .   Dry lake . . . .

Foreshore flat . . . .   Rock or coral reef . . . .

Sounding, depth curve .   10   Piling or dolphin . . . . .   o

Exposed wreck . . . . . .   Sunken wreck . . . .

Rock, bare or awash; dangerous to navigation . . . . *

Marsh (swamp) . . . .   Submerged marsh . . . .

Wooded marsh . . . .   Mangrove . . . .

Woods or brushwood . .   Orchard . . . .

Vineyard . . . .   Scrub . . . .

Inundation area . . . .   Urban area . . . .

Pipe Fittings and Valves

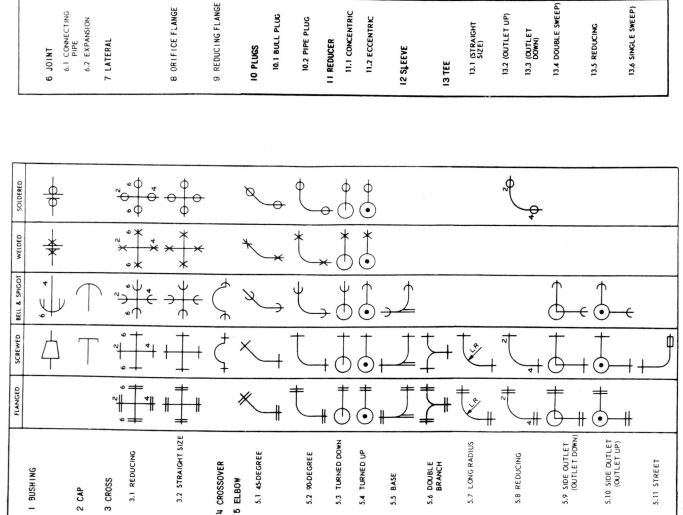

(ANSI)

Drawing Sheet Layouts

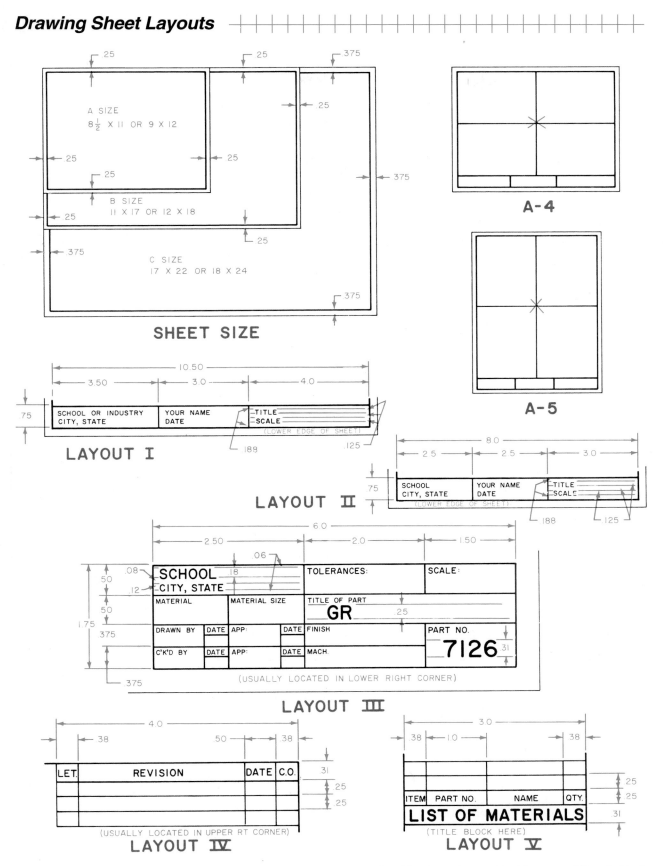

SHEET SIZE

A SIZE
8½ X 11 OR 9 X 12

B SIZE
11 X 17 OR 12 X 18

C SIZE
17 X 22 OR 18 X 24

.25   .25   .375   .25   .25   .375   .25   .25   .375   .375

A-4

A-5

LAYOUT I

10.50   3.50   3.0   4.0   .75

SCHOOL OR INDUSTRY CITY, STATE	YOUR NAME DATE	TITLE SCALE

(LOWER EDGE OF SHEET)
.188   .125

LAYOUT II

8.0   2.5   2.5   3.0   .75

SCHOOL CITY, STATE	YOUR NAME DATE	TITLE SCALE

(LOWER EDGE OF SHEET)
.188   .125

LAYOUT III

6.0   2.50   2.0   1.50
.06   .08   .50   .12   .50   .375   1.75   .18   .375

SCHOOL CITY, STATE		TOLERANCES:	SCALE:
MATERIAL	MATERIAL SIZE	TITLE OF PART GR .25	
DRAWN BY	DATE	APP:	DATE	FINISH	PART NO.
C'K'D BY	DATE	APP:	DATE	MACH.	7126 .31

(USUALLY LOCATED IN LOWER RIGHT CORNER)

LAYOUT IV

4.0   .38   .50   .38   .31   .25   .25

LET.	REVISION	DATE	C.O.

(USUALLY LOCATED IN UPPER RT CORNER)

LAYOUT V

3.0   .38   .10   .38   .25   .25   .31

ITEM	PART NO.	NAME	QTY.

LIST OF MATERIALS

(TITLE BLOCK HERE)

These suggested sheet layouts are recommended for use with the various sizes of drafting sheets. Layouts I and II are especially suited to A size sheets in the horizontal position or vertical. Layouts III, IV and V are suggested title block, change block and materials block for larger size sheets. A-4 and A-5 illustrate the manner of dividing an "A" size sheet into four sections horizontally and vertically.

Glossary

A

Abrasive: A substance such as emery, aluminum oxide, or diamond that cuts material softer than itself. It may be used in loose form, mounted on cloth, paper, or bonded in a wheel.

Abscissa: The "X"-coordinate of a point. In other words, its horizontal distance from the "Y"-axis measured parallel to the "X"-axis. Also, the horizontal axis of a graph or chart.

Absolute Coordinates: Refers to exact point locations measured from the origin.

Absolute System: A system of numerically controlled machining that measures all coordinates from a fixed point of origin or "zero point." Also known as "point-to-point" CNC machining.

Active Device: An electronic element capable of gain or control, such as a transistor or vacuum tube.

Acute Angle: An angle less than 90°.

Addendum: The radial distance between the pitch circle and the top of the tooth.

Addendum Circle: The diameter of the pitch circle plus twice the addendum (same as outside diameter).

AEC: An acronym for architecture, engineering, and construction. Used to refer to anything that can be applied to these industries.

Allowance: The intentional difference in the dimensions of mating parts to provide for different classes of fits.

Alloy: A mixture of two or more metals fused or melted together to form a new metal.

Alphanumeric: A general term used to describe a set of characters that includes letters of the alphabet (alpha), punctuation elements, and numbers (numeric). This is data that is distinct from graphic lines, points, and curves.

Ammonia: A colorless gas used in the development process of diazo and sepia prints. The molecular formula for ammonia is NH_3.

Angle of Thread: The included angle between the sides, or flanks, of the thread measured in an axial plane.

Annealing: A form of heat treatment that reduces the hardness of a metal to make it machine or form more easily.

Anodize: The process of protecting aluminum by oxidizing in an acid bath and using a direct current.

ANSI: The acronym for American National Standards Institute. A national group setting performance standards for various fields. Formerly known as U.S.S.I. and A.S.A.

Aperture Card: A card with a rectangular hole, or holes, prepared specifically for the mounting or insertion of microfilm.

Approximate: Describes a value that is nearly but not exactly correct or accurate.

Arbor: A shaft or spindle for holding cutting tools.

Arc: A part of a circle.

Architectural Drafting: The area of design and drafting that specializes in the preparation of drawings for the building of structures, such as houses, churches, schools, shopping centers, airports, commercial buildings, and factories.

Archival Quality: How well a processed print or film, or an electronic file, will retain its integrity and characteristics during a period of use and storage.

Area: The amount of surface enclosed by a circle, ellipse, polygon, or irregular shape composed of several entities.

Array: A pattern of copies placed in a rectangular or circular design.

Artificial Intelligence (AI): Information placed in the computer's memory that would cause the computer to make decisions normally associated with human intelligence.

As-built Drawings: Since construction drawings do not show exactly how something was built, it is preferable to use the term "Record Drawings."

ASCII Code: Acronym for American Standard Code for Information Interchange. Binary code (digital number) assigned to alphanumeric characters and graphic reference positions.

Aspect Value: The ratio of text width to text height. For example, an aspect value of "2" means the width of a text character will be twice its height.

Assembly Drawing: A drawing showing the working relationship of the various parts of a machine or structure as they fit together.

Asymptote: A straight line that is the limit of a tangent to a curve as the point of contact moves off to infinity.

Automated Guided Vehicles (AGVs): Typically a small, wheeled vehicle that follows a preprogramed path. The AGV can follow a taped path or a wire laid below the floor surface. These machines can be used to deliver parts, assemblies, or intra-building mail.

Autopositive: A print made on paper or film by means of a positive-to-positive silver type emulsion.

Auxiliary View: An additional view of an object, usually of a surface inclined to the principal surfaces of the object to provide a true size and shape view.

Axes: Plural of axis.

Axial Pitch: The distance between corresponding sides of adjacent threads in a worm.

Axis: An imaginary line that parts rotate around or are regularly arranged around.

Axis, "X," "Y," "Z": The three linear axes used with two-dimensional ("X," "Y") and three-dimensional ("X," "Y," "Z") drafting and modeling.

Axonometric: One of several forms of one-plane projection. Axonometric projection gives the pictorial effect of perspective with the possibility of measuring the principal planes directly.

Azimuth: The angle that a line makes with a north-south line, measured clockwise from the north.

B

Backing: The distance from the back of the gear hub to the base of the pitch cone measured parallel to the gear axis.

Backlash: The play (lost motion) between moving parts, such as a threaded shaft and nut, or the teeth of meshing gears.

Back-up: A copy of electronic data or of a hardcopy is stored in a safe place. This allows the material to be retrieved in case the original material is lost.

Bar Code: Automatic identification technology that encodes information with an array of varying-width parallel lines or rectangular bars and spaces. Bar codes may also include alphanumeric coding. Bar codes can be found on almost any item purchased. Bar codes are what is "scanned" at the checkout counter.

Base Circle: The circle from where the involute profile is generated.

Base-line Dimensioning: A system of dimensioning where features of a part are located from a common set of datums. As many features as practical are located from the common datums.

Basic Dimension: A theoretically exact value used to describe the size, shape, or location of a feature.

Basic Size: The size that the limits of size are derived from by the application of allowances and tolerances.

Baud Rate: The number of digital elements, bits-per-second, serially transmitted via a communications line or network. For example, a 9600 baud modem will transfer 9600 bits/second.

Bearing: An angle having a value of 0° to 90°, measured from either the north or south. A line with a bearing of 20° to the west of south would be stated as South 20° West.

Bend Allowance: The amount of material added to compensate for the changes incurred by bending the metal. As metal is bent, the thickness of the metal changes at the bend.

Bend Angle: Full angle that the sheet metal is bent through. Not to be confused with the angle between the flange and the adjacent leg.

Bend Line: Tangent line where the bend changes to a flat surface. Each bend has two bend lines.

Bevel Gears: Used to transmit motion and power between two or more shafts whose axes are at an angle (usually 90°) and would intersect if extended.

Binary: A characterization of data as having only two states: "1" or "0," "yes" or "no," "on" or "off." This is the primary functional mode of computers.

Bisect: To divide something into two equal parts.

Bit: A unit of computer information. This is the smallest unit of information in the computer. Eight bits make up a byte. A byte represents one character, such as a letter, number, or symbol.

Bit Map: A representation of a CRT display where each resolution pixel is addressed by individual bits for enhanced graphic control.

Blank: A flat sheet metal piece of approximately the correct size, that a pattern has been laid out on and is ready for machining and forming.

Blanking: A stamping operation where a press uses a die to cut blanks from flat sheets of metal.

Blind Rivets: Rivets that can be installed in a joint that is accessible from only one side.

Block: Part of a stream of data that is being transferred or processed. This is sometimes used interchangeably with "sector." A block is 52 bytes of data.

Block Diagrams: Closely related to single-line diagrams: each contains basic information that presents an overview of a system in its simplest form. Squares and rectangles are primarily used on block diagrams, but an occasional triangle or circle may be used for emphasis. Graphic symbols are rarely used, except to represent input and output devices.

Blowback: General term for an enlargement from a reduced-size (intermediate) camera negative.

Blow Molding: A tube or cylinder of heated thermoplastic resin, called a parison, is extruded and placed between halves of the split mold. The mold closes, pinching the ends of the plastic tube, then a blast of air forces the thermoplastic against the mold.

Blueline Print: A paper reproduction having blue lines on a white background. Also known as a whiteprint or a diazo print.

Blueprint: A reproduction using iron-sensitized papers having white lines on a blue background (negative image). A blueprint is made from an original drawing or positive intermediate.

Bolt: A bolt has a head on one end and is threaded on the other end to receive a nut.

Boring: Enlarging a hole to a specified dimension by use of a boring bar. Boring may be done on a lathe, jig bore, boring machine, or mill.

Boss: A small local thickening of the body of a casting or forging. A boss is designed to allow more thickness for a bearing area or to support threads.

bps: Abbreviation for bits-per-second (or bits/sec).

Brazing: The process of joining metals by adhesion with a low melting point alloy. This process does not melt the parent metal. A copper base with tin, zinc, and/or lead is commonly used.

Broach: A tool for removing metal by pulling or pushing it across the work. The most common use is producing irregular hole shapes such as squares, hexagons, ovals, or splines.

Broaching: The process of pulling or pushing a tool over or through the workpiece to form irregular or unusual shapes. The broach is a long tapered tool with cutting teeth that get progressively larger so that at the completion of a single stroke, the work is finished.

Brownprint: A reproduction method using light-sensitive iron and silver salts that produces a negative brown image from a positive master. Prints are often referred to as "van dykes."

Burnish: To smooth or polish metal by rolling or sliding a tool over surface under pressure.

Burr: The ragged edge or ridge left on metal after a cutting operation.

Bushing: A metal lining that acts as a bearing between rotating parts such as a shaft and a case. Also used on jigs to guide the cutting tool.

Butt Welding: Resistance welding of an entire joint or seam.

Byte: A byte is made up of eight bits. A byte represents one letter, number, or symbol. There are 1024 bytes in a kilobyte.

C

Cadastral Maps: Maps drawn to a scale large enough to accurately show locations of streets, property lines, and buildings.

CADD: Acronym for computer-assisted design and drafting.

CAE: Acronym for computer-assisted engineering.

Calendering: The process of making plastic sheets. The plastic resin is fed from a hopper, heated, and passed through a series of rollers. The rollers reduce the resin sheet to the desired thickness.

Callout: A note on a drawing that gives a dimension specification or machine process.

Cam: A rotating or sliding device used to convert rotary motion into intermittent or reciprocating motion.

CAM: Acronym for computer-aided manufacturing. A plan for utilizing numerical controlled machines to perform a wide variety of manufacturing processes.

Cap Nut: Used for appearance.

Cap Screw: Similar to a bolt with a head on one end, but usually a greater length of thread on the other. It is screwed into a part with mating internal threads for greater strength and rigidity

Captive: Multiple-threaded nuts that are held in place by a clamp or binding device of light gage metal.

Carburize: Heating of low-carbon steel for a period of time to a temperature below its melting point in carbonaceous solids, liquids, or gases, then cooling slowly in preparation for heat treating.

Cartesian Coordinate System: A two-dimensional system of locating points using "X" and "Y" axes.

Cartography: The science of mapmaking.

Case-hardening: Forms a hard outer layer on a piece, leaving the inner core more ductile. This produces a part that is very hard, long wearing, and resistant to breaking under impact.

Casting: The process of pouring molten metal into a mold where it hardens into the desired form as it cools.

Cathode Ray Tube (CRT): An output screen that makes an image by propelling an electron beam against a phosphor-coated screen surface.

Centerline of Bend: A radial line, passing through the bend radius, that bisects the included angle between bend lines.

Chain Dimensioning: Successive dimensions that extend from one feature to another, rather than each originating at a datum. Tolerances accumulate with chain dimensions unless the note "TOLERANCES DO NOT ACCUMULATE" is placed on the drawing.

Chamfers: Small bevels usually cut on the ends of holes, shafts, and threaded fasteners to facilitate assembly.

Change Block: The place on a drawing where changes that have been approved and made on the drawing are recorded.

Chemical Milling: A method of removing material by etching with a chemical.

Chordal Addendum: The radial distance from the top of the tooth to the chord of the pitch circle.

Chordal Thickness: The length of the chord along the pitch circle between the two sides of the tooth.

CIM: Acronym for computer-integrated manufacturing. The full automation of all facets of industry to produce a product. CIM includes design, documentation, materials handling, machine processes, quality assurance, storing/shipping, management, and marketing.

Circuit Diagram: A line drawing using graphic symbols or pictorial views to show the complete path of flow in a hydraulic or electronic system.

Circuit: The various connections and conductors of a specific device. Also, the path of electron flow from the source through components and connections and back to the source.

Circular Pitch: The length of the arc along the pitch circle between similar points on adjacent teeth.

Circular Thickness: The length of the arc along the pitch circle between the two sides of the tooth.

Clearance: The radial distance between the top of a tooth and the bottom of the tooth space of a mating gear.

Clockwise: Rotation in the same direction as hands of a clock.

Closed Traverse: A traverse that returns to its point of origin in a previously identified point.

CNC: Acronym for computer numerical control. The programming and control of a numerical control machine by a computer for production machine tools. A numerical code can be produced from a CAD drawing or it can be manually "written" and "entered."

Cold Heading: The process of applying tremendous pressure to metal, forcing it to flow upward into a narrow space around a punch and die.

COM: Acronym for computer output to microfilm. Microfilm images containing data produced by an imager direct from computer-generated electrical signals.

Command: A short word that achieves some function from the computer when entered.

Command Syntax: The correct order that commands must be entered into the computer. This can be thought of as "computer grammar."

Common Nuts: Nuts used on bolts for assemblies.

Compression Molding: One of the most common processes used in forming thermosetting plastics. A measured amount of plastic resin is placed in the open heated mold, then the mold is closed. Pressure is applied to force the plastic into the shape of the mold cavity.

Computer: An electronic machine capable of making logical decisions under the control of programs.

Computer Graphics: The process of designing industrial products and the production of graphic documents with the aid of a computer and related input and output devices.

Concentric: Having a common center.

Conjugate Diameters: Two diameters are conjugate when each is parallel to the tangents at the extremities of the other.

Connection and Interconnection Wiring Diagrams: Drawings that supplement schematic diagrams. These drawings contain information used in the manufacture, installation, and maintenance of electrical and electronic equipment.

Construction Documents: The material specifications, working drawings, and contracts required to build a facility.

Contact Frame: A device for making same-size reproductions from translucent materials that are held in position by vacuum. Items are exposed to either an internal or external light source through a glass cover.

Contact Print: A print produced by contact exposure (the original physically touches the reproduction) with the original positive or negative. These prints are always the same size as the original.

Contemporary: Belonging to the present time period.

Contour: The irregularly shaped lines found on topographic and other plan-view maps to indicate changes in terrain elevation. On any single contour line, every point is at the same elevation.

Control Unit: The part of the computer that directs the flow of data.

Coordinate Dimensioning: A type of rectangular datum dimensioning where all dimensions are measured from two or three mutually perpendicular datum planes. All dimensions originate at a datum and include regular extension and dimension lines and arrowheads.

Coordinate Pair: The "X" and "Y" values of a point in the Cartesian coordinate system.

Copy: An imitation, duplicate, or reproduction of an original. Also, a pattern.

Counterboring: Involves cutting deeper than spotfacing to allow fillister and socket head screws to be seated below the surface.

Counterdrilling: Smaller holes that are partially drilled with a larger diameter drill to allow room for a fastener or feature of a mating part.

Countersinking: Cutting a beveled edge (chamfer) in a hole so that a flat head screw will seat flush with the surface.

CPU: Acronym for central processing unit. The CPU is the primary working center of the computer. This is where all of the controlling functions and calculations take place.

Crest: The top surface of thread joining two sides (or flanks).

Crown Backing: The distance from the back of the gear hub to the crown of the gear, measured parallel to the gear axis.

Crown Height: The distance from the cone apex to the crown of the gear tooth measured parallel to the gear axis.

Cursor Control Device: Moves the cursor across the screen and is used to enter commands (however cannot be used to trace an existing drawing).

Cut and Tape: A systems drafting where reusable data images are cut out from a copy, relocated, and taped in place. This is also known as "scissors drafting."

D

Dash Number: A number preceded by a dash after the drawing number that indicates right-hand or left-hand parts. This also indicates neutral parts and/or detail and assembly drawings. The coding is usually special to a particular industry.

Database: A collection of information that can be recalled by a computer from an electronic storage device. A database can be graphics, text, or numerical information.

Data Storage Device: A data storage device holds electronic data such as a CADD program or drawing data.

Datum: Points, lines, planes, cylinders, and similar objects assumed to be exact size and shape, and to be in exact location. Datums are used to locate or establish the geometric relationship (form) of features of a part.

Dedendum Circle: The circle that coincides with the bottom of the gear teeth.

Default Values: Parameters built into the software that the computer assigns until the user changes them.

Delineation: To represent pictorially a chart, a diagram, or a sketch.

Design Size: The size of a feature after an allowance for clearance has been applied and tolerances have been assigned.

Detail Drawing: A drawing of a single part that provides all the information necessary for the production of that part.

Developed Length: Length of the flat pattern layout. This length is always shorter than the sum of mold line dimensions on part.

Development: Refers to the layout of a pattern on flat sheet stock.

Deviation: The variance from a specified dimension or design requirement.

Diagram: A figure or drawing that is marked out by lines. Also, a chart or outline.

Diameter: The length of a straight line passing through the center of a circle and terminating at the circumference on each end.

Diametral: Of a diameter or forming a diameter.

Diametral Pitch: The number of teeth in a gear per inch of pitch diameter.

Diazo: Diazo material is either a film or paper sensitized by means of azo dyes. This film or paper is used for reproducing drawings.

Diazo Print: A reproduction made by using diazo paper. These reproductions are known as whiteprints. Also called blueline prints, blackline prints, brownline prints, depending on the color of the reproduced lines.

Die: A tool used to cut external threads by hand or machine. Also, a tool used to cut a desired shape from a piece of metal, plastic, or other typically flat material.

Die-casting: A method of casting metal dies. Also the part formed by die-casting.

Die Stamping: A piece cut out by a die.

Digitizer: A device that provides input coordinate data by scanning or pointing. Digitizers are used in the process of converting a drawing or image to digital form in CADD operations.

Digitizing Tablets: Used to move the cursor across the screen, enter commands, and trace existing drawings.

Dihedral Angle: The true angle between two planes.

Dimension: Measurements given on a drawing such as size and location.

Displacement: Refers to the distance a cam follower moves in relation to the rotation of the cam.

Displacement Diagram: A graph or drawing of the displacement (travel) pattern of the cam follower caused by one rotation of the cam.

Display Screen: Also known as the monitor. The display screen allows the drafter to view the CADD drawing being created.

Distribution Prints: Final reproductions circulated to various locations.

DNC: Acronym for distributive numerical control. The control of machine tools by a main host computer that controls several intermediate computers coupled to certain machine tools, robots, or inspection stations.

Document, Documentation: Those pieces of paper, film, or tapes for CNC machining that describe the engineer's idea in physical terms and tell manufacturing or construction personnel what to make.

DOD Standards: Specifications and standards for operation set by the U.S. Department of Defense.

DOS: Acronym for disk operating system. One of the software operating systems available for controlling the functions of a computer.

Double-curved Geometrical Surfaces: Surfaces generated by a curved line revolving around a straight line in the plane of the curve.

Double-threaded Screw: This screw has two threads side by side and moves forward into its mating part a distance equal to its lead, or 2P. (Refer to Fig. 24-5B.)

Dowel Pin: A pin that fits into a hole in a mating part to prevent motion or slipping, or to ensure accurate location of assembly.

Draft: The angle or taper on a pattern or casting that permits easy removal from the mold or forming die.

Drafting Machine: A machine that combines all functions of a T-square or straightedge, triangle, scales, and protractor.

E

Eccentric, Eccentricity: Not having the same center or off center.

Effective Thread: The complete thread. Also, the portion of the incomplete thread having fully formed roots, but having crests not fully formed.

Effectivity: The serial number(s) of an aircraft, machine, assembly, or part that a drawing change applies to. The change can be indicated by an effective date and would apply from that date forward.

Electrical and Electronics Drafting: Involves the same basic principles used in other types of drawings. The difference is in the special symbols that have been developed to represent electrical circuits and wiring devices.

Electrical Discharge Machining (EDM): The working of metals by eroding the material away with an electric spark.

Electrochemical Machining (ECM): The reverse of electroplating. An electric current in used to remove metal from a piece that is suspended in a chemical solution.

Electron Beam Welding: A high-intensity beam of electrons focused in a small area at the surface to be welded.

Electrostatic Printer: A hardcopy output device that produces images in dot matrix or raster format using heat and various forms of graphite toners.

Ellipse: A circle viewed at an angle.

Expert Systems: A branch of artificial intelligence placed in the computer's memory expressed as "rules" of human expertise.

Exploded Assembly: A drawing where components, usually drawn in pictorial form, are shown with an axis line showing the sequence of assembly.

External Thread: The thread on the outside of a cylinder such as a machine bolt.

Extrusion: Metal that has been shaped by forcing it through dies. Extrusion can be done to either hot or cold metal.

F

Face Angle: The angle between an element of the face cone and the axis of the gear or pinion.

Face Width: The width of the tooth measured parallel to the gear axis.

Fastener: A mechanical device for holding two or more bodies in a definite position with respect to each other.

FEA: Acronym for finite element analysis. A CAD method for analyzing engineering problems involving static, dynamic, and thermal stressing of parts.

Feature: A portion of a part such as a diameter, hole, keyway, or flat surface.

Ferrous: Metals that have iron as their base material.

File: A group of related data. In the case of a CADD drawing file, the data in the file describes a drawing.

Fillet: A concave intersection between two surfaces.

Film Positive: A positive reproduction made on photographic film.

Finish: General finish requirements such as paint, chemical, or electroplating. Does not indicate surface texture or roughness. (See also "surface texture.")

Finished Nuts: Used for close tolerances.

Finishing Washer: These washers distribute the load and eliminates the need for a countersunk hole.

Finite Element Analysis: Since solid models are made of small elements, like a real object is made of atoms, a solid model can be cut, pressed, pulled, and twisted much like a real object. If you put force on one area of the model, you can analyze the effect of that force in all areas of the model. This is finite element analysis.

Fit: The clearance or interference between two mating parts.

Fixed Characteristics: Characteristics of a part or object that remain the same for all parts involved.

Fixture: A device used to position and hold a part in a machine. It does not guide the cutting tool.

Flame Hardening: Similar to case-hardening. It is a method of producing surface or localized hardening by directly heating a surface, then immediately quenching the piece before the heat has had a chance to penetrate far below the surface. This process is widely used to harden gears, splines, and ratchets.

Flange: An edge or collar fixed at an angle to the main part or web.

Flash Welding: Resistance welding where the ends of two metal parts are brought together under pressure and resistance welded.

Flat Pattern: Pattern used to lay out a sheet metal part on a blank.

Flat Washers: Used primarily for load distribution.

Flow Charts: A graphic means of depicting the sequence of technical processes that would be difficult to describe in narrative form.

FMS: Acronym for flexible manufacturing system. A production system of highly automated machines, assembly cells, robots, inspection equipment, materials, and storage systems. This type of system is capable of processing a wide variety of similar products.

Foamed Plastics: Plastic resins that air or gas are added to, forming a sponge-like substance.

Follower: The follower makes contact with the surface or groove of the cam. The follower is held against the cam by gravity, spring action, or by a groove in the groove cam. Reciprocating motion is input into or taken off of the follower.

Font: Refers to the appearance, or typeface, of text.

Forging: The forming of heated metal by a hammering or squeezing action.

Form Tolerancing: The permitted variation of a feature from the basic shape indicated on the drawing.

Frisket: A special paper with an adhesive back used to shield a drawing during airbrushing. The frisket is laid over the entire drawing, then "windows" are cut out of the material over any areas to be airbrushed.

Function Keys: When pressed, function keys perform a series of commands that otherwise would have to be entered separately.

Functional Drafting: Making a drawing that includes just those lines, views, symbols, notes, and dimensions needed to completely clarify the construction of an object or part.

Functional Drawing: A drawing that uses a minimum number of views, details, and dimensions, and yet maintains accuracy.

Fusion Welding: This type of welding includes oxyacetylene, arc, TIG, and MIG welding. The heat generated by a flame or arc causes the parent metal and a feeder metal rod to melt and "fuse" into one piece. This type of welding joins the metals by cohesion.

G

Gage: A number that represents the thickness of sheet metal. This number does not represent an actual physical measurement, but "places" the metal on a scale in relation to other thicknesses of metal.

Gas-plasma Displays: Made of a thin glass enclosure containing a low-pressure neon or neon/argon gas. A grid of wires runs through the enclosure. When two crossing wires are energized, the gas at their intersection glows.

Gears: Machine parts used to transmit motion and power by means of successively engaging teeth.

Generation: Each succeeding stage in the reproduction of an original. For example, the blowback made from a microfilm of an original drawing is a first-generation print. The reproduction made from the first-generation print is a second-generation print.

Geology: The study of the earth's surface, its outer crust and interior structure, and the changes that have taken place and are taking place.

Geometric Dimensioning and Tolerancing: A means of dimensioning and tolerancing a drawing with respect to the actual function or relationship of part features that can be most economically produced. It includes positional and form dimensioning and tolerancing.

Ghosting: A smudged area on a reproduction copy of a drawing caused by damage to the drawing sheet due to erasing or mishandling.

Graphics: Data in the form of pictorial communications such as drawings, charts, cartographic manuscripts, engineering designs, and plotted data.

Grid: Intersecting lines that are displayed as lines or intersecting points. These act as references or as limits to movement.

Grinding: The process of removing metal by means of abrasives.

Group Technology (GT): A manufacturing philosophy that consists of organizing components into families of parts to be produced in machining cells.

Gusset: A small plate used in reinforcing assemblies.

H

Hard Copy: An end-use copy that is usually on paper, vellum, or film.

Hardcopy Device: The hardcopy device makes a paper or film copy of the drawing held in computer memory or stored on a data storage device.

Hardness Test: Techniques used to measure the degree of hardness of heat-treated materials.

Hardware: The CPU, monitor, keyboard, digitizer, disks, plotter, and other *physical components* of a computer system.

Harmonic Motion: This motion will move a cam follower in a smooth continuous motion.

Heat Treating: Heating metal to a high temperature, then cooling it at various rates to produce qualities of hardness, ductility, and strength.

Heavy Nuts: Used for a looser fit, for large-clearance holes, and for high loads.

Hexagon: A polygon having six angles and six sides.

Hobbing: A special gear cutting process. The gear blank and hob rotate together as in mesh during the cutting operation.

Honing: Done with blocks of very fine abrasive materials under light pressure against the work surface (such as inside of a cylinder). They are rotated rather slowly and are moved backwards and laterally. This process provides a very smooth surface.

Horizontal: Parallel to the horizon.

Horizontal Curve: A change of direction in the horizontal or plan view that is achieved by means of a curve.

Human Factors: Characteristic dimension of humans such as reach, body form, body size, and comfort factors. Used in the design of products.

I

Icons: Small pictures that represent computer commands. When the cursor is moved over an icon and the icon is selected, a command is given to the computer.

Identification Code: A number assigned by the federal government to industries doing contractual work for the government.

Image: A representation of an object, or information sources produced by light rays, drafting, reproduction, or digital means.

Inclined: A line or plane at an angle to a horizontal line or plane (not at 90°).

Incremental System: A system of numerically controlled machining that always refers to the preceding point when making the next movement. Also known as continuous path or contouring method of CNC machining.

Indicator: A precision measuring instrument for checking the trueness of work.

Induction Welding: Similar to resistance welding. However, in induction welding the heat generated for the weld is produced by the resistance of the metal parts to the flow of an induced electric current. The welding action may occur with or without pressure.

Injection Molding: Plastic granules are loaded into the hopper of the injection-molding machine, then softened by heating. The softened plastic is forced through a nozzle into the mold to form and cool.

Input Device: Used to enter, or "input," commands. Also used to locate positions on a CADD drawing when constructing and editing lines, circles, and other entities.

Input/Output (I/O): The process of transferring data to or from a computer system.

Inseparable Assembly: A component permanently assembled by welding, bonding, riveting, potting, or similar permanent process.

Insoluble: A material that cannot be dissolved.

Integrated Circuit: A complete electronic circuit composed of various electronic devices fabricated on a common substrate. Usually very small in size.

Interactive Graphics: The ability to communicate with the computer. Icons are a good example of interactive graphics.

Interchangeability: The ability of units or parts of a mechanism or an assembly to be exchanged with another part manufactured to the same specifications. For example, an 8UNC thread nut should fit on an 8UNC bolt no matter where the nut and bolt are manufactured.

Interlaced Monitor: Uses the raster scan method. It will scan every line, from left to right and top to bottom. However, two passes are required for every line. On the first pass, the monitor will scan every other pixel. On the second scan, the remaining pixels are scanned.

Intermediate: A translucent reproduction from an original drawing, typically reverse reading. Also, a print used in place of the original for making other copies.

Internal Thread: The thread on the inside of a cylinder such as a nut.

Interpolation: A technique used to locate, by proportion, intermediate points between grid data given in contour plotting problems.

Intersecting Lines: Lines that have a common point lying at the exact point of intersection.

Intersections: The lines formed at the junction when two or more objects (such as two planes or a cylinder and a square prism) join or pass through each other.

Involute: A spiral curve generated by a point on a chord as the chord "unwinds" from a circle or a polygon.

Isometric Drawing: A pictorial drawing of an object positioned so that all three axes make equal angles with the picture plane. Measurements on all three axes are made to the same scale.

J

Jig: A device used to hold a part to be machined. A jig positions and guides the cutting tool. (This is how a jig differs from a fixture.)

JIT: Acronym for just-in-time manufacturing. A concept of manufacturing where the aim is to reduce work-in-progress by reducing lead times, inventories (raw materials, parts, and finished products to be stored), and setup times to an absolute minimum.

Joggle: An offset in the face of a part that has an adjacent flange.

Justification: Refers to how a text string will be placed in relation to a selected location point. Left justified means that the text will start at the point and continue to the right. Right justified means that the text will end at the point and start to the left of the point. Center justified means that the text will be centered about the point.

K

Kerf: The slit or channel left by a saw or other cutting tool.

Key: A small piece of metal (usually a pin or bar) used to prevent rotation of a gear or pulley on a shaft.

Keyboard: The most common input device. Used to type in text and enter commands into the computer.

Keyseat: The slot machined in a shaft for square and flat keys.

Keyslot: The slot machined in a shaft for Woodruff type keys.

Keyway: The slot machined in a hub for all types of keys.

Knurled Nuts: Allow for hand tightening.

Knurling: The process of forming straight-line or diagonal-line (diamond) serrations on a part to provide a better handgrip or interference fit.

L

Ladder Logic: The "logic" used in PLC programming. The printout of a ladder logic program looks like rungs of a ladder. The PLC will "look" along one "leg" of the ladder, examining each rung, until a condition that can be met with the current set of inputs is found.

Laminate: The product resulting from bonding two or more layers of material together. The individual layers making up the laminate may be sheet plastic, cloth, paper, or wood.

LAN: Acronym for local area network. A network that links several computers in close proximity.

Landscape: A printing orientation where the text and graphics is printed along the longest horizontal paper dimension. This would be as if you turned a standard sheet of notebook paper sideways and printed on it that way.

Lapping: Similar to honing, except a lapping plate or block is used with a very fine paste or liquid abrasive between the metal lap and work surface.

Laser: An acronym for light amplification by stimulated emission of radiation. A device for producing light by emission of energy stored in a molecular or atomic system when stimulated by an input signal.

Lay: The direction of the predominant surface pattern.

Layering Commands: Allow the CADD drafter to separate different parts of a drawing into levels, much like using layers of film in overlay drafting.

Layout Drawing: Often the original concept for a machine design or for placement of units. It is not a production drawing, but rather serves to record developing design concepts.

Lead Angle: The angle between a tangent to the helix of the thread at the pitch diameter and a plane perpendicular to the axis of the worm.

Lead: The axial advance of the worm in one complete revolution.

Left-hand Thread: A thread, when viewed in the end view, winds counterclockwise to assemble. Left-handed (LH) threads are indicated.

Limits: The extreme permissible dimensions of a part resulting from the application of a tolerance.

Locknuts: A type of special nut that prevents the nut from loosening up once it is properly tightened.

Logic Testing: Simulates the operation of the circuit board.

Logon Procedure: The process of "entering" or starting a computer program. However, usually refers to connecting to a network.

M

Machine Screw: Similar to a cap screw except it is smaller and has a slotted head.

Machining Center: A grouping of high precision, multi-functional machines for performing specific operations without scheduling secondary operations.

Magnaflux: A nondestructive inspection technique that uses a magnetic field and magnetic particles to locate internal flaws in ferrous metal parts.

Magnetic North: Indicated by a magnetic compass.

Major Diameter: The largest diameter on an external or internal screw thread.

Maximum Material Condition (MMC): When a feature contains the maximum amount of material as allowed by the tolerances. For example, MMC exists when a hole is at the minimum diameter and when a shaft is at the maximum diameter.

Median Line: A line along the interior of sheet metal that remains true length throughout the bending process. The metal on the outside of this line is stretched and the metal on the inside of the bend is compressed.

Megabyte (MB): A megabyte is 1,048,576 bytes, or 1024 kilobytes. Most storage devices are rated in megabytes.

Menu: A list of computer commands displayed on the screen or printed on a digitizing tablet overlay.

Metricize: To convert a unit other than metric to its metric (SI) equivalent.

Microfiche: Transparent film approximately 105mm x 148mm (4″ x 6″) containing micro images in a group pattern with a title heading large enough to be read by the naked eye.

Microfilm: A fine-grain, high-resolution film typically 35mm or 16mm wide containing an image greatly reduced in size from the original.

MIG: Acronym for metal inert gas welding. It is a gas-shielded arc welding process similar to TIG welding. In MIG welding, the electrode is a filler wire that is fed into the weld automatically.

Mill: To remove metal with a rotating cutting tool on a milling machine.

Minor Diameter: The smallest diameter on an external or internal screw thread.

Mismatch: The variance between depths of machine cuts on a given surface.

Miter Gears: Bevel gears of the same size and at right angles.

Modeling: The process of creating a three-dimensional view of an object.

Modem: Acronym for modulator demodulator. A telecommunications device used to send serial data between nodes in a network or over phone lines.

Mold Line: Line of intersection formed by projection of two flat surfaces.

Monomasters: Semi-complete drawing of simple parts that show a common shape and/or combination of features ready for the addition of dimensions and other specific data by the user. For example, the drawing of a compression spring with a form for completion of data for a specific size is a monomaster.

Mosaic: A series of aerial photographs or radar images of adjacent land areas, taken with intentional overlaps and fitted together to produce a larger picture.

Motion: Refers to a cam follower's rate of speed or movement in relation to the uniform rotation speed of the cam.

Mounting Distance: The distance from a locating surface of a gear (such as end of hub) to the centerline of its mating gear. It is used for proper assembling of bevel gears.

MS-DOS: Acronym for Microsoft disk operating system. An operating system for personal computers.

Multitasking: Multitasking is the computer's ability to run more than one program at the same time.

Multiview Projection: Two or more views of an object as projected upon the picture plane in orthographic projection.

N

Neck: A groove on a shaft provided to terminate a thread.

Negative Print: A print, usually on opaque material, that is opposite to the original drawing. In other words, light lines appear on a dark background.

Nesting: Grouping of related commands under a main head. This allows many commands to be used, but not displayed on the computer screen at one time.

Next Assembly: The next object or machine that the part or subassembly is to be used on.

Nominal Size: A general classification term used to designate size of a commercial product.

Nonferrous: Metals not derived from an iron base or an iron alloy base, such as aluminum, magnesium, and copper.

Non-interlaced Monitor: Uses the raster scan method. The difference between interlaced and non-interlaced is that non-interlaced monitors scan every pixel in a line in one pass. This means that each line is scanned only once. Since only one pass is made, "flicker" is eliminated.

Normalizing: A process where ferrous alloys are heated and then cooled in a still-air to room temperature to restore the uniform grain structure free of strains caused by cold working or welding.

North: The direction normally indicated on the top of a map.

Numerical Control: A system of controlling a machine or tool by means of numeric codes. These codes control devices attached to, or built into, the machine or tool.

O

Oblique Drawing: A pictorial drawing of an object drawn so that one of its principal faces is parallel to the plane of projection, and is projected in its true size and shape. The third set of edges is oblique to the plane of projection at some convenient angle.

Obtuse Angle: An angle larger than 90°.

Octagon: A polygon having eight angles and eight sides.

On-line: Condition when device or user is directly connected, or actively working with the computer or CPU. Commonly used to indicate that the computer is connected to some remote terminal or system.

Open Traverse: A traverse that neither returns nor ends at a previously identified point.

Ordinate Dimensioning: A type of rectangular datum dimensioning where all dimensions are measured from two or three mutually perpendicular datum planes. Datum planes are indicated as zero coordinates and dimensions are shown on extension lines without the use of dimension lines or arrowheads. Sometimes called "arrowless dimensioning."

Ordinate: The "Y"-coordinate of a point. In other words, the vertical distance from the "X"-axis measured parallel to the "Y"-axis. Also, the vertical axis of a graph or chart.

Original: The material that copies are made from, such as handwritten copy, typed copy, printed or plotted matter, tracings, drawings, databases, and photographs.

Origin: The "zero" point of the Cartesian coordinate system.

Orthographic Projection: A projection on a picture plane formed by perpendicular projectors from the object to the picture plane. Third angle projection is used in the United States. First angle projection is used in most countries outside the United States.

Outline Assembly: These drawings are used for the installation of units and provide overall dimensions to show size and location of points necessary in locating and fastening each unit in place. The outline assembly drawing provides the dimensions essential to making electrical, air, and other connections, as well as the amount of clearance required to operate and service the unit.

Output Devices: Allow the CADD drafter to view the created drawing. Some output devices include display screens and hardcopy devices.

Outside Diameter: The diameter of the crown circle of the gear teeth.

Outside Diameter of the Worm: The pitch diameter of the worm plus twice the addendum.

Overlay: Transparent or translucent prints that form a composite picture when suitably registered on top of one another.

P

Parallel: Having the same direction, such as two lines that would never meet if extended.

Parallel Circuit: A circuit that contains two or more paths for electrons supplied by a common voltage source.

Parametric Programming: A listing of a series of CADD commands used to create a certain object. Parametric programs automate the drafting process. Only certain values (such as the radius and center point of a circle) need to be supplied to the computer, and it will do the rest.

Parison: A tube or cylinder of heated thermoplastic resin that is used in blow molding.

Part Programming: The description of a part that is to be machined by numerical control or for the development of precision artwork for printed circuit boards.

Passive Device: An electronic element incapable of gain or control, such as a resistor or capacitor.

Pentagon: A polygon having five angles and five sides.

Perpendicular: A line or plane at a right angle to a given line or plane.

Perspective Drawing: A pictorial drawing where receding lines converge at vanishing points on the horizon. It is the most "natural" of all pictorial drawings.

Photo Drafting: The drafting process of combining actual photographs and sections of several drawings onto one new or revised drawing using photographic techniques.

Photodrawing: A photograph of a drawing or an object that additions, changes, or dimensions have been added to.

Photogrammetry: The use of photography, either aerial or land-based, to produce useful data for the preparation of contour topographic, planimetric, and orthophoto cross-section profile maps.

Pickle: The removal of stains and oxide scales from parts by immersion in an acid solution.

Piercing Point: The point of intersection between a plane and a line inclined to that plane.

Pilot: A protruding diameter on the end of a cutting tool designed to fit in a hole and guide the cutter in machining the area around the hole.

Pilot Hole: A small hole used to guide a cutting tool for making a larger hole. Also used to guide a drill of larger size.

Pinion: The smaller of two mating gears.

Pins: A fastening device used where the load is "primarily" shear.

Piping Drawings: A type of assembly drawing that represents a piping layout using symbols and either double-line or single-line drawings.

Pitch: The distance from a point on one thread to a corresponding point on the next thread.

Pitch Angle: The angle between an element of the pitch cone and its axis.

Pitch Circle: An imaginary circle located approximately half the distance from the roots and tops of the gear teeth.

Pitch Diameter: The diameter of an imaginary cylinder passing through the thread profiles at the point where the widths of the thread and groove are equal.

Pitch Diameter of the Worm: The diameter of the pitch circle of a worm thread.

Pixel: A single point on a raster display (monitor).

Plan View: The top view of an object.

Plasticizers: Flexibility agents that are added to plastic resins to improve the workability of the mixture and reduce brittleness after curing.

Plastisols: A mixture of plastic resins with plasticizers (flexibility agents) that improve the workability of the mixture and reduce brittleness after curing.

Plat: A plan that shows land ownership, boundaries, and subdivisions.

Plotter, Raster: A hardcopy output drafting device in a computerized design system that converts digital data into a graphical representation made up of a finite series of high resolution dots.

Plotter, Vector: A hardcopy output drafting device in a computerized design system that converts digital data into a graphical representation using pens to form a series of straight lines and curves.

Plotting: The process of making a hardcopy of a drawing.

Polar Coordinate Entry: Points are located by an angle and a distance from a previous point.

Polygon: A plane geometric figure with three or more sides.

Portrait: A printing orientation where text and graphics are printed along the longest vertical paper dimension. When you write in a standard notebook, your printing is aligned in portrait orientation.

Positional Tolerancing: The permitted variation of a feature from the exact or true position indicated on the drawing.

Positive Print: A print that is similar to the original drawing. In other words, dark lines appear on a light background.

Potassium Dichromate: An orange-red crystalline compound used in developing with a dark blue background (blueprints).

Precision: The quality or state of being precise or accurate.

Precision Sheet Metal: Working light-gage metal to machine shop tolerances.

Pressure Angle: The angle of pressure between contacting teeth of meshing gears.

Primary Auxiliary View: Projected from orthographic views—horizontal, frontal, or profile.

Primary Revolution: Drawn perpendicular to one of the principal planes of projection.

Primitives: Form the building blocks of a computer model. Primitives include cubes, spheres, cones, pyramids, and other shapes.

Print: Refers to any hard copy.

Prism: A solid whose bases or ends are any congruent and parallel polygons, and whose sides are parallelograms.

Prismatic: Pertaining to, or like a prism.

Process Drawings: Also called operation drawings. These drawings usually provide information for only one step or operation in the making of a part.

Process Specification: A description of the exact procedures, materials, and equipment to be used in performing a particular operation.

Profilometer: A device that measures the smoothness of a surface or finish.

Program, Computer: A set of step-by-step instructions telling the computer how to react to data inputs.

Programmable Logic Controller (PLC): A microprocessor that is programmed and typically used to control a machine. PLCs typically have no monitor attached.

Project: To extend from one point to another.

Proportion: Proper relation between objects or parts.

Punch Cards: A way of controlling NC machines. The program is "punched" into the cards and the cards are run through the machine. This type of machine control has been replaced by CNC equipment.

Q

Quenching: Cooling metals rapidly by immersing them in liquids or gases.

Queue: A temporary storage location, buffer, or list.

R

Rack: A spur gear with its teeth spaced along a straight pitch line.

Radius: The straight-line distance from the center of a circle or arc to its circumference.

Random Access Memory (RAM): Holds information while the computer is being used. When the power is turned off, all information stored in RAM is lost. RAM is not to be confused with disk storage space.

Rasterization: The converting of drawing data into series of dots that can be reproduced by a plotter or printer.

Read Only Memory (ROM): Contains data and instructions that cannot be changed by the user. This data is typically used when the computer is first turned on.

Reaming: To finish a drilled hole to a close tolerance.

Reference Dimension: Used only for information purposes and does not govern production or inspection operations.

Refresh Displays: These displays constantly "redraw" the image to prevent it from becoming dim. This is similar to scanning, yet the electron beam refreshes the image by drawing the same line pattern of the drawing (similar to how a human would draw the line).

Regardless of Feature Size (RFS): The condition where tolerance of position or form must be met irrespective of where the feature lies within its size tolerance.

Relative Coordinates: Define a point using the distance from a previous point, not the origin.

Release Notice: Authorization indicating the drawing has been cleared for use in production.

Relief: A notch cut in sheet metal to provide a clearance for a bend.

Rendering: Finishing a drawing to give it a realistic appearance. Also, a rendering is a representation.

Reproducible: Capable of being used as a master for making prints by the action of radiant energy (light) on chemically treated media.

Resistance Welding: Using the resistance of metals to the flow of electricity to produce heat for fusing the metals to one permanent piece.

Resistor: An electronic component that resists flow of an electric current.

Resolution: A measure of sharpness of an image. Typically expressed as the number of dots-per-inch discernible in an image.

Retaining Rings: Inexpensive devices used to provide a shoulder for holding, locking, or positioning components on shafts, pins, studs, or in bores.

Revolution: A method available to the drafter/designer in defining spatial relationships of rotating or revolving parts.

Right-hand Thread: A thread, when viewed in the end view, winds clockwise to assemble. A thread is considered to be right-handed (RH) unless otherwise stated.

Robots: Devices that are controlled by a computer and perform tasks that might otherwise be performed by a human.

ROM: Acronym for read only memory. This memory is static memory. It is retained even when the power is shut off to the computer.

Root: The bottom surface of thread joining two sides (or flanks).

Root Angle: The angle between an element of the root cone and its axis.

Root Diameter: The diameter of the root circle. It is equal to the pitch diameter minus twice the dedendum.

Rotational Molding: Plastic resin in the form of powder or liquid is placed in a mold of the desired form. The mold rotates, spreading the powder or liquid evenly over its interior surface.

Roughness: Refers to the finer irregularities in a surface. Included are those irregularities that result from the machine production process, such as traverse feed marks.

Roughness Width: The distance between successive peaks or ridges of the predominant pattern of roughness.

Ruled Geometrical Surfaces: Surfaces generated by moving a straight line.

S

Sandblast: The process of removing surface scale from metal by blowing a grit material against it at very high air pressure.

Scale: A measuring device with graduations for laying off distances. Used to draw objects to full, reduced, or enlarged size. Also refers to size an object is drawn, such as full-size, half-size, or twice-size.

Schematic Diagram: The most frequently used drawing in the electronics field. It serves as the master drawing for production drawings, parts lists, and component specifications.

Scissors Drafting: When part (or all) of one drawing is used to create part (or all) of a "second-original" drawing.

Secondary Auxiliary View: Projected from a primary auxiliary and a principal view.

Section View: A view of an object obtained by "cutting away" the front portion of the object to show the interior detail.

Security Copies: Exact duplications of the original drawings.

Self-retaining Nuts: Multiple-threaded nuts that are held in place by a clamp or binding device of light gage metal.

Sems: A generic term for screw and washer assemblies.

Sensitized: A reproduction material coated with a light-sensitive emulsion.

Sepia: A yellow-brown print, sometimes called a "van dyke," made directly from the original tracing. These are the most common negatives used for making positive prints or duplicate intermediates. The negative is printed in reverse. From the negative the positive print is made. The negative is contacted face to face with the positive to be printed.

Series Circuit: A circuit which contains only one possible path for electrons through the circuit.

Serrations: Condition of a surface or edge having notches or sharp teeth.

Server: A main computer that controls the functions of other computers.

Set-back: Amount of deduction in length resulting from a bend developed in a flat pattern.

Setscrew: Used to prevent motion between two parts, such as rotation of a collar on a shaft.

Shaft Angle: The angle between the shaft of the two gears, usually 90°.

Shim: A piece of thin metal used between mating parts to adjust their fit.

Single-line Diagrams: Simplified representations of complex circuits or entire systems.

Single-thread Engaging Nuts: Nuts formed by stamping a thread engaging impression in a flat piece of metal.

Single-threaded Screw: This screw will move forward into its mating part a distance equal to its pitch in one complete revolution (360°).

Slope: The angle that a line makes with the horizontal plane.

Software: The program used to instruct the computer to perform CADD functions.

Solenoid: A coil of wire carrying an electric current possessing the characteristics of a magnet.

Solids Modeling: The most refined type of computer modeling. Instead of recognizing only surfaces, the computer "thinks" of the object as solid material. In wireframe modeling, the computer recognizes only exterior surfaces. However, in solids modeling the computer recognizes both exterior and interior features of the object.

Solution: The answer to a problem. Also, a homogeneous molecular mixture of two or more substances.

Special Nuts: Used where an application requires features not found on common nuts.

Specifications: A written set of instructions with a proposed set of plans, giving all necessary information not shown on the prints such as quality, manufacturer's name, and how work is to be conducted.

Splines: Similar to multiple keys on a shaft. Splines prevent rotation between the shaft and its related member.

Spotface: A machined circular spot on the surface of a part to provide a flat bearing surface for a screw, bolt, nut, washer, or rivet head.

Spot Welding: Resistance welding where the metal is fluxed only in the contact spots.

Spur Gears: Used to transmit rotary motion between two or more parallel shafts.

Stable: Refers to drafting media whose dimensional characteristics remain constant (or relatively so) with changes of temperature and humidity.

Staggered: To arrange in an offset fashion.

Standard Rivets: A small cylinder of metal that is inserted into a clearance hole in two mating parts. The rivet is then peened on both ends providing a permanent fastener.

Stations: The turning points in a map traverse. In highway construction surveys, points at 100 foot intervals on the center line in the plan view are also called stations. They are located by stakes with station numbers on them.

Stress Relieving: To heat a metal part to a suitable temperature and hold that temperature for a determined time, then gradually cool it in air. This treatment reduces the internal stresses induced by casting, quenching, machining, cold working, or welding.

Stretchout: A flat pattern development for use in laying out, cutting, and folding lines on flat stock, such as paper or sheet metal, to be formed into a useful object (a container, air duct, or funnel).

Stud: A rod threaded on both ends to be screwed into a part with mating internal threads.

Substrate: A base material.

Successive Auxiliary Views: Views projected after a secondary auxiliary.

Successive Revolutions: This method is similar to finding the true size of an oblique surface through successive auxiliary views.

Superheterodyne: A radio receiver where the incoming signal is converted to a fixed intermediate frequency before detecting the audio signal component.

Supersede: The replacing of one part by another. A part that has been replaced is said to be superseded.

Surface Models: Computer models constructed by connecting edges. Basically, a surface model begins with a wireframe model. Then each plane is converted to a surface as if a sheet of plastic covered the wireframe.

Surface Texture: The roughness, waviness, lay, and flaws of a surface.

Surveying: The process or occupation of determining the location of boundaries for construction, roads, and other purposes.

Symbol: A letter, character, or schematic design representing a unit or component that can be inserted into a drawing in any position.

Symbol Library: A special directory on the data storage device where similar symbol types–mechanical, electrical, architectural, welding, and others–are stored together.

T

Tabular Dimension: A type of rectangular datum dimensioning where dimensions from mutually perpendicular datum planes are listed in a table on the drawing instead of on the pictorial portion.

Tabulated Drawing: A type of detailed working drawing that provides information needed to fabricate two or more items that are basically identical but vary in a few characteristics. These variable characteristics typically involve dimensions, material, or finish.

Tangent: A line drawn to the surface of an arc or circle so that it contacts the arc or circle at only one point. (A circle or arc can also be constructed tangent to a line or another circle or arc.)

Tap: A rotating tool used to produce internal threads by hand or machine.

Technician: One who works in a technical field as a member of the engineering support team.

Tempering: Creating ductility and toughness in metal by heat treatment process.

Template: A pattern or guide.

Tensile Strength: The maximum load (pull) a piece supports without breaking or failure.

Thermoplastic Materials: These materials become soft when heated and harden as they cool. They can be softened repeatedly by heating. Included in this group are the styrenes, vinyls, acrylics, polyethylenes, and nylons.

Thermosetting Plastics: These plastics cannot be softened by reheating, since they change chemically during the curing process. Included in the thermosets are the phenolics, epoxies, ureas, melamines, and polyesters.

Thread Class: The fit between two mating thread parts with respect to the amount of clearance or interference which is present when they are assembled. "Class 1" represents a loose fit and "Class 3" a tight fit.

Thread Form: The profile of the thread as viewed on the axial plane. (Refer to Fig. 24-4 for standard thread forms.)

Thread Series: The groups of diameter-pitch combinations distinguished from each other by the number of threads per inch applied to a specific diameter.

Thread-cutting Screws: These screws act like a tap. They cut away material as they enter the hole. These screws have flutes or slots in the point to form a cutting edge.

Thread-forming Screws: These screws form threads by displacing the material rather than cutting it.

TIG: Acronym for tungsten inert gas welding. It is a gas-shielded arc welding process.

Tolerance: The total amount of variation permitted from the design size of a part.

Toner: The "powder" that creates the image in electrostatic printers.

Topography: The detailed description and analysis of the features of a relatively small area, district, or locality of the earth's surface.

Torque: The rotational or twisting force in a turning shaft.

Traditional: The customs and style from an earlier period.

Trammel: An instrument consisting of a straightedge with two adjustable fixed points for drawing curves and ellipses.

Transfer Molding: The plastic in this type of molding is not fed directly into the mold as it is in compression molding. Instead, the resin is placed in a separate chamber and heated under the pressure of a plunger until molten. Higher pressures are then exerted, forcing the softened resin through runners and gates into the mold cavities

Translucent: A material that permits the passage of light (partially transparent).

Traverse: Measuring or laying out a line, such as a property line, by means of angular and linear measurements.

Triple-threaded Screw: This screw has three individual threads. A triple-threaded screw moves forward a distance equal to its lead, or 3P. (Refer to Fig. 24-5 (c).)

True North: Determined by sighting on Polaris (the "North Star"). This is slightly different than magnetic north, but is considered to be the most accurate.

True Position: The "basic" or theoretically exact position of a feature.

Truncated: Having the apex, vertex, or end cut off by a plane.

Tumbling: The process of removing rough edges from parts by placing them in a rotating drum that contains abrasive stones, liquid, and a detergent.

Turnkey Systems: CADD workstations that include all of the hardware (computer and related equipment) and software needed to run the system. The user simply needs to "turn the key" to operate the system.

Typical (TYP): This term, when associated with any dimension or feature, means the dimension or feature applies to the locations that appear to be identical in size and configuration unless otherwise noted.

U

Ultrasonic Machining: Using an electrode that is placed in a strong magnetic field that fluctuates rapidly. An alternating current is used to reverse the magnetic field about 25,000 to 30,000 times per second. The "cutting tool," a rod of brass or soft steel of the desired shape, is attached to the nickel rod.

Undercut: A recess at a point where the shaft changes size and mating parts such as a pulley must fit flush against a shoulder.

Unified Coarse (UNC): This series is used for bolts, screws, nuts, and threads in cast iron, soft metals, or plastic where fast assembly or disassembly is required.

Unified Extra-fine (UNEF): This series is used for very short lengths of thread engagements. It is also used for thin-wall tubes, nuts, ferrules, and couplings. It is also used for applications requiring high stress resistance.

Unified Fine (UNF): This series is used for bolts, screws, and nuts where a higher tightening force between parts is required. The Fine series is also used where the length of the thread engagement is short and where a small lead angle is desired.

Unified Thread Series: The American standard for fastening types of screw threads.

Uniform: Having the same form or character. Also, unvarying.

V

Variable Characteristics: Characteristics of a part or object, such as dimensions, that change from part to part.

Vernier Scale: A small movable scale attached to a larger fixed scale, for obtaining fractional subdivisions of the fixed scale.

Vertex: The highest point of something, the top, or the summit.

Vertical Curve: A change of direction in the grade, shown in a profile view, that is achieved by means of a curve (usually a parabolic curve).

Vertical: Perpendicular to the horizon.

Vertices: Plural of vertex.

W

Waviness: The widest-spaced component of the surface; it covers a greater horizontal distance than the roughness-width cutoff.

Waviness Width: The spacing from one wave peak to the next, and waviness height is the distance from peak to valley, measured in inches or millimeters.

Welding Drawings: A type of assembly drawing that shows the components of an assembly in position to be welded, rather than as separate parts. Specification of the type of welds to be used on various joints has become standard procedure on welding drawings.

Whiteprints: Prints made by the diazo process.

Whole Depth: The total depth of a tooth (addendum plus dedendum).

Wing Nuts: Allow for hand tightening.

WIP: Acronym for work-in-progress. This term indicates that a product has not yet been completed, and that more processes have to be performed to it.

Wireframe Models: Computer models created by connecting "points" of an object. "Points" refers to intersections of lines of the object. Connecting these points makes the object appear three-dimensional.

Working Depth: The sum of the addendums of two mating gears.

Working Drawings: A set of drawings that provide details for the production of each part and information for the correct assembly of the finished product.

Workstation: The equipment included in a CADD system. A typical workstation includes a computer, CADD software, a display screen, an input device, and a hardcopy device.

Worm Mesh: A gear type used for transmitting motion and power between nonintersecting shafts usually at 90° to each other. The worm mesh consists of the worm and the worm gear.

Z

Zero Point: The point that is defined as the origin on a CNC machine. All features are located from this point.

Zoom: The enlargement or reduction of display image as an apparent scale change for increased clarity.

Index